Human Biology

Human Biology

Joseph A. Mannino, Ph.D.

Associate Professor
College of Human Biology
University of Wisconsin, Green Bay

St. Louis Baltimore Berlin Boston Carlsbad Chicago London Madrid
Naples New York Philadelphia Sydney Tokyo Toronto

Publisher: James M. Smith
Editor: Robert J. Callanan
Assistant Editor: Kristin M. Shahane
Project Manager: Carol Sullivan Weis
Production: Publishers Services, Inc.
Design: Matthew Doherty Design

Printed in the United States of America.

Composition by Black Dot Graphics
Printing/binding by Von Hoffman

Mosby-Year Book, Inc.
11830 Westline Industrial Drive
St. Louis, MO 63146

International Standard Book Number 0-8016-5850-0

*This book is dedicated to my daughter Sophia
and to the memories of Mark Skubick
and Richard Lytie.*

Preface

Humans are interested in learning about human biology. Television programs, movies, and radio reports cover a wide range of human biology topics and issues and vividly attest to our continued interest in matters of human biology. There are dozens of professional journals and magazines published each year whose primary subject is human biology. These publications cover topics ranging from the findings of continued research in the areas of human evolution and genetics to topics concerned with the maintenance and improvement of human health including nutrition, weight loss, physical fitness, cancer prevention, and aging. Many US medical schools and public health centers publish newsletters and bulletins with readerships over 100,000 that survey current research and important discoveries made in the health and medical sciences. Furthermore, it is not unusual, in any given month, to find that the *New York Times* best-seller list includes one or even two books concerning some topic on human biology.

One reason biology is and always has been the focus of human interest and the media is that many of today's important social problems and political debates often involve fundamental human biological issues. Some of the current problems and debates involving human biology include the establishment of public health policies regarding control of infectious diseases, such as AIDS and tuberculosis; what types of basic and applied medical research should be government funded; what diagnostic procedures and types of diseases and injuries should medical insurance companies cover; the ethics of genetic counseling as a means of preventing the conception of offspring with genetic disease or disability; whether or not biological evolution should be taught in public schools; drug testing in the workplace; and the ongoing problems of environmental degradation and pollution. These issues may affect us personally or as a society, now or in the near future, and all of these issues involve biology. While we are naturally curious about human biology, it is also necessary today to be knowledgeable about basic biological principles to be informed and responsible citizens.

When learning basic biology in the classroom, students are introduced to a body of knowledge they perceive as having a language of its own made up of unpronounceable and intimidating words. Many students look at biology as a science that follows a logic of inquiry obsessed with the discovery of the smallest and most obscure facts. Students often feel that an overwhelming number of facts must be learned to understand the simplest biological structure or function. Furthermore, many come away from the experience unclear or unconvinced that much of what was learned has anything to do with understanding human biology.

Despite the need for a basic understanding of biology and the student's apprehension at tackling the subject, a solution does suggest itself. Perhaps students are best served if they begin by exploring concepts and principles or the *ideas* of biology. By first developing a conceptual framework, or perspective, the facts of biology and human biology can be learned and understood.

This book is about the biology of humans, from the functions of our bodies to our relationships with other living organisms. HUMAN BIOLOGY provides students with the tools necessary to understand current biological facts, as well as those yet to be discovered.

HUMAN BIOLOGY is intended to educate an important audience, the nonscience major in a one-quarter or one-semester course. The important topics of HUMAN BIOLOGY are presented clearly and concisely, allowing the student to grasp the major concepts without being overwhelmed by unnecessary detail. Essential concepts and principles are reinforced with key illustrations and human applications and analogies, assisting the reader in visualizing how human biology *works.*

ORGANIZATION

HUMAN BIOLOGY guides the student from the level of basic cellular biochemical processes to the basics of inheritance and evolution, the structure and function of tissues, and the integration of the various body systems. Chapter 1 introduces human biology and gives an overview of the characteristics shared by living organisms. Also included in this chapter is a description of the scientific method, including examples illustrating essential terms and concepts. An emphasis on scientific methodology is found throughout the text and provides students with a method of investigation useful during their study of human biology and beyond. Chapters 2 and 3 cover basic chemistry and cell structure and function, respectively. Chapters 4 and 5 build on the student's knowledge of cellular chemistry and provide thorough coverage of the patterns and biochemical basis of inheritance. Chapter 6 focuses on the process of human evolution. Unlike other human biology textbooks, this text presents genetics and evolution (Chapters 4, 5, and 6) early in the text. These topics, along with the introductory chapters, provide the student with a biological perspective and the historical context with which to proceed in the study of human biology. Biological principles and concepts introduced in the beginning chapters are emphasized and reinforced throughout the text and where appropriate, expanded upon in the context of human biology. Chapters 7 through 18 describe the systems of the human body. Human development through adolescence is discussed in Chapter 19, while Chapter 20 is dedicated to adulthood and the aging process. Chapter 21 explores human ecology and human impact on the environment. The nature of human biological variability is covered in Chapter 22. This chapter documents the range of human variability of several traits and objectively interprets differences among individuals and populations. A major portion of the chapter is devoted to the question of human race and the assessment of its biological validity.

PEDAGOGY

The following pedagogical elements of HUMAN BIOLOGY focus students' attention on the important concepts of the chapter and help them organize the material and review for exams.

For Review and Objectives

Each chapter begins with a list of terms and concepts described in previous chapters and the page numbers from those chapters. The **For Review** terms allow students to quickly review terms that they should be familiar with before progressing into the current chapter. A number of learning **Objectives** begin each chapter providing the student with a quick preview of the chapter's main points. The **Objectives** also aid the student in organizing the important concepts of the chapter.

Chapter Introductions

Each chapter opens with a brief preview of the topic to be covered. Many **Chapter Introductions** begin with a question about the chapter topic. This approach not only helps students gain a perspective of the chapter but also actively encourages them to find connections between the question posed and the chapter content.

Vocabulary Aids

Important terms are presented in **Boldface Print** throughout the text, allowing the student to quickly identify new or important vocabulary. The boldface terms are defined when first used in the chapter. Phonetic spellings are provided for most key terms, and the Greek or Latin derivation is given where appropriate. Essential terms and concepts and their page numbers are listed in the Chapter Review section at the end of each chapter in a **Vocabulary Review.** Students can easily test their understanding of key terms and concepts and assess their performance by easily referring to the page where the term is defined and described. For easy reference, the **Glossary** contains the most important terms from the text as well as phonetic spellings and page references.

Colored Key Concepts

Key concepts in the text are highlighted in blue type for the student. Highlighted Key Concepts provide immediate visualization of important concepts and assist students in studying and reviewing material for exams.

Illustrations/Design

A **Full Color Art Program** helps students visualize important concepts and processes, enhancing the text description. Some **Unnumbered Figures** are run right into the text, allowing immediate visualization of topics just described in the narrative. The unnumbered figures allow students to focus on the reading instead of searching for figures. The unique, eye-catching design and layout is user-friendly and will actively engage the reader.

Biology in Action Boxes

Biology in Action boxes are scattered throughout each chapter and discuss clinical, historical, and topics of general human interest. These asides are immediately relevant to the content of the chapter, illustrating chapter material in a way that stimulates student interest.

Chapter Review

Each chapter ends with a **Chapter Review** which includes a narrative **Summary** highlighting the essential concepts of the chapter. The Chapter

Review also includes **Fill-in-the-Blank Questions** which allow students to master chapter content and to prepare for exams. Answers to the Fill-in-the-Blank Questions are found in Appendix A. **Short Answer Questions,** also found in the Chapter Review, challenge the student to go beyond identification of facts and simple recall. These questions assume a particular understanding of the chapter content and require the student to synthesize the material in several different ways.

ANCILLARIES

This text is accompanied by several carefully developed ancillaries that support and enhance its conceptual approach. The ancillary package is designed to increase the text's effectiveness for both students and instructors.

Student Learning Guide

Developed by Carl J. Thurman of the University of Northern Iowa, the Student Learning Guide is designed to provide the reinforcement and practice essential for the student's success in learning human biology.

Human Biology Laboratory Manual

The laboratory manual by Randall Oelerich, M.D., consists of 38 labs of varying lengths, giving instructors alternatives as to amount of time needed to perform labs and as alternatives to dissection.

Instructor's Manual and Test Bank

The instructor's manual written by Michael Stearney of the University of Wisconsin, Green Bay, assists instructors in the preparation and development of a well-organized course. The instructor's manual contains a chapter overview, lecture outline, and teaching tips for each chapter. The test bank consists of approximately 1,000 questions ranging in level of difficulty.

Computerized Test Bank

The test bank questions are also available on Computest, a computerized test generating system for IBM users. It has many features that make it easy for the instructor to design tests and quizzes. The instructor can browse and select questions for inclusion on an examination using several different criteria, including question type and level of difficulty. A similar test-generating system is available for Macintosh and Apple users.

Transparency Acetates

A set of full-color transparencies emphasizes the most important concepts and processes from the text. All of the acetates were selected from key illustrations in the text, providing the student with an invaluable connection between the text and classroom lecture.

Acknowledgments

I would like to thank the reviewers of HUMAN BIOLOGY for their insightful comments and constructive suggestions. Their comments were invaluable in the development of the text. I would like to extend my gratitude to Dr. Linda S. Kollett of Massasoit Community College (Massachusetts) who spent considerable time and effort searching for errors and ambiguities in the text in the final stages of production.

I have been fortunate in working with some extremely talented and enthusiastic professionals at Mosby-Year Book, Inc. I extend my appreciation to Kristin Shahane, Assistant Editor, whose excellent editing and organizational skills are evident throughout this text. Carol Sullivan Weis, Project Manager, kept the production of this book on course from its initial development to final printing. I also wish to thank Robert Callanan, Editor, for his well-timed encouragement and consistent perspective throughout the writing of this text. Lori Toscano of Publishers Services, Inc., whose responsibilities included coordinating many aspects of production, was consistently professional and also extended her skills and insights in assisting with many key editorial decisions. I also extend my gratitude to the artists at Publishers Services, Inc. and the production staff at Black Dot Graphics who created the excellent artwork and executed the innovative design for the textbook.

Joseph A. Mannino

Reviewers

Andy Anderson
Utah State University

James Carrel
University of Missouri - Columbia

Vic Chow
City College of San Francisco

Joe Colosi
Allentown College

Dale Erskine
Lebanon Valley College

Robert Elgart
State University of New York - Technological College

Sheldon Gordon
Oakland University

Stephen Grebe
The American University

Martin Hahn
William Peterson College

Laszlo Hanzely
Northern Illinois University

Clare Hays
Metropolitan State College

Ron Hoham
Colgate University

Michael Hudecki
State University of New York - Buffalo

Patricia Humphrey
Ohio University

Vincent Johnson
Saint Cloud State University

Linda Kollett
Massasoit Community College

John McClure
Saint Cloud State University

Donald Nash
Colorado State University

Lee Peachy
University of Pennsylvania

Charles Schauf
School of Science at Indianapolis

John Sherman
Erie Community College - North

Richard Sjolund
University of Iowa

Robin Tyser
University of Wisconsin - Lacrosse

F. William Vockell
Florida Community College

Jean Wagner
Mohawk Valley Community College

Contents in Brief

Contents

CHAPTER 8:

The Nervous System

CHAPTER 9:

The Senses

CHAPTER 10:

The Endocrine System

CHAPTER 11:
The Musculoskeletal System

CHAPTER 12:
The Circulatory System

CHAPTER 13:
Blood

CHAPTER 14:
Immunity

CHAPTER 19:
Human Development

CHAPTER 20:
Adulthood and Aging

CHAPTER 21:
The Human Environment

CHAPTER 22:
Human Variability

Human Biology

Chapter One

The Unity and Diversity of Life

Objectives

After reading this chapter you should be able to:

1. Explain the five general characteristics common to all organisms.
2. List and describe each of the three basic scientific principles.
3. Define hypothesis and explain the four activities involved in the scientific method.
4. Explain the three ways to determine causation.

A tremendous diversity of living organisms inhabits the world. All of them, including humans, share a number of characteristics that define the process called life. Organisms are composed of cells, are able to acquire energy and materials from their surroundings and use them to grow and maintain themselves, are able to reproduce, and are capable of adapting to ever changing environmental conditions. Some organisms are also able to capture the energy of the sun directly. This captured energy then continues to flow through and sustain all other forms of life on earth.

Unifying Concepts of Life Processes

When a human infant is only several months old, it spends a great deal of time simply observing the movement of its fingers and hands. A mixture of wonder and delight shines on the infant's face as, day by day, it discovers each new aspect of its body

and how it behaves. As the child grows and develops, observations extend beyond the self and other humans to include a larger world of living organisms. The child's new vocabulary begins to include categories and classifications for the surrounding world. Dogs are not the same as cats. All sorts of plants, trees, bushes, and flowers with interesting names and an almost endless variety of insects, some named and others not, march through the child's expanding world. Crawling and flying creatures appear and disappear from sight in an increasingly diverse parade of life. If taken to the zoo, on a nature walk, or through a museum, the child asks an almost unending stream of questions. Why were dinosaurs so big? Why aren't brothers and sisters the same? Is that what *I* look like inside?

As the child's awareness continues to grow, so does biological self-awareness. For example, the infant's primary response to a changing physical and physiological state was reactive. When hungry, tired, or uncomfortable, the infant cried until the uncomfortable condition was eliminated. Adults can more accurately determine the infant's state of physical being at this stage. For example, adults are usually able to diagnose certain illnesses and injuries, such as the flu or an injured knee, or what causes fatigue or hunger. From the very beginning of life, then, humans explore, compare, question, and learn about their own biology, as well as the biology of other organisms around them.

From all the diverse knowledge a person accumulates through living come some basic insights into life. Each of us recognizes some of the basic similarities we share with other living organisms. But how do we define life? A member of the clergy, a physician, a chemist, or a person with a relative on mechanical life-support systems may have very different answers to this fundamental question. We all have some insight into what it means to be alive. People around us are alive. Plants and animals are alive. In contrast, a desk and a personal computer are not alive. However, determining in what ways you are alive and your desk is not is difficult in part because living organisms are so varied and nonliving things sometimes possess some lifelike characteristics.

Human biology is the study of the structural, functional, and behavioral bases of the human organism's ability to adapt to and survive in its environment. Human biology also provides practical knowledge about ourselves, including information about nutrition, exercise, birth control, fertility, development, aging, and disease. Basic knowledge about human biology can assist you in dealing with such matters as communicating with health-care professionals or making an informed judgment about the potential effects of toxic waste in the environment.

Human biology also includes understanding our relationship to our environment and other living organisms. Our view of nature has changed over time. For example, one common Western European belief during the eighteenth century was that God created the world and all it contained exclusively for humans. During the nineteenth century Europeans developed economies not concerned with the long-term effects of depleting nonrenewable natural resources such as coal and oil or polluting and contaminating renewable resources such as air, water, plants, and animals.

By the mid–twentieth century it became apparent that many human activities were profoundly disturbing the environment. For example, oil was being consumed at rates that, if continued, would lead to severe shortages by the beginning of the twenty-first century. Forests were being depleted faster than they could be regrown. By 1988 scientists began to notice unusual world weather patterns. During the summer of 1988, for instance, many U.S. cities regularly recorded record-high daily temperatures of over 100° F. These high temperatures helped reduce U.S. agricultural production by 30%. Scientists detected holes in the protective ozone layer of Earth's atmosphere. Such evidence sparked a debate about global warming caused by human activities that came to be called the Greenhouse Effect. This debate is often complicated, controversial, and confusing. However, with a basic understanding of biological and ecological principles, one can, as an informed citizen, make rational decisions concerning such issues as the value of recycling, pollution control, and energy-conservation policies.

We begin here by revisiting our attempt to define life. Although no one has yet distilled the properties of life into a single equation or sentence, all organisms share certain general properties, ones we think must ultimately have been derived from the first organisms on Earth. We recognize other living things, and also what we mean by the process of life, by these general characteristics, five of which are common to all organisms on Earth.

Cellular Complexity and Organization

Compared with nonliving matter of the same size, living organisms are highly organized and complex. For example, a crystal of table salt consists of just two elements, sodium and chloride, organized geometrically. Salt is organized but simple. Earth's oceans contain atoms of all of the naturally occurring elements, randomly distributed. Oceans are complex but not organized. In contrast a single-celled organism, such as a paramecium, is much more complex and highly organized than an equal amount of salt or ocean water. The paramecium contains many different elements combined into larger chemical substances. These are further organized into subcellular structures designed to carry out specific tasks.

The living world also has a hierarchical structure, each level based on the one below and providing the foundation for the one above. The simplest

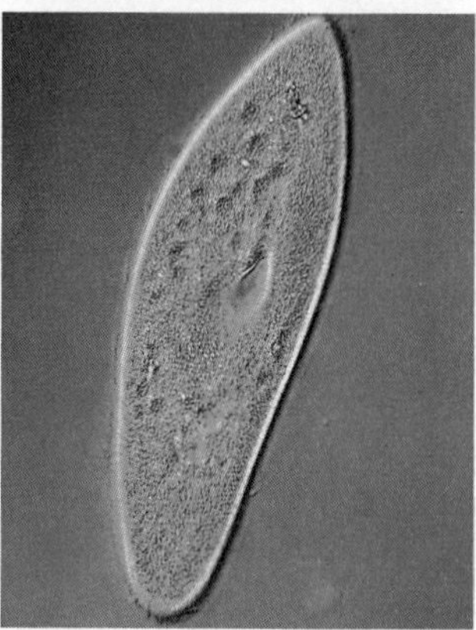

level is the **atom,** the smallest particle of an element that retains the properties of that element. All physical matter is composed of atoms. For example, a diamond is composed of carbon atoms. If it were possible to cut up a diamond into ever smaller pieces, each piece would still be carbon until the diamond had been separated into isolated carbon atoms. Any further subdivision would produce subatomic particles that would no longer be carbon.

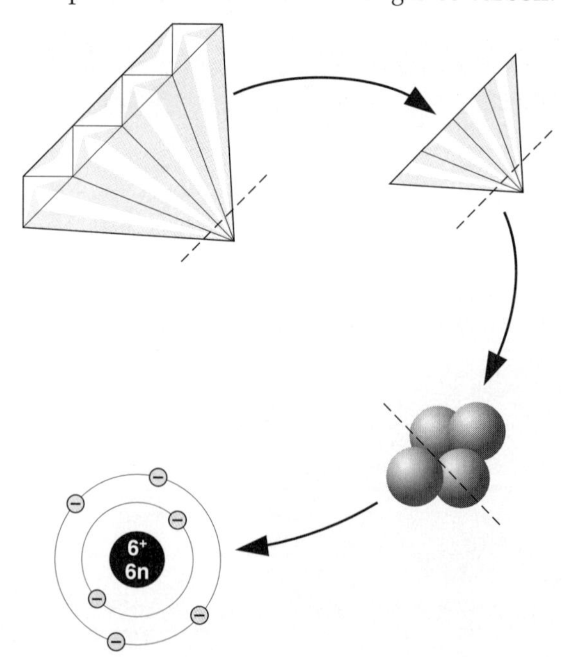

Atoms may combine in specific ways to form larger structures called **molecules,** the next level. For example, one oxygen atom and two atoms of hydrogen produce a molecule of water. Although many simple molecules form spontaneously in the environment, only living organisms produce large complex molecules.

As an atom is the smallest unit of all physical matter, the cell is the smallest unit of life. A **cell** has several components, including hereditary material with the information needed to control the cell's life and subcellular structures that use this encoded information to carry out essential life activities. These subcellular structures are surrounded by and suspended in **cytoplasm,** a watery solution inside of all cells. Cells also have a thin membrane separating them from the outside world. The simplest organisms consist of a single cell and are usually microscopically small. Larger organisms are **multicellular,** composed of many cells. Your body, for instance, is composed of approximately 100 trillion cells.

Multicellular organisms also have a hierarchical structure and organization. In humans, for example, cells of similar type form **tissues,** such as bone or muscles. A variety of tissues, such as nervous and connective tissues, combine to form a structure, having a specific function, called an **organ.** The human brain, for example, has both nervous and connective tissue. Several organs collectively performing a single function, such as the brain, spinal cord, and sense organs, create an **organ system.** All of your body's organ systems function cooperatively and collectively, making up an individual living entity called an **organism.**

A hierarchical organization also includes and extends beyond the individual organism. A group of organisms of the same kind living in the same area is called a **population.** Populations of different kinds of organisms living and interacting in the same area make up a **community.** Living communities and their nonliving surroundings, such as rocks, water, and soil, make up an **ecosystem,** or **biome.** Finally, Earth's surface and all living organisms on it constitute the **biosphere** (Figure 1-1). These concepts are more fully explored in Chapter 21.

Metabolism

Organisms are able to acquire energy and materials from their external environments, convert them into new forms, and use these forms to grow, in a process called **metabolism** (me-tab′o-lizm). Plants and some other organisms use the sun's energy to build complex molecules from simple molecules of carbon dioxide (CO_2) and water (H_2O) through a process called **photosynthesis** (fo-to-sin′the-sis). The complex molecules produced by photosynthesis are then used to build and maintain the organism's structure and organization. All other organisms obtain their energy by consuming photosynthetic plants or organisms directly or by consuming one another, forming a many-layered hierarchy through which energy constantly flows. Organisms, including humans, maintain their life processes by using chemical energy ultimately

produced by photosynthesis. In organisms' cells, through the process of metabolism, this energy is systematically released and then transferred from place to place by means of a small energy carrier molecule called **ATP,** which we discuss in Chapter 3.

Homeostasis

Maintaining relatively stable internal conditions under changing external environmental conditions is called **homeostasis** (ho′-me-o-sta′sis). Homeostasis operates at all levels of a multicellular organism. For example, each human body cell maintains a particular concentration of water and other molecules within itself. The cell's structural and functional mechanisms maintain these vital conditions. The human body in turn maintains an internal temperature of 98.6° F, regardless of the external environmental temperature. At or very close to this temperature, individual cells, as well as vital organs such as the brain and heart, can function properly. Temperature balance is accomplished through biological mechanisms involving the interactions of cells, tissues, organs, and organ systems. Homeostasis and the body's organization are described in Chapter 7.

Reproduction

Although some organisms may live many years and others only minutes, all organisms eventually die. For life to continue organisms must be able to reproduce new organisms similar to themselves. How is it, for example, that bacteria produce only bacteria or that a geranium seed germinates to produce a similar geranium? This transmission of characteristics from parent to offspring is called **heredity** (he-red′i-ti). Heredity is possible because most organisms are capable of replicating a complex mole-

Figure 1-1 Hierarchical organization of life

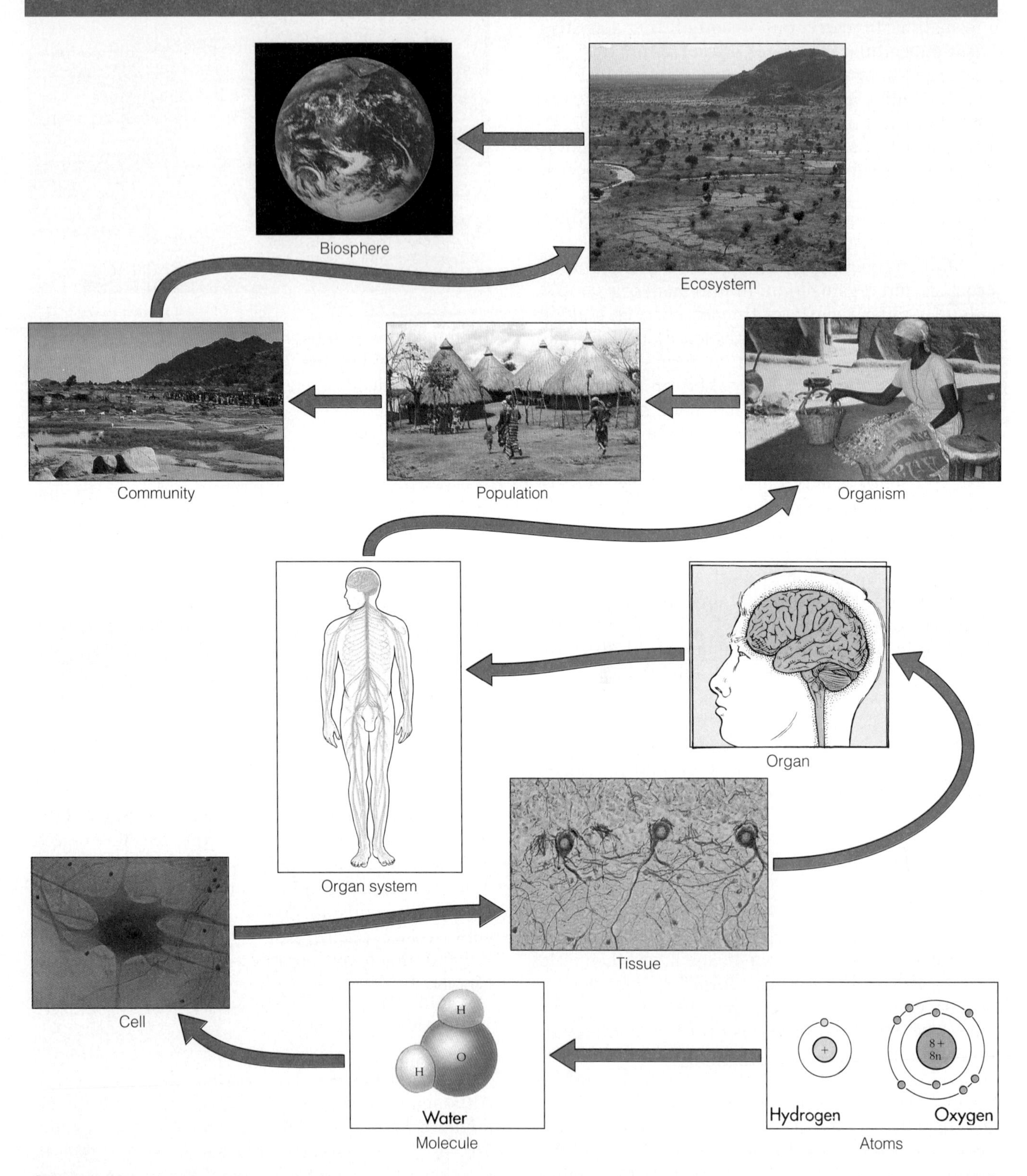

There is a hierarchy of organization within organisms, beginning with atoms that form molecules which in turn form the basic building block of life called the cell. Cells form tissues which, in combination with other tissues, form organs. Groups of organs form organ systems that compose an individual organism. A hierarchical organization also includes and extends beyond the organism. Organisms belong to a population within a community of different types of organisms which make up an ecosystem that belongs to Earth's biosphere.

cule called **DNA.** The order of the subunits, called nucleotides, that make up DNA determines what an individual organism will be like. We discuss the nature of the hereditary material in Chapter 2.

Adaptation, Change, and Evolution

Adaptation is any variety of structure or behavior that increases the likelihood of an organism's survival and the number of offspring it is able to produce relative to other individuals in its environment. Adaptation is the basis for the ability of all life to change and diversify through time. Adaptation is the driving force of **evolution,** the permanent heritable changes accumulated in a population of organisms through time.

For any trait in a population of organisms there will be a range of variation. For example, in a population of humans each person is a particular height but not all people are the *same* height (Figure 1-2). If a particular variety of a trait enables an organism to survive and reproduce in greater numbers relative to other organisms in the population with other varieties of that trait, the number of individuals in the next generation with the favorable variety will increase. If this process continues for several generations, the frequency of the advantageous trait in the population increases. More individuals in succeeding generations will thus come to have a particular variety of the trait as opposed to other varieties. Although each organism in a population may not evolve, the frequency or number of organisms with a particular variety of a trait changes through time. Eventually the descendant population may differ substantially from the ancestral populations of many generations ago.

All organisms require a constant supply of energy and materials from their environments. However, the energy and materials any organism requires are limited and often in short supply relative to the numbers of other individuals in the same popula-

Figure 1-2 Variation among individuals of a population

There is variation for many human characteristics. For example, within a group, individuals will be different heights.

tion. If a particular variety of a trait allows an individual organism access to greater portions of or more efficient use of the limited energy and material, then that individual is more likely to survive and reproduce more offspring relative to other organisms in the population with other varieties of that trait. This process, first described by Charles Darwin in the nineteenth century, is called **natural selection.** Natural selection results in producing organisms adapted to the particular conditions of their environment.

Principles and Methods of Science

Human biology is a science because it shares the same principles and methods of investigation as those of any other science, such as chemistry, physics, or astronomy. What, then, is science?

Every means of explaining the world around us begins with a set of assumptions or statements. Science is based on three general assumptions, called

Biology in Action: VIRUSES—LIVING EXCEPTIONS

As is often the case, nature does not adhere to human definitions, categories, or systems of classifications. It is what it is. For example, consider our attempts to define life. As you have just learned, all living things, except viruses, share several common characteristics. Viruses are not cellular, and they do not possess their own means of metabolism. This means that they cannot release energy or synthesize molecules and cannot replicate or grow on their own. Instead they are **parasites** (Gr. *parasitos* = guest), living in another life-form or **host.** Cells are the hosts of viruses, providing the energy, materials, and means to synthesize new viruses. Viruses are extremely host-specific. For example, viruses that infect multicellular organisms require not only a specific organism but also a specific tissue or cell type in the organism for their survival.

Although viruses exist in many shapes and sizes, they are relatively simply organized and all share many characteristics. For example, all viruses consist of at least one strand of hereditary material. This material is contained in a core surrounded by a protein coat and is usually surrounded by an outer envelope. Most viruses are geometrically shaped and are commonly threadlike or spherical with multiple faces or sides. How do viruses reproduce and carry on their life cycle in a particular host? First they must attach to a particular cell. They then enter the cell where they disassemble or uncoat, exposing their hereditary material. Depending on the viral hereditary material, two different alternatives are possible. In one case the viral hereditary material combines with and becomes part of the host cell's hereditary material. When the host replicates its hereditary material it will also replicate the viral hereditary material and reproduce the virus along with itself. The other alternative is that once inside the cell, the viral hereditary material simply directs the construction of new viruses using the metabolic machinery of the host cell. In either case, after the viruses reproduce they leave the host cell to infect other cells of the same type.

Viruses cause numerous diseases in humans, including the common cold, measles, smallpox, warts, and, of particular concern today, acquired immunodeficiency syndrome (AIDS). Viruses are common to many diverse life-forms, infecting bacteria, plants, fish, mammals, and even insects.

The origins of viruses are uncertain. However, it is unlikely that viruses are the forerunners of life because they cannot reproduce without infecting more complex cells. Some biologists think viruses originated from simple parasitic cells that over time evolved such complete dependence on their hosts they lost the ability to perform basic processes of life. Whatever their origin their success today continually challenges all living organisms.

scientific principles (Table 1-1). The three basic scientific principles are natural causality, uniformity in space and time, and common perception.

Historically people have used two different approaches to explain causality, or cause and effect.

Table 1-1 Principles of Science

PRINCIPLE	EXPLANATION
Natural causality	All events can be traced to preceding natural causes. Every event has at least one discoverable cause.
Uniformity in time and space	For example, explanations for the present biological diversity on Earth are also used to explain life-forms no longer present on Earth.
Common perception	All humans perceive natural events through their senses in the same way.

One approach assumes that events occur through the intervention of supernatural forces, usually thought to be beyond human understanding. For example, the ancient Romans believed that when a woman gave birth to identical twins it was because she had been visited by one of the gods.

In contrast to this supernatural causality is the assumption made in science called **natural causality**, which states that all events can be traced to preceding natural causes, every event having at least one discoverable cause. Today we know one cause of twinning in humans is that the fertilized cell divides into two separate cells, which develop independently to form two identical individuals.

Our confidence in the natural causality principle is reinforced by the number of causes scientists have isolated for events formerly thought to be supernatural, such as earthquakes, many diseases, the movement of celestial bodies, and pregnancy.

The second basic scientific principle is that natural explanations or laws do not change with time or distance. Astronomy, for example, provides a valid explanation of the apparent movement of the sun across the sky. The astronomical laws of planetary motion, first noted more than 350 years ago, were valid then, are now, and presumably will be in the future. These laws are true in London, New York, or Mexico City. The calculations required to send people to the moon or satellites around Venus were successfully based on these astronomical explanations. The assumption of uniformity in time and space is essential to biology because many events, such as the present diversity of life forms on Earth, happened before humans were around to observe them.

The third basic scientific principle is that we all share **common perception.** All humans perceive natural events through their senses in the same way. If a giraffe were to join students during a lecture, all of the students still awake, able to see, and looking would see the giraffe. Furthermore, the students' other senses would also agree with what they saw, and they could feel, hear, and smell the giraffe.

Common perception is a concrete, specific scientific assumption. Value or moral systems are not so exact. For example, we may each perceive the notes of a musical score in a similar way, but we do not perceive the aesthetic value of the music identically. However, the *kinds* of scientific questions we ask are guided by moral and ethical values. Today much of science is focused on improving technologies that increase financial wealth and power, often at the cost of harming the environment. How we deal with these problems depends on what we believe to be right or important, which reflects a particular moral system.

Science has also produced results that many find disturbing. How best to use the results of science is also a decision based on moral values. Moral systems influence, for example, what kinds of scientific research get funding from the federal government and what kinds from private sources. Thus is it more important to fund weapons research, the NASA space program, or a cure for cancer?

Scientific Method

Applying these three scientific principles, how do biologists study human life? Ideally any science uses the **scientific method,** which is the experimental testing of a hypothesis. The scientific method

consists of the four related activities of observation, the formulation of a hypothesis, experiment, and conclusion.

Every scientific endeavor begins with an observation. Then a **hypothesis** (hi-poth'e-sis) is proposed. A **hypothesis** states that a particular preceding event is the cause of a particular subsequent observation, that event A causes event B. A hypothesis may be proven true or false. To be proven true a hypothesis must be able to predict the outcome of further observations or experiments. Failure to correctly predict an outcome would indicate the hypothesis is false. Then based on the results of the experiments or on further observations, a conclusion is made. The simplest way to test a hypothesis is to state that a single preceding event is the cause of a single subsequent observation, usually as an "if . . . then" statement. The most difficult, time-consuming, and challenging part of the endeavor, however, is designing an experiment to demonstrate this.

The steps of the scientific method may seem to be regimented, but they are really little more than organized common sense, much like what you routinely use every day. Here we use two examples to illustrate the scientific method. First is a simple and common experience. Second is an important historical experiment.

One day you go to your car and discover it will not start (an observation). You would then ask, why? (identify the problem). By observing and identifying the problem, you might propose a testable hypothesis, such as the car is out of gas. If the car is out of gas, then the gas gauge should read empty (a hypothesis). You would then perform a simple experiment. In this case you simply look at the gas gauge. If the gas gauge reads empty, then you have confirmed your hypothesis. If, however, it reads half full, your hypothesis is false. You would then discard the "empty gas tank" hypothesis and ask new questions to create a new hypothesis.

Our second example is an experiment originally conducted by Louis Pasteur (1822–1895). Even though as early as 1688 some experiments cast strong doubts on the notion of **spontaneous generation,** the production of living organisms from nonliving matter, some in the mid–nineteenth century still strongly adhered to this notion. People were able to demonstrate, for example, that even when organic broths were boiled for a prolonged time and then left to cool in the open air, within 48 hours molds or fungi would appear in the liquid. Its cause was said to be gases in the air. Oxygen was thought to organize certain kinds of nonliving matter in such a way as to produce living organisms. In 1862 Pasteur conducted a series of experiments that finally refuted the notion of spontaneous generation.

Using his observations of mold growth in organic broths, Pasteur proposed a testable hypothesis. He thought that living airborne particles or "germs" caused molds to grow in organic broths, not oxygen or some other atmospheric gas. Pasteur proposed that if airborne particles were kept away from the liquids, molds would not appear.

In scientific experiments a potentially important factor is called a **variable.** In experiments there are two types of variables. The first one is called the **independent variable,** which is the variable controlled and manipulated by the experimenter. In this experiment, the airborne particles that Pasteur believed to be the causative agent of mold growth comprise the independent variable. The other type of variable is called the **dependent variable,** which is influenced by or dependent on the independent variable. The dependent variable in this experiment is the appearance of mold in the broth.

To test a hypothesis about a variable, it is important to hold all other known variables constant so the experiment does not produce confusing or misleading results. One method that eliminates this potential problem is to carry out two experiments in parallel. The **experimental group** is the group in which the variable of interest is to be altered or manipulated to test the stated hypothesis. The **control group** is identical to the experimental group, except for the single variable being tested. In every other way the experiments are identical. In this way any difference observed between the two experiments must result from the influence of the altered variable because all of the other known variables are identical (Figure 1-3).

Pasteur first prepared an organic broth containing a quantity of sugar, water, and protein and then separated it into two identical flasks. He took both flasks and heated them to the same temperature for the same amount of time. One was the **experimental** flask. In it the broth was exposed for 5 minutes to air that had first been forced through a cotton filter to remove any airborne particles. The flask was then sealed. The other flask was the **control** flask. In it the broth was exposed to unfiltered air for 5 minutes and then sealed. Except for the quality of the air, the variables were the same.

After 72 hours the broth exposed to filtered air was unchanged. No mold or fungus had grown in it. The control flask, however, was covered with mold. Pasteur concluded his hypothesis was correct and that organic broths did not spontaneously generate life because they were exposed to oxygen. Controlled experiments such as this finally refuted the incorrect belief in spontaneous generation.

Scientists cannot always be certain all variables other than the one being studied have been controlled. For example, although Pasteur's cotton filter

Figure 1-3 Controlled experiment

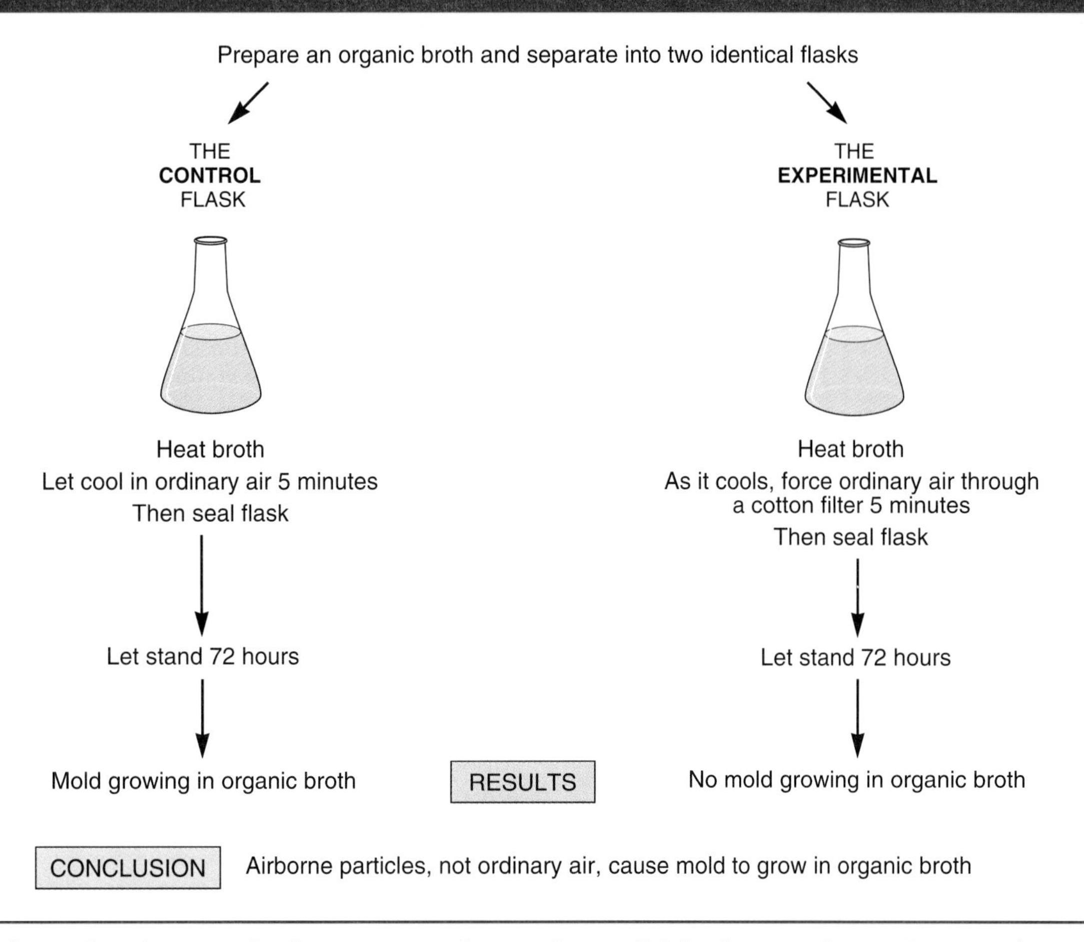

This experiment involves conducting two experiments in parallel. In the experimental group, the independent variable, in this case, the airborne particles thought to be the cause of mold growth, is tested. The control group is identical to the experimental group, except for the single variable being tested.

kept away any airborne particles, perhaps some other unknown gas in the air was also filtered out that was responsible for growth or particles smaller than what the cotton filter could trap got through. Although the experiment supported Pasteur's hypothesis, it did not rule out all other possible hypotheses. Because of this kind of uncertainty all scientific conclusions must remain tentative and subject to revision if new observations arise or other experiments are tried.

Sometimes, controlled experiments are more complicated than in this case. We can look at how a drug was discovered to be effective in prolonging the lives of people infected with the human immunodeficiency virus (HIV) and having acquired immunodeficiency syndrome (AIDS). In testing the effectiveness of this drug on HIV the experiment was designed as a double-blind controlled study.

AIDS is thought to be caused by a virus that affects and dramatically reduces the body's ability to fight disease and infections (see Chapter 14). Most people infected with this virus eventually die of **opportunistic infections,** diseases the human body can normally fight off but infect people with AIDS. In laboratory experiments using isolated human cells infected with HIV, a drug called azidothymidine, or AZT, was discovered to interfere with the virus's reproduction. The next step was to test the drug's effectiveness in infected humans.

People already infected were asked to volunteer for the study. A code number was assigned to each, and then a computer randomly picked who was to receive either AZT or a **placebo** (pla-se'bo), a substance that looked and tasted just like AZT but was inactive. Placebos are often used in controlled medical experiments because subjects and sometimes researchers will report an improvement in health simply because the subject took something for the problem. Using placebos helps avoid this psychological bias.

Figure 1-4 Double-blind controlled experiment

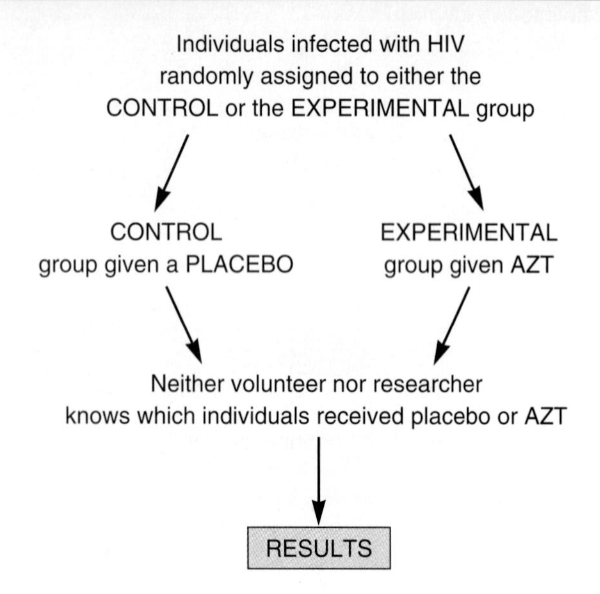

Individuals infected with HIV
randomly assigned to either the
CONTROL or the EXPERIMENTAL group
CONTROL
group given a PLACEBO
EXPERIMENTAL
group given AZT
Neither volunteer nor researcher
knows which individuals received placebo or AZT
RESULTS
Individuals given AZT had significantly lower rates of
opportunistic infections compared to individuals
given the placebo
CONCLUSION
AZT significantly lowers replication rate
of HIV in humans

This experiment is similar to the controlled experiment except that neither the investigators nor the subjects know who belongs to the experimental group and the control group.

In our example neither the subjects nor the researchers knew which volunteers received the AZT or the placebo. Hence we use the term *double-blind experiment.* Both the experimental and the control groups were otherwise identical (Figure 1-4).

The results of this experiment showed that subjects who received AZT had significantly lower rates of opportunistic infections than did those taking the placebo. This was because AZT interfered with the reproduction of HIV, reducing its spread through the immune system. With these findings the experiment was discontinued and the control group was also given AZT.

Association, Causation, and Proof

When the scientific method is used to discover or explain relationships between events, sometimes a simple association of two events can be confused with causation. Causation may operate in either direction in the relationship, or it may not directly link the two events at all. For example, it would be a mistake for a farmer to think a rooster crowing every morning causes the sun to rise. It is sometimes difficult to unravel causation in the real world, especially if many variables are involved in causing a certain observation or event. In spite of overwhelming evidence that smoking tobacco causes lung cancer, some tobacco companies claim that cigarettes do not promote lung cancer, but that people susceptible to lung cancer are more likely to smoke cigarettes. This statement at first may appear nonsensical, but because many variables are involved in the actual development of lung cancer, it has not been possible until recently to confirm a simple hypothesis stating that smoking tobacco is *the cause* of lung cancer.

When there are problems in which several variables are suspected to cause a particular outcome, such as lung cancer, causation may be determined in three ways. The first way to determine causation is to establish a relationship in time. Did tobacco use occur early enough to cause or increase the likelihood of lung cancer? The second way to determine causation is to establish the consistency of the association. Does the association of tobacco smoking and lung cancer exist consistently among many different populations? The third way to determine causation is to use animal studies. Does the evidence from animal studies support observations concerning human populations of smokers? Many apparently contradictory scientific statements people read or hear about the causes of many human diseases come largely from experiments testing either single variables or different combinations of variables under different conditions, each leading to somewhat different conclusions. Scientific theories are developed to try to untangle such situations.

Scientific Theories

The meaning of the term *theory* as it is used in science differs from its meaning as most people might use it in daily life. For example, you might have a "theory" about why your favorite sports team did not have a winning season or why you did not win the state lottery last week. You are actually saying something such as "I have a hunch" or an opinion about those events. In science as knowledge accumulates and scientific verification increases, certain hypotheses are tested many times. Each time, for example, an object falls to Earth rather than up into the sky, the hypothesis concerning the gravitational attraction between masses is supported. When many independent experiments or observations all support a hypothesis or a group of hypotheses that explain a large number of events, such as gravity, these hypotheses are called a **theory.**

In science a theory is a hypothesis supported by so many observations or experiments that few scientists doubt its validity. For instance, Charles Darwin's original observations and hypotheses about the forces that determine evolution have been confirmed and have grown to become the theory of evolution by means of natural selection (see Chapter 6). When scientists talk about evolutionary theory, they are referring to a well-established body of observations and experimental proofs that explain much about life on this planet. Since Darwin's day the theory of evolution has been supported by such evidence as fossils, geological studies, radioactive dating, genetics, biochemistry, and population breeding experiments. The explanatory power of this theory is so great that as Theodosius Dobzhansky, an evolutionary biologist, once stated, nothing in biology makes sense except in the light of evolution.

Your Study of Human Biology

The topics introduced in this chapter and their order of presentation provide a framework for organizing the material you are about to learn. We begin by describing the basic units of matter called atoms and learn how and under what conditions larger organic molecules are formed. We investigate the basic unit of life, the cell, and outline its structure and functions. We then discuss the pattern of heredity, the manner in which traits are passed from parent to offspring. Following this we describe the biochemical structure and function of the hereditary material that ultimately determines traits and their expression in an individual.

We then present the process of evolution and the forces known to cause change in populations of organisms through time. Several important aspects of human biology are also specifically viewed from an evolutionary perspective.

We discuss the specific tissues, organs, and organ systems that make up the human body and how human variability is determined. Finally we conclude by studying the interactions of humans with their communities, ecosystems, and Earth's biosphere.

We emphasize the methods and logic of science and an evolutionary perspective of human biology throughout this text. The scientific method is a powerful tool in helping us to understand ourselves and our surroundings. Whereas, for example, it is one thing to recognize a relationship such as "sexual intercourse may result in pregnancy," it is quite another to understand the biological basis of that relationship. Recognition allows us to predict an outcome, but understanding the *relationship* enables us to have more control over the world around and within us. This is true whether we are trying to understand the probability of passing on a particular trait to future offspring, the cause of lung or breast cancer, the impact of cutting down tropical rain forests, or the behavior of children in reaction to parents' behavior.

A Biological Perspective

We cannot understand either ourselves or any other biological organisms in isolation from their surroundings or their biological history. Knowledge of evolution allows us to view ourselves and other organisms in an environmental and historical context. A variety of biological traits make us distinct from our closest biological relatives, the chimpanzees and gorillas. Two of those traits include upright posture and bipedalism (walking on two legs), and having a relatively large brain. Why are we bipedal? To answer this question we must first determine the structure and function of bipedalism and then attempt to reconstruct the biological history of its development. While doing so we also come closer to discovering the environmental conditions that selected for increasing brain size and the development of human intelligence. By reconstructing the evolutionary history of bipedalism, we may also find answers to other questions concerning human biology.

An issue of more immediate concern also illustrates this approach. What is the origin of AIDS? Why is it spreading now in human populations? We can begin to answer these questions by viewing the virus that causes the disease and the human response to it in a way similar to answering the question of bipedalism. First we describe the biology and evolution of the disease-producing agent, HIV. Does it or a similar virus exist in other primates or mammals? If a similar virus exists in other organisms, in what ways is their biological response similar to or different from ours and why? By understanding the underlying principles shared by all living organisms, we can discover and understand the uniqueness of human biology at many different levels.

Biology in Action THE GAIA HYPOTHESIS

James Lovelock (1979), a British biochemist, and Lynn Margulis, an American molecular biologist, proposed the intriguing idea that "the physical and chemical condition of the Earth's surface, of the atmosphere, and of the oceans has been and is actively made fit . . . by the presence of life itself." This idea has been called the **Gaia** (Ga′uh) hypothesis, named after the Greek earth goddess, Gaia, from whom all other Greek gods were born.

To support the Gaia hypothesis these two scientists cite evidence that living organisms interact with and significantly affect Earth's atmosphere and chemical cycles. Lovelock in particular also maintains that all life has evolved into a kind of "global superorganism," called Gaia, whose parts affect and adjust to various environmental conditions, such as carbon dioxide and oxygen levels. In other words, a system of checks and balances among all living organisms enables life to continue.

Other evidence used to support the Gaia hypothesis is that Earth's atmosphere and oceans act like global circulatory systems, carrying chemical substances across the planet and distributing them as needed. Both living and dead plants and animals process and temporarily store such important substances as oxygen and carbon dioxide, releasing them and affecting the Earth's temperature and atmosphere.

Homeostasis is the term used to describe the ability of an organism to maintain stable internal conditions under changing external environmental conditions. Lovelock and Margulis propose that Gaia has actively maintained planetary homeostasis since the first appearance of life on Earth.

The Gaia hypothesis is controversial and is actively debated among scientists. Regardless of the debate's final outcome, the Gaia hypothesis provides a useful alternative perspective. It reminds us that organisms interact with and alter the environment in many ways. The changing composition of Earth's atmosphere and changes in weather have, for example, been partly a consequence of human activities. What will come of these changes and how they will affect our lives remain unknown.

CHAPTER REVIEW

SUMMARY

1. Living organisms are highly organized and complex structures compared to nonliving matter of comparable size. Within multicellular organisms there is a hierarchy of structural organization beginning with the basic unit of life called the cell. Cells of similar type are organized into tissues, organs, organ systems, and, finally, the organism.
2. The basic principles defining matter, life, and the hierarchy of its organization are based on the ability of some organisms to capture and utilize energy first derived from the sun through the chemical process of photosynthesis. Through the process of cellular metabolism, organisms are able to release the stored energy of molecules and use it to grow and maintain the organisms' own structural organization.
3. The diversity of living things comes about as a result of the variability in populations of organisms, the limited resources upon which they rely, and the capacity of organisms to reproduce. The diversity of form, function, and behavior in living organisms represents specialized ways of obtaining and utilizing energy and materials.
4. The mechanism explaining the diversity of living things and how they change through time, or evolve, is called natural selection. Any variety of a characteristic allowing an organism to gain a greater share of limited resources or to utilize resources more efficiently increases the likelihood of that organism's survival and its production of offspring compared with other organisms in the population.
5. The outcome of the process of natural selection is adaptation; populations of organisms

are better suited to survive in the environment they inhabit. Populations of organisms are always in the process of adapting to their environments because no environment remains constant through time but is always changing in some way.

6. Human beings share with other living things the same fundamental characteristics that define life processes and are subject to the same forces governing evolutionary change. We too have a biological history and continue to change through time in response to a changing environment.
7. Human biology is a science and therefore determines the nature of causal relationships between naturally occurring events through use of the scientific method. This method rests on three assumptions regarding the world: natural causality, natural explanations that do not change with time or distance, and humans' sharing a common perception.

FILL-IN-THE-BLANK QUESTIONS

1. The smallest unit of life is called a ______.
2. Many communities and their nonliving surroundings constitute an ______, or ______.
3. The ability of living organisms to actively maintain relatively stable internal conditions is ______.
4. The assumption that all events can be traced to preceding natural causes is referred to as ______ ______.
5. Every scientific inquiry begins with an ______ of the world.
6. An ______ variable is the causal agent in an experiment.
7. A statement that a particular preceding event is the cause of a particular subsequent observation is called a ______.
8. ______ ______ is the belief that living organisms can be produced from nonliving matter.
9. For any characteristic in a population there will be a range of ______ present.
10. The process that uses sunlight to build complex molecules from simple molecules of carbon dioxide and water is ______.

SHORT-ANSWER QUESTIONS

1. What is the relationship between adaptation and evolution?
2. How does the diversity of living things come about?
3. What distinguishes a hypothesis from other kinds of questions?
4. In Pasteur's experiment what was the independent variable? The dependent variable?
5. Why is it difficult to sometimes determine causation in science?
6. Compare and contrast supernatural causality and natural causality.

VOCABULARY REVIEW

control group—*p. 10*
experimental group—*p. 10*
homeostasis—*p. 5*
hypothesis—*p. 10*
metabolism—*p. 5*
natural causality—*p. 9*
placebo—*p. 11*
scientific method—*p. 9*
theory—*p. 13*
variable—*p. 10*

Chapter Two

CHEMISTRY AND LIFE

OBJECTIVES

After reading this chapter you should be able to:

1. *Discuss the relationship between elements, atoms, molecules, and compounds.*
2. *Differentiate between the subatomic particles of an atom and among their organization within the atom.*
3. *Explain what it means for an atom to be electrically balanced and describe the electrical charge of ions.*
4. *Define and give examples of isotopes.*
5. *Describe the three types of chemical bonds.*
6. *List the two properties of water that make it important for living organisms.*
7. *Compare and contrast acids and bases and describe them in terms of the pH scale.*
8. *Describe the importance of buffers in chemical reactions.*
9. *Explain how a carbon atom is important as a basic building block of life and differentiate between organic and inorganic molecules.*
10. *Differentiate between synthesis and hydrolysis reactions.*
11. *Define energy of activation and describe how catalysts affect it.*
12. *Name the four categories of organic molecules.*
13. *Describe amino acids and their role in the formation of proteins.*

As the atom is the basic unit of matter, so too are small organic molecules the basic chemical units of life. These molecules are structurally similar and simple, usually having a backbone of no more than 20 carbon atoms to which are bonded either single atoms or small groups of atoms. Through the action of biological catalysts called enzymes, small organic molecules are modified and rearranged to construct four basic groups of macromolecules. These in turn make possible the structure and activities of the basic biological unit of life called the cell.

The Structure of Matter

The advertising slogan of a well-known U.S. chemical manufacturer is "Better Living Through Chemistry." If you were to develop an advertising campaign promoting the unique properties common to life on Earth, your slogan might read, "We're living because of chemistry." This may strike you as odd, but on consideration you will realize it is a fundamental statement about your own life. For example, how is it you can perceive and then understand these words? Why do you age? How do you digest food? Why are some substances in the air, water, or your food thought to cause cancer? Is genetic engineering of simple organisms dangerous? The answers to these and many other questions can be better answered if you understand some basic principles of chemistry.

General Composition of Matter

All matter, whether it is a flower, the book you are reading, or a distant planet, is composed of elements. An **element** is a substance that cannot be divided or decomposed by chemical change into simpler substances. There are 92 known elements in the universe. All 92 occur naturally on Earth, and 26 different elements are found in the human body (Table 2-1). Each element has been named and given a one- or two-letter symbol. For example, the letter H represents the element hydrogen, whereas O and C stand for oxygen and carbon, respectively. However, in chemistry as in life, things are not always what they seem. For example, Na is the symbol for sodium. The symbol Na comes from *natrium,* the Latin word for sodium. Similarly, the element potassium is represented by the letter K, which comes from *kalium,* the Latin word for potassium.

Each element is composed of like atoms. (For example, oxygen is an element composed only of oxygen atoms.) An **atom** is defined as the smallest unit of an element that still retains the properties of that element. Each type of atom has its own particular structure and chemical behavior. Although nonchemical means can be used to divide an atom into smaller units, when this occurs the atom loses its particular properties and behaves differently.

Table 2-1 The Most Common Elements on Earth and Their Distribution in the Human Body

ELEMENT	SYMBOL	ATOMIC NUMBER	PERCENT OF HUMAN BODY BY WEIGHT
Oxygen	O	8	65.0
Silicon	Si	14	Trace
Aluminum	Al	13	Trace
Iron	Fe	26	Trace
Calcium	Ca	20	1.5
Sodium	Na	11	0.2
Potassium	K	19	0.4
Magnesium	Mg	12	0.1
Hydrogen	H	1	9.5
Manganese	Mn	25	Trace
Fluorine	F	9	Trace
Phosphorus	P	15	1.0
Carbon	C	6	18.5
Sulfur	S	16	0.3
Chlorine	Cl	17	0.2
Vanadium	V	23	Trace
Chromium	Cr	24	Trace
Copper	Cu	29	Trace
Nitrogen	N	7	3.3
Boron	B	5	Trace
Cobalt	Co	27	Trace
Zinc	Zn	30	Trace
Selenium	Se	34	Trace
Molybdenum	Mo	42	Trace
Tin	Sn	50	Trace
Iodine	I	53	Trace

The Structure of an Atom

Every atom has a central mass known as the **atomic nucleus,** with a number of smaller particles orbiting at some distance around it. The atomic nucleus is composed of two types of subatomic particles called protons and neutrons. A **proton** has a positive electrical charge represented by a plus sign (+). A **neutron** has no electrical charge. In orbit around the atomic nucleus are subatomic particles called **elec-**

trons. Each electron has a negative electrical charge, represented by a minus sign (–). The electron is by far the smallest subatomic particle.

The total number of protons in the nucleus of an atom is its **atomic number.** The total number of protons *and* neutrons in the nucleus is its **atomic mass number.** The element hydrogen is the simplest atom, with only one proton in its nucleus and one electron orbiting it. Helium has the next simplest atom, having two protons and two neutrons in its nucleus with two electrons orbiting it. Therefore the atomic number of hydrogen is one and the atomic number of helium is two. However, although the atomic mass number for hydrogen is also one, the atomic mass number for helium is four. When the number of protons in the nucleus equals the number of electrons orbiting around it, the atom is said to be **electrically balanced.**

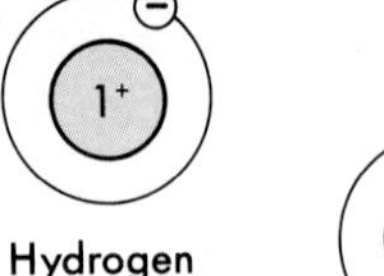

Helium

Ions

Electrons stay in their orbits because they are attracted to the positive charge of the protons in the nucleus. Sometimes because of other forces, atoms may lose or gain electrons. If this occurs, the atom is no longer electrically balanced. When an atom loses an electron, it has more positively charged protons in its nucleus than negatively charged electrons orbiting it. Therefore the atom has a net positive charge (+). In contrast, when an atom gains an electron, there are more negatively charged electrons orbiting the nucleus than there are positively charged protons in it, so the atom has a net negative charge (–). **Ions** are atoms that carry a positive or negative electrical charge.

Isotopes

All atoms of the same element have the same number of protons but do not necessarily contain the same number of neutrons in their nuclei. Consequently, atomic mass numbers among atoms of the same element can vary. Atoms with the same number of protons but different numbers of neutrons are called **isotopes** of that element. For example, among every 5000 hydrogen atoms one isotope, called deuterium (du-ter'i-um), has one proton *and* one neutron in its nucleus. Another isotope of hydrogen, called tritium, has one proton and two neutrons in its nucleus. Naturally, adding one more neutron to the tritium's nucleus makes its atomic mass number greater than each of the other two hydrogen atoms. Isotopes of an element are still atoms of that element, despite their different mass, because they contain the same number of *protons.*

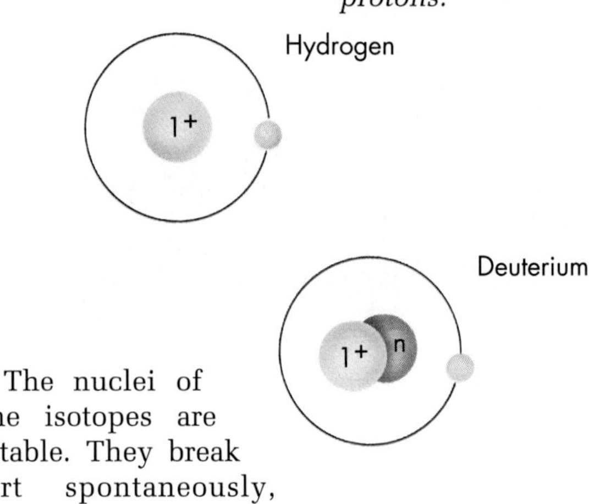

The nuclei of some isotopes are unstable. They break apart spontaneously, emitting energy, subatomic particles, or both. These isotopes are known as **radioisotopes,** and their particular properties make them important for research in both medicine and science.

Arrangement of Electrons Within Atoms

The key to understanding the particular properties and chemical behavior of atoms lies in the arrangement of the electrons orbiting the nucleus. Electrons spin around the atom's nucleus at specific distances from it in shells of space, or **energy levels.** Each shell, or energy level, in an atom can accommodate a certain number of electrons. The innermost shell can hold two. The second and third shells can each hold eight.

Regardless of how many electron shells an atom has, its chemical behavior is determined by the outermost shell. Atoms tend to react (form bonds) with other atoms in ways that keep their outer shells full. Only atoms whose outermost shells are maximally filled can exist as free atoms. These are known as **inert elements.** For example, refer to the helium and hydrogen figures on this page. Notice that helium has the maximum number of two electrons in its first and only shell. The hydrogen atom has only one electron in the first and only shell. Consequently helium is an inert element that exists freely in nature. Hydrogen atoms do not exist free-

Biology in Action: ISOTOPES AS RESEARCH TOOLS

Each radioactive isotope has its own rate of decay called a *half-life,* which is the time required for one half of a given amount of isotope to decay. For example, cobalt-60 has a half-life of 5.3 years. If we started with 10 g of cobalt-60, only 5.0 g of that isotope would remain after 5.3 years. After another 5.3 years only 2.5 g would remain.

Because the half-life of any isotope is constant, it can be a "molecular clock" to help determine the age of organic and inorganic substances. Carbon-14, for example, allows us to estimate when a plant or animal was last alive. This *carbon-dating* procedure is based on knowing that the isotope carbon-14 is formed at a constant rate and its known half-life is 5730 years.

On Earth every plant, while taking in carbon dioxide, also absorbs some carbon-14 into its system. Plants are eaten by animals, which in turn are eaten by other animals. Thus carbon-14, in small but constant amounts, is incorporated into all living things. As long as a plant or animal is alive and taking in carbon-14, there is a constant known proportion of carbon-14 in its body. However, once the organism dies the carbon-14 is no longer replenished.

By analyzing the changed proportion of carbon-14 in such organic remains, we can calculate when the organism was last alive. For example, if an ancient bone contains only 50% of the proportionate amount of carbon-14 presently in the atmosphere (meaning a half-life had passed), the animal from which the bone came must have died approximately 5730 years ago. If only 25% of the expected carbon-14 is found in the bone (meaning two half-lives had passed), the animal must have died about 11,460 years ago. This molecular clock can be used to date remains accurately within the past 70,000 years. Beyond that point other isotopes with longer half-lives can be used.

Isotopes can also be used in other ways. Suppose you wanted to know where iodine is concentrated in the body. An isotope of iodine could be injected into the bloodstream to be circulated throughout the body. Then by using a device called a Geiger counter, which detects the amount of energy emitted from radioactive isotopes, it would be possible to detect the greatest concentration of energy being emitted from the body. That concentration would mark the location of the iodine isotope, which would have made its way to the thyroid gland in the neck. Many facts presented in this book were discovered through the use of isotopes.

---

ly but combine with other atoms to form molecules.

Combinations of atoms are called **molecules.** A molecule, although composed of multiple atoms, behaves in many ways as a single unit. (You might compare it to a television, which can be recognized as a single functional object even though it has many parts.) The atoms in a molecule may be the same or different. For example, a molecule of oxygen has two oxygen atoms, whereas a molecule of water has two atoms of hydrogen and one atom of oxygen. The molecular composition of these and all other substances may be represented in several different ways. One way, called a **molecular formula,** indicates the actual numbers and types of atoms in a molecule. In the molecular formula for water, H_2O, the subscript 2 indicates there are two atoms of hydrogen (H). No subscript after the O indicates there is only one atom of oxygen (O). Another way to represent molecular composition is a **structural formula,** which shows the relative arrangements of the atoms in the molecule.

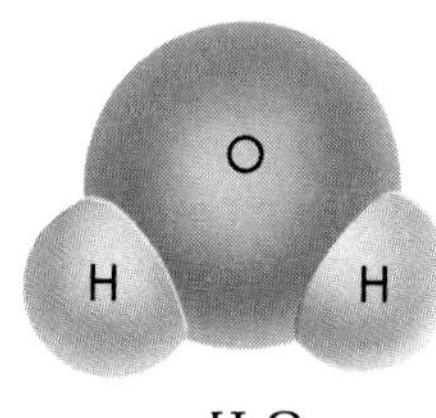

H_2O

Molecules containing two or more different types of atoms are called **compounds.** Water (H_2O) is a compound whose molecules have two hydrogen atoms and one oxygen atom. Table salt (NaCl) is a compound whose molecules each have one sodium atom and one chlorine atom.

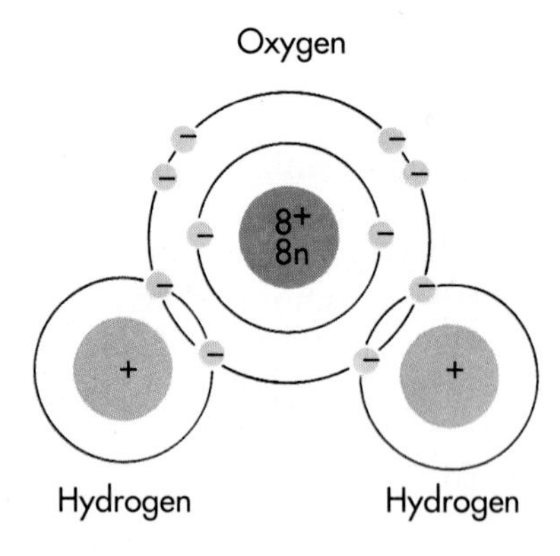

Chemical Bonds

The attractive force that holds two atoms together in a molecule is known as a **chemical bond.** This force is created by interactions between the electrons in the outermost shells of the two atoms. A chemical bond usually requires energy to form, and breaking a chemical bond usually releases energy. (This concept will be important in Chapter 16, where we discuss how the body obtains energy by breaking the bonds in food molecules.)

Chemical bonds allow an atom to fill its outer shell with the maximum number of electrons. An atom can gain one or more electrons from another atom, it can lose one or more electrons to another atom, or it can share one or more electrons with another atom. Based on these potential interactions, the three types of chemical bonds are ionic, covalent, and hydrogen.

Ionic Bonds

Each molecule of table salt (sodium chloride, or NaCl) contains one atom of sodium and one atom of chlorine. A sodium atom has one electron in its outermost shell. A chlorine atom has seven electrons in its outermost shell. Because of their incomplete outer shells, both sodium and chlorine are chemically reactive atoms.

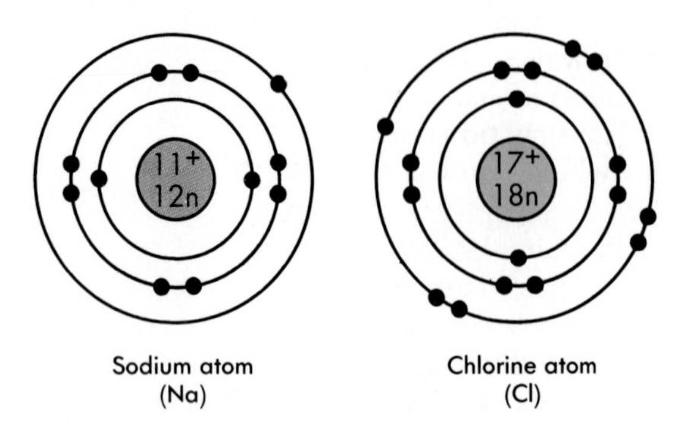

Sodium atom (Na) | Chlorine atom (Cl)

The chlorine atom requires one electron to complete its outermost shell. Conversely, if the sodium atom were to lose the single electron in its outer shell, it would also become stable. Not surprisingly then, sodium donates the single electron of its outer shell to chlorine's outer shell. Thus both atoms lose their electrical balance. The chlorine ion now has a net *negative* electrical charge of one, while the sodium ion now has a net *positive* charge of one (Figure 2-1). As a result of this electron transfer the newly formed sodium and chloride ions are oppositely charged and attracted to each other. This type of chemical bond, in which electrons are *transferred* from one atom to another, is called an **ionic bond.**

Covalent Bonds

A **covalent bond** forms when two atoms *share* one or more pairs of electrons. For example, a hydrogen atom has only one electron in its first and only shell. Therefore the atom is chemically active. When two hydrogen atoms form a chemical bond, they share their orbiting electrons and complete their shells. The result is a molecule of hydrogen (H_2). This chemical bond, in which *one* pair of electrons is shared between atoms, is called a **single**

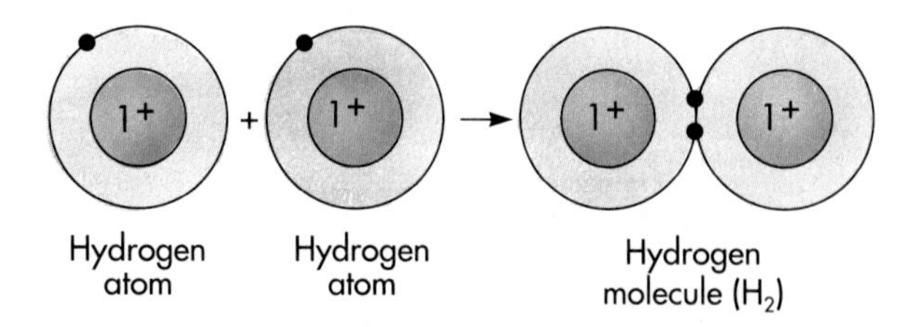

covalent bond. If *two* pairs of electrons are shared between atoms, a **double covalent bond** results. For example, a carbon dioxide molecule (CO_2) results from a double covalent bond between a carbon atom and each of the oxygen atoms. Triple covalent bonds are also possible.

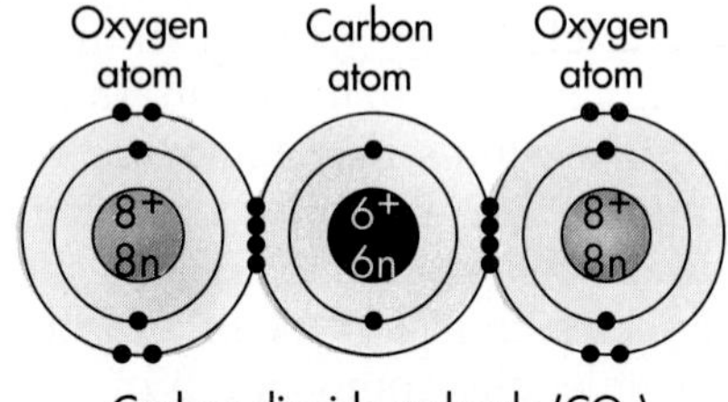

Carbon dioxide molecule (CO_2)

Electron pairs shared between two different atoms are not necessarily shared equally. Because

Figure 2-1 Ionic bond

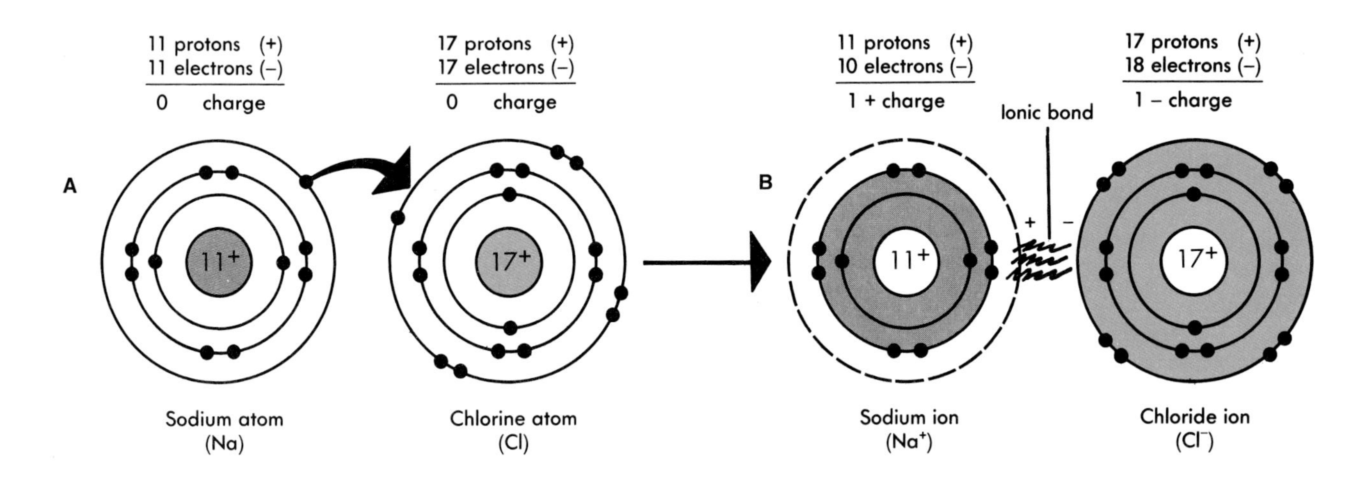

An ionic bond forms between sodium and chlorine after an electron transfer from the sodium atom to the chlorine atom occurs (*A*), resulting in full outer shells for both atoms. The new oppositely charged ions attract each other, forming an ionic bond (B).

atoms of different elements do not have the same number of protons in their nuclei, their force of attraction for electrons is different. In a **polar covalent bond,** a greater attraction for the electrons by one of the atoms results in an *unequal sharing* of the electrons involved. For example, a molecule of water has one oxygen atom and two hydrogen atoms. Each hydrogen atom requires an additional electron to complete its outer shell, and the oxygen atom requires two electrons to complete its outer shell. The oxygen atom has a stronger pull on the shared electrons than do the hydrogen atoms. Thus the oxygen side of the molecule acquires a partial negative charge, while the hydrogen ends of the molecule acquire a partial positive charge. This produces a **polar molecule.** The word **polar** refers to the fact that this type of covalent bond produces a molecule with weak opposite electrical charges. A polar molecule's structure is similar to that of a magnet, in which one pole is positive and the other pole is negative. Most polar molecules usually result from a covalent bond between a hydrogen atom and either an oxygen or a nitrogen atom.

Hydrogen Bonds

Atoms involved in polar covalent bonds may also come under the influence of other atoms and molecules. This allows some polar covalent molecules to participate in a third type of bond—the **hydrogen bond.** A hydrogen bond occurs when the partially positive (delta +) hydrogen atom in a polar molecule is attracted to the partially negative (delta –) atom of another polar molecule (usually the second atom involved is an oxygen or a nitrogen atom) (Figure 2-2). Although hydrogen bonds are relatively weak, they are significant in compounds as simple as water. Hydrogen bonds may also be in compounds as complex as DNA (see p. **33**). The composition of water is the focus of the next section.

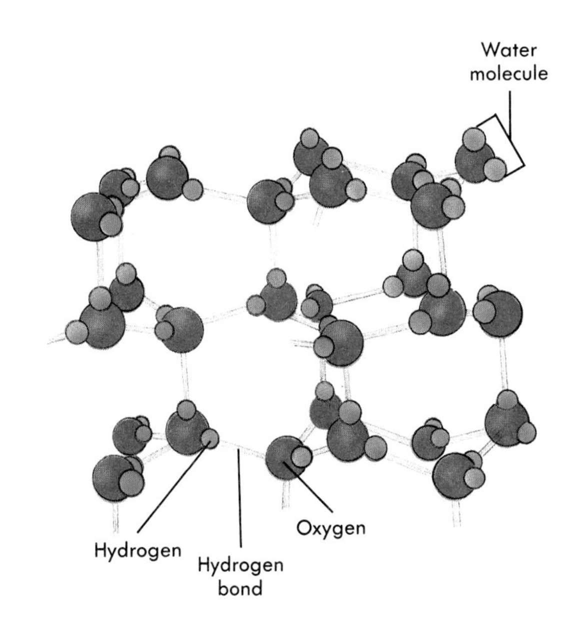

Figure 2-2 Water is a polar molecule

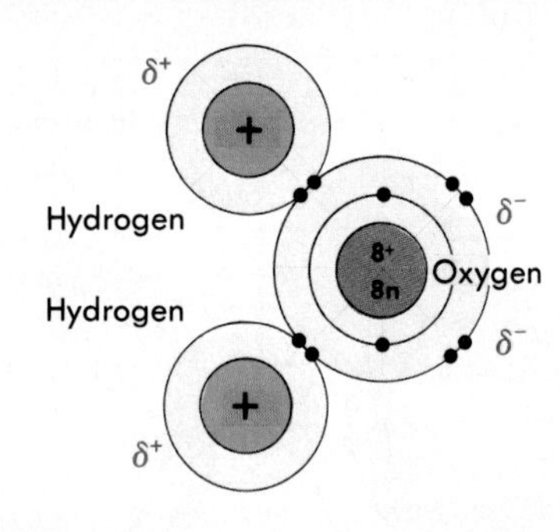

The oxygen side of the molecule acquires a weak partial negative charge, and the hydrogen ends of the molecule acquire a weak partial positive charge, resulting in a polar molecule. The Greek letter delta (δ) designates the weak charge located at a site within a polar molecule.

The Particular Properties of Water

Life first arose in water, and all life depends on its properties. Water is the medium in which all chemical activities of the cell and also between cells occur. Water accounts for 75% to 85% of the weight of an active cell and is also found between cells. About 60% of your body weight is water.

A water molecule is polar because the electron pairs shared between the hydrogen and oxygen atoms spend more time near the oxygen atom than to the hydrogen atoms. The water molecule is shaped like an isosceles triangle with an H-O-H angle of 104.5°. This angle puts the weak opposite charges at almost the maximum distance from one another. (If the H-O-H angle formed a straight line, there would be no weak positive end to the molecule.) The combination of its molecular geometry

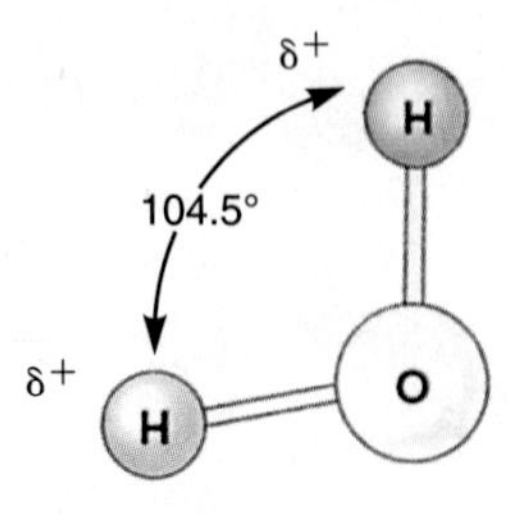

and the polar covalent bonds between its atoms makes water one of nature's most polar molecules. These polar characteristics of water give it two important properties: (1) it is liquid at the temperature range in which chemical reactions common to life occur, and much energy is required to alter its temperature; and (2) many substances dissolve in it.

Temperature Stability

The chemical reactions of human life can occur only within a narrow range of temperatures. You may know from personal experience that a change in body temperature of just a few degrees can be a serious health threat (Figure 2-3). The temperature of a substance is a consequence of molecular movement. Molecules within a substance are not motionless. They constantly move and bounce off one another. A substance's *temperature* is a measure of the velocity of its molecules. The greater the average speed of the molecules, the higher the temperature.

Figure 2-3 Temperature range within which the human body normally functions

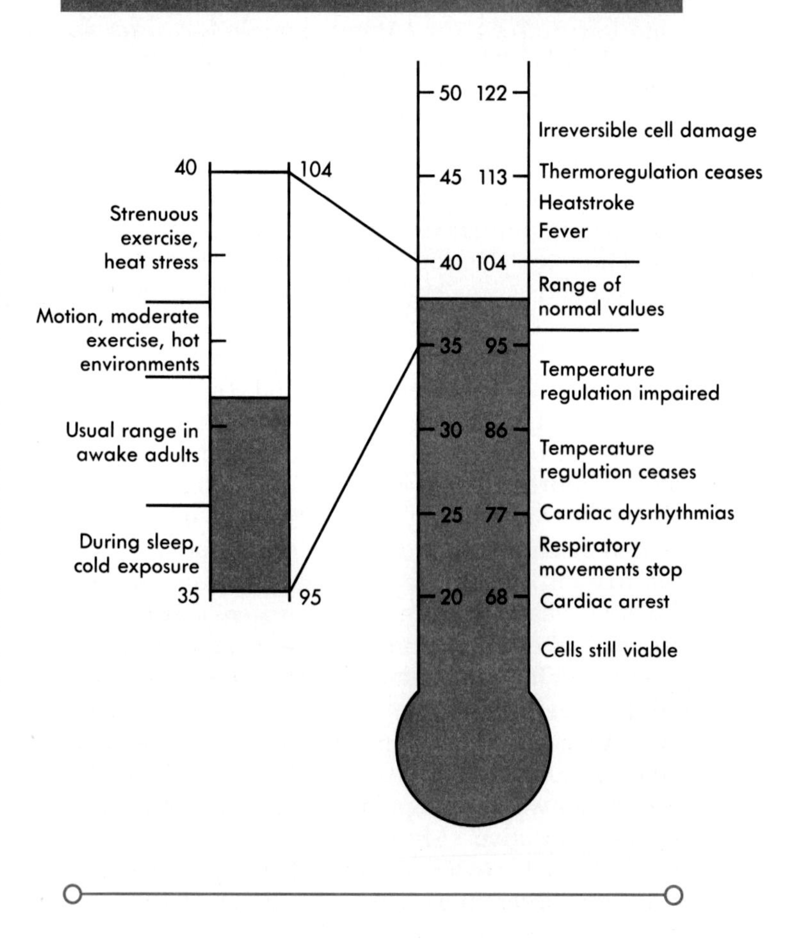

Temperature is related to energy. In biology the basic unit of energy is the *calorie,* defined as the energy required to raise the temperature of 1 g of water 1° C. The *specific heat* of a substance is the amount of energy (calories) needed to raise the temperature of 1 g of that substance 1° C. The specific heat of water is 1.0 calorie/gram. Compared with other substances water has a high specific heat. In other words a relatively large amount of energy is required to raise its temperature. (For comparison, alcohol has a specific heat of 0.6 calories per gram, and only 0.02 calories will raise the temperature of 1 g of granite.) This is because individual water molecules are weakly linked to one another by hydrogen bonds. When heat enters a watery system such as a living cell, much of the heat energy goes into breaking hydrogen bonds between water molecules rather than into speeding up individual molecules and raising the temperature of the system. Because of water's high specific heat, however, sunbathers can pick up a great deal of heat energy without sending their body temperatures soaring.

Another property of water that helps us survive high temperatures is its high **heat of vaporization,** the heat required to convert liquid to vapor. Water has one of the highest heats of vaporization known: 539 calories per gram. This is also due to its hydrogen bonds. To evaporate, a water molecule must move fast enough to break all of the hydrogen bonds holding it to the other water molecules. For example, as a sunbather's body temperature begins to rise, perspiration covers it with a film of water. Heat energy is transferred from the skin to the water. Evaporating just 1 g of this water uses up 539 calories, which means the bather's body has cooled by 1° C. In other words a great loss of heat occurs without much loss of water.

Water as a Solvent

Movement of molecules between cells as well as within them occurs in a watery medium. Water is also a suitable medium for the chemistry of life because it is an excellent **solvent.** This means it is capable of dissolving many different substances. Water molecules are polar. Therefore, **hydrophilic** ("water-loving"), or both polar and ionic molecules, are **soluble** (able to be dissolved) in it. For example, when salt is placed in water, it dissolves (Figure 2-4). Water also similarly dissolves polar molecules because its positive and negative ends attract oppositely charged regions of such molecules. Many biologically important substances exist as either ionic compounds or polar molecules.

In contrast, nonpolar molecules do not dissolve in water because they possess no electrical charge. They are therefore incapable of forming hydrogen bonds with water molecules. Instead the water molecules will form hydrogen bonds among themselves and exclude the nonpolar molecules (Figure 2-5). The excluded nonpolar molecules tend to gather together, forming droplets that eventually fuse to form a separate layer "outside" the water. Hydrophilic ("water-fearing"), or nonpolar molecules, *do not* interact with water and are insoluble, such as oil in water.

Acids, Bases, and Buffers

Although water molecules are stable because of their covalent bonds, at any given time a few are ionized. **Ionization** occurs when a hydrogen atom (H) separates from the water molecule, leaving its electron behind. This is represented by the following equation:

$$H_2O \rightarrow OH^- + H^+$$

The hydrogen atom leaves its electron with the –OH portion of the molecule because the positive attraction of the oxygen atom is stronger than the positive attraction of the lone hydrogen proton. Notice that the separated hydrogen atom is no longer in electrical balance. It is now a positively charged hydrogen ion (H^+). The remaining covalently bonded oxygen and hydrogen atoms now have an additional electron and become negatively charged (OH^-). This negatively charged pair is called a *hydroxide ion.* Such spontaneous ionization of water is important for life chemistry.

As you will learn in the following chapters, hydrogen atoms take part in many chemical events, being donated by one molecule and accepted by another when such molecules are put together or pulled apart. The chemical behavior of molecules and the cellular events in which they take part can be significantly altered by changes in the concentration of hydrogen ions in a solution.

Acids and Bases

Any molecule or compound, when dissolved in water, releases hydrogen ions in solution is called an **acid.** In contrast, any molecule or compound that, when dissolved in water, releases hydroxide ions is called a **base.** The following illustrates this:

$$HCL \rightarrow H^+ + Cl^-$$

(hydrochloric acid)

$$NaOH \rightarrow Na^+ + OH^-$$

(sodium hydroxide)

The effect of acids is to increase the number of hydrogen ions in a solution. Bases reduce the number of hydrogen ions in a solution because the

Figure 2-4 Water as solvent

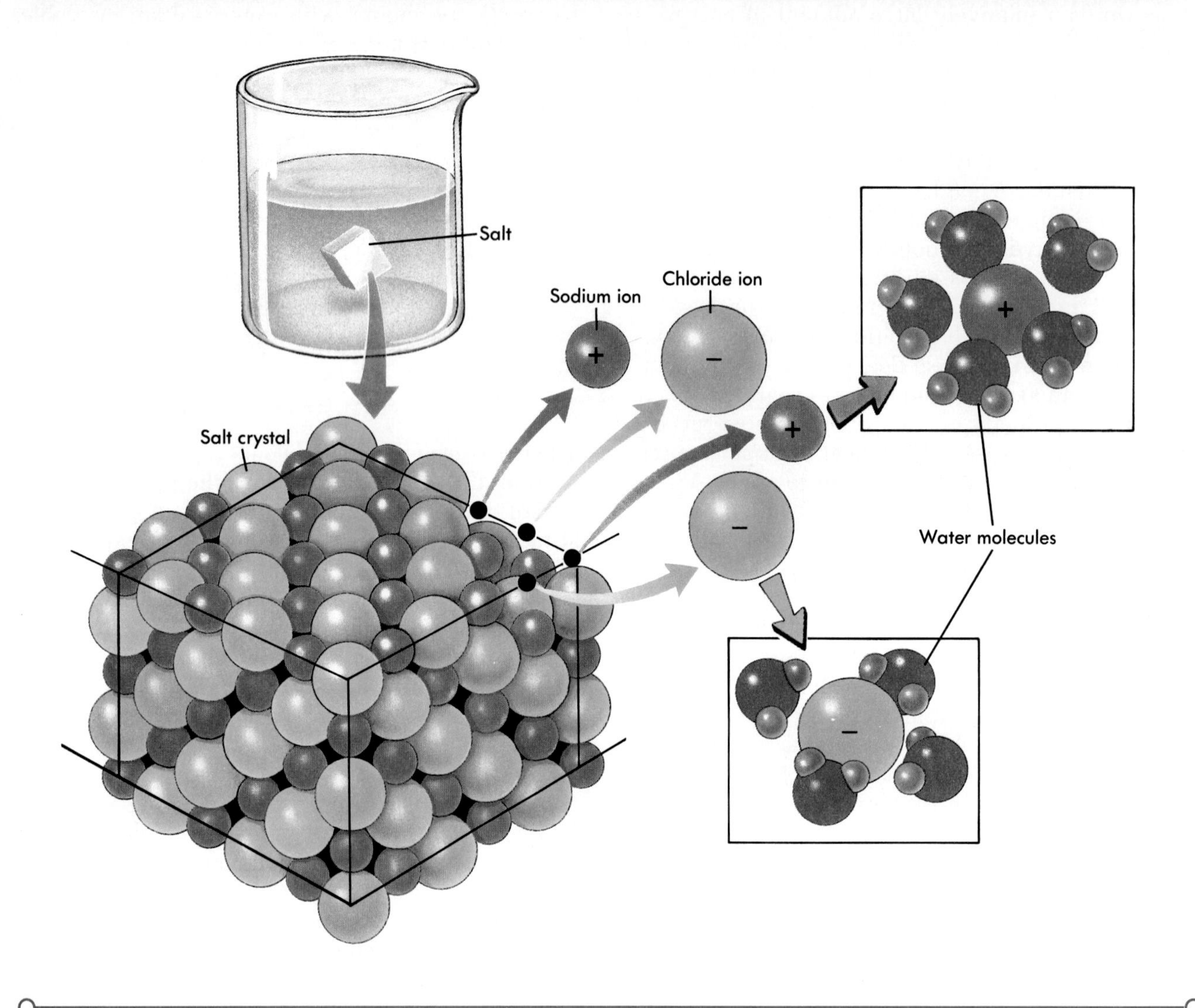

Ionic compounds dissociate in water. For example, sodium chloride (salt) dissociates to Na^+ and Cl^-. The positive sodium ions are surrounded by the negative parts of the water molecules, and the negative chloride ions are surrounded by the positive part of the water molecules. This prevents the ions from reforming into salt molecules.

hydroxide ions released by the base combine with the hydrogen ions to form water molecules.

The pH Scale

A solution's degree of acidity is measured on a **pH scale** (*p* = power, *H* = hydrogen ion concentration). The **pH scale,** which ranges from 0 to 14, represents the hydrogen ion concentration in a solution (Figure 2-6). Pure water is neutral, neither acidic nor basic, because equal numbers of hydrogen and hydroxide ions are being formed through spontaneous ionization only to reform again into water molecules. The reference point on this scale is 7, which is the pH of pure water. Each number on the pH scale represents a hydrogen ion concentration 10 times greater than the next higher number. Any solution with a pH lower than 7 contains a higher concentration of hydrogen ions than of hydroxide ions when compared to pure water, and is acidic. Any solution with a pH higher than 7 contains a lower concentration of hydrogen ions than hydroxide ions when compared to pure water, and is basic.

Figure 2-5 Hydrophobic interactions

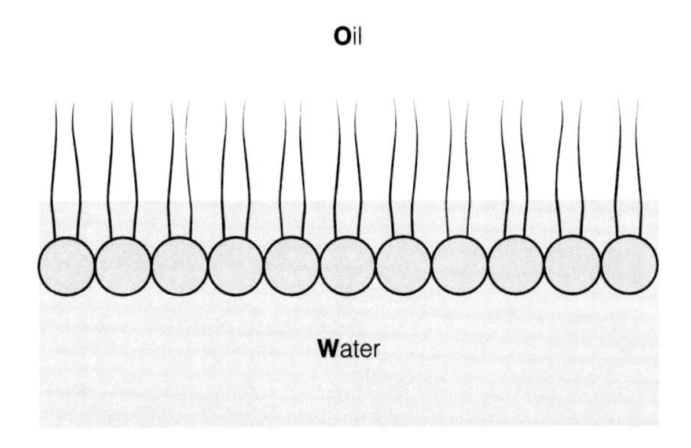

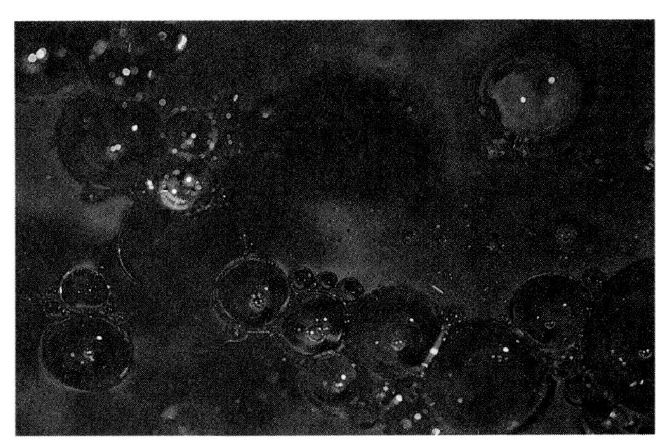

Nonpolar molecules such as oil are not attracted to polar water molecules. As a result the nonpolar molecules gather together and form a layer separate from the water.

Buffers

In living organisms ongoing chemical events and processes are sensitive to changes in pH and, with only minor exceptions, function best in nearly neutral conditions. For example, the interior portions of most human cells have a pH of about 6.8, whereas the fluid portion of human blood, as well as other fluids surrounding the cells, have a pH of 7.3 to 7.4.

If hydrogen ions are transferred during chemical events and the pH of solutions in and around cells must be nearly neutral to function properly, how is the correct pH maintained in the body? Special mechanisms aid in stabilizing body fluids so cells will not be subject to significant changes in pH. The chemical substances that aid in maintaining nearly neutral pH conditions are called **buffers.** Buffers have the ability to combine with hydrogen ions and remove them from solution whenever their concentrations begin to rise. Conversely, buffers can release hydrogen ions back into solution when their concentrations begin to fall.

For example, bicarbonate (HCO_3^-) is a buffer found in human blood. Any excess hydrogen ions formed in the blood will react with bicarbonate molecules to form carbonic acid (H_2CO_3). Carbonic

Figure 2-6 The pH scale

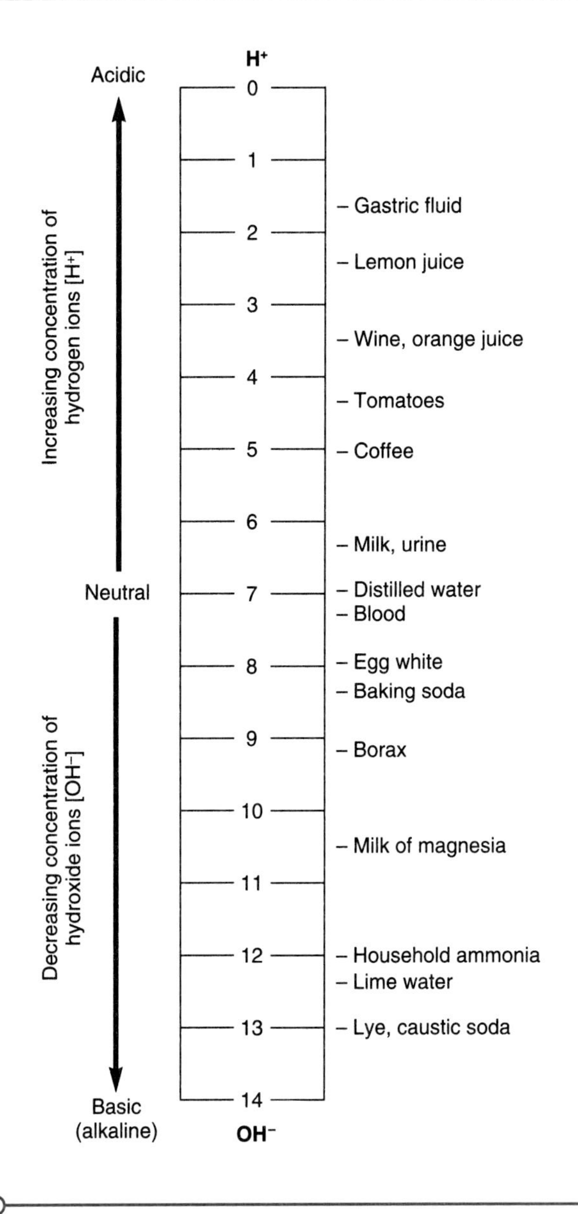

The pH scale is based on the number of hydrogen ions in a solution. If the corresponding concentration of hydrogen and hydroxide ions is equal, the solution is described as neutral, which is 7 on the scale.

acid in turn will separate into water and carbon dioxide, which is removed from the blood by the lungs and exhaled, as follows:

$$H^+ + HCO_3^- \rightarrow H_2CO_3$$

$$H_2CO_3 \rightarrow H_2O + CO_2$$

In later chapters you will learn what "drives" this particular reaction and also about buffers found in other body organ systems.

The Importance of Carbon

Approximately 93% of the atoms in your body are oxygen, hydrogen, and carbon. Much of the oxygen and hydrogen exists as water molecules. As was the case with water, learning about the atomic structure of carbon will enable you to understand its fundamental functions in life chemistry.

Carbon has six protons in its nucleus and six electrons in orbit around it. Two electrons are located in the first shell, and four reside in the second shell. If carbon either gained or lost four electrons by forming ionic bonds, its electrical charge would be thrown too far out of balance. Therefore carbon can complete its outermost shell only through covalent bonding with other atoms. For example, a single carbon atom can form four single, two single and one double, one single and one triple, or two double covalent bonds with other atoms. Carbon atoms can covalently bond to one another to form long chain or ringlike structures. Linked carbon atoms act as the backbone or skeleton in the construction of larger molecules.

To early chemists molecules were classified as either organic or inorganic. The word *organic* originally indicated carbon-containing molecules existed only within living organisms and could be manufactured only by them. Today, however, we know that some of the same molecules exist in both living and nonliving structures. The current definition of an **organic molecule** is any molecule that contains both carbon and hydrogen atoms. Most organic molecules important in the chemistry of life usually have no more than about 20 carbon atoms and are also referred to as **small organic molecules.**

Small organic molecules can exist either separately or as part of much larger complex structures called **macromolecules.** Some types of macromolecules are also called **polymers** when they consist of similar organic molecules linked together. Just as trains are made by coupling cars together, organic molecules can be coupled to form polymers and macromolecules. Also like a train, which uses standard connectors to attach cars to one another, organic molecules almost always use the same type of chemical reaction (a covalent bond) to join to one another.

Synthesis Reactions and Hydrolysis

Macromolecules and polymers are made by **synthesis reactions.** Synthesis reactions covalently bond small organic molecules together and produce a water molecule for each created bond. A synthesis reaction proceeds as follows: A hydrogen ion (H^+) is removed from one organic molecule, and a hydroxide ion (OH^-) is removed from the second organic molecule. The two ions bond, forming a water molecule as the two organic molecules are joined by a covalent bond. The synthesis reaction process uses energy, which is stored in the resulting chemical bond, and takes place with the help of enzymes (see the next section).

Macromolecules and polymers can also be broken down into their component organic molecules by means of **hydrolysis** (hi-drol′-i-sis) **reactions.** Hydrolysis reactions occur when the covalent bond between two organic molecules is broken and a hydrogen ion is added to one molecule and a hydroxide ion is added to the other. In other words the water molecule atoms are added back to the organic molecules. The energy held in the chemical bond is released.

How Chemical Reactions Between Molecules Occur

For atoms or molecules to react chemically they must collide with enough energy to overcome the repulsion between their electrons and to produce interactions between the outermost shells of electrons. The amount of energy required to overcome the repulsion between electrons and initiate a chemical reaction is called the **energy of activation.**

Four factors affect the energy with which atoms and molecules collide. The first is particle size. The smaller the atom or molecule the faster it moves and therefore the more frequently it collides with other atoms or molecules. A second factor is temperature. Raising the temperature of a substance increases the velocity of its atoms or molecules.

This higher velocity increases their rate of collision with other atoms or molecules. The third factor is concentration. Chemical reactions proceed more rapidly when the interacting atoms or molecules are present in high concentrations. The larger the number of atoms or molecules in a given space, the greater the probability of collision. However, if these were the only factors involved in generating a sufficient energy of activation, chemical reactions in our bodies would proceed far too slowly to sustain life. The chemistry of life occurs at a much faster rate. Most chemical reactions in biological systems usually occur one *million* times faster than they would occur otherwise.

The fourth factor that allows atoms or molecules in cells to effectively overcome the energy of activation necessary to form chemical bonds involves **catalysts. Catalysts** are substances that increase the rate of chemical reactions without themselves becoming a part of the outcome. Catalysts work by lowering the energy of activation required to initiate chemical reactions. **Enzymes** (en′-zimz) are catalysts in biological systems. They are the most important factor determining the rate of chemical reactions occurring in our bodies, the means by which chemical reactions can occur at sufficiently high rates to sustain life processes. How enzymes lower the energy of activation necessary to create or break chemical bonds will be discussed later in the chapter.

Organic Molecules, Macromolecules, and Polymers

Nearly all the organic molecules found in our bodies fall into one of four categories: carbohydrates, lipids, amino acids, or nucleotides.

Carbohydrates and Polysaccharides

Carbohydrates are molecules that contain carbon, hydrogen, and oxygen atoms. The ratio of hydrogen atoms to oxygen atoms is 2:1. Because this ratio is the same as the ratio of atoms found in water molecules, *carbohydrates,* meaning hydrates (water) of carbon, is an appropriate name. Carbohydrates are also an important energy source for the body's cells because they contain many carbon-to-hydrogen bonds.

Among the simplest carbohydrates are the so-called simple sugars, or **monosaccharides** (mon-o-sak′a-rids). Simple sugars may have as few as three carbon atoms, but the primary molecule used for the body's energy is a six-carbon sugar called **glucose.**

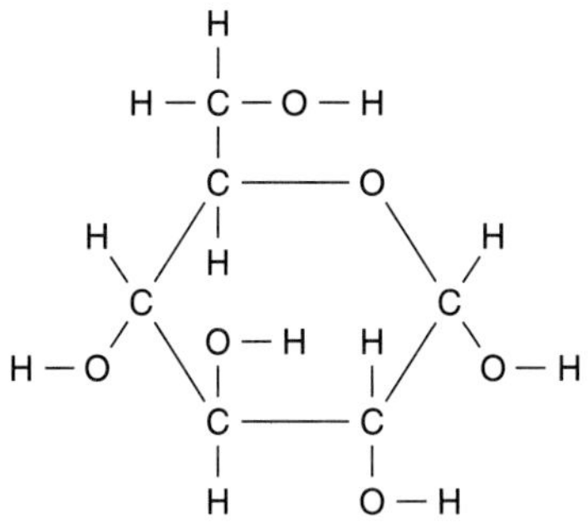

```
                H
                |
            H — C — O — H
                |
                C ——————— O
         H    / |          \      H
          \  /  H           \    /
           C                  C
        /   \   O — H    H  /  \
  H — O      \  |        | /     O — H
               C ———————— C
               |          |
               H          O — H
```

Our bodies are able to store surplus glucose molecules by covalently linking them into long polymers called **polysaccharides,** also called **glycogen** (gli′ko-jen) (Figure 2-7). Most glycogen is manufactured and stored in the liver and some muscle tissue. Between meals as glucose is taken up from the blood by the body's cells, glycogen is hydrolyzed into its component glucose molecules and released into circulation for further distribution in the body.

Glycerol, Fatty Acids, and Lipids

Lipids include a diverse assortment of molecules each sharing a common characteristic. **Lipids** are insoluble in some polar solvents such as water. They also have few oxygen atoms. Almost all of the chemical bonds in lipids are nonpolar, occurring between carbon and carbon atoms or between carbon and hydrogen atoms. As a result lipid molecules cannot interact with polar molecules, so most lipids do not dissolve in water. They are hydrophobically excluded from water and cluster together. Another consequence of having many hydrogen-to-carbon bonds is that lipids have a high concentration of stored energy, about twice that of simple sugars. As a result some types of lipids are used for semipermanent or long-term energy storage. Some lipids also function as components of cell membranes, as components of other cellular structures, and as chemical messengers in the growth, development, and maintenance of human sexual reproductive structures.

There are three categories of lipid molecules: (1) oils and fats, also called triglycerides, which are similar in structure and contain only carbon, hydrogen, and oxygen atoms; (2) phospholipids, which are structurally similar to oils but also contain phosphorus and nitrogen atoms; and (3) lipid molecules composed of carbon rings and known as steroids.

Triglycerides consist of two types of molecules. One molecule is a short chain of three carbon atoms, called **glycerol.** The other molecules, known as **fatty acids** or **glycerides,** are longer chains of carbon atoms and attached hydrogen atoms with a carbon-oxygen-oxygen-hydrogen group at one end called a carboxyl group. **Triglycerides** (*tri* = three, glycerides) are formed from the attachment of three fatty acid chains to a glycerol molecule. Each fatty acid chain bonds to the glycerol molecule by a synthesis reaction. A hydrogen atom is removed from the carboxyl group of the fatty acid chain, and a hydroxyl group is removed from the glycerol molecule. This results in a covalent chemical bond between the glycerol and fatty acid chain and a water molecule.

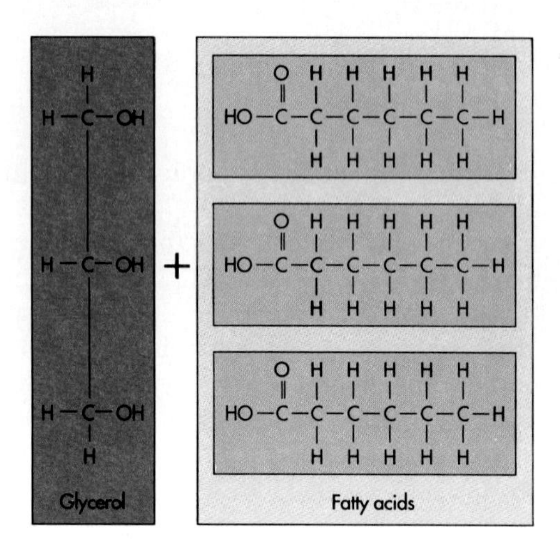

The number of carbon atoms in the fatty acid chains may vary, usually from 14 to 20. Fatty acids whose carbon atoms are each bonded to two hydrogen atoms are said to be **saturated** because they contain the maximum possible number of hydrogen atoms. Other fatty acids may contain double covalent bonds between one or more pairs of carbon atoms. Because the double bonds replace some of the hydrogen atoms, these fatty acids are called **unsaturated** (Figure 2-8). If a fatty acid has more than one double bond in the carbon chain, it is called **polyunsaturated.**

Unsaturated fatty acid chains kink or bend at the double bonds between the carbon atoms. This prevents them from packing close together the way saturated fatty acids can. Lipids produced by safflower, olive, and peanut plants are unsaturated and liquid at room temperature. They are called oils. Saturated chains are common to animal fats found in butter fat and meats, which are solid at room temperature. A vegetable oil, however, can be converted to a fat by breaking the double bonds between carbons, replacing them with single bonds and adding hydrogen atoms to the remaining bond positions. This is the "hydrogenated oil" listed in the ingredients on a package of margarine made from corn oil.

The second type of lipid, called a **phospholipid,** consists of a glycerol molecule with only two fatty acid chains attached to it. In place of the third fatty acid chain, normally found in triglycerides, there is a phosphate-containing group of atoms, usually with a nitrogen-containing group attached at the end. Both of these molecules are polar. One end of a phospholipid, containing the two fatty acid chains, is nonpolar and hydrophobic. The other end of the phospholipid, containing the polar phosphate and nitrogen groups, is polar and hydrophilic. The two nonpolar fatty acids extend in one direction approximately parallel to each other. The polar region extends in the other direction. Because of this orientation phospholipids are often drawn as a polar "ball" with two trailing nonpolar "tails." Cell membranes are largely composed of phospholipid molecules (see the next chapter).

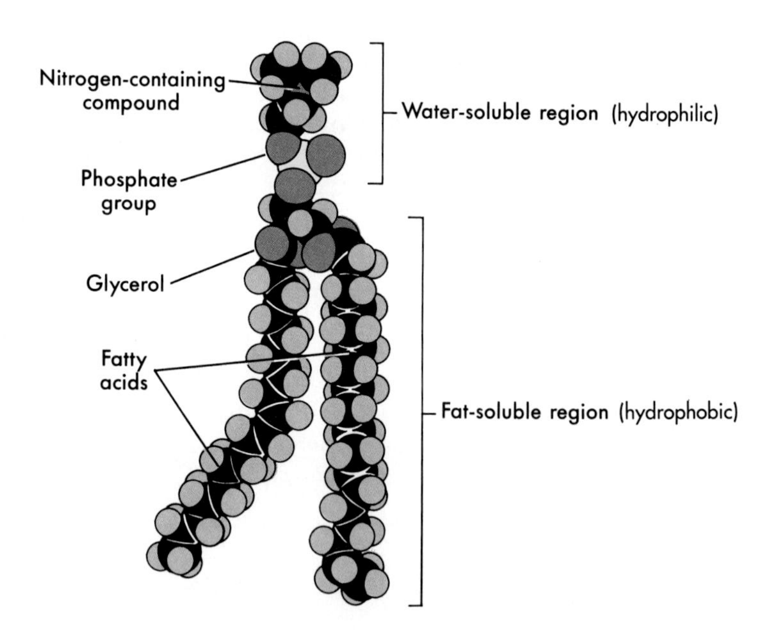

Except for being insoluble in water our third category of lipids, called **steroids,** are unlike other lipids. **Steroids** are composed of four connected carbon rings with a number of oxygen, hydrogen, and other carbon atoms attached to the rings. One steroid, called **cholesterol** (kuh-les′-ter-ol), is in many cell membranes along with phospholipids (Figure 2-9). Cholesterol molecules are also modified by chemical reactions to become hormones, such as estrogen and testosterone. The role of steroid hormones is discussed in Chapter 18.

Figure 2-7 Glucose storage

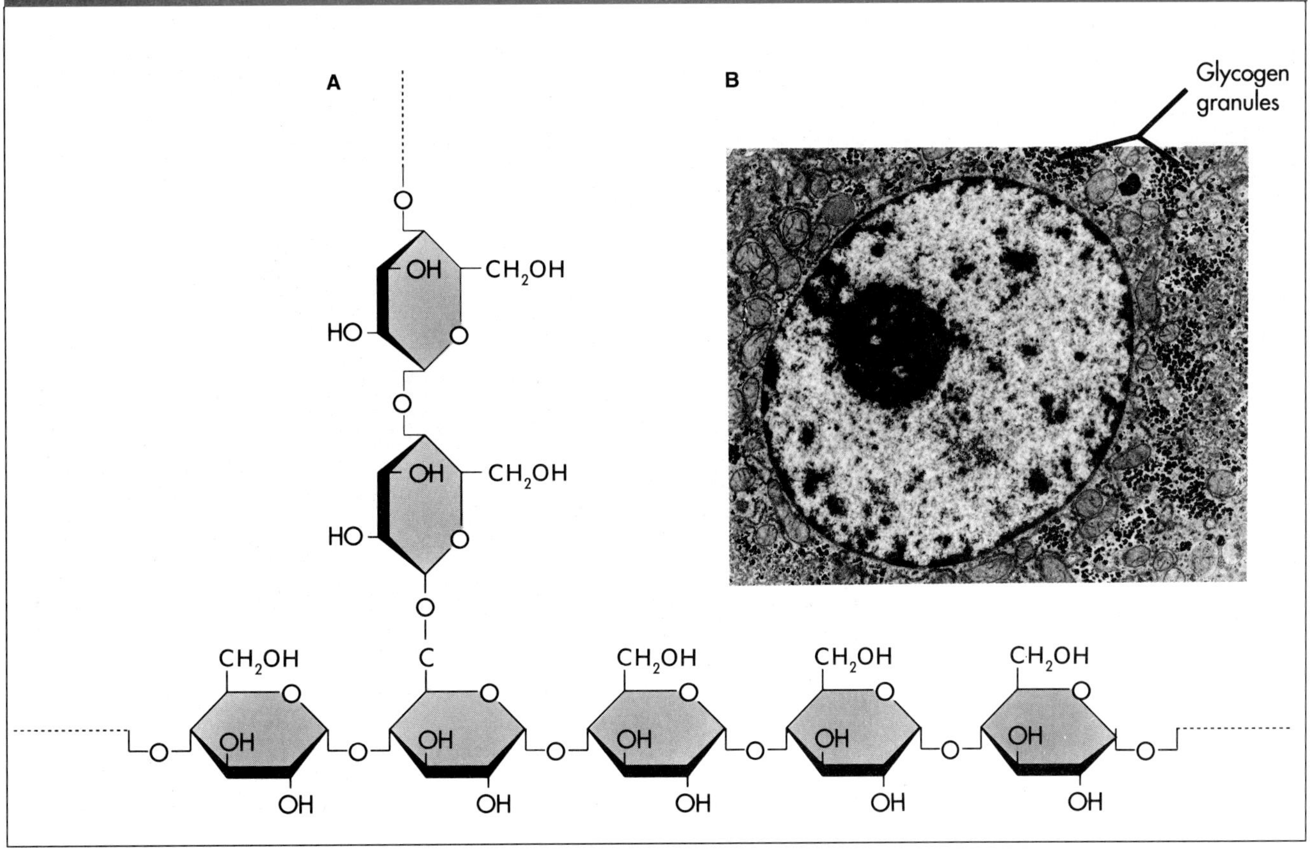

A

B

Glycogen granules

The immediate storage form of glucose molecules in the human body is a long highly branched polysaccharide called glycogen (A). Most glycogen is manufactured and stored in the liver. This micrograph shows the nucleus of a liver cell surrounded by dense granules of glycogen (B).

Figure 2-8 Saturated and unsaturated fatty acids

A

Palmitic acid (saturated)

B

Linolenic acid (unsaturated)

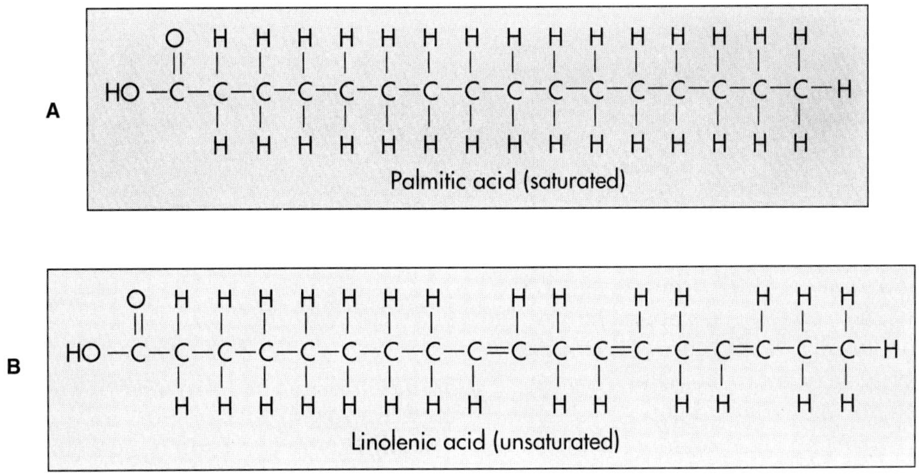

Saturated fatty acids contain only single covalent bonds between carbon atoms (A). Unsaturated fatty acids have at least one double-covalent bond between carbon atoms (B).

Figure 2-9 Steroid lipid structure

A Cholesterol

B Estrogen (estradiol)

C Testosterone

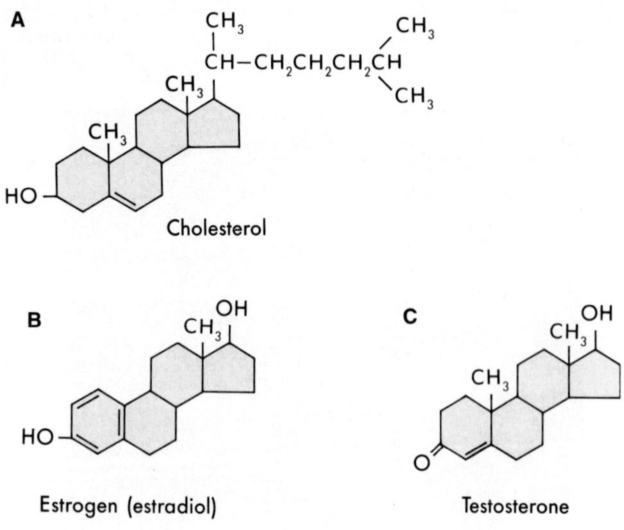

The carbon atoms of a steroid molecule form linked rings. Steroid lipid structure is the basis of other important compounds, such as cholesterol (A) and sex hormones (B, C).

Amino Acids and Proteins

Of all the major macromolecules, proteins are the most diverse and most numerous organic compounds in the human body. Some proteins are used for structural purposes. A protein called **collagen** (kol′la′jen), for example, is responsible for giving organs and many cells their characteristic shapes and bones the ability to resist twisting and bending forces. Other proteins transport molecules and compounds through the body. For example, hemoglobin, a protein in all red blood cells, transports oxygen from the lungs to all other cells. Proteins also make cell movement possible. The contractile proteins in muscle tissue make it possible for them to shorten and lengthen and thereby generate movement. Still other proteins are chemical messengers, assisting in coordinating many bodily activities. The most diverse and important proteins, however, are enzymes, which are catalysts that speed up chemical reactions by lowering the energy required to activate or start the reaction. How enzymes accomplish these amazing tasks is discussed in the description of protein structure on p. 31.

Despite the diversity of protein function, proteins all have the same basic structure. Each **protein** is composed of a long chain of amino acids linked end to end.

Amino acids are simple organic molecules with four components, each covalently bonded to one central carbon atom. The parts of every amino acid are these: an amino group ($-NH_2$), a carboxyl group ($-COOH$), a hydrogen atom, and a variable group of atoms designated R. Simply think of the letter *R* as an abbreviation for "remainder of the molecule." The general form of all amino acids can be represented as follows:

$$H_2N-\underset{\displaystyle H}{\overset{\displaystyle R}{C}}-COOH$$

All amino acids have identical components except for the R group. Differences in the numbers and arrangement of the atoms making up the R group make each amino acid unique in its behavior.

The amino acids are joined by synthesis reactions, with the amino end ($-NH_2$) of one amino acid bonding to the carboxyl end ($-COOH$) of the next amino acid. A water (H_2O) molecule is produced, and the resulting covalent bond between the two amino acids is called a **peptide bond.** Two bonded amino acids form a **dipeptide,** and three or more are called **polypeptides.**

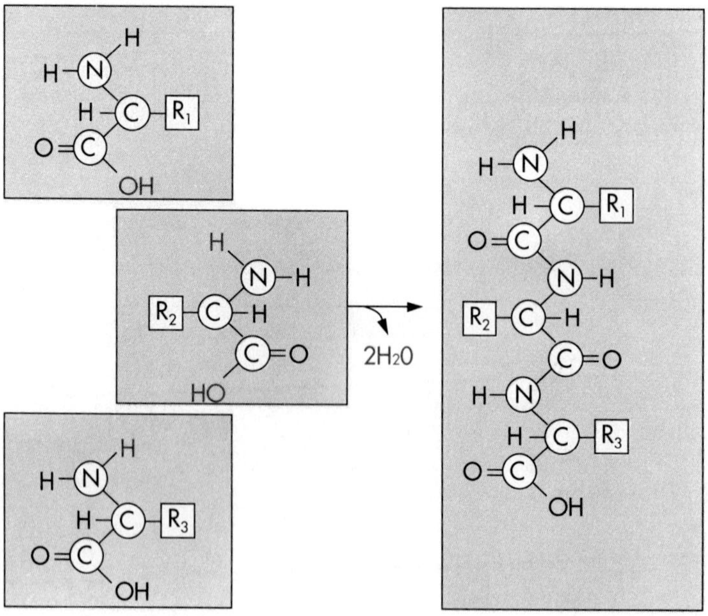

All the body's proteins can be made using various combinations of just 20 amino acids. The number and types of amino acids and the sequence in which they occur determine the unique properties of each protein. Although our bodies are able to generate most of the amino acids we require, there are seven or eight amino acids referred to as **essential amino acids,** meaning that humans cannot manufacture them from individual components. Instead we must ingest them already formed within the food we eat.

You might think of the 20 amino acids as a 20-letter "alphabet" used in specific combinations to form "words" (proteins). Just as a change in one letter can produce a word with an entirely different meaning (*bat, cat*), a change in the type or position of an amino acid can produce a different function.

The particular sequence of amino acids determines not only the function of a protein and how it will interact with other substances, but also the shape it can assume. There are as many as four distinct structural levels of protein structures. First all proteins have a particular linear sequence of amino acids, called the **primary structure** of the protein (Figure 2-10, *A*). The R groups of different amino acids can interact. Under cellular conditions amino acids are ionized and can also interact with their neighbors to form hydrogen bonds. Because of these bonds regions of a polypeptide chain, or sometimes the whole chain, fold into sheets or wrap into coils. This is the **secondary structure** (Figure 2-10, *B*).

The folded sheets or helical shape of a protein further fold or bend into more complicated shapes, called the **tertiary structure** (Figure 2-10, *C*).

When two or more protein chains come together to form a functional unit, the chains are referred to as subunits. Hemoglobin, for example, is composed of four subunits. The way these subunits are assembled into a functioning whole is called the **quaternary** (kwah'ter-ner'i) **structure** (Figure 2-10, *D*).

After this introduction to protein structure we can turn our attention back to protein enzymes and see how they function as biological catalysts.

Enzymes

Enzymes increase the rate of chemical reactions between atoms or molecules by lowering the activation energy needed for the reaction to occur. The tertiary structure of most enzyme proteins is globular, containing "grooves" called **active sites.** The molecules an enzyme interacts with are called **substrates.** When the enzyme alters substrates, the resulting molecular arrangement or combination is referred to as **product.** The substrates first chemically bond to the active site of the enzyme. Facing into the active sites are the R groups of specific amino acids that make up the protein enzyme. The R groups chemically interact with the substrate by stressing or distorting a particular bond. This consequently lowers the amount of activation energy needed to break the bond. The **enzyme-substrate complex,** formed temporarily in the course of the reaction, dissociates to yield a product. The free, unchanged enzyme can now bond to another substrate molecule. Enzymes are not altered or used up by the chemical reactions they control.

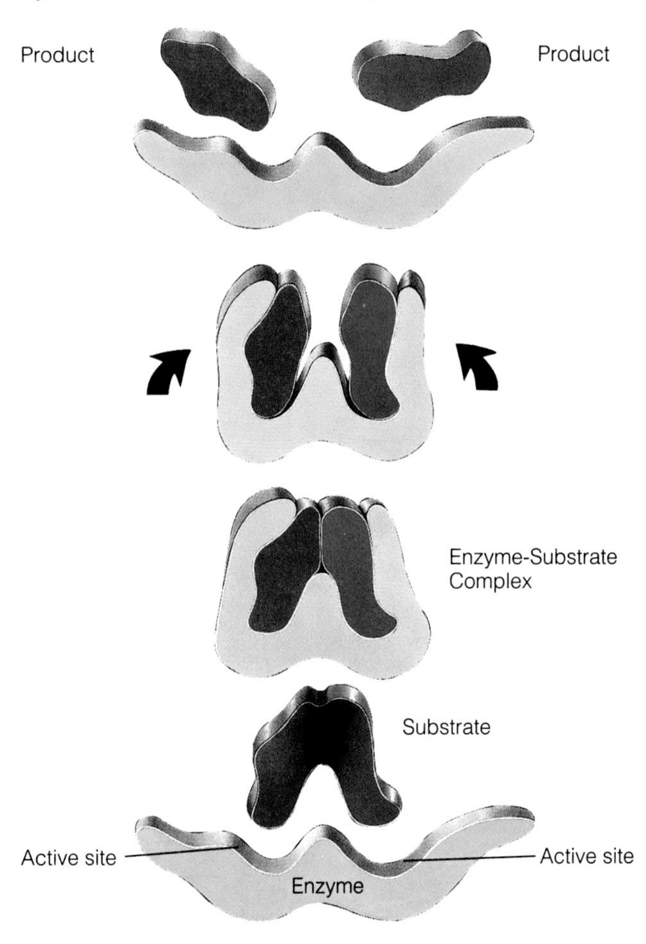

Each enzyme is specific to a substrate and can control only a single chemical reaction or a small group of related reactions. Consequently the presence of specific enzymes determines not only which reactions will be speeded up but also which reactions will occur at all.

Nucleotides and Nucleic Acids

The last of the major groups of biologically important macromolecules are called **nucleic acids.** **Nucleic acids** are composed of small organic molecules called **nucleotides.** Each **nucleotide** is in turn composed of one five-carbon-

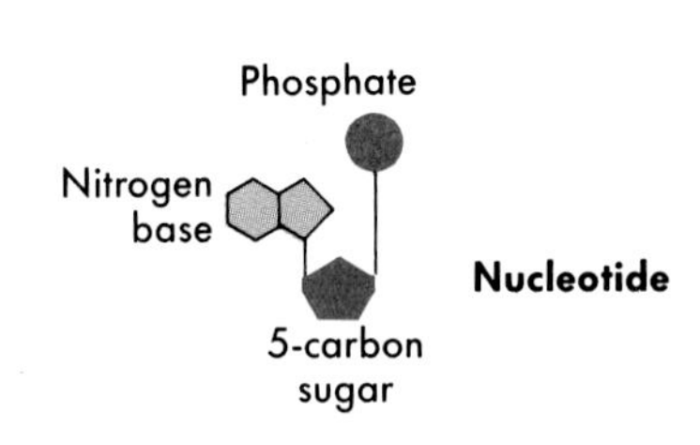

Figure 2-10 Structural levels of protein

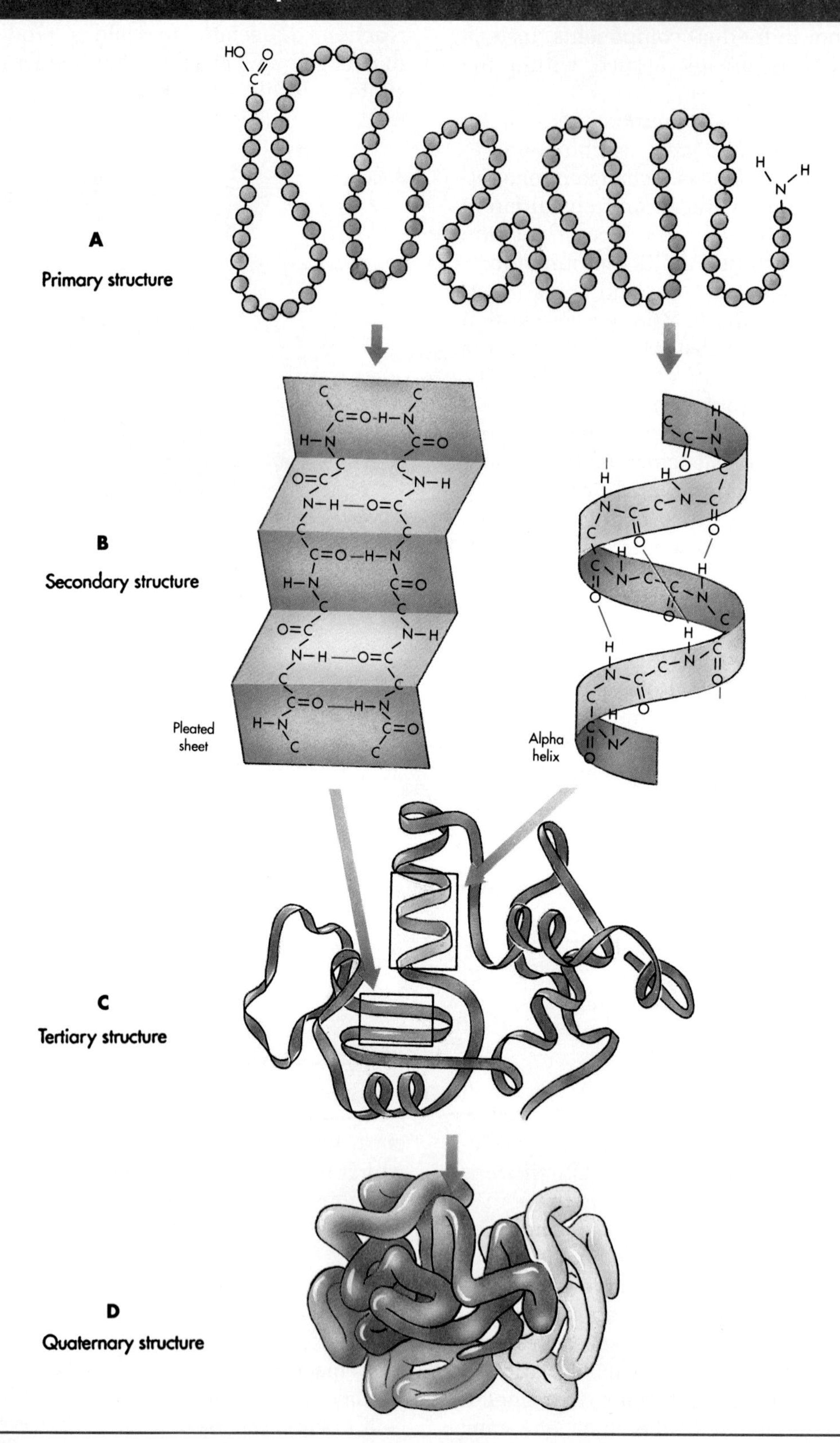

The *primary structure* is determined by the number, kind, and sequence of amino acids in the chain (A). The *secondary structure* is formed when hydrogen bonds (red lines) form between amino acids, usually forming folded sheet or helical shapes (B). The overall three-dimensional shape of the protein is known as its *tertiary structure* (C). The *quaternary structure* results from the bonding of two or more amino acid chains (D).

Figure 2-11 Types of nucleotides

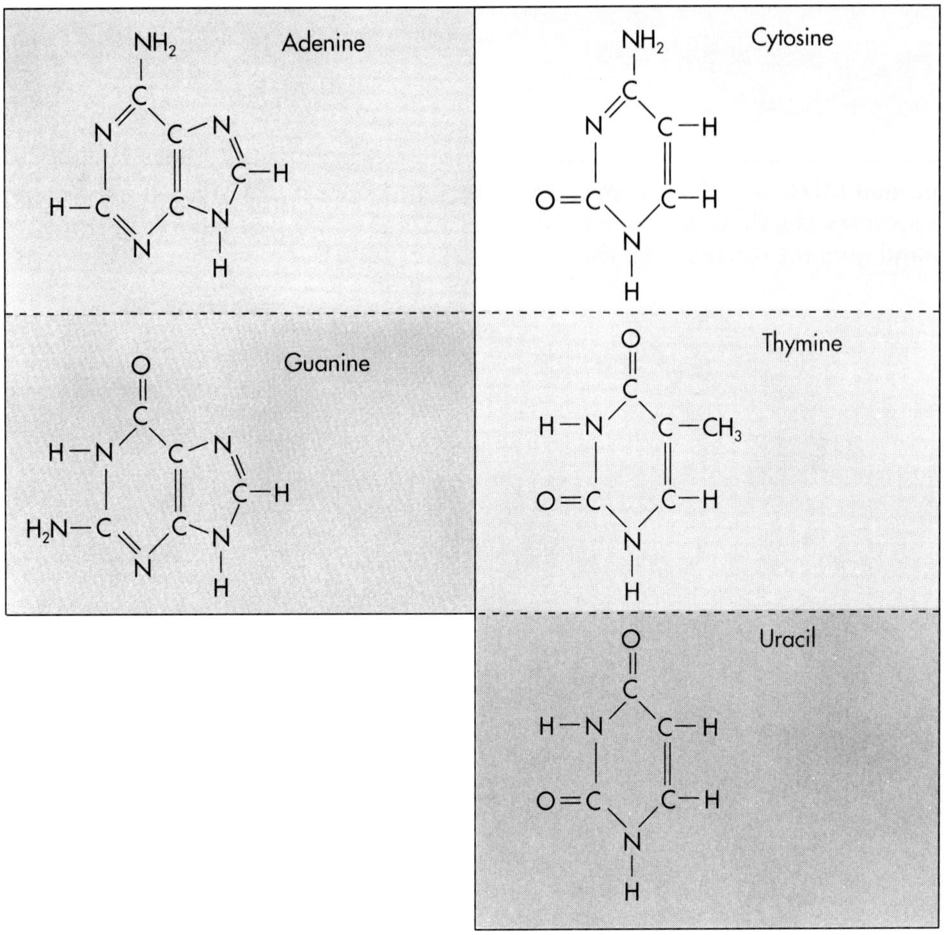

Adenine

Guanine

Cytosine

Thymine

Uracil

Each of the five nucleotides has a different nitrogen-containing base. Thymine, cytosine, and uracil each have a single nitrogen-containing carbon ring base. Adenine and guanine each have two nitrogen-containing carbon ring bases.

ringed sugar called either **ribose** or **deoxyribose,** a phosphate group, and a nitrogen-containing carbon ring compound known as a **base.** Five different nitrogen-containing bases are found in nucleotides. Three bases, thymine, cytosine, and uracil, each have a single nitrogen-containing carbon ring structure. The other two, adenine and guanine, each have two nitrogen-containing carbon rings (Figure 2-11). The components of each nucleotide are joined by synthesis reactions.

Different combinations of nucleotides form two different nucleic acids that make up the basic heredity material of the cell. The first of these nucleic acids is a double-stranded macromolecule called **deoxyribonucleic** (de-ox'si-ri'bo-nu-kle'ik) **acid (DNA).** The second is single stranded and called **ribonucleic acid (RNA)** (Figure 2-12). For now you might think of DNA as the basic molecular "cookbook" whose "recipes" determine your biological structure and function and are passed down from generation to generation. RNA helps "translate" and assemble what the "recipes" (DNA) specify. The role of both DNA and RNA and the specific mechanisms of inheritance are discussed in Chapter 5.

Some nucleotides also combine with other molecules. **Adenosine** (ah-den'o-seen) **triphosphate,** or **ATP,** contains the nucleotide adenine. ATP consists of the base adenine and one ribose sugar. Together they are called adenosine, to which three phosphate groups are attached. The wavy lines between the phosphate groups represent high-energy chemical

Figure 2-12 DNA and RNA

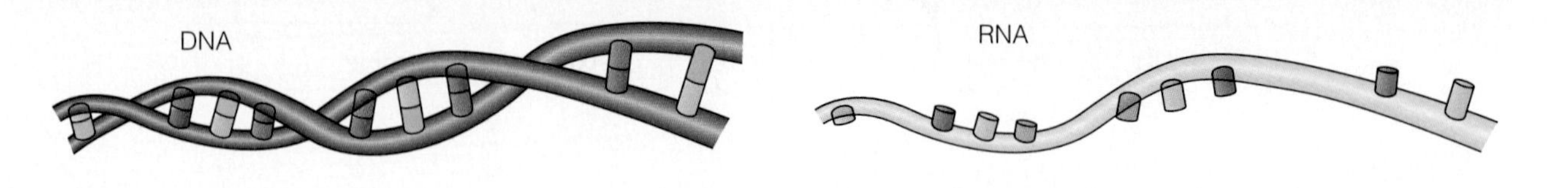

Deoxyribonucleic acid (DNA) is a double-stranded macromolecule made of adenine, thymine, cytosine, and guanine nucleotides (A). Ribonucleic acid (RNA) is a single-stranded macromolecule made of adenine, uracil, cytosine, and guanine nucleotides (B).

bonds. When these bonds are broken, a large amount of energy is released. Conversely, the creation of these chemical bonds also requires a large amount of energy. More about the role of ATP in cell structure and function is discussed in the next chapter.

With this introduction to the primary organic molecules and the larger polymers and macromolecules constructed from them, we can next look at how they interact to build the basic biological building block of life called the cell.

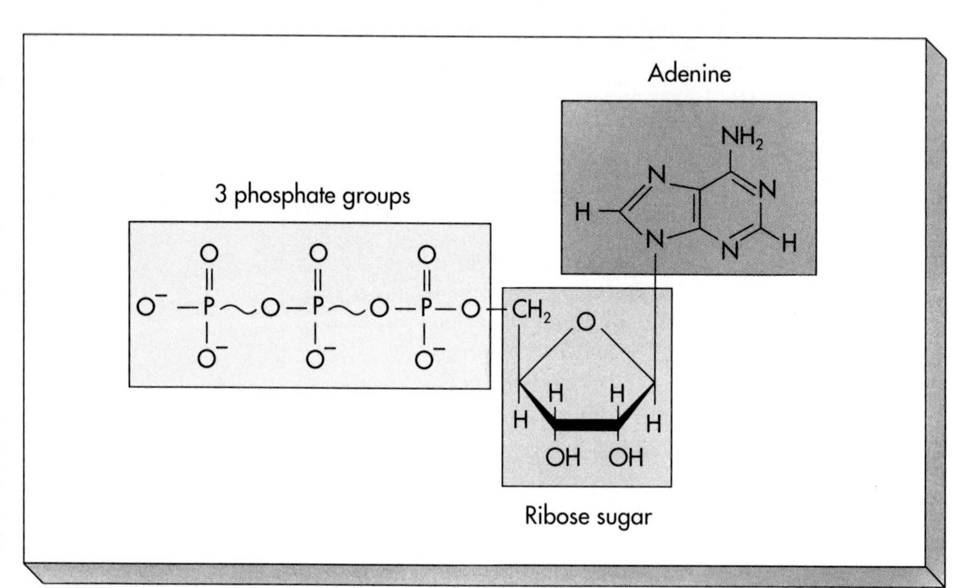

CHAPTER REVIEW

SUMMARY

1. The smallest unit of matter nondivisable by chemical means is the atom. An atom has a nucleus composed of protons and neutrons. Orbiting around the nucleus and occupying specific energy shells are negatively charged particles called electrons.
2. The number of electrons in the outer shell of an atom determines its chemical properties. If the outer shell is not filled, an atom can enter into one of several kinds of chemical bonds with other atoms. Atoms can form ionic, covalent, and hydrogen bonds.
3. Atoms are joined to form molecules. Combinations of two or more different types of atoms are called compounds.
4. Because water consists of polar molecules it has special properties essential to the biochemical processes of life. The weak hydrogen bonds that form between polar water molecules give water great temperature stability. It is also an excellent solvent capable of dissolving many different substances.
5. Acids, bases, and buffers are important components of the changing conditions brought about by chemical reactions during cellular processes. Because hydrogen ions are frequently transferred during chemical events, the nearly neutral pH conditions required by cells is maintained by buffers. Buffers are able to bond to hydrogen ions whenever their concentrations begin to rise and also release them back into solution when their concentrations begin to fall.
6. Carbon is the most important structural atom in the human body. Carbon atoms form the "skeleton" or "backbone" of small organic molecules most important to life. Because of its outer electron shell configuration, carbon forms only covalent bonds with other atoms.

7. Small organic molecules bond to build larger complex macromolecules. If the sequence of small organic molecules, such as simple sugars, is similar, the macromolecule is also called a polymer.
8. Biological catalysts, called enzymes, speed up the rate of specific chemical reactions and determine which reactions occur in the body. Enzymes are a class of proteins that allow the chemistry of life to proceed; without them chemical reactions would proceed too slowly to maintain life.

FILL-IN-THE-BLANK QUESTIONS

1. Every atom has a nucleus composed of ______ and ______ with a number of ______ orbiting the nucleus.
2. When an atom gains ______, it becomes ______ charged, and when an atom loses ______, it becomes ______ charged.
3. The chemical behavior of an atom is determined by the number of ______ found in its outermost shell.
4. The forces holding atoms in combination are called ______ ______.
5. A ______ ______ ______ forms when there is an unequal sharing of electrons between two atoms.
6. With polar compounds, such as water, the partially positive end of the molecules is attracted to the partially negative end of another molecule, forming a weak ______ ______.
7. Water is able to absorb relatively more heat compared with other common liquids because of its high ______ ______ ______.
8. ______ is a measure of hydrogen ion concentration in a solution. The higher the concentration, the more ______ the solution; and conversely the lower the concentration, the more ______ the solution.
9. If there is a transfer of an electron from one atom to another, the union of the now oppositely charged ions is called an ______ bond.
10. Enzymes increase the ______ of chemical reactions by lowering the ______ ______ required to initiate the reaction.
11. Dehydration synthesis always involves the formation of a chemical bond and the production of one ______ molecule.
12. Amino acids have identical structures except for the ______ groups.
13. Enzymes interact with and alter ______, resulting in a ______.

SHORT-ANSWER QUESTIONS

1. Write the molecular formula and structural formula of water.
2. What are the structural differences between a hydrogen atom and a hydrogen ion?
3. What are the differences between an atom of oxygen and a molecule of oxygen?
4. What is the primary difference between a nonpolar covalent bond and a polar covalent bond?
5. What kinds of molecules are hydrophobic and why?
6. Describe the functions of a buffer.
7. List and describe each of the four small organic molecules important to life.
8. List each of the structural levels of a protein.
9. What occurs when a macromolecule undergoes a hydrolysis reaction?
10. What occurs when two amino acids form a peptide bond?
11. Describe the similarities and differences between the different categories of lipid molecules.
12. Describe how an enzyme lowers the activation energy required for a chemical reaction to occur.

VOCABULARY REVIEW

acid—*p. 23*
atom—*p. 17*
ATP—*p. 33*
base—*p. 23*
buffer—*p. 25*
carbohydrate—*p. 27*
DNA—*p. 33*
electron—*p. 17*
enzyme—*p. 27*
lipid—*p. 27*
neutron—*p. 17*
nucleic acid—*p. 31*
pH scale—*p. 24*
protein—*p. 30*
proton—*p. 17*
RNA—*p. 33*

Chapter Three

CELL STRUCTURE AND FUNCTION

For Review

Here are some important terms and concepts that you will encounter in this chapter. If you are not familiar with them, you should review them before proceeding.

Hydrolysis and synthesis reactions (page 26)

Lipids (page 27)

Polar and nonpolar molecules (page 21)

Proteins (page 30)

OBJECTIVES

After reading this chapter you should be able to:

1. *List and describe the three characteristics common to all cells.*
2. *Describe the structure of the cell membrane and name its three functions.*
3. *Explain the arrangement of the phospholipid bilayer.*
4. *Define selective permeability and list the five most important transmembrane proteins and the function of each.*
5. *Describe the processes of simple diffusion and osmosis.*
6. *Define tonicity and differentiate between isotonic, hypotonic, and hypertonic solutions.*
7. *Compare and contrast passive and active transport.*
8. *Explain and differentiate between the processes of endocytosis and exocytosis.*
9. *Discuss the two important functions of cell organelles and describe the function of the major cell organelles.*
10. *Define energy and explain the role of ATP in the transfer of energy.*
11. *Differentiate between hydrolysis and synthesis pathways.*
12. *List the three metabolic pathways of cellular respiration.*
13. *Describe chromosomes and define homologue and sister chromatids.*
14. *Describe what happens in each of the four phases of mitosis.*
15. *Define cytokinesis.*

The biological building block of the human body is the cell. Cells are highly organized, dynamic systems possessing both permanent and temporary structures. Through the enzyme-controlled steps of cellular respiration, cells are able to release the energy necessary for the growth, maintenance, and reproduction of their biological structures and for driving all chemical reactions and processes in the cell.

The Generalized Cell

All organisms are composed of cells. Some, such as bacteria, are composed of a single cell, whereas others, such as human beings, are composed of many cells. Cells are dynamic, living systems. Structures within cells are constantly growing, extending, moving, sometimes multiplying, and then disappearing. Many molecules pass into and out of cells, and cells may respond to many different environmental signals. All this activity occurs in fractions of a second. Imagine that what you are about to read is equivalent to viewing a single frame from a movie called *The Life of a Cell.* From the single frame you would discover the basic components of cell structure and, possibly, how they function. However, all of the details of the cell, the activities of each component, and the cell's astounding flexibility would not be completely known until you saw the cell in "action." Using your imagination will help create the action to assist in learning some important facts about the cell. The following describes the basic structures and functions of a cell.

General Characteristics of a Cell

Cells that make up the human body vary in size, shape, and complexity, but they share at least three characteristics. First, a **membrane** surrounds the cell, separating it from its surroundings. There are passages and communication channels through the membrane, allowing the cell to *selectively* interact with the outside world. Second, cells contain **cytoplasm** (si'-to-plaz-zim), a semifluid material that contains chemical substances used by the cell, such as sugars, amino acids, proteins, and ions, necessary for cell maintenance, growth, and reproduction. Third, cells also contain membrane-bound structures, called **organelles,** with specific cellular functions.

The Cell Membrane

A cell is surrounded by a **cell membrane,** or **plasma membrane.** It is only several molecules thick and consists primarily of two layers of phospholipids stabilized by specifically oriented proteins. Depending on their function these proteins either partially or completely penetrate the bilayer. Through this thin membrane, approximately 10 nm thick, the cell selectively interacts with its surroundings.

The cell membrane has three functions. First the cell's life processes must be separated from the cell's surroundings. Molecules such as nutrients and enzymes must stay inside the cell. Consequently the cell membrane isolates the cytoplasm from the exterior environment. However, a cell must also acquire nutrients and building materials from its surroundings and eliminate waste products. The cell membrane's second function is to regulate the flow of molecules and substances into and out of the cell. Finally the third function is to allow communication with other cells. The millions of human body cells must be able to coordinate and integrate their diverse activities. Structural components of the cell membrane make highly specific communication between cells and their surroundings possible.

Phospholipid Bilayer

Recall that a phospholipid molecule consists of a glycerol molecule, with two fatty acid chains and a phosphorus- and nitrogen-containing group, each

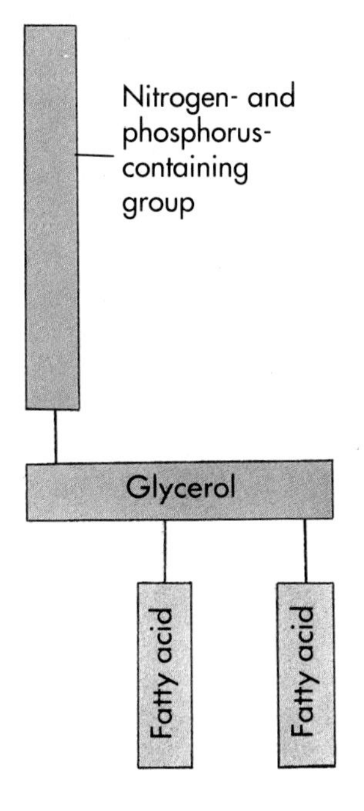

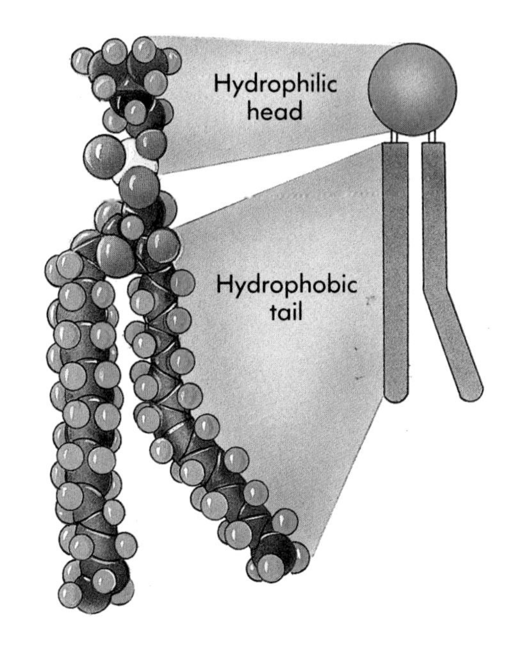

covalently bonded to the glycerol molecule. The nitrogen- and phosphorus-containing group is polar, making one end, the "head," of the phospholipid molecule water-soluble, or hydrophilic. The two nonpolar fatty acid chains make the other end, the "tail," of the phospholipid water insoluble, or hydrophobic. When phosopholipids are placed in water, the polar region of the phospholipid molecules form hydrogen bonds with the surrounding water molecules. The nonpolar tails, or fatty acid chains, will be "pushed away" from the water molecules. Water spontaneously organizes the phospholipids into two layers. The hydrophilic heads are toward the water, and the hydrophobic tails are toward each other and therefore away from the water. The result of this organization is called a **phospholipid bilayer.** Phospholipid bilayers are the foundation of all biological membranes. Because the interior portion of the bilayer is nonpolar, it repels water-soluble molecules, polar molecules, or electrically charged ions that attempt to pass through it. However, if a cell membrane were a pure phospholipid bilayer, the cell would be completely isolated from such water-soluble molecules as simple sugars and amino acids, which are necessary for the cell's survival.

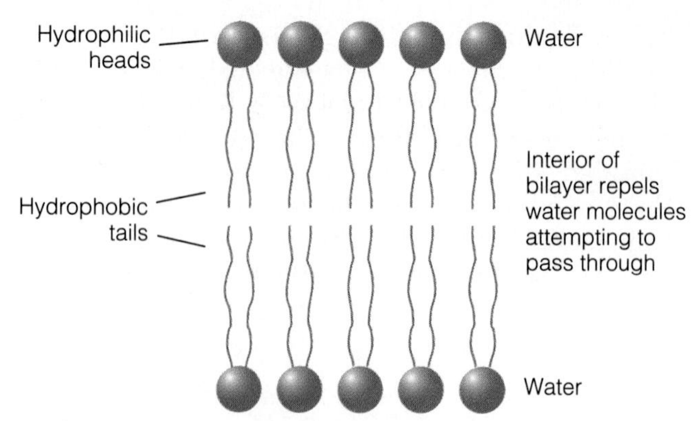

Membrane Proteins

The second structural component of the cell membrane, proteins, make the cell membrane **selectively permeable.** **Selective permeability** means that some molecules are allowed to pass through the cell

Figure 3-1 Cell membrane structure according to the fluid mosaic model

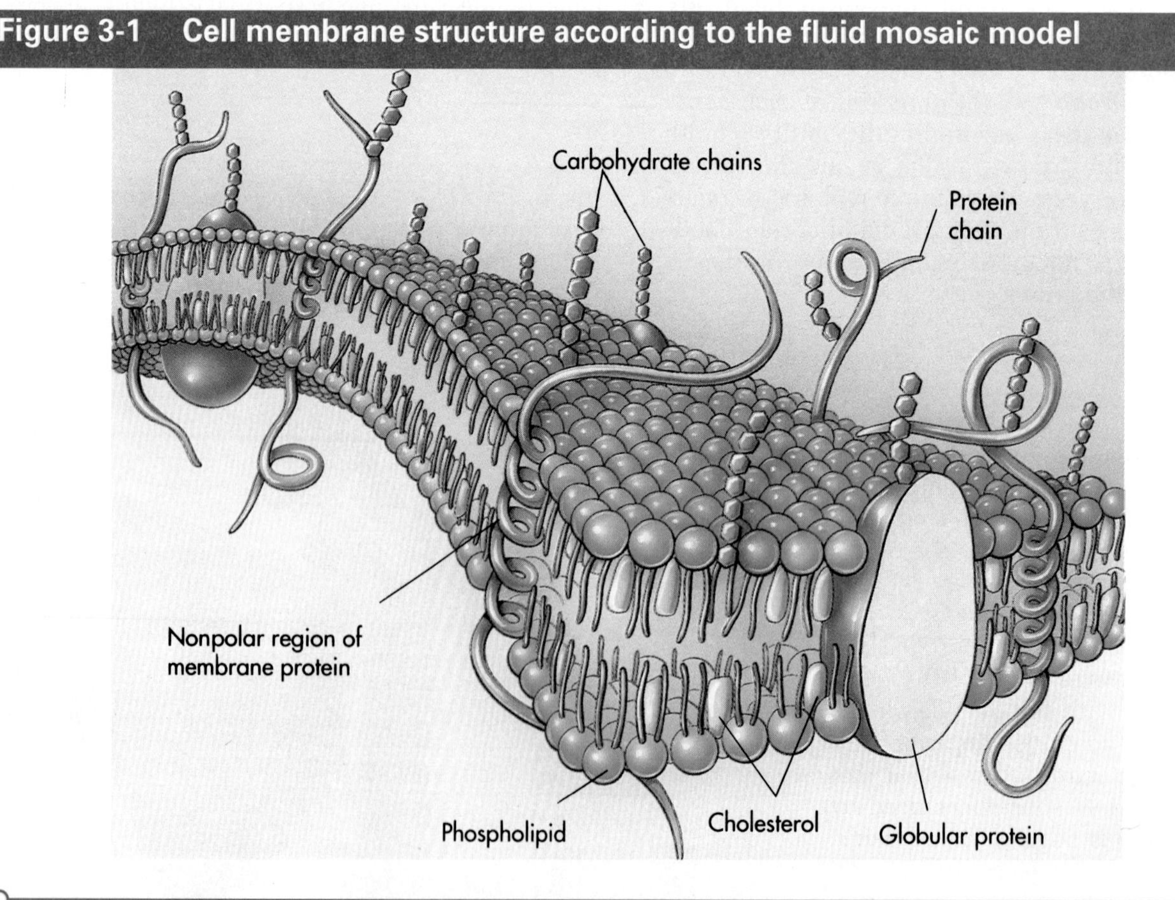

The cell membrane has a phospholipid bilayer with a variety of transmembrane proteins embedded in it. Some transmembrane proteins make possible the selective permeability of the cell membrane.

membrane while others cannot.

Different types of proteins are also embedded in the phospholipid bilayer. Thus the cell membrane is described according to the **fluid mosaic model.** The phospholipid molecules and proteins are "fluid" because they move around and jostle one another in reaction to the surrounding water molecules. The word *mosaic* refers to the assemblage of proteins in the membrane that help to stabilize the lipid bilayer and create various passages and channels through it (Figure 3-1). The most important transmembrane proteins are the channel; transport, or carrier; receptor; recognition; and attachment proteins (Figure 3-2).

Channel proteins allow water molecules and small water-soluble molecules to pass across the lipid bilayer. Channel proteins are also selective.

Figure 3-2 Transmembrane proteins

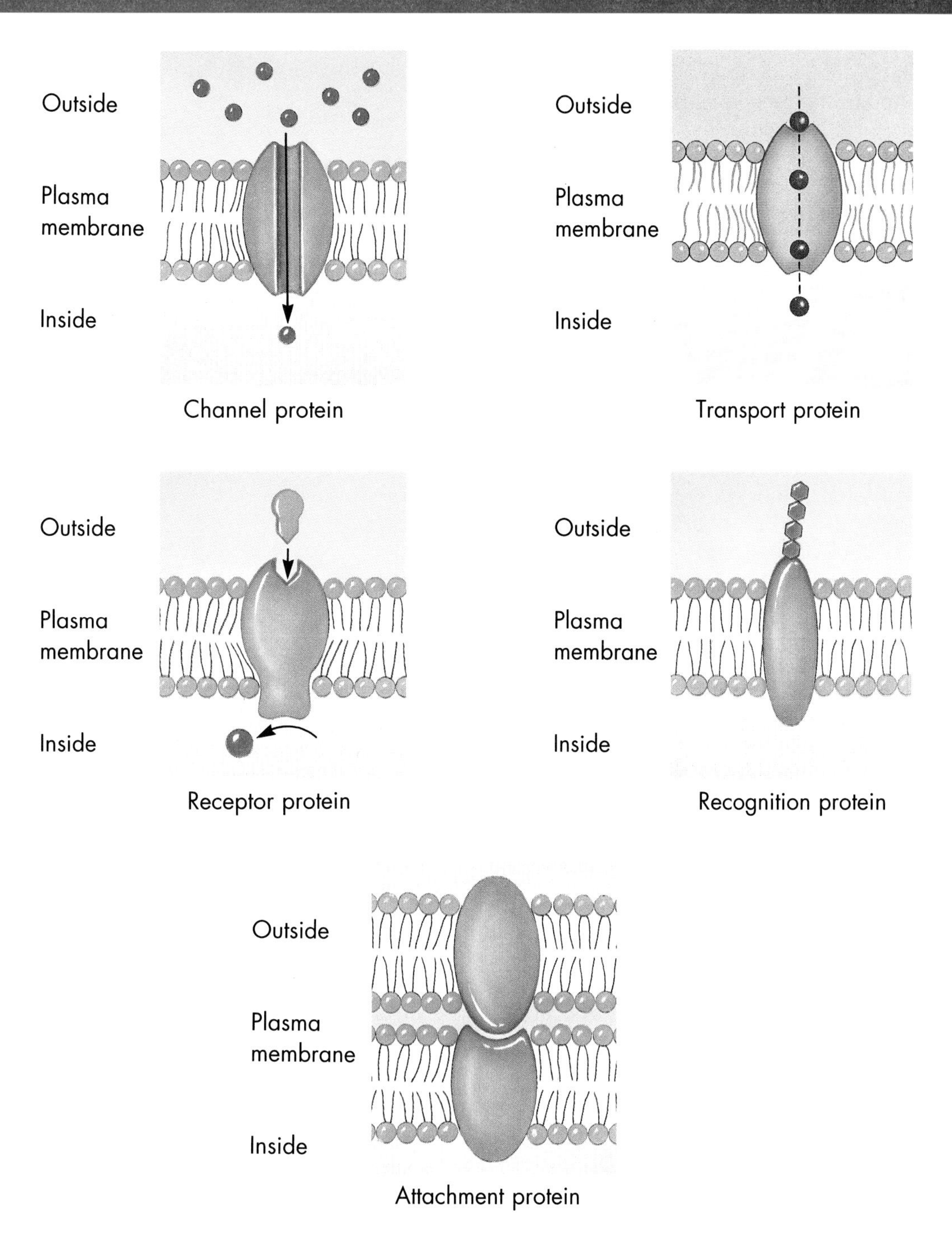

For example, some channel proteins can attract or repel particular ions. Other channel proteins are always open, and some open only under certain conditions.

Transport, or carrier, proteins assist molecules across the lipid bilayer. With some transport proteins, the cell must expend energy to move specific molecules in a particular direction across the membrane.

Receptor proteins, also known as cell-surface receptors, are embedded in the cell membrane and transmit information into the cell. The end of the receptor protein exposed to the cell's surroundings fits specific signal molecules. When a particular molecule encounters the receptor protein, it chemically bonds to it. This produces a change in the shape of the end of the receptor protein protruding into the cell. This shape change causes a change in some aspect of cell activity.

Recognition proteins embedded in the phospholipid bilayer indicate to other body cells that the cell belongs to the same body, that the cell is not foreign. All the cells of a particular person have the same "self" marker. Some recognition proteins also have lipid and carbohydrate molecules attached to them called **glycolipids.** The specific role of self-recognition proteins is described in Chapter 14.

Attachment proteins protrude from the exterior surface of the cell membrane, allowing cells to form connections with one another. Many body cells are in contact with other cells, usually as members of tissues in such organs as the lungs, heart, or skin. Attachment proteins help make this organization possible. Other proteins are arranged on the inner surface of the cell membrane and are secured to certain internal proteins, constituting a weblike network that helps to maintain cell shape and in some cells makes movement possible.

Movement Across the Cell Membrane

Molecules are in constant, random motion. As a consequence collisions between molecules can occur, causing them to bounce off one another in different directions. If there are many molecules in a given volume of space, the molecules will collide with one another more frequently and move away from each other toward areas where there are fewer molecules. Although random collisions send the molecules back and forth, the **net movement** is outward from the area of higher concentration. As a result molecules move down their **concentration gradient.** This means they move from an area of higher concentration to areas of lower concentration. For example, if a cube of sugar is placed in a beaker of water, the sugar molecules will dissolve and move from the area of highest concentration (the cube) to the area of lesser concentration (the water in the beaker), eventually resulting in an even distribution of sugar molecules throughout the water. The random movement of molecules or ions down their concentration gradients from an area of higher concentration to an area of lower concentration is called **simple diffusion.** Some of the more important lipid-soluble molecules that move by simple diffusion across the cell membrane include oxygen and carbon dioxide.

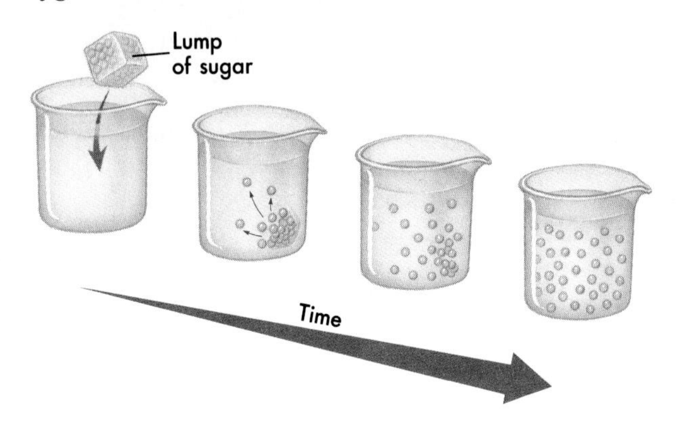

Osmosis

The cytoplasm of the cell is a semifluid watery medium containing many substances. The mixture of these or any substances in water is called a **solution.** The fluid portion of this solution is called the **solvent.** The substances dissolved in the solvent make up the **solute.** Some simple experiments demonstrate an important relationship between the solvent and solute portions of a solution and also illustrate how substances move through the cell membrane.

A container is divided into two equal sections by a selectively permeable membrane that allows water, but not the larger sugar molecules, to pass through freely. Side *A* holds pure, distilled water, meaning there are no solutes in it. Side *B* holds a solution of 90% water and 10% sugar. Which side has the greater concentration of water molecules? Side *A* has a greater concentration of water molecules per unit of volume compared with side *B*. As a result water will flow down its concentration gradient, moving from side *A* to side *B* until the concentration of water molecules on both sides of the membrane is equal (Figure 3-3). At equilibrium, side *B* contains *93%* water and *7%* sugar. The diffusion of water across a selectively permeable membrane is called **osmosis.**

Figure 3-3 Osmosis

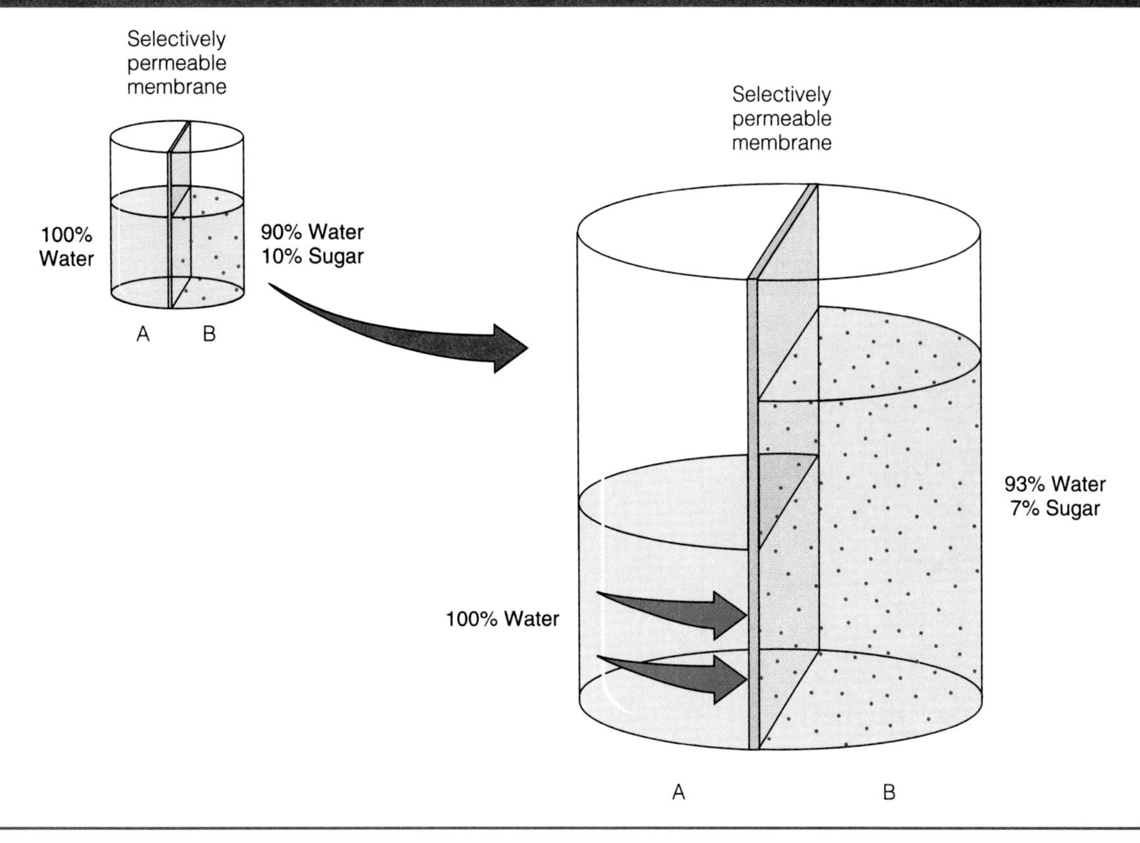

Osmosis is the diffusion of water across a selectively permeable membrane.

Tonicity

The osmotic movement of water across a selectively permeable membrane is in turn affected by **tonicity.** **Tonicity** is the relative concentration of solutes in two fluids on either side of the membrane.

Another experiment illustrates tonicity. A solution of 90% water and 10% sugar is placed into a bag made of a selectively permeable membrane. The bag is then placed into a larger container of 90% water and 10% sugar (Figure 3-4). When solute concentration is equal on both sides of the membrane, it is called **isotonic.** There is no net osmotic movement of water in either direction. If the bag is removed and placed into a container of 100% pure water containing no solutes, then the concentration of solute is *relatively* less outside of the bag compared to inside. The solution is said to be **hypotonic** (*hypo* = below) when solute concentration is *less* outside the membrane. Water moves from the container into the bag, resulting in the bag swelling up. Finally if the bag with the original solution of 90% water and 10% sugar is placed into a container of solution made up of 80% water and 20% sugar, the solution in the container is **hypertonic** (*hyper* = above) *relative* to the solution in the bag. The solution is said to be **hypertonic** when solute concentration is *greater* outside the membrane. The net osmotic movement of water will be from the inside of the bag into the container, and the bag will shrink.

Because of the channel proteins and molecular size, water easily moves through the cell membrane in response to solute concentration. (About 100 times the volume of water in a cell crosses the cell membrane every second.) Even though water is rapidly entering and leaving cells, the cells normally do not swell up or shrink because the solute concentration in the solution surrounding the cells is maintained at the same level as the solute concentration in the cell.

Selective Transport of Molecules Across the Cell Membrane

With the exception of water molecules, soluble polar molecules, such as amino acids or glucose, cannot cross the cell membrane on their own, regardless of their concentration gradients. Although the selectively permeable membrane ensures that the large polar proteins and other molecules essential to its survival stay within the cell, mechanisms for transporting essential sub-

Figure 3-4 Tonicity

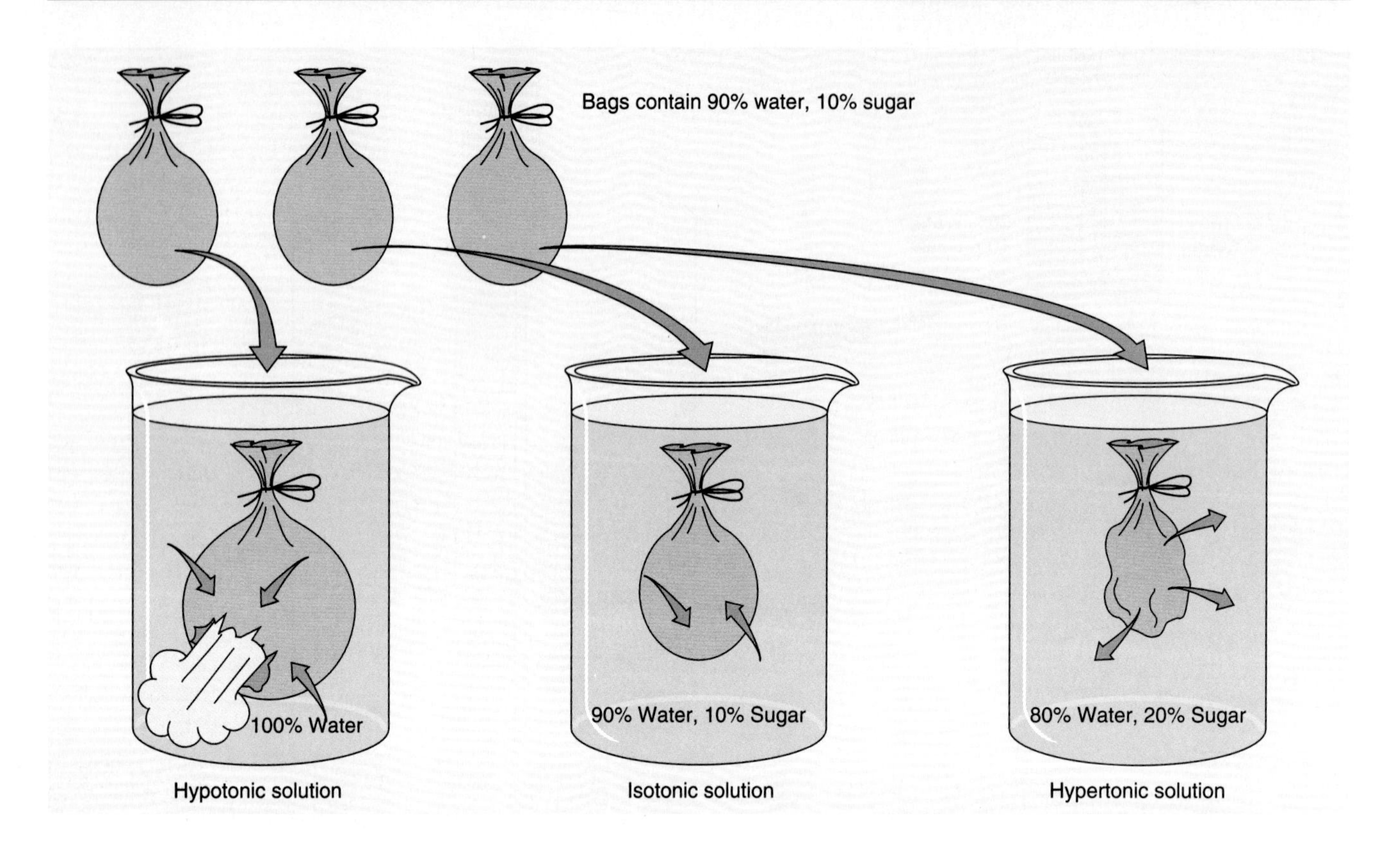

When solute concentration is equal on both sides of a selectively permeable membrane-bound bag, there is no net movement of water. Therefore the solution is *isotonic.* When solute concentration is relatively less outside the bag, the solution is *hypotonic.* The movement of water is into the bag. When, however, the concentration of solute is relatively higher outside of the bag, or *hypertonic,* the movement of water is out of the bag and into the surrounding solution.

stances into the interior of the cell must also exist. The two different methods of transporting solutes across the cell membrane are passive and active transport.

In **passive transport** there is a carrier protein embedded in the cell membrane specific to the molecule to be transported. On contact the molecule chemically bonds to the carrier protein. This

Extracellular fluid
Transported molecule
Carrier molecule
Cytoplasm

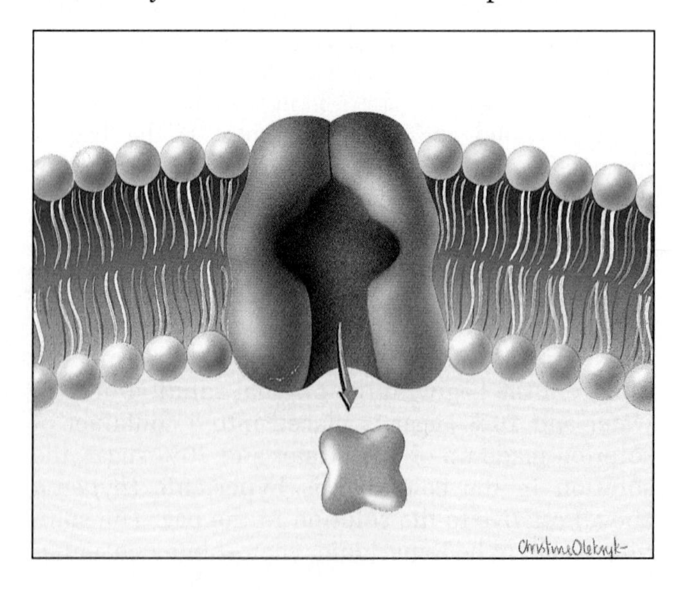

enables the molecule to pass through the cell membrane in the direction of its concentration gradient, either into or out of the cell. The carrier protein facilitates, or helps, the molecule pass into or out of the cell. This transporting method is thus also called **facilitated diffusion.** In passive transport the carrier protein is specific to a particular molecule, and the process is passive. The direction of movement is determined by the concentration gradient of the transported molecule.

Active Transport

Many molecules can cross the cell membrane against their concentration gradients. For example, amino acids and nucleotides cross the cell membrane into the cytoplasm even though they are maintained at higher concentrations in the cell compared with its surroundings. On the other hand, some cells continue to expel ions into their surroundings even though they have a much lower concentration of ions in their cytoplasm compared with their surroundings.

To move molecules and ions against their concentration gradients, the cell must expend energy. This type of transport, called **active transport,** requires a carrier protein and the expenditure of energy in the form of ATP, which enables the cell to move molecules across the cell membrane against their concentration gradients. The molecule to be transported first chemically bonds to a specific carrier protein of the cell membrane, causing the protein to change shape slightly. In the cell an ATP molecule chemically bonds to the carrier protein, then breaks down and releases its stored energy. This causes the protein to change shape again, which allows the molecule to move to the high concentration side of the membrane. The expended ATP molecule breaks away from the carrier protein, allowing it to regain its original shape.

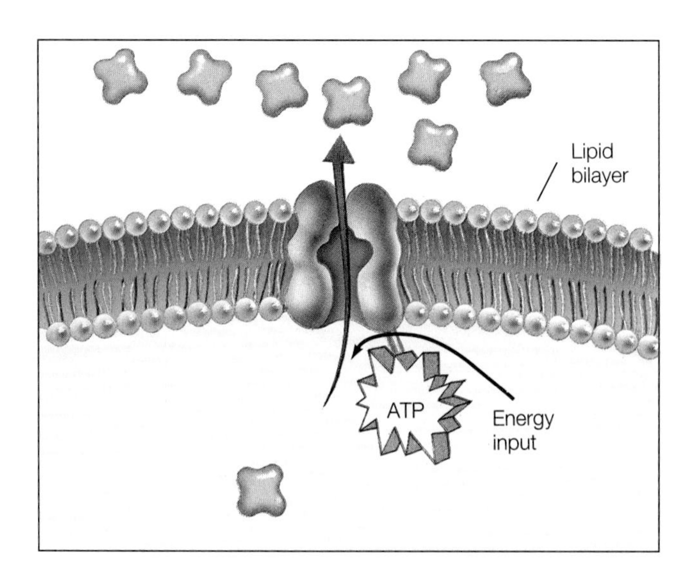

Generalized Endocytosis

Cells of the body are also capable of bringing in large polar molecules, particles, and dissolved molecules by engulfing and incorporating them into their interior by a process called **generalized endocytosis.** During **endocytosis** the cell membrane first completely surrounds the substance to be taken into its interior and then pinches off. This creates a **vesicle** (ves′i-kl), meaning "small vessel," that encloses the engulfed material within the cell.

If the material brought into the cell is a particle, such as a bacterium or some cell fragment, the process is called **phagocytosis** (fag′o-si′to′sis). If the material is liquid and contains dissolved molecules, the process is called **pinocytosis** (pin′o-si-to′sis) (Figure 3-5).

Cells can also expel material from their cytoplasm into the surrounding extracellular fluid through a process called **exocytosis.** In **exocytosis** organelles in the cell create a vesicle around the material to be removed. After fusing with the cell membrane, the vesicle opens to the exterior, and its contents diffuse out (Figure 3-6).

Cell Organelles

Organelles are membrane-enclosed, subcellular structures that perform specialized tasks. Organelles have two important functions. First they physically separate different chemical reactions in the space of the cytoplasm. Second they also separate different chemical reactions in time. The following section describes the structure and function of each organelle illustrated in Figure 3-7.

Nucleus

The **nucleus** is the largest single organelle in the cytoplasm of the cell and is usually located near the center of the cell. The nucleus is surrounded by a double-layered membrane called the **nuclear envelope,** which separates the nucleus from the rest of the cell. Each layer of the envelope is a phospholipid bilayer. Scattered over the surface of the nuclear envelope are holes called **nuclear pores.** Nuclear pores are points at which the two membrane layers fuse to form openings through the membrane. The pores are lined with channel proteins that permit the contents of the nucleus to

Figure 3-5 Phagocytosis and pinocytosis

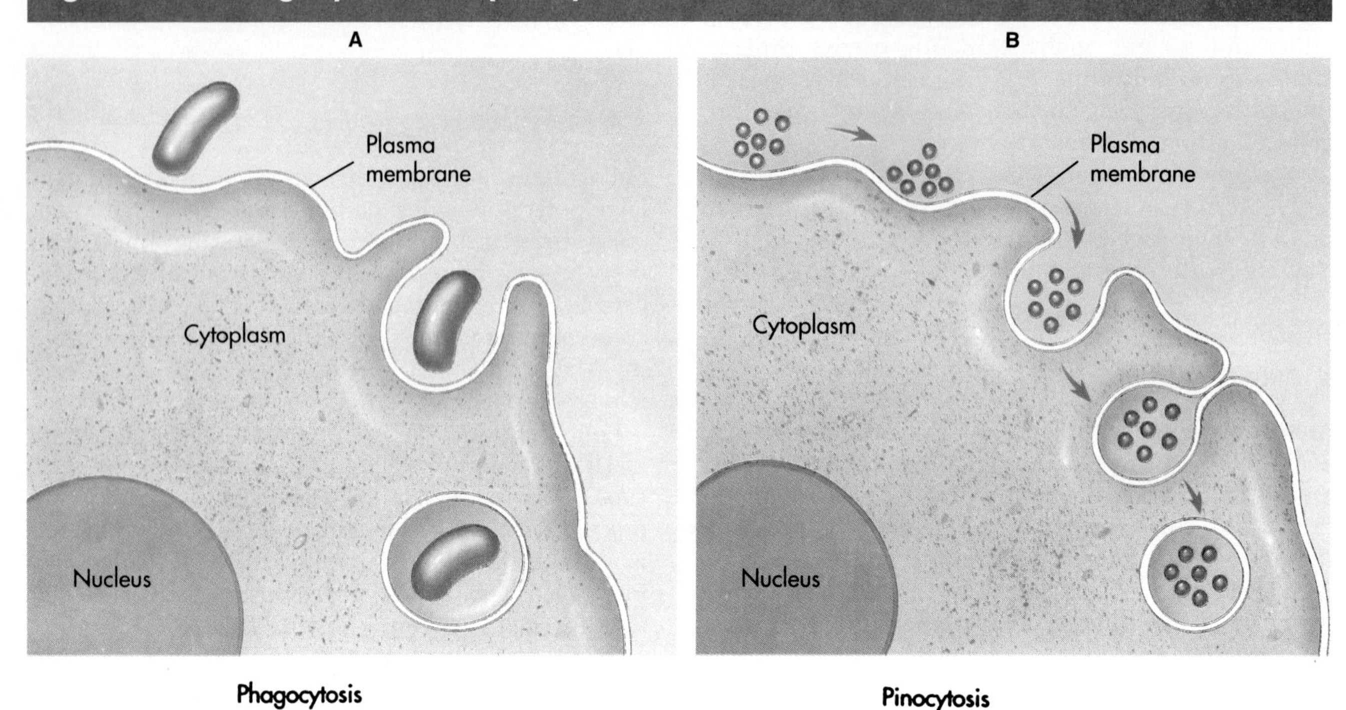

Most cells move particles through the cell membrane and into the cytoplasm by a process called phagocytosis (A). If the material to be brought into the cell is a liquid containing dissolved molecules, the process is called pinocytosis (B). Both phagocytosis and pinocytosis are special types of endocytosis.

Figure 3-6 Exocytosis

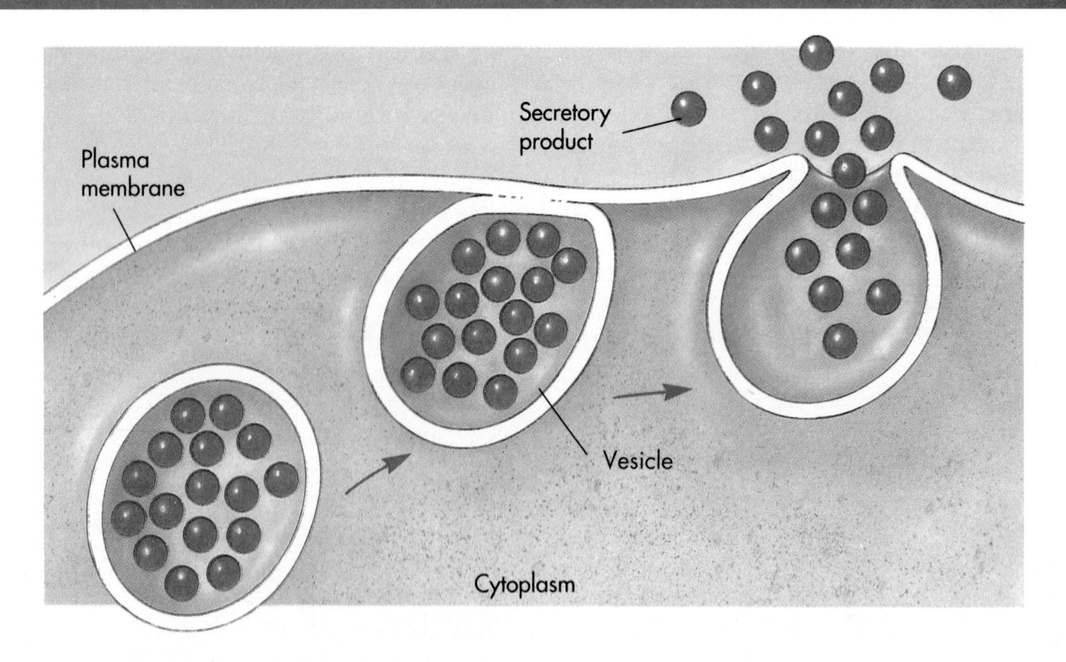

A cell can expel material from its interior through the process of exocytosis. The material to be removed is first surrounded by a membrane-bound vesicle which then moves to the cell membrane and fuses with it, releasing the material to the cell's surroundings.

selectively interact with the rest of the cell.

The nucleus contains the cell's genetic material, DNA. The nucleus also contains small spherical bodies called **nucleoli** (noo-kle-uh-li). Nucleoli produce small structures, called **ribosomes,** needed by the cell to assemble proteins.

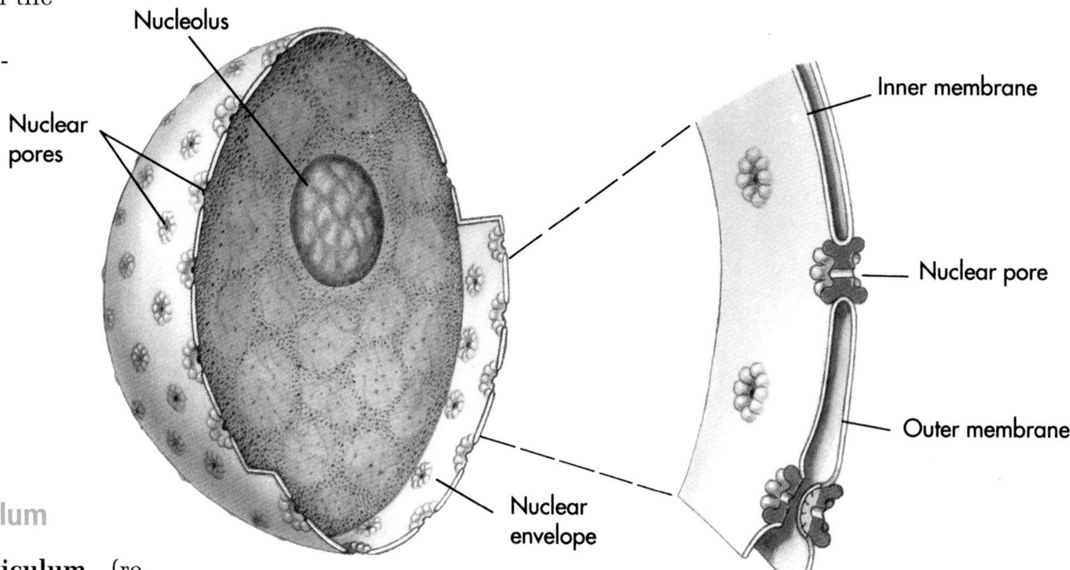

Endoplasmic Reticulum

The **endoplasmic reticulum** (re-tik'u-lum) (**ER**) is an extensive membranous system of interconnected fluid-filled tubules and flattened sacs that coils and twists through the cytoplasm. It is continuous with the outer membrane of the nuclear envelope.

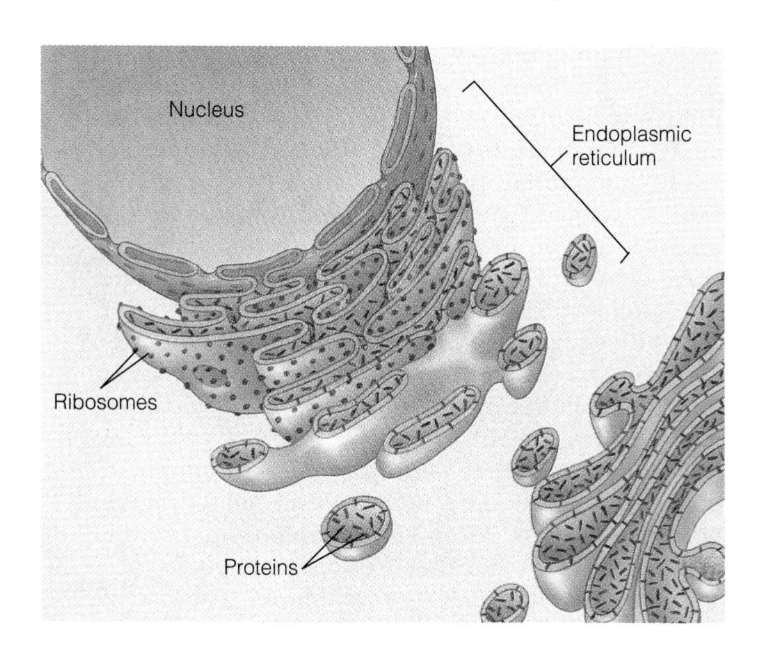

Like the cell membrane the ER is composed of a phospholipid bilayer, with various protein enzymes attached to its surface. When viewed under an electron microscope, part of the ER appears grainy. The grains are actually ribosomes attached to the membrane surface of the ER. This part of the ER is referred to as the **rough endoplasmic reticulum.** The surface of the **rough endoplasmic reticulum** is where the cell manufactures proteins destined for export from the cell.

Other portions of the ER membrane do not have ribosomes and are called the **smooth endoplasmic reticulum.** The **smooth ER** is involved in the synthesis, storage, and intracellular transport of lipid-containing substances.

In some cells the enzymes of the smooth ER are responsible for steroid synthesis. In the other cells, such as those of the liver, the smooth ER contains enzymes that detoxify a variety of drugs and toxic substances.

Golgi Complex

The **Golgi complex** is a set of flattened, slightly curved, membrane-bound sacs called **Golgi bodies.** They are thin in the middle but have enlarged edges. Golgi bodies collect, package, and distribute molecules synthesized elsewhere in the cell. The ER and Golgi complex are functionally related. They produce, package, and determine the destination of protein and lipid substances produced in the cell. Most of the newly synthesized molecules produced by the ER migrate, via vesicles, to the Golgi complex (Figure 3-8).

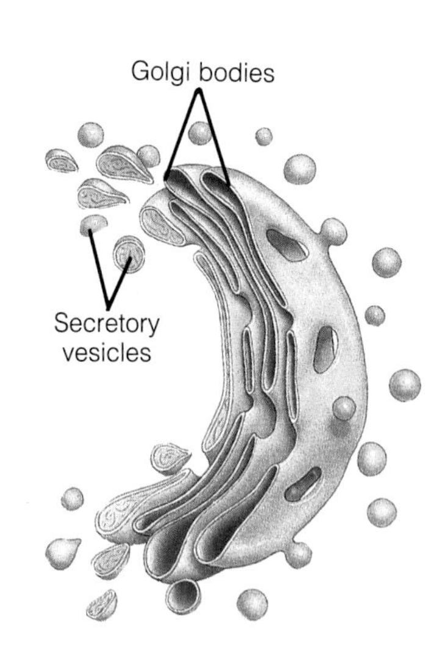

Lysosomes

Within the cell some vesicles that bud from Golgi bodies become **lysosomes,** which are membrane-bound organelles containing digestive enzymes synthesized by the cell and packaged in the Golgi complex. Unlike the secretory vesicles, lysosomes stay in the cell's cytoplasm. Lysosomes are the primary organelles of intracellular digestion.

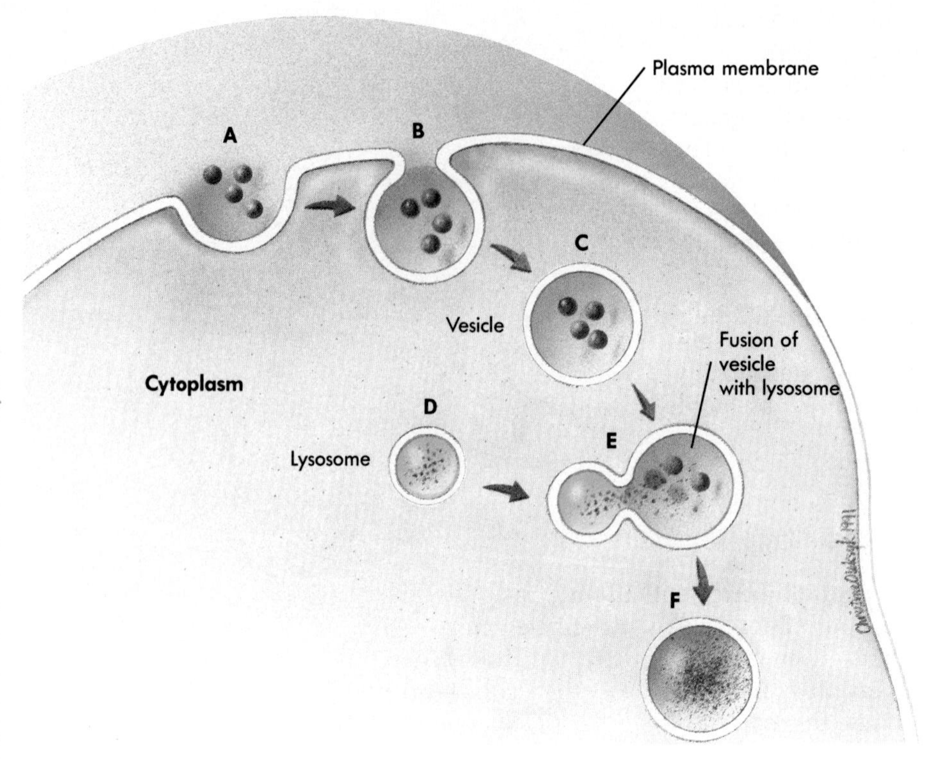

The digestive enzymes in lysosomes are able to rapidly break down proteins, nucleic acids, lipids, and carbohydrates. Lysosomes digest worn-out cellular components, making space for newly formed ones while recycling the materials "locked up" in the old ones. If needed, lysosomes are also able to digest material brought into the cell through endocytosis. The lysosomes carry out their tasks by first fusing with the vesicle containing the newly introduced material and then releasing its digestive enzymes into it. In the same way, lysosomes also fuse with aged or damaged organelles. If a whole cell is damaged or dies, the lysosomes rupture and release their enzymes into the cytoplasm, ultimately digesting the whole cell.

Mitochondria

Mitochondria (mi-to-kon-dre-uh, sing. mitochondrion) are the energy organelles of the cell. The mitochondria, often called the "powerhouses" of the cell, extract usable energy from nutrients required to maintain the cell's activities. The number of mitochondria per cell varies, depending on the energy requirements of each particular cell type.

Mitochondria appear under the microscope as sausage-shaped structures. Each mitochondrion has an outer and an inner membrane, each a phospholipid bilayer. The outermost membrane surrounds the mitochondrion itself, and the inner membrane forms a series of walls creating mazelike compartments called **cristae** (kris-te). The interior spaces of the mitochondria are filled with a semifluid substance called **matrix.** The matrix itself is a mixture of enzymes necessary to prepare the nutrient molecules for the final extraction of usable energy by the cristae's enzymes. Many enzymes embedded in the cristae are also responsible for converting molecules into a usable form of energy for the cell. The role of mitochondria in energy production is described later in this chapter.

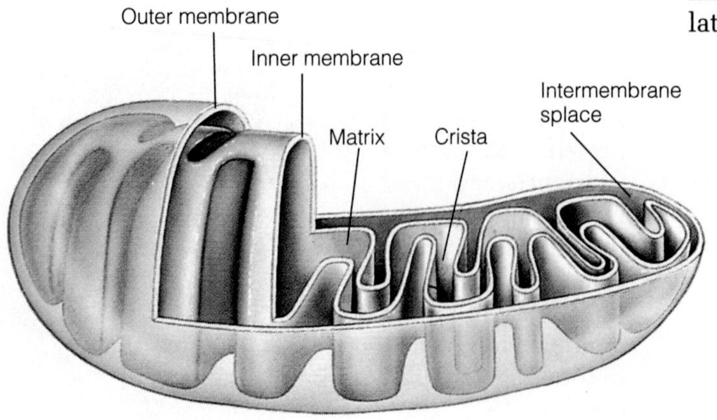

Other Cellular Components

Human body cells are of many different shapes and sizes. In addition to these differences, some cells are capable of self-propelled movement while othes are able to move substances across their surfaces.

Figure 3-7 A generalized cell and its organelles

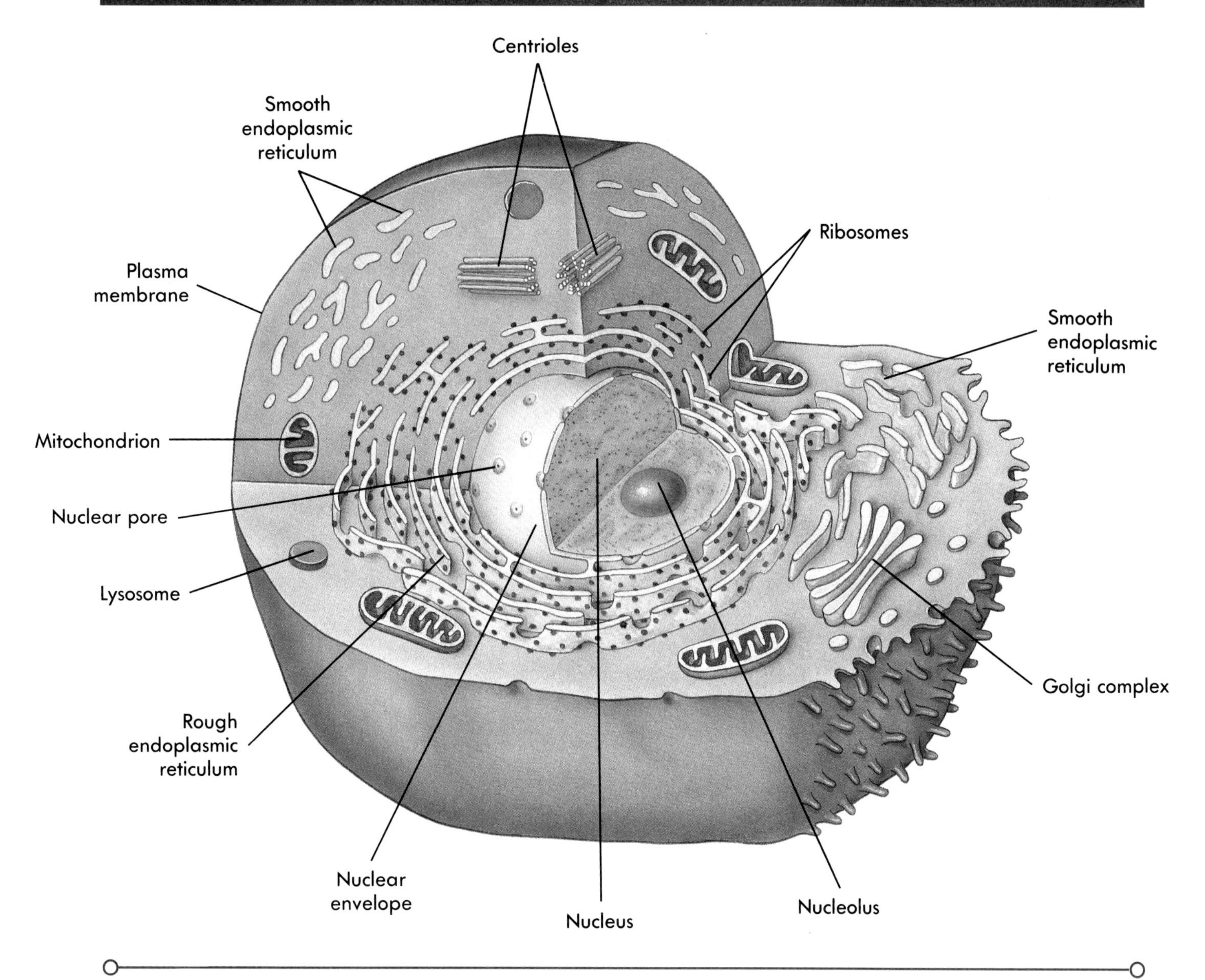

There are approximately 100 different cell types that make up the human body. Depending on its shape and function, each cell type has different numbers of each organelle in its cytoplasm.

Centrioles

Centrioles are two short rod-like structures that lie at right angles to one another near the cell's nucleus. Each centriole is composed of triplets of protein fibers organized in a circle. The centrioles act as anchoring points for spindle fibers which move the hereditary material during cell division. The role of centrioles in cell division is described in Chapter 5.

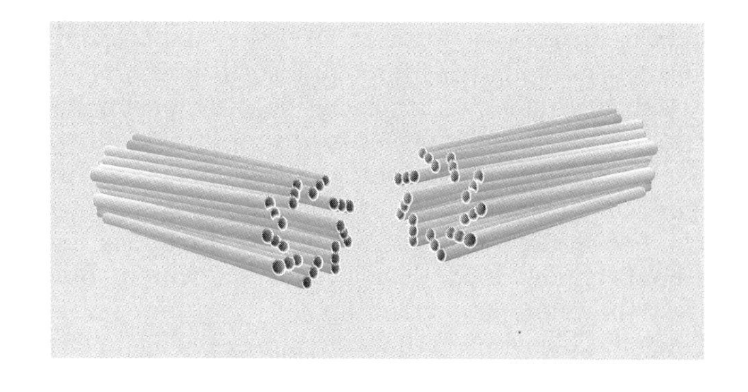

Cytoskeleton

The cytoplasm is crisscrossed by a network of protein fibers that assist in maintaining the shape of the cell and in anchoring organelles in the cytoplasm. This protein network is called the **cytoskeleton.** The **cytoskeleton** is responsible for maintaining cell shape and anchoring the organelles found in the cytoplasm and, depending on the structure and function of the cell, is composed of several different protein fibers.

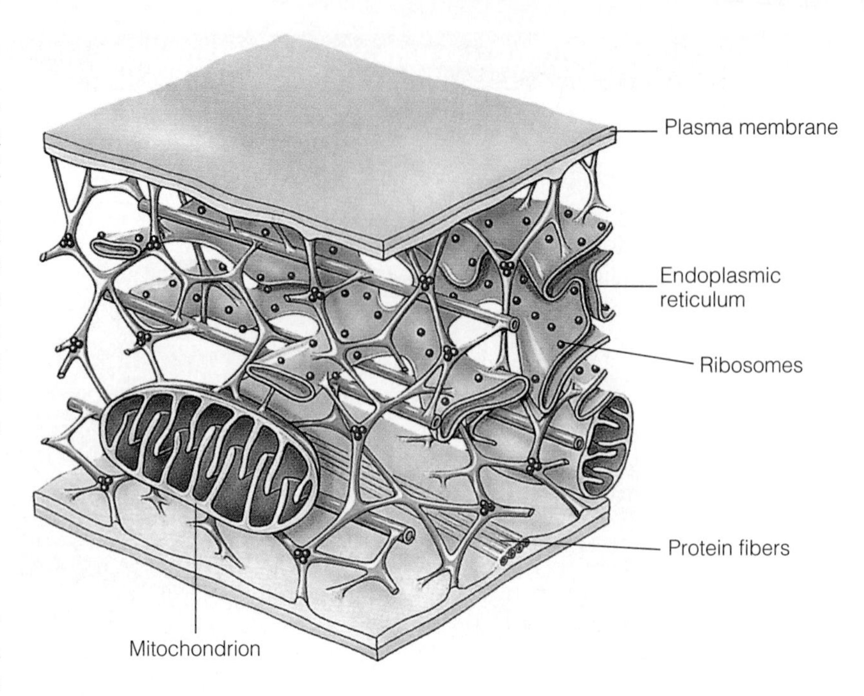

Cilia and Flagella

Cilia and flagella are specialized protrusions of the cell surface that allow a cell to either move materials across its surface or propel itself through its environment.

Cilia (sih′-le-uh) are numerous tiny, hairlike protrusions, whereas a **flagellum** (fluh-jeh′-lum) is a single, long, whiplike appendage. They differ only in length. Even though they project from the cell surface, both cilia and flagella are intracellular structures covered by the cell membrane (Figure 3-9).

Approximately 100 different cell types make up the human body. Depending on its function, each cell type has different numbers of organelles in its cytoplasm. How cells extract energy from molecules is the focus of the next section.

How Cells Acquire Energy

Imagine that you are walking down a street, watching a bird flying overhead and noticing your neighbor waving to you from her front porch. All these behaviors are made possible because of the highly coordinated action of the millions of cells that make up each of the organisms in this scene. Each cell is alive with activity. Proteins are being synthesized, substances are passing through the cell membranes, cellular components are being constructed, modified, and replaced, and organelles are functioning. Each activity requires energy. All the organisms in this scene obtain energy from foods. If any of them stopped eating, their bodies would first use the stored energy, but then eventually would die. Growth, reproduction, and heredity require energy. Once this energy is used it cannot be used again. To understand the energy production process we must first understand the nature of energy itself.

Energy

Energy is defined as the ability to bring about change or the capacity to do work. For example, the movement of a car, the force of a pencil falling off a desk, or the swift movement of a bird's flight each require energy. Energy is also "heat," such as the warmth of a fire, the pull of a magnet, or the flow of an electrical current. Each form of energy can bring about change.

Energy exists in two states. Some energy is actively doing work. Reading this book, turning pages occasionally, blinking the eyes, and breathing are all possible because of working energy. Other energy, not actively engaged in change or work, is called **potential energy,** or stored energy. For example, before the pencil fell from the top of the desk it had the potential energy stored from someone placing it there originally.

Although energy can take many forms, there is a simple and uniform way to measure it. The unit of measurement is based on the basic principle that all forms of energy can be converted to heat. Recall from Chapter 2 that the basic unit of measurement is the **calorie.** Because the calorie is such a small measure and organisms use a great deal of energy, the **kilocalorie,** equal to 1000 calories, is used. To avoid confusion *Calorie,* spelled with a capital *C,* has been defined to represent a kilocalorie (1000 calories). For example, food charts indicating the

Figure 3-8 Function of the Golgi complex and endoplasmic reticulum

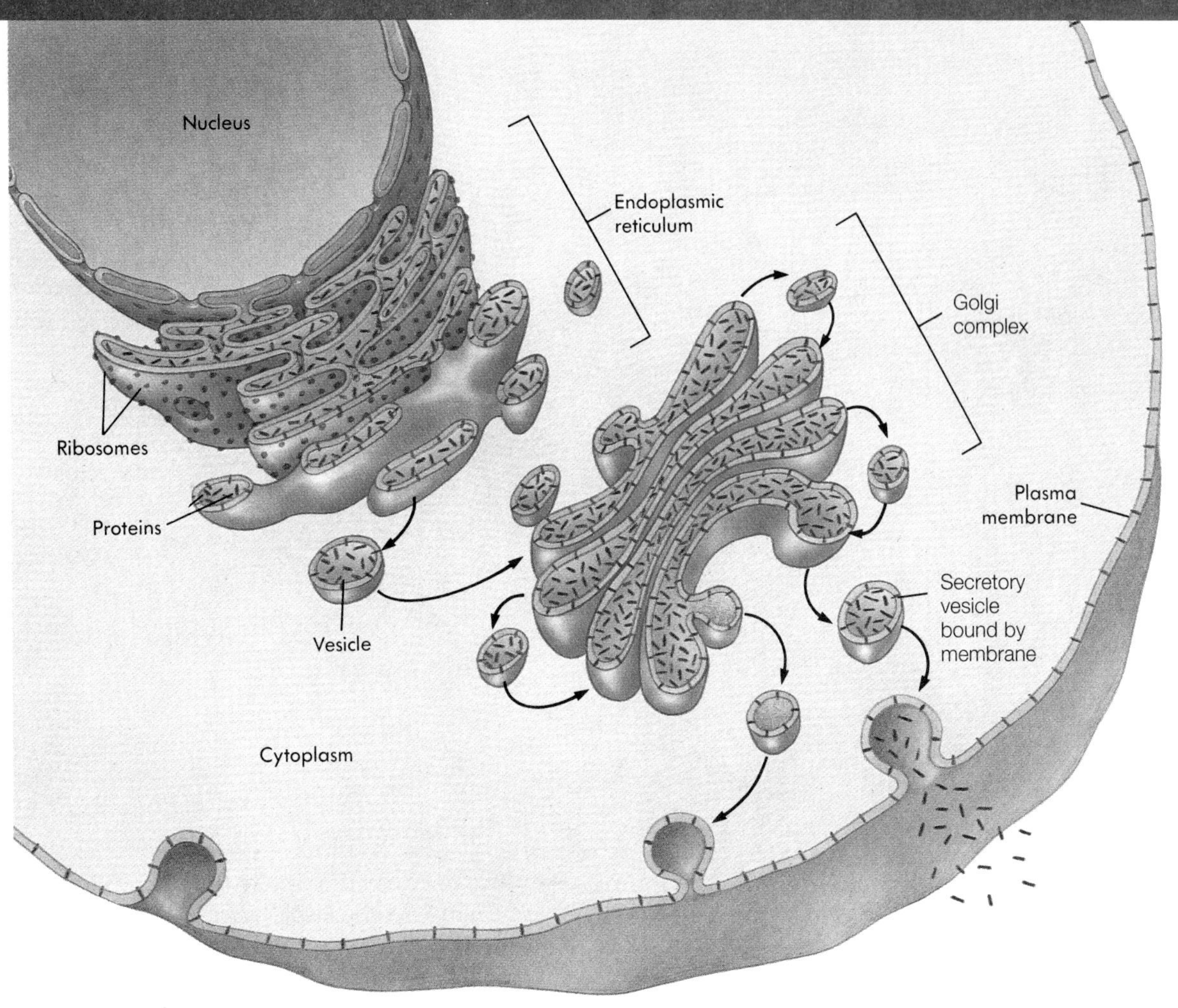

Proteins destined for export from the cell are manufactured in the rough endoplasmic reticulum and then packaged into vesicles that move to the Golgi bodies of the Golgi complex. The vesicle fuses with a Golgi body, releasing its contents into its interior where the protein is further processed. The processed protein moves to the outer margin of the Golgi body where another vesicle forms around the material. The newly created secretory vesicle moves to the cell membrane, fuses with it, and releases the protein into the cell's surroundings.

number of "calories" of various foods are actually using the term *Calorie* as used in biology.

Hydrolysis and synthesis reactions are constantly occurring in cells. Synthesis reactions require energy, and hydrolysis reactions release energy to the cell. However, synthesis and hydrolysis reactions usually occur in different places in the cell. There are two reasons why cells cannot directly use the energy released from hydrolysis to drive other synthesis reactions. First the hydrolysis reaction would have to occur at the same time and the same place as the synthesis reaction. Second the amount of energy released in the hydrolysis reaction would have to equal the amount of energy required by the synthesis reaction. This would be an inflexible and inefficient form of cellular chemistry.

ATP: The Universal Energy Currency

The cell uses an energy-carrier molecule to temporarily store the energy released from hydrolysis reactions and transfer it to the site of all synthesis reactions and other energy-requiring reactions, such as actively transporting substances across cell membranes. The energy carrier molecule is **adenosine triphosphate,** or **ATP.**

When the cell requires energy, the high-energy chemical bond of the last phosphate group in ATP is broken down as follows:

$$ATP \rightarrow ADP + P_1 + ENERGY$$

The result is **adenosine diphosphate,** or **ADP,** a free inorganic phosphate molecule (P_1), and some energy. On the other hand, by using the energy

Figure 3-9 Cilia and a flagellum

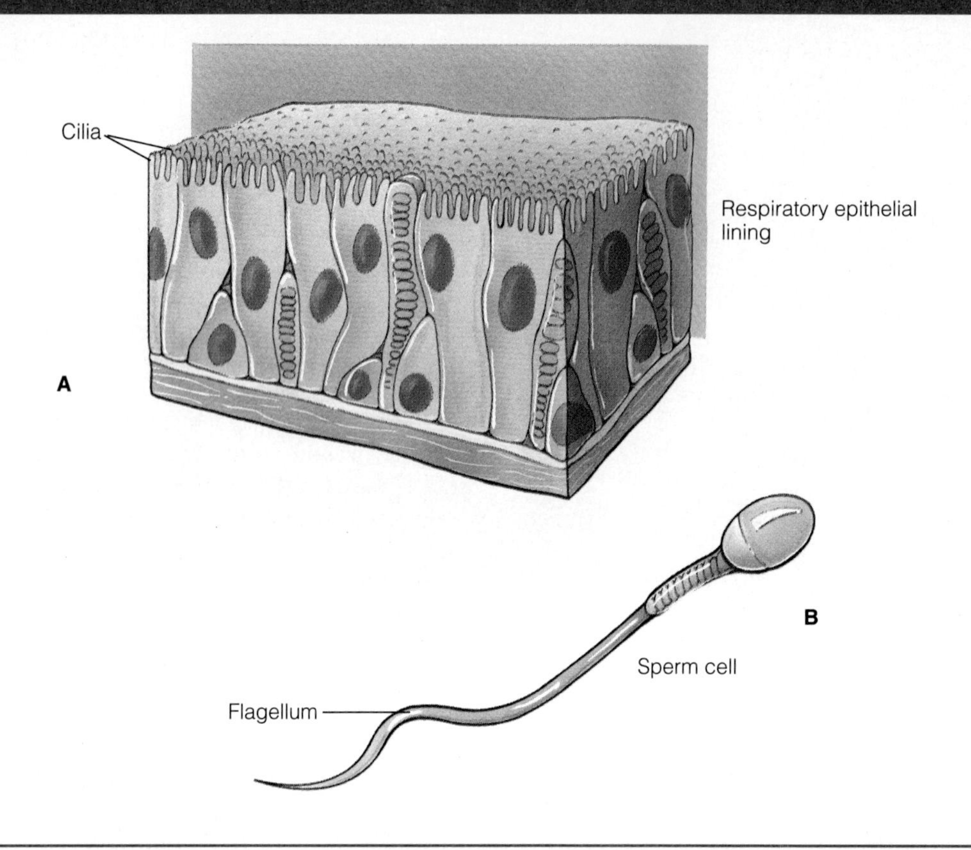

Cilia are tiny, hairlike protrusions from the cell membrane. Cilia are found on cells lining the respiratory airways of the lung. The wavelike motion of the cilia move inhaled debris back toward the entrance of the respiratory system (A). A flagellum is a single, large, long whiplike appendage protruding from the cell membrane. A male sperm cell moves by means of a flagellum (B).

released from hydrolysis reactions the cell can also manufacture ATP, as follows:

$$ADP + P_1 + ENERGY \rightarrow ATP$$

You might think of ATP as the energy "currency" of the cell. It "pays" the energy costs of synthesis and other energy-requiring reactions by breaking down ATP into ADP and releasing energy for energy-requiring activity. It also "saves" the energy released in hydrolysis reactions by synthesizing ATP from ADP and phosphate molecules.

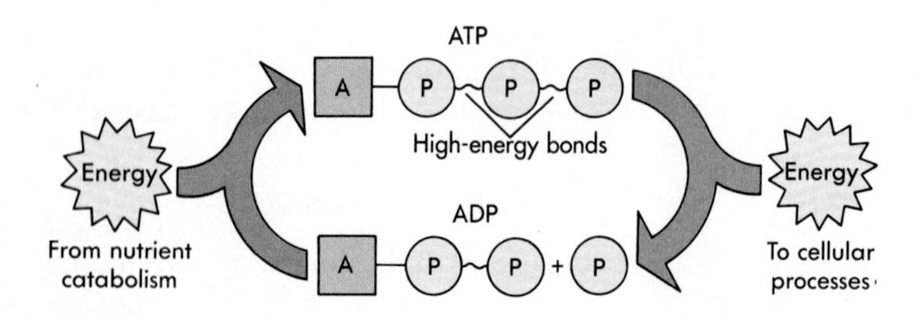

Metabolic Pathways

All the chemical reactions and transformations in living cells are known as **cellular metabolism.** All cellular metabolic chemical reactions occur in orderly sequences of steps called **metabolic pathways.** Each step of a metabolic pathway is controlled by a specific enzyme. Each usually produces a small change in a substance, such as an electron, atom, or molecule becoming attached to or detached from the substance.

Metabolic pathways have many distinctive parts. The organic and inorganic substances that enter into a reaction are called **reactants.** They are also referred to as substrates. **Metabolites** are substances being passed through a metabolic pathway. They are the intermediate forms in the assembly or breakdown of substances. **Cofactors** are substances that assist enzymes, usually carrying atoms or electrons

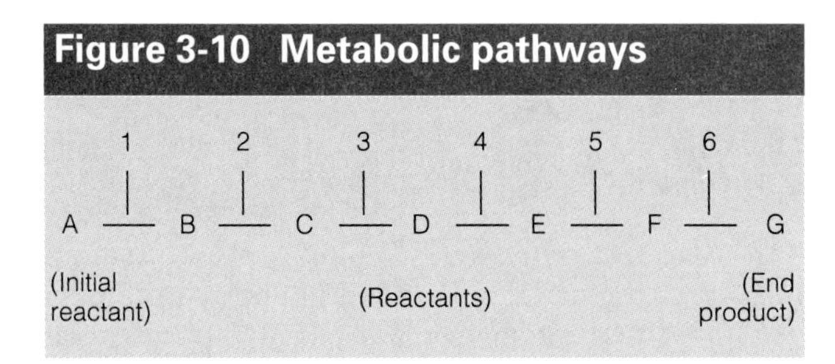

Figure 3-10 Metabolic pathways

The letter A represents the entering substance, called the reactant or substrate. The letters B through F are the metabolites and G, the end product. The numbers in the metabolic pathway represent different enzymes, each controlling a specific chemical reaction. If enzyme number 3 is missing or inoperative, the pathway cannot be completed beyond that point. It will stop at C.

stripped from a substance to another reaction site. When a cofactor is a nonprotein organic molecule, it is called a **coenzyme.** For example, many of the vitamins that your body requires, such as thiamine (vitamin B_1) or riboflavin (vitamin B_2), are important coenzymes. Finally there are **end products,** which are substances at the end of a metabolic pathway (Figure 3-10).

There are two types of metabolic pathways in cellular metabolism. **Hydrolysis pathways** involve the breakdown of chemical substances such as macromolecules into simpler substances, releasing some energy. In contrast, **synthesis pathways** involve the synthesis of complex chemical substances from simpler substances and require energy to proceed. The next process to consider is how the cell harnesses the energy released from hydrolysis pathways to produce ATP molecules.

Energy Release: Cellular Respiration

Generally cells release and capture the potential energy of glucose to manufacture ATP in a highly organized series of three linked metabolic pathways. This series allows the gradual and controlled release of energy. If all of the chemical bonds of a glucose molecule were completely broken down in one step, the cell would not be able to control the released energy. Instead ATP molecules are synthesized from the energy released by the hydrolysis of glucose in a sequence of three related metabolic pathways in the cell, collectively known as **cellular respiration.** Each glucose molecule hydrolyzed in cellular respiration yields enough energy to synthesize as many as 36 ATP molecules.

Glycolysis

The first metabolic pathway in cellular respiration, called **glycolysis** (*glyco* = sugar, *lysis* = breakdown), is in the cell's cytoplasm.

As the glucose molecule enters the glycolytic pathway, two ATP molecules are first broken down to ADP and their energy released. This is the activation energy required to begin glycolysis. Along the way hydrogen atoms are removed from the intermediate metabolites and accepted by a carrier molecule, **nicotinamide adenine dinucleotide,** or **NAD^+.** Two NAD^+ ions each accept a hydrogen atom and an electron to become **NADH** (Figure 3-11). Glycoly-

Figure 3-11 Glycolysis

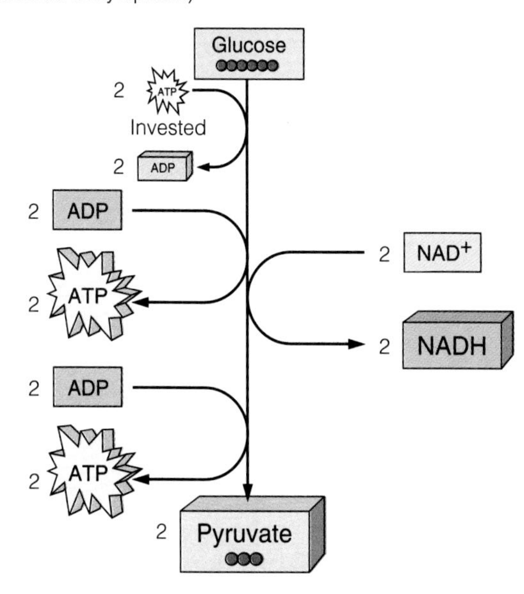

This is the first metabolic pathway in cellular respiration. This 10-step pathway hydrolyzes a glucose molecule into two 3-carbon pyruvate molecules, releasing enough energy along the way to produce 4 ATP molecules. However, because 2 ATP molecules were used to initiate glycolysis, there is a *net* gain of only two ATP molecules.

sis breaks down each glucose molecule into two molecules of pyruvate. This releases enough energy along its 10-step enzyme-controlled pathway to produce a total of four ATP molecules and two molecules of NADH. However, because two ATP molecules were used to initiate glycolysis, there is a *net* gain of only two ATP molecules.

The two pyruvate molecules produced from the breakdown of a single glucose molecule during glycolysis are further modified before passing into the second metabolic pathway of cellular respiration. First, one of the three carbons of each pyruvate is removed, departing as carbon dioxide. (The carbon dioxide released from this chemical reaction is what you exhale with each breath.) Left behind is a two-carbon molecule, called an **acetyl group,** and a hydrogen atom. The hydrogen atom and electron attach to an NAD^+ molecule, which becomes NADH (Figure 3-12). The acetyl group is then added to a carrier molecule called **coenzyme A,** or **CoA,** producing **acetyl-CoA.** Each of the two acetyl-CoA molecules is then able to enter the second metabolic pathway of cellular respiration.

Figure 3-12 Production of acetyl-CoA

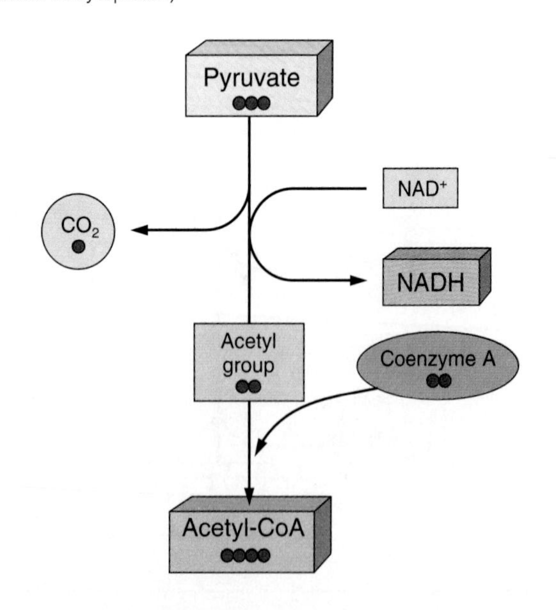

Each pyruvate molecule must be further modified before it enters the second metabolic pathway of cellular respiration. As a carbon atom is removed from pyruvate (converting the pyruvate to an acetyl group), enough energy is released to convert the carrier molecule NAD^+ to NADH. The 2-carbon acetyl group then combines with coenzyme A, producing acetyl-CoA.

Citric Acid Cycle

The other two pathways of cellular respiration are located in the cell's mitochondria and require oxygen molecules to function. They are therefore called **aerobic** (with oxygen) **metabolic pathways.**

When acetyl-CoA moves into the mitochondrion, the molecule enters the **citric acid cycle** and is released from coenzyme A to combine with a four-carbon molecule to become a six-carbon molecule of citric acid. Coenzyme A is free again to go back and pick up another acetyl group.

As the citric acid molecule proceeds through the steps of the cycle, two molecules of carbon dioxide are produced, accounting for all three carbon atoms of the original pyruvate molecule. Three molecules of NADH and one ATP molecule are also formed. However, in one step of the citric acid cycle, the hydrogen atom and electron are not energetic enough to attach to NAD^+. Thus a different carrier molecule, called **flavin adenine dinucleotide,** or **FAD^+**, picks them up instead, becoming **$FADH_2$.** Left is the original four-carbon compound that is able again to pick up another acetyl group brought to the citric acid cycle by coenzyme A (Figure 3-13). Because two molecules of pyruvate are produced from the breakdown of each glucose molecule, the citric acid cycle is completed twice for each glucose molecule broken down in glycolysis. All that is left of the original glucose molecule at the end of the citric acid cycle is its energy, which is captured in 4 ATP molecules and 12 electron carriers. The last metabolic pathway then uses the energy captured by NADH and $FADH_2$ to produce ATP molecules.

Electron Transport Chain

The NADH and $FADH_2$ molecules formed during glycolysis and the citric acid cycle move to the third pathway. Embedded in the cristae of the mitochondrion, the **electron transport chain** pathway is like the citric acid cycle in that it is an aerobic metabolic pathway. The **electron transport chain** is made up of a series of carrier proteins. The NADH and $FADH_2$ molecules transfer their hydrogen atoms and electrons to the first of the carrier proteins (Figure 3-14).

After accepting the hydrogen atoms and electrons the carrier proteins release hydrogen ions and pass their electrons along to the next protein. The hydrogen ions remain on either side of the inner membrane as the electrons continue to be passed

Figure 3-13 Citric acid cycle

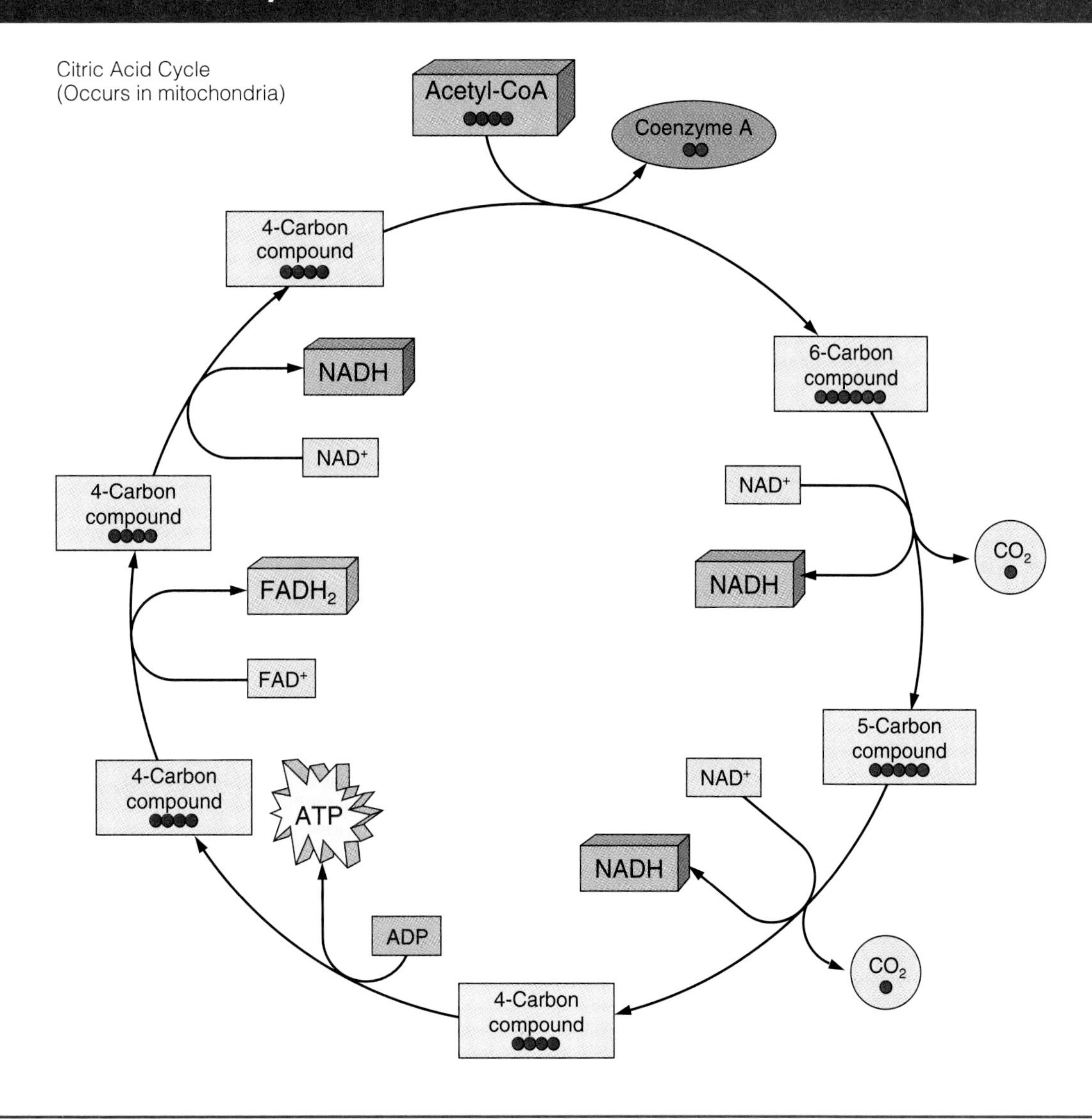

Acetyl-CoA enters this cycle by first combining with a 4-carbon compound to create a citric acid molecule. As it follows the cycle carbon atoms are removed, releasing energy to form 3 carrier molecules of NADH, one carrier molecule of $FADH_2$, and one molecule of ATP. For each glucose molecule that enters cellular respiration, the citric acid cycle accepts two acetyl-CoA molecules.

along a sequence of carrier proteins, moving "downhill" in a series of chemical reactions along the electron transport chain. As the electrons move from one protein to the next, they are moving to a lower energy state. This results in the controlled release of enough energy to produce as many as 32 ATP molecules.

At the end of the electron transport chain, the electrons join again with the hydrogen ions to form hydrogen atoms. The hydrogen atoms then combine with oxygen molecules (O_2), acquired by breathing, to produce water molecules. The primary function of oxygen in the cell is as a final acceptor of hydrogen atoms in cellular respiration. Without oxygen as the final acceptor, the electrons would build up in the pathway and prevent NADH and $FADH_2$ molecules from entering the electron transport chain. This would ultimately stop ATP production in both the electron transport chain and the citric acid cycle.

The release of the stored energy of a single glucose molecule produces a net yield of 2 ATP molecules from glycolysis, 2 ATP molecules from the citric acid cycle, and as many as 32 ATP molecules from the electron transport system. This results in 36 ATP molecules that can be used by the cell for energy-requiring activities (Figure 3-15).

Figure 3-14 Electron transport chain

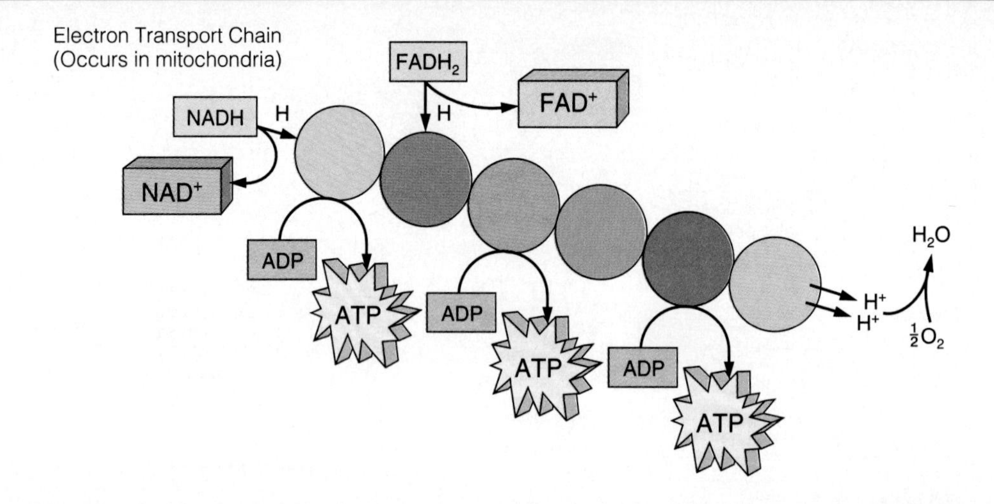

This last metabolic pathway of cellular respiration is a series of carrier proteins that accept hydrogen atoms and pass along their electrons to lower energy states. As the electrons move "downhill," enough energy is released to produce as many as 32 ATP molecules. Oxygen molecules at the end of the pathway combine with the reformed hydrogen atoms to produce water molecules (H_2O).

Figure 3-15 Number of ATP molecules and energy carriers produced during cellular respiration

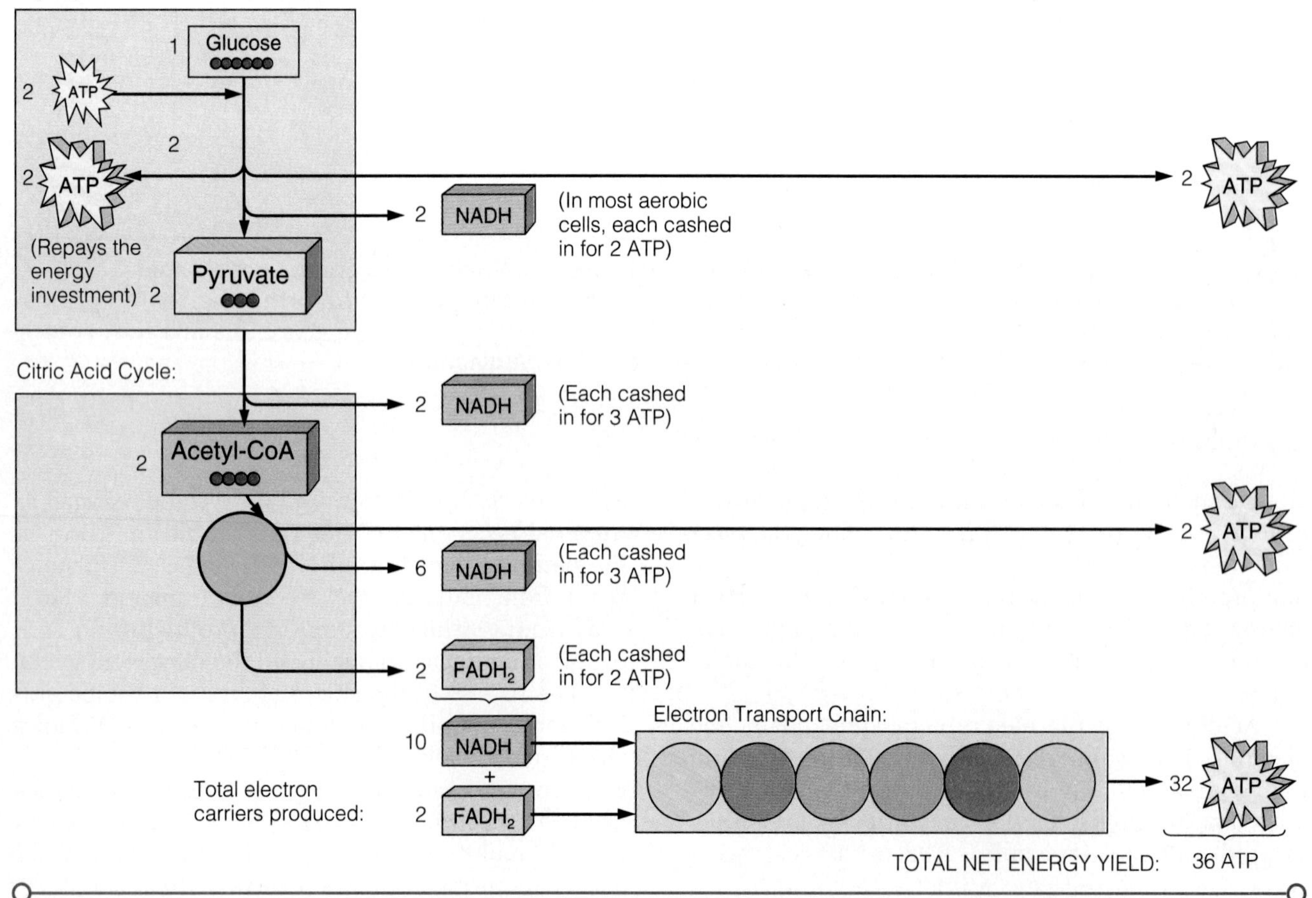

Biology in Action LACTIC ACID FERMENTATION

Cells do not have large stores of NAD^+; therefore, as NAD^+ is converted to NADH, it will eventually be depleted. In order for glycolysis to continue, NADH must be recycled back to NAD^+. If oxygen is present, the NADH produced during glycolysis is later used in the electron transport chain to produce ATP. However, if oxygen is not present, pyruvate and NADH enter another pathway that is anaerobic and also found in the cytoplasm called **lactic acid fermentation.**

In lactic acid fermentation NADH passes its hydrogen atom to pyruvate, converting it to lactic acid. The reformed NAD^+ can then be recycled and used in glycolysis. The demands of vigorous exercise may exceed the ability of the body's respiratory and circulatory systems to deliver enough oxygen molecules to the tissues to maintain the level of cellular respiration required by the activity. When this occurs, glycolysis can be anaerobic and still continue to function. However, the lactic acid that is produced is toxic to the cell. The fatigue and muscle cramps you feel during a bout of strenuous exercise are consequences of lactic acid buildup in muscle cells being exercised.

After exercise stops, your breathing rate remains elevated for some time because you have acquired an "oxygen debt." The additional oxygen being inspired is used in cellular respiration in the following way. The accumulated lactic acid is first carried away by the blood to the liver. In the liver some lactic acid is converted to glycogen. The remaining lactic acid is reconverted to pyruvate. It then reenters the cell's aerobic pathways of cellular respiration, where it is completely broken down into carbon dioxide and water, and the remaining energy is used to synthesize ATP or converted to glycogen.

The Metabolism of Other Molecules

Glucose molecules are the main source of cellular energy. Our cells use free glucose for as long as it is available, then they break down stored polysaccharides, or glycogen. If glycogen is not available, lipids are then degraded, and finally proteins are hydrolized as a last resort. These other molecules are metabolized by the following process.

Fats must first be split into their component fatty acid chains and glycerol. The fatty acid chains are broken down and converted to acetyl CoA and fed into the citric acid cycle. Glycerol, a three-carbon compound, also enters the citric acid cycle after being modified slightly. For protein to participate in cellular respiration, it must be hydrolized into its amino acids. These are then stripped of their amine groups (NH_2) and converted into fatty acids. The fatty acids are converted to acetyl-CoA and enter the citric acid cycle (Figure 3-16). All types of molecules used in cellular respiration must first be converted to acetyl-CoA to enter the mitochondrion and complete cellular respiration.

Cells are also able to control ATP production. When a cell has sufficient energy, ATP concentration is high. The ATP molecules inhibit a key enzyme of glycolysis, which slows down the production of pyruvate and cellular respiration. When more energy is required, ATP molecules decrease in concentration, the key enzyme becomes active, and

Figure 3-16 Other molecules used in the production of ATP

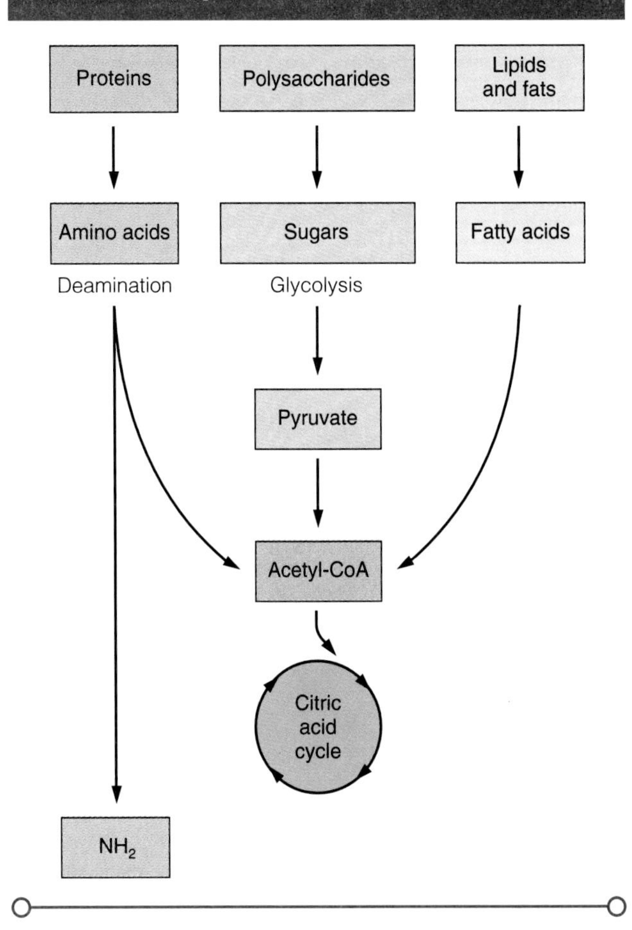

the cellular respiration of glucose and other molecules can again be broken down, rapidly supplying the cell with additional ATP molecules.

To put cellular respiration into perspective, consider a candy bar once it has been eaten. It consists primarily of carbohydrates, fat, and some protein. The candy bar enters the digestive system, where its macromolecules are broken down into their component simple organic molecules, carbohydrates into simple sugars, fat into glycerol and fatty acid chains, and protein into individual amino acids. The simple organic molecules then pass out of the digestive system and into the bloodstream, where they are distributed through the body and absorbed by the cells. The fate of glucose, fatty acids, glycerol, and amino acids, once inside the cells, varies depending on the metabolic needs of the particular cell type. For example, if the energy requirements of the cell are particularly high, all four substances may be broken down in cellular respiration. On the other hand, if the energy requirements of the cell are modest, only the glucose may be broken down, and the glycerol, fatty acids, and amino acids might be used to construct cellular components, such as phospholipid molecules, structural proteins, or enzymes, which are each produced from energy-requiring synthesizing pathways.

How Cells Replicate

Cells are very active. They reproduce, grow, specialize, and eventually die. Some 50 million body cells die every second. During that same period 50 million new cells are produced to take their place. As with all living things cells vary in the duration of their life span. For example, the cells lining the stomach live for about 36 hours, whereas red blood cells live for approximately 120 days. At the other end of the life-span spectrum, a nerve cell lives for 60 years or longer. The complete sequence of activities in a cell from one cell division to the next constitutes the **cell cycle.**

When a cell reproduces, it cannot simply divide its DNA in half and give each new cell, called **daughter cells,** half. The cell must first duplicate its DNA and then pass a complete set of DNA on to its daughter cells. However, even a full set of DNA is useless without materials with which to work. Each newly formed cell must also receive the molecules it needs to live long enough to acquire new materials from the environment. When a cell divides, its cytoplasm is divided about equally between the two daughter cells. This includes the organelles, nutrients, enzymes, and other needed molecules. A description of the behavior of the chromosomes

Figure 3-17 Chromosomes of a human male

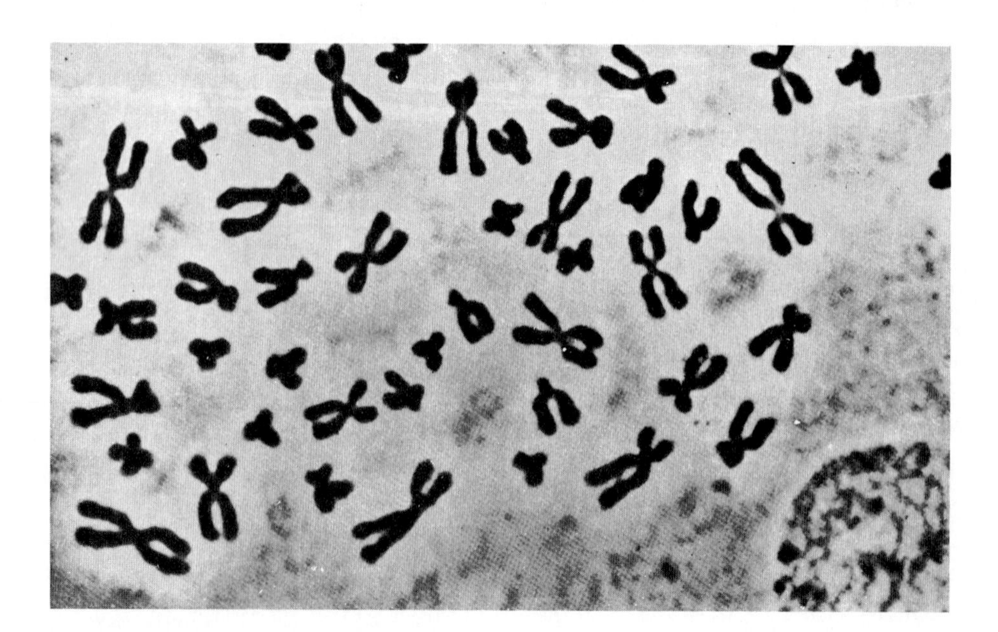

There are 46 chromosomes, or 23 homologous pairs of chromosomes, in somatic cells. This photograph was taken right before cell division, so each chromosome has duplicated itself. As long as a chromosome and its duplicated copy remain attached at the centromere, they are called sister chromatids.

during cell replication is next, followed by a description of cytoplasmic division.

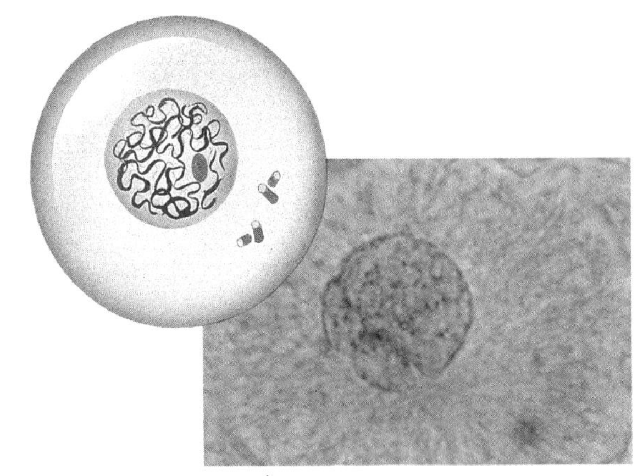
Interphase

Chromosomes

Toward the end of the nineteenth century, biologists noticed that during cell division the nucleus disappeared and in its place dark, threadlike structures appeared. These threads were called **chromosomes,** meaning "colored bodies." Scientists discovered later that the darkly stained material within the nucleus and chromosomes are actually the same thing. The chromosomes were the condensed form of the darkly stained material. **Chromosomes** contain the hereditary material of the cell and consist primarily of DNA.

The condensed chromosomes of each species assume characteristic shapes, sizes, and, when special stains are applied, particular color patterns. In the nonsex cells, called **somatic cells,** there are pairs of chromosomes of the same length. When stained, they appear similar. Each pair also has similar hereditary information controlling individual traits. The members of a pair of chromosomes are called **homologues.** All cells with pairs of homologous chromosomes are referred to as **diploid** (dip′loyd). There are 46 chromosomes, or 23 homologous pairs of chromosomes, in human somatic cells (Figure 3-17).

Each chromosome has a specialized region along its length, usually toward the center, called the **centromere.** During cell division each chromosome duplicates itself. As long as a chromosome and its duplicated copy remain attached to one another at the centromere, they are called **sister chromatids.** During one phase of cell division the two sister chromatids separate and move into the new daughter cells. Each chromatid is then an independent chromosome.

The cell cycle consists of two major activity periods. The period between cell divisions is called **interphase,** and nuclear division itself is called **mitosis.**

Interphase

Most of the life of a cell is spent in interphase. The cell grows, maintains itself, and performs other functions specific to its cell type. A portion of interphase is devoted to preparing for cell division. During this time each of the 46 chromosomes in the nucleus replicates itself, producing sister chromatids that remain attached at the centromere. The chromosomes then begin to condense. The cell also begins to assemble the intracellular components that will later move the chromosomes to opposite points or poles of the cell. Two **centrioles** begin to produce protein fibers called microtubules. Finally the replication of organelles also occurs. On average, cells spend about 10 hours replicating their DNA and 2 hours preparing for cell division. Cell division itself takes approximately 1 hour.

Nuclear Division: Mitosis

Nuclear division, called **mitosis,** is subdivided into four phases. The prefix of each name of these phases is derived from Greek and describes their relative positions in the sequence of division. Mitosis begins with prophase (*pro* = first), then metaphase (*meta* = between), anaphase (*ana* = later), and finally telophase (*telo* = end) (Figure 3-18).

Prophase begins when chromosome condensation reaches the point at which individual condensed sister chromatids first become evident. Early in prophase the two centriole pairs start to move apart, forming between them microtubules referred to as **spindle fibers.** The centrioles continue to move apart until they reach the opposite ends or poles of the cell. As the centrioles are moving away

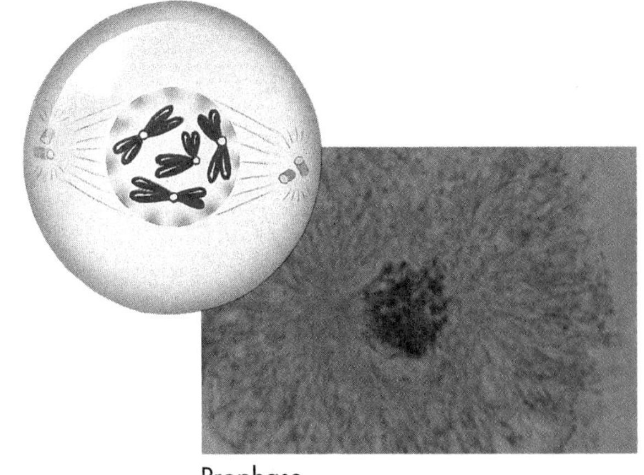
Prophase

Figure 3-18 Overview of mitosis

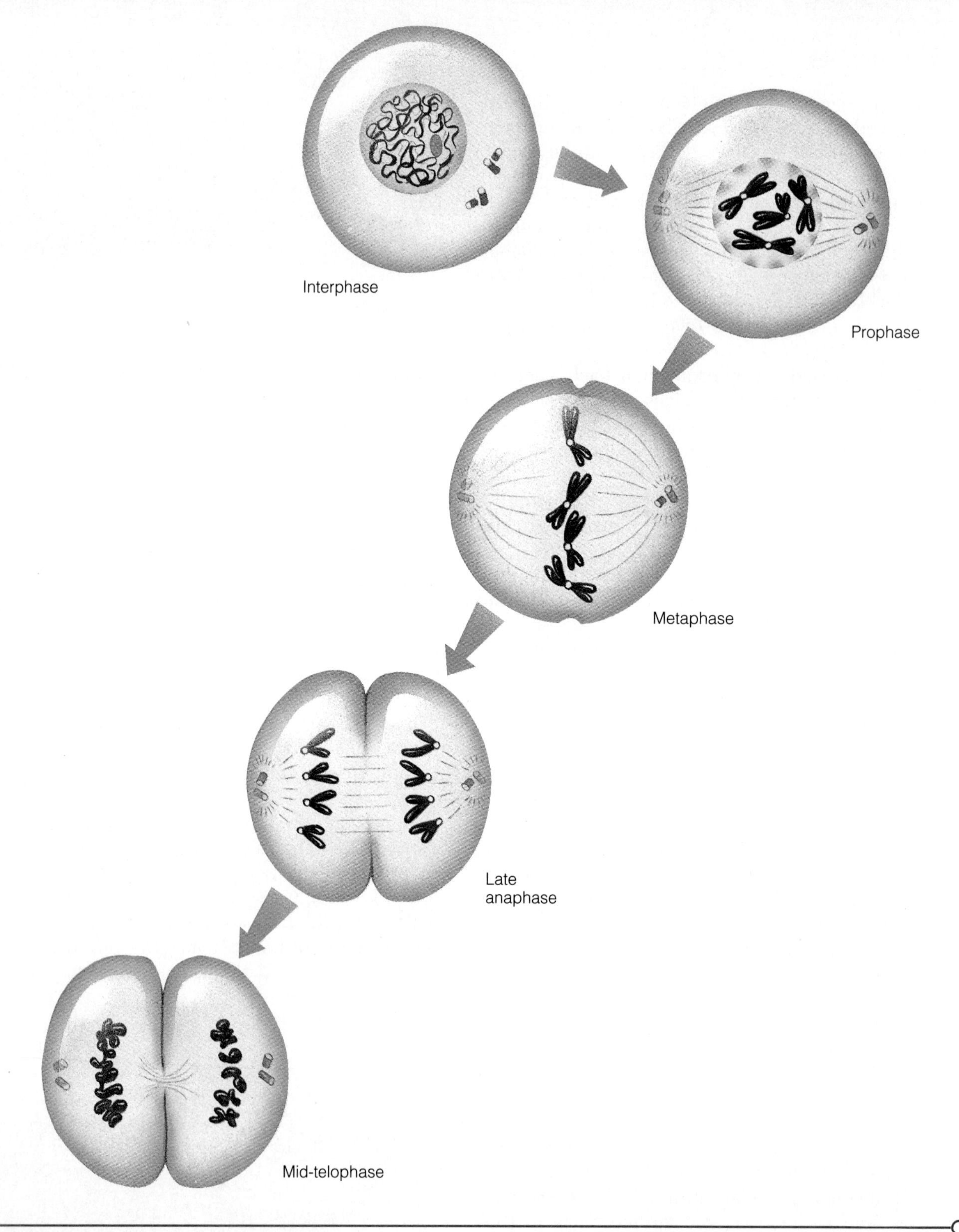

Toward the end of interphase, each chromosome has duplicated itself. Prophase begins when sister chromatids first become visible. Metaphase begins as the sister chromatids align along an imaginary equator in the center of the cell. In anaphase, each sister chromatid separates into two separate chromosomes. Telophase occurs when the chromosomes have reached opposite ends of the cell and a nuclear membrane begins to form around each group of chromosomes.

from each other the nuclear envelope breaks down and disappears.

Metaphase begins as the spindle fibers align the sister chromatids along an imaginary equator in the center of the cell perpendicular to the spindles. The sister chromatids of each chromosome face opposite poles of the cell, with spindle fibers running from each centromere to the centrioles.

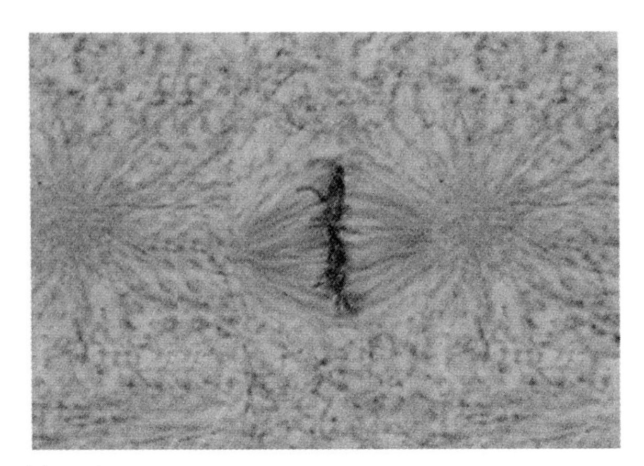

Metaphase

At the beginning of anaphase the centromere of each chromosome divides, and the sister chromatids separate into two independent chromosomes. The chromosomes move toward opposite poles of the cell. One sister chromatid of each chromosome moves toward each pole of the cell. Because the sister chromatids are identical copies of the original chromosomes, the two groups of chromosomes each contain one copy of every chromosome.

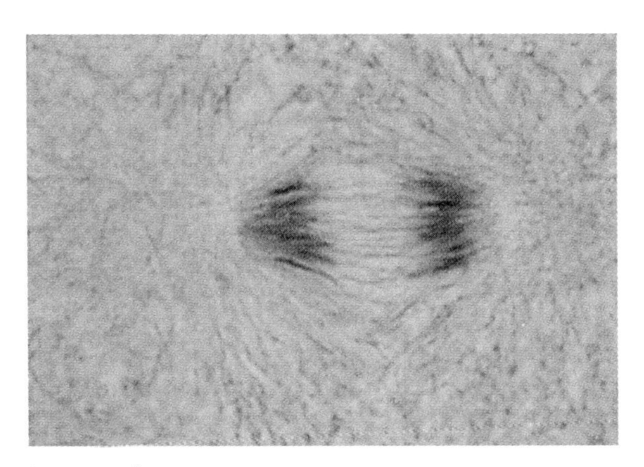

Late anaphase

Telophase begins when the chromosomes reach the poles of the spindle. The spindle fibers then disintegrate, a nuclear membrane begins to form around each group of chromosomes, and the chromosomes begin to extend and appear threadlike.

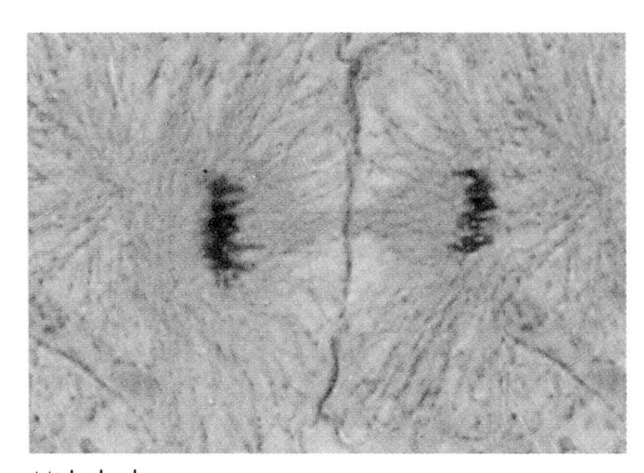

Mid-telophase

Cytokinesis

During telophase, changes are also occurring in the cytoplasm of the cell outside of the nucleus. The division of the cytoplasm into equal halves is called **cytokinesis** (si-to-kin-e′sis). Small protein fibers, attached to the cell membrane, first form a ring around the equator of the cell. The ring then contracts around the equator, much like tightening a belt on a pair of pants. Eventually the ring completely contracts, dividing the cytoplasm into two new daughter cells.

CHAPTER REVIEW

SUMMARY

1. All cells of the human body share at least three characteristics: a selectively permeable membrane, specialized membrane-bound structures within it called organelles, and a semifluid substance called cytoplasm that fills the space between the organelles and the cell membrane. Although there are structural differences between cell lines, functional differences are due largely to numbers of organelles in the cytoplasm.
2. The cell membrane is selectively permeable because it is composed of two layers of phospholipid molecules into which a variety of proteins are embedded. Some of the proteins allow large insoluble molecules and substances to pass through the phospholipid

bilayer. Other proteins transport substances against their concentration gradients into and out of the cell.

3. Selected substances can cross the cell membrane passively without the cell expending energy, or actively, with a cellular expenditure of energy. The driving force in passive transport is diffusion, while active transport requires that the cell expend some of its ATP energy molecules.
4. The functions of organelles include physically separating chemical reactions in the cell's interior and separating different chemical reactions in time enabling the cell to control sequential biochemical reactions. Organelles also function together, as in the case of the endoplasmic reticulum and the Golgi complex, to synthesize, modify, and package various substances produced by the cell.
5. The nucleus is the largest intracellular organelle and ultimately directs all cellular activities. DNA, forming structures called chromosomes, is the specific substance responsible for the control of cellular activities.
6. All chemical reactions and transformations occurring within the living cell are referred to as cellular metabolism. Cellular metabolism consists of a number of orderly sequences of steps called metabolic pathways. Each chemical reaction or step in a metabolic pathway is controlled by a specific enzyme.
7. Metabolic reactions that release energy are called hydrolysis pathways, and metabolic reactions that require energy to proceed are called synthesis pathways. Both types of metabolic reactions occur simultaneously within the cell.
8. The cell uses an energy-carrier molecule called ATP to transfer energy released from hydrolysis pathways to synthesis pathways that require energy. Free cellular energy is used to produce ATP from ADP and a free phosphate molecule. When energy is needed, ATP is broken down into ADP, a free phosphate molecule, with some energy released.
9. There are three hydrolysis pathways, collectively known as cellular respiration, used by the cell to release energy from molecules and manufacture ATP. The first pathway, called glycolysis, occurs in the cytoplasm. The other two, called citric acid cycle and the electron transport chain, occur within mitochondria.
10. Cells are able to replicate themselves, a process called **mitosis.** During mitosis each chromosome duplicates itself, forming sister chromatids. Toward the end of mitosis, cytokinesis occurs, which is a division of the cell's cytoplasm into two equal halves, resulting in the formation of two new daughter cells.

FILL-IN-THE-BLANK QUESTIONS

1. Cells have a cell membrane, a number of membrane-bound intracellular structures called ______, and a semifluid substance within the cell called ______.
2. The cell membrane is composed of a ______ bilayer with a variety of transmembrane ______ embedded in it.
3. A solution consists of a liquid portion called the ______ and substances dissolved in it referred to as the ______.
4. The movement of water across a selectively permeable membrane in response to ______ concentration is called ______.
5. Movement of a molecule across the cell membrane that requires a transport or carrier protein but no expenditure of energy is called ______ ______.
6. The organelle responsible for extracting energy from food and transforming it into a usable form of energy is the ______.
7. The cell uses a particular molecule called ______ as its energy source.
8. ATP molecules are synthesized in a series of metabolic pathways known as ______ ______.
9. One metabolic pathway in cellular respiration, called the ______ ______ ______, is able to synthesize as many as 32 ______ molecules.
10. Glycolysis does not require oxygen to proceed and is therefore called ______ ______ ______.
11. Glucose, glycerol, fatty acids, and amino acids must be hydrolyzed into ______ before entering the citric acid cycle.
12. Molecular ______ is the final hydrogen acceptor in the electron transport chain.

SHORT-ANSWER QUESTIONS

1. Why is ATP considered to be a high-energy molecule?
2. Describe the events of lactic acid fermentation.
3. Describe the cooperative activities of the endoplasmic reticulum and the Golgi complex.
4. Describe the different functions of the proteins embedded in the cell membrane.
5. Compare and contrast active and passive transport.
6. What are the functions of lysosomes?

VOCABULARY REVIEW

chromosome—*p. 57*
citric acid cycle—*p. 52*
cytoplasm—*p. 37*
cytoskeleton—*p. 48*
diffusion—*p. 40*
electron transport chain—*p. 52*
endoplasmic reticulum—*p. 45*
glycolysis—*p. 51*
Golgi complex—*p. 45*
mitochondria—*p. 46*
mitosis—*p. 57*
nucleus—*p. 43*
osmosis—*p. 40*
ribosome—*p. 45*

Chapter Four

PATTERNS OF INHERITANCE

For Review

Here are some important terms and concepts that you will encounter in this chapter. If you are not familiar with them, you should review them before proceeding.

Diploid
(page 57)

DNA
(page 33)

Homologous chromosomes
(page 57)

Mitosis
(page 57)

Sister chromatids
(page 57)

OBJECTIVES

After reading this chapter you should be able to:

1. *Differentiate between dominant and recessive varieties in terms of Mendel's experiments.*
2. *Define Mendel's law of segregation and law of independent assortment.*
3. *Define meiosis.*
4. *Explain what happens during prophase I, metaphase I, and telophase I of meiosis.*
5. *Describe what happens during crossing over.*
6. *Explain the two important differences between meiosis I and mitosis.*
7. *Define gene and allele.*
8. *Differentiate between genotype and phenotype.*
9. *Define homozygous and heterozygous.*

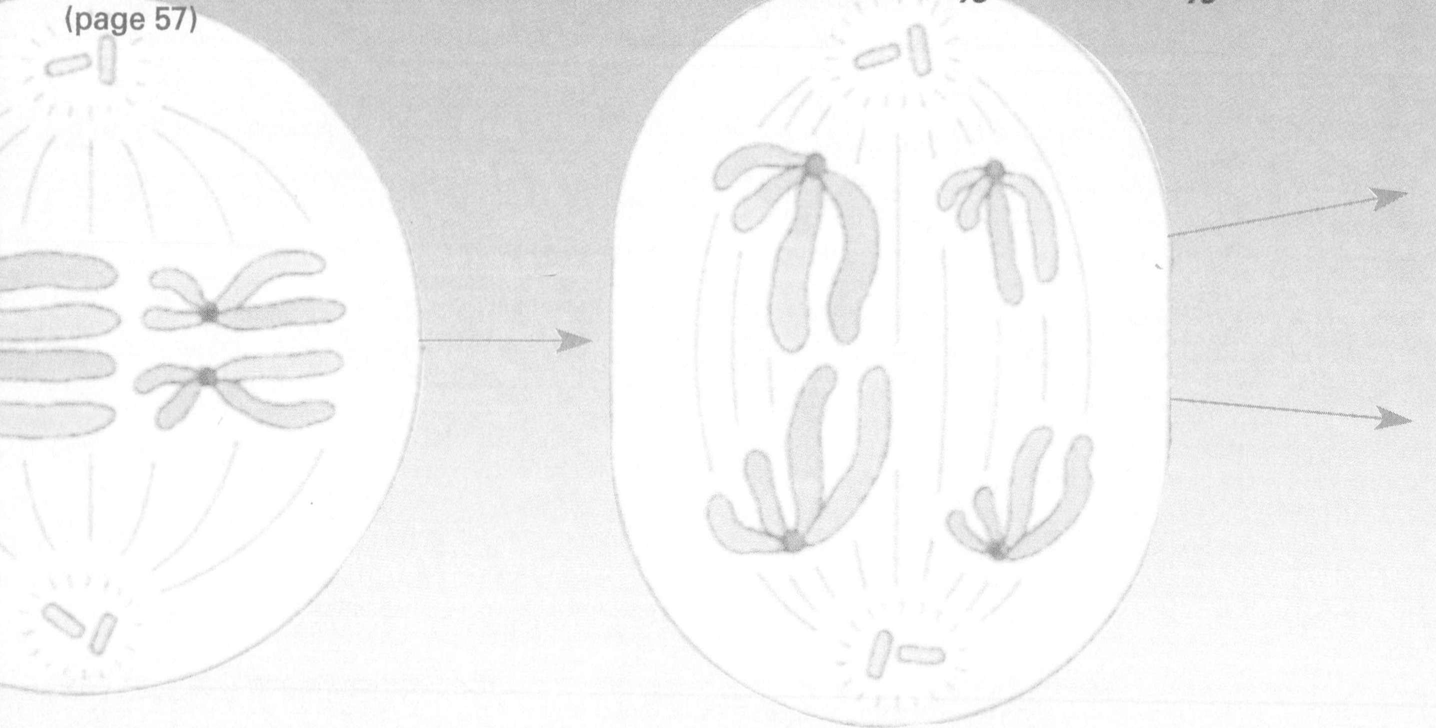

"All cells come from cells." The nineteenth-century biologist Rudolf Virchow summarized both cell theory and the crucial importance of cellular reproduction for both unicellular and multicellular organisms. Each of us began life as a single cell that then underwent cellular division and specialization. Each of those cellular divisions first involved replication of the full complement of hereditary material in the cell nucleus. Although there is cellular specialization, every body cell, with the exception of reproductive cells, has a complete copy of the hereditary material.

The Problem of Inheritance

A characteristic common to all organisms is the capacity to reproduce offspring, to create a new generation of similar organisms. People have known for centuries several important facts about reproduction. Within a population of organisms variability (or "varieties") usually exists for the characteristics of the organism. For example, all human beings have a blood type, but not all have the same blood type. People have also known that among sexually reproducing organisms, both parents contribute in some way to the production of offspring. For centuries it has been known that offspring usually resemble their parents and each other. Also, many traits in the offspring are not exactly like those of either parent. Children usually resemble their parents and each other but display varying combinations of parental traits.

With these and other observations of nature, humans have been able to manipulate both animal and plant breeding to develop particular characteristics. Over 4000 years ago in Egypt, farmers realized they could improve their crops and animals by selective breeding. For example, artificial pollination of date palms was common by 2000 B.C.

By 1727 the French government began a seed breeding program to improve food and wine production. They were able to develop the sugar beet, used to produce table sugar. This had great commercial value because sugar usually had to be imported. By 1760 the British government began a 30-year program to improve livestock herds. Many of the sheep and dairy and beef cattle common today were first developed then.

Although the results were impressive, selective breeding was based as much on trial and error as it was on luck because no one knew exactly what determined the pattern of inheritance. Many plant and animal breeders had noticed that some characteristics passed from one generation to the next unchanged, some appeared to be a mixture of parental characteristics, and others appeared and disappeared from generation to generation.

Gregor Mendel

Gregor Mendel began to unravel this confusing knot of heredity. He was born in 1822 to peasant parents in a small Austrian farming village with a long tradition of supplying skilled gardeners to the great landowning families. Mendel's father was himself a skilled fruit grower. Mendel attended grammar school and was an extremely gifted mathematics student. After grammar school Mendel entered a monastery at Brunn, Czechoslovakia, as a monk to continue his studies. Five years later he was sent to study mathematics and science at the University of Vienna. After graduation he planned to become a science teacher at the monastery. Largely because of his agricultural background Mendel knew much about botany and basic biology. However, he failed his examination in botany because he disagreed with the examiners over several questions about plant breeding and consequently did not obtain his teaching certificate.

Having given up on being a teacher, Mendel returned to Brunn, where he did administrative work in the monastery. However, still convinced his answers on the examination were correct, he quietly carried out some plant breeding experiments in his spare time. Mendel began his experiments in 1856 behind monastery walls, in a modest garden plot measuring 120 feet by 20 feet. These experiments occupied him for the next 8 years and resulted in solving the riddle of inheritance.

Mendel chose the common garden pea plant for his breeding experiments for several important reasons. First, pea plants are small, easy to grow, and they produce large numbers of offspring. A generation is completed in a single growing season. Second, pea plant pollination is easily controlled because plants contain both male and female repro-

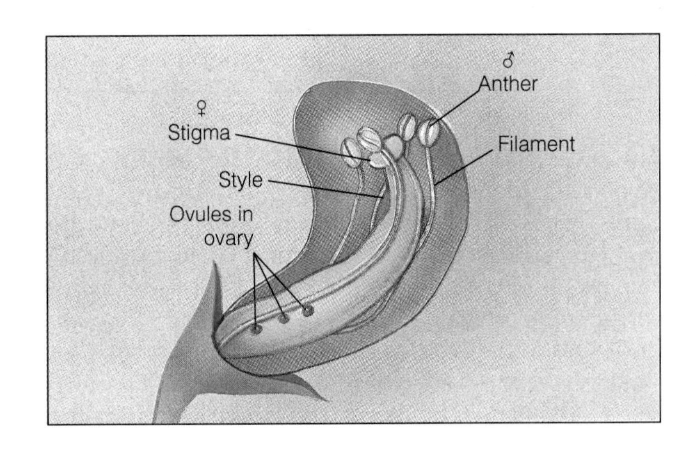

ductive structures in the flower. Normally the plants undergo **self-pollination.** The female reproductive cells in each flower are pollinated by the male reproductive cells from the pollen of the same flower. Offspring are produced from a single pea plant. However, different plants can be mated manually. In **cross-pollination** the flowers are carefully picked apart. Pollen is collected from one plant and transferred to a flower of another plant. In this way a plant breeder could mate two different plants and see what varieties of offspring resulted. The offspring produced from cross-pollination are called **hybrids.**

Third, Mendel chose the pea plant because many varieties of pea plants were readily available. He was able to obtain 34 varieties of pea plants from several commercial growers. These differed in such characteristics as flower color, seed shape, and height. Each variety of pea plant had been cultivated for many generations and was considered **true breeding.** **True breeding** means that all offspring produced through self-pollination are identical to the parent. For example, if a particular pea plant produced purple flowers, all its offspring, generation after generation, also produced only purple flowers.

Mendel chose seven different characteristics to observe and manipulate. He then selected two contrasting varieties for each of these characteristics. For example, for flower color, he used purple or white, for seed shape he chose smooth or wrinkled, and for plant height he selected either tall or short (Figure 4-1). These contrasting varieties were distinctly different and therefore would be easily identified from generation to generation. Mendel thus initially made only **monohybrid crosses.** A **monohybrid cross** is breeding for only one characteristic at a time.

Mendel's Experiments and Results

Mendel conducted his monohybrid breeding experiments in two steps. First he cross-pollinated true-breeding plants having alternative varieties for a particular characteristic. He removed the male parts from flowers of the pea plants to prevent self-pollination. Next he manually removed the pollen from one plant and carefully pollinated the female parts of another plant. For example, pollen from a white flower was placed on the female parts of a purple flower. He also made a reciprocal cross by taking the pollen from a purple-flowered pea plant and introducing it to a white-flowered plant. The second step was to allow the hybrid offspring produced by cross-pollination to self-pollinate and produce offspring. To keep track of his experiments

Figure 4-1 Seven characteristics of pea plants with their contrasting varieties

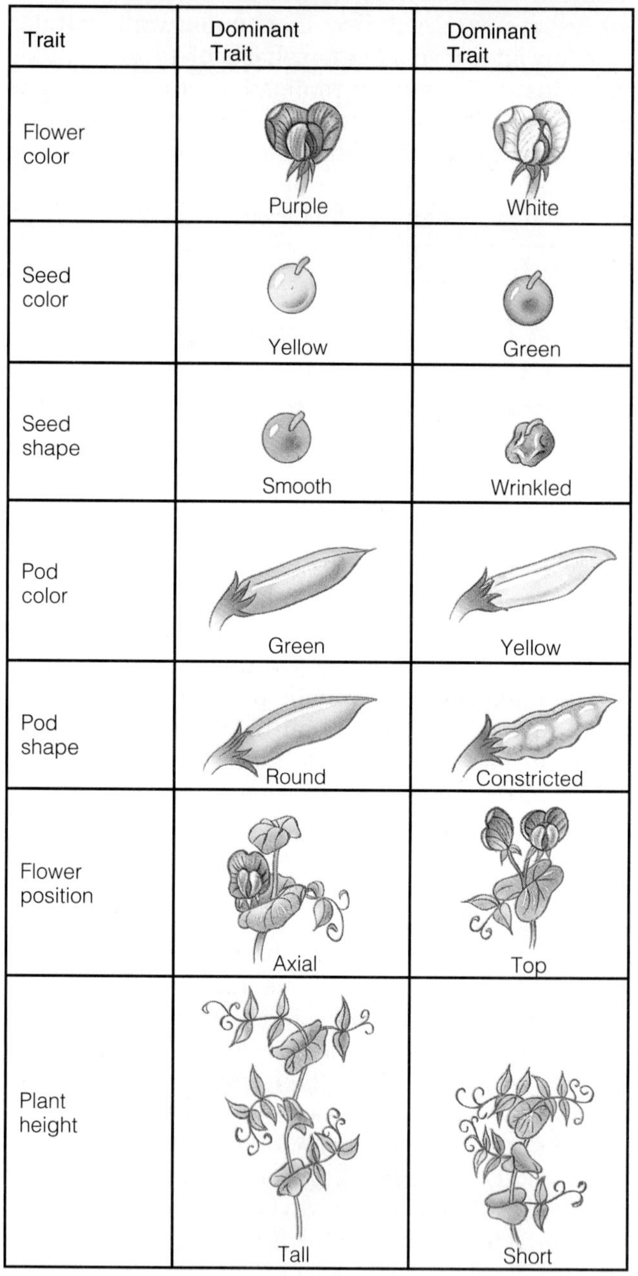

Trait	Dominant Trait	Dominant Trait
Flower color	Purple	White
Seed color	Yellow	Green
Seed shape	Smooth	Wrinkled
Pod color	Green	Yellow
Pod shape	Round	Constricted
Flower position	Axial	Top
Plant height	Tall	Short

Mendel selected seven different characteristics of the pea plant, each with two contrasting varieties, for his breeding experiments.

Mendel referred to the first cross as the **parental generation,** or **P generation.** The first offspring generation was called the **first filial** (fil′ial) or $\mathbf{F_1}$ **generation.** The next generation he designated as the **second filial** or $\mathbf{F_2}$ **generation.**

When Mendel crossed two contrasting varieties,

such as purple-flowered pea plants with white-flowered plants (P generation), the hybrid offspring (F_1 generation) produced always resembled one of the parental varieties. The flowers of the F_1 generation hybrids were as purple as the flowers of the purple parent, and there were no white-flowered offspring (Figure 4-2, *A*). The variety expressed in the F_1 plants, purple, was called **dominant.** The alternative variety, white, not expressed in the F_1 plants, was called **recessive.** For the contrasting

Figure 4-2 Monohybrid cross

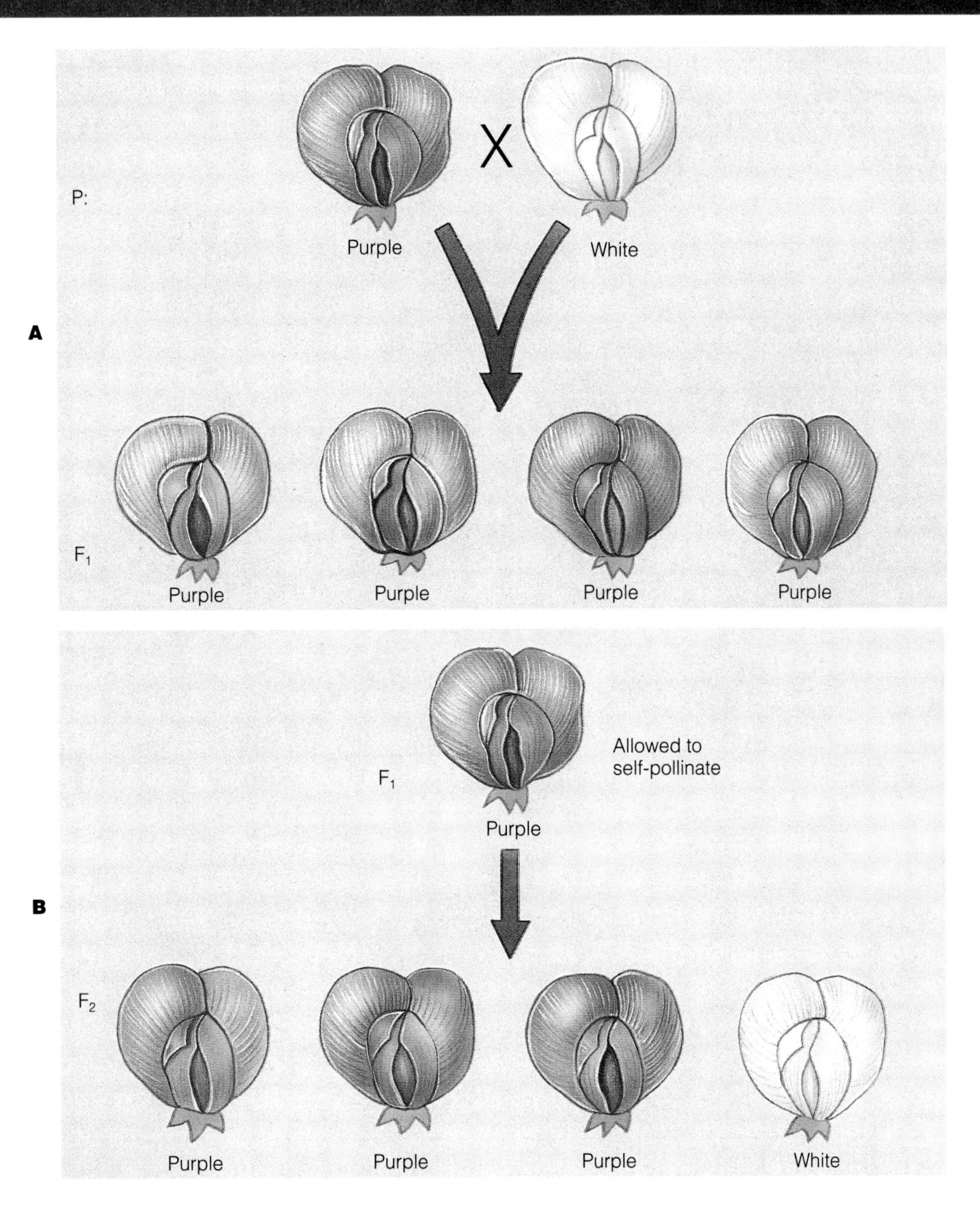

In his initial breeding experiments Mendel cross-pollinated two true-breeding pea plants, each with a different variety of the same characteristic. For example, he took pollen from a purple-flowered plant and introduced it into a white-flowered plant. He also made a reciprocal cross in which he took pollen from a white-flowered plant and introduced it into a purple-flowered plant. The flowers of all of the F_1 pea plants were purple (A). Mendel allowed F_1 hybrids to self-pollinate. Both purple and white flowers were produced in the F_2 generation. He also discovered that for every three purple flowers produced there was 1 white flower produced (B).

Figure 4-3 Mendel's experiment to determine the number of hereditary factors controlling a characteristic

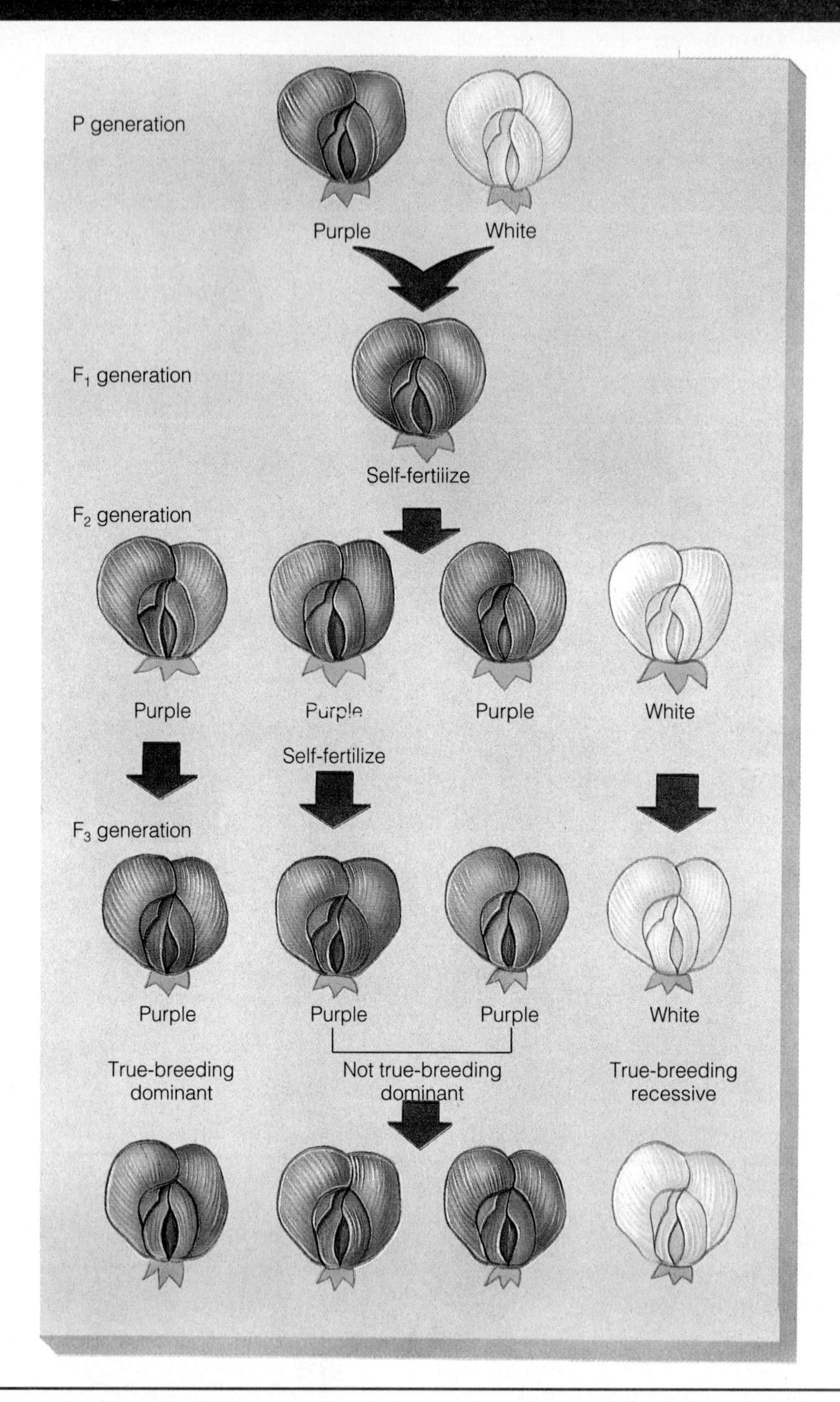

To determine the number of hereditary factors responsible for a characteristic, Mendel let the F_2 generation plants self-pollinate. He found that only one third of the purple-flowered plants produced only white-flowered offspring.

varieties of each of the seven characteristics Mendel tested, one variety was dominant and the other was recessive. But was that an accurate conclusion, and if so, what happened to the recessive variety?

Mendel began the second step of his experiment to find this out. After the individual F_1 plants matured and were allowed to self-pollinate, Mendel collected and planted the seeds from each plant to observe what the F_2 would look like. He found that F_1 plants produced white flowers as well as purple flowers in the F_2 generation. The recessive variety was not lost in the F_1 generation but remained "hidden" and then expressed in the F_2 plants.

Mendel counted the numbers of each variety among the F_2 offspring. For example, of 929 F_2 individuals produced, 705 had purple flowers and 224 had white flowers. Almost one quarter of the F_2 offspring had white flowers, the recessive variety (Figure 4-2, *B*). Mendel then examined each of the seven characteristics by counting the varieties present among the offspring produced from the F_1 generation. He discovered that for every three plants displaying the dominant variety, one plant displayed the recessive variety. The ratio was always the same.

Mendel called the unknown agent responsible for each characteristic a **hereditary factor.** However, it was still not clear how many factors were involved. He reasoned that the purple flower F_1 hybrid must contain at least one hereditary factor for purple because it is purple, but it must also contain at least one hereditary factor for white because when self-fertilized, it produced some white flowered plants in the F_2 generation. To find out how many factors were present for each characteristic, he allowed the F_2 to self-pollinate. The white-flowered F_2 plants produced only white-flowered offspring. However, only one third of the purple-flowered dominant F_2 plants produced only purple-flowered offspring. The remaining two thirds were hybrids and produced both purple- and white-flowered offspring in a ratio of 3:1. The F_2 generation consisted of one-fourth true-breeding dominant purple plants, one-half hybrid purple plants that produced both purple- and white-flowered offspring, and one-fourth recessive true-breeding white-flowered plants (Figure 4-3). Mendel's background in mathematics helped him at this point to solve the problem of inheritance and explain what these ratios meant.

Mendel saw that the F_1 hybrid purple-flowered plants producing both purple- and white-flowered offspring in the F_2 generation must contain hereditary factors for both varieties. The simplest conclusion was that a hybrid plant for flower color contains two hereditary factors, which he labeled *P* and *p*. The capital letter *P* represents the dominant hereditary factor for purple. The lowercase letter *p* represents the recessive hereditary factor for white. The true-breeding purple-flowered and white-flowered plants of the parental generation must also contain PP and pp, respectively. If this is correct, then clearly PP and pp plants breed true because when they self-pollinate, only true breeds could be produced with no other factors present. The Pp hybrids, however, do not necessarily breed true because they are carrying both P and p factors. Mendel designed a final set of experiments to determine if his reasoning was correct.

The Law of Segregation

Mendel proposed the hypothesis that two hereditary factors determine a particular characteristic and they must separate from each other when reproductive cells are formed. For example, the hybrid plant must produce two kinds of pollen and two kinds of eggs, either P-bearing or p-bearing. If this were true, then the three possible pollen-and-egg combinations were PP, Pp, and pp. However, observation alone could not confirm this hypothesis because of the dominant and recessive relationship between the two varieties. Purple flowers could thus be the PP or Pp combination.

To test this hypothesis, Mendel devised a simple breeding technique called a **testcross.** In the **testcross** a hybrid plant is cross-pollinated with a true-breeding recessive plant. To learn which hereditary factors a self-pollinated purple-flowered plant from the F_1 generation possessed, Mendel crossed it with a white-flowered pea plant. Two alternatives from a testcross were possible. If the test plant from the F_1 generation contained identical factors, PP for example, and was crossed with a white-flowered recessive plant, pp, all of the offspring would have purple flowers. The other possibility was that if the test plant contained two different factors, Pp, one half of the offspring would have white flowers and the other half would have purple flowers. The dominant and recessive varieties would then appear in a 1:1 ratio (Figure 4-4). The results were that for each pair of contrasting varieties Mendel investigated, each testcross produced a 1:1 ratio, just as his hypothesis predicted.

Using conclusions from his experiments, Mendel proposed the **law of segregation.** The **law of segregation** states that pairs of hereditary factors segregate or separate from one another during the formation of reproductive cells and each reproductive cell receives only one of the pair of hereditary factors.

The Law of Independent Assortment

Mendel then went on to ask whether the factors for *different* characteristics also segregated independently of one another. Would, for example, the particular factor that determined seed shape influence which factors a plant inherited for seed or flower color?

Mendel conducted breeding experiments similar to his original monohybrid crosses but this time made **dihybrid crosses.** A **dihybrid cross** involves breeding for two characteristics at a time. He again started with true-breeding lines of peas that differed from one another in two of his seven chosen charac-

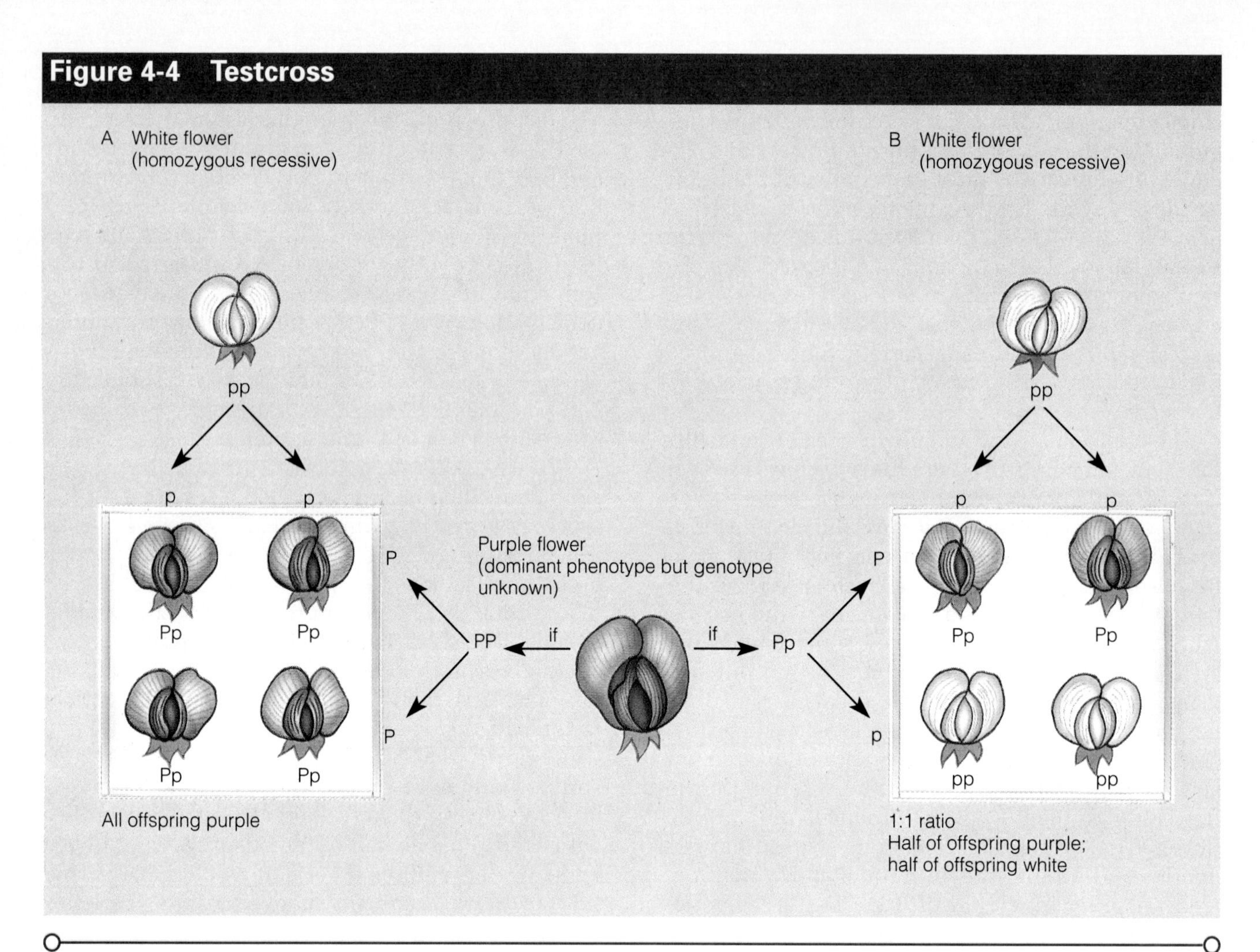

To test the hypothesis that two hereditary factors were responsible for a characteristic, Mendel devised a testcross in which a hybrid plant is cross-pollinated with a true-breeding recessive plant. If the hybrid contained two identical factors, all the offspring would have purple flowers (A). If the hybrid contained two different factors, however, one half of the offspring would have white flowers and the other half would have purple flowers (B). For each pair of contrasting varieties Mendel investigated, each testcross produced a 1:1 ratio, confirming his hypothesis that two factors determined a characteristic.

teristics. Then he cross-pollinated contrasting pairs of the true-breeding lines. In a cross involving different seed-shape varieties, smooth (S) and wrinkled (s), and different seed-color varieties, yellow (Y) and green (y), all of the F_1 offspring were identical. Each plant has smooth and yellow seeds. From his monohybrid experiments Mendel knew each plant contained different hereditary factors for both seed shape, Ss, and seed color, Yy. The F_1 plants were then allowed to self-pollinate. Mendel examined 556 plants produced from a dihybrid cross that had been allowed to self-pollinate and obtained the following F_2 generation results:

$\frac{315}{556}$ $(\frac{9}{16})$ smooth and yellow
$\frac{108}{556}$ $(\frac{3}{16})$ smooth and green
$\frac{101}{556}$ $(\frac{3}{16})$ wrinkled and yellow
$\frac{32}{556}$ $(\frac{1}{16})$ wrinkled and green

These results express a 9:3:3:1 ratio. From this, Mendel concluded that the chances for a seed to be smooth or wrinkled are independent of its chances to be yellow or green. The characteristics assort independently of one another. How did Mendel reach this conclusion?

If a seed has a three-fourths chance of being smooth and a three-fourths chance of being yellow, and if the characteristics assort independently of one another, its chances of being both smooth and yellow at the same time would be $\frac{3}{4} \times \frac{3}{4} = \frac{9}{16}$ $(\frac{315}{556})$ (Figure 4-5). This is the ratio Mendel observed. If the characteristics were assorting independently of one another, he should have been able to predict the expected ratio of a particular combination in the F_2 dihybrid cross by multiplying the chances of occurrence of the individual varieties of the two characteristics together. For example, what are the chances of a seed being both smooth and green?

There is a three-fourths chance of a seed being smooth. There is a one-quarter chance of a seed being green. So the chance of being both smooth and green would be $\frac{3}{4} \times \frac{1}{4} = \frac{3}{16}$. Each of his predicted ratios matched the observed ratio in the F_2 generation. From these dihybrid crosses Mendel reached a second conclusion about the pattern of inheritance, called the **law of independent assortment.** The **law of independent assortment** states that the distribution of hereditary factors for one characteristic into the reproductive cells does not affect the distribution of hereditary factors for other characteristics.

Rules of Probability

The numerical ratios observed among the offspring of Mendel's controlled breeding experiments led him to discover the pattern of inheritance. The significance of these ratios relies on the rules of probability, illustrated by the flip of a coin.

In mathematics **probability** is defined as the number of times one outcome or event will occur divided by all possible outcomes or events. For probability, or chance, to apply to a group of events, two conditions must be true. First the individual events cannot influence one another. They must be independent events. Flipping heads on the first coin toss does not in any way influence the outcome of the second toss of the coin. Second the events must be mutually exclusive events. If one event occurs, the other event or events are excluded. For example, when tossed the coin cannot show both heads and tails at the same time. When these conditions are true about a group of events, then the probability, or chance, of a particular event

Figure 4-5 Dihybrid cross

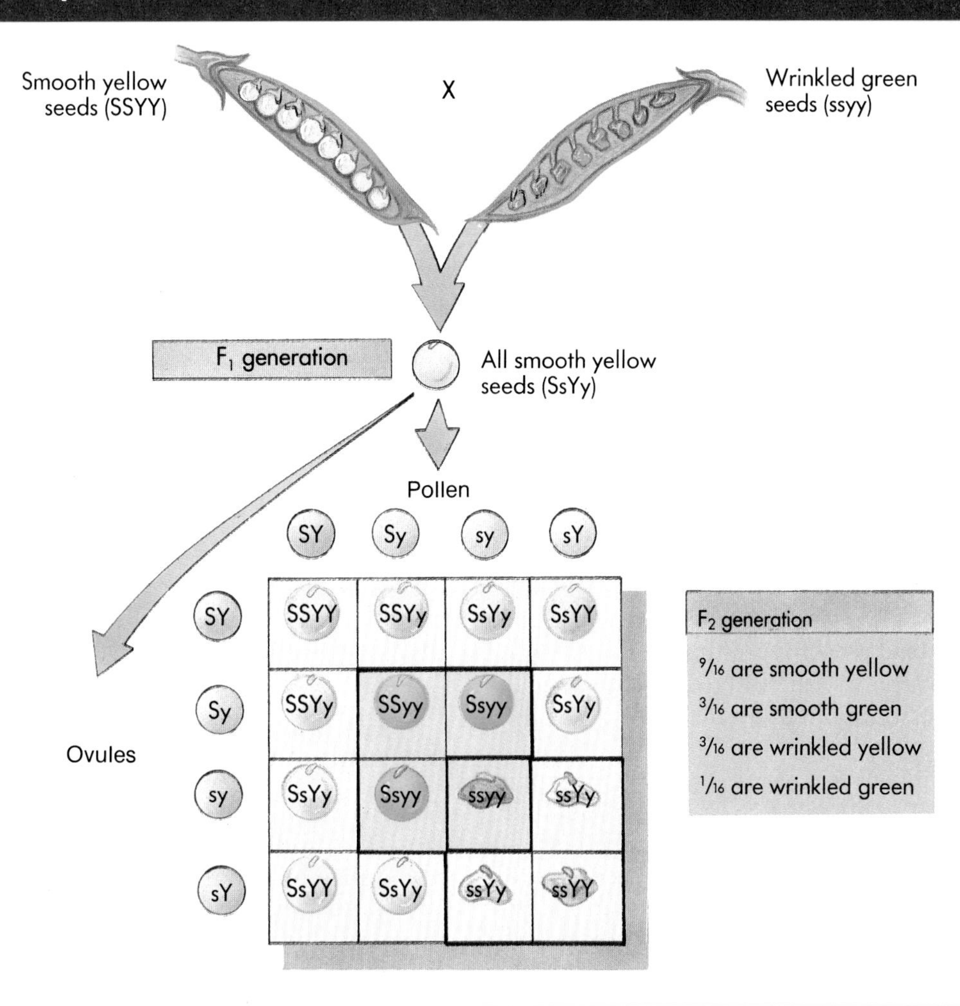

Mendel wanted to know whether the factors for different characteristics assorted independently. After he allowed the F_1 hybrids to self-pollinate, the different combinations of varieties expressed a 9:3:3:1 ratio. This ratio could occur only if the hereditary factors for different characteristics assorted independently in reproductive cells.

occurring is its individual occurrence divided by the total number of events possible.

When both conditions are met, it is also possible to calculate the chance, or probability, of particular combinations of events occurring. For example, taking two coins, what is the probability of both landing heads side up? We know there is a one-in-two chance of flipping heads on each coin. Therefore both their individual probabilities of occurrence are multiplied together, $\frac{1}{2} \times \frac{1}{2} = \frac{1}{4}$. Thus there is a one-in-four chance of both coins coming up heads. What is the probability of flipping a heads/tails or tails/heads combination? Again for each combination the individual probabilities of occurrence are first multiplied together, $\frac{1}{2} \times \frac{1}{2} = \frac{1}{4}$. However, because there are two ways to get an unlike combination, the probability of each combination must then be added together: $\frac{1}{4}$ (heads/tails combination) + $\frac{1}{4}$ (tails/heads combination) = $\frac{1}{2}$, the probability of the unlike combination occurring.

Remember that expected ratios or frequencies of occurrences of an event or combinations of events are more likely to appear with larger numbers of trials. For example, a single coin may have to be tossed 100 times before the expected probability of heads, $\frac{1}{2}$, or 50% is realized. Mendel was also aware of this potential problem. He used hundreds of pea plants and their offspring in his breeding experiments to improve the chance of observing any pattern of inheritance. Although Mendel knew mathematics, unfortunately, for science, his colleagues did not.

Mendel published his results in 1866. His explanations were concise, and the logic of his experiments was clearly presented. The paper was published in a local natural history society journal, and only 115 copies were sent out. Scientists read it and promptly forgot about it because the study of mathematical ratios was not part of their training. In 1900, 16 years after Mendel's death, three different investigators independently rediscovered his original work while reviewing the literature to prepare to publish their own findings. Each had reproduced results similar to those Mendel had quietly presented more than 30 years earlier.

The ratios Mendel observed in both his monohybrid and dihybrid crosses occurred because the individual hereditary factors acted as independent and mutually exclusive events. However, Mendel did not know what the hereditary factors were. He knew only their pattern of behavior as they were transmitted from one generation to the next.

Sexual Reproduction and Meiosis

Now we can compare Mendel's pattern of inheritance with the behavior of chromosomes in the cell.

Human reproductive cells are **haploid** (hap'loyd), meaning that they possess only one copy of each pair of homologous chromosomes. These haploid cells are the **gametes** (gam'-ets), called sperm in the male and egg in the female. At fertilization male and female gametes fuse to form a diploid cell, called a **zygote** (zi'got), with a full complement of 23 pairs of chromosomes. The zygote then divides, and specialized cell lines develop, which eventually result in offspring. Each parental gamete has contributed one chromosome of each of the 23 pairs of chromosomes of the offspring. A special type of cell division produces gametes with one half of the number of chromosomes found in somatic cells. We begin with an overview of this process.

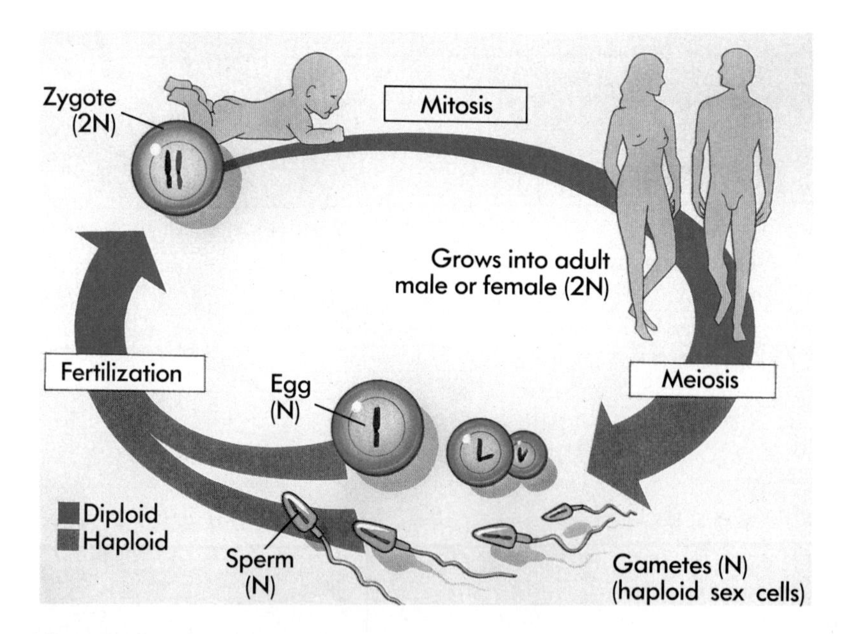

Overview of Meiosis

Meiosis (mi-o'-sis) is a special type of nuclear division that produces haploid gametes with one half of the number of chromosomes found in somatic cells. The word *meiosis* is derived from the Greek word meaning to diminish because that is what meiosis does, it diminishes or reduces the number of chromosomes by one half.

Meiosis involves two cell divisions, called meiosis I and meiosis II. As in mitosis the chromosomes are first duplicated during interphase before meiosis I. Sister chromatids remain attached at the centromere. During meiosis I homologous pairs of chromosomes separate, each cell receiving one homologue. However, each chromosome still consists of two joined sister chromatids at the end of meiosis I. In meiosis II the sister chromatids split into two independent chromosomes, each one going into a different daughter cell. Each division is accompanied by cytokinesis, so a cell undergoing both meiosis I and meiosis II can produce as many as *four haploid cells.* The phases of meiosis are given the same names as the phases in mitosis, although there are important differences, especially in meiosis I (Figure 4-6).

Meiosis I

In prophase I of meiosis I individual chromosomes first become visible under the light microscope, as their DNA coils during condensation. Because the DNA has already replicated before the onset of meiosis, each of these chromosomes actually consists of two sister chromatids joined at their centromeres. During prophase I the homologous chromosomes line up side by side. This configuration is referred to as a **tetrad** (meaning "groups of four"). While chromosomes are in a tetrad configuration, **crossing-over** occurs, meaning that the *nonsister* chromatids of the homologous chromosome pairs exchange segments of DNA (Figure 4-7). Once crossing-over is complete, each tetrad moves to the center of the cell, lining up along an imaginary equator. Spindle fibers from the centrioles form and attach to the centromeres of the chromosomes.

Two important differences between meiosis I and mitosis now become apparent. In mitosis the chromosomes, each consisting of two sister chromatids attached at the centromere, line up *individually* along the equator of the cell. However, in meiosis I chromosomes line up in *homologous pairs* along the equator. Second in mitosis chromatids separate, becoming chromosomes, and move away from the equator, resulting in the original number of homologous pairs of chromosomes in each daughter cell. In contrast, in meiosis I both sister chromatids of one homologue are attached to spindle fibers extending toward one pole of the cell. The two sister chromatids of the other homologue attach to spindle fibers extending toward the opposite pole. Movement away from the equator separates homologous pairs of chromosomes but keeps intact the attached sister chromatids. Which member of a pair of homologous chromosomes faces which pole of the cell is based only on chance. For example, for some homologous pairs the chromosome derived from the male parent may face one direction. For other pairs the chromosomes derived from the male parent may face the opposite pole.

The reassortment of maternally and paternally derived chromosomes makes possible much of human diversity. For example, each human gamete receives one of the two copies of each of the 23 different chromosomes, but which copy of a particular chromosome it receives does not influence which copy of a different pair of chromosomes it receives. This means there are 2^{23}, or more than 8 million, different combinations of 23 chromosomes possible for each gamete formed.

Crossing-over also generates extensive variation by reshuffling the genetic material. For example, suppose a homologous pair of chromosomes contains four different traits, labeled A, B, C, and D. One homologue possesses the dominant variety for each characteristic, and the other homologue contains the recessive variety for each of the same characteristics. Each homologue duplicates, and as meiosis I proceeds, homologous pairs come together to form a tetrad, and crossing-over occurs. As a result one chromatid might still contain the original combination of dominant varieties and another chromatid might contain the original combination of recessive varieties. However, the two nonsister chromatids that crossed over will each be a new and unique combination of both dominant and recessive varieties. The fusion of two gametes creates a unique individual with a new combination of 23 pairs of chromosomes.

In metaphase I, spindle fibers attach to the centromeres and homologous pairs of chromosomes line up along an imaginary equator.

In anaphase I, sister chromatids move to one pole of the dividing cell while their homologue moves to the opposite pole. This separation results in each pole containing one member of each pair of homologous chromosomes. Although each chromosome contains its duplicated copy, these sister chromatids are not necessarily identical because of the crossing over in prophase I.

In telophase I the chromosomes are gathered around their poles and the spindle fibers disappear. The cell then divides into two daughter cells. Then meiosis II begins.

Meiosis II

Meiosis II occurs in the nuclei of both daughter cells formed by meiosis I and closely resembles mitosis. Within each daughter cell the spindle fibers reform and attach to the centromeres of the sister chromatids. The sister chromatids then move to the center of each cell. The sister chromatids then separate from one another, the spindle fibers

Figure 4-6 Meiosis

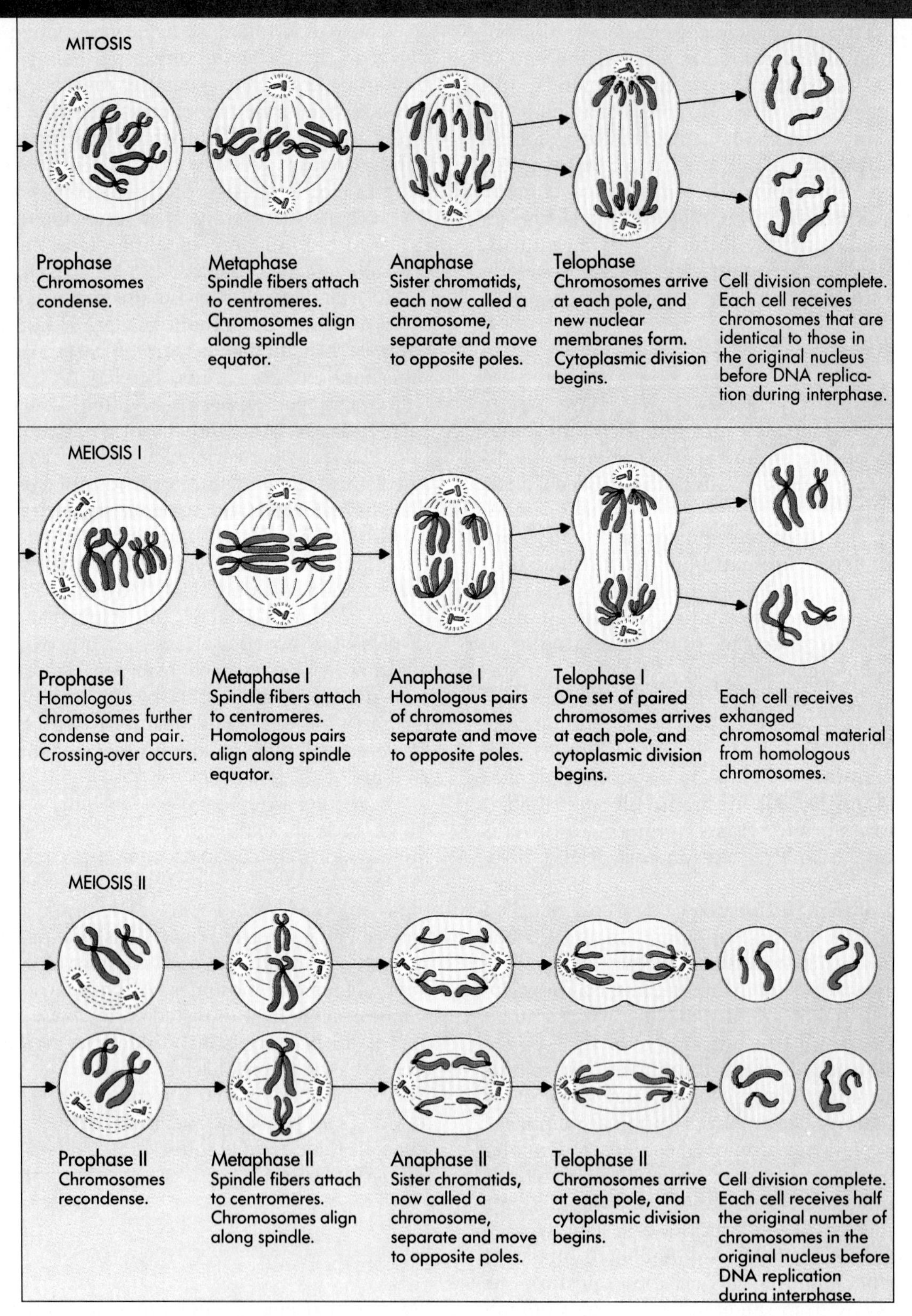

There are two important differences between meiosis and mitosis. First, in meiosis chromosomes, each consisting of two sister chromatids, line up in *homologous pairs* along the equator, forming tetrads. In mitosis chromosomes, each consisting of two sister chromatids, line up *individually* along the equator. Second, in meiosis I both sister chromatids of one homologue move toward the same pole. In mitosis, chromatids become separate chromosomes and move toward opposite poles.

Figure 4-7 Crossing-over

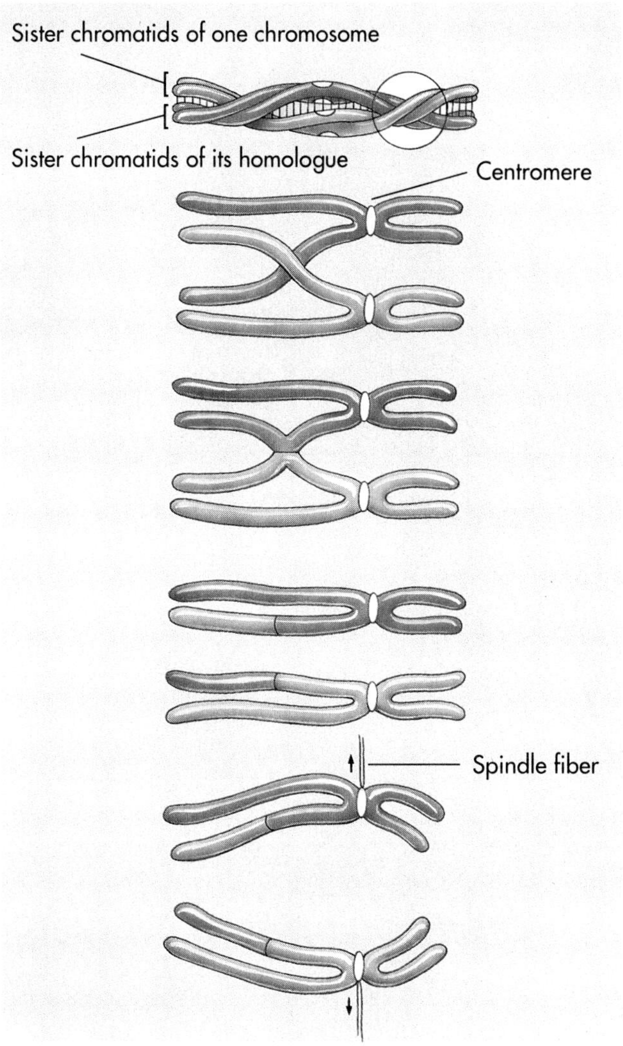

Sister chromatids of one chromosome
Sister chromatids of its homologue
Centromere
Spindle fiber

In meiosis I, homologous chromosomes, each consisting of two sister chromatids, line up along the equator, forming tetrads. Nonsister chromatids of the homologous chromosomes then exchange segments of DNA.

pulling each chromosome to opposite ends of the cell. Cytokinesis occurs, the nuclear membrane reforms, and the chromosomes relax into their threadlike extended state. As many as four haploid gametes are produced.

Mendel and Meiosis

Mendel discovered the pattern of inheritance without knowing the nature of the hereditary material. In light of knowledge gained since his initial discoveries, new terms that combine the concepts of Mendelian inheritance and the behavior of chromosomes have developed.

Today Mendel's hereditary factors are called **genes.** Although there are several related definitions for a gene, for now we define a **gene** as a sequence of DNA nucleotides on a chromosome that determines an inherited characteristic. The gene's location on a chromosome is referred to as a **locus.** Most genes exist in two or more alternative forms called **alleles,** with differences at one or more nucleotide positions in the DNA. Different alleles of a gene are usually recognized by the change in appearance or function resulting from the nucleotide differences. For example, in pea plants there are two alternative alleles governing flower color, giving rise to either purple or white.

Similarly, as Mendel hypothesized, each of us also has two genes for many different characteristics. One copy is inherited from each parent. For example, humans possess one gene with two alternative alleles that determines the shape of the earlobe. One allele is dominant (U) and produces unattached lobes. The other allele is recessive (u) and produces attached earlobes. Each of us has either two unattached-earlobe genes (UU), two attached-earlobe genes (uu), or one of each (Uu). If you inherit the attached-earlobe gene from one parent and the unattached-earlobe gene from the other, you will not have half-attached lobes, or one attached and the other unattached. Because of gene dominance both earlobes will be unattached (Figure 4-8). Other human traits with similar patterns of inheritance extend from the ability to roll one's tongue, the ability to taste a bitter substance called PTC, short fingers, or life-threatening disorders such as cystic fibrosis (Table 4-1).

The total set of genes in an individual is called a **genotype** (je'no-tip). Genotype also refers to the set of genes for a particular characteristic. For example, the genotype for flower color may be PP, Pp, or pp. The genotype for earlobes may be UU, Uu, or uu. The outward physical appearance or function of the genotype of an organism is called a **phenotype** (fe'no-tip). Phenotype may also apply to a particular characteristic. For example, the phenotype for flower color may be either purple or white. For earlobe type in humans, the phenotype may be attached or unattached.

When a pair of genes for a characteristic are identical, the organism is said to be **homozygous** (ho-mo-zi'gus) for that characteristic. For example, the homozygous genotype for unattached earlobes would be UU. On the other hand, if the genes differ, the organism is **heterozygous** (het-er-o-zi'gus) for that characteristic. The heterozygous genotype for unattached earlobes is Uu. However, because of the dominant and recessive relationship of the alleles,

Figure 4-8 Earlobe shape is determined by dominant and recessive alleles.

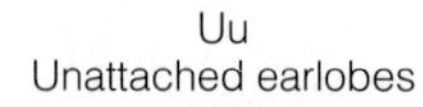

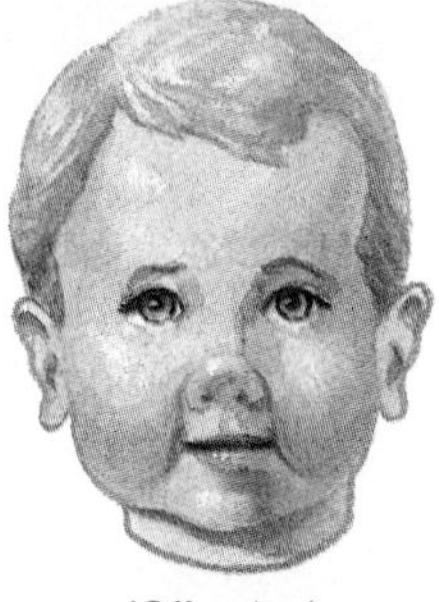

Table 4-1 Some Dominant and Recessive Human Traits

TRAIT		COMMENTS
Little fingers	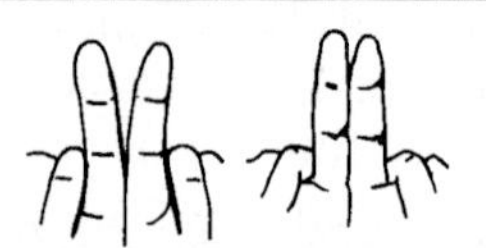 	Hold your hands with the palms toward you and press the little fingers together. Bent finger tips are dominant over parallel fingers.
Fingers and toes		Having extra fingers and toes is dominant over the normal number of digits. Short fingers are dominant over those of normal length.
Nose	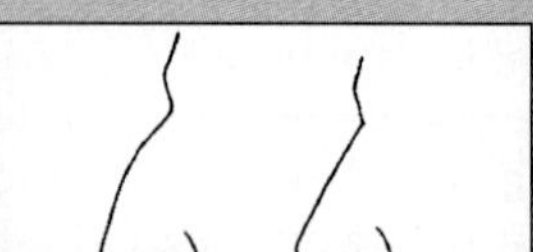 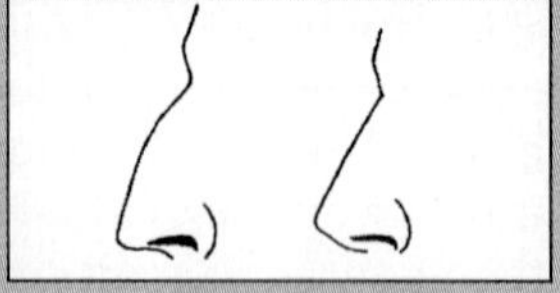	A large, convex nose is dominant over a smaller, straight one.
Chin dimples or fissures		Chin dimples or fissures are dominant over no chin dimples or fissures.
Freckles	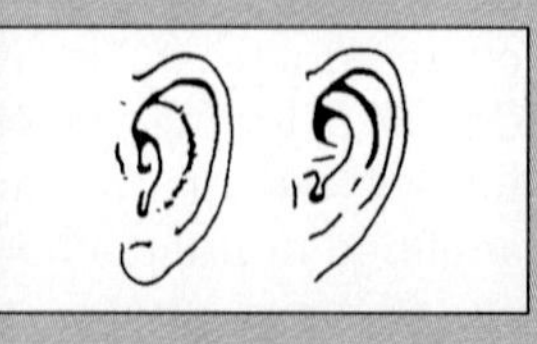	Freckles appear to be dominant over no freckles.
Ear traits		Free earlobes are dominant over attached ones. Shallow ear pits (one or both ears) are dominant over nonpitted ears. A small tubercle on the upper rim of the ear is also a dominant trait.
Tongue-rolling	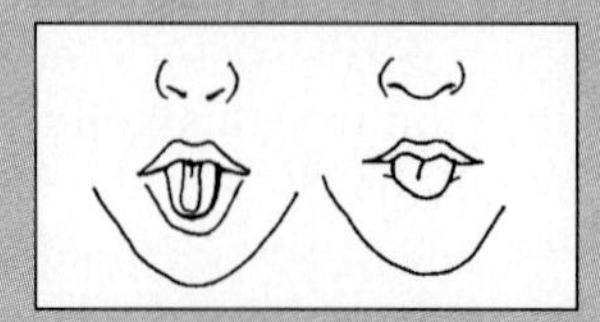	Tongue-rolling is dominant over the lack of this ability.
Common baldness		The M-shaped hairline that recedes with age is recessive to a normal hairline.

Biology in Action WHY SEX?

Organisms reproduce either sexually or asexually. In sexual reproduction a plant or animal produces haploid gametes through meiosis. In humans two gametes from a male and a female fuse to form a diploid offspring. Because an offspring receives genes from two parents it is genetically different from either one. In asexual reproduction a single-celled organism produces offspring by splitting into two cells, a process called **binary fission**. The simplest form of asexual reproduction occurs in some forms of bacteria. A single bacterium simply undergoes binary fission, producing identical copies of itself. Offspring are genetically identical to the parent.

Comparing these two general modes of reproduction in the context of biological evolution raises an interesting question. If evolutionary success is measured by how many offspring an organism reproduces, why not reproduce asexually and generate many identical copies of oneself? After all, sexual reproduction is biologically "expensive." Individuals have to find a mate, court, possibly fend off rivals, and usually waste thousands or even millions of sperm and eggs that never unite to form an offspring. Even though less efficient than asexual reproduction, sexual reproduction's advantage is that it allows for new gene combinations that increase the probability of survival and the reproduction of the offspring. The combined processes of crossing-over during meiosis and gamete fusion to form a zygote produce genetically diverse offspring. Because all environments are subject to change through time, sexual reproduction usually generates sufficient diversity each generation, thus enabling a species to adapt to changing conditions over time.

the underlying genotype for unattached earlobes could be homozygous UU *or* heterozygous Uu.

Human beings possess many thousands of genetically determined traits, all packaged into 23 pairs of homologous chromosomes. This means that many genes are contained on one chromosome and must travel together during meiosis. Genes found on the same chromosome are said to be **linked genes.** Why then do we say that Mendel's second law of independent assortment is valid? The farther apart two genes are on a chromosome, the more likely they are to be separated during crossing-over. Usually when two genes are close to one another on the same chromosome, they assort independently *less frequently* than would otherwise be expected. For example, imagine you have a rope 2 feet long with two knots tied in it 18 inches apart. Theoretically there are many places between the knots where the rope may be cut to separate the knots.

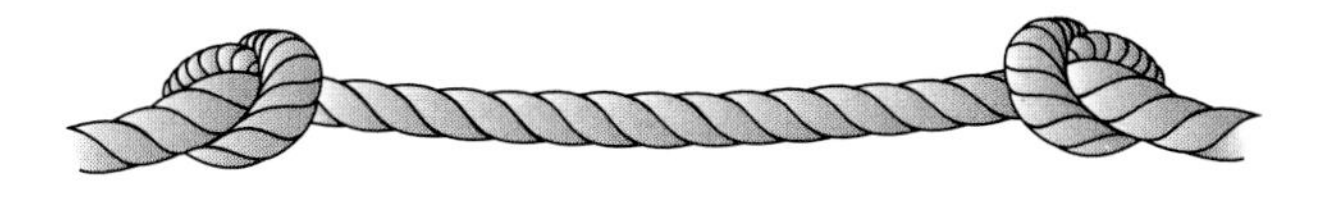

However, if the two knots are only 1 inch apart, there are few places to cut the rope. Similarly if genes are far apart on a chromosome, during meiosis I there are many places that crossing-over can occur. There is a good chance the genes will assort independently. On the other hand, if the genes are close together, there are far fewer places for crossing-over to occur. The genes are then less likely to assort independently.

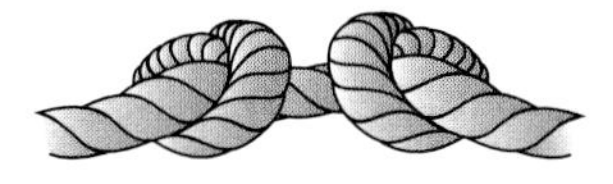

In the next chapter we describe the gene and the message encoded in its DNA structure.

CHAPTER REVIEW

SUMMARY

1. Gregor Mendel discovered that pairs of hereditary factors determined the characteristics of his experimental pea plants and that these factors behaved according to the rules of probability. Although earlier biologists had conducted similar experiments, they did not discover the pattern of inheritance because they did not recognize the significance of numerical ratios.

2. From plant breeding experiments Mendel formulated the law of segregation, which says that hereditary factors segregate from each other during the formation of gametes and that each gamete receives only one of the pair of hereditary factors. Mendel confirmed this hypothesis by making a testcross of F_2 plants with recessive homozygous plants.
3. Mendel's second conclusion, called the law of independent assortment, stated that the distribution of a hereditary factor for one characteristic into a gamete did not affect the distribution of hereditary factors for other characteristics. The probability, or chance, of different combinations of particular varieties of characteristics appearing in an offspring is the product of their individual probabilities of occurrence.
4. Today Gregor Mendel's hereditary factors are called genes, which are sequences of DNA nucleotides along the length of a chromosome. Alternative forms of a gene, which give rise to different varieties of a characteristic, are called alleles.
5. Each gene of a gene pair is located on a chromosome, called a homologue, of a homologous pair of chromosomes. All cells that are not gametes contain homologous pairs of chromosomes and are therefore diploid cells. During both mitosis and meiosis each homologue and its copy are attached to one another at the centromere. As long as they remain attached they are called sister chromatids.
6. When each gene of a pair is identical for a characteristic, the individual is homozygous for that characteristic; when each gene is different, the individual is heterozygous for the characteristic. Because of the dominant and recessive relationship between most alleles, only the dominant allele is expressed in the heterozygous state.
7. Meiosis produces gametes that contain only one homologue of each homologous pair of chromosomes and are referred to as haploid cells. There are as many as four haploid gametes produced from the meiotic division of a single cell.
8. During fertilization the haploid male sperm and female egg unite to form a single diploid cell called a zygote. Each gamete contributes one homologue to each homologous pair of chromosomes in the zygote. Ultimately from this single cell come all the specialized cells that make up the offspring.

FILL-IN-THE-BLANK QUESTIONS

1. Plants produced by cross-pollination are called ______.
2. A particular variety of a characteristic expressed in the F_1 generation is ______ over the other variety not expressed.
3. Cross-pollinating an F_2 plant with a true-breeding recessive variety is called a ______ ______.
4. Mendel discovered that a monohybrid cross produces a ______:______ phenotypic ratio in the F_1 generation.
5. Mendel also discovered that dihybrid crosses produced a ______:______:______:______ phenotypic ratio in the F_1 generation.
6. When homologues replicate and remain attached to their copies, they are called ______ ______.
7. There are ______ chromosomes in a human gamete.
8. During nuclear division ______ ______ pull the chromosomes toward the opposite poles of the cell.
9. Meiosis reduces the number of chromosomes by ______ ______.

SHORT-ANSWER QUESTIONS

1. Compare and contrast meiosis and mitosis.
2. What conditions must be present for the rules of probability to operate in a group of events or outcomes?
3. Why did Mendel obtain a 1:1 phenotypic ratio in his testcrosses?
4. What is an allele?
5. What are the chances of parents having two male children in a row? A male and female child in a row?

VOCABULARY REVIEW

allele—*p. 73*
dominant allele—*p. 65*
gene—*p. 73*
genotype—*p. 73*
heterozygous—*p. 73*
homozygous—*p. 73*
phenotype—*p. 73*
recessive allele—*p. 65*
testcross—*p. 67*

Chapter Five

MOLECULAR GENETICS: THE BIOCHEMICAL BASIS OF HUMAN INHERITANCE

OBJECTIVES

After reading this chapter you should be able to:

For Review

Here are some important terms and concepts that you will encounter in this chapter. If you are not familiar with them, you should review them before proceeding.

Dominance and recessiveness
(pages 65–66)

Hydrogen bonds
(page 21)

Protein synthesis
(page 30)

1. *Discuss the relationship between the complementary base pairs of DNA.*
2. *Define semiconservative replication.*
3. *Explain the processes of transcription and translation.*
4. *Describe the role of mRNA in transcription and tRNA in translation.*
5. *Define mutation and differentiate between point mutations, frameshift mutations, and chromosomal abnormalities.*
6. *Discuss the implications of sex-linked inheritance for males.*
7. *Define pleiotropy.*
8. *Name and discuss the two stages in the development of cancer.*
9. *Define recombinant DNA and discuss its importance in laboratory technology.*
10. *Explain the action of restriction enzymes and discuss their importance in recombinant DNA technology.*

The hereditary material passed from generation to generation of humans, as well as between generations of cells, is composed of a biochemical compound called DNA. DNA encodes the information necessary to construct and maintain the biological complexity of humans. Ultimately, the information encoded in DNA determines the construction of specific polypeptides necessary for life.

The Composition of Chromosomes

In 1952 Cambridge University scientists James Watson and Francis Crick proposed that DNA has a **double helix** (he′-licks) structure. A double-stranded helical structure is like a circular staircase that always maintains the same diameter and width of the steps, with a connecting railing on either side. The "railings of the staircase" are composed of the chemical bonds between the phosphate and sugar molecules of the nucleotides, repeating without change the entire length of the structure. Watson and Crick proposed that each step contains two nucleotides held together by hydrogen bonds, adenine pairing with thymine and guanine pairing with cytosine. This arrangement made sense in light of earlier observations. The amount of adenine (A) in DNA equals the amount of thymine (T), and the amount of guanine (G) equals the amount of cytosine (C). The nucleotide bases that pair via hydrogen bonds are called **complementary base pairs.**

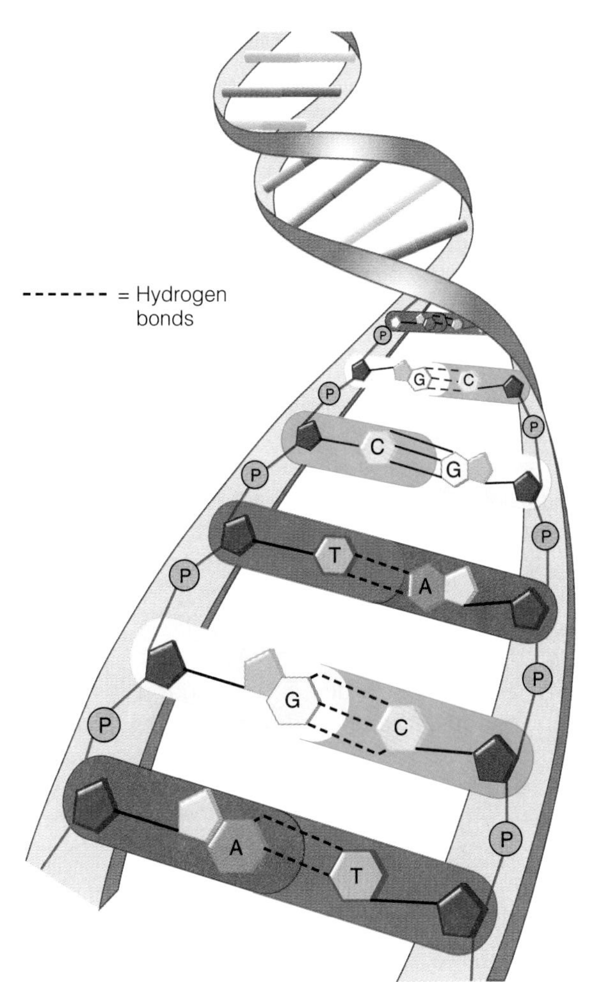

Replication of DNA

DNA is replicated by means of **semiconservative replication** in which a complementary strand is

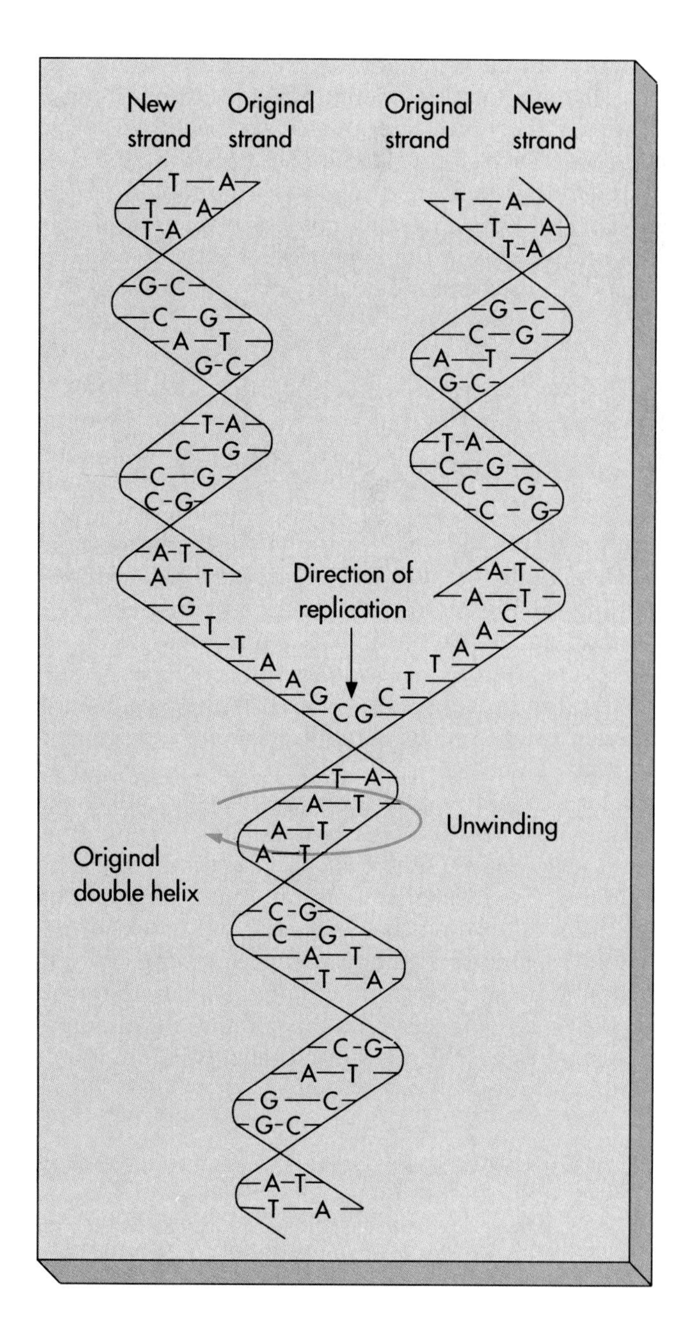

formed from each of the original double strands. This results in two new double strands, each made of one original single strand and one newly formed single strand. **Replication** begins as the hydrogen bonds between the complementary base pairs break, allowing the two strands to unzip or unwind. An enzyme called **DNA polymerase** (pol'i-mer-as) then binds to each strand. Each separated strand is a template for the synthesis of a new complementary strand. The DNA polymerase moves along each single strand, adding the complementary nucleotide base to each exposed base of the original strand. For example, if one exposed nucleotide of a strand is adenine, in the presence of DNA polymerase its complementary base pair thymine forms a hydrogen bond with it. As hydrogen bonds form, the sugar and phosphate molecules of the adjacent nucleotides chemically bond, producing a complementary strand from the original DNA template.

DNA replication occurs during interphase, just before either mitosis or meiosis begins. During prophase each chromosome of a pair contains two chromatids and each chromatid is a double helix of DNA. Each chromatid is composed of one of the original strands of the parent DNA helix and a new strand that is a complementary copy of that strand.

Genetic Code and Protein Synthesis

Each somatic body cell contains a full complement of DNA in its nucleus. DNA is like a cookbook containing all the recipes necessary to construct you. Depending on its particular needs throughout its life cycle, the cell is able to open up the cookbook to the appropriate recipe. Because the recipes are valuable and used repeatedly, the cookbook is kept in a safe place (the nucleus). If the cell needs a particular recipe, it copies a recipe from the cookbook. The copied recipe is brought to the kitchen (the cell's cytoplasm), where all the necessary ingredients are. The ingredients are put together by a chef (an organelle called the ribosome) in the order and amount specified by the recipe. The result is a product with a particular function, to be used either inside or outside the cell. What then is the nature and language of the recipe, and what is its product?

Genetic Code

DNA encodes information specifying the construction of proteins or parts of proteins. The information stored in DNA is similar in many ways to Morse code. In Morse code a set of symbols, dots and dashes can be translated into another set of symbols, alphabetical letters. In the genetic code nucleotide bases can be translated into another set of structures, amino acids in proteins. Because there are 4 different bases in DNA and RNA and 20 different amino acids in proteins, the bases cannot be a one-to-one code for amino acids because there are not enough of them. Therefore, as in Morse code, there are short sequences of bases to encode each amino acid. In a sequence of two bases coded for an amino acid, there would be 4^2, or 16, possible combinations of bases. This is not enough to code for all amino acids. However, three bases per amino acid produces 4^3, or 64, possible combinations. This is more than enough to code for all amino acids. The units of information, or "words," of the DNA genetic code are **triplets**, each three nucleotides long. To understand how information encoded by DNA becomes a protein, we must first describe the messenger that makes cellular protein production possible.

RNA

In Chapter 2 you learned that in addition to DNA there was also another macromolecule called **ribonucleic acid**, or **RNA**, involved in heredity. There are three important differences between RNA and DNA macromolecules. First, RNA nucleotides contain ribose sugar, which retains all its oxygen atoms. Second, RNA macromolecules contain the single carbon-nitrogen ringed nucleotide base **uracil (U)** instead of thymine (T). Third, when RNA nucleotides bond they form a single-stranded chain rather than the double helical structure of DNA.
Triplets of RNA nucleotides, each coding for a specific amino acid, are called **codons.** As in any other code or language, it must be clear where sentences start and stop. How does the protein synthesizing apparatus of the cell recognize where codons start and stop and where entire protein codes start and stop? The codon AUG signals "start," the beginning of a protein. Three codons, UAG, UAA, and UGA, signal "stop," the end of a protein.

In addition to the "start" codon and three "stop" codons, there are 60 remaining codons to code for 20 amino acids. Because several codons may specify the same amino acid, the genetic code is called **redundant.** For example, there are six different codons for the amino acid leucine. Thus, a **gene** is a sequence of DNA nucleotide triplets on a chromosome that codes for a polypeptide chain.

Transcription

RNA controls protein synthesis in two stages. The first step in protein synthesis is called **transcription.** Transcription is a cellular process that produces a single strand of RNA complementary to a segment of DNA. The newly formed single-stranded RNA, called **messenger RNA (mRNA),** then separates from the DNA strand and moves from the nucleus into the cytoplasm.

The double-stranded helical DNA first unwinds or unzips, as it also does during DNA replication. However, only certain sections along the double strand containing the genes to be expressed unwind. Recalling the cookbook analogy, we can see it is as if we've turned to a particular page for a recipe, temporarily ignoring the other recipes. As the double-stranded DNA unwinds, the hydrogen bonds between complementary base pairs break, exposing nucleotide bases. A special enzyme called **RNA polymerase** bonds to one end of the exposed DNA nucleotides (Figure 5-1). How does the polymerase know which exposed strand of DNA to bind to and where to begin? Each gene begins with a **promoter site,** which is a specific sequence of nucleotide bases recognized by the RNA polymerase.

Free RNA nucleotides enter the nucleus from the cytoplasm. Through the action of RNA polymerase, they begin forming complementary base pairs with the corresponding bases exposed along one of the DNA strands. Cytosine attaches to exposed guanine, guanine attaches to exposed cytosine, and adenine attaches to exposed thymine. No thymine is present in RNA to attach to exposed adenine, however, but uracil is. As a result where adenine lies exposed along a DNA strand, the RNA nucleotide uracil is attached to it, forming a complementary base pair. The RNA nucleotides chemically bond to one another by way of the sugar-phosphate portion of their structures. This results in a long single strand of RNA complementary to the exposed portion of DNA. The newly formed messenger RNA, or mRNA, then separates from the DNA strand and moves from the nucleus into the cytoplasm. The RNA has transcribed (that is, "copied") the exposed segment of DNA. A particular recipe has been copied from the cookbook. Next comes the assembly of the ingredients located in the cell's cytoplasm.

Figure 5-1 Transcription

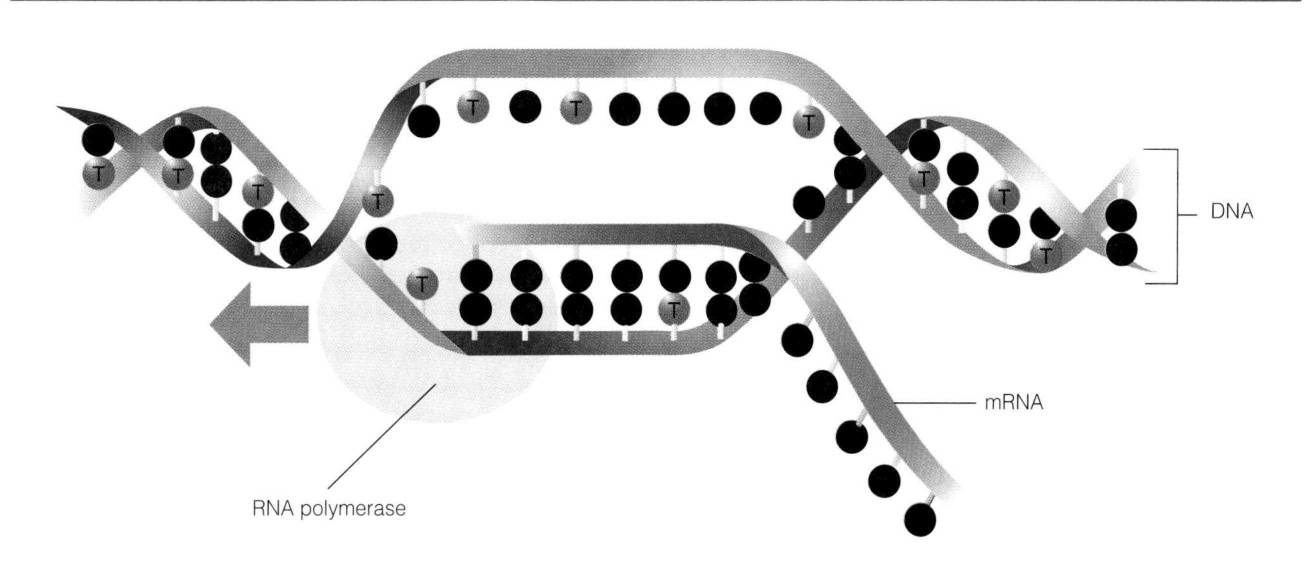

After a segment of the double-stranded DNA unwinds and exposes its nucleotide bases, RNA polymerase bonds to the promoter site, which is the beginning of the gene to be copied. In the presence of RNA polymerase, free RNA nucleotides form temporary complementary base pairs with the exposed DNA nucleotides. This results in a complementary single strand of mRNA.

Translation

Translation is the cellular process that results in the synthesis of a polypeptide chain having an amino acid sequence specified by the sequence of codons in mRNA.

After the newly formed mRNA leaves the nucleus, it moves into the cytoplasm, where it encounters one of the cell's ribosomes. It joins with a ribosome to form a ribosome-mRNA complex. During transcription another series of events also takes place in the cytoplasm. Another type of RNA, located in the cytoplasm and called **transfer RNA,** or **tRNA,** has formed and bonded to amino acids. There are as many types of tRNAs as there are amino acids. It might help to think of tRNA as "code books." They are the only molecules in the cell that can decipher the particular sequences of mRNA nucleotides and translate them into a polypeptide chain.

Each tRNA is a single RNA strand. The strand is folded back onto itself, forming three loops much like a three-leafed clover growing from one stem. The loops are formed by hydrogen bonds between complementary bases of the RNA nucleotides. Enzymes in the cytoplasm recognize each tRNA strand and attach the correct amino acid to the stem. Each type of tRNA bonds to a specific amino acid. One loop of the tRNA has three nucleotide bases called the **anticodon.** The **anticodon** is a three-nucleotide sequence complementary to the codon of mRNA that specifies the amino acid attached to that tRNA. The tRNA molecules delivering their specific amino acids to the mRNA-ribosome complex begin the translation process.

OH Amino acid attaches here

Anticodon loop

GAU Anticodon

CUA Codon

mRNA

The ribosome contains two bonding sites, called the **P site** and the **A site,** that can accept tRNA carrying amino acids. The ribosome first bonds to the start codon (AUG) of the mRNA so it is positioned at the P site. The tRNA-amino acid molecule bearing the anticodon UAC forms a complementary base pair with the start codon of the mRNA. The second

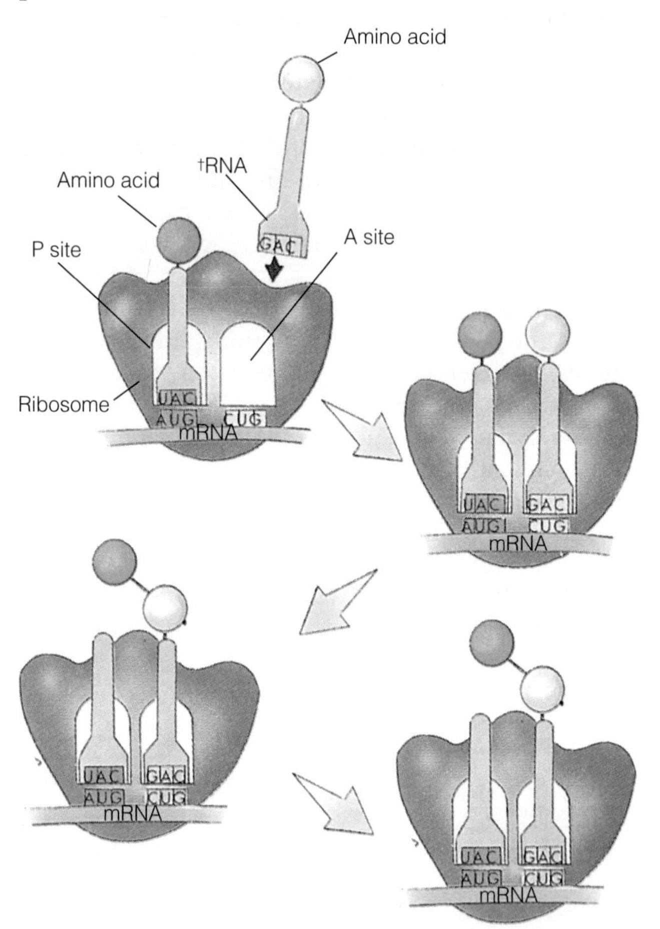

codon in the sequence is at the A site. At the A site an amino acid–bearing tRNA molecule with the appropriate anticodon will form a complementary base pair with the second mRNA codon. A peptide bond forms between the two amino acids by the transfer of the amino acid on the tRNA at the P site to the amino acid attached to the tRNA at the A site. Having lost its amino acid, the tRNA at the P site breaks its bond with the codon of the mRNA and drifts off.

The tRNA at the A site, now carrying the first two amino acids of the protein being synthesized, moves to the empty P site as the ribosome moves one codon down the mRNA. This now positions a new mRNA codon at the A site, to which the appropriate amino acid–carrying tRNA can bond. The growing amino acid chain, attached to the tRNA at the P site, is transferred to the amino acid carried by the tRNA at the A site, forming a peptide bond. The process continues until the ribosome encounters a stop codon, UAA, UGA, or UAG. Enzymes in the ribosome then split the chemical bond between the last amino acid of the completed polypeptide chain and the tRNA to which it is attached, freeing the newly synthesized protein.

The newly formed polypeptide chain, which is either a protein or a portion of one, may be a structural material, an enzyme, a transport protein, or a hormone. For example, an enzyme formed in this way will participate in the biochemical activities of that cell and therefore help to determine the cell's specific structure and function. If each somatic cell possesses a full complement of DNA, why are there cells of different types and functions?

Different Cells, Different Proteins

Geneticists have found that in bacteria and some simple multicellular organisms, gene promoter sites can be blocked or uncovered, depending on the presence of intracellular substances. Promoter sites act as "on/off" switches regulating gene transcription. But geneticists do not know how regulation occurs or what cellular events control the timing of gene transcription in more complex multicellular organisms.

Chromosomes are good at replicating themselves. For example, you developed from a single-celled zygote to about 100 trillion cells as an adult. By adding the approximately 50 million cells replaced in your body each day during adulthood, you can understand how the accuracy of DNA replication is staggering. However, mistakes do occasionally occur.

The Source of Genetic Variability

Organisms face a biological dilemma of sorts. Chapter 1 noted that environments change through time and organisms must also be able to adapt to environmental changes or become extinct. Adaptation to changing conditions is possible if individuals vary in a population. We have already discussed two methods of changing gene combinations and maintaining population variation. The first method is through sexual reproduction. Each parent contributes different alleles to the offspring via its contribution of one half of each pair of the offspring's chromosomes. The offspring represents a unique combination of parental genes.

The second method is crossing-over, or genetic recombination, in which new combinations of alleles are produced during the exchange of chromosome segments. However, both sexual reproduction and crossing-over can produce new combinations of genes only if a number of different alleles *already* exist.

Despite the accuracy of chromosomal replication, random mistakes occur, altering nucleotide sequences of DNA. Such changes in the genetic message are rare. For example, in a human male the probability that a particular gene is altered is about one out of every million gametes (sperm) he produces. Limited as this appears to be, this steady trickle of change is the ultimate cause of genetic variation and the basis for biological evolution.

Mutations

A change in the genetic message of a cell is called a **mutation** (mu-ta′shun). When a mutation occurs within a gamete that fuses with another gamete to produce a zygote, the mutational change is passed along as part of the hereditary information to future generations. This introduces new variability. In contrast, when a mutation occurs in a somatic cell, it is not passed along to later generations because the mutation does not directly affect the gametes.

Point Mutations

The most common type of mutation occuring in DNA replication is a **point mutation.** A **point mutation** is the substitution or change in one nucleotide base pair in a gene. Three different effects may result from a single base substitution in a nucleotide sequence. We will consider the possible effects of a single base substitution of the DNA sequence CTC, which codes for the amino acid glutamic acid.

The first possibility is that a mutation may not change the amino acid sequence of the encoded protein. The genetic code is redundant in that one amino acid may be encoded by several different codons. If a mutation changes the codon CTC to

DNA triplet changed from

CTC ——► CTT

The mRNA would transcribe

GAG ——► GUU

Originally GAG coded for glutamic acid. Although the mutation has changed the sequence, it still specifies the amino acid glutamic acid.

CTT, the new sequence still codes for glutamic acid and the protein synthesized from the mutated gene does not change. A second possibility is that the mutation may code for an amino acid functionally equivalent to the original amino acid. Many proteins contain sequences of amino acids that can vary. For example, in hemoglobin the amino acids that face outward in the three-dimensional protein must be hydrophilic to keep the protein dissolved in the red blood cell's cytoplasm. Which hydrophilic amino acids are on the outside is not significant. For example, a point mutation causing a change of the codon CTC to CTA replaces glutamic acid, which is hydrophilic, with aspartic acid, which is also hydrophilic. This substitution does not change the solubility of hemoglobin. Mutations that do not change the function of an encoded protein are called **neutral mutations.**

DNA changed from

CTC ⟶ CTA

The mRNA would transcribe

GUG ⟶ GAU

Originally GAG coded for glutamic acid. The mutation has changed the sequence so that it now specifies the amino acid aspartic acid.

A third possible outcome is that the mutation may encode for a functionally different amino acid. A mutation from the codon CTC to CAC replaces glutamic acid, which is hydrophilic, with valine, a hydrophobic amino acid. In human hemoglobin this substitution results in the genetic defect of sickle-cell anemia. This hemoglobin variant sticks to itself, causing the red blood cells to become distorted and malfunction.

DNA triplet changed from

CTC ⟶ CAC

The mRNA would transcribe

GAG ⟶ GUG

Originally GAG coded for glutamic acid. The mutation has changed the sequence so that it now specifies the amino acid valine.

Mutations can also involve adding or deleting a single nucleotide base pair. In these **frameshift mutations** one nucleotide base pair in a gene is added or deleted. Because the DNA code is read as triplets of nucleotides, frameshift mutations cause all triplets after the deletion or addition to be read differently.

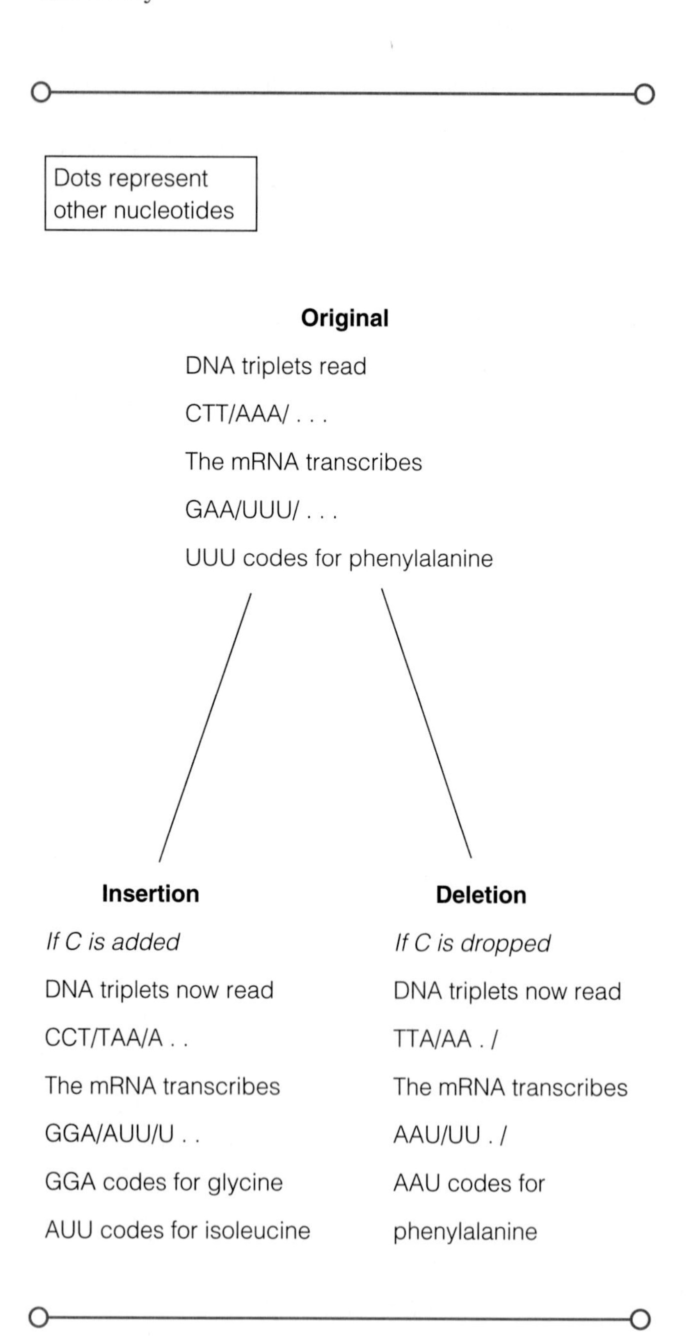

Dots represent other nucleotides
Original
DNA triplets read
CTT/AAA/ . . .
The mRNA transcribes
GAA/UUU/ . . .
UUU codes for phenylalanine
Insertion
If C is added
DNA triplets now read
CCT/TAA/A . .
The mRNA transcribes
GGA/AUU/U . .
GGA codes for glycine
AUU codes for isoleucine
Deletion
If C is dropped
DNA triplets now read
TTA/AA . /
The mRNA transcribes
AAU/UU . /
AAU codes for phenylalanine

Random changes in nucleotides usually result in changed proteins that function less effectively than the original protein. The probability of a random change improving the function of this structure is thus small. It is similar to a blindfolded person randomly changing a single part of a watch hoping to improve its ability to keep time. It probably will not. However, on the rare occasion when an altered protein functions better, or functions in a way adaptive in a new environment, the individual possessing this mutation will be at a selective advantage. The individual will have a relatively greater probability, compared with other individuals, of surviving to reproduce. In this way new alleles are tested by the environment and, if neutral or adaptive, contribute to evolutionary change and diversity.

Chromosomal Abnormalities

Chromosomal abnormalities are another category of mutations and involve changes in either the number or structural arrangement of chromosomes in the gametes.

Occasionally during meiosis an event called **nondisjunction** of chromosomes occurs. **Nondisjunction** occurs when two homologous chromosomes do not separate. Instead of one member of a pair of homologous chromosomes going to each of the two daughter cells, both go into the same daughter cell, and the other gamete gets none. If this occurs, for example, in an ovum, and it is fertilized by a sperm with one of these chromosomes, the resulting zygote will have *three* homologous chromosomes instead of the normal complement of *two*.

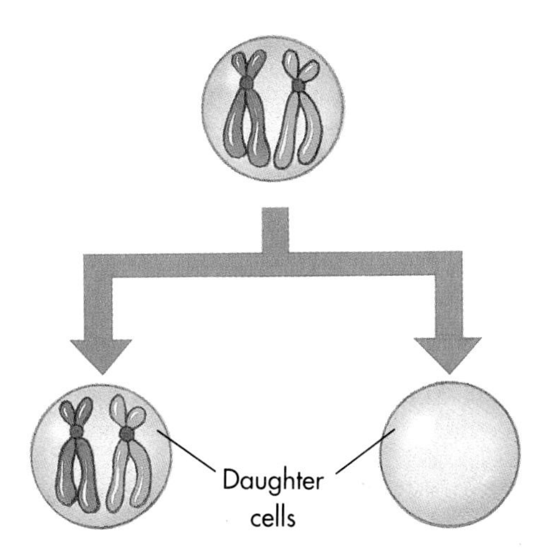

A well-known example of nondisjunction in humans is **Down syndrome.** An individual with this condition has three instead of two chromosomes number 21. The condition is thus also known as **trisomy** (*tri* = three, *soma* = body) **21.** People with Down syndrome usually have a shortened life span and some mental retardation. The average incidence of trisomy 21 is one out of 750 births.

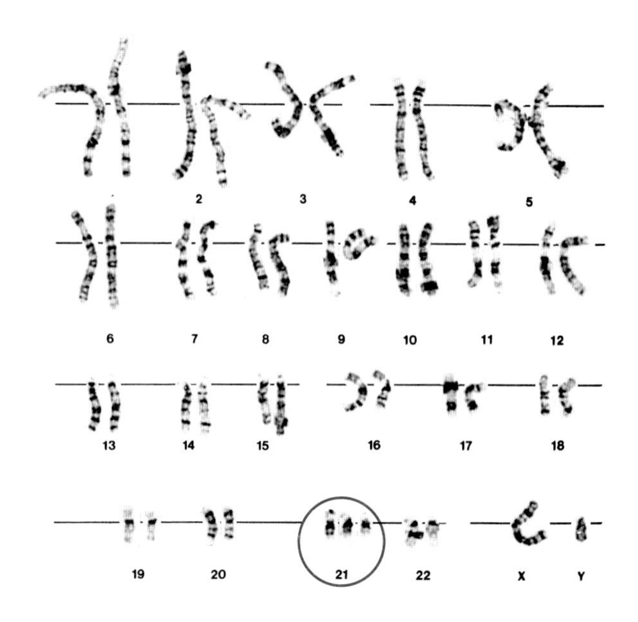

Three important classes of chromosomal rearrangements are **deletions, translocations,** and **inversions.** Deletions occur when a portion of a chromosome is lost. Translocations occur when a segment of one chromosome is moved to another nonhomologous chromosome. Inversions occur when the orientation of a portion of a chromosome becomes reversed.

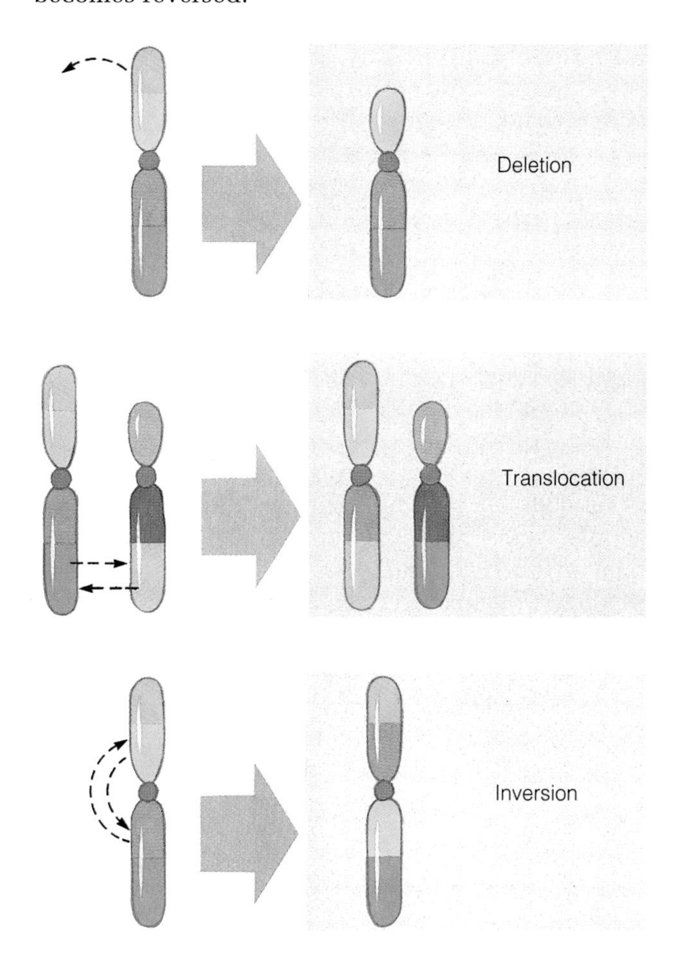

Deletions are generally harmful to the individual. The larger the deletion, the greater the harm. Recently many physical and mental symptoms known as **Prader-Willi syndrome** have been associated with a deletion of part of chromosome number 15. As infants these individuals do not feed well because of a poor sucking reflex. However, by the age of five, they develop an uncontrollable urge to eat that results in obesity and related health problems, such as diabetes.

Causes of Mutation

Aside from their spontaneous occurrence, mutations can also be produced by external factors, such as radiation or chemicals. Any external factor or agent capable of damaging DNA is called a **mutagen.** When mutations are caused by a mutagen, the mutation is said to be **induced,** rather than spontaneous. Examples of mutagens and their effects are also described later.

Human Heredity

Mendel's laws of inheritance also apply to human beings. Genetic knowledge gained from studying plants and fruit flies often tells us a great deal about ourselves. As you know, DNA is common to all forms of life. The number and sequence of individual nucleotides, called genes, specify polypeptides that ultimately determine whether the organism is a human being, or a pea plant, or any other organism.

Aside from the shared common biochemical structure of DNA, there are other practical reasons why biologists study other organisms in trying to understand human heredity. Mendel studied pea plants because they had a short life span, produced large numbers of offspring, and matings between selected plants and their environment were easily controlled. Unlike pea plants, however, humans have relatively long life spans and few offspring per couple, their matings cannot be controlled, and human beings interact complexly with their environments. Compared with the numbers of offspring produced in a single cross with peas, humans produce few offspring and these usually represent only a small portion of the existing genetic combinations. For example, if two heterozygous pea plants, Aa × Aa, are crossed, the expected phenotypic frequency is that three fourths of the offspring will express the dominant allele A, while only one fourth will express the recessive allele, a. Mendel was able to count hundreds of offspring from such a cross and record the offspring in all expected classes, establishing a phenotypic numerical ratio of 3:1.

Suppose a human couple has two children, one of whom is a son with a genetic disorder. The ratio of phenotypes in this case is 1:1. What is the underlying pattern of inheritance? Without large numbers of offspring it is difficult to decide. Simple techniques of investigating family histories and lineages are used to accurately answer these and other problems regarding human inheritance patterns.

Human Chromosomes

Human chromosomes are examined by first collecting a sample of an individual's cells, usually as a small blood sample. Chemicals are added that cause the cells to begin mitosis. At metaphase I when the duplicated chromatids of each homologous pair of chromosomes are aligned along the equator of the cell, another chemical is added that stops further division. At metaphase I the chromosomes are most condensed and therefore most easily distinguished from one another. So at this point the cells are broken open and the chromosomes are separated from the rest of the cell and stained. The chromosomes are photographed, and each chromosome is cut out and arranged according to size. The result of this procedure is called a **karyotype** (kari-o-tip) (Figure 5-2).

Of the 23 pairs of human chromosomes, 22 are perfectly matched in both males and females and are called **autosomal** chromosomes. The remaining pair of chromosomes is not always homologous. The chromosomes of this pair are involved in sex determination and are called **sex chromosomes.** Females have two homologous chromosomes designated as **X** chromosomes, and males have a nonhomologous pair of one X chromosome and one **Y** chromosome.

Pedigree Analysis

Next we need to describe the methods used to determine the patterns of human heredity, several genetic disorders that occur in human populations, and how some human cancers are genetically influenced.

The basic methods of human genetics are observational and not experimental. Rather than designing and conducting experiments to directly test a hypothesis, biologists and geneticists reconstruct events that have already taken place. Extensive documentation of human families and lineages from around the world is available. In some cultures this

Figure 5-2 Human karyotype

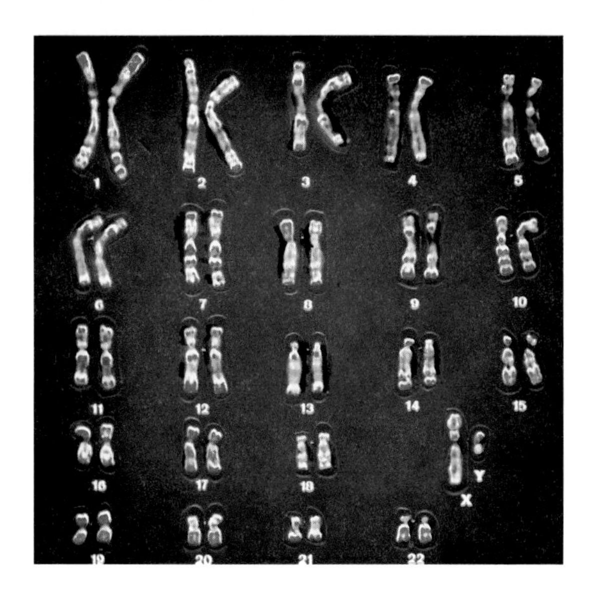

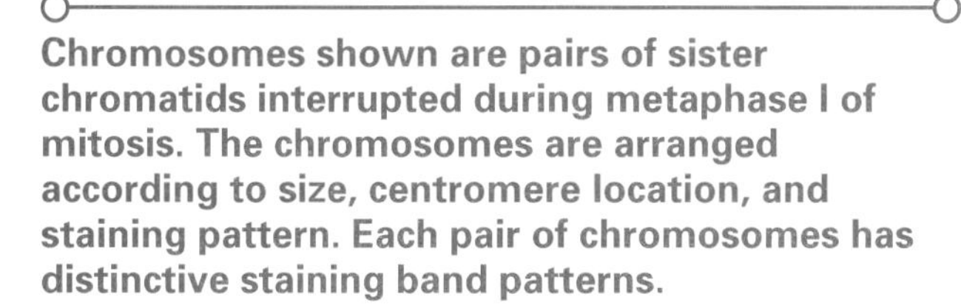

Chromosomes shown are pairs of sister chromatids interrupted during metaphase I of mitosis. The chromosomes are arranged according to size, centromere location, and staining pattern. Each pair of chromosomes has distinctive staining band patterns.

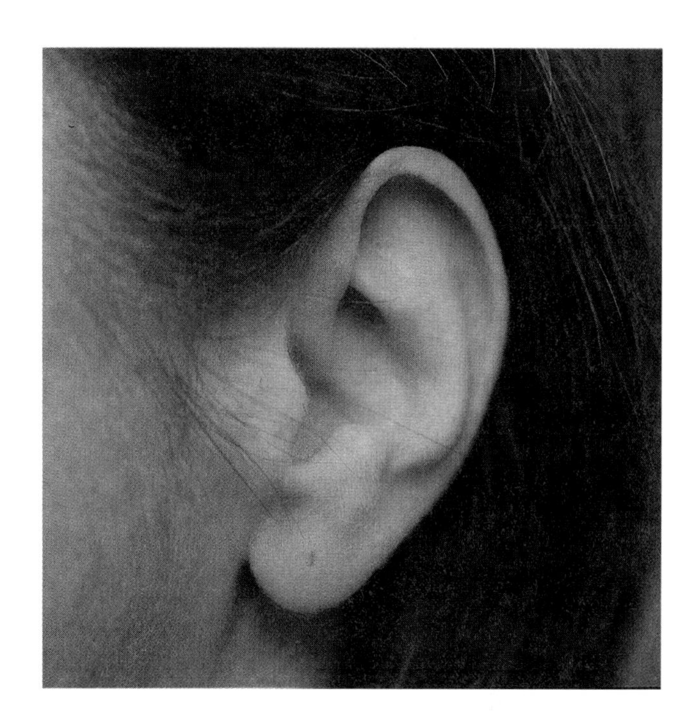

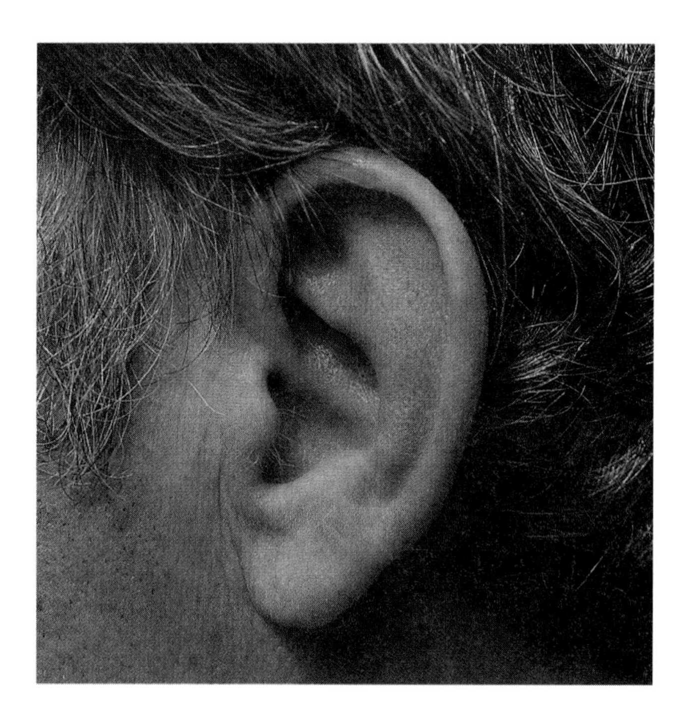

documentation is recorded in writing, whereas in other cultures it is passed along as oral tradition and history. Also, the human population is quite large. Even though families are usually small, examining many family histories that display the same genetic trait of interest may make patterns of inheritance for a particular trait apparent.

Almost 4000 genetic traits have been identified in humans. The chromosome location of a few of these traits is known, and the molecular basis of traits associated with detrimental or harmful phenotypes is known in a small percentage of cases. Other genetically determined traits are of general interest and illustrate human patterns of heredity. Earlobe shape or whether a hairline is straight across the forehead or forms a widow's peak are just two examples. However, many people are concerned that various health problems and diseases might also have a hereditary basis. For example, if a member of your family has had cancer or a stroke, it is difficult not to worry about your own future state of health, knowing that some forms of cancer and tendencies for strokes may run in families.

To study heredity in humans, biologists often look at the results of crosses that have already occurred, studies called **pedigrees.** A pedigree chart is an orderly, readable documentation of family information. By studying which relative exhibits a trait, it is often possible to determine the trait's pattern of inheritance and whether it is dominant or recessive. After determining the pattern of inheritance, it is often possible to infer which individuals in the pedigree are homozygous and which are heterozygous for the gene specifying the characteristic. Pedigrees use a set of standard symbols similar to those shown here.

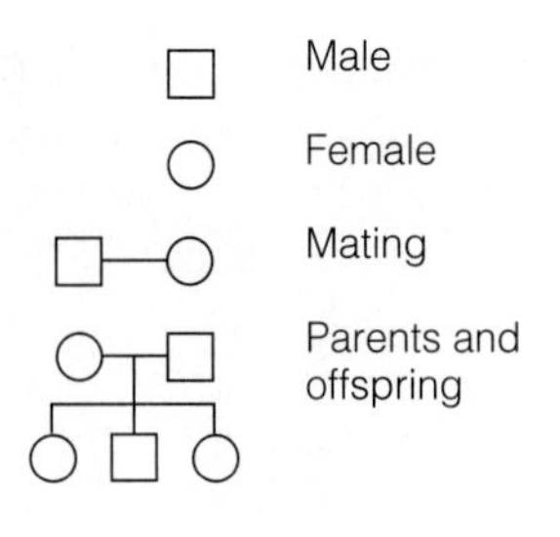

Although human families are relatively small, examining several generations usually provides enough information to determine whether the trait in question is recessive. In the analysis of a pedigree several factors indicate whether a trait is recessive.

1. The trait usually appears in the offspring of unaffected parents.
2. Affected individuals are usually members of alternate generations, unless both parents are affected.
3. Approximately one fourth of the members of the pedigree will be affected.
4. Two affected individuals cannot have an unaffected child.

These statements are simply a review of Mendel's laws of segregation and independent assortment.

When a detrimental allele occurs at a significant frequency in human populations, the harmful effect it produces is called a **genetic disorder.** One category of genetic disorders found in humans is known as **autosomal recessive traits.** Autosomal recessive traits are caused by a recessive gene on an autosomal chromosome.

Autosomal Recessive Traits

Albinism (al'bi-nizm) is a genetic disorder associated with a lack of pigmentation in the skin, eyes, and hair. Cells called melanocytes produce a granular brown-colored protein called melanin that gives skin, hair, and eyes their color. In albinism the melanocytes do not produce melanin. Pedigree analysis of albinism confirms expectations that this is an autosomal recessive trait (Figure 5-3).

The frequency of albinism varies widely among populations. In the United States, in individuals of European ancestry, the frequency of albinism is 1 individual in 37,000 live births; in individuals of West African ancestry the frequency is 1 in 15,000 live births.

Cystic fibrosis is the most common fatal autosomal recessive genetic disorder in the United States among individuals of European and northern European descent. Its main effect is on the mucous and sweat glands. Because these glands are defective they secrete a thick mucus. The results of this disease are far-reaching. In the lungs, for instance, the mucus clogs the airways. The pancreas and liver both produce several digestive enzymes that enter the intestines through a duct system. In cystic fibro-

Figure 5-3 A pedigree of albinism

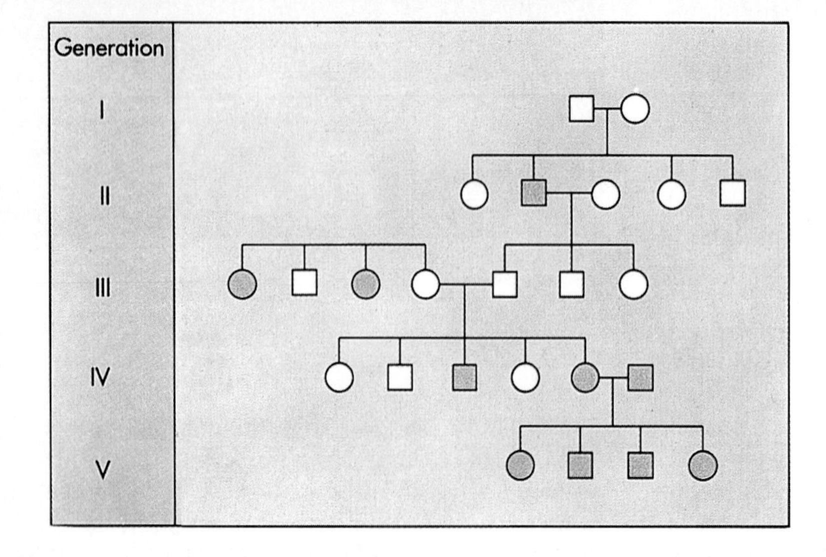

Albinism is an autosomal recessive genetic disorder. In this pedigree, the affected individuals expressing the disorder are shown in blue and unaffected individuals are shown in white.

sis the duct becomes clogged by the thick mucus, preventing the enzymes from assisting in the digestion of food. As a result the affected child sometimes is malnourished despite a normal food intake. Eventually the blocked ducts cause the pancreas to form fluid-filled sacs, or cysts. Later the pancreas degenerates and becomes fibrous. These symptoms originally gave rise to the disorder's name. A similar pattern of events also affects the liver. Death results either from a lung disease caused indirectly by the mucus clogging the airways or from pancreas or liver failure.

Among Americans of European and northern European ancestry, about 1 in 20 individuals has a copy of the defective allele but shows no symptoms. The homozygous recessive individuals occur about 1 in 1,800 live births.

Sickle-cell anemia is an inherited disorder in which individuals homozygous for the sickle-cell allele are unable to transport oxygen to their tissues properly because the protein molecules in the red blood cells carrying oxygen, called hemoglobin, are defective. When oxygen is released by the red blood cell and moves into the surrounding body tissues, the hemoglobin in sickle-cell anemia causes the red blood cells to change shape and become crescent or sickle-shaped. The sickling occurs because the hemoglobin molecules stick together instead of remaining free in the cell. Their clumping together causes the red blood cell to become distorted. In this state the deformed blood cells are fragile and easily broken apart. Because new blood cells are not produced fast enough to replace those lost, the oxygen-carrying capacity of the blood is reduced, resulting in anemia. The deformed blood cells also clog small blood vessels, causing additional problems of blood circulation and further reducing oxygen transport.

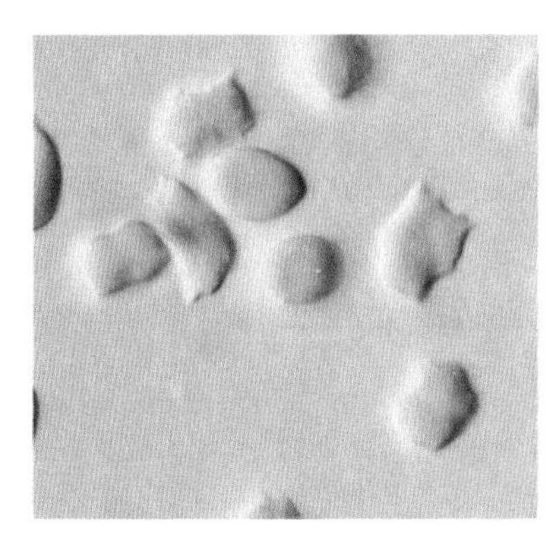

Because of the number of systems in the body affected and the severity of the effects, sickle-cell anemia is usually lethal. In contrast to those with the homozygous recessive condition, individuals who are heterozygous for the allele are generally indistinguishable from unaffected individuals. However, some of the red cells show the sickling characteristic when they are exposed to low levels of oxygen, such as high altitude environments. The allele responsible for the sickle-cell trait is common among Americans whose ancestors are from specific areas of western Africa or around the Mediterranean Sea basin. In the United States about 1 in every 500 Americans of West African descent is homozygous recessive and about 1 in every 12 is heterozygous for the sickle cell allele.

Tay-Sachs disease, named after the two physicians who first described the symptoms, is an incurable recessive hereditary disorder in which the brain of the homozygous individual deteriorates. The brain cells fail to synthesize an enzyme located in the lysosomes necessary to break down a special class of lipid called **gangliosides** (gang'gli-o-sidz). Because of the defective enzyme the lysosomes fill with gangliosides, swell, and eventually burst, releasing their enzymes into the brain cells and destroying them. The developing child who is homozygous for the Tay-Sachs allele has progressive mental retardation, blindness, and lack of motor control and will usually die before five years of age.

Tay-Sachs disease is a rare genetic disorder in most human populations, occurring in 1 in 360,000 live births. However, it has a high incidence among Jews of Eastern and Central Europe and among the American Jewish population, 90% of whom are descendants of these European populations. Approximately 1 in 3600 babies born among American Jewish populations has Tay-Sachs disease. Most people carrying the allele do not develop the symptoms because it is a recessive condition. An estimated 1 in 28 individuals in American Jewish populations is a heterozygous carrier of the allele. A blood test for Tay-Sachs disease exists, which involves taking a sample of white blood cells to determine the enzyme's presence. Affected individuals have no detectable enzyme activity in the cell sample.

Autosomal Dominant Traits

Many human physical traits, including dimpled chins and freckles, are inherited as **autosomal dominant traits.** Autosomal dominant traits are caused by a dominant gene on an autosomal chromosome. Similarly not all genetic disorders are caused by recessive alleles.

Huntington's disease causes a slow progressive deterioration of parts of the brain, resulting in loss of bodily coordination, behavioral disturbances, and eventually death. About 1 in 10,000 individuals in the United States has Huntington's disease. Because Huntington's disease is controlled by a dominant allele, every individual carrying the allele expresses the trait. If this is the case, why will the allele not die out? The first symptoms of Huntington's disease do not usually appear until the affected individual is at least 30 years old, and by that age most individuals have already reproduced. In other words the allele is transmitted before its lethal condition develops.

Metabolic Disorders

Many proteins function as enzymes in diverse metabolic pathways, making our lives possible. A block in any single reaction results in a shutdown of the metabolic pathway beyond that point and of substrate accumulation before that point. Usually these metabolic pathway disorders are caused by recessive autosomal genes. They were first identified by Archibald Garrod and are called **inborn errors of metabolism.**

Have you ever noticed warning labels on cans of diet soda that read "Phenylketonurics: Contains phenylalanine" and wondered what it means? An inborn error of metabolism known as **phenylketonuria** (fen'il-ke'to-nu'ri-ah), or **PKU,** is the reason. The affected individuals inherit two recessive alleles and are unable to break down the amino acid phenylalanine and convert it to another amino acid called tyrosine. Phenylalanine is instead converted to other metabolic products that accumulate in the bloodstream. Although not harmful to adults, these substances are harmful to the developing brain cells of infants. An infant with PKU has severe mental retardation and rarely lives beyond early adulthood. Many states, however, routinely take blood samples from newborn infants and screen them for this problem. The affected infant who is homozygous for PKU can then be treated nutritionally simply by restricting the intake of phenylalanine in the diet. This is why there is a warning label on that can of diet soda. A low-calorie sweetener called **aspartame** (as-par-tame') is widely used in diet soft drinks and other foods. Although it is safe, it contains large amounts of phenylalanine, and therefore phenylketonurics should avoid it.

Multiple Alleles

Mendel studied pairs of contrasting varieties of traits in pea plants. The plants he cross-bred were, for example, either tall or short, the flowers either white or purple, and so on. Many human genes also exhibit two alternate alleles, including the genetic disorders just described. Many genes, however, possess more than two alleles. For example, the human ABO blood type system has three alleles. In an individual heterozygous for the A and B alleles, both are expressed at the same time. Therefore the A and B alleles are said to be **codominant.** Either allele, however, is dominant over the O allele.

Different combinations of the three alleles occur in different individuals because each possesses two copies of the chromosome bearing the gene and may be homozygous for any allele or heterozygous for any two alleles. Because of the particular dominance relationships between the three alleles, four phenotypes (A, B, AB, and O) are possible.

The **Rh system** is determined by a gene encoding a protein found on the surface of the red blood cell. The Rh blood group system produces two phenotypes, Rh^- and Rh^+. Rh^+ is dominant over Rh^-. The Rh designation is used because a similar protein was first identified in the blood of Rhesus monkeys. Rh^+ individuals have red blood cells with the protein marker, and Rh^- individuals do not. An individual who is phenotypically Rh^+ may have the genotype + + or + –. An individual who is Rh^- must have the genotype – –. Our immune systems are able to recognize the differences between the Rh phenotypes. Transfusing an Rh^- individual with Rh^+ blood causes **agglutination,** or clumping, of the infused blood cells.

The Rh blood group is also important in a condition known as **erythroblastosis** (e-rith'ro-blas-to'sis) **fetalis,** or **hemolytic disease of the newborn (HDN).** This condition will be discussed in Chapter 14.

Codominance and Incomplete Dominance

The human ABO blood type has three different alleles, A, B, and O. Two of the alleles, A and B, are **codominant,** meaning that both are phenotypically detectable in heterozygotes. If an offspring inherited the A allele from one parent and the B allele from the other parent, the blood would be type AB.

Incomplete dominance involves two alleles, neither of which is dominant. In humans sickle-cell anemia is considered a trait of incomplete dominance. Sickle-cell anemia is caused by a type of hemoglobin molecules that stick together, distorting the shape of the red blood cell. The allele encoding for nonsickling adult hemoglobin is designated as Hb_a, and the allele encoding for the hemoglobin causing sickling of the red blood cell is designated as Hb_s. Although the red blood cells of heterozygous individuals appear to be normal, they contain a mixture of 50% normal and 50% abnormal hemoglobin. Heterozygous individuals have none of the symptoms of sickle-cell anemia and are phenotypically normal. However as mentioned earlier, under low-oxygen conditions some sickling occurs. Based on this, their blood cells can be distinguished from those of either homozygous individuals by a blood test. The test involves subjecting a blood sample to low oxygen levels. This causes the heterozygous blood cells to deform *slightly*. The recessive homozygous blood sample shows severe sickling, and the dominant homozygous blood shows no sickling (Figure 5-4). Consequently at the cellular level the heterozygote phenotype can be distinguished from the phenotypes of either homozygote phenotype and shows incomplete dominance.

The sickling trait corresponds to the genotypic ratio of $\frac{1}{4}$ Hb_aHb_a: $\frac{1}{2}$ Hb_aHb_s: $\frac{1}{4}$ Hb_sHb_s, which agrees with Mendel's principles of inheritance. Although incomplete dominance confirms Mendelian principles, it increases the difficulty of inferring genotypes from phenotype.

Sex-Linked Inheritance

The human X and Y chromosomes determine an individual's genetic sex. The X chromosome carries at least 139 genes, with another 160 suspected of also being there. Some of these genes determine nonsexual traits, such as eye color. Any gene on the X chromosome is called an **X-linked** gene. Any gene on the Y chromosome is called a **Y-linked** gene.

Figure 5-4 Incomplete dominance of sickle-cell allele

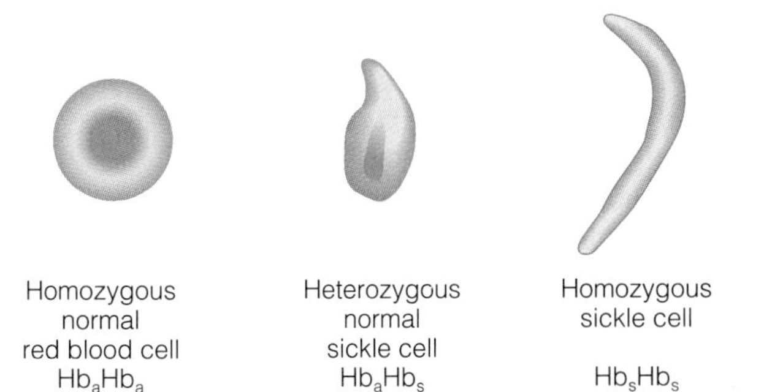

The allele for sickle cell is shown as Hb_s and the allele for nonsickling hemoglobin as Hb_a. In the homozygous nonsickling state, the red blood cells do not deform under low oxygen conditions. In the heterozygous state, there is only a slight sickling of the red blood cells under low oxygen conditions. In the homozygous sickle-cell state, severe sickling occurs under the same conditions.

As mentioned earlier, females have a homologous pair of sex chromosomes designated as XX while the male's sex chromosomes are nonhomologous and designated as X and Y. Recessive alleles on the single X chromosome of males thus have no active counterpart on the Y chromosome. Because of this, recessive alleles on the X chromosome are always expressed in males, a phenomenon called **sex-linked inheritance.** As you may already suspect, the genes determining whether a zygote will develop into a male are located on the Y chromosome. Any individual with at least one Y chromosome is a male, and any individual lacking a Y chromosome is a female.

A son receives his X chromosome from his mother and his Y chromosome from his father. A man must therefore inherit the X chromosome genes from his mother, and can pass them on only to his daughter. For recessive alleles found on the X chromosome, there is usually a particular pattern of inheritance. These recessive traits appear most frequently in males and can skip generations when an affected man passes the trait on to a phenotypically normal daughter, called a **carrier.** She in turn may bear sons expressing the trait. One familiar genetic disorder caused by recessive alleles of the X chromosome genes includes several forms of color blindness (Figure 5-5).

Three different genes are involved in color perception. Each gene encodes for a particular protein found in light-sensitive nerve cells in the back of the eye. Each protein is sensitive to only a part of the visible light spectrum, and each nerve cell contains only one type of protein. Two of these genes are X-linked. The most common form of this disorder is called **red-green color blindness** because usually either the red- or the green-sensitive protein is missing. If the red-sensitive protein is missing, red and green appear the same. If the green-sensitive protein is missing, green appears red. Red-green color blindness shows a pattern typical of a recessive X-linked gene; males express the trait and females pass the allele to their offspring.

Hemophilia (he'mo-fil'i-ah) is a hereditary condition in which the blood is slow to clot or does not clot. Several forms of hemophilia are also sex-linked.

One reason you do not keep bleeding from a cut is that the blood in the immediate area of the cut is converted to a solid gel that seals the cut. Many proteins are involved in this clotting process, and all of them must function properly for clotting. Mutations causing the loss of activity of any of these factors lead to different forms of hemophilia.

Hemophilias are recessive disorders, expressed only when an individual does not possess at least

Figure 5-5 Color blindness is a sex-linked trait

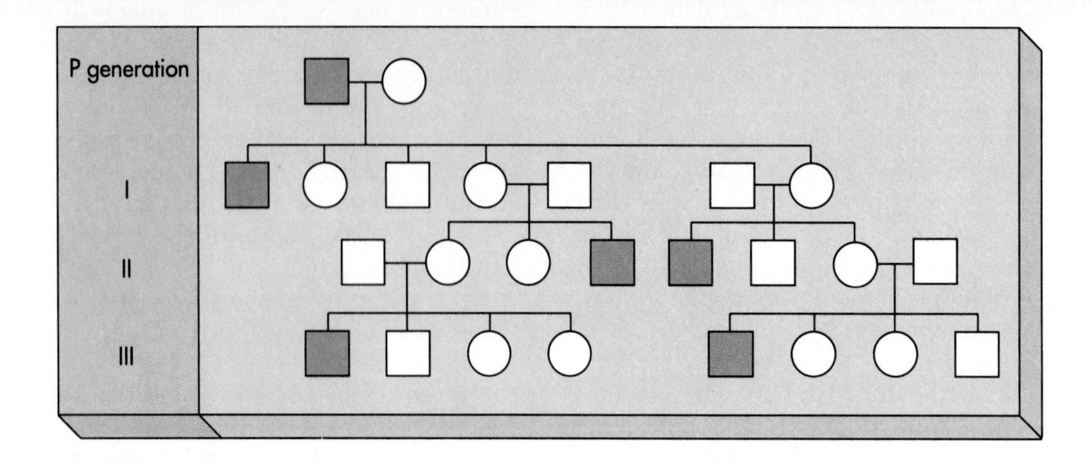

Males have one X chromosome. If that chromosome carries the recessive allele for color blindness, it will be expressed. Because females have two X chromosomes, if one of a female's X chromosomes carries the recessive allele, she is phenotypically normal. However, there is a 50 percent chance she may pass on this chromosome, and thus color blindness, to any son. Also, she may pass the chromosome to any daughter, who may continue to pass along the chromosome to future generations. In this pedigree, affected individuals are blue and unaffected individuals are white.

one copy of the normal gene and so cannot produce one of the proteins necessary for clotting. Most of the genes encoding for the proteins important in clotting blood are located on autosomes, but two are found on the X chromosome. With these protein-clotting genes any male who inherits a defective allele will develop hemophilia because his other sex chromosome is the Y chromosome, which lacks an allele of the gene.

A mutation in an X chromosome protein-clotting gene occurred in one of the parents of Queen Victoria. The British royal family escaped the disorder because Queen Victoria's son King Edward VII did not inherit the defective allele. Three of Victoria's nine children received the defective allele, however, and carried it by marriage into many of the royal families of Europe (Figure 5-6). It is still being transmitted to future generations along these family lines.

Some autosomal genes may also be expressed differently in males and females. These are known as **sex-influenced genes.** One example is pattern baldness, most often seen in men but also seen in women. Pattern baldness is influenced by the male hormone testosterone. However, evidence also suggests there is an X-linked gene involved. If, for example, a male inherits an X chromosome with the gene for baldness, it will be expressed and he will eventually go bald. In contrast, if a woman inherits one X chromosome with the gene for baldness, her hair may only thin. Only if a woman inherits both X chromosomes containing the gene will she go bald (Figure 5-7).

Abnormal Numbers of Chromosomes

The intricate mechanisms of meiosis ensure that each gamete receives one homologue from each homologous pair of chromosomes. Occasionally this elaborate dance of the chromosomes misses a step, resulting in gametes with either too many chromosomes or too few chromosomes. Such errors, called **nondisjunction,** can affect both autosomal and sex chromosomes. Nondisjunctions of either chromosomes or chromatids usually occur during meiosis. For example, during sperm cell formation when the duplicated pairs of chromatids first line up along the cell's equator, a duplicated pair might pass into only one cell instead of equally dividing into both cells. If meiosis I is normal, nondisjunction can still occur during meiosis II if duplicated chromatids fail to divide equally into gametes. We will first examine some examples of nondisjunction of autosomal chromosomes then discuss nondisjunction of sex chromosomes.

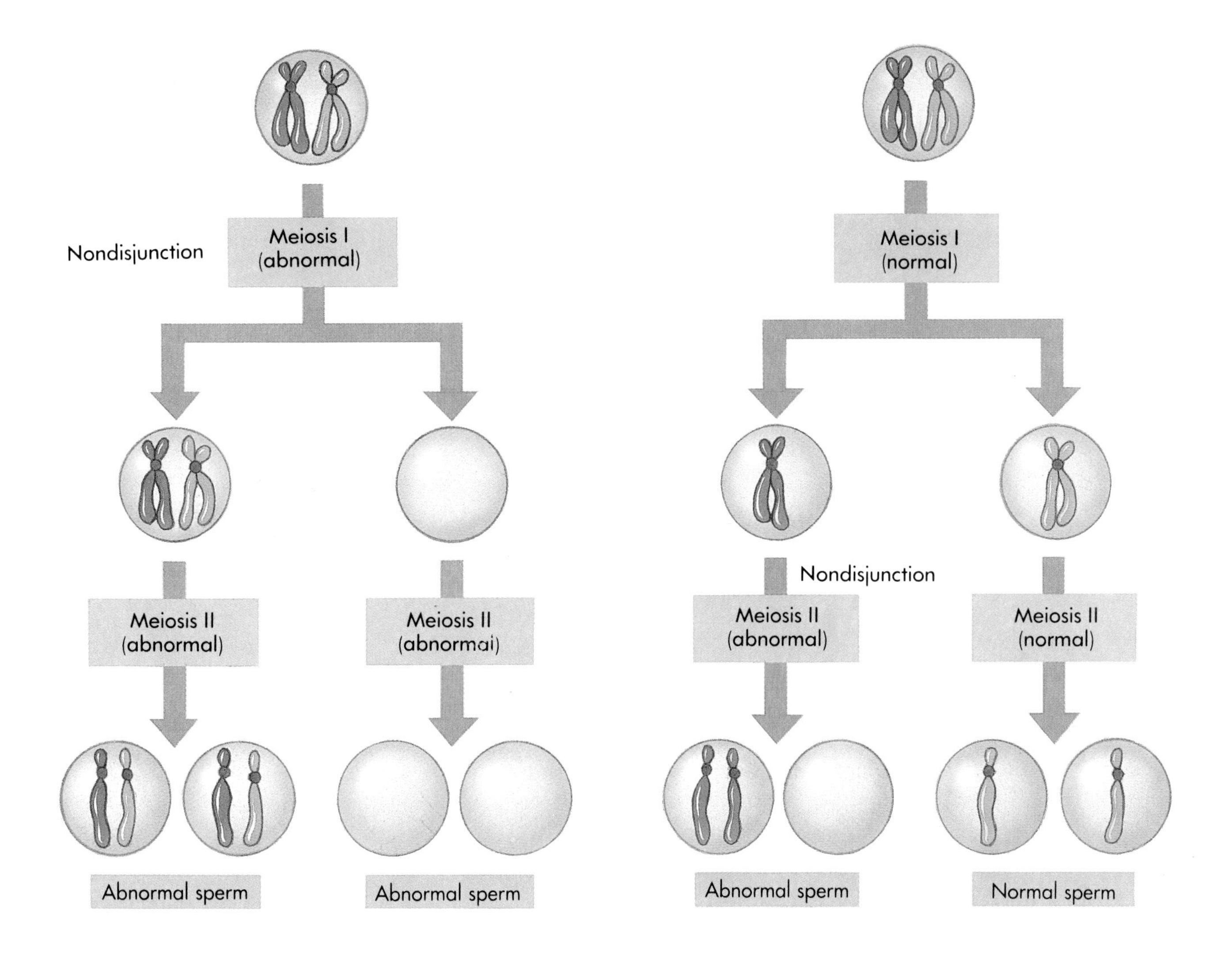

Abnormal Numbers of Autosomal Chromosomes

Recall from page 85 of this chapter that Down syndrome (trisomy 21) is caused by the nondisjunction of chromosome 21. When the gamete having two chromosomes 21 fuses with a gamete with one copy of the chromosome, the resulting zygote has three chromosomes of the same number (21), causing a complex of abnormal characteristics. Down syndrome children have several distinctive physical characteristics, including lack of muscle tone, a small mouth held partially open because it cannot accommodate the tongue, and distinctively shaped eyelids. Much more serious defects include low resistance to infectious diseases, heart malformations, and mental retardation.

Although we do not know the cause of nondisjunction of chromosome 21, we do know its incidence increases with the mother's age. In mothers 20 to 30 years old the incidence of Down syndrome is only about 1 in 1400 births. However, in mothers 30 to 35 years old the risk doubles, to 1 in 750 births. In mothers older than 45 the risk is as high as 1 in 16 births (Figure 5-8).

Nondisjunctions are believed to occur much more often in women than in men because of sex differences in gamete production. All the eggs a woman will ever produce have begun their development, to the point of prophase of meiosis I, by birth. The eggs are present before and during a woman's entire reproductive life span and are thus exposed to many potentially harmful external environmental factors. In males after reaching reproductive maturity, the development of sperm is initiated continuously in large numbers. With the large numbers of sperm produced and their short life span, sperm cells have a low probability of being affected by external environmental factors. A much greater chance therefore exists in women for nondisjunction to occur and to accumulate over time in the gametes than in men. This is probably why the mother's age is more critical than that of the father when a couple contemplates having children.

Figure 5-6 Hemophilia and British royalty

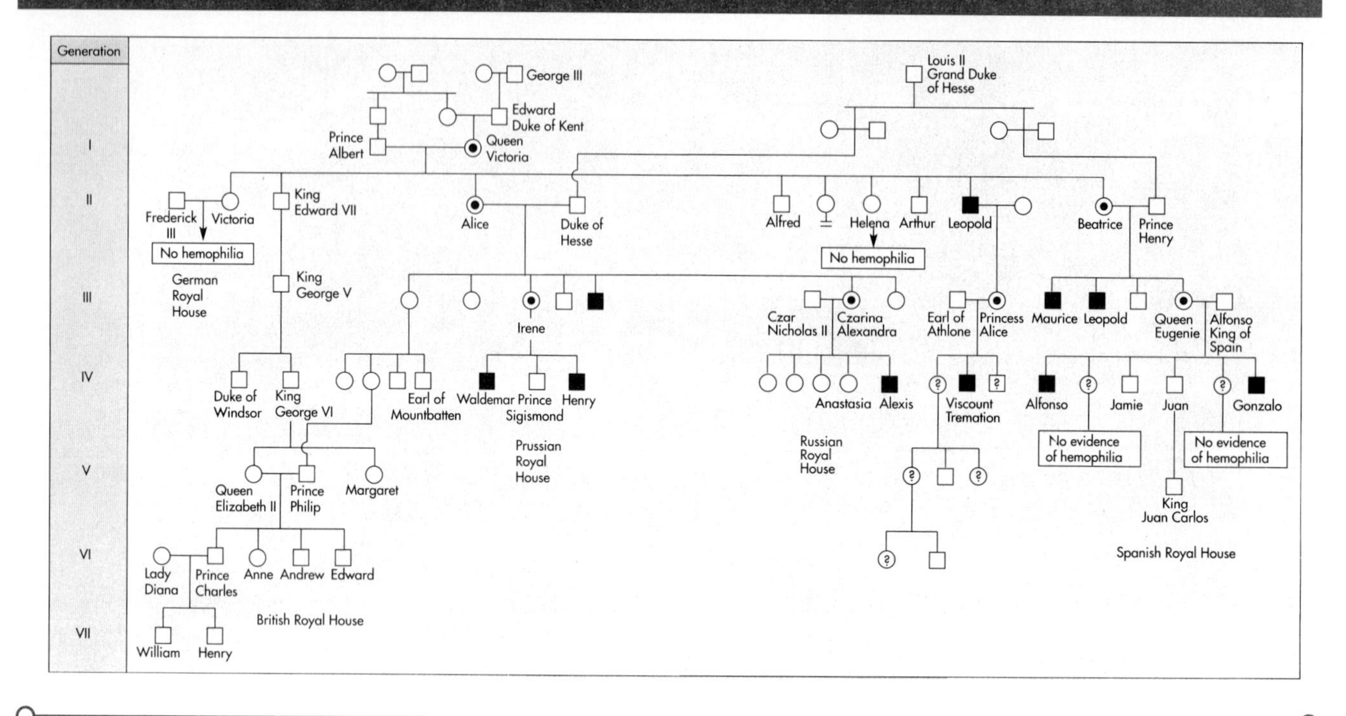

From Queen Victoria's daughter Alice, hemophilia was introduced into Russian and Austrian royal families. From her daughter Beatrice, it was introduced into the Spanish royal family. Victoria's son Leopold, himself a hemophiliac, also transmitted the disorder into a third line of descent.

Abnormal Numbers of Sex Chromosomes

Failure of the X chromosome to separate during meiosis leads to some XX gametes and others with no sex chromosome, designated O. If the XX gamete joins an X gamete to form an XXX zygote, the zygote develops into a phenotypically normal female. In contrast, if the XX gamete joins a Y gamete, the effects are more serious. The resulting XXY zygote develops into a sterile male, a condition called **Kleinfelter's syndrome.** This occurs in about 1 of every 500 live male births. Most Kleinfelter's syndrome males have longer than average limbs and underdeveloped sex organs and are sterile. About one half of these males also develop femalelike breasts.

The other gamete produced when the X chromosomes fail to separate lacks any X chromosome. If this O gamete fuses with a Y gamete, the resulting OY zygote fails to develop and is lost. Apparently, human beings cannot survive without any of the genes on the X chromosome. If, however, the O gamete fuses with an X to form an XO zygote, a condition called **Turner's syndrome,** the result is a sterile female who has immature sex organs. The physical growth of XO individuals is usually retarded, so they are short, and they often have a large fold of skin along the sides of the neck. Turner's syndrome occurs in about 1 of every 2000 live female births.

The Y chromosome may also fail to separate during meiosis. This creates a YY gamete and produces viable XYY zygotes, which develop into fertile males. Most of these males are more than 6 feet tall and on average reach sexual maturity earlier than other males. They are otherwise phenotypically normal. The frequency of XYY males is about 1 per 1000 live male births.

Variations on Mendelian Patterns of Inheritance

Recall from Chapter 4 that some of Mendel's pea plant experiments also consisted of dihybrid crosses in which two traits were followed simultaneously from one generation to the next. These dihybrid crosses confirmed the law of independent assortment. But through our understanding of genes and chromosomes, a problem becomes apparent. Humans possess perhaps 100,000 genes, which are contained on 23 pairs of homologous chromosomes. Thousands of genes must be located on each chromosome. You also know that during meiosis, chromosomes assort independently, so the genes on a

Figure 5-7 Male pattern baldness

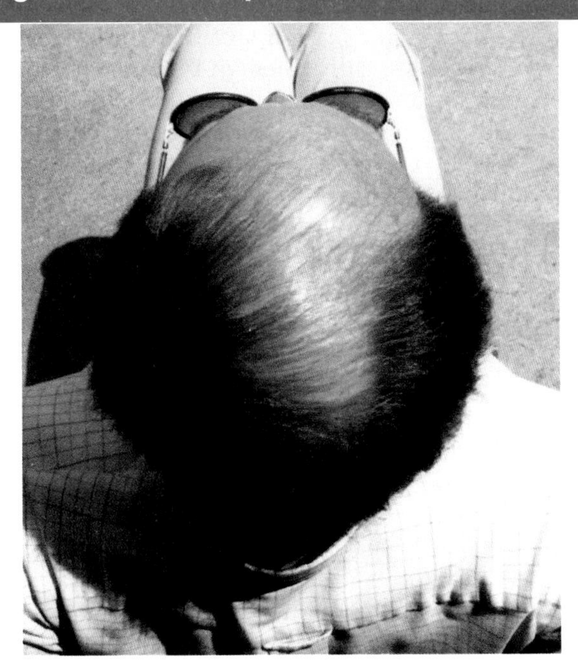

Because baldness is expressed differently in males and females, it is called a sex-influenced trait. If a male inherits an X chromosome with the gene for baldness, in the presence of the male sex hormone (testosterone), baldness will occur. For a woman to go bald, both X chromosomes must have the gene for baldness.

Figure 5-8 Incidence of Down syndrome and age of mother during pregnancy

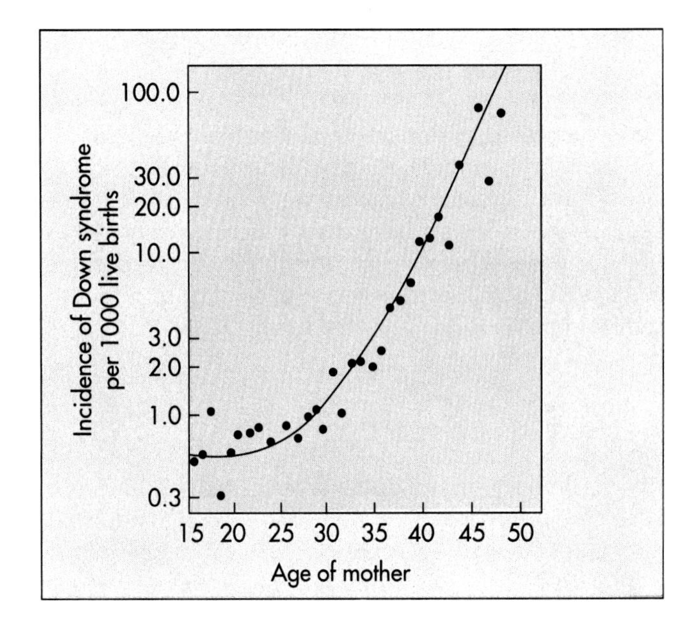

The incidence of Down syndrome babies increases as the age of the pregnant mother increases. The greatest increase occurs in pregnant mothers over the age of 45 years.

chromosome must also tend to assort together. Suppose, for example, you conducted a dihybrid cross and both of the traits were on the same chromosome. In the F_2 generation the 9:3:3:1 phenotypic ratio predicted by Mendel's law of independent assortment would not occur. Does this mean the law of independent assortment is invalid? No, but it must be modified slightly.

Gene Linkage

Linkage is a condition in which two or more genes are found on the same chromosome and tend to be inherited together. Because of linkage the law of independent assortment must be modified as follows:

> Genes located on *different chromosomes* assort independently during meiosis.

Notice the phrase "tend to" in this discussion of gene linkage and independent assortment. We say "tend to be inherited" because linked genes do not *always* stay together. During meiosis parts of homologous chromosomes are often exchanged between the chromosomes in a process called crossing-over. The farther apart the linked genes are, the more likely crossing-over will occur. Conversely, the closer together, the less likely a crossing-over event will separate them. The frequency, or how often linked genes travel together, is a function of how close together they are on the chromosome.

One example of gene linkage in humans is a group of genes that make up the **major histocompatibility complex,** or **MHC** (*histo* = tissue), which is responsible for encoding protein-recognition markers found on all cell membrane surfaces. These recognition proteins are markers of our biological individuality. They are the molecules our immune system learns to recognize as "self." Other biological tissues or cells with different cell-surface markers are "not self" and are attacked by the alert cells of our immune system. The MHC system is made up of four closely linked genes, designated as A, B, C, and D. Each gene has several alleles. The B gene, for example, has at least 20 different alleles. To date, no study has revealed that these linked genes have ever been separated because of crossing-over during meiosis.

Pleiotropy

A gene can also indirectly affect the phenotypic expression of many different traits in an organism. Genes that produce several often apparently unrelated effects are called **pleiotropic** (pli-ot'ro-pik) **genes.** Several of the genetic disorders described in this chapter exhibit phenotypic effects that appear to be unrelated to one another. For example, in cystic fibrosis a change in one gene produces many effects in the major organ systems of the body. The lungs, pancreas, and liver are all affected because of the unusually thick mucus produced by the faulty gene. PKU also produces many defects in affected individuals.

Polygenic Inheritance

Many of the genetic traits determined by a single gene are inherited in a relatively simple either/or pattern. You are either Rh^+ or Rh^-. You are blood type A, AB, B, or O. Either you have PKU or you do not. These traits of single-gene inheritance are referred to as discontinuous or discrete because the phenotypes do not overlap. Most human traits, however, are determined by the interaction of two or more genes. This is called **polygenic inheritance.** A characteristic determined by a number of genes shows **continuous variation,** meaning that small degrees of phenotypic variation occur over a more or less continuous range. For example, human height is a polygenic trait. Suppose you measured the height of all male students in your human biology class. You would get a range of heights from the shortest to the tallest male in the sample. As you included more individuals in your sample, say from the next classroom, you would eventually have individuals at every height within a certain range, say individuals 59, 60, 61 centimeters tall and so on. With enough individuals the data would graph as a smooth bell-shaped curve showing continuous variation for height. Most individuals would cluster around some particular height, whereas fewer individuals would be found toward both the taller and shorter end of the height range.

Cancer

Cancer is a complex group of diseases that can affect many different body cells and tissues. All cancers are characterized by the uncontrolled growth and division of cells. Cancer usually begins when an apparently normal cell begins to grow in an uncontrolled fashion. The result is a mass of cells called a **tumor.** Some cells may form small growths that reach a certain size then stop growing. These growths are called **benign tumors.** Generally, benign tumors are not a significant medical problem unless they put pressure on nerves or perhaps block blood vessels. Of much greater concern are **malignant tumors,** which continue growing and often spread to other parts of the body. If cells leave the mass and spread, forming new tumors at other sites, the spreading cells are called **metastases** (me-tas'-ta-sez).

Although improved medical care has reduced deaths from infectious disease and led to increased human longevity, the benefits have also helped make cancer a major cause of illness and death in the United States. Because the risk of many cancers is age related and more people are living longer, people are at greater risk of developing cancer. According to the American Cancer Society, 30% of all Americans now living will develop cancer. Each year about 500,000 individuals die of cancer, and over 1,000,000 new cases are diagnosed annually in the United States. Currently more than 10,000,000 individuals are being medically treated for cancer in the United States. In this section we limit our discussion to the genetic changes and mechanisms thought to cause cancer. Specific types of cancers, such as lung, breast, and skin cancer, and the external environmental factors known to initiate cellular changes causing them are described in later chapters devoted to that particular cell line and organ system.

To understand the causes of cancer, we must answer two questions. First, what changes occur in a cancerous cell that allow it to escape the normal controls of cell division and growth? Second, what agents or factors initiate these cellular changes?

Oncogenes

In 1981 biologists discovered genes in cancerous cells called **oncogenes,** genes that cause cancer. A potential oncogene may exist in all cells but causes cancer only when activated by some external event. There are two principal mechanisms by which oncogenes can cause cancer.

All cells have genes that stimulate cell division and growth. These genes are active during embryonic growth and development, but they are either turned off or transcribed more slowly as we mature. Biologists suspect some oncogenes may be growth genes that have mistakenly been either reactivated or turned on "full speed." Experiments have shown that chromosomes in some cancerous cells have been rearranged, so an embryonic growth gene is transferred to a part of a chromosome normally transcribed rapidly. The protein synthesized under the direction of the growth gene then stimulates cell division and growth. The daughter cells produced from mitosis also have identical copies of the rearranged chromosome, resulting in an ever-accelerating cancerous growth of this cell line. The other mechanism is that a harmless gene may mutate into an oncogene. For example, a gene normally directs the synthesis of a protein that governs cell division and growth at levels sufficient to replace cells lost through normal body activity. A mutation in this gene, referred to as a **proto-oncogene,** may alter the protein so it greatly increases the rate of cell division. Cancer would result from this increase in cells.

There are two stages in cancer development. The first stage, called **initiation,** occurs when one or more mutations occur in the cellular DNA. Perhaps 70% to 90% of these mutations are caused by an external chemical, physical, or biological substance. Cancer-causing mutations do not immediately produce cancer. Many tumors may begin 20 or more years after the actual mutation. This time lag is referred to as the **latent period,** following initiation, and is one important reason why it is sometimes difficult for medical researchers to determine the specific causes of cancer.

The second stage in cancer development is called **promotion,** which refers to any influence that causes a cancer cell to grow in an uncontrolled manner. Promoters are varied and abundant. For example, some are environmental factors, such as a diet high in cholesterol, that promote certain kinds of cancer. Other promoters involve genes directly. A promoter region, for example, somewhere in the chromosome that is necessary to turn on the oncogene may mutate and as a consequence be reactivated. This reactivation switches the oncogene "on," producing cancerous growth of the cell and its descendants.

In the past 50 years evidence has accumulated that many forms of cancer are caused by particular environmental agents called **carcinogens.** The common characteristic of all carcinogens is that they produce mutations. Some substances are harmless, but after entering the body they are converted to carcinogens, which can produce mutations in cellular DNA. For example, **nitrites,** chemical compounds created when red meat is grilled, are harmless, but once nitrites are ingested, liver enzymes convert them to carcinogenic nitrosamines, thought to be responsible for several forms of digestive-system cancers.

Recombining DNA in the Laboratory

Organisms within the same species have been recombining their DNA through sexual reproduction and crossing-over and producing new genotypes for millions of years. By the 1970s biologists were beginning to recombine DNA in the laboratory in an entirely new way. The term **recombinant DNA** refers to the artificial association of DNA fragments or genes, usually from two different organisms, that are not found together naturally. **Recombinant DNA technology** has developed techniques for cutting and splicing DNA so genes from one organism can be inserted into and used by another organism.

Recombinant DNA technology usually involves using microorganisms, such as bacteria, and inserting genes from other organisms into them to produce large quantities of a gene product. The results of this recombinant DNA technology are widely discussed in the media. Cures for genetic diseases, drought-resistant corn and wheat, bacteria designed to consume oil spills, and genetic monsters that threaten to destroy all life are a few of the more dramatic public images associated with this technology. Although fascinating, what DNA recombinant technology is and what it is capable of producing are not nearly that dramatic.

We can return to our earlier analogy of DNA as a cookbook. The cells from two different organisms, such as a human being and a bacterium, each contain different DNA cookbooks. Suppose there was a particular recipe in the human DNA cookbook you were interested in preparing in great quantities. Knowing that bacteria reproduce in far greater numbers and at a faster rate than do human beings, if you could cut the desired recipe out of the human DNA cookbook and insert it into the DNA cookbook of the bacteria, the gene and its product would also be produced in much greater quantities compared with the human rate. With these ideas in mind, we can discuss recombinant DNA technology by describing how genes can be clipped out of the DNA of one organism and spliced into the DNA of another organism. We can also describe how genes are transferred from one organism's DNA to another.

Restriction Enzymes

Beginning in the late 1960s molecular biologists discovered a type of enzyme with peculiar properties found in some kinds of bacteria. These enzymes, called **restriction endonucleases,** or

Biology in Action SCIENTISTS AND THE PUBLIC GOOD

The first recombinant DNA experiments were conducted in 1973. Most often these experiments involved modifying the genetic information in a bacterium, called *Escherichia coli,* that lives in the human intestinal system. However, scientists immediately recognized that these modified intestinal bacteria could be potentially dangerous to humans. For example, after insertion of foreign genes into the *Escherichia coli,* might these new instructions unknowingly turn this beneficial bacterium, which helps us to absorb several vitamins from our food, into a genetic "monster" producing substances lethal to humans?

A group of scientists called on the National Academy of Sciences requesting a panel of scientists be appointed to study the risks and possible control of recombinant DNA research and technology. Shortly after, a second group of scientists published a letter in the journals *Science* and *Nature* calling for a temporary halt to recombinant DNA experiments until the potential hazards could be assessed.

In 1975 a set of strict guidelines resulting from this conference was published by the U.S. government under the direction of the National Institutes of Health (NIH), an agency that funds much of the biomedical research in this country. Legislation was also proposed in the U.S. Congress and at the state levels of government to regulate recombinant DNA technology use.

By 1978 research had shown that the most common laboratory strain of *Escherichia coli,* called K12, was much safer for use in recombinant DNA research than originally thought. It was shown that K12 could not survive in the human intestinal system and was unable to survive outside of the specific conditions of the laboratory.

In 1982 the NIH issued a new set of guidelines that have eliminated most of the original constraints on recombinant DNA research. No experiments are currently prohibited, although notification must be made to the NIH if certain experiments are to be conducted.

The important issue, however, is that the scientists who developed the methods were the first to call attention to the possible dangers of recombinant DNA research. They did so only because of its potential for harm. Scientists voluntarily stopped some of their research until the situation could be properly and objectively assessed. Only when independent investigation demonstrated there was no danger did these experiments resume. Many scientists care deeply about the consequences of their discoveries and their potential impact, both positive and negative, on the human community.

restriction enzymes, were capable of breaking the chemical bonds of DNA by cutting them within the DNA strand. Bacteria use restriction enzymes to restrict the activities of invading viruses. Just as you occasionally come down with a viral infection, so can bacteria become infected with viruses. Viruses invading a bacterium do so by first attaching themselves to the membrane of the bacterium and then injecting their DNA into the cytoplasm of the bacterium. The bacterium then incorporates the viral DNA into its own genetic material. As the bacterium replicates itself it also makes copies of the virus. After the virus assembles inside the bacterium, it breaks out of the bacterium and the new copies of the virus are released into the environment to reinfect other bacterial cells. Bacteria use restriction enzymes as part of their defense against viruses, cutting up the invading viral DNA as it enters the bacterial cells and thus rendering it nonfunctional.

The names given restriction enzymes are based on the names of the types of bacteria that originally produced them. For example, the restriction enzyme called Eco RI comes from the bacterium *Escherichia coli.* The RI (Restriction I) means this was the first restriction enzyme isolated from this bacterium. The enzyme known as Hind III comes from the bacterium *Hemophilus influenzae* strain D (Restriction III). These and other similar enzymes are the basic tools of recombinant DNA technology.

Restriction enzymes, derived from bacteria, cut DNA by recognizing a sequence of four to six nucleotides that reads the same on both sides of the strands in different directions. Eco RI, for example, recognizes this nucleotide sequence in DNA:

↓

" G" A" A" T" T" C"

" C" T" T" A" A" G"

↑

The enzyme breaks the chain at the same point in the sequence on each side, as indicated by the arrows. The strands look like this:

" G"   " A" A" T" T" C"

" C" T" T" A" A"   " G"

Because the nucleotide sequence at the cutting site is complementary, the cut ends, or "tails," of the strands can reassociate with each other or with other DNA fragments cut by the same enzyme because they, too, have similarly cut ends. Recombinant DNA technology uses restriction enzymes to cut open DNA from two different organisms, for example, a human gene and a bacterial chromosome.

Plasmids

Unlike the double-stranded DNA common in multicellular organisms, some types of bacteria contain DNA organized in two slightly different ways. A large primary strand of DNA forms a circle, and much smaller pieces of circular DNA also exist, called **plasmids.** Although bacterial plasmids do not contain genes essential for their growth and development, they act as a reservoir of genetic variability for the bacteria. For example, plasmids may contain genes that allow the bacterium to metabolize many nutrients from its environment or give it resistance against environmental toxins it may encounter.

The circular plasmids are removed from the bacterium and opened up by a restriction enzyme (Figure 5-9). The human DNA is also cut up into smaller **restriction sequences** by the same restriction enzyme. Because both the plasmid and the human DNA were cut by the same restriction enzyme, they will also have the same sequence of nucleotides exposed. Consequently both of the opened ends of the plasmid and the human DNA fragment are complementary. The restriction sequence containing the particular gene of interest is then introduced to the opened plasmid. The ends line up, and hydrogen bonds form between the exposed nucleotides. In the presence of a DNA

Figure 5-9 Plasmids

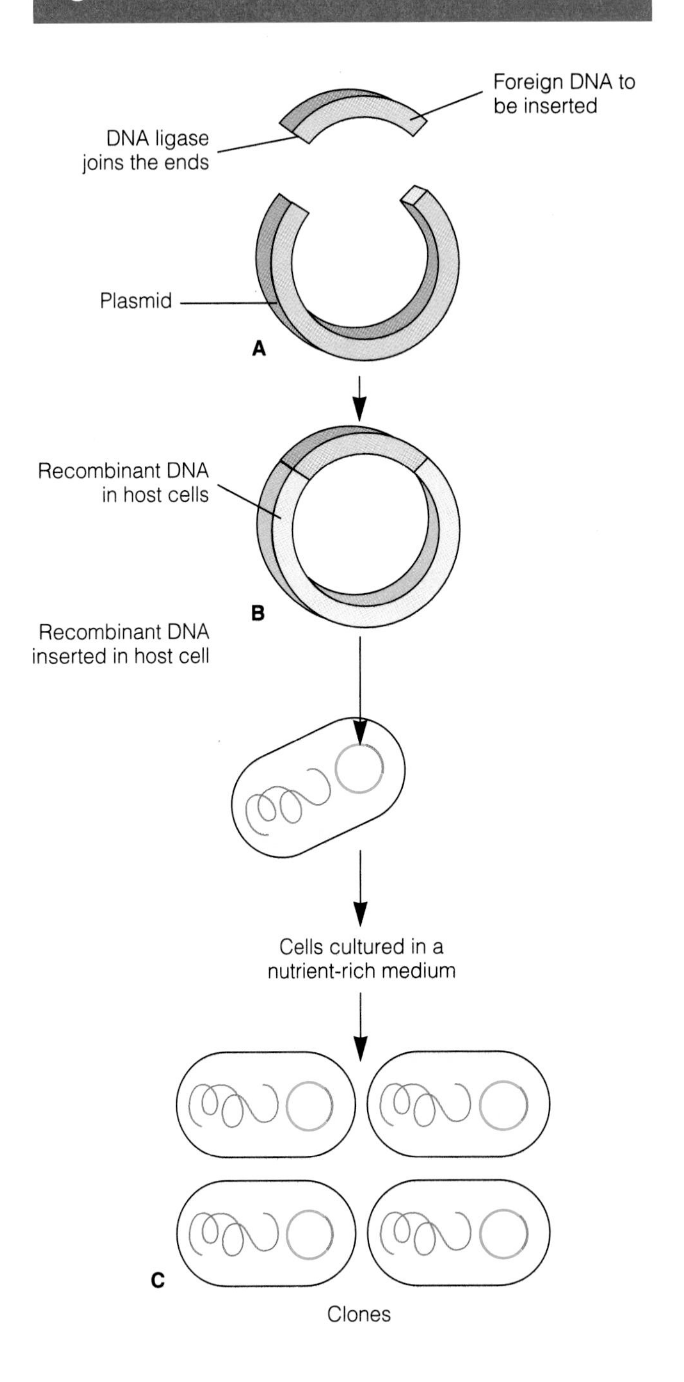

A bacterial plasmid is cut open by the same restriction enzyme used to cut out the foreign DNA from its original chromosome. The foreign DNA is then inserted into the bacterial plasmid (A). The recombinant DNA is inserted into a bacterial cell (B). The bacterium is grown in a nutrient-rich medium, asexually reproducing identical copies, or clones, of itself (C). The foreign gene is transcribed and translated along with the bacterial DNA and the product of the foreign gene is then removed from the medium.

repair enzyme called **DNA ligase,** covalent bonds form between the deoxyribose sugars and phosphate groups, so the two pieces are permanently joined and the plasmid again forms a circle.

The plasmid containing the human DNA fragment must be returned to the bacterium. The plasmids are incubated with specially treated bacteria, which will take up the plasmids. Once inside the bacterium the plasmid will replicate along with the primary DNA of the bacterium.

The bacterium containing the newly incorporated plasmids is allowed to reproduce asexually producing many genetically identical bacteria, known as **clones.** When the human gene is expressed, the bacteria become tiny factories to manufacture a product that may be extremely useful for medical and technological applications. We will focus on one of these medical applications here.

Recently we have seen an explosion of interest in applying recombinant DNA technology to practical human problems. Because they can be grown cheaply in bulk, bacteria incorporating human genes can produce the human proteins the genes specify in large amounts. This method has been used to produce human insulin, used by people with diabetes, and human growth hormone, used to treat some forms of human dwarfism. It has also produced several forms of human interferon, a protein used to fight the spread of cancer in the body.

In the next chapter we discuss the forces that change DNA and genes and ultimately account for the tremendous diversity of life on this planet, from plants to fleas to humans.

CHAPTER REVIEW

SUMMARY

1. Chromosomes are made of DNA, a double-stranded helical molecule composed of four different nucleotides. Each DNA strand is formed through the bonding of the sugar molecule of one nucleotide to the phosphate group of another nucleotide.
2. Humans possess 22 pairs of homologous chromosomes and one pair of nonhomologous chromosomes that for the purposes of analysis can be arranged according to chromosome size, centromere location, and staining pattern. A cell is stopped during mitosis, and the chromosomes are stained and then spread out and photographed. The chromosomes are then cut out and arranged; this process results in a karyotype of the cell.
3. The double strands are held together throughout their length by hydrogen bonds between nucleotide bases on opposite strands. The hydrogen-bonding nucleotides form complementary base pairs; adenine always bonds with thymine and cytosine with guanine.
4. The genetic code of the DNA consists of sequences of three nucleotide bases called triplets. Each triplet codes for a specific amino acid. However, because several different triplets can code for a single amino acid, the genetic code is called redundant.
5. Genes are sequences of triplets encoding for a sequence of amino acids. Each gene along the length of a chromosome has a promoter site that signals the RNA polymerase where to begin transcription.
6. Proteins are constructed from the genetic code of the DNA by transcription of the gene; the sequence of triplets is first copied by a messenger. The molecules transcribing DNA are mRNA. The specified sequence of amino acids is then assembled in the cell's cytoplasm through a process called translation. The molecules bringing the amino acids to the mRNA-ribosome complex are tRNA.
7. Mistakes occasionally occur during DNA replication, producing mutations. Mutations can either occur spontaneously or be induced by external factors or agents capable of damaging the DNA, called mutagens. The sequence of genes on a chromosome can also be rearranged by accident, called chromosomal rearrangements.
8. Determination of patterns of human heredity rely on pedigree analysis, records of family histories for several generations. From these family histories it is possible to determine the pattern of a particular trait and infer who is heterozygous for the trait.
9. In humans most genetically determined defects, called genetic disorders, display a recessive pattern of inheritance. Dominant genetic disorders can survive in humans if the disorders are expressed late in the reproductive life of the individual.
10. Human cancers are primarily the result of mutations caused by environmental carcinogens. Many human cancers occur in older individuals, and since we are living longer, cancer has become a major cause of death in the United States.

11. If accidentally reactivated or mutated, genes that normally control cell division and growth become oncogenes, genes that cause cancer. Every cell in your body, then, has a potential cancer gene that is subject to change under certain conditions.
12. DNA from different organisms can be combined in the laboratory using methods developed by recombinant DNA technology. Restriction enzymes are used to clip segments of DNA from one organism and insert them into another organism. Recombinant technology is used primarily to produce large quantities of a particular gene product by using rapidly growing bacteria-plasmid hybrids.

FILL-IN-THE-BLANK QUESTIONS

1. DNA is a ______ stranded ______ structure, and RNA is ______ stranded.
2. Bacteria use ______ ______ to protect themselves against invading viruses.
3. The nucleotide called ______ is found only in RNA and not in DNA.
4. The ribosome contains ______ bonding sites called the ______ site and the ______ site.
5. A change in the genetic message of a cell is referred to as a ______.
6. By making a ______, human chromosomes can be examined and analyzed.
7. The incidence of ______ ______ increases with mother's age.
8. Most genetic disorders have a ______ pattern of inheritance.
9. Some forms of hemophilia show a pattern of ______ ______ inheritance.
10. Genes that produce several often apparently unrelated effects are called ______ genes.
11. Traits of polygenic inheritance display ______ variation in their phenotype.
12. It appears that oncogenes were once genes that normally controlled cell ______ and ______.

SHORT-ANSWER QUESTIONS

1. Explain semiconservative replication.
2. What are the possible consequences of a point mutation occurring in a gene?
3. Describe the main features of translation.
4. Describe sex-linked inheritance.
5. Describe the two steps by which oncogenes can cause cancer.
6. Explain how incomplete dominance agrees with Mendel's principle of independent assortment.
7. Explain how, if a female who is Rh^- becomes pregnant by an Rh^+ male, there is a 50% chance that the fetus will also be Rh^+.

VOCABULARY REVIEW

anticodon—*p. 82*
carcinogen—*p. 97*
codon—*p. 80*
mutation—*p. 83*
oncogene—*p. 96*
plasmid—*p. 99*
recombinant DNA—*p. 97*
replication—*p. 79*
sex chromosome—*p. 86*
transcription—*p. 81*
translation—*p. 82*

Chapter Six

THE PROCESS OF EVOLUTION

For Review

Here are some important terms and concepts that you will encounter in this chapter. If you are not familiar with them, you should review them before proceeding.

Alleles
(page 73)

Mendelian principles of inheritance
(page 67)

OBJECTIVES

After reading this chapter you should be able to:

1. *Differentiate between artificial and natural selection.*
2. *Describe the two implications of natural selection proposed by Darwin.*
3. *Define breeding population and discuss its importance in Darwin's ideas of evolution.*
4. *List the three categories of evidence for evolution.*
5. *Differentiate between homologous and analogous structures.*
6. *Define and give examples of a biological species.*
7. *Define adaptive radiation and discuss when it occurs.*
8. *Describe the two related causes of species extinction.*
9. *Define taxonomy.*
10. *List the characteristics that define human beings.*

The theory of biological evolution seeks to explain both the unity and the diversity of life on Earth. It provides a framework for understanding the structure and function of all organisms, including humans. Through the mechanism of natural selection, the differential reproduction of alternative varieties of biological characteristics over many generations often results in descendants that differ greatly from their ancestors.

Biological Evolution

All life, including humans, shares a set of underlying properties. For example, from Chapter 2 you recall that one common property of life is cellular organization. How can each organism, from bacteria and plants to elephants and humans, be constructed from similar building blocks and yet each be so dramatically different? From Chapter 5 you learned that the hereditary material, called DNA, encodes instructions for constructing sequences of amino acids. Ultimately the results of the complex interaction of these different proteins and polypeptides determines whether a particular organism is a frog or a human. But we still need to ask what are the forces that have come to modify the instructions encoded by DNA into such dramatically different life-forms?

The key to understanding both life's diversity and its underlying unity resides in understanding another property shared by all life-forms, the ability to adapt to and change with their environments. The changes that groups of organisms undergo through time, called **biological evolution,** are the foundation on which biology or any life science is built. Biological evolution is the framework within which any living entity, including ourselves, is understood.

We begin by describing the mechanism that allows plants and animals to adapt to their environments. First we will reconstruct Charles Darwin's argument and his evidence for proposing natural selection as the primary force of change in the adaptation of organisms to their environments.

Charles Darwin and the Voyage of the *Beagle*

That Charles Darwin was aboard a sailing ship destined to circumnavigate the globe was the result as much of luck as of his own efforts (Figure 6-1). At age 16 Darwin entered the university at Edinburgh,

Figure 6-1 The route of the H.M.S. *Beagle*

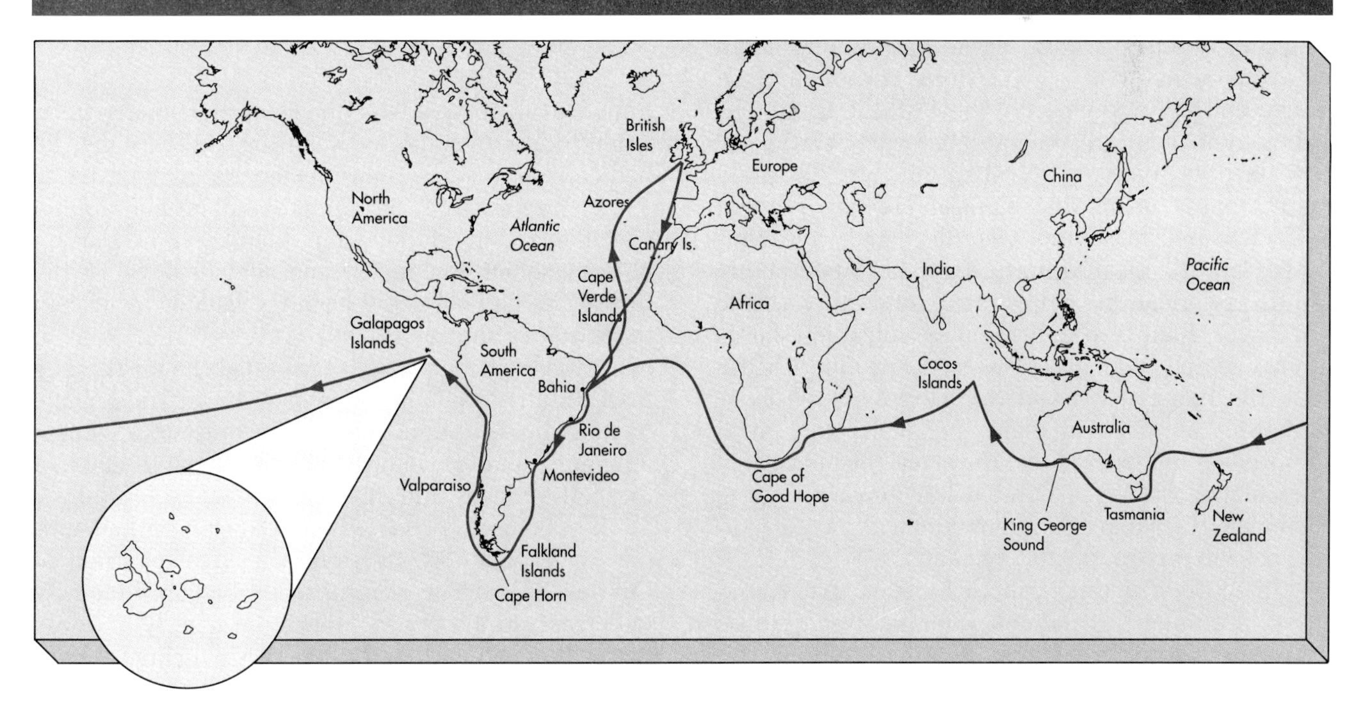

The purpose of the voyage of the *Beagle* was to chart both the east and west coastlines of South America. As was typical of nineteenth-century British ships of exploration, a naturalist was aboard to collect animal and plant specimens and make other recordings of interest. Charles Darwin was the naturalist on board the *Beagle.*

Scotland, to learn medicine, as had his father and grandfather. However, after he witnessed his first surgery, without benefit of anaesthesia for the patient, he quickly persuaded his father that he had no talent or stomach for a medical career. He later enrolled at Cambridge University in England and earned a theology degree. He also pursued his passion for collecting plants and animals. While at Cambridge he developed friendships with two important scientists of the day, John Henslow, a noted botanist, and Adam Sedgwick, a professor of geology. After completing his studies at Cambridge, Darwin went home and reluctantly began sending out inquiries for a position as a parish minister.

During the summer John Henslow wrote to Darwin informing him of a survey ship called the *Beagle* that needed a naturalist. Darwin's father agreed to let his 22-year-old son take the position as naturalist on board. On December 27, 1831, the *Beagle* left Plymouth Harbour and headed southwest, toward the Atlantic coast of South America. The *Beagle* would not return to England until October 2, 1836, nearly five years later. During the voyage Darwin developed the foundations of a theory destined to shake the foundations of European notions about life on Earth and the role of humans in its ecosystem.

The *Beagle* was to chart the lands and coastline of South America, allowing Darwin to spend a great deal of time on land. In 1832 he discovered bones of extinct fossil animals in a cliff face on the coast just south of Buenos Aires, Argentina. Among them were the remains of a giant sloth. Although it no longer existed in any part of the world, Darwin was struck by its similarity to the smaller tree sloths living further inland. This situation was repeated many times during the voyage. He encountered extinct animal forms that were in many ways similar but not identical to living animals. After leaving South America the ship's next landfall was the Galápagos Islands, a collection of volcanic islands approximately 600 miles west of Ecuador (Figure 6-2). By then Darwin had collected and preserved numerous animal and plant specimens and taken hundreds of pages of notes. However, the animals of the Galápagos Islands pointed him toward finding the mechanism explaining how life could change in response to environmental demands.

After Darwin went ashore he soon discovered unusual animal populations scattered among these otherwise desolate islands. Reptiles were everywhere, including several forms of iguana that browsed on seaweed along the shoreline, and an inland form that ate low bushes and cacti. Huge tortoises also lived on each of the islands he visited. Birds, many of which Darwin thought were unnamed, also lived on the islands. Among them were finches in some ways similar to the finches he had collected on the South American mainland.

Figure 6-2 The Galápagos Islands

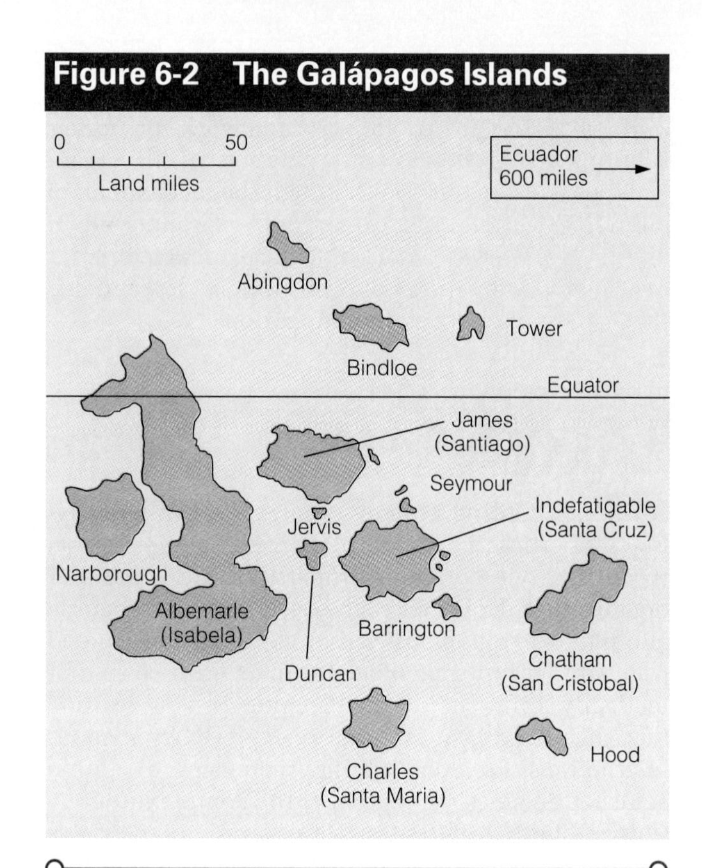

The animals of the Galápagos Islands prompted Darwin to propose the mechanism to explain how organisms adapt to their environment.

As he prepared the numerous plant and animal specimens for storage on board ship, Darwin noticed something interesting about several of the finches he had collected from the islands. Although the bird's plumage is dull, varying from gray to black to green, Darwin observed that the birds taken from different islands differed in the size and shape of their beaks. Some were short and stout, others long and slender. He did not know what to make of these differences (Figure 6-3).

Just before the *Beagle* set sail, Darwin had dinner with Nicholas Lawson, the vice-governor of Charles Island. The governor said he could identify which island a tortoise came from simply by looking at the size and shape of the shell (Figure 6-4). Darwin wondered if each island had somehow modified its own forms of tortoise, finch, or iguana. Because of that conversation Darwin carefully sorted his animal and plant specimens according to their islands of origin.

Figure 6-3 Differences among animals in the Galápagos Islands

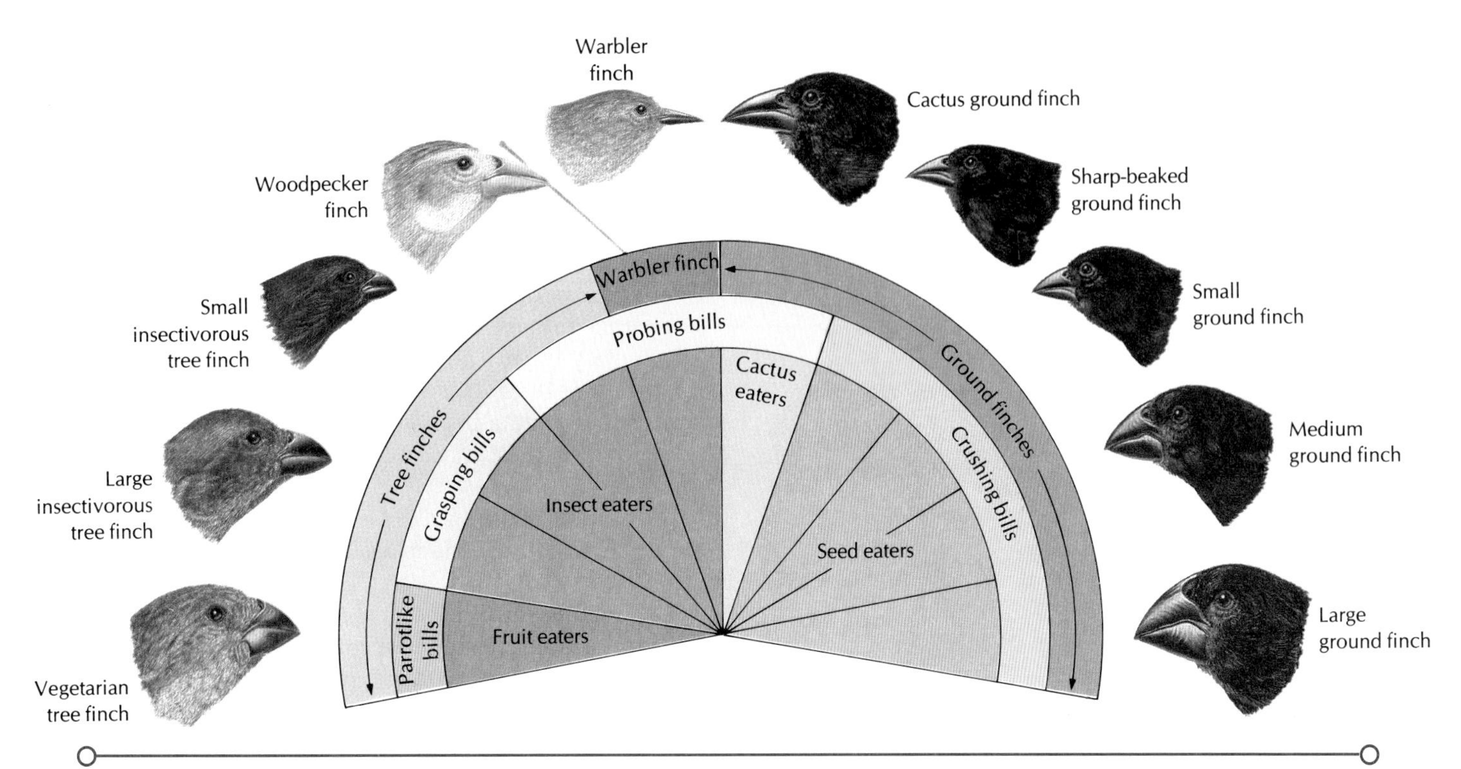

A different species of finch inhabits each island of the Galápagos. Although the birds are similar in appearance, their beaks differ according to the particular environment they exploit on the islands. Darwin later discovered that the different beaks of the finches were specialized characteristics used to exploit different environments.

Darwin speculated that the finches must have come originally from the South American mainland because the volcanic islands of the Galápagos must have formed later than the continent. Darwin asked, "Why were these birds so different from those on the mainland and why did the assortment of animals on each island differ so much from one to the next?" After making stops in Tahiti, Australia, and South Africa, the *Beagle* finally sailed into the English port of Falmouth Harbour on October 2, 1836, where Darwin left the ship, never to leave England again. However, during his five-year voyage Darwin had witnessed diverse fossils, animals, plants, and geography, which convinced him that organisms were able to change in response to environmental demands.

Natural Selection

Convinced that the environment shaped the physical features of animals, Darwin set about to discover the mechanism by which this occurred. He began by examining domesticated animals from pigeons to dairy cattle and noting the kinds of **artificial selection** used by farmers to enhance or diminish a particular trait (Figure 6-5).

Darwin observed that individuals in every animal population vary. The differences may be slight, but they exist. From both his knowledge of animal breeding and his observations of nature, Darwin wondered whether individual variability might be somehow connected to the ways in which animals adapt to their environment. The answer to this question came from an unexpected source.

In 1838 about 15 months after he had returned, Darwin came across an essay entitled "Population," written by the Reverend Thomas Malthus (1766–1834) in 1798. The essay was about human overpopulation. Malthus began by describing the reproductive rates of animal populations. He observed that many more individuals are born into animal populations than ever survive to maturity. Although reproductive potential is high, the number of adult animals remains constant or grows relatively slowly through time because the essential resources of animal populations are finite. Only a limited amount of food and shelter is found in any environment.

Could an organism's survival be related to particular physical variations? Darwin reasoned that

Figure 6-4 Variations in tortoise populations inhabiting the Galápagos Islands

Although not divided into separate species, different breeding populations of tortoise inhabit each of the Galápagos Islands. The populations show slight but noticeable differences with respect to shell shape and limb length.

if all animals in a population varied in physical features, some variations might permit better or more efficient access to essential environmental resources. The environment then might favor those individuals whose variations made them better able to survive in that environment. These animals stand a better chance of surviving to maturity and reproducing. The variations that permitted these animals to survive would be passed on to the next generation. In contrast, variations that were not as adaptive would be eliminated by the environment. Darwin called this process natural selection. **Natural selection** means that individuals with favorable varieties of physical characteristics are more likely to survive and reproduce and pass those varieties on to the next generation. In contrast, those individuals with unfavorable varieties of physical characteristics were less likely to survive and reproduce.

Charles Darwin and Alfred Russel Wallace

Early in 1858 an unexpected event occurred, prompting Darwin to act on his hypothesis. Darwin received the first draft of a paper from a young English naturalist named Alfred Russel Wallace (1823–1913), working in Indonesia. Like Darwin, Wallace had traveled widely, studying animal and plant life and collecting specimens for the zoos and museums of Europe. The paper was an outline of the principles of natural selection, and Wallace wanted Darwin's opinion.

Darwin had originally planned to present his hypothesis of natural selection to the Linnaean Society of London in late 1858. After he received Wallace's paper, however, he and Wallace jointly announced their findings in a paper presented to the society in July and published in August 1858. In November 1859 Darwin published much of his accumulated data in his book *The Origin of Species.* The first edition sold out by late afternoon of the same day.

To briefly summarize Darwin and Wallace's evidence for natural selection, consider the following points:

- **Organisms reproduce**
- **Populations are stable or grow slowly**
- **Individuals vary within a population**
- **Resources on which organisms rely are limited**

Figure 6-5 Artificial selection

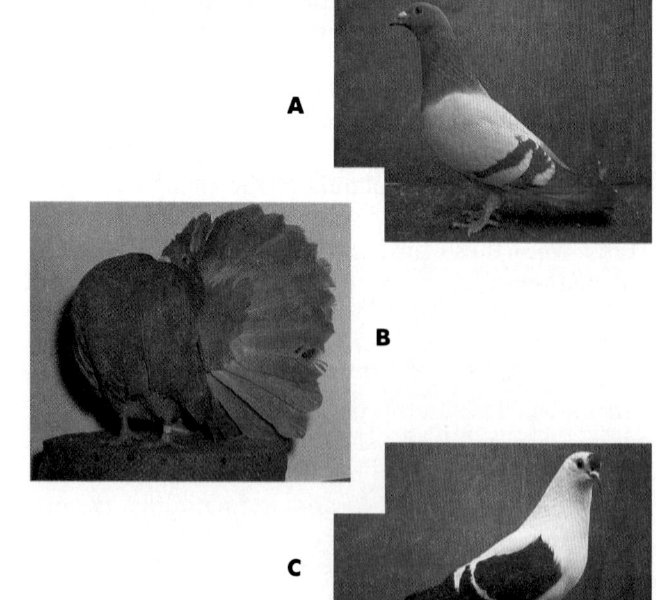

These are some of the differences that have been obtained by artificial selection of the wild European rock pigeon (A). Compared to the original group, the differences seen in domestic groups, such as the red fantail (B) and the fairy swallow (C), are very great.

Biology in Action DARWIN AND THE PROBLEM OF INHERITANCE

Charles Darwin's theory of evolution rested on the selection in populations of small variations that he showed to be numerous and believed to be heritable. However, in the preface to *The Descent of Man,* published in 1871, he stated, "I now admit . . . that in earlier editions of my *Origin of Species* I perhaps attributed too much to the action of natural selection." What had occurred since 1859 to cause Darwin to seriously question a lifetime of work? It was the problem that had plagued him privately while writing *The Origin of Species:* the origin of variability or, more precisely, the problem of inheritance.

Although biologists of the mid–nineteenth century believed inheritance to be a blending of parental characteristics, Darwin thought inheritance to be hard in the sense that characteristics were passed intact from parent to offspring. To support his argument Darwin relied upon his keen observations of nature and the practical knowledge of artificial selection used by plant and animal breeders. The evidence initially supported Darwin's argument, but within eight years of the publication of *The Origin of Species,* the same evidence was used to argue against natural selection. Darwin's confidence in his theory began to erode.

In 1867 Fleeming Jenkin, an engineer and mathematician, wrote an article challenging Darwin's theory of evolution. In the article Jenkin stated that a "newly emergent character" possessed by one or a few individuals would rapidly disappear or be swamped by mating with the majority of individuals in the population normal for the character. He reasoned that the only way a population could significantly change through time was if large numbers of individuals simultaneously possessed the new character. But didn't Darwin's theory rest on individual variation of biological characteristics already present in populations and not upon "newly emergent characters"?

Jenkin used Darwin's evidence of artificial selection to support the necessity of new variability, not present variability. He argued that it is not possible to change one type of animal into another by either artificial or natural selection because animals cannot change beyond a certain point. For example, an animal breeder cannot change a pig into a cow on the basis of the range of variability in pigs. That is why, Jenkin argued, the same newly emergent character in a number of individuals at the same time would be necessary for one kind of animal to evolve into another. Jenkin believed his argument put nature back into the hands of a supernatural being and also implied a direction to the evolution of life.

Darwin's response was the theory of **pangenesis.** According to Darwin, every organ of the body produced minute hereditary particles called gemmules. These particles collected in the sperm and egg of each individual. Through their respective gametes each parent contributed the particles used to create an offspring. Darwin could now account for environmental influences such as the use and disuse of organs by an organism. Enlarged or diminished organs produced either more or fewer gemmules that would create the appropriately sized tissue in the offspring. He could also account for the noninheritance of mutilations, such as in polled cattle whose horns had been removed, by saying that the horns had already produced their horn gemmules before their removal.

With the rediscovery of Mendel's work in 1901, the inheritance of characteristics was proven to be particulate. With the discovery of mutations and the potential variability they produce, Darwin's original theory of evolution based on the constant selection of small variations was proven to be valid.

Because resources are limited and organisms produce more offspring than needed to replenish the population, there must be, as Malthus stated in his essay, **a struggle for survival** in which some individuals are eliminated. Therefore the population remains stable or grows slowly. This process of elimination is selective because organisms vary in their suitability for particular environments. Those individuals best suited for the environment will survive and produce offspring similar to themselves. Over many generations this differential reproduction among individuals with different varieties of physical characteristics changes the population's composition.

Determining When Natural Selection Occurs

Darwin was able to test natural selection by determining the **relative fitness** of an individual, or its *relative* capacity for leaving offspring. For example, imagine there are 100 individuals in a population and one half of the population possesses one variety of a characteristic, called A, and the other half of the population possesses variety B. We can determine if natural selection is occurring at this trait by calculating the relative fitness of each variety. First we count the total number of offspring produced by individuals with each variety. We then determine the average number of offspring for each variety. For example, suppose we calculate that the average number of offspring for individuals with variety A is 2.0 and the average number of offspring for individuals with variety B is 1.5. Variety A is increasing in frequency, while variety B is decreasing. Thus natural selection at this trait is occurring for variety A and against variety B. With these facts in mind, Charles Darwin defined biological evolution as descent with modification by means of natural selection.

The Implications of Natural Selection

Darwin stressed that the concepts of "favorable" and "unfavorable" varieties of traits, as used in the definition of natural selection, are relative terms. First whether a particular variety is favorable depends on the environment in which it occurs. Favorable variations of a characteristic help an individual survive in a given environment, and the most fit individuals are those producing the most offspring. For example, if a particular mammal lives in a desert, any variety of a characteristic that helps the animal gain or retain water in this dry environment will be considered favorable because individuals with these varieties are more likely to survive and reproduce relative to animals with other varieties of the same characteristics. On the other hand, if the same animal lives in a tropical rain forest where water is abundant, any variety of a characteristic that assists in gaining or retaining water would not confer a greater chance of survival because water is plentiful and this variety would not be favorable compared with other varieties of the characteristic.

Another implication of natural selection concerns the range of variability for biological characteristics. Natural selection can select only from the range of variability in a population. For example, if the response of a population to predation is to run away from the predator, the fastest individuals in that population are probably going to escape predation, and relatively slower individuals are probably going to be the predator's next meal. Within the range of variation in the population for speed, the fastest will survive. This does not mean, however, that this is the fastest speed that individuals in this population can obtain. In a future generation there may be individuals even faster than their ancestors.

A Question Darwin Could Not Answer

Although Darwin described the role and importance of variability in populations, he was nonetheless laboring under a great burden. He had no heredity theory to support his evolutionary theory. In *The Origins of Species* he cited numerous cases of artificial selection, used by farmers to breed better milking cows, better egg-laying hens, and higher-yielding wheat and corn. This suggested to Darwin that variation had to be heritable. In the same way nature also selected for or against certain varieties of a characteristic. However, if heritable variation was constant from one generation to the next, how did variation arise at all? Darwin did not know the source of variation, but from his close study of nature he was sure of two things: (1) variation was abundantly available, and (2) it was heritable. Answers to questions about inheritance and the origin of variability had to await the rediscovery of Gregor Mendel's breeding experiments and other research that followed during the first decade of the twentieth century.

Biological Evolution and Population Genetics

When the word *evolution* is mentioned, many people think of dramatic alterations and large-scale changes that have occurred in the history of life on Earth. Such changes would be from water to land animals, from reptiles to mammals, or from apelike creatures to humans. Decades after Darwin presented his persuasive argument for natural selection, many scientists agreed with this dramatic view of evolution. Yet these impressive reorganizations in physical characteristics are only the visible evidence of small alterations accumulated over millions of years. However, Darwin stated that it is the accumulation through time of favorable varieties of a trait or traits "however slight" that eventually produces dramatic differences in descendant populations compared with their remote ancestors. But what exactly is the nature and extent of these changes occurring in a population through time? Based on Darwin's original theory of evolution, Mendelian principles of inheritance, and biochemical genetics, we are able to answer these important questions. We can begin by first defining the basic unit of study used in population genetics, the breeding population.

The Breeding Population

A **breeding population** is defined as a group of potentially interbreeding individuals inhabiting a defined area. For example, the tortoises on each of the Galápagos Islands can each be defined as a breeding population. For many different species of organisms breeding populations are relatively easy to distinguish. With humans, because of their great mobility and cultural biases, it is often difficult to determine the boundaries of a particular breeding population. Two general ways—geographical and social—are used to divide the human species into breeding populations.

Geography often determines the likelihood of matings. People living in Africa are not likely to encounter and mate with people from Alaska. However, our social and cultural behavior often complicates attempts to define human breeding populations on the basis of physical proximity alone. For example, in Franklin County, Pennsylvania, are approximately 350 people belonging to a religious group known as Dunkers. They are descendants of a group that originally emigrated from Germany in the early 1700s. Because of their religious beliefs they have maintained a strict isolation from the surrounding people. Since 1850 most of their marriages have been with other members of this group. Maintaining a cultural practice thus also qualifies the Dunkers as a biological breeding population. They do not intermarry with people not belonging to their religion, even though they may be neighbors. However, religious beliefs alone do not necessarily define breeding populations. American Catholics are not a breeding population to the extent that the Dunkers are because interreligious marriages involving Catholics are not rare.

In general the solution to such complex problems of defining human breeding populations depends largely on the research project and the questions being asked.

Using our understanding of breeding populations, it is possible to define more precisely Darwin's original thoughts on evolution and natural selection. Biological evolution is now understood as changes in allele frequencies in breeding populations through time. Modern biology also defines the process of natural selection as the differential reproduction of alternative genotypes.

Evidence for Evolution

What is the evidence that evolution has occurred in the past, and is it still occurring today, and how do new kinds of organisms come about? In other words, how do different types of organisms evolve from the same ancestral population? The evidence supporting our understanding of evolution falls into three general categories. Fossils, data from several comparative sciences, and direct observational studies are the three categories of evidence for evolution.

Fossils

The most direct evidence for large-scale evolutionary changes is the **fossil** (fos'el, L. *fossilis* = to dig up). **Fossils** are the mineralized parts of an organism's skeleton or impressions of its anatomy left in sedimentary rock. Some fossils are created when organisms become buried in sediment and the bone

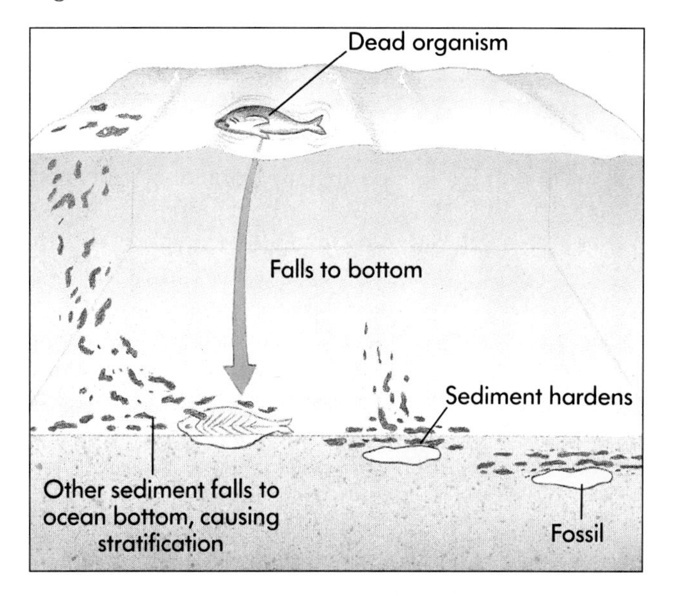

and other hard tissue is mineralized. In other instances the organism completely breaks down, leaving only an impression or outline in the sedimentary material. In both situations, however, the sediment eventually converts to rock. The sedimentary layers of rock reveal a history of life on Earth in the fossils they contain.

By dating the rock strata in which fossils occur, we can make an accurate estimate of the fossils' age. For example, by using the potassium-argon radiometric technique (see Biology in Action in Chapter 2) on a layer of volcanic dust and debris found in association with a fossil, the oldest known human fossil remains have been dated to 3.4 million years ago.

Fossils also document the large-scale changes in the evolution of life. They provide us with a history of life on Earth. Biologists know, for example, that all land-dwelling organisms evolved from a particular group of primitive fish. These early terrestrial forms in turn gave rise to reptiles. Birds and mammals evolved from two early groups of reptiles.

Comparative Evidence

Comparative evidence from such disciplines as comparative anatomy and chemistry is important in helping to determine the evolutionary relationships between different types of organisms. The more similar the characteristics of two groups are, the more recently they shared a common ancestor. As an example of comparative evidence, used with fossils as evidence of evolution, consider primates, especially the great apes, which are similar to humans both in anatomy and in behavior. We can also recognize that other mammals, such as dogs or horses, share fewer similarities with us, whereas reptiles are quite different from us.

Although contemporary organisms are adapted to a variety of environments and interact with them in different ways, many underlying anatomical structures are similar. Detailed comparison of body structure is called **comparative anatomy.** For example, the forelimbs of birds are used primarily for flying, whereas the forelimbs of mammals may be used for flying, swimming, running, or grasping objects. Despite this enormous diversity of function, the skeletal anatomy of all bird and mammal forelimbs is similar (Figure 6-6). Characteristics with a similar internal structure are called **homologous structures.** Such similarity is expected if bird and mammal forelimbs evolved from a common reptilian ancestor. Sometimes, because of similar environmental demands, organisms evolve analogous structures. **Analogous structures** are superficially similar structures that evolve in dissimilar and distantly related organisms. For example, the wing of a bird and that of a fly serve the same function, flight. Although superficially similar, the evolution of these two structures started from two different points. Since bird and fly wings have different underlying structures, they are only distantly related (Figure 6-7).

Observational Studies

The third category of evidence for evolution is based on observational studies of the environment. Many deal with insects and microorganisms because they have short life spans relative to us, the observers, and therefore are able to change rapidly.

One of the better-known examples of the direct observation of evolution in action is the common peppered moth found in Great Britain. During the mid–nineteenth century, a speckled light-gray form was common in the forest and the dark-gray form was rare. Between 1848 and the turn of the century, the dark form increased in frequency because of a change in natural selection.

Peppered moths are active at night and rest during the day on tree trunks. Before the Industrial Revolution in Great Britain, most tree trunks were covered with light-gray lichens. Light-gray speckled moths resting on lichen-covered trees blended in with the background and were not apparent to their natural predators, birds. Soot and other pollutants from surrounding factories began killing the lichen and covering the tree bark. The light-gray speckled moths were now easily spotted by the birds and

Figure 6-6 Homologous structures

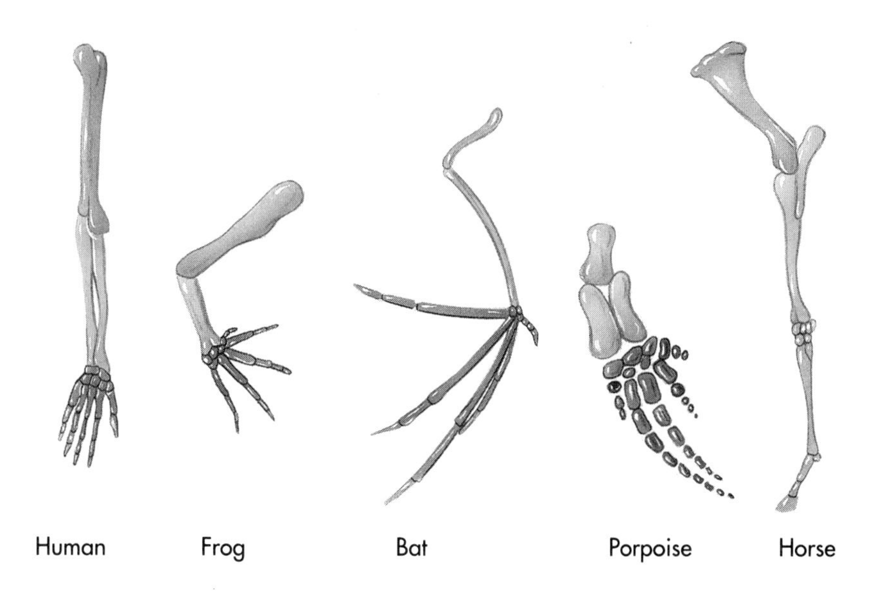

The bones in the forelimbs of amphibians, reptiles, birds, and mammals all resemble one another despite functional differences. The underlying resemblance of limb structure is caused by the fact that they all share an early reptilian common ancestor.

quickly eaten. In contrast, the dark forms were now camouflaged when resting on the soot-darkened bark and were not preyed on. Dark moths survived and reproduced relatively greater numbers of offspring compared with light moths. The dark moths increased in frequency while the light moths decreased in frequency from generation to generation (Figure 6-8).

Over the last 50 years one of the greatest advances in medicine has been the development of many antibiotics, chemicals such as penicillin and streptomycin that can kill or retard the growth of particular microorganisms. Although these drugs are a valuable and indispensable part of medicine, they are, to a degree, self-defeating. The widespread use of antibiotics leads to the selection and evolution of populations of microorganisms resistant to those drugs. Like all organisms, bacteria are variable. Some bacteria contain mutations that may allow them to survive in an environment containing a particular antibiotic. When this antibiotic is first used, the mutation becomes favorable to the bacteria's survival. The bacteria without the mutation die, but the bacteria containing the mutation survive the antibiotic and treatment and reproduce. With the evolution of bacteria resistant to common antibiotics, it becomes necessary to discover and produce new antibiotics capable of controlling the evolved bacteria.

A similar process has resulted in insects becoming resistant to the effects of DDT and other insecticides. After World War II, DDT was widely used to eradicate many insects ranging from disease-carrying mosquitoes to crop-damaging locusts. Before DDT was banned because of its negative impact on the environment, more than 300 insect species were resistant to DDT. This is also a common occurrence with many insecticides used today. Crop damage

Figure 6-7 Analogous structures

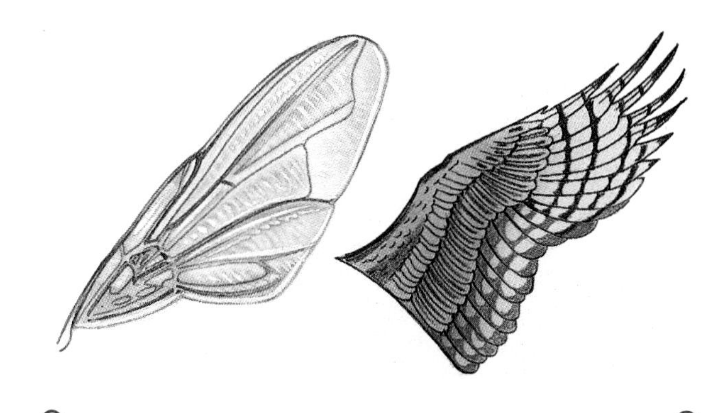

Similar selective pressures acting on organisms only distantly related may result in the evolution of outwardly similar structures. The wings of both flies and birds are designed for the same function, flight, but the wings' underlying structures are different.

Figure 6-8 Pepper moths

A

B

In Great Britain during the mid–nineteenth century, a speckled light-gray form of pepper moth was most common in forests (A). With increased pollution from local industry, the light-gray tree bark was covered with dark-gray soot, making the light moth more visible to predators. The dark-gray form of the moth was now favored and increased in frequency through the generations.

from insects is actually greater now than it was before the widespread use of insecticides 50 years ago.

The Origin of Species

Each of us recognizes that cats, dogs, raccoons, and humans are different kinds of organisms and that members of these populations do not interbreed. From comparative evidence we also know that many organisms, such as humans and apes, shared a common ancestor at one time. Finally, natural selection implies certain populations may eventually emerge that are radically different from their ancestors. How then do adaptive changes in populations lead to the origin of new types of organisms such as dogs, cats, raccoons, or apes and humans?

Each of these different types of organisms is a **biological species.** A **biological species** consists of individuals capable of potentially interbreeding and producing viable and fertile offspring. For example, all humans are members of a single species, since viable (living) and fertile (able to reproduce) offspring result from matings between members of any human population. In contrast, the horse and donkey are not classified as members of the same species. Although they can produce viable offspring, these offspring, called mules, are sterile and cannot themselves reproduce.

Geographical Isolation

One important factor in creating new species is **geographical isolation. Geographical isolation** occurs when two populations of a single species live in different locations and are separated by a physical barrier. Depending on the kind of animal under consideration, separation might follow a change in the course of a river, the appearance of a mountain or desert, or the building of a highway or a city, anything that will effectively prevent mating between members of the two populations.

Because no two places on the earth are exactly the same, the inhabitants of the two separated areas will be subject to slightly different selection pressures. They may have slightly different types or amounts of food, be exposed to different amounts of rainfall, or be preyed on by predators with different habits. From these differences in natural selective pressures, we can expect group members to become more and more different over time. It is also unlikely that at the time of separation, the same range and frequencies of alleles occurred in both populations or the mutations occuring in one population will be duplicated exactly in the other. If separation between the two groups is maintained long enough, the groups will ultimately become so different that even if the geographical barrier were removed the animals could no longer produce fertile offspring. When this occurs, the result is speciation (Figure 6-9).

Speciation does not occur at a constant rate producing a given number of species per time period. Sometimes the rate of divergence can be rapid and the number of new species fairly large. Sometimes, however, a species undergoes **adaptive radiation,** giving rise to many new species specialized for survival in diverse environments. This process occurs when a species enters an environment that offers many unused new ways of life. In response to the

Figure 6-9 Geographical isolation

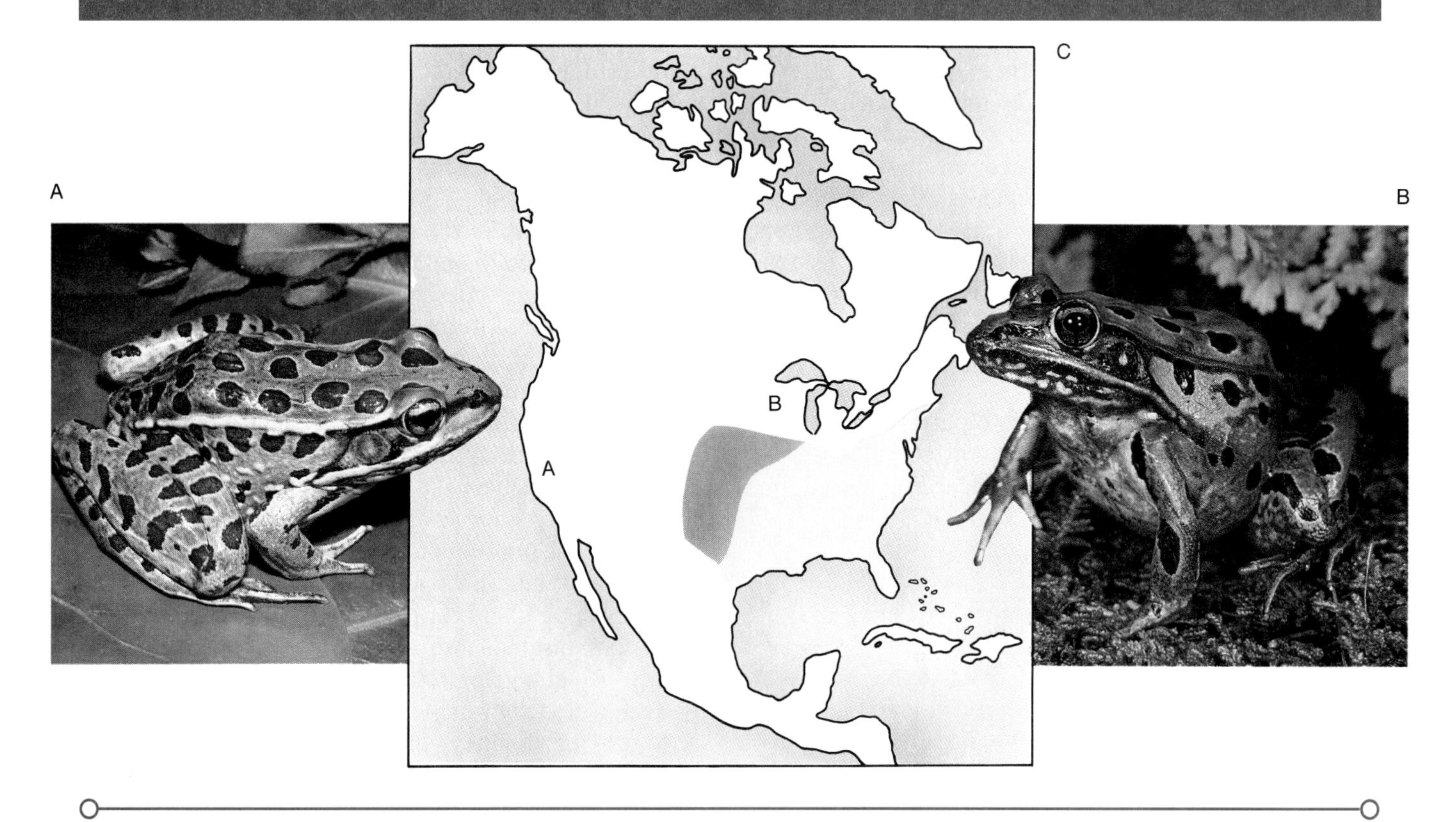

The southern leopard frog, *Rana berlandieri,* is found in California (A). The leopard frog, *Rana pipiens,* is common in Wisconsin (B). Both of these species have come about as a consequence of geographical isolation (C).

different selective pressures in those particular ways of life, different species may eventually evolve.

Adaptive radiation has occurred many times during animal evolution. An example would be the finches Darwin discovered on the Galápagos Islands. When the ancestors of these finches first came to the islands there were no other birds inhabiting them. The finches radiated into many of the available niches in the new island habitat, eventually resulting in at least 14 different species.

Extinction

Dinosaurs, saber-toothed tigers, and wooly mastodons no longer roam the earth. They are known to us only from fossils. Biologists today estimate that at least 99% of all the species ever in existence are now extinct. If natural selection results in organisms becoming *better adapted* to their environments, how does extinction occur?

There are two related causes of species extinction. First, as environmental conditions change, organisms must change with them, migrate from the area in search of more favorable conditions, or become extinct. If a species cannot migrate and fails to adapt rapidly enough to changing conditions, it will become extinct. Many environmental alterations cause animal extinction. The saber-toothed tiger, a predator that lived in northern climates during the last ice age, preyed on slow-moving, thick-hided grazing animals, such as mastodons and giant sloths. As the climate warmed they disappeared and as a result, so did the saber-toothed tigers.

The second major cause of extinction is extreme specialization. The Everglades kite, a type of hawk, for example, feeds only on a certain species of freshwater snail. As humans drain the swamps of Florida, the snail population shrinks. If the snails become extinct, the kite will almost certainly also disappear. Each species evolves a set of genetic adaptations in response to selective pressures from its particular environment. Occasionally this produces specializations to a narrow portion of its environment. The Everglades kite is well adapted to a specific way of life. If this environment vanishes, the kite cannot "disassemble" its genotype quickly enough to take another evolutionary path to avoid extinction.

Human Evolution

Humans are subject to the same evolutionary forces as other living organisms. Thus many human characteristics have emerged as modifications of what went before. Although we possess unique characteristics, many of these are a reorganization and elaboration of structures present in other organisms. For example, the biological basis for such traits as the capacity to learn and modify behavior and manipulate the environment by creating and using tools is also present, to some degree, in our nearest biological relatives, the great apes.

Understanding the basic biological characters that define humans and their adaptive values provides an essential perspective for learning about our present human biology.

Taxonomy

Taxonomy is the science by which organisms are classified and placed into categories based on their evolutionary relationships. The naming of organisms is based on binomial nomenclature, which is Latin for "two names". The first part of the organism's name represents the genus, which usually consists of several closely related species. For example, dogs and wolves are placed in the genus *Canis.* The second name designates the species. Dogs are put into *Canis familiaris.* Similarly, the taxonomic name for humans is *Homo,* meaning man, *sapiens,* meaning wise. We are the only living species in the genus *Homo.* Taxonomic categories form a hierarchy or a series of levels each more inclusive than the last, like a set of nesting bowls. Working down the hierarchical categories, the common biological characteristics become more and more specific until the level of the most specific category of comparison is reached, the biological species. The most inclusive category, a **kingdom,** includes all animals, including humans. The most specific category is the species, such as *Homo sapiens,* which contains only modern humans.

Four major types of information are used to place organisms in taxonomic groupings. These are anatomy, embryological and developmental stages, biochemical similarities, and behavior. The most important and frequently used is anatomy. In addition to outward similarities in external body structure, details such as skeleton and tooth structure are also carefully compared. As indicated earlier, the presence of homologous structures provides evidence of a common ancestry. Clues to common ancestry may also be provided by the developmental stages animals undergo on their way to adulthood. Biochemical similarities may also exist between closely related species. For example, the structure of the oxygen-carrying protein found in red blood cells, called hemoglobin, is identical in both chimpanzees and humans.

The final category used in taxonomic comparisons is behavior. As with each of the other categories, the more similar the behavior between different species, the more recently they shared a common ancestor. For example, chimpanzee and gorilla behavior is similar. They last shared a common ancestor perhaps as recently as 9 million years ago.

Figure 6-10 Trends in human evolution

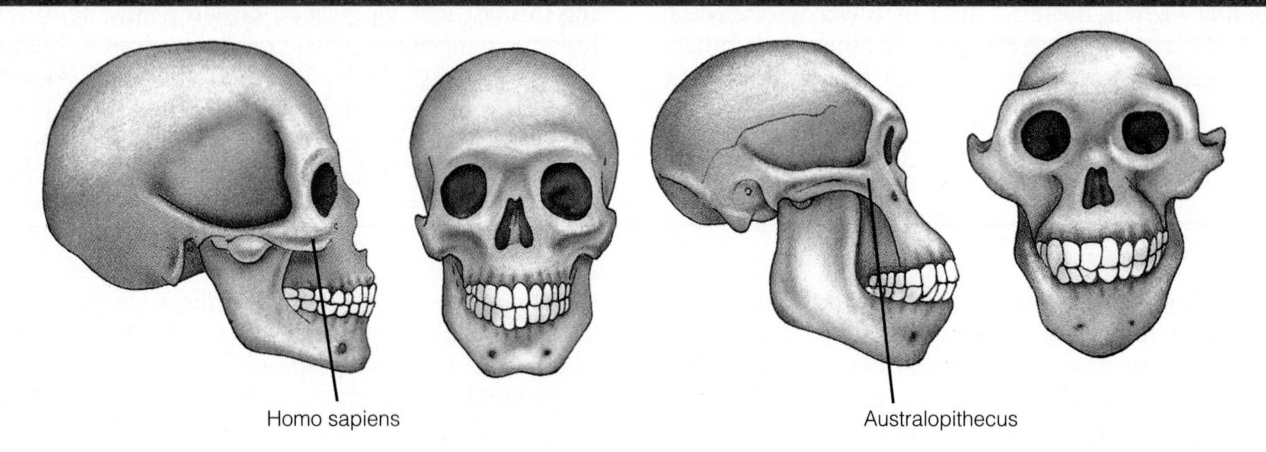

Two trends in human evolution are a reduction in the size of the teeth, jaw, and supporting structures of the face and an increase in brain size. A vertical line drawn at the origin of the cheekbone shows the differences in the shape of the oldest hominid skull, on the left, and a modern human skull, on the right. This makes these two evolutionary trends readily apparent.

Defining Humans

Humans possess particular characteristics, that define them as a distinct species. It is these traits, some of which are shared by other species, that as a suite define humans. Human characteristics include bipedalism (*bi* = two, *ped* = footed), or walking upright on two legs; complex tool making and social behavior; a large brain; language; and small, nonprojecting canine teeth. Although we do not know how bipedalism first came about, from the fossil record we have found it is the earliest hominid characteristic. Fossil remains indicate that the first humanlike creatures, or **hominids,** entered into an open grasslands environment bipedally, with a tooth and jaw structure allowing them to be **omnivorous** (*omni* = all, *voros* = to eat), that is, able to eat both plants and animals and not be dependent on only one food source.

Two evolutionary trends emerge during the course of human evolution. They are the increasing size of the cerebrum and the decreasing size of the molars and those facial features affected by tooth size, such as the jaw and points of muscle attachment controlling chewing. Comparison between the earliest-known hominid fossils and modern human skulls illustrates these two trends. If a vertical line is drawn at the origin of the cheekbone in both skulls, the differences in the size of braincases in relation to the face become apparent (Figure 6-10).

Human Fossils

The earliest hominids are assigned to the genus **Australopithecus** (os-tral-oh-pith′-e-kus, L. *austro* = southern, Gr. *pithekos* = ape). Among their human features were bipedal locomotion and tooth structure.

The australopithecines lived on the ground in the open grasslands of eastern and southern Africa. All australopithecine groups were bipedal, had the same tooth and jaw structure, and had relatively large brains. These and other characteristics varied, however. There was a wide range of variation in body size, from approximately 3 feet 6 inches to almost 5 feet 0 inches in height; variation in weight and muscular development, from 55 pounds to nearly 150 pounds; and molar tooth size, an almost 50% increase from the smallest to the largest australopithecines. Although the australopithecines were smaller in overall size compared with modern humans, their molars were two to two and one half times larger than ours. The size of australopithecine brains ranged from 400 to 800 cc. As a basis of comparison, the cranial capacity of gorillas is 450 cc and the current average human brain size is 1450 cc.

Figure 6-11 Fossil footprints

These footprints, found at the Laetoli site in eastern Africa, were made by a fully erect bipedal hominid. The impression of the big toe is in line with the other toes, and there is also an impression of a well-developed arch.

Noting this anatomical variation, scientists believe there were at least two and possibly four or more species of australopithecines. Some species were living at the same time, whereas others were time-successive species, meaning that one species gave rise to another descendant species. Here we will describe the different australopithecine groups, their relationships to one another, and their most recent descendants, the *Homo sapiens.*

The earliest clearly identified australopithecine fossils, consisting of both cranial and skeletal material, are radiometrically dated from 3.4 million years ago. These australopithecines are known as *A. afarensis* (*A.* is the abbreviation of the genus *Australopithecus*) and are found at a number of places in eastern Africa. Also of importance are hominid footprints dated to 3.7 million years ago (Figure 6-11). They were made shortly after a volcanic eruption covered the area in a layer of ash several inches thick. It then rained, turning the ash into mud. Many animals, including several hominids, then moved through it. The sun baked the mud, and the imprints were soon covered by a much deeper layer of volcanic ash from another eruption. These fossil footprints were made by a bipedal hominid. The big toe is in line with the other toes, and there is evidence of a well-developed arch.

A. afarensis was bipedal, but there is some debate as to *how* bipedal it was. *Afarensis* had relatively long arms and slightly curved finger bones, suggesting both considerable climbing ability and its apelike ancestry. *Afarensis* stood between 3.5 and 4.0 feet in height and weighed approximately 50 pounds (Figure 6-12, *A*).

In existence from approximately 3 to 2 million years ago in both eastern and southern Africa was a hominid form called *A. africanus.* (Figure 6-12, *B*). It had an average cranial capacity of 440 cc, and its body was somewhat larger than that of *A. afarensis.* Other australopithecines found in southern Africa and dated from 2 to 1.5 million years ago had much larger molars and facial skeletons and somewhat larger skeletons compared with *A. africanus.* This group is called *A. robustus* because it was more heavily built and larger than *A. africanus* (Figure 6-12, *C*).

About 2.5 to 1.0 million years ago were a group of australopithecines even larger than *A. robustus.* They had large back teeth and a massive supporting facial skeleton. The species name is *A. boisei.* The species name comes from the Boise Fund, which sponsored the research leading to this discovery. *Boisei* was approximately 5 feet tall and weighed between 100 and 150 pounds. Like other australopithecines, it too had a relatively small brain, ranging in size from 410 to 530 cc (Figure 6-12, *D*).

In 1986 a fossil skull was found in eastern Africa and dated to 2.5 million years ago using the potassium-argon dating method. This skull shows characteristics similar to those of *A. boisei,* such as a large facial skeleton. Other features, however, are more primitive than those of *A. boisei* and are more typical of *A. africanus.* Two such traits include the shape of the joint where the jaw joins with the skull. This specimen has been given the species designation *A. aethiopicus* (Figure 6-12, *E*).

During the early 1970s many hominid specimens, radiometrically dated from 2 to 1.5 million years ago, were discovered in eastern Africa. These hominids had smaller molars and premolars and a larger brain than any of the other australopithecines discovered previously. As determined by four specimens, cranial capacity ranged from 500 to more than 760 cc. Because of these differences, the fossils were assigned to the genus *Homo* and given the species name *habilis* (Figure 6-12, *F*).

Early Humans and Tools

At a site in eastern Africa called Oldowan and dated to 2.5 million years ago, researchers found crude stone tools. These tools were made by knocking several chips off of a rounded stone to give it a rough cutting edge. The tools, as well as the manufacturing technique, are known as the **Oldowan tradition.** At several sites in eastern Africa dated to roughly 2.0 million years ago stone tools were found at sites of both *A. boisei* and *H. habilis.* Consequently we do not know if only one or if several different australopithecine species manufactured and used tools. It is reasonable to hypothesize that they may have used other more perishable materi-

Figure 6-12 Comparison of australopithecine populations

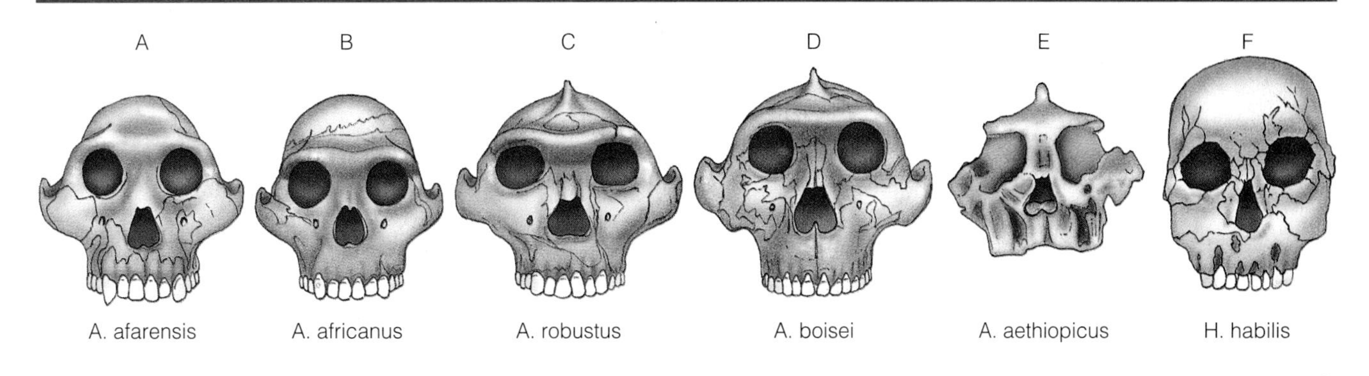

Australopithecine populations varied in brain size, body size, weight, muscular development, and molar tooth size. The skulls show relative differences in brain size and facial structures related to jaw size and molar tooth size.

als, such as wood and bone, as tools, much as some modern human populations do.

Given the often confusing proliferation of names and dates of our earliest ancestors, what is really known of these different populations with respect to later human evolution? Three things are known with reasonable certainty. The fossil record indicates *A. afarensis* is the oldest known hominid, *H. habilis* is a direct ancestor of later hominids, and *A. boisei* was a side-branch, a "biological cousin" that became extinct. We know this because *A. boisei* was also a contemporary of *H. erectus,* which is the next time-successive species in human evolution. The placement and presumed relationships between the other populations of australopithecines give rise to at least four different possible early hominid evolutionary trees.

Even though they understand the general anatomical traits of the australopithecines and *H. habilis,* human biologists lack basic information on the range of variation for many traits. For example, the range of variation of several characteristics between australopithecine populations exceeds the known range of variation of any known living or extinct primate species. This fact suggests at least two species of hominids living at the same time. However, what about the variation between southern and eastern African forms? Do these differences represent different species or only geographical variation? Another problem is that the same information is not present for the same time periods for both geographical regions. Because *A. africanus* has been found in southern Africa, does this simply reflect the dearth of fossils found for the same time period from eastern Africa? Or does it represent different species in different regions? These are only a few of the questions remaining. As evidence continues to accumulate, some of these questions may be answered.

The early evolution of the genus *Homo* appears to have occurred in Africa. There fossils belonging to the second and only other extinct species of *Homo, Homo erectus,* are widespread and abundant from 1.7 million until about 500,000 years ago. The cranial capacity of *H. erectus* is about 44% larger than *H. habilis* and more than 114% larger than *A. africanus.* The range in cranial capacity for *H. erectus,* from 727 to 1125 cc, overlaps the high end of the range of *H. habilis* and the low end of the range for modern humans. Jaws and teeth are large compared with modern humans but smaller than those of the australopithecines. By approximately 1 million years ago, *H. erectus* had migrated into Asia, Europe, and Indonesia.

The stone tool tradition used by *H. erectus* was much more sophisticated compared with the Oldowan tradition of earlier hominids. It is called the **Acheulian tradition,** the name being derived from the fossil site in which it was first discovered. Acheulian tools are worked on both sides of the stone. They are flatter and have straighter and sharper edges than do Oldowan tools. The basic tool is the hand ax, which probably had many uses. Other tools were also produced, presumably for other activities. This increased tool specialization allows for more efficient tool use and also implies greater mental complexity in tool design and manufacture. The wide distribution of tools made by the same technique suggests the widely dispersed populations of *H. erectus* communicated with one another. The first evidence of the controlled use of fire by humans occurs at the campsites of this species in the Rift Valley of Kenya at least 1.4 million years ago, and fire was associated with populations of *H. erectus* from that time.

The first appearance of our own species, *Homo sapiens,* may have occurred as early as 250,000 years ago in Europe. These early populations are sometimes called "archaic" (old) *H. sapiens.* Their primary differences include a larger brain, reduced tooth size, and enlargement of the rear of the skulls compared with those of *H. erectus.* Of all the variant populations of archaic *H. sapiens,* probably the best-known group, spanning the time range from 125,000 to 35,000 years ago, are the Neanderthals. The first evidence of them was discovered in the Neander River Valley in Germany in 1856. The name given to this species is derived from this site. The Neanderthals lived in the area surrounding the Mediterranean Sea, including western and eastern Europe, part of the Middle East, and North Africa.

Compared with modern *H. sapiens,* Neanderthals were powerfully built, short, and stocky. Compared with other archaic *H. sapiens,* Neanderthal skulls were longer, with protruding faces and noses and rather heavy bony ridges above their eyes. Their brains were larger than those of modern humans, possibly as a consequence of their heavy, large bodies. The fossil record reveals that Neanderthals made diverse tools and built hutlike structures or lived in caves. Neanderthals took care of their injured and sick and commonly buried their dead, sometimes placing animal bones and weapons with the bodies. Analysis of pollen grains found in the soil of Neanderthal burial sites also reveals that flowers were placed at the burial site. Such attention to the dead strongly suggests they believed in an afterlife. For the first time the kinds of thought processes characteristic of modern *H. sapiens,* including symbolic thought, are evident in these acts.

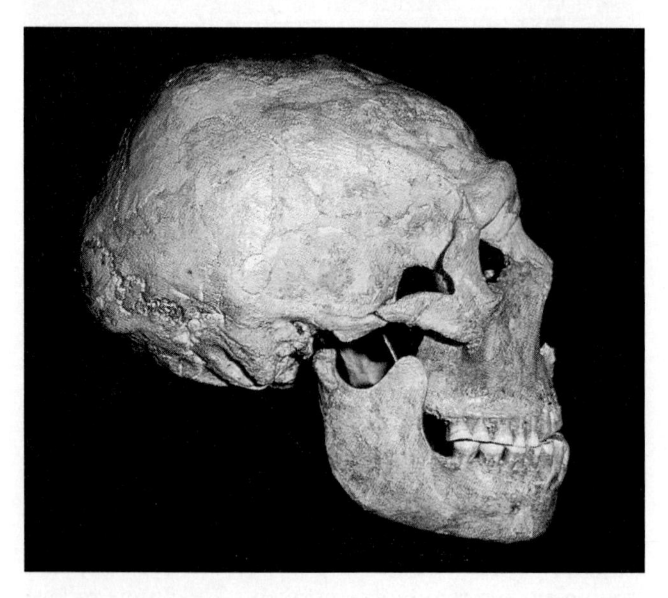

About 34,000 years ago the European Neanderthals were rapidly replaced by modern *H. sapiens.* Evidence indicates these populations originally came from Africa, where almost identical fossils, but perhaps 100,000 years old, have been found. They seem to have replaced the Neanderthals completely in southwest Asia by 40,000 years ago and then spread across Europe during the next 5000 to 8000 years.

These modern humans used sophisticated stone

Figure 6-13 Bering Land Bridge

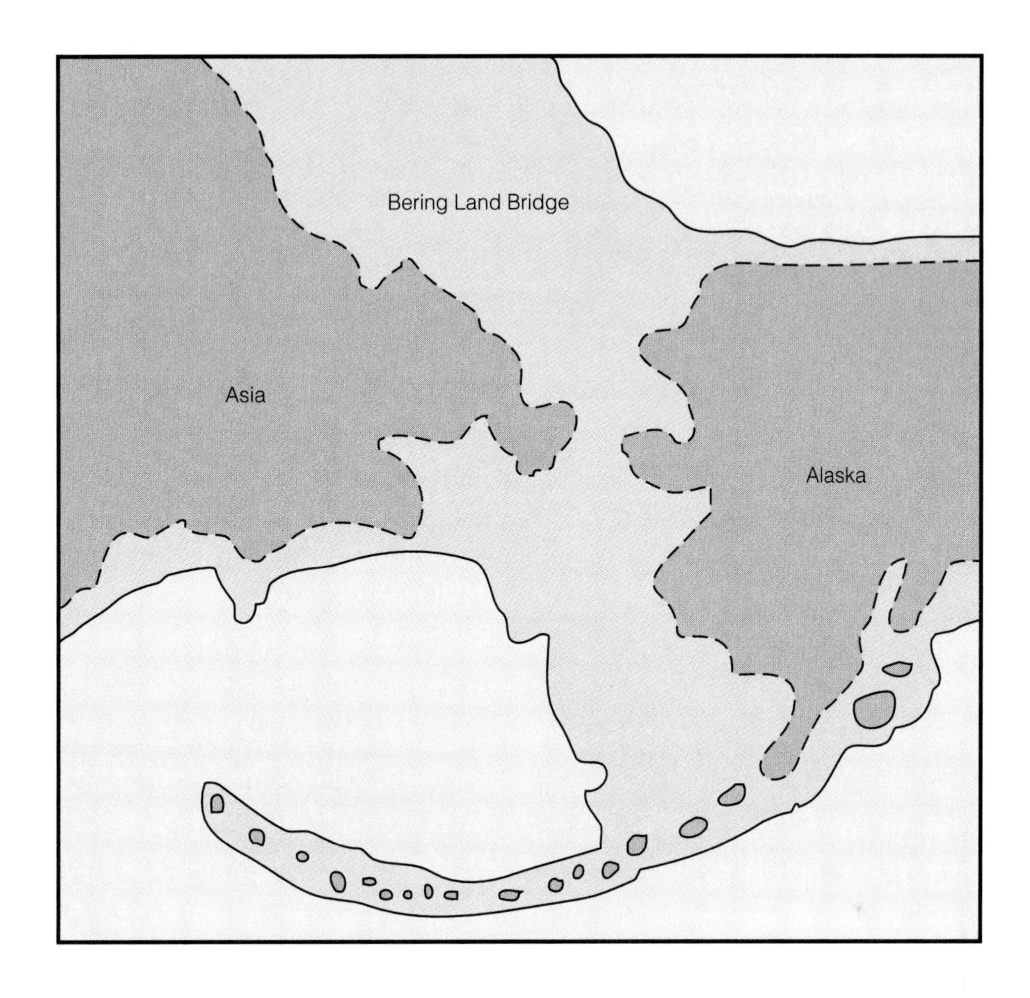

Modern *H. sapiens* migrated from Asia to the North American continent by crossing the Bering Land Bridge. It appeared as glaciers formed in the northern continent, temporarily locking up water as ice, lowering sea levels and exposing the land bridge. As the glaciers melted, sea levels rose, covering the land bridge and creating the Bering Sea between Asia and North America.

tools that rapidly became more diverse and specialized. They also made tools out of bone, ivory, and antler. They were hunters who killed game by using complex tools. Soon after the appearance of modern *H. sapiens,* cave paintings also frequently appeared.

Modern *H. sapiens* eventually spread across Siberia to the Americas, which they reached perhaps 30,000 years ago. Their route from eastern Asia to the North American continent was over the Bering Land Bridge. Today Siberia and Alaska are separated by the Bering Sea. But during the last 40,000 years there have been several ice ages in which large areas of northern Europe, Central Asia, and northern North America were covered by giant glaciers. When this occurs, water is temporarily locked up in the form of ice, producing dramatic drops in sea level. During these periodic cycles of glaciation, the Bering Land Bridge emerged (Figure 6-13). It connected Asia and North America and was as much as 1000 miles wide. Early modern humans migrated from Asia to the North American continent and south into South America.

Soon after people began to spread throughout North America, they caused the extinction of many animal and plant populations, a process that has recently accelerated. Human remains are often associated with the bones of the large animals they hunted, and these animals often disappeared rapidly from different parts of the world shortly after the first appearance of humans.

CHAPTER REVIEW

SUMMARY

1. Charles Darwin proposed a mechanism, called natural selection, that explained how organisms adapted to their environment and changed through time. His argument was based on evidence accumulated from his five-year sea voyage around the world, artificial breeding techniques of farmers, and geology.
2. Since the rediscovery of Mendel's principles of inheritance, evolution is understood as changes in allele frequencies in a population through time and natural selection as differential reproduction of alternative genotypes. Darwinian evolution, Mendelian principles of inheritance, and the biochemistry of the genetic material are all used to understand the evolutionary process.
3. Human evolution is characterized by two major evolutionary trends: an increase in brain size and a decrease in tooth size and those facial characteristics associated with them. The increase in the human brain has occurred primarily in the cerebral cortex.
4. Some biological characteristics defining humans include bipedalism, or walking exclusively in an upright manner on two feet; a relatively large brain; complex social behavior; and language. Of these characteristics it appears from the fossil record that bipedalism was the first distinct hominid characteristic to emerge.
5. Biological evolution is an ongoing process because environments change and the life-forms in them must also adapt to those changes to survive. If organisms cannot keep pace with the rate of environmental change, they face extinction.
6. The most direct evidence for evolution comes from the mineralized remains of past life-forms, called fossils. Comparative evidence supports the fact that through time natural selection has modified similar structures for different functions. Observational studies also demonstrate that evolution is ongoing. As is true of all science, as new information is gathered more evolutionary hypotheses can be tested, confirming or disconfirming what we presently know.
7. For one species to evolve into two new species, at least two populations within the species must become reproductively isolated from one another. The most common way is through the appearance of a physical barrier, referred to as geographical isolation. As differences accumulate the populations will eventually be unable to interbreed even if they come back into contact with one another.
8. Organisms usually become extinct because they cannot keep pace with environmental change or because of extreme specialization. Environmental change refers to any aspect of the species environment, from the sudden destruction of the environment to competition from other species for the same or similar resources.

FILL-IN-THE-BLANK QUESTIONS

1. Charles Darwin's reading of ______ ______ essay entitled "Population" provided him with the insight necessary to solve the problem of how organisms change and adapt to their environments.
2. Darwin defined biological evolution as ______ with modification by means of ______ ______.
3. Fitness is defined in terms of the relative ______ for leaving ______.
4. In light of genetics evolution is understood as a change in ______ frequencies through time.
5. The modern definition of natural selection is differential ______ of alternative ______.
6. According to the fossil record, the earliest hominid characteristic appears to be ______.
7. The use of fire is first associated with *Homo* ______ approximately one million years ago.
8. ______ are the mineralized parts of an organism's skeleton.

9. Although they might have different functions, structures that are internally similar are referred to as ______ structures.

10. In nature two individuals that interbreed and produce viable and fertile offspring belong to the same ______ ______.

SHORT-ANSWER QUESTIONS

1. What is the importance of individual variation of biological characteristics in Darwin's theory of evolution?
2. Describe one important implication of natural selection as first defined by Darwin.
3. Why do scientists state there were several species of *Australopithecus* living at the same time?
4. Compare and contrast the features of *Homo sapiens* with those of *Neanderthal* populations.
5. Compare and contrast homologous and analogous structures.

VOCABULARY REVIEW

adaptive radiation—*p. 112*
analogous structures—*p. 110*
Australopithecus—*p. 115*
geographical isolation—*p. 112*
hominid—*p. 115*
homologous structures—*p. 110*
natural selection—*p. 106*
taxonomy—*p. 114*

Chapter Seven

ORGANIZATION OF THE HUMAN BODY

For Review

Here are some important terms and concepts that you will encounter in this chapter. If you are not familiar with them, you should review them before proceeding.

Cilia
(page 48)

Homeostasis
(page 5)

OBJECTIVES

After reading this chapter you should be able to:

1. *Describe the ways in which body cells function both individually and collectively.*
2. *Discuss the importance of interstitial fluid.*
3. *Define homeostasis and explain how negative feedback helps a system maintain homeostasis.*
4. *Define tissue, organ, and organ system.*
5. *Name and describe the four classifications of primary tissue types.*
6. *Classify the major types of epithelium.*
7. *Name the three major types of specialized connective tissue.*
8. *List and describe each of the three subtypes of muscle tissue.*
9. *Define membrane and differentiate between serous and mucous membranes.*
10. *Describe the six directional terms used when describing the body.*
11. *List the eleven organ systems and give the major functions and components of each.*

Although the human body consists of diverse cell types, these cell types are organized into four kinds of tissues that in turn combine to form eleven organ systems. Cells thus function on two interrelated levels, both as individual cells and as integrated parts of the body. Both individually and collectively, cells maintain the optimum internal environment needed to sustain themselves and ultimately the body itself.

From Cell to Body

In Chapter 3 you learned that all cells regularly exchange energy and materials with their surroundings. This exchange is precisely controlled to sustain the biochemical pathways governing cellular growth, development, maintenance, and reproduction. A disruption in the regular exchange of materials and energy with the environment may result in disturbance and death of the cell or even the organism itself. Cells also *actively* maintain their structure and organization by maintaining an internal environment usually different from their surroundings. Living organisms, then, must maintain optimum internal conditions or balance within narrow limits, often under the conditions of wide fluctuations in environmental conditions.

It is easy to imagine how a single-celled organism might be able to exchange energy and materials with its environment and precisely maintain its internal environment. It is a single cell completely surrounded by the external environment it inhabits. But what about a multicellular organism such as ourselves? A muscle cell buried deep within your leg has no way of obtaining oxygen or other essential materials directly from your body's external environment, nor can it eliminate wastes directly into that environment.

Our body cells must function on two interrelated levels, both as individual cells and as integrated parts of the body. The individual cells perform functions the organism itself performs. They must obtain oxygen, water, and nutrients from their environment. They must metabolize these nutrients to release energy needed for energy-requiring activities, or to provide raw materials to manufacture cell parts. They must eliminate metabolic wastes, and they usually must reproduce. Ultimately the body is the cell's link with the external environment. The food and water you ingest is digested by your body, supplying the essential nutrients and other required materials to all body cells; and breathing supplies oxygen to all body cells; and excretion eliminates the cumulative wastes of body cells. Similarly, body cells, organized into particular systems and acting in coordination, maintain the delicate internal environment necessary for existence. How then does the cell function both individually and collectively to ensure its survival?

Cells not only contain watery cytoplasm within their membranes but they also are surrounded by a watery fluid called **interstitial fluid.** **Interstitial fluid,** or extracellular fluid, is the medium through which substances are exchanged between the cells and the body's circulatory system. Nutrients and oxygen, carried in the blood, pass from the smallest blood vessels of the circulatory system, called **capillaries,** into the interstitial fluid. The nutrients and oxygen then pass through the cell membrane and into the cell. Conversely, cell wastes cross the cell membrane, enter the interstitial fluid, pass into the capillaries, and are ultimately eliminated from the body. The interstitial fluid is an intermediary in the exchange of materials between individual cells and the body's external environment.

This chapter describes the basic principles and mechanisms of homeostasis, the levels of cellular organization and integration from tissues to organ systems, and the body cavities the organs occupy.

Homeostasis

Self-regulating mechanisms in all organisms are able to sense changes in the internal environment and make appropriate adjustments to minimize changes and restore balance. The ability for this self-regulated control of the internal environment is called **homeostasis** (ho′-me-o-sta′-sis). Every single body cell must maintain its own internal homeostasis, and it also contributes to the internal homeostasis of the whole body. The external environment of the body's cells is the body itself. The cell must also help maintain this environment for its continued existence. Our component body parts work together to maintain a stable internal fluid environment required by the body's living cells.

All homeostatic control mechanisms of the body have at least three interdependent components. First is the **control center,** which analyzes the incoming information it receives, comparing it with an internal **set point,** the normal value or range for that particular variable. Second is the **receptor,** a type of sensor that monitors the body's environment and responds to changes, or to stimuli, by sending information to the control center. Third is the **effector,** providing the means by which the control center can cause a response to the incoming information, or stimulus, it has received. The results of the response then "feed back" to influence the stimulus. The effect is to decrease the original effect of the stimulus, slowing it down or shutting it off

Figure 7-1 A negative feedback system

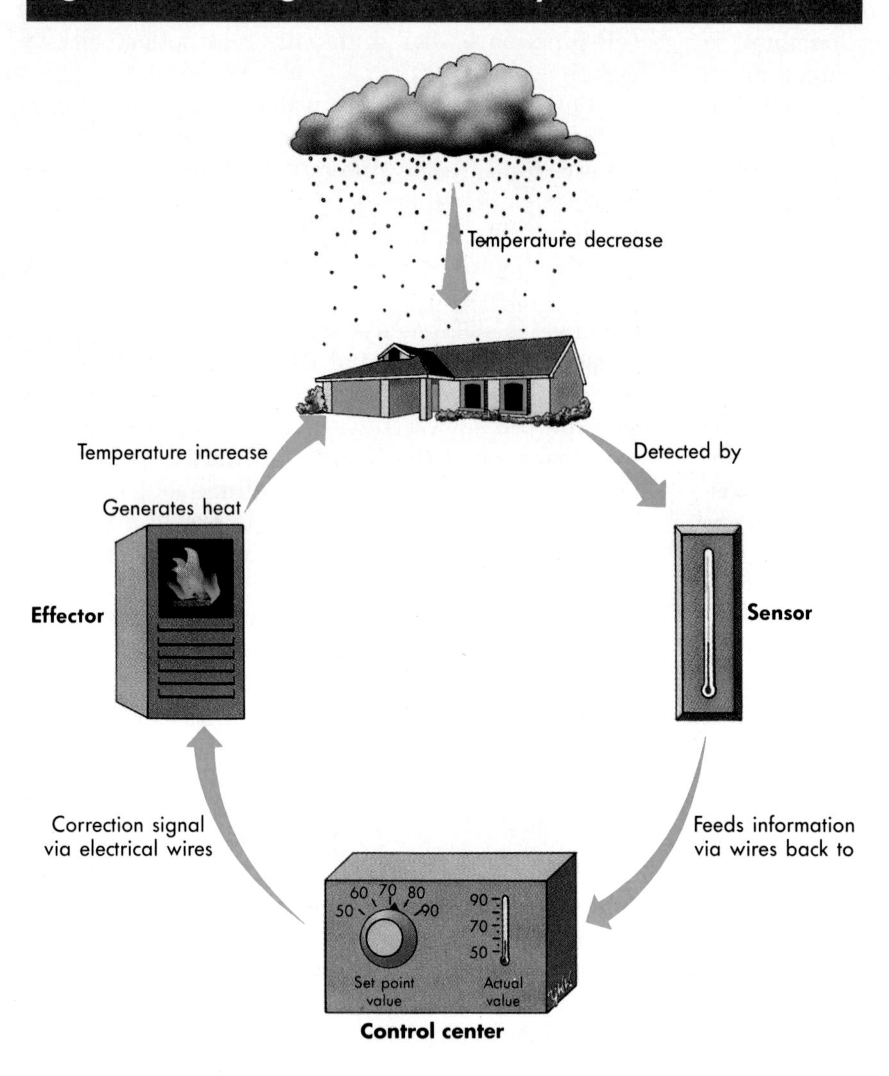

As the temperature in the house decreases, the sensor, in this case a thermostat, sends a signal to the control center. The control center compares the incoming information with the normal set point of temperature. If the temperature in the house is below the normal value, the control center sends a signal to the appropriate effector, in this case the furnace. The furnace goes on, warming the house until the set point temperature is reached, and then turns off.

completely. The response causes the variable to change in the opposite direction of the initial change. Thus it is called a *negative*-feedback mechanism (Figure 7-1). Following is an example of negative feedback as it assists in maintaining homeostasis in the body.

Negative Feedback

Negative feedback occurs when an increase in the output of the system decreases input into the system. Most body systems maintain homeostasis through negative feedback. Negative feedback does not prevent variation but maintains variation within a normal range.

An example of homeostasis by means of negative feedback is water-balance maintenance in the human body. Each body cell is surrounded by a small amount of fluid, and the normal function of that cell depends on maintaining its fluid environment within a narrow range of conditions, including volume, temperature, chemical content, and water concentration. If the concentration of water in the fluid surrounding cells deviates from the normal range, the cells and possibly the whole body may die. Any loss or gain of water from optimum levels sets into motion homeostatic mechanisms to counteract the change. If the amount of body water lost causes a drop below the minimum normal range, body receptors detect this condition and signal specific homeostatic mechanisms to restore water levels back to within the normal range. For example, increased thirst leading to greater intake of water and decreased water elimination by the kidneys are two ways to help restore water balance. In contrast, if the body's water content exceeds the normal range, this is also detected, and homeostatic mechanisms are triggered to decrease the amount of water until it drops back to the normal range. Increased elimination of water by the kidneys is the primary means of restoring water balance (Figure 7-2).

Homeostasis is not simply the maintenance of all conditions within a narrow range of values at all times. For example, during exercise blood pressure and heart rate increase dramatically. The elevated blood pressure is required to supply extra nutrients and oxygen to muscles to maintain their increased rate of activity. In other words, increased blood pressure and heart rate are required to maintain homeostasis of increased cellular metabolism. This range is higher and the variation broader than the resting range. After exercise the range returns to the earlier resting condition. As a result, blood pressure and heart rate are maintained under continuously changing, or **dynamic,** conditions.

Systems of Cells

The human body can be understood at seven related structural levels: chemical, organelle, cellular, tissue, organ, organ system, and finally the entire body. The chemical, organelle, and cellular levels of structure and function were described in Chapter 3. Now we move to tissue, organ, and organ system levels of organization. A **tissue** is a group of cells with similar structure and function plus the extracellular substances located between them. **Organs** are composed of two or more tissue types that perform one or more common functions. An **organ system** is a group of organs classified as a unit because of a common function or set of functions.

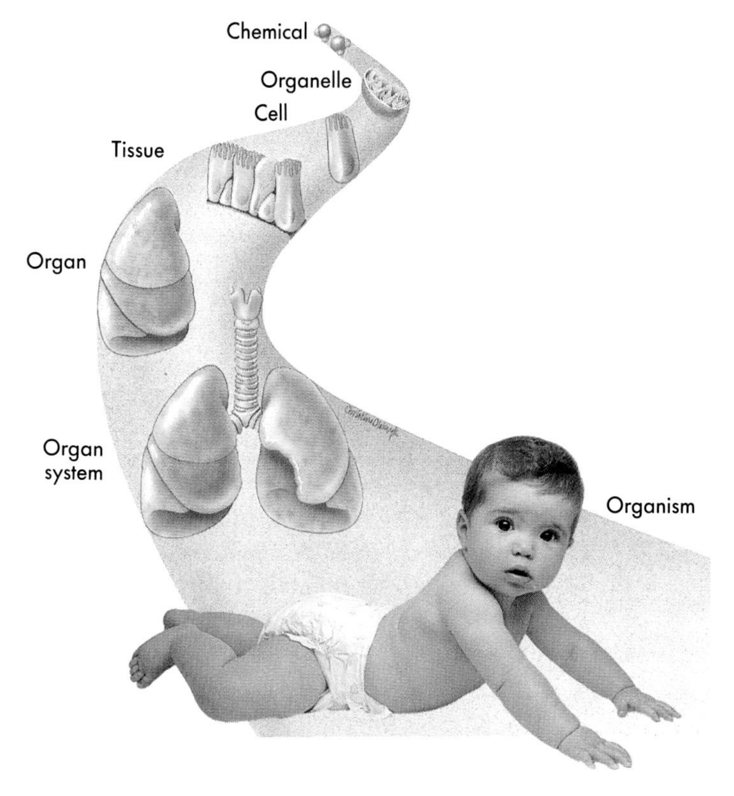

Types of Tissues

Cell structure and composition of the extracellular substance are the primary characteristics used to classify types of body tissue. Although many tissues make up the human body, they can be classified into four primary tissue types. The **four primary tissue types** are epithelial, connective, muscular, and nervous. Each primary tissue type has several subtypes (Table 7-1). Epithelial and connective tissues are the most abundant and diverse of the four tissue types and are components of every organ in the human body.

Table 7-1 The Primary Tissue Types and Their Subtypes

Epithelial Tissue	**Muscle Tissue**
Membranous	Skeletal
Glandular	Cardiac
Connective Tissue	Smooth
Connective tissue proper	**Nervous Tissue**
Dense connective tissue	
Loose connective Tissue	
Specialized connective tissue	
Cartilage	
Bone	
Blood	

Epithelial Tissue

Epithelial (ep-i-the-le-ul) **tissue** consists almost entirely of cells with little extracellular fluid between them. In general, epithelium covers surfaces or forms structures, such as glands, that embryologically were part of the body surfaces. The epithelium forms the outer covering of the body and of individual organs and the inner lining of various body cavities. It also forms tubes, such as in the digestive and respiratory systems, ducts of glands, and the inner lining of the heart and blood vessels.

The two basic subtypes of epithelium are membranous epithelium and glandular epithelium. A common characteristic of **membranous epithelium** is one free surface not associated with other cells. The free surface might be toward the external environment, as is the case of skin, or it might be the inner lining of a tube, such as the windpipe or digestive system. Because membranous epithelium forms the boundaries that separate us from the outside world, nearly all substances received or given off by the body must pass through epithelium. On the opposite side of the epithelium is the **basement membrane,** which binds it to the underlying tissue.

Membranous epithelial tissues are further classified according to the number of cell layers and cell shape. **Simple** epithelium consists of a single layer of cells, each cell extending from the basement membrane to the free surface of the tissue.

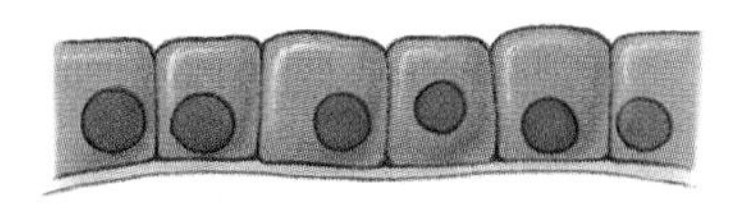

Figure 7-2 Negative feedback controlling water balance

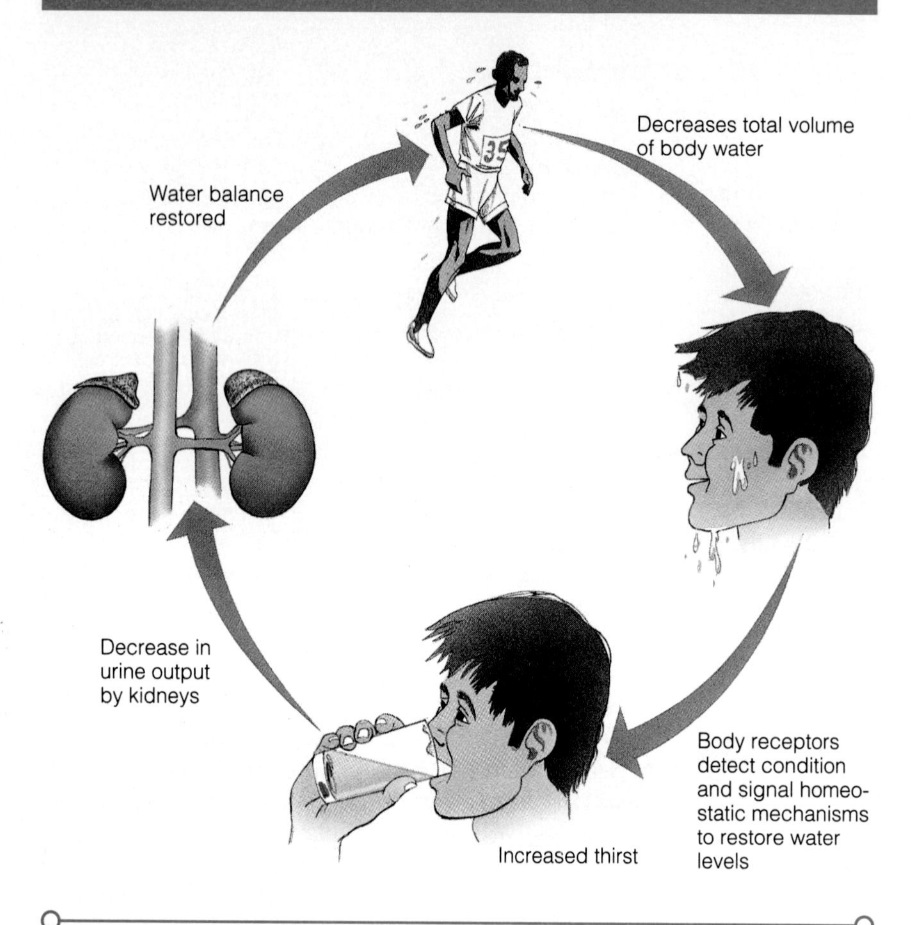

Any loss or gain of water from optimum levels sets into motion homeostatic mechanisms to counteract the change. For example, if excessive sweating occurs because of exercise, the total volume of body water decreases below the normal range. Sensory receptors in the body detect this condition and set into motion homeostatic mechanisms, such as increased thirst and decreased urine output by the kidneys, to restore water balance.

Stratified epithelium (*strata* means layers) is composed of more than one layer of cells, but only one layer touches the basement membrane. **Pseudostratified** epithelium (*pseudo* means false) consists of a single cell layer but appears multilayered. This type of epithelium contains a combination of cells, some of which extend from the basement membrane to the free surface.

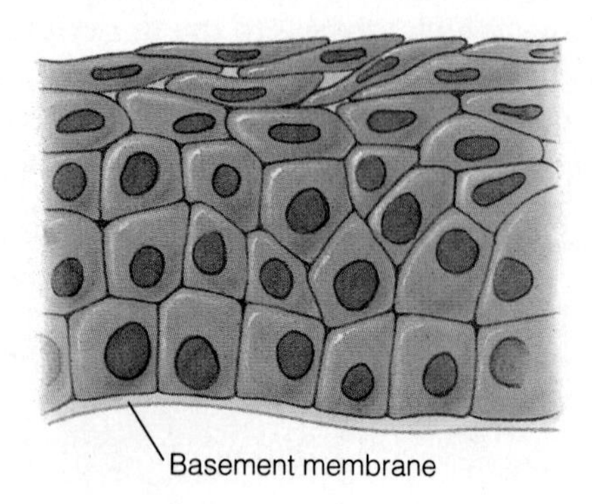

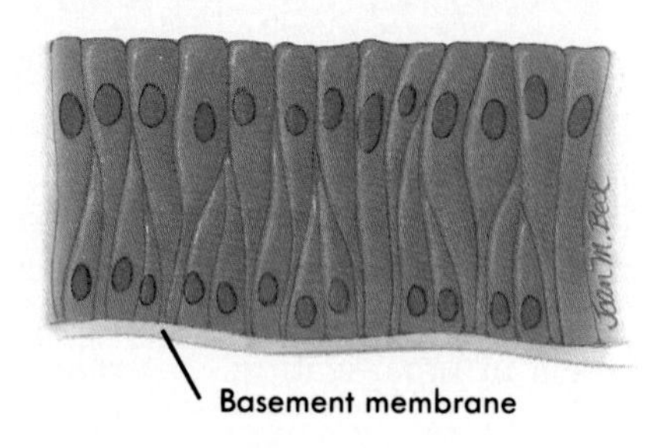

There are three distinct cell shapes found in epithelium: **squamous** (skwa′mus) cells are flat and thin, **cuboidal** are cube-like, and **columnar** cells are tall and thin like a column (Figure 7-3). Most membranous epithelia are given two names. The first name indicates the number of layers, and the second describes cell shape. For example, the walls of the microscopic air sacs in the lungs are composed of simple squamous epithelium. In contrast, pseudostratified ciliated columnar epithelium lines part of the air passageway which connects the mouth and nose to the lungs.

Glandular epithelium comprises glands. A gland consists of one or more cells that produce and secrete a particular product. This product, called a secretion, is a water-based fluid which can contain a variety of substances. Two kinds of glands are composed of glandular epithelium: endocrine glands and exocrine glands.

Endocrine glands (*endo* = within) secrete their substances directly into the bloodstream where they are distributed throughout the body. **Exocrine glands** (*exo* = outside) secrete their products through a duct onto body surfaces or into hollow body organs. Exocrine glands are much more numerous than endocrine glands. They include single-celled glands, called **goblet cells,** that secrete a thick, watery substance called mucus; multicellular sweat and oil glands that secrete their products onto the surface of the skin to cool and lubricate it;

Figure 7-3 Epithelium cell shapes

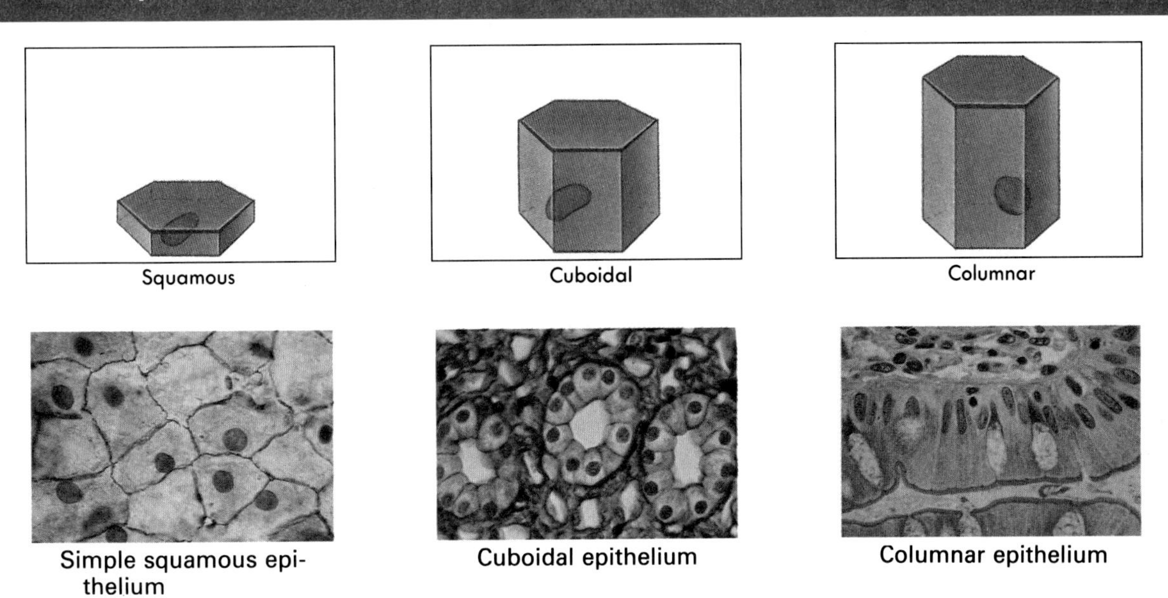

glands that secrete enzymes into the digestive system to aid in the breakdown of food; and female mammary glands that produce milk for a newborn baby (Figure 7-4).

Cell Surfaces and Cell Connections

To better understand how membranous epithelium functions, we must also discuss its specialized cell-membrane surfaces, of which there are three types. The first, described earlier, is the free surface that faces away from the underlying tissue. The second is a surface that faces other cells. The third faces the basement membrane. Free surfaces can be smooth, have fingerlike projections called microvilli, or be ciliated. Smooth free surfaces reduce friction. For example, epithelium with smooth free surfaces lines blood vessels, enabling blood to flow easily through the vessels. As you recall from Chapter 3, microvilli and cilia are extensions from the cell membrane. Microvilli greatly increase surface area and are found in cells involved in absorption or secretion, such as in the lining of the small intestine. Cilia propel materials along the cell's surface. The epithelium lining the airways of the lungs are ciliated and also contain goblet cells. The cilia help to move foreign particles, trapped by the mucus, from the deeper airway passages to the upper airways, where they can be removed.

Cell surfaces other than free surfaces have modifications that tightly hold epithelial cells to each other or to the basement membrane. One type is called an **adhering junction,** where the membranes of adjacent cells are attached to one another by protein filaments. Small circular areas of protein, called **desmosomes** (dez-mo-soms), form the sites of adhesion between epithelial cells. Additional proteins extend into the cell's cytoskeleton and also cross the extracellular space to link to the desmosome of the adjoining cell. Adhering junctions are

Figure 7-4 Example of exocrine glands

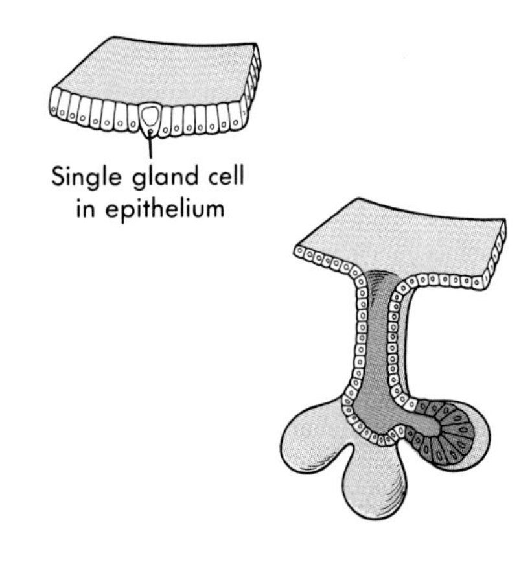

Exocrine glands vary from single-cell goblet cells that secrete mucus, to multicellular sweat glands that secrete a watery substance onto the skin's surface to cool it.

Figure 7-5 Cell connections

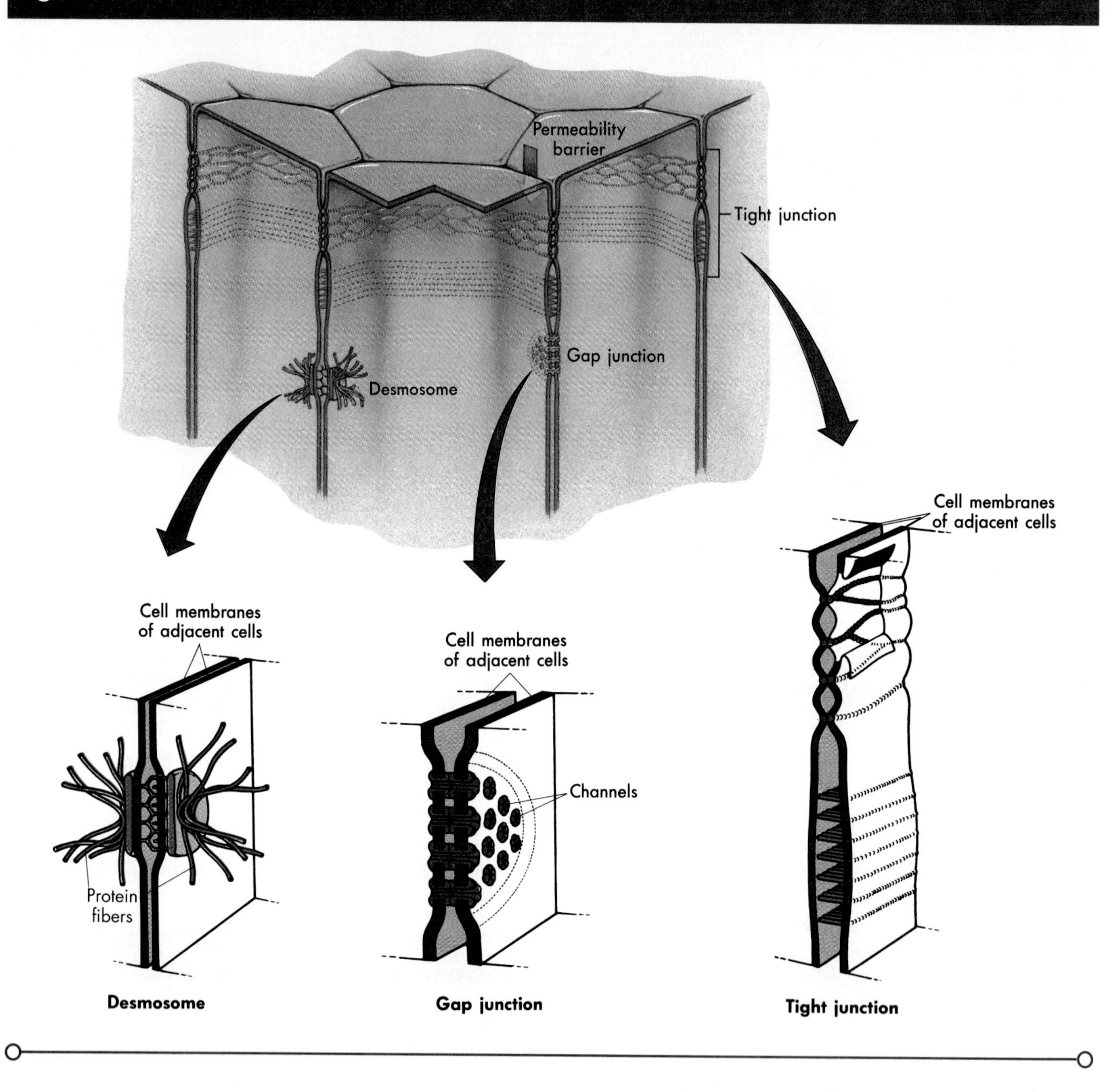

abundant in the skin and other tissues that must withstand movement in many different directions without separating.

Tight junctions seal off spaces between adjacent cells of the epithelium and prevent leakage across it. Substances can cross the epithelial tissue only by moving through the cytoplasm of its individual cells. Tight junctions are often found between cells of glands and hollow organs where they seal the contacting surfaces of neighboring cells. These junctions prevent the diffusion of substances between the cells in the organ wall, requiring instead that materials pass through the cells. For example, the simple molecules derived from the breakdown of food in the digestive system pass through the cells of the digestive organs into the bloodstream and are carried throughout the body. **Gap junctions** are clusters of small channels that cross the membrane of one cell and match up with similar channels of an adjacent cell. These channels allow passage of ions and small molecules between cells, so individual cells can coordinate activities throughout a tissue. Each channel is composed of protein molecules embedded in the phospholipid bilayer of the cell membrane. Gap junctions occur between muscle cells of the heart, making precise activity coordination possible. (Figure 7-5.)

Connective Tissue

The second body tissue type, called **connective tissue,** binds structures together, provides support and protection, fills spaces, stores fat, and forms blood cells. Connective tissue is found everywhere in the body and is the most abundant of the primary tissues. However, the amount of connective tissue in particular structures or organs varies greatly. Bone and skin, for example, consist primarily of connective tissue, whereas the brain contains very little. The essential characteristic distinguishing connective tissue from the other three tissue types is that it consists of cells separated from each other by an **extracellular matrix** surrounding the individual cells. The extracellular matrix of connective tissue has three major components. The first is protein fibers, which may be of two major types. One is called **collagen,** long protein strands that look like microscopic ropes and are strong and flexible but do not stretch. Collagen accounts for about one

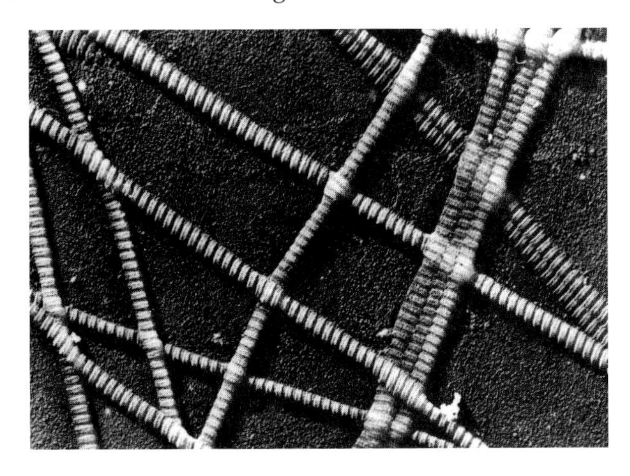

third of the total body protein or about 6% of total body weight. The second component is a type of protein called **elastin.** It gives the tissue in which it is found elastic properties. Although not as strong

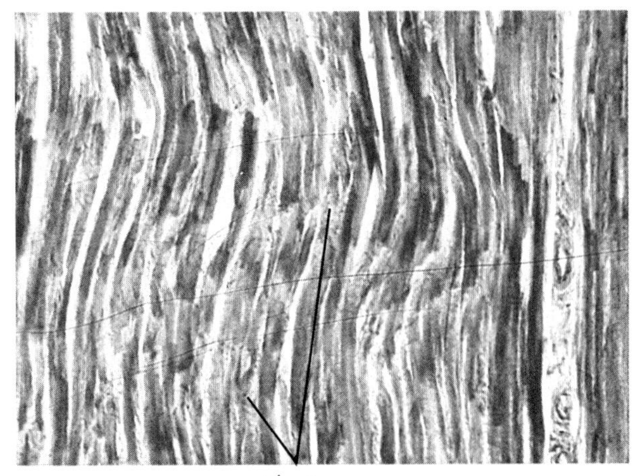

as collagen fibers, elastin, when stretched, will recoil back to its original shape. The third component is a watery fluid. The relative amounts and types of these three components form the basis of connective tissue classification that follows.

The two major types of connective tissue are **connective tissue proper** and **specialized connective tissue,** each having several subtypes. **Connective tissue proper** is an important structural component of the body and has two subtypes. The first subtype, known as **dense connective tissue,** contains a large amount of collagen closely packed and may be bundled together into fibers oriented in the same direction. Dense connective tissue makes up tendons, which attach muscle to bone, and ligaments, which connect bone to bone at joints. Both tendons and ligaments encounter stress in only one direction, and the collagen fibers are aligned in this direction of stress. For example, at the back of the ankle just above the heel is the Achilles tendon. If you put your thumb and index finger on either side of it and flex your foot up and down, you will feel the action of the tendon. It shortens and lengthens in only one direction. The cells that produce collagen fibers found in connective tissue are called **fibroblasts** (*blast* means to create).

The second subtype of connective tissue proper, called **loose connective tissue,** supports most epithelial tissue and many organs and also surrounds blood vessels and nerves. Think of loose connective tissue as packing material. For example, in addition to forming around blood vessels, it binds the muscle cells of skeletal muscle and forms a soft internal skeleton supporting the cells that make up such organs as the liver and spleen.

Some loose connective tissue is specialized for fat storage. This is called **adipose tissue** (ad′i-pos) and functions as an energy reserve, padding for some body organs, and insulation against heat loss. Lipid molecules accumulate in a single vacuole in each adipocyte, or fat cell. As the vacuole increases in volume, it pushes the cytoplasm and nucleus to the cell periphery. Under the microscope a fat droplet dominates the cell's appearance.

The second major type of connective tissue is specialized connective tissue. The three subtypes of **specialized connective tissue** are cartilage, bone, and blood. **Cartilage** (kar′ti-lij) is made of cells called **chondrocytes** (*chrondros* = cartilage), which lie in small chambers called **lacunae** (la-ku′ne, the latin word for pit). They are separated by an extracellular matrix containing both collagen and elastic fibers, along with water. Cartilage is both tough and flexible, providing a slightly resilient rigidity to any structure it supports. One type of cartilage, called **hyaline** (hi′a-lin) **cartilage,** contains large amounts of collagen that gives it a very smooth, shiny

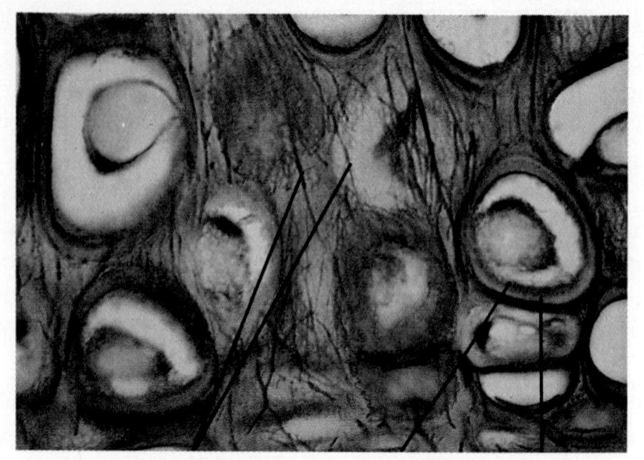

appearance. The word *hyaline* is derived from the Greek word meaning glass. Hyaline cartilage is the most common cartilage type, found in the nose, at the ends of many bones and ribs, and in the supporting rings of major airway passages of the lungs. It is also found in the walls of the major airway passage connecting the oral cavity to the lungs. If you

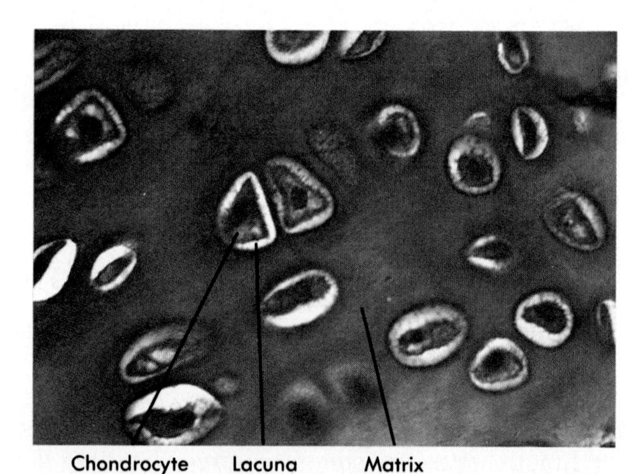

tilt your head back slightly and gently run a finger down the front of your neck, you will feel the rings of hyaline cartilage that help protect and keep open the airway passage that connects your mouth and nose to your lungs.

Another type of cartilage, known as **fibrocartilage,** is found in areas of the body subjected to body weight, such as the pads between the vertebrae of the spinal column and the knee joint.

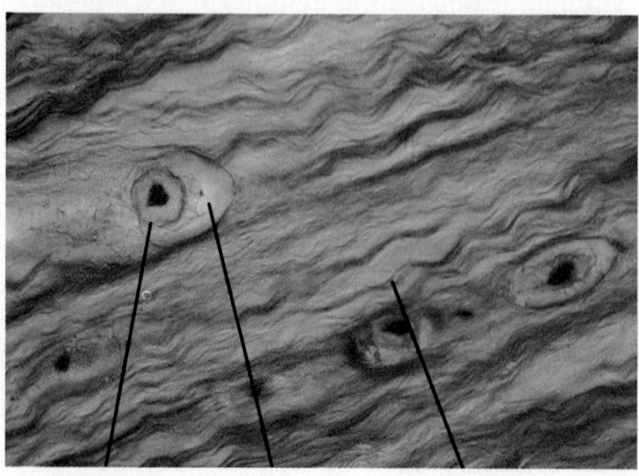

Biology in Action RHEUMATOID ARTHRITIS

In your grandparent's generation doctors diagnosed most indefinite aches and pains of the joints as rheumatism, a word derived from Greek meaning a flux, or periodic episode of pain. Today the most common connective tissue disease associated with joint pain is rheumatoid arthritis.

Rheumatoid arthritis is an inflammation of the connective tissue membrane that covers a joint. The longer the tissue remains inflamed, the greater the damage to other parts of the joint, including the bones. Rheumatoid arthritis first affects the smaller joints in the hands and feet and usually spreads to the larger joints, such as the ankles, wrists, elbows, knees, and shoulders. Its incidence is higher among women than men and frequently afflicts people between the ages of 40 and 60 years of age. Approximately 80% of the U.S. population has some rheumatoid arthritis pain sometime in their lives.

Although we do not know the cause of rheumatoid arthritis, we do know it is an autoimmune disease (*auto* means self). It occurs when the body's immune system, designed to recognize and destroy foreign invaders, begins to malfunction and attacks the body's own tissues. Recent research indicates that rheumatoid arthritis may occur in individuals when a virus or some other environmental agent enters joints and elicits an immune response not only against the foreign virus but also against the joint tissue.

Whatever the cause turns out to be, rheumatoid arthritis has plagued human populations for a long time. A human fossil found in southern France in the early part of the twentieth century and dated to 80,000 years ago showed signs of rheumatoid arthritis in both knees.

Bone is the most rigid of the specialized connective tissues. It consists of living cells and an extremely hard extracellular matrix containing both organic and inorganic substances. The organic portion is primarily made of collagen fibers. The inorganic portion consists of modified calcium, which gives bone its stiffness and rigidity. Bone cells called **osteocytes** (os′-te-o-sitz) are located in lacunae, in the matrix.

The two types of bone are called **spongy bone** and **compact bone. Spongy bone** contains a latticework of bars separated by irregular spaces. The latticework of little bars, called **trabeculae** (tra-bek′ule), gives this bone its spongelike appearance. In contrast, **compact bone** is dense and highly organized. The osteocytes, located in lacunae, are more numerous than in spongy bone and are regularly arranged in concentric rings around canals containing blood vessels and nerves. These nutrient canals allow bone to grow, maintain, and remodel itself (Figure 7-6).

Blood is unique among connective tissues because the extracellular matrix is liquid. The liquid matrix of blood, called **plasma,** allows it to flow rapidly through the body, carrying nutrients, oxygen, and waste products. Plasma is also unique in that most of it is produced by cells contained in other body tissues.

Figure 7-6 Bone

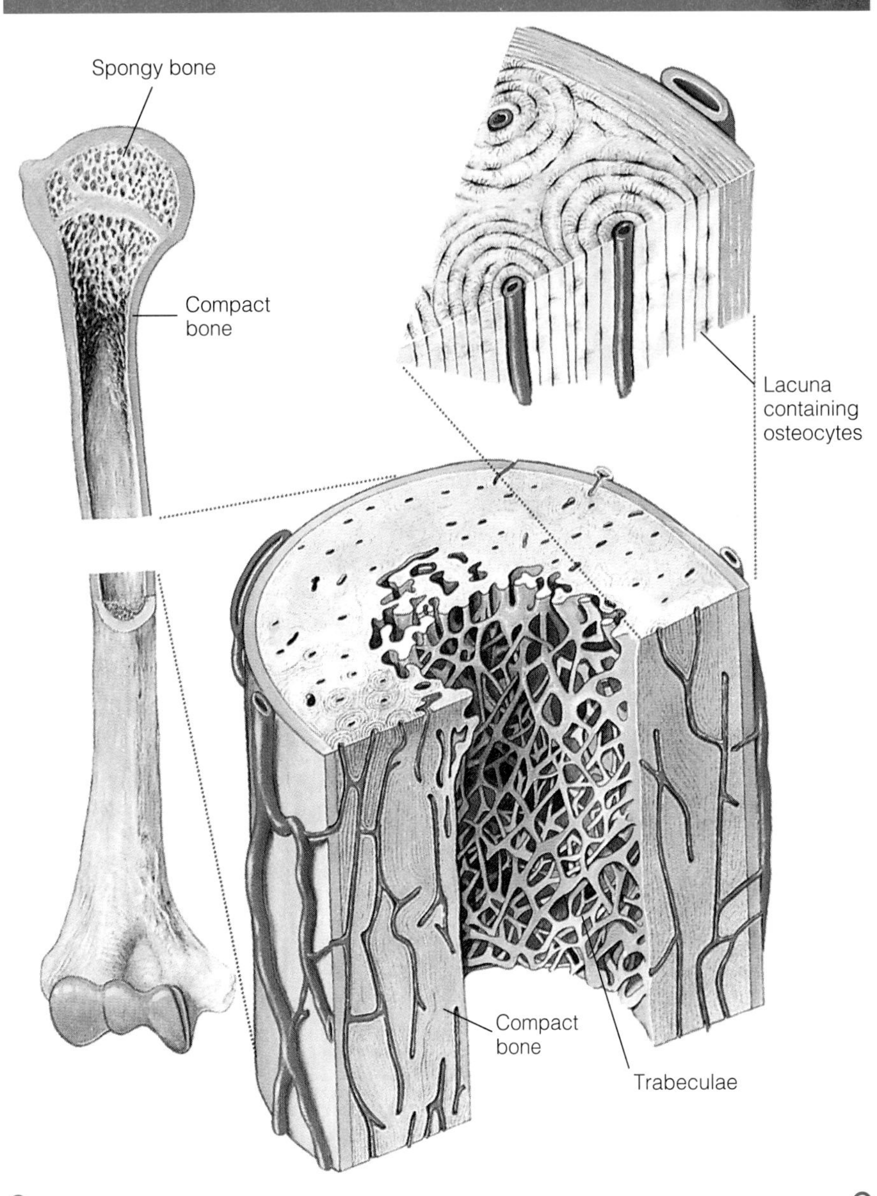

Spongy bone is made of a latticework of trabeculae produced by osteocytes and separated by irregular spaces. Compact bone is dense and contains many more osteocytes than does spongy bone. The osteocytes are arranged in concentric rings around canals containing blood vessels and nerves.

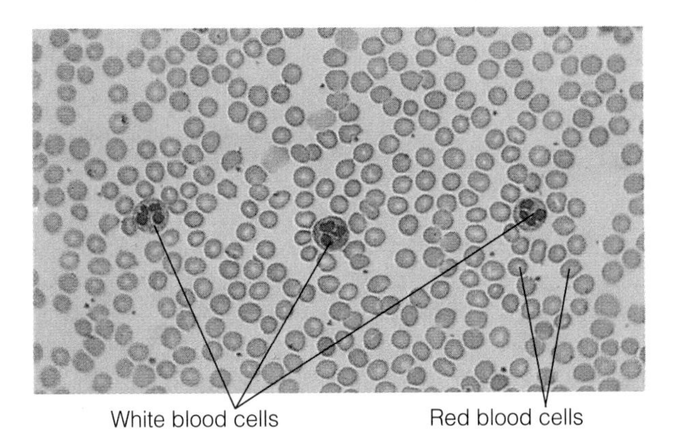

Muscle Tissue

The third primary tissue type is **muscle tissue.** The main characteristic of **muscle tissue** is that it contracts and is therefore responsible for movement. Muscle contraction occurs because of the interaction between two proteins, called **actin** and **myosin,** found in each muscle cell. Muscle is in nearly every organ of the body. When stimulated, muscle cells contract, causing movement or a change in the shape of some body part. For example, muscles contract to move the body, to pump blood from the

Figure 7-7 Muscle types

A

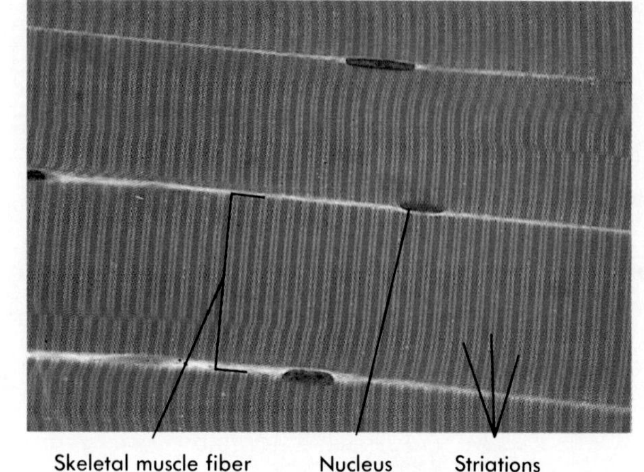

B

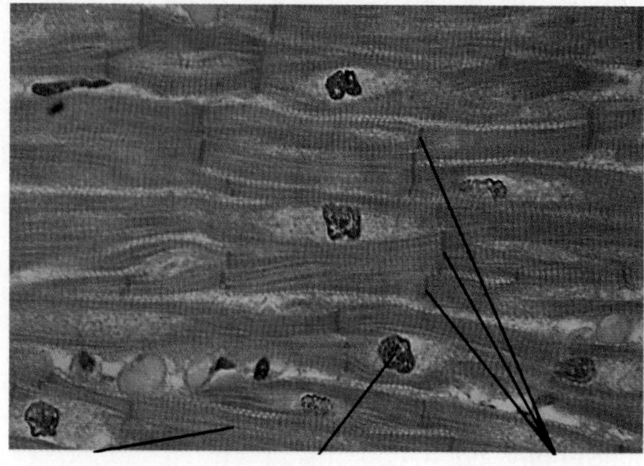

C

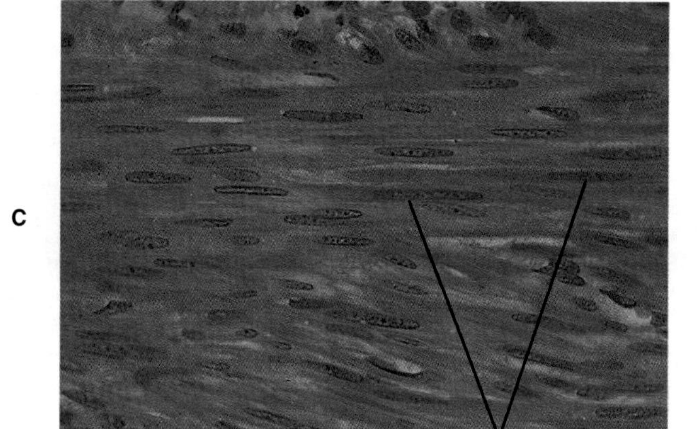

Skeletal muscle is also described as striated because of the presence of microscopic bands (A). It is also referred to as voluntary, meaning they are consciously controlled. Cardiac muscle is also striated but is referred to as involuntary because it is not normally under conscious control (B). Smooth muscle is nonstriated and involuntary (C).

heart through blood vessels, and to decrease the size of hollow organs such as the stomach. What many people think of as muscle, such as the biceps, is really a group of many individual muscle cells held together with connective tissue. Muscles are also responsible for more intricate movements, such as eye movement, smiling, or handwriting.

The three subtypes of muscle tissue are skeletal, cardiac, and smooth muscle. Skeletal muscle is what is normally thought of as muscle and represents much of the body's total weight. **Skeletal muscle** attaches to the bones of the skeleton and by contracting causes major body movements (Figure 7-7, *A*). **Cardiac muscle** makes up the walls of the heart and by contracting is responsible for pumping blood (Figure 7-7, *B*). **Smooth muscle** is in the walls of hollow organs, such as the digestive system and blood vessels. Smooth muscle is widespread and is responsible for many organ functions (Figure 7-7, *C*).

Nervous Tissue

The final primary type of tissue is **nervous tissue,** which is characterized by the ability to conduct electrical signals called action potentials. The tissue consists of **neurons,** which are responsible for producing and conducting action potentials, and support cells called **neuroglia** (nu-rog'li-ah).

Neurons are composed of three major parts: the cell body, dendrite, and axon. The **cell body** contains the nucleus and is the site of cell functions.

Dendrites and axons are projections of cytoplasm surrounded by cell membrane. **Dendrites** usually receive electrical impulses and conduct them toward the cell body. The **axon** usually conducts impulses away from the cell body. Each neuron has only one axon but may have several dendrites. Neuroglia are the support cells of the nervous tissue, which nourish, protect, and insulate the neurons.

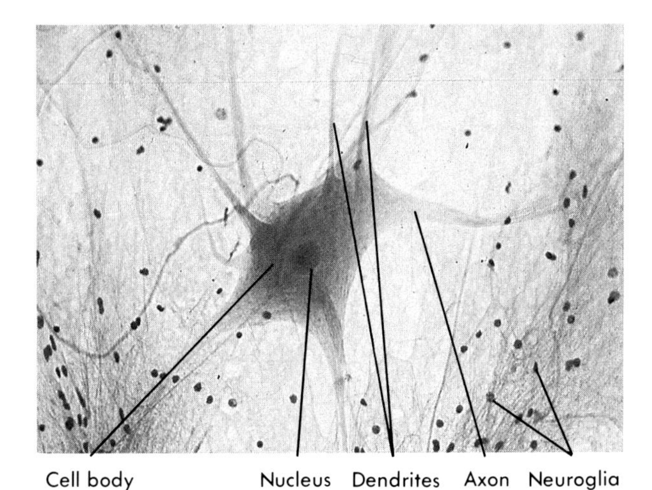

Membranes

A **membrane** is a thin sheet or layers of tissue that cover a structure or lines a cavity. Most of the body's membranes are formed from epithelium and the connective tissue to which it is attached. The two major types of membranes are serous and mucous membranes.

Serous (ser'us) **membranes** consist of epithelium and its basement membrane, which is attached to a thin layer of loose connective tissue. Serous membranes line cavities and do not contain glands. The membrane produces a small amount of fluid that helps to protect the internal organs from friction and to hold them in place.

Mucous membranes consist of epithelial cells and their basement membrane, which rests on a thick layer of loose connective tissue. Mucous membranes line cavities that open to the outside of the body, such as the digestive, respiratory, and reproductive passages. They contain goblet cells that secrete mucus to protect the membrane and trap debris.

Describing Body Structure

When describing the body, it is essential to know where one structure is in relation to another. For example, you know a home may have several

Biology in Action: WHEN YOU CUT YOUR FINGER

Epithelial tissue repairs itself by **regeneration** and by a process called **fibrosis.** Which of these two processes occurs depends on the type of epithelial tissue damaged and the severity of the injury. Regeneration is replacement of destroyed tissue by proliferation of the same kind of cells. Fibrosis involves producing fibrous connective tissue, forming scar tissue. However, in most epithelial tissues repair usually involves both processes. For an example, we consider what happens when you accidently cut your finger.

The process of tissue repair begins while the inflammatory reaction is going on (Chapter 14). Briefly stated, the inflammatory reaction is a process whereby the body isolates the damaged area, preventing any foreign bacteria or other infective agents from spreading to surrounding tissues. During the inflammatory reaction the tissue immediately surrounding the damage site begins to swell, the watery portion of the blood, called plasma, "walling off" the damaged area. Blood and proteins leak into the injured area forming a clot, which stops blood loss and holds the edges of the wound together. When the clot dries, it becomes a scab.

The next phase is called **organization,** a process during which the blood clot is eventually replaced by the growth, beginning at the edges of the wound, of a pink tissue composed of several elements. This tissue is called **granulation tissue** and consists of fibroblasts, collagen, and the smallest blood vessels of the circulatory system, the capillaries. Eventually, normal connective tissue replaces the granulation tissue. Sometimes if the wound is large enough, granulation tissue forms and persists as a visible scar. At first the scar may be pink to bright red because of the capillaries present. Eventually, however, the scar turns from red to white as more collagen fibers accumulate and the blood vessels are compressed beneath the skin. Occasionally an abnormal proliferation of connective tissue during skin healing results in a large mass of surface scar tissue. This is known as a **keloid** scar, derived from the Greek word meaning a spot.

rooms. If you are told to take the hallway to the left of the living room and go to the second door on the left, you are able to easily locate the specific room. So it is with the body, and there are six general terms used when describing relative positions.

1. **Superior** and **inferior.** When standing upright, *superior* means toward the head and *inferior* means toward the feet. For example, the heart is superior to the diaphragm, whereas the stomach is inferior to it.
2. **Anterior** and **posterior.** *Anterior* means front or in front of, and *posterior* means back or in back of. Because humans walk in an upright position, *ventral,* meaning toward the belly, can be used in place of *anterior;* and *dorsal,* meaning toward the back, can be used for *posterior.* For example, the eyes are on the body's anterior (ventral) surface, and the shoulder blades are on its posterior (dorsal) surface.
3. **Medial** and **lateral.** *Medial* means toward the midline of the body, and *lateral* means toward the sides of the body, or away from its midline. For example, the heart lies medial to the lungs, and the lungs lie lateral to the heart.
4. **Proximal** and **distal.** *Proximal* means toward or nearest the trunk of the body, and *distal* means away from or farthest from the trunk. For example, the elbow lies at the proximal end of the lower arm, and the hand lies at its distal end.
5. **Superficial** and **deep.** *Superficial* means nearer the surface, and *deep* means farther away from the body surface. For example, the skin of the leg is superficial to the muscles below it, and the bone of the leg is deep to the muscles that surround and cover it.
6. **Supine** and **prone.** *Supine* and *prone* are terms used to describe the body in a horizontal position. In the supine position the body is lying face upward, and in the prone position the body is lying face downward.

Body Cavities

The human body contains two major cavities, the **dorsal body cavity** and the **ventral body cavity.** Most organs are in these two cavities and are attached to them by connective tissues and membranes. The dorsal body cavity is located in the head region and extends down the back of the body as a hollow tube. It is further divided into the **cranial cavity,** which contains the brain, and the **spinal cavity,** which contains the spinal cord and the nerves extending from it (Figure 7-8).

The ventral body cavity is subdivided by a sheet of muscle called the diaphragm, which assists in breathing. The cavity is divided into an upper region called the **thoracic** (tho-ras'ik) **cavity** and a lower region called the **abdominopelvic cavity.** The **thoracic cavity** is subdivided into three compartments separated by serous membranes. The central compartment, called the **pericardial cavity,** contains the heart. The compartments on either side of it, called the **pleural cavities,** each contain a lung. The **abdominal cavity** contains the organs of digestion, excretion, and reproduction.

Figure 7-8 Body cavities

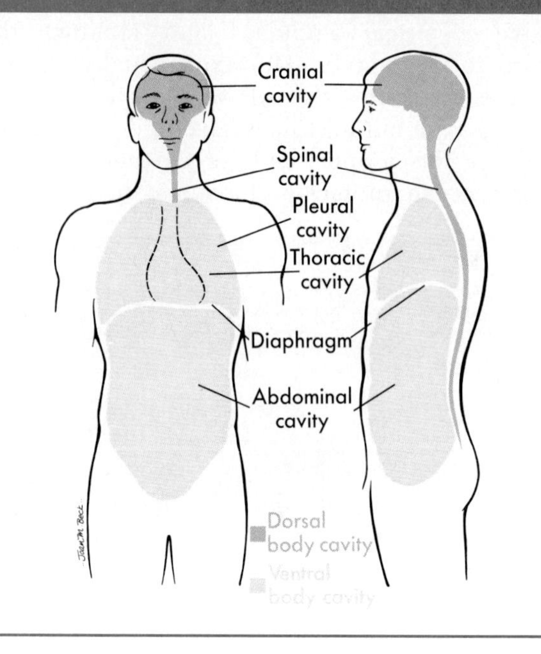

Organ Systems

Much of this textbook describes the major organ systems of the human body. The following is an overview to provide some perspective into how the various systems of the body integrate with one another to maintain life.

The **integumentary system** (in-teg'u-men-tari), consists of the skin and its accessory structures, which include hair, nails, and exocrine glands. It protects the body from the loss of water and other critical materials and from invasion by bacteria and viruses, and it is important in regulating temperature and in sensing changes in the external environment (Figure 7-9).

The **skeletal system** includes the bones, cartilage, and other dense connective tissues. It provides support for the body and protects critical parts. Closely associated with the skeletal system is the **muscular system.** The **muscular system** is formed by the skeletal muscles that attach to the bones of the skeletal system, allowing coordinated body movements (Figure 7-10).

Figure 7-9 Integumentary system

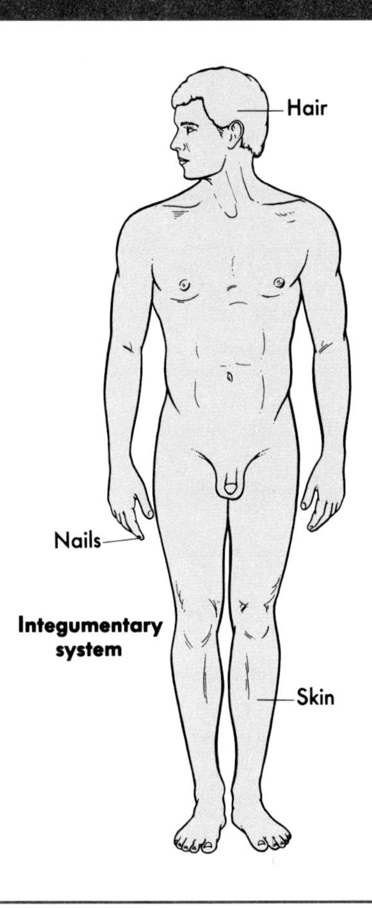

The integumentary system covers and protects the body.

Figure 7-10 Skeletal and muscular systems

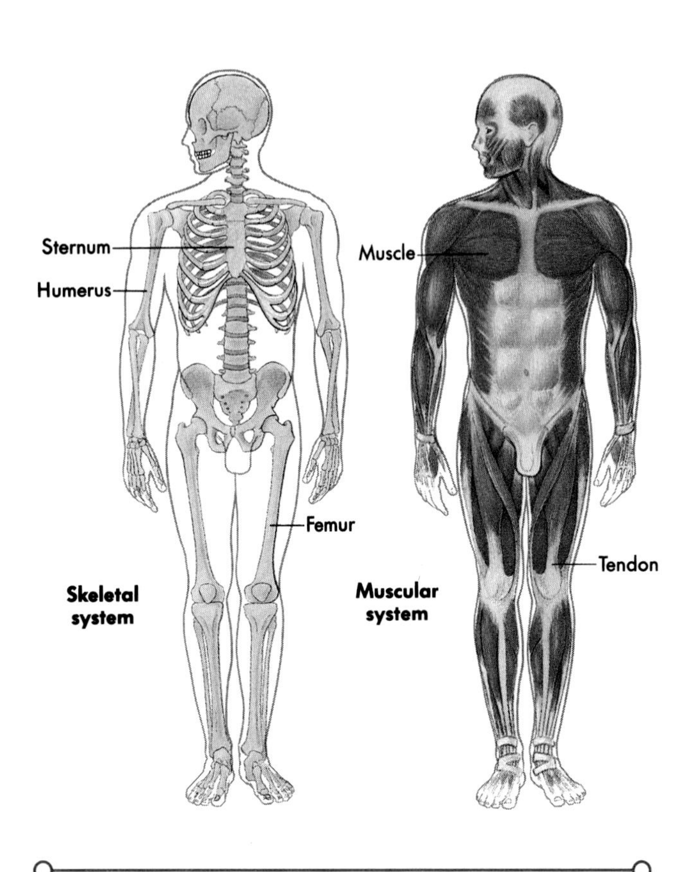

The skeletal and muscular systems provide support and movement.

The primary means by which bodily activities are controlled is through the **nervous system.** The **nervous system** consists of the brain, spinal cord, nerves outside of the spinal cord, and special sense organs.

All the endocrine glands make up the **endocrine system.** The secretions of the many endocrine glands enter the bloodstream to be carried to specific organs or tissues on which they have a particular effect. The endocrine system coordinates the metabolic activities of many organs and is critically important in growth and development (Figure 7-11).

The **circulatory system** consists of blood, the heart, and all vessels that circulate blood. The **lymphatic system** is a network of vessels, ducts, nodes, and organs that helps to protect, maintain, and circulate a watery fluid called lymph in the body. It is also important in defending against foreign invasion and infection and returning some of the extracellular fluid to the blood (Figure 7-12).

The **respiratory system** includes the lungs and the hollow passages through which air reaches them. It is involved in supplying the blood with oxygen and removing carbon dioxide from the blood (Figure 7-13).

The **digestive system** consists of a long muscular tube, beginning at the mouth and ending at the anus. The digestive system also includes many associated glandlike structures, such as the salivary glands, pancreas, liver, and gall bladder, that empty their secretions into the digestive tube. The kidneys, bladder, and tubes involved in water regulation and in evacuating urine constitute the **urinary system** (see Figure 7-13).

The **reproductive system** is involved with producing offspring. In the male the primary reproductive organs are the testes, which produce sperm cells, and the penis. The female reproductive organs include the ovaries, which produce egg cells, the Fallopian tubes, and the uterus. Other external structures, such as the mammary glands, provide nourishment for the dependent offspring after birth (Figure 7-14).

How does the body coordinate its tissues, organs, and organ systems to respond appropriately

Figure 7-11 Nervous and endocrine systems

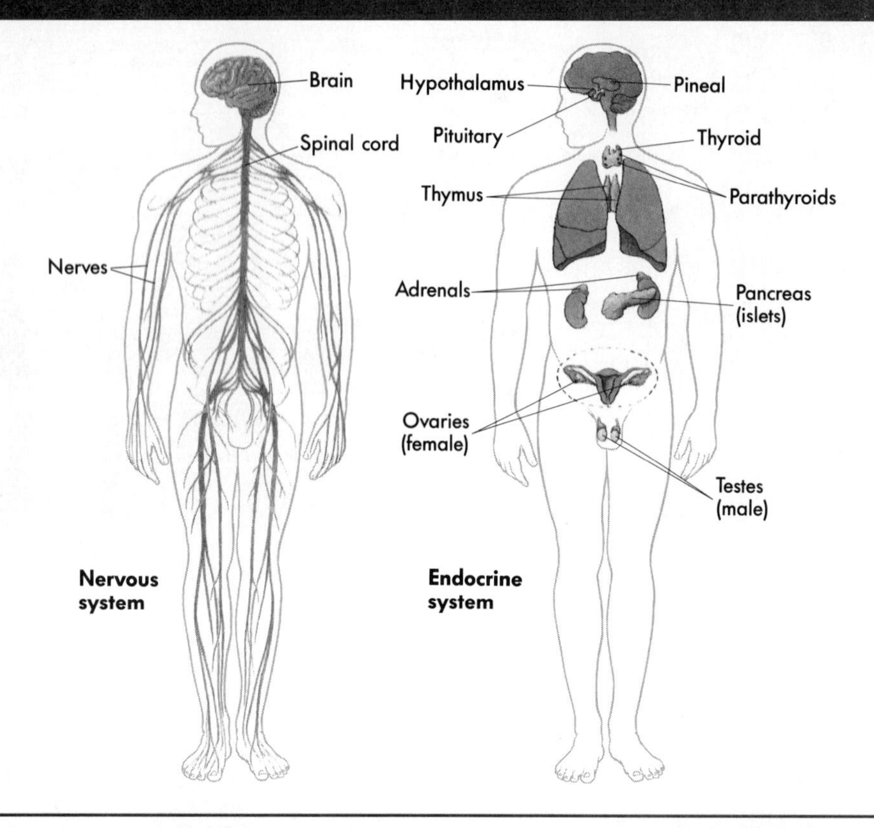

The nervous and endocrine systems provide communication, control, and integration.

Figure 7-12 Circulatory and lymphatic systems

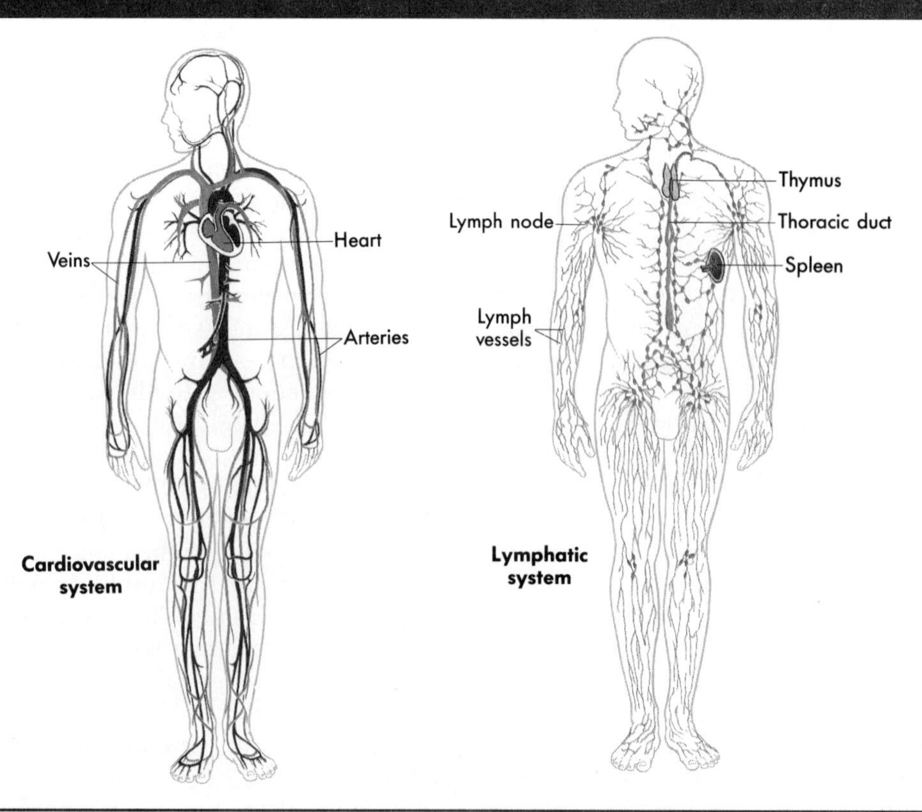

The circulatory and lymphatic systems provide transportation and defense.

to internal and external environmental conditions? The next two chapters describe the endocrine system and the nervous system, two organ systems responsible for controlling and coordinating bodily functions.

Figure 7-13 Respiratory, digestive, and urinary systems

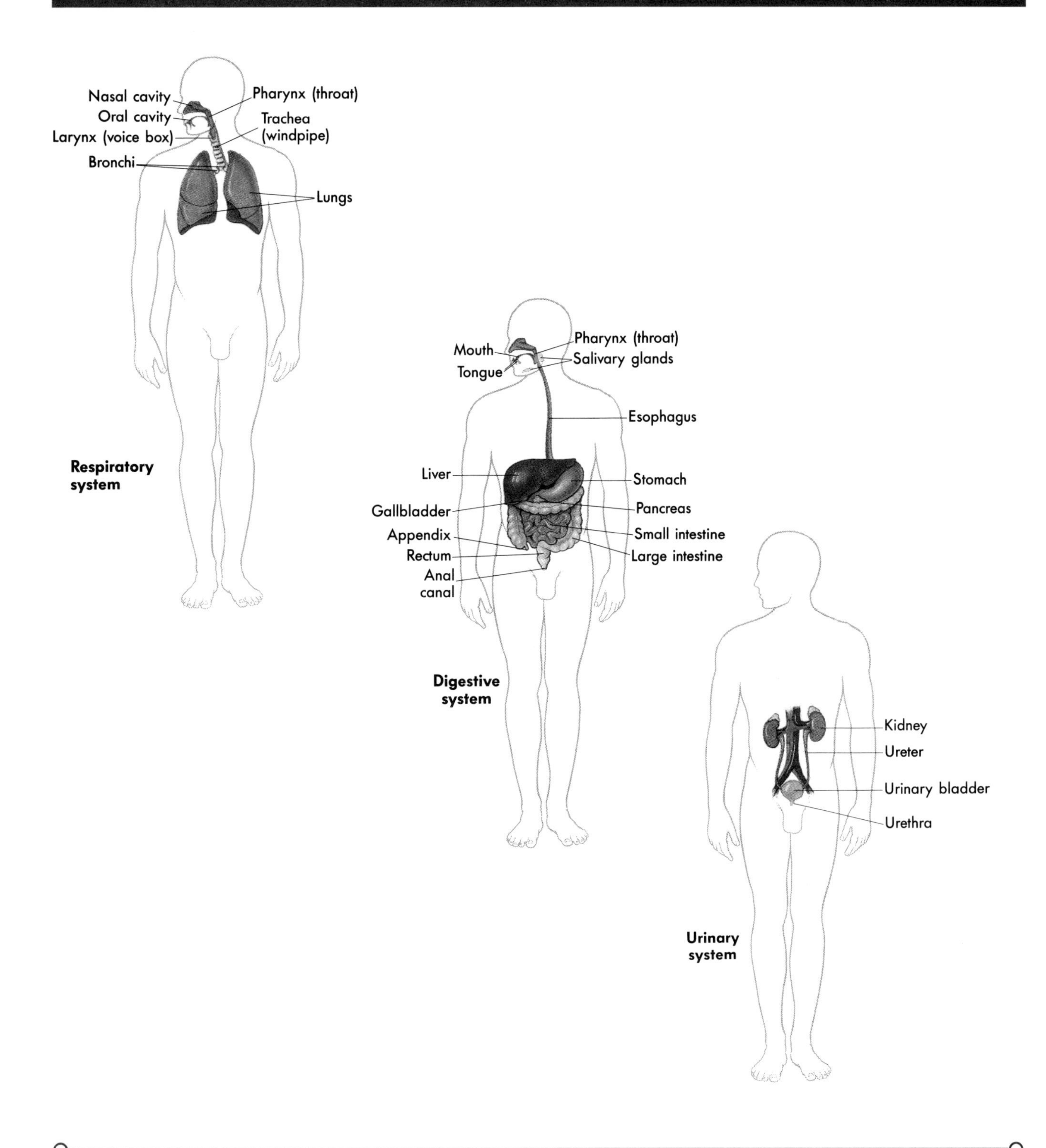

The respiratory, digestive, and urinary systems are involved with processing, regulating, and maintaining the body.

Figure 7-14 Male and female reproductive systems

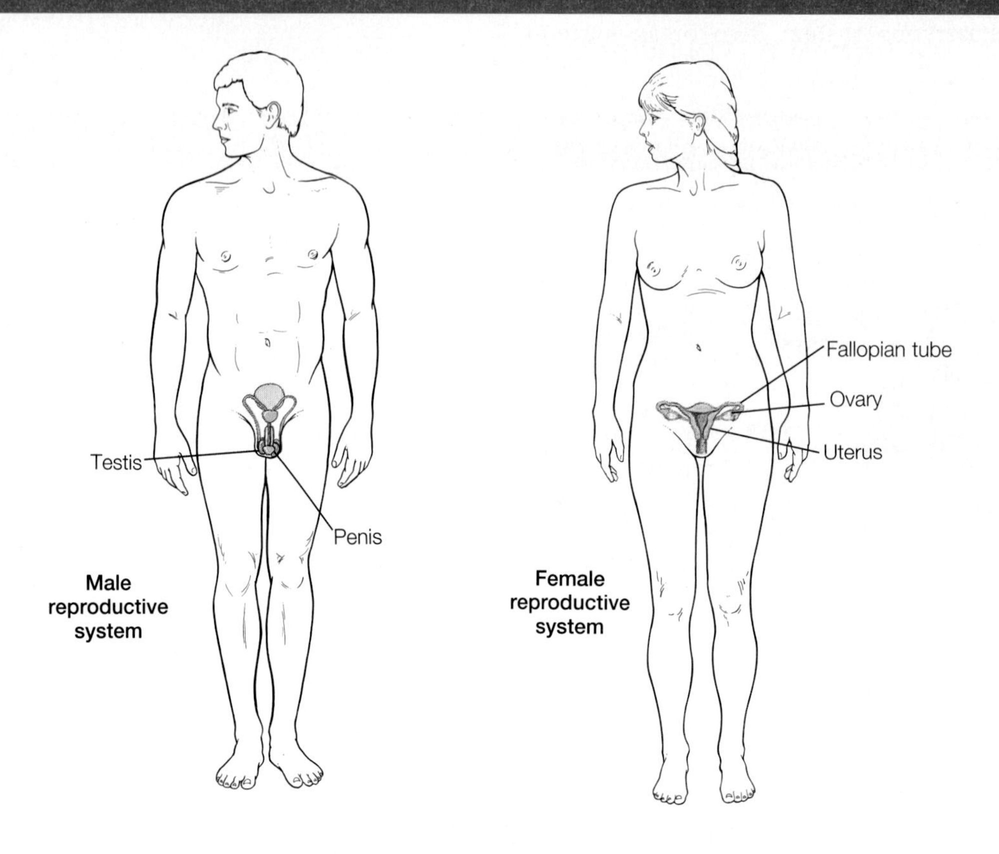

The male and female reproductive systems are involved with the reproduction and development of offspring.

CHAPTER REVIEW

SUMMARY

1. In humans cells function on two interrelated levels: as individual cells and as integrated parts of the body that they compose. This is accomplished in large part through cellular specialization and organization.
2. All body cells are surrounded by a watery fluid called interstitial fluid. This fluid serves as the intermediary in the exchange of materials between individual cells and ultimately the body's external environment. Capillaries bring nutrients and oxygen to the interstitial fluid as well as removing the accumulated cellular waste products from the fluid.
3. All parts of the body work together to maintain a stable internal environment required to sustain each living cell. Cells are organized into larger systems and act to maintain a constant internal environment to ensure their survival.
4. The ability for self-regulating control of the internal environment is called homeostasis. All homeostatic mechanisms are similarly organized and possess a control center, a receptor, and an effector.
5. Most systems of the body maintain homeostasis through negative feedback, a process that reduces or eliminates the stimulus that set the system in motion. Through the process of negative feedback, the internal environment can be maintained within an acceptable range of variation under both internally and externally changing conditions.
6. The body is organized into groups of cells separated by extracellular substances with similar structures and functions called tissues. Two or more tissue types that perform one or more common functions are referred to as organs. An organ system is a group of organs classified as a unit because of a common function or set of functions.

FILL-IN-THE-BLANK QUESTIONS

1. Each cell of the human body is surrounded by a watery fluid called ______ ______.
2. Each homeostatic control mechanism has some type of sensor called a ______ that monitors the body's environment.
3. Negative feedback ______ the stimulus so that the control mechanism is shut off.
4. ______ tissue consists of cells that have very little extracellular matrix between them.
5. ______ ______ seal off spaces between adjacent cells of the epithelium and prevent leakage across it.
6. The extracellular matrix of connective tissues is composed of two types of protein fibers, ______ and ______, and fluid.
7. Dense regular connective tissue makes up ______ and ligaments.
8. An organ system is a group of ______ classified as a unit because of a common function.

SHORT-ANSWER QUESTIONS

1. What are the primary differences between exocrine and endocrine glands?
2. Compare and contrast the extracellular matrix of cartilage and bone.
3. What characteristics make blood unique among connective tissues?
4. Describe the basic components of a homeostatic control mechanism.
5. What is the importance of interstitial fluid to the cells of the body?

VOCABULARY REVIEW

cartilage—*p. 129*
connective tissue—*p. 129*
epithelial tissue—*p. 125*
interstitial fluid—*p. 123*
mucous membrane—*p. 133*
muscle tissue—*p. 131*
negative feedback—*p. 124*
nervous tissue—*p. 132*
serous membrane—*p. 133*

Chapter Eight

The Nervous System

For Review

Here are some important terms and concepts that you will encounter in this chapter. If you are not familiar with them, you should review them before proceeding.

Active transport
(page 43)

Cell membrane
(page 37)

Ions
(page 18)

Tight junctions
(page 128)

Objectives

After reading this chapter you should be able to:

1. *List the three classes of neurons and describe the function of each.*
2. *Describe the structure of a nerve.*
3. *Define membrane potential and explain a resting membrane potential.*
4. *Explain how an action potential is generated and describe the two phases of an action potential.*
5. *Describe the path of transmission of an action potential.*
6. *Describe the two ways in which neurotransmitters modify the postsynaptic membrane.*
7. *Describe the organization of the nervous system and differentiate between its two functional groups.*
8. *Name the two divisions of the autonomic nervous system and describe the differences between them.*
9. *Describe the function of the blood-brain barrier.*
10. *Describe the function of the brain stem and list its three regions.*
11. *Describe the functions of the cerebellum.*
12. *List and describe the most important structures of the limbic system.*
13. *Describe the structure and function of the cerebral cortex.*
14. *Name and describe the function of each of the cortical areas of the cerebral cortex.*
15. *Define memory and differentiate between short- and long-term memory.*

Whether you are asleep, reading these words, or about to say hello to a friend, your nervous system is directing and coordinating diverse parts of your body to respond quickly and precisely to the momentary changes occurring both inside your body and in the environment. The nervous system processes a tremendous amount and variety of information and consists entirely of nerve cells capable of conducting simple electrical impulses along with some supporting cells. The association and patterning of these electrical impulses enables you both to read these words and to think about how your body allows you to read these words.

The Neuron

Your nervous system directs and coordinates many different parts of your body to respond rapidly and precisely to information from both inside and outside your body. The activities of each part of your body are continually monitored and evaluated to regulate their individual performance and their contribution to maintaining homeostasis. Homeostasis is maintained largely by the nervous system's regulatory and coordinating activities, which depend on its ability to detect, interpret, and respond to internal and external changes. Sensory receptors monitor a wide array of internal body activities, such as blood pressure, body fluid pH, temperature, and external stimuli such as light, sound, and smell. This information is transmitted from the different sensory receptors to integration centers in the brain and the spinal cord as electrical impulses. Using this information, the brain and the spinal cord assess the conditions inside and outside the body. The information may produce a response, stored for later use or ignored.

This chapter considers the structure and function of the specialized cell of nervous tissue, the neuron; the nerve impulse; how neurons communicate with one another; the centers of integration and interpretation of information entering and leaving the nervous system; and its overall organization.

The human body contains many different kinds of neurons (Figure 8-1). Some of them are very small with a few projections, others are bushy with many projections, and still others have extensions as much as 1 m long. Despite specific differences in the number of cell projections a neuron may have, they all share a common basic structure. Each neuron has three major parts, which include a **cell body**, containing the nucleus, the site of cellular functions, and two types of cytoplasmic projections surrounded by the cell membrane. The first type of

Figure 8-1 Generalized neurons

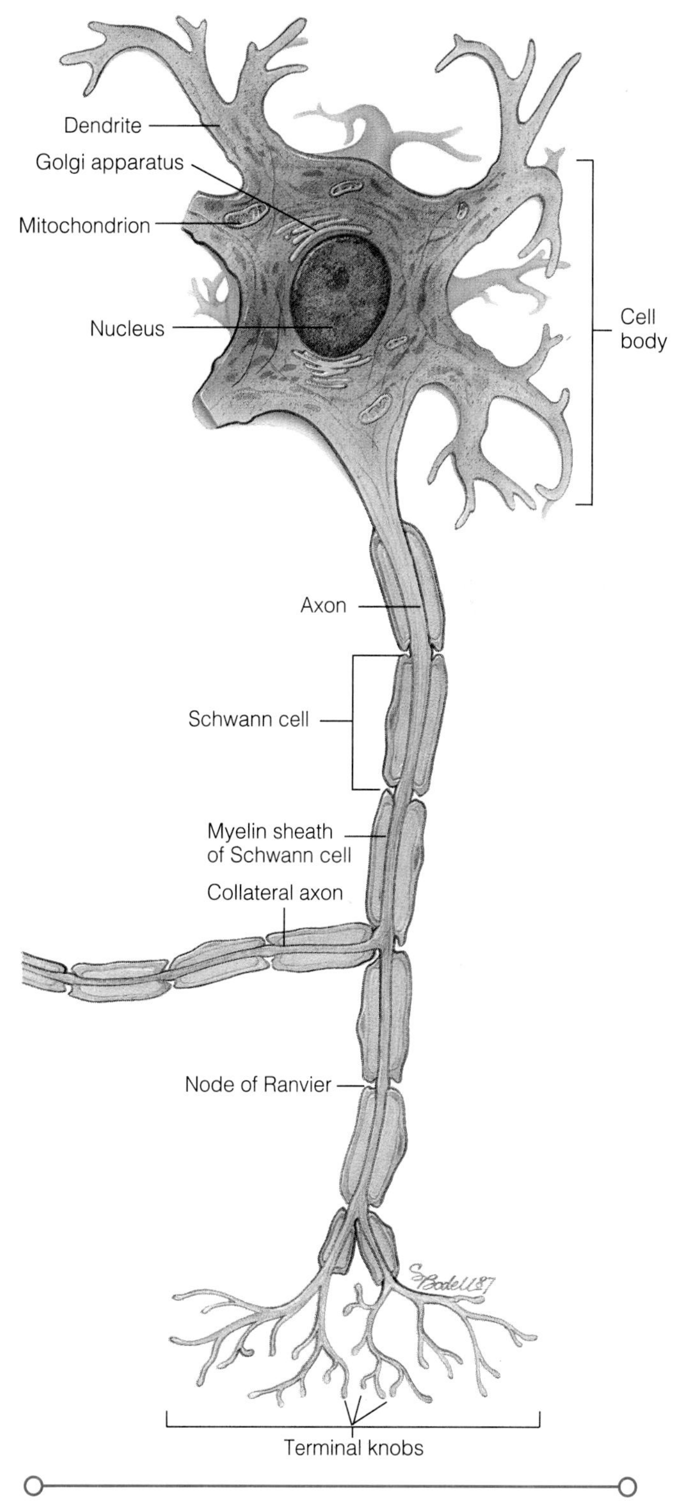

A neuron consists of three parts: two types of membrane-bound cellular extensions, called dendrites and the axon, and a cell body containing the nucleus and other organelles essential for cellular functions. Neurons may have many dendrites but have only one axon, which may occasionally have sidebranches called collaterals.

Figure 8-2 Types of neurons

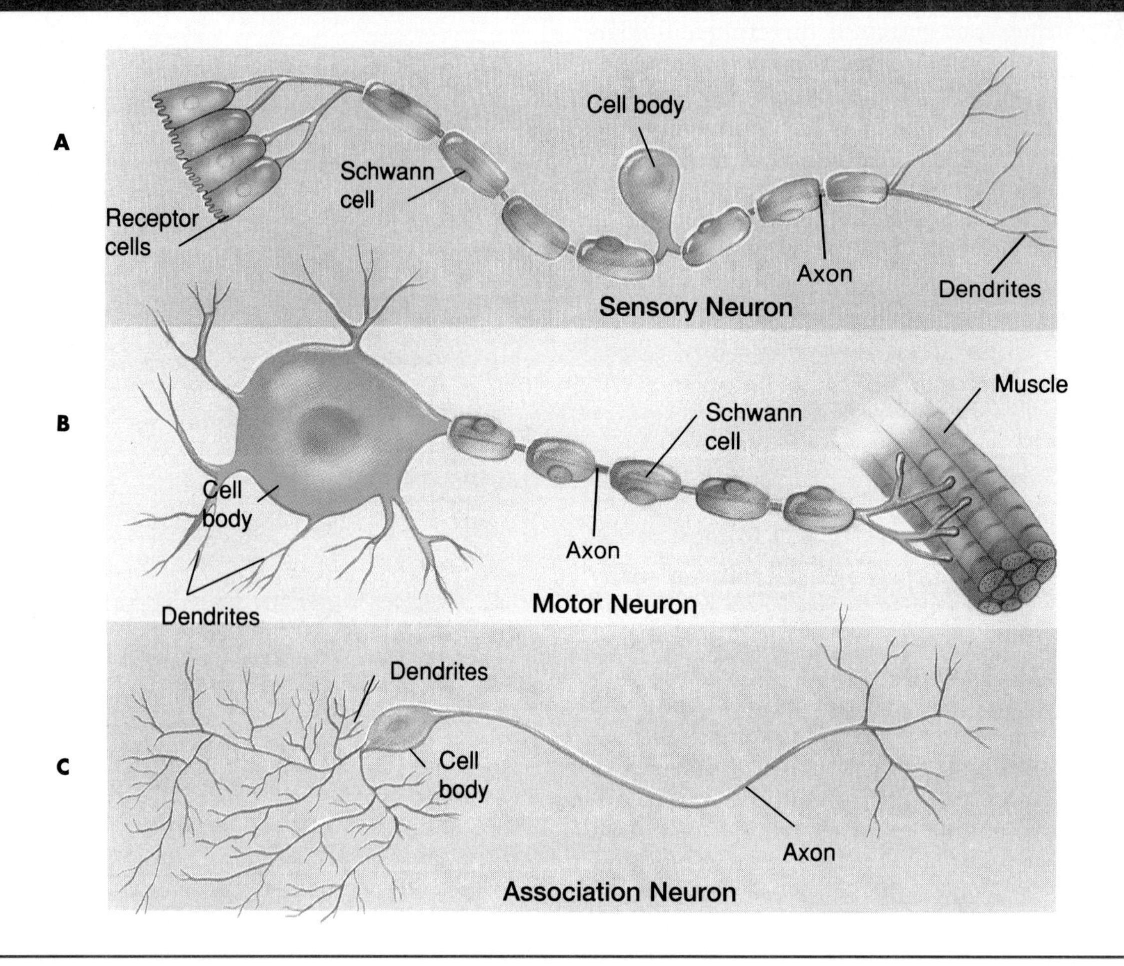

There are three classes of neurons found in the human nervous system. Sensory (afferent) neurons are receptors for specific stimuli from either the internal or external environment (A). Motor (efferent) neurons relay nerve impulses away from integrating centers to muscle or gland cells (B). The third class of neurons are association neurons, which are found mainly in the human nervous system's areas of integration such as the brain and spinal cord (C). Association neurons combine the information arriving from sensory neurons and then influence other neurons.

projection is the **dendrite,** which receives electrical impulses and conducts them toward the cell body. The second type is an **axon,** which carries the electrical impulse away from the cell body. Although each neuron has only one axon, it sometimes gives off side branches, called **collaterals.** At the end of an axon is a small swelling called a **terminal knob,** which allows a neuron to chemically communicate with other neurons. In contrast to its single axon, a neuron may have as many as 1000 dendrites (Figure 8-1).

Three classes of neurons are in the human nervous system. **Sensory neurons** are the body's receptors for specific stimuli from either the internal or external environment. They are also called **afferent neurons.** The term *afferent* is derived from the Latin word meaning to bring to. For example, sensory neurons in your eyes respond to light energy, sensory neurons in your skin respond to pressure or temperature change, and sensory neurons in your circulatory system respond to changes in carbon dioxide levels in the blood. Each sensory neuron sends information toward the spinal column and the brain. Most sensory neurons have long dendrites and a short axon.

The second type of neuron, the **motor neuron,** relays nerve impulses away from integrating centers to muscle or gland cells. Motor neurons are also called **efferent neurons.** The term *efferent* is derived from the Latin word meaning to bring out or away from. Motor neurons usually have short dendrites and a long axon. **Association neurons** are the third type of neuron found in the human nervous system, and are mainly in the body's areas of integration, such as the brain and the spinal cord. **Association neurons** combine the information arriving from sensory neurons and then influence other neurons (Figure 8-2).

Most neurons in the human body cannot survive alone for long. They require the nutritional support and protection provided by companion **neuroglial cells.** *Neuroglia* literally means nerve glue. More than one half of the nervous tissue in the human nervous system consist of neuroglial cells. Outside the brain and spinal column a modified neuroglial cell, called a Schwann cell, plays a passive but important role in rapidly transmitting electrical impulses through a neuron.

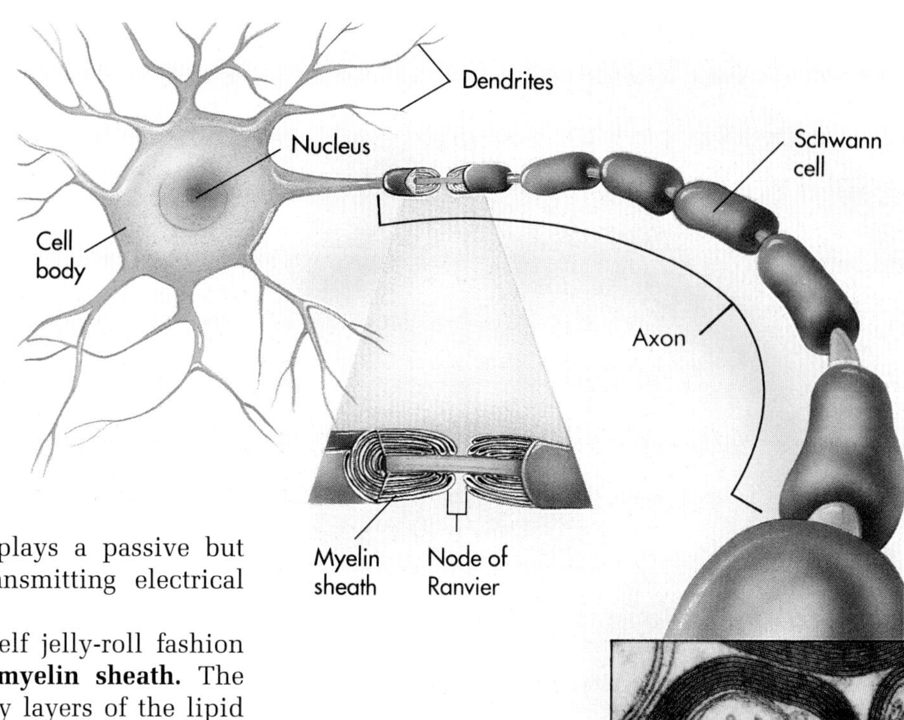

A **Schwann cell** wraps itself jelly-roll fashion around the axon, forming a **myelin sheath.** The myelin sheath consists of many layers of the lipid bilayer that composes the Schwann cell's cell membrane. This myelin sheath insulates the axon. There are gaps between each Schwann cell called **nodes of Ranvier** (ron′ve-a), where the axon is in direct contact with the surrounding extracellular fluid. (The nodes are named after a nineteenth-century French physiologist who first described their structure and function in 1871.) An axon and its associated Schwann cells form a **myelinated fiber.** Bundles of nonmyelinated and myelinated fibers or bundles of axons are called **nerves** (Figure 8-3).

Figure 8-3 A nerve

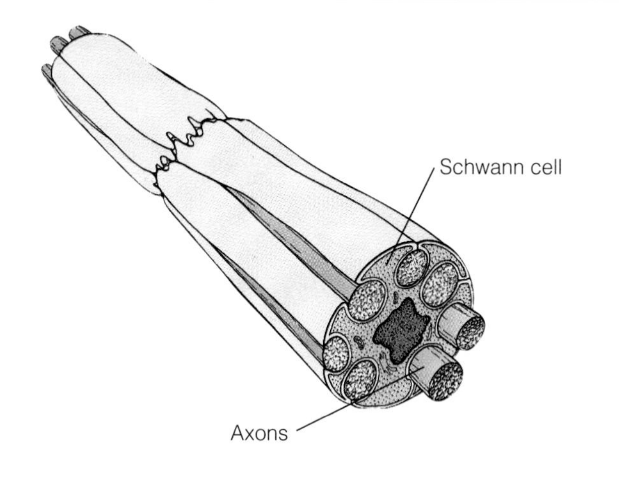

A nerve may consist of either bundles of both myelinated and nonmylinated fibers or bundles of axons.

Membrane Potential

How do neurons produce and conduct electrical impulses? To answer this question we need to remember some basic characteristics shared by all human body cells. The cell membrane separates two chemically distinct areas, the cytoplasm and the surrounding extracellular fluid. Cells have a slight excess of positively charged substances on the outside and a correspondingly slight excess of negatively charged substances on the inside. This occurs because many proteins in the cytoplasm possess a negative charge and are kept in by the cell membrane. There is also a greater concentration of Na^+ ions outside the cell. In combination these differentially distributed substances make the inside of a cell more negative than the outside. The separation of oppositely charged substances across a cell membrane is called a **membrane potential** (Figure 8-4).

A neuron also has a membrane potential. However, neurons have a unique property regarding their membrane potentials. These cells are able to undergo rapid reversals in their membrane potentials. These reversals are the electrical impulses transmitted by the neurons. We will consider this property more closely.

The membrane of a neuron is able to distinguish between sodium (Na^+) and potassium (K^+) ions and control the relative permeability of each. Active transport proteins, called **sodium-potassium**

Figure 8-4 Membrane potential

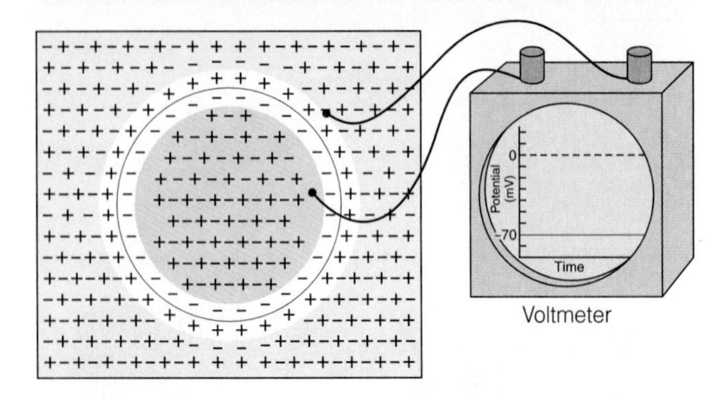

The separation of oppositely charged substances across a cell membrane is called a membrane potential. The word *potential* refers to the potential energy available to do work. The voltmeter measures the amount of energy between the two oppositely charged areas available to do work.

pumps, in the neuron membrane *actively* transport ions across the membrane. These pumps require energy, in the form of ATP, because they are moving ions against their concentration gradients.

The sodium-potassium pumps move an average of three sodium ions out for every two potassium ions brought into the cell. Some potassium ions are passively diffusing out of the cell, but the sodium-potassium pump brings more potassium ions in than diffuse back out. The membrane is not, however, as permeable to sodium ions. Once sodium ions are pumped out of the cell they cannot easily move back in. This net difference in positive ions along with the negatively charged proteins permanently located inside the neuron results in a net electrical difference across the cell membrane. The interior is negatively charged compared with the surrounding extracellular fluid, which is positively charged. The attraction between these separated charged substances will cause them to accumulate in a thin layer along the cell membrane's outer and inner surfaces (Figure 8-5). These separated charges represent only a small fraction of the total charges in the extracellular fluid and the neuron's cytoplasm. The rest of the positive and negative charges in the extracellular fluid and the cytoplasm cancel one another out and result in an electrically balanced condition. Only a few charged ions immediately surrounding the cell membrane are responsible for the membrane potential. In addition to sodium-potassium pumps, there are also separate gated sodium-ion channels and gated potassium-ion channels in the neuron's membrane. *Gated* means the channels can be opened to let ions pass through the membrane or closed to prevent ion passage.

Resting Membrane Potential

A neuron is said to be at rest when its cell membrane is not responding to a stimulus. During **resting potential** opposite charges separate across the neuron's cell membrane, and the neuron is said to be **polarized.** The sodium-potassium pump is actively transporting three Na^+ ions out of the cell for every two K^+ ions it admits to the cell. The gated ion channels are also closed. This differential distribution of positive ions along with negatively charged proteins kept inside of the cell results in a resting potential of –70 mV. A millivolt is $\frac{1}{1000}$ of a volt, which is a measure of energy between two oppositely charged areas available to do work.

Action Potential

An **action potential** is a brief reversal of membrane potential brought about by rapid changes in cell

Figure 8-5 Sodium-potassium pump

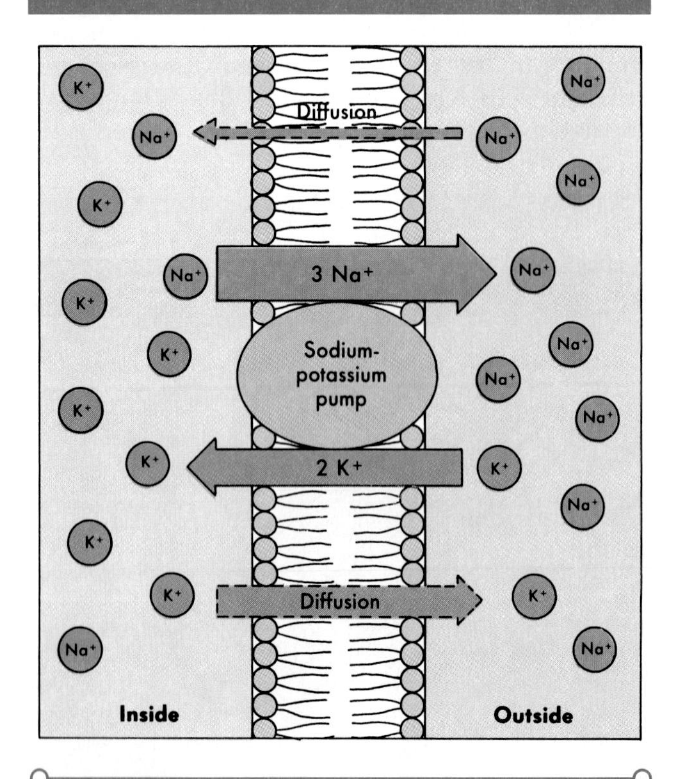

This is a transmembrane protein that uses energy from ATP to pump sodium ions out of the neuron and potassium ions into the cell. On average, for every three sodium ions pumped out, there are two potassium ions pumped into the neuron.

Figure 8-6 The action potential

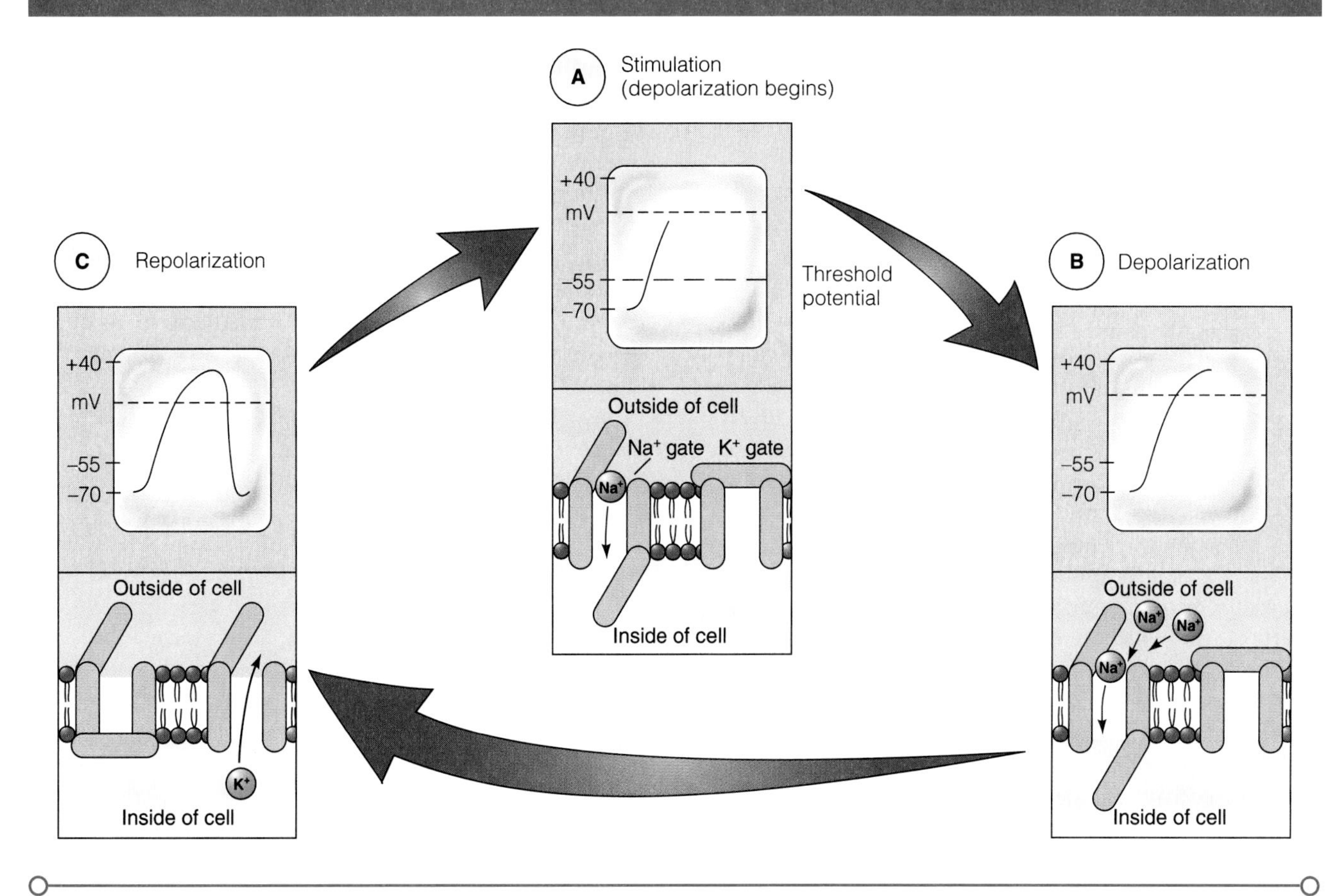

An action potential is a brief reversal of membrane potential brought about by rapid changes in cell membrane permeability. Stimulation of the neuron begins the phase called depolarization (A). Once depolarization reaches threshold potential, an action potential is generated as the sodium ion channels open and sodium ions rush down its concentrations gradient into the neuron. As soon as the maximum interior charge is reached, about + 40 mV, the sodium ion channels close (B). During repolarization, sodium ion channels close and potassium ion channels open, allowing potassium ions to rush out of the neuron. Along with the sodium-potassium pump, the membrane potential is restored to its original resting potential (C).

membrane permeability. An action potential occurs when the neuron is stimulated, for example, by pressure, temperature, or chemical activity. The two sequential phases in an action potential involve changes in membrane permeability occurring during each phase.

Stimulation of the neuron begins the phase called **depolarization.** When stimulated, gated sodium-ion channels begin opening, allowing Na^+ ions to move down their concentration gradient into the neuron. Depolarization begins slowly at first until it reaches about –55 mV, a point referred to as **threshold potential.** Any stimulus that brings a neuron to threshold potential will generate an action potential. A neuron membrane either responds to an action potential or it does not. This is called the **all-or-none response** (Figure 8-6, *A*).

At threshold potential, rapid depolarization takes place because all the gated sodium-ion channels are opened, allowing Na^+ ions to rush in and eliminate the internal negativity and actually make the inside of the cell slightly more *positive* compared with the outside. The interior of the axon is now about +40 mV compared with the outside. As soon as this maximum interior positive charge is reached, the sodium-ion channels close and the potassium-ion channels open. The membrane immediately becomes impermeable to Na^+ ions and permeable to K^+ ions, which rush out of the cell down their concentration gradient. Along with the continuous action of the sodium-potassium pumps and the outward movement of K^+ ions, the interior of the neuron becomes negative (Figure 8-6, *B*).

The final phase, called **repolarization,** restores the membrane potential to its resting potential. The sodium-potassium pump maintains the original distribution of ions across the membrane and is responsible for maintaining the resting potential (Figure 8-6, *C*).

Transmission of the Action Potential

An action potential occurs at the stimulation point on the neuron, typically at a dendrite or a specialized part of the cell body, called the axon hillock. The inward rush of Na^+ ions will spread to the region immediately next to it, which still maintains a resting potential of –70 mV. The spread of Na^+ ions causes the membrane of this adjacent region also to become permeable to Na^+ ions. Rushing across the membrane into the neuron, Na^+ ions produce an action potential in the adjacent region. The result is that each action potential acts as a stimulus for generating another action potential at the adjacent region of the neuron membrane. Once an action potential is initiated in one part of the membrane, a self-perpetuating cycle begins so that the nerve impulse continues throughout the rest of the axon automatically. It is not the original action potential that travels down the length of the axon. Instead it triggers an identical new action potential in the adjacent area of the membrane, the process being repeated down the axon. It is similar to setting dominoes on end in a row. Once the first domino is tipped over, it hits the next, which in turn hits the next one down the line until the last domino falls. The last action potential is identical to the first action potential, regardless of the length of the axon.

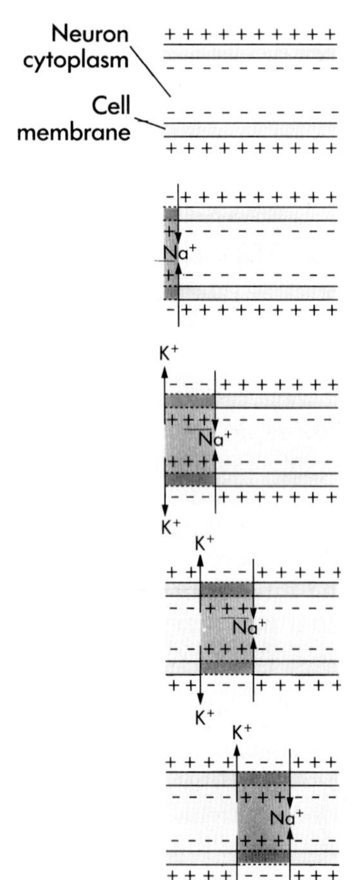

Why does the action potential travel only in one direction? During the time a particular region of the membrane is generating an action potential, it is incapable of initiating another action potential. By the time the adjacent region of the membrane is undergoing an action potential, the first region is becoming repolarized. The time required for the sodium-potassium pumps to restore the ion concentrations in the original region is called the **refractory period.** The original region cannot be depolarized again until the original charge difference is reestablished by the sodium-potassium pumps. By the time the original region is repolarized, the action potential of the adjacent region has caused the region next to it to undergo depolarization. By the time the recovery process has been completed, however, the signal has moved too far away for the original site of depolarization to be influenced by it. This process results in the action potential moving away from the stimulation point in only one direction.

Saltatory Conduction

In nonmyelinated neurons the action potential at one point on the membrane generates the action potential at the very next point on the membrane so that the nerve impulse is propagated along the entire length of the neuron. In myelinated neurons action potential conduction is different. The myelin sheath acts as an insulator, like rubber on an electrical wire, to prevent ions from passing across the cell membrane. However, at the nodes of Ranvier the neuron's membrane is in direct contact with the extracellular fluid. When an action potential occurs at a node, the change in membrane permeability allows an inflow of Na^+ ions, generating an action potential. The nodes are usually about 1 mm apart, a distance short enough that the first action potential influences the exposed membrane of the next node. This causes depolarization to reach threshold potential and generates an action potential. Then depolarization occurs at the next node, and so on. This form of nerve impulse conduction is called **saltatory conduction,** from the Latin word *saltare,* meaning to jump. Saltatory conduction in myelinated neurons propagates action potentials more rapidly than local propagation that occurs in nonmyelinated neurons, approximately 120 meters per second (270 miles per hour) for larger neurons. Nonmyelinated neurons, on the other hand, conduct at about 0.5 meters per second. Saltatory conduction also requires less energy to propagate because there is less membrane depolarization and fewer sodium-potassium pumps needed to restore to its resting potential when only nodes are depolarized, compared with the entire nerve membrane in nonmyelinated neurons.

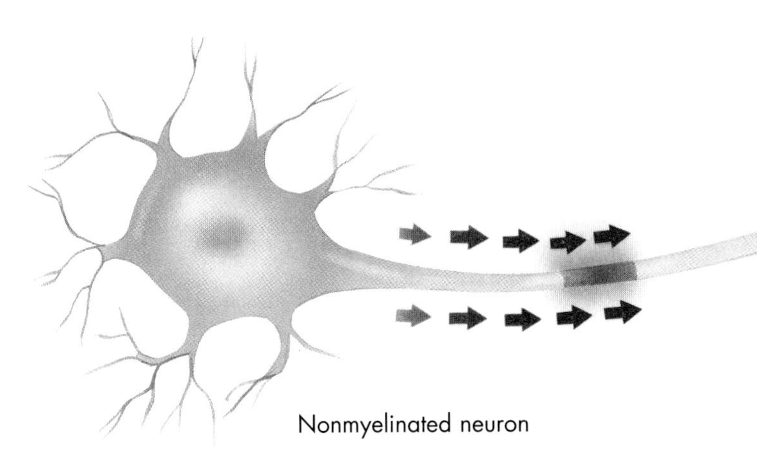

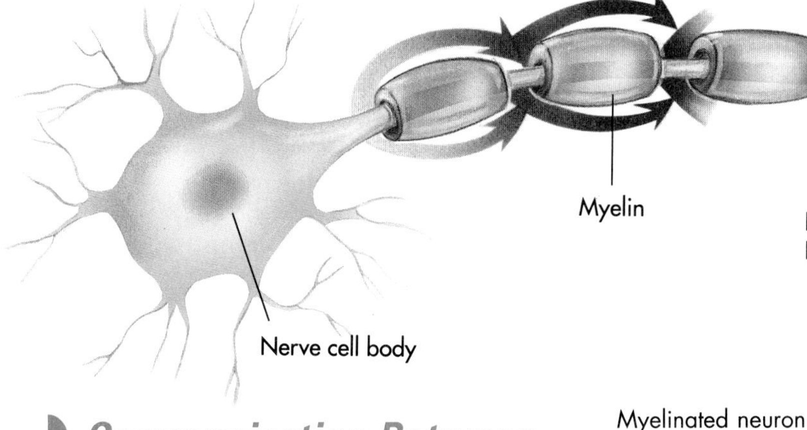

Communication Between Neurons

What happens to the action potential when it reaches the end of an axon? The axon of a neuron may terminate at one of three cell types: another neuron, a muscle cell, or a gland's secretory cell. The region that incorporates the terminal knob of the axon and the membrane of another cell is called a **synapse** (si′-naps). All synapses share two general characteristics. First the terminal knob does not actually make contact with the other cell it approaches. A very small gap separates the end of the axon from the next cell. This gap is called a **synaptic cleft.** Second the means by which an action potential passes from one neuron to the next is different from the propagation of the action potential along the length of a neuron. When an action potential arrives at the synaptic cleft, it is converted to a chemical signal and passes across the gap *chemically* by way of **neurotransmitter substances.** The membrane on the axonal side of the synaptic cleft covering the terminal knob is called the **presynaptic membrane.** When the action potential reaches the terminal knob, it stimulates the release of a neurotransmitter substance into the cleft. The chemical rapidly diffuses to the other side of the gap. Once there it combines with receptor sites located on the membrane surface of the target cell called the **postsynaptic membrane.** The advantage of a chemical junction such as this compared with a direct electrical contact is that the chemical transmitter can be different in different junctions, permitting different kinds of responses.

The events that occur in the synaptic cleft when an action potential arrives depend on the type of neurotransmitter chemical released into the cleft and on the type of receptors of the postsynaptic membrane. To understand this we will discuss nerve-to-nerve conduction. We will discuss nerve-to-muscle conduction and nerve-to-gland conduction in later chapters.

Nerve-to-Nerve Connections

Within the terminal knob are a number of **presynaptic vesicles** containing neurotransmitter substances. When an action potential reaches the terminal knob at the end of the axon, a series of events occur in rapid succession. First the presynaptic membrane becomes permeable to calcium (Ca^{++}) ions. Because Ca^{++} ions are in much higher concentration in the extracellular fluid, they diffuse into the terminal

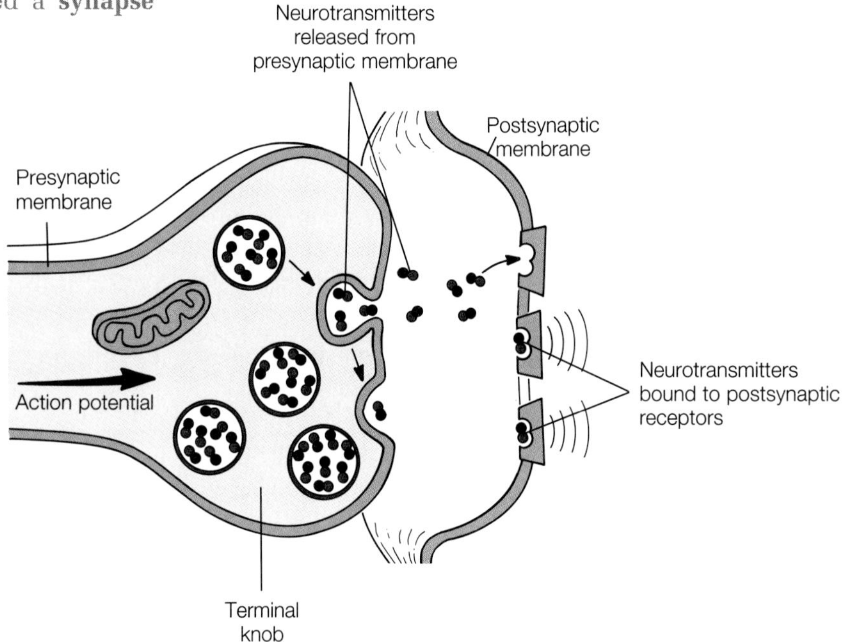

Figure 8-7 Transmission of an action potential between two neurons

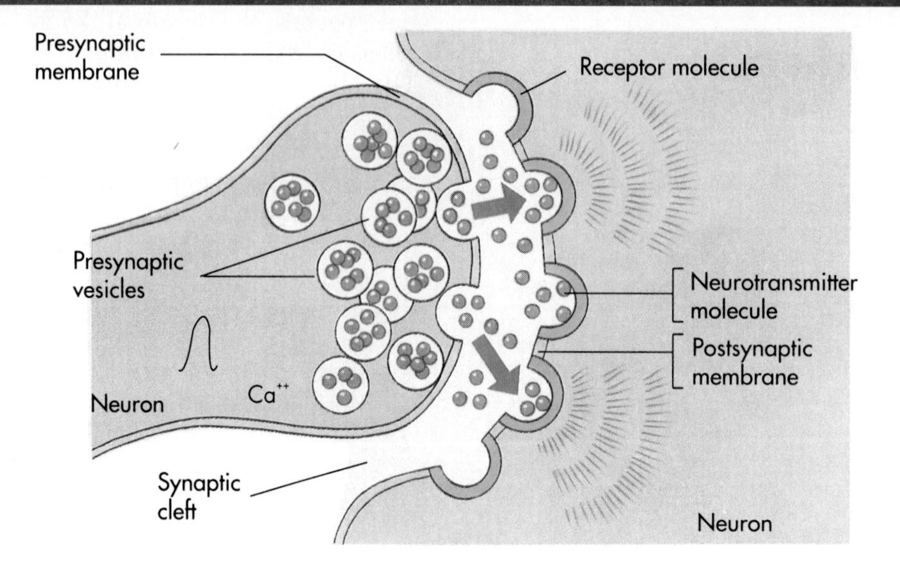

When an action potential reaches the terminal knob of an axon it causes the presynaptic membrane surrounding it to become permeable to calcium ions. Calcium ions move to the inside of the knob causing presynaptic vesicles to move to the presynaptic membrane, fuse with it, and release the neurotransmitters they contain into the synaptic cleft. The neurotransmitter diffuses across the cleft and binds to receptor sites on the postsynaptic membrane, causing an action potential to begin in the neuron.

knob. Once inside, the ions cause the presynaptic vesicles to move to the membrane and, through exocytosis, release their contents into the synaptic cleft. The released neurotransmitter substance diffuses across the cleft and chemically binds with specific protein receptor sites on the postsynaptic membrane. It takes about 0.5 to 1.0 milliseconds for the presynaptic action potential to release the neurotransmitter substance into the synaptic cleft, diffuse across, and bind to the receptor sites of the postsynaptic membrane (Figure 8-7).

Neurotransmitters modify the postsynaptic membrane in two different ways. The first way the postsynaptic membrane is modified is called an **excitatory postsynaptic potential,** where the neurotransmitter produces a small depolarization of the postsynaptic membrane. This brings the postsynaptic membrane closer to threshold potential, increasing the likelihood for an action potential. The second is called an **inhibitory postsynaptic potential,** where the neurotransmitter causes hyperpolarization of the postsynaptic membrane, which means greater negativity inside the neuron. Hyperpolarization moves the membrane potential even farther away from threshold potential, lessening the likelihood the postsynaptic neuron will generate an action potential.

Because a single axon may branch at its end and have many terminal knobs, one neuron may transmit impulses to many other neurons. The neuron may also receive impulses from many other neurons. However, a given synapse is either always excitatory or always inhibitory because a particular presynaptic neuron always releases the same neurotransmitter from its terminal knobs. An individual neuron may have both excitatory and inhibitory synapses from other neurons. When signals from both excitatory and inhibitory synapses reach the neuron's cell body as the input from the dendrites, the various excitatory and inhibitory electrical effects tend to cancel or reinforce each other. If the postsynaptic neuron is brought to threshold potential, it generates an action potential. If threshold potential is not reached, the neuron will not generate an action potential.

Regardless of the particular effect a neurotransmitter has on the postsynaptic membrane, as long as it remains combined with the receptors of the postsynaptic membrane the neuron continues in that state, either excited or inhibited. The separation of the neurotransmitter from postsynaptic membrane receptors is accomplished in various ways, depending on the particular synapse and the type of neurotransmitter involved. The three methods of neurotransmitter termination are enzymatic degradation of the neurotransmitter substance; uptake of the neurotransmitter by the presynaptic cell, either to be reused or degraded; and simple diffusion of the neurotransmitter substance away from the membrane receptors.

Figure 8-8 Transmission of an action potential from a neuron to a skeletal muscle

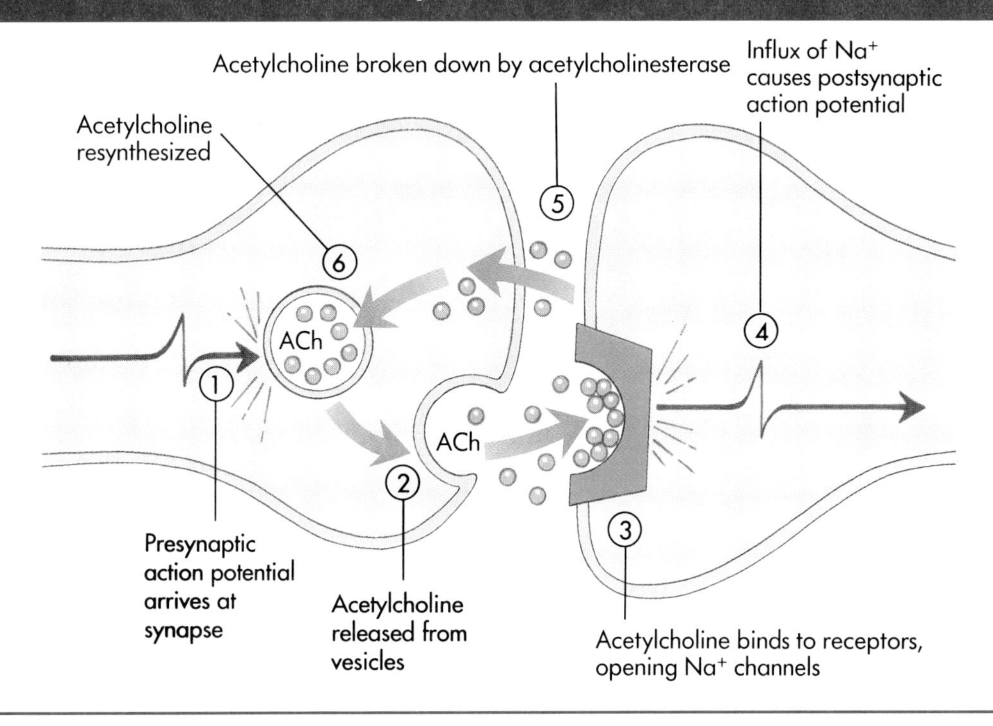

Neurotransmitter Substances

One neurotransmitter commonly released by most neurons is **acetylcholine** (a′-sih-til-ko-len), or **ACh.** Acetylcholine is released by all neurons that stimulate skeletal muscles. After ACh separates from the receptors, it is immediately broken down by the enzyme **acetylcholinesterase** (a′-sih-tel-ko-lih-neh′-ster-as), or **AChE,** which is mainly in the postsynaptic membrane. The breakdown products are taken up by the presynaptic membrane, where acetylcholine is resynthesized to be used again (Figure 8-8).

Circuit Organization

Action potentials spread away from the cell body. The direction in which an action potential flows through the nervous system depends on the way neighboring neurons and their cellular processes are oriented relative to one another. With synapses as the connections, neurons form circuits for transmitting and processing action potentials. These circuits may involve only a few neurons or many thousands.

One of the simplest examples of a neuron circuit is the **reflex arc.** A **reflex** is an automatic, involuntary response to specific stimuli occurring inside or outside of the body. Some reflexes require the brain to process the information, such as blinking the eye. Others, such as withdrawing the hand from a hot object, involve mainly the spinal cord. An example of the simplest type of reflex arc is the knee jerk (Figure 8-9).

The knee-jerk reflex and similar stretch-contraction reflexes are the simplest type because they involve only the sensory (afferent) neuron and motor (efferent) neuron. Most reflexes, however, involve circuits that include many association neurons located in the spinal column. For example, if you step on a piece of glass with your left foot, the left leg flexes and the opposite leg extends. Other reflex arcs are also activated to help to maintain body balance by bringing about specific movement of the arms and trunk. Reflex behavior in response to specific stimuli is of great value to us because it occurs automatically and rapidly. If the brain were involved in processing the information, these reactions would be significantly slower.

Most reflex arcs are not fully isolated from conscious or brain control. For example, if you are walking barefoot across hot asphalt pavement, you can do so consciously, without withdrawing your foot from the surface. In this case the reflex arc is overridden by conscious decision making. How does your nervous system perform such remarkable and sometimes mysterious information sorting and processing?

Figure 8-9 Knee jerk reflex

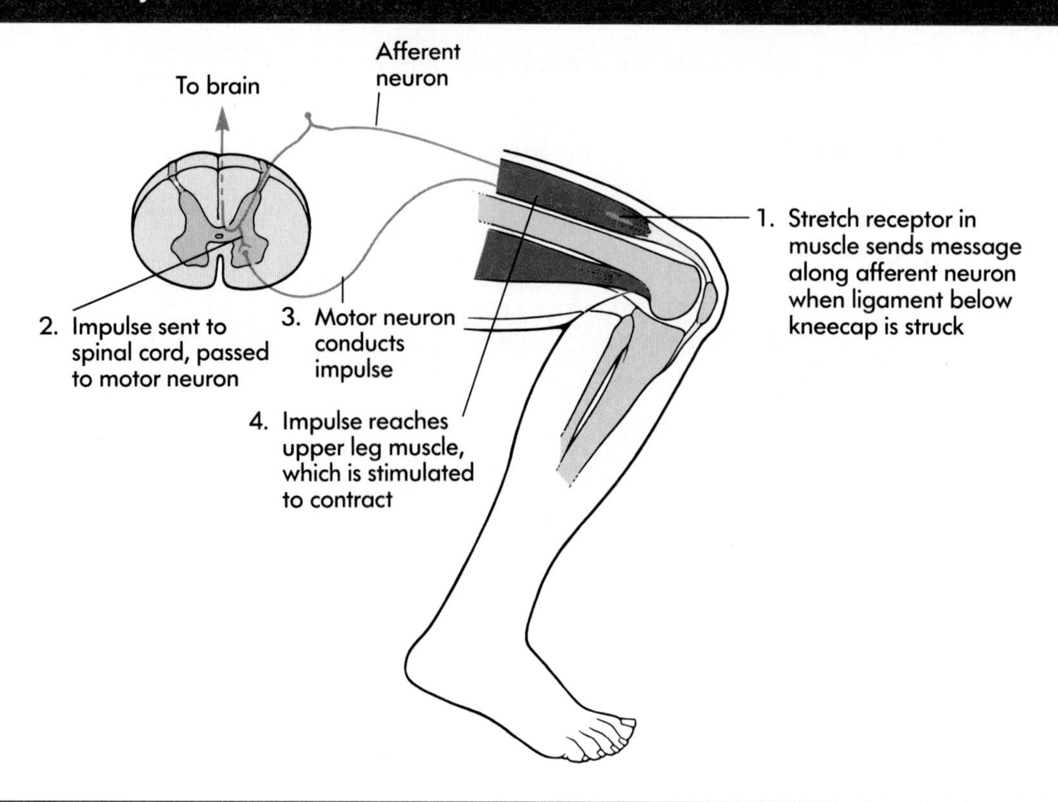

Organization of the Nervous System

The overall organization of the nervous system consists of (1) nerves that bring information into the nervous system, called **sensory input;** (2) a central processing region, or **brain** and **spinal cord,** that interprets sensory input through a process called **integration;** and (3) nerves that transmit commands from the central nervous system to muscles and glands, effecting responses known as **motor output.**

The human nervous system is organized into two functional groups. The **central nervous system (CNS)** is composed of the brain and the spinal cord and is the site of information processing in the nervous system. The **peripheral nervous system (PNS)** includes all the nerves of the body outside the brain and the spinal cord. These nerves are commonly divided into **sensory,** or **afferent pathways,** which transmit information to the CNS, and **motor,** or **efferent pathways,** which transmit commands from it. The PNS is divided into two functionally distinct groups of nerves called the somatic nervous system and the autonomic nervous system (Figure 8-10).

Somatic Nervous System

The voluntary, or **somatic nervous system,** contains afferent pathways that relay sensory information from the skin, skeletal muscles, and special senses, such as the sense organs, to the CNS and efferent pathways carrying action potentials from the CNS to the body's skeletal muscles. The sensory receptors that send impulses through the somatic nervous system are the topic of the next chapter, and the somatic nervous system control of skeletal muscle activity is discussed in Chapter 11.

As you learned previously nerves consist of individual neurons. They may be afferent or efferent neurons, or both, and some of the neurons may be myelinated. The neurons are bundled together like the strands of a telephone cable. Within the CNS these bundles of neurons are called **tracts.** In the PNS they are simply called **nerves.** The cell bodies from which nerves extend are often clustered into groups, called **nuclei** if they are in the CNS and **ganglia** (gang′-gli-ah) if they are in the PNS (Figure 8-11).

Autonomic Nervous System

The second major division of the PNS, called the **autonomic nervous system,** consists of efferent pathways that control the involuntary, or automatic, activities of the smooth and cardiac muscles and glands.

The nerves of the autonomic nervous system fall

Figure 8-10 Organization of the human nervous system

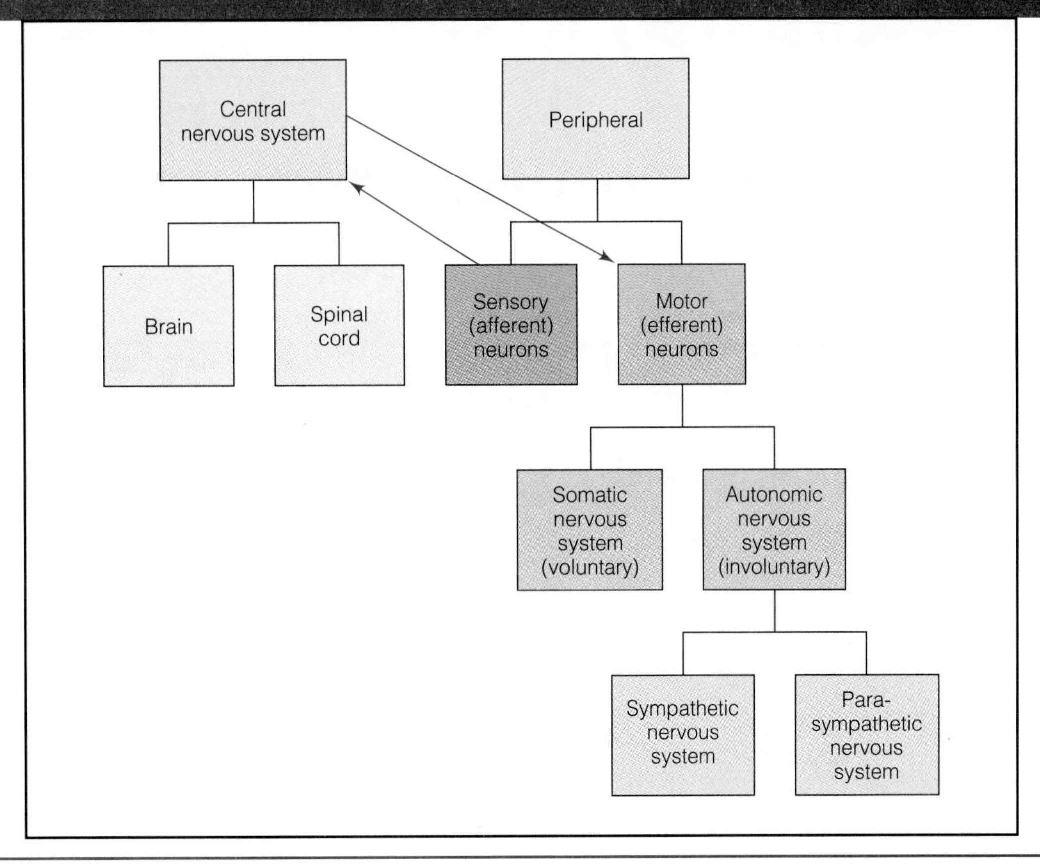

The human nervous system is divided into two major divisions. The central nervous system (CNS) includes the brain and spinal cord. The peripheral nervous system (PNS) includes all the nerves of the body outside the CNS. The peripheral nervous system is further subdivided into sensory pathways and motor pathways. The motor pathways are in turn divided into the somatic nervous system, which includes motor neurons controlling skeletal muscles, and the autonomic nervous system, which includes motor neurons controlling glands and smooth and cardiac muscle activities. The autonomic nervous system is further divided into the sympathetic system and the parasympathetic system. The sympathetic system, when stimulated, increases body activities that prepare the body for intense activity, usually of short duration. Parasympathetic system stimulation maintains normal organ activities of the body, such as digestion and respiration.

into two divisions, called sympathetic and parasympathetic systems. Both divisions are organized in the same way. There are always two efferent neurons extending from the CNS to the organ. The cell body and dendrites of the first of these two neurons are in the CNS. The axon of the first neuron extends from the cell body to a ganglion, where it synapses with the dendrites and cell body of a second neuron. The axon of the second neuron then carries action potentials to a particular organ or tissue. The first neuron of the two is called a **preganglionic neuron,** and its axon is called a **preganglionic fiber.** The second neuron is called a **postganglionic neuron,** and its axon is called a **postganglionic fiber.**

The sympathetic and parasympathetic divisions of the autonomic nervous system differ in where they connect with the CNS, where their ganglia are located, and the effects they have on the tissues they innervate. There is also an important difference in the organization of the preganglionic and postganglionic fibers of the two autonomic nervous system divisions. The preganglionic fibers of the sympathetic system are able to communicate with one another. This particular organization enables the sympathetic system to undergo massive and widespread discharge. The **sympathetic division** promotes responses that prepare the entire body for strenuous physical activity, as during an emergency. The preganglionic fibers of the parasympathetic system do not directly communicate with one another. The **parasympathetic division** individually controls diverse targets. It dominates in quiet and relaxed situations, controlling "housekeeping" activities of the body, such as digestion.

The preganglionic neurons of the sympathetic division arise from the middle portions of the

Figure 8-11 Structure of a nerve

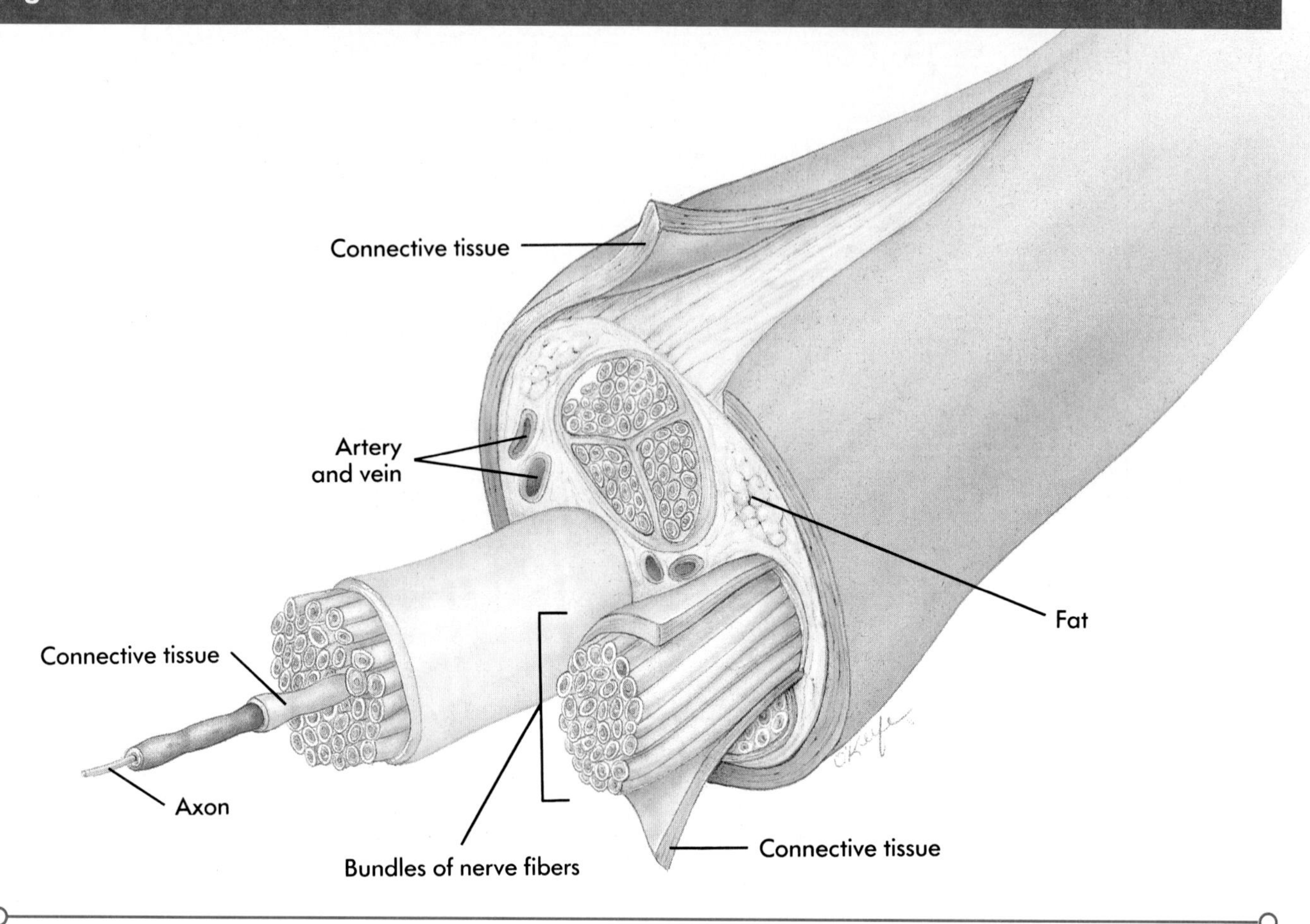

Neurons are bundled together like the strands of a telephone cable into nerves. Within the central nervous system, nerves are called tracts. In the peripheral nervous system they are called nerves.

spinal cord, called the thoracic and lumbar sections (Figure 8-12). After leaving the spinal cord most preganglionic fibers synapse with postganglionic fibers in the **sympathetic chain ganglia,** which are chains of 22 ganglia that lie on either side of the vertebral column. The postganglionic fiber leaves the sympathetic chain ganglia and synapses with a particular organ or tissue. In this division the

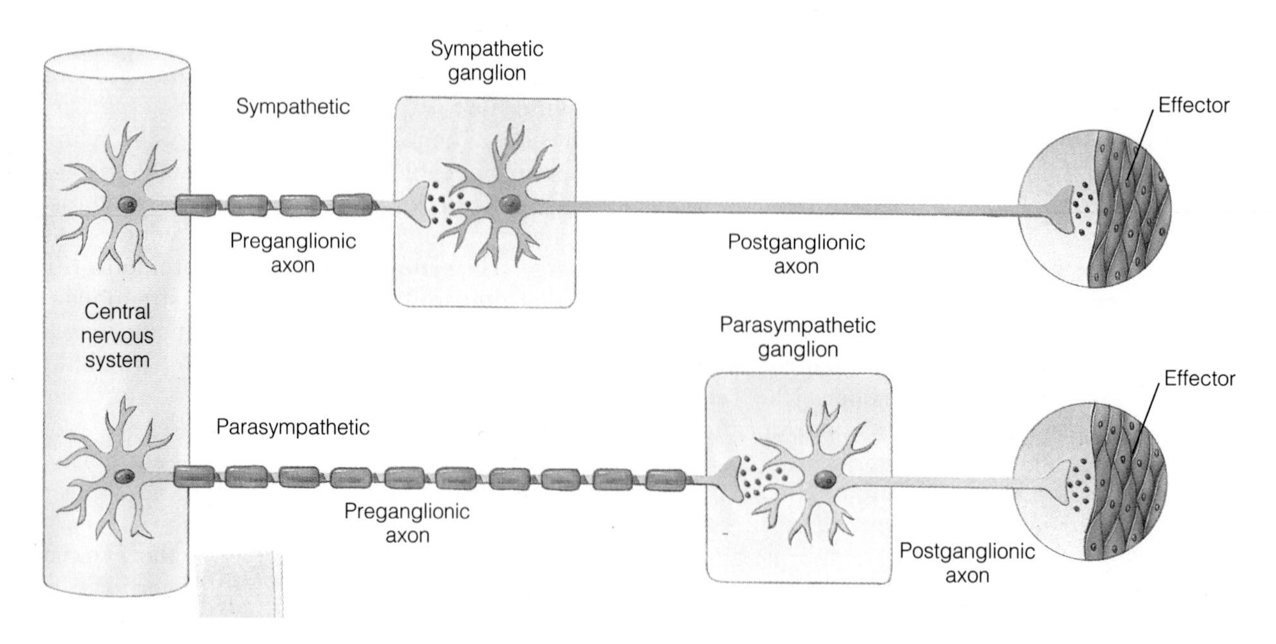

Figure 8-12 Sympathetic and parasympathetic divisions of the autonomic nervous system

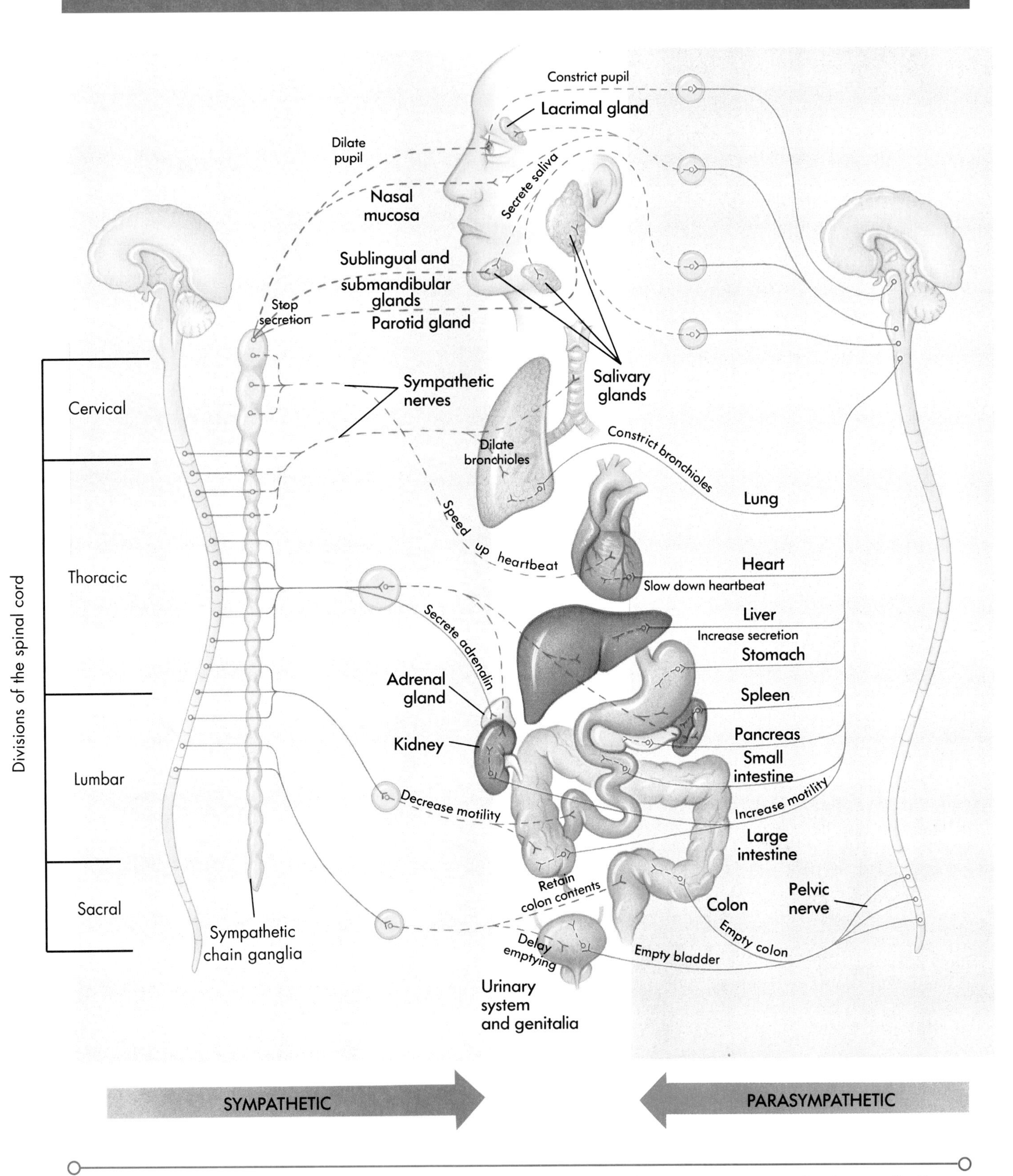

Sympathetic chain ganglia lie near the spinal cord on either side of it. The parasympathetic ganglia lie near or on the organs they affect.

preganglionic fibers are short, but the postganglionic fibers that make contact with the organ or tissue are long.

The cell bodies and dendrites of the parasympathetic division lie in the lower portion of the brain and the lower end of the spinal cord. The efferent pathways from these two parts of the CNS are called cranial nerves and sacral nerves, respectively. In contrast to the ganglia of the sympathetic division, those of the parasympathetic division lie near or within the organs and tissues they innervate. In the parasympathetic division the preganglionic fibers are long, and the postganglionic fibers are short.

Most organs of the body are innervated by both parasympathetic and sympathetic divisions of the autonomic nervous system. In general the effect of stimulating sympathetic fibers to a particular organ is usually antagonistic to the effect of stimulating parasympathetic fibers to the same organ. When your body is not receiving much outside stimulation, as in resting or recovering from some physiological stress, parasympathetic nerve action tends to slow down overall body activity and divert energy to basic maintenance tasks. For example, parasympathetic stimulation reduces the frequency and strength of the heartbeat and decreases the blood pressure and breathing rate. This decreases the supply of oxygen and nutrients going to the cells, a decrease which is appropriate because the body is at rest. Parasympathetic stimulation also increases the process of digestion and urine elimination from the bladder.

During times of elevated stimulation, excitement, or danger, sympathetic stimulation tends to slow down maintenance functions and increase overall body activities that prepare the body for intense activity, usually of short duration. For example, under sympathetic stimulation heart rate and the strength of muscle contraction increase, glycogen is rapidly broken down in the liver, raising blood glucose levels, and breathing rate increases. Because blood is circulating faster, greater amounts of oxygen and glucose are distributed throughout the body. Consequently metabolic output is increased dramatically. In addition sympathetic nerves to the medulla of the adrenal gland (Chapter 10) stimulate secretion of the hormone epinephrine, which increases and sustains all these physiological events.

The Central Nervous System

The central nervous system consists of the spinal cord and the brain. Because nervous tissue is soft and delicate, both the brain and the spinal cord are protected by hard bony material; a durable membranous covering of dura mater, the arachnoid layer, and pia mater, collectively called the **meninges;** and a water cushion of fluid called the **cerebrospinal fluid** (Figure 8-13).

Figure 8-13 Coverings of the brain

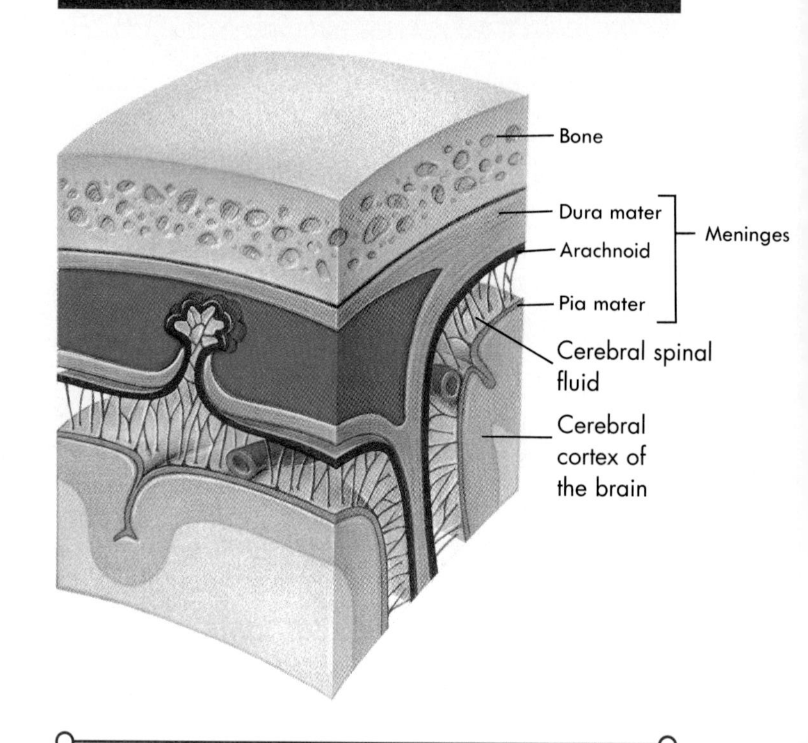

Both the brain and the spinal cord are protected by hard, bony material; a durable membranous covering of dura mater, the arachnoid layer, and pia mater (collectively called the meninges); and a watery cushion of fluid called cerebrospinal fluid.

Blood-Brain Barrier

No other body tissue is so absolutely dependent on a constant internal environment as is the brain. In other body tissues the extracellular concentrations of hormones, amino acids, glucose, and ions continuously undergo small fluctuations, particularly after eating or exercise. Because some hormones and amino acids are neurotransmitters in the brain, and some ions, such as potassium, modify the threshold potential for brain-cell firing, if the brain were exposed to even the slightest chemical variations, uncontrolled neural activity might result.

The **blood-brain barrier** helps ensure that the brain's environment remains stable. The smallest blood vessels of the body are formed by a single layer of cells. Usually pores between the cells making up the blood vessel permit free exchange of many blood components with the surrounding interstitial fluid. In the blood vessels of the brain, however, the cells are joined by tight junctions that completely seal the blood vessel wall so nothing can be exchanged across the wall by passing between the cells. The only possible exchanges are through the blood vessel cells themselves. Water molecules and lipid-soluble substances, such as oxygen, carbon dioxide, alcohol, anesthetics, and steroid hormones, penetrate these cells easily by diffusing through the lipid cell membrane. All other substances exchanged between the blood and the brain's interstitial fluid, including such essential materials as glucose, amino acids, and ions, are transported by highly selective membrane-bound carriers.

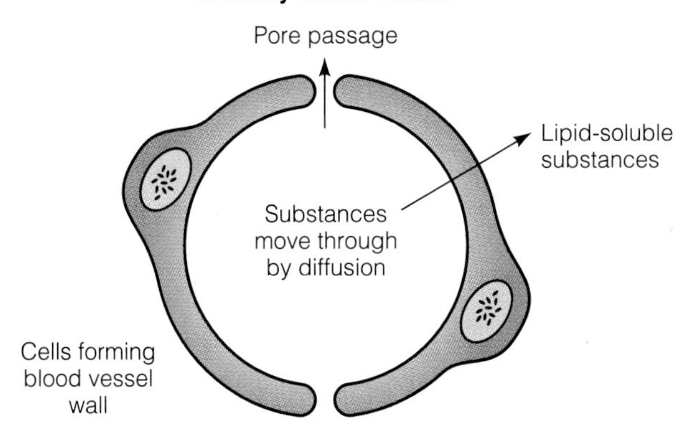

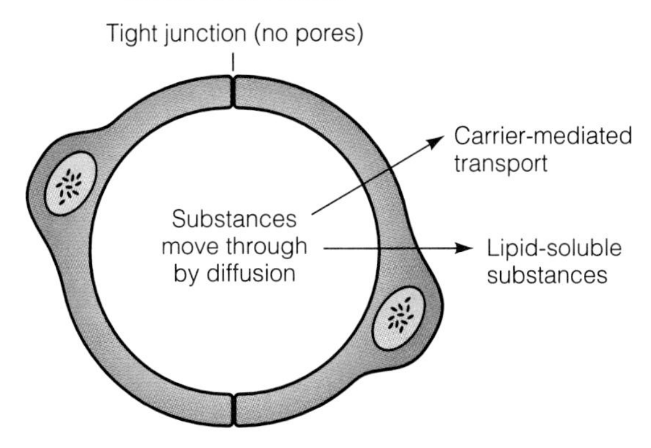

The brain is highly dependent on a constant blood supply, more than any other tissue. Unlike most tissues that can resort to anaerobic metabolism to produce ATP for at least short periods, the brain cannot produce ATP without oxygen. Also in contrast to most tissues, which can use other sources of fuel for energy production in the absence of glucose, the brain uses *only* glucose but does not store any of it. Therefore the brain is absolutely dependent on a continuous, adequate blood supply of oxygen and glucose. Brain damage results if the brain is deprived of its critical oxygen supply for more than 2 to 4 minutes or if its glucose supply is cut off for more than 10 to 15 minutes.

Spinal Cord

Like the brain the **spinal cord** is protected by bone, cerebrospinal fluid, and meninges. All the bones protecting the spinal cord, called **vertebrae,** have a common structural pattern, which is described in Chapter 11. The spinal cord resides in a central canal formed by the walls of 26 vertebrae stacked and connected in such a way that a flexible, curved structure called the **vertebral column** results.

The spinal cord, extending from the brain stem, is about 42 cm, or 17 inches, long and about the thickness of your thumb, approximately 2.0 cm, or $\frac{3}{4}$ inches in diameter. The **spinal cord** is a two-way conduction pathway to and from the brain. We begin our description of the spinal cord by looking first at its cross-sectional anatomy (Figure 8-14).

Gray matter in the spinal cord forms a butterfly-shaped region on the inside surrounded by white matter on the outside. Cell bodies and dendrites form the **gray matter.** The **white matter** consists of long myelinated association neurons that form tracts connecting the spinal cord and the brain.

Afferent neurons carrying action potentials from the peripheral sensory receptors, such as from the skin, enter through the **dorsal roots** of the spinal cord. The cell bodies of the sensory neurons are in an enlarged region of the dorsal root called the **dorsal root ganglion.** The axons from these cell bodies may take many different routes after entering the spinal cord. Some axons enter the posterior white matter of the cord directly and travel to synapse at the level of the brain. Other axons synapse with neurons in the gray matter at the level they enter. Recall our earlier description of the simple reflex action resulting from someone accidentally stepping on a

Figure 8-14 Section through the spinal cord

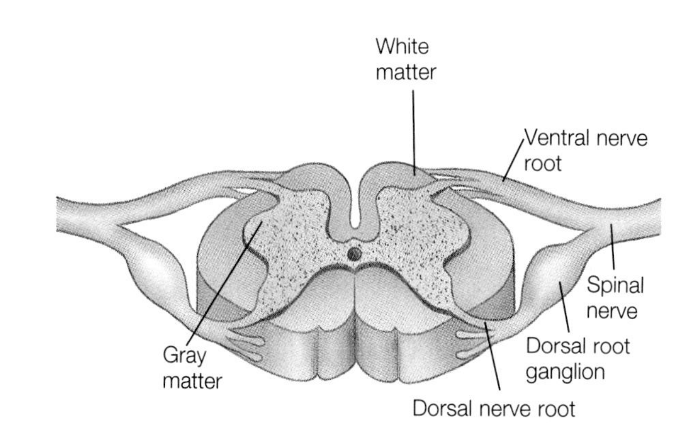

Figure 8-15 Coordination of reflex actions

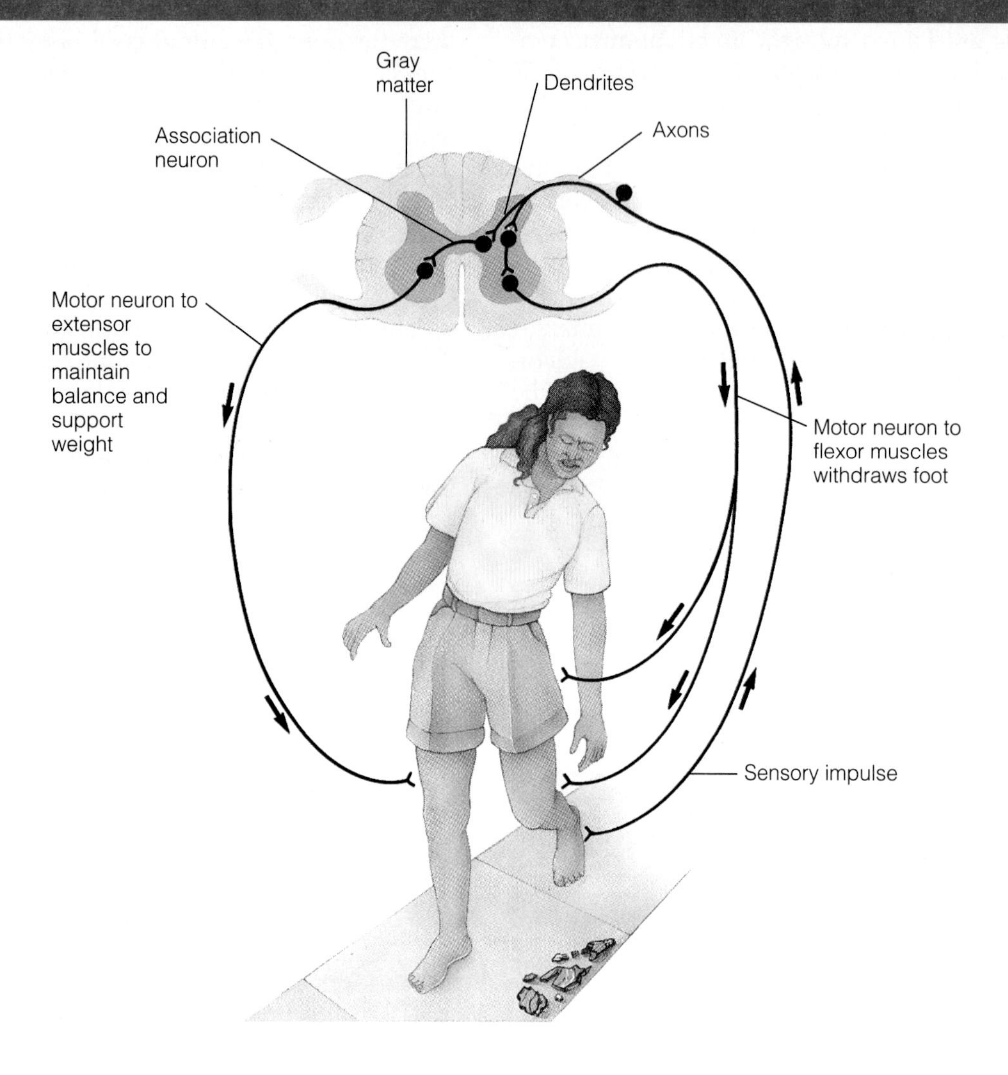

piece of glass. The sensory neurons of the foot send their impulses via dendrites toward the cell bodies located in the dorsal root ganglion. The axons of these cell bodies synapse with motor neurons, which send their impulses back to the leg muscles, causing them to contract and quickly move the leg away from the pain stimulus. At the same time sensory axons synapse with association neurons in the gray matter of the spinal cord and relay their impulses to motor-neuron cell bodies at the same level but on the opposite side of the spinal cord. The axons of the motor neurons leave the **ventral root** of the spinal cord to synapse with the appropriate muscles of the opposite leg. Finally the sensory impulses also synapse with neurons in the white matter, sending the pain stimulus to the brain, where it is consciously perceived (Figure 8-15).

In addition to simple reflexes, entire programs for operating complex activities also reside in the spinal cord. All the neurons needed for you to walk and run, for example, are in the cord. In these situations, the brain's role is to initiate and guide the spinal cord activity. The advantage of this arrangement is an increase in speed and coordination. Messages do not have to travel all the way up the cord to the brain and back down again merely to swing one of your legs forward as you walk or run.

The white matter of the cord is composed of myelinated tracts running in three directions: (1) ascending to higher centers of the CNS, (2) descending through the cord from the brain, and (3) running from one side of the cord to the other.

The ascending tracts conduct afferent signals entering the spinal cord up toward various parts of the brain. Our ability to identify and appreciate the kind of sensation being transmitted, whether hot or cold, pain, or pressure, depends on the specific areas of the brain where the ascending spinal cord tracts synapse. The nature of the message is always action potentials.

The descending tracts deliver efferent impulses from the brain to the spinal cord. These tracts are motor neurons concerned with regulating muscles involved in posture, in the more coarse movements of the limbs, and in coordinating head and neck, and eye movements that allow you to follow objects in your field of vision. Most of these movements depend on reflex activity. These tracts descend, without synapsing, from the upper regions of the brain all the way to the spinal cord. There they synapse with motor neurons in the spinal cord, which controls limb muscles.

General Organization of the Brain

The outward appearance of the human brain reveals few clues to its remarkable abilities. It is about two good handfuls of pinkish gray tissue with a wrinkled surface that looks remarkably like a cauliflower (Figure 8-16). The average adult male's brain weighs about 1600 g, or 3.5 lbs, and the average adult female's brain weighs about 1450 g. However, in terms of brain weight per body weight, males and females have equivalent brain sizes. The brain is composed of approximately 100 billion neurons and an equal number of neuroglial cells that support, protect, and nourish the neurons. We discuss the organization of the brain by breaking it down into general anatomical and functional specializations. We begin with the brain stem.

Brain Stem

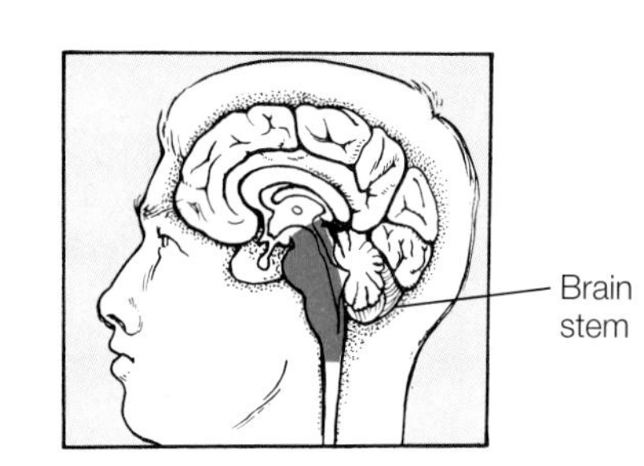

The brain stem includes portions of the brain that are continuous with the spinal cord. The **brain stem** generally conducts all efferent and afferent signals to and from the brain. It sorts and transmits sensory input to higher brain centers and controls vital bodily functions. The brain stem consists of the medulla oblongata, the pons, and the midbrain regions (Figure 8-17).

Figure 8-16 The human brain in cross-section

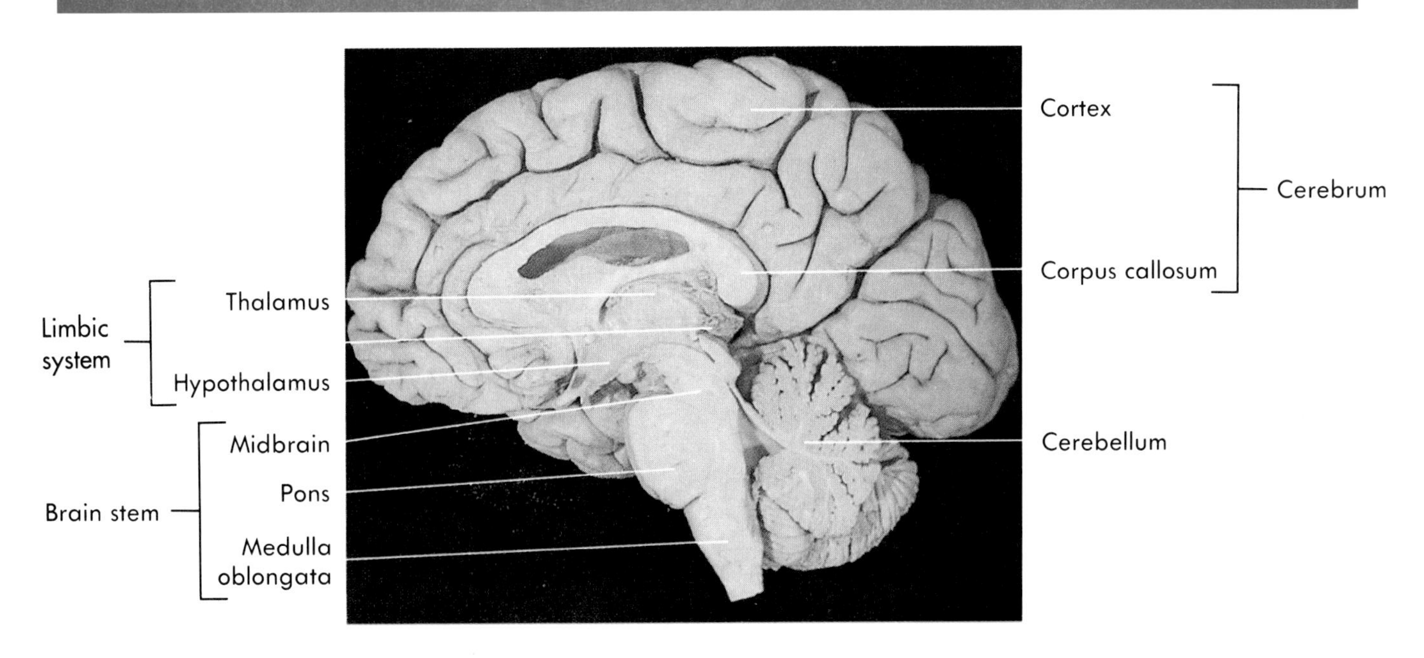

The largest area of the human brain is its outermost covering called the cerebrum. The limbic system, a ring of structures connected by nerve tracts, controls and influences emotional states. The part of the brain connected to the spinal cord and called the brain stem is responsible for many essential reflex activities, such as breathing and swallowing. Parts of the brain stem also filter and sort out incoming sensory information and direct it to appropriate higher brain areas. The cerebellum controls the timing and appropriate patterns of skeletal muscle contraction needed for coordinated movements, such as walking or writing.

Figure 8-17 The brain stem

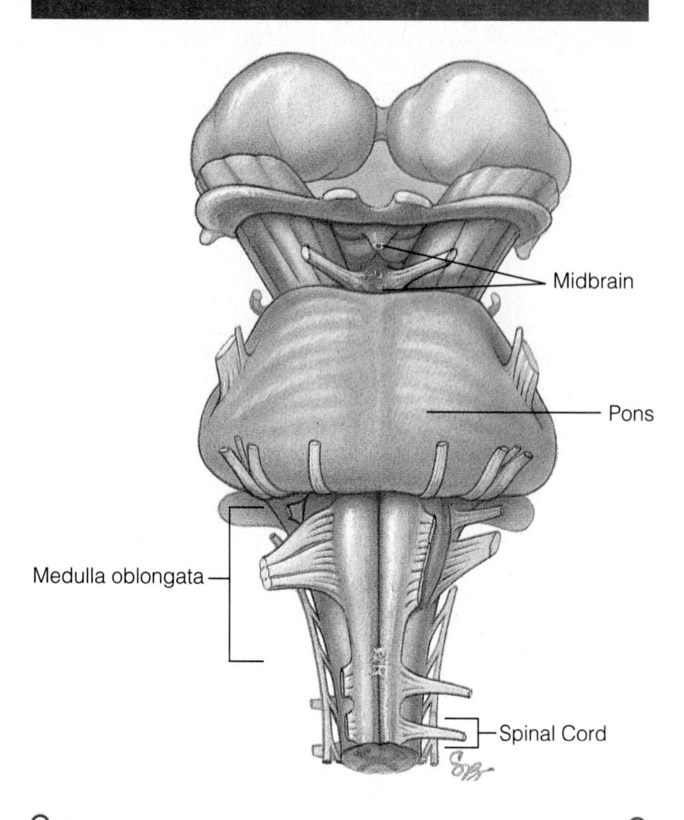

An anterior view of the brain stem shows that it consists of the medulla oblongata, pons, and midbrain. Sitting on top of the brain stem are the hypothalamus and thalamus, parts of the limbic system of the brain.

The **medulla oblongata** is about 3 cm long and is continuous with the spinal cord. Unlike the spinal cord it has no distinct regions of gray and white matter. Instead the gray and white areas are intermingled. Many of the descending motor neurons passing through this region cross over, or **decussate** (decus'sate), to the opposite side of the spinal cord. The decussation of these motor fibers explains why skeletal muscles on one side of the body are controlled by the opposite side of the brain. Many ascending tracts also cross over in the medulla oblongata.

Three reflex centers that control vital bodily functions are also in the medulla oblongata. These are the **cardiac center,** which adjusts the force and rate of heart contraction to meet the body's needs; the **vasomotor center,** which regulates blood pressure by acting through constriction or dilation of the walls of blood vessels to change their diameters; and the **respiratory centers,** which control the rate and depth of breathing and help to maintain respiratory rhythm. Other reflex centers control such actions as swallowing, coughing, sneezing, and vomiting.

On the ventral surface and slightly above the medulla oblongata is a bulging region of the brain stem called the **pons.** In Latin *pons* means bridge, which suggests its function. The pons relays impulses from the spinal cord to higher brain centers. Nuclei in the pons control various facial muscles and muscles to control head movement.

The **midbrain** is the upper portion of the brain stem lying just above the pons. Like the pons its primary function is to pass action potentials between the spinal cord and the higher brain centers. In addition, the midbrain contains nuclei that act as reflex centers for movements of the eyeballs, head, and trunk in response to sights, sounds, and other stimuli.

Reticular Activating System

The reticular activating system, located in the center of the medulla oblongata, pons, and midbrain, is a cluster of ascending sensory tracts and nuclei. Generally, the RAS is responsible for sorting and filtering all incoming sensory information and transmitting it to higher brain centers. It is also essential for wakefulness, attention, and concentration.

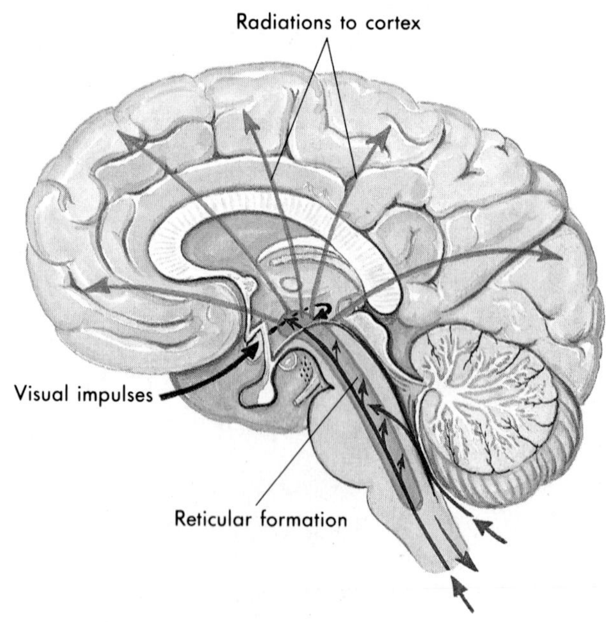

Cerebellum

The **cerebellum** (seh'-rih'beh'lum), meaning small brain, is located behind the pons and medulla and protrudes dorsally from the brain stem (see Figure

8-16). The **cerebellum** continuously processes inputs received from the higher brain centers, various brain stem areas, and sensory receptors from the body to provide precise timing and appropriate patterns of skeletal muscle contraction needed for smooth, coordinated movements. Cerebellar activity occurs subconsciously. We have no awareness of its functioning. The cerebellum is similar to the automatic pilot of an airplane. It continually compares the higher brain's intention with the body's actual performance and sends out signals to initiate the appropriate corrections. It thus helps the body in producing smooth voluntary movements, such as a smooth stride as you walk down the street, regardless of an uneven or broken sidewalk. The cerebellum is also responsible for maintaining normal muscle tone and proper posture and balance. It receives signals from sensory neurons in the inner ear, which are able to detect changes in body position and movement relative to the force of gravity. This information is sent to the cerebellum, which in turn sends signals to those muscles whose contraction maintains or restores body balance.

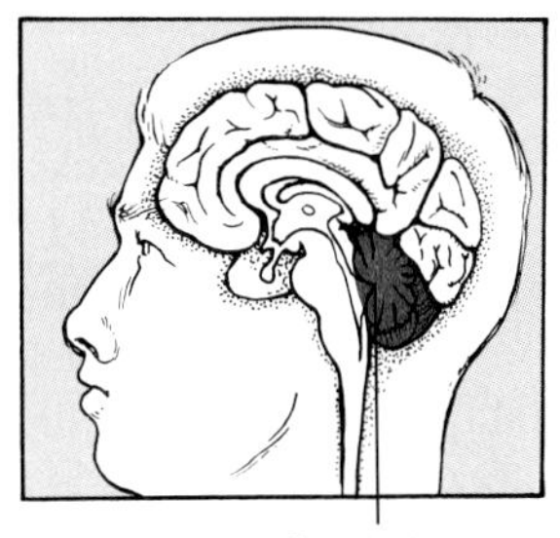

Limbic System

The **limbic system** is a ring of structures surrounding the brain stem and interconnected by nerve tracts. Some of the important structures found in the limbic system include the thalamus, hypothalamus, hippocampus, and amygdala. The limbic system is often called the emotional brain (Figure 8-18).

The **thalamus** is located just above the midbrain and is almost in the exact center of the brain. The **thalamus** acts like a filter, sorting out the flood of sensory information coming into the brain. Familiar, repetitive, and weak signals are filtered out, while unusual, important, and strong signals are allowed to pass up to consciousness. For example, you are probably unaware of the pressure exerted by the belt around your waist, but if the buckle suddenly released, it would immediately gain your attention.

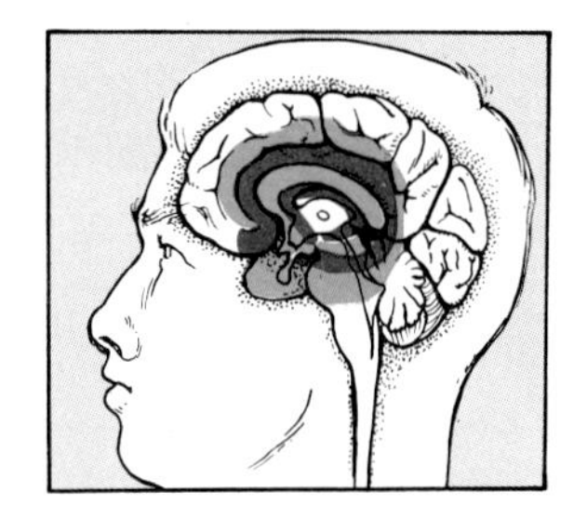

Figure 8-18 The limbic system

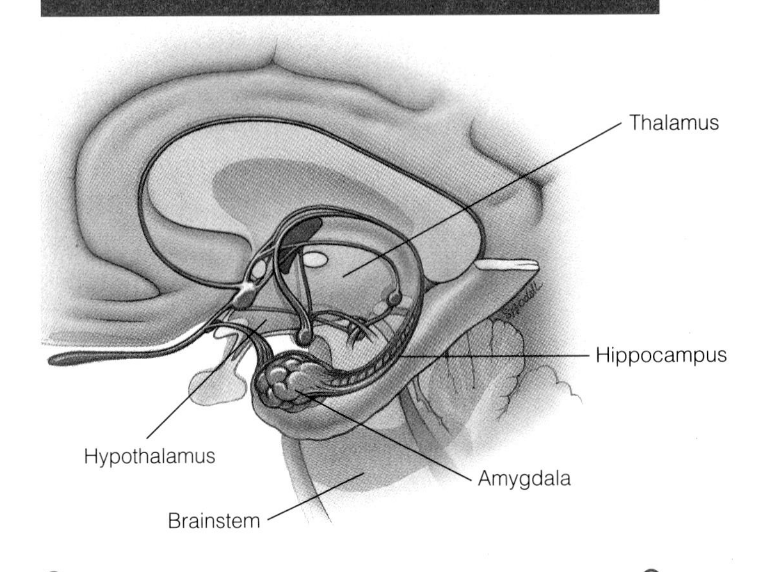

The limbic system consists of a ring of structures surrounding the brain stem and interconnected by nerve tracts. Parts of the limbic system control and influence body homeostasis and emotional states.

The hypothalamus is located below the thalamus. The **hypothalamus** (hi′po-thal′a-mus) is concerned with homeostasis. As you will learn in Chapter 10, the hypothalamus is a direct link between the hormone-secreting endocrine system and the nervous systems.

The hypothalamus is also responsible for these aspects of homeostatic balance:

1. It regulates autonomic nervous system activity by controlling the autonomic centers in the brain stem. For example, it can influence blood pressure and the rate and force of heart contraction.
2. The hypothalamus has numerous connections with higher brain centers. Nuclei involved in pain, pleasure, fear, and rage sensation have been localized in the hypothalamus. The hypothalamus acts through the autonomic nervous system to initiate most physical expressions of emotion. For example, the physical manifestations of the emotion of fear, such as a pounding heart, elevated blood pressure, sweating, and a dry mouth, are initiated by the hypothalamus.

The **hippocampus** is located near the ventral surface of the pons. The **hippocampus** is a channel through which current incoming sensory signals generate particular limbic system responses, based on previous encounters with the same information, found in memory storage. For example, there is no objective reason to fear a human biology examination, but you may have learned that not doing well on the last examination resulted in being excluded from the family will. You now associate dire consequences with taking the next exam, a fear you can blame on your hippocampus. The role of the hippocampus in the memory process will be discussed later in this chapter.

The **amygdala** (a-mig'da-lah) is an almond-shaped complex of nuclei just ventral to the hippocampus. The **amygdala** receives impulses from all portions of the limbic system and from the higher brain centers. In experimental animals electrical stimulation of the amygdala produces aggressive behavior. It appears also to control human aggression.

Cerebrum

The **cerebrum** (se-re'-brum), or **cerebral cortex,** is the center for association, integration, and learning in the brain. It is the outermost part of the brain and accounts for more than 60% of total brain weight. Imagine how a mushroom cap covers the top of its stalk and you have a fairly good image of how the cerebrum covers and obscures other specialized brain regions (Figure 8-19). The large size of the human brain is a result of the elaboration of the cerebrum. Its extraordinary size is responsible for our ability to initiate voluntary motor activity, control many emotional reactions, and display unique functions, such as speech, language, creativity, moral judgments, abstract reasoning, memory, and learning.

The cerebrum consists of two halves, called **cerebral hemispheres.** These halves are connected by a nerve tract called the **corpus callosum** through which they communicate.

The cerebral cortex appears gray because it consists of nerve cell bodies, unmyelinated nerves, and neuroglial cells. This outermost layer is approximately 4 mm, or $\frac{1}{16}$ inch, thick. Almost the entire surface of the cerebral hemispheres is marked by convolutions called **gyri** (ji-ri, singular *gyrus*), separated by shallow grooves called **sulci** (sul'ki, singular *sulcus*) or by deeper grooves called **fissures.** Only one third of the entire surface area of the cerebral cortex is visible because the remaining two thirds is in the sulci and the fissures. The cerebral cortex is similar to a large, thin piece of paper that has been crumpled up to fit into a much smaller container.

Figure 8-19 Cerebral cortex

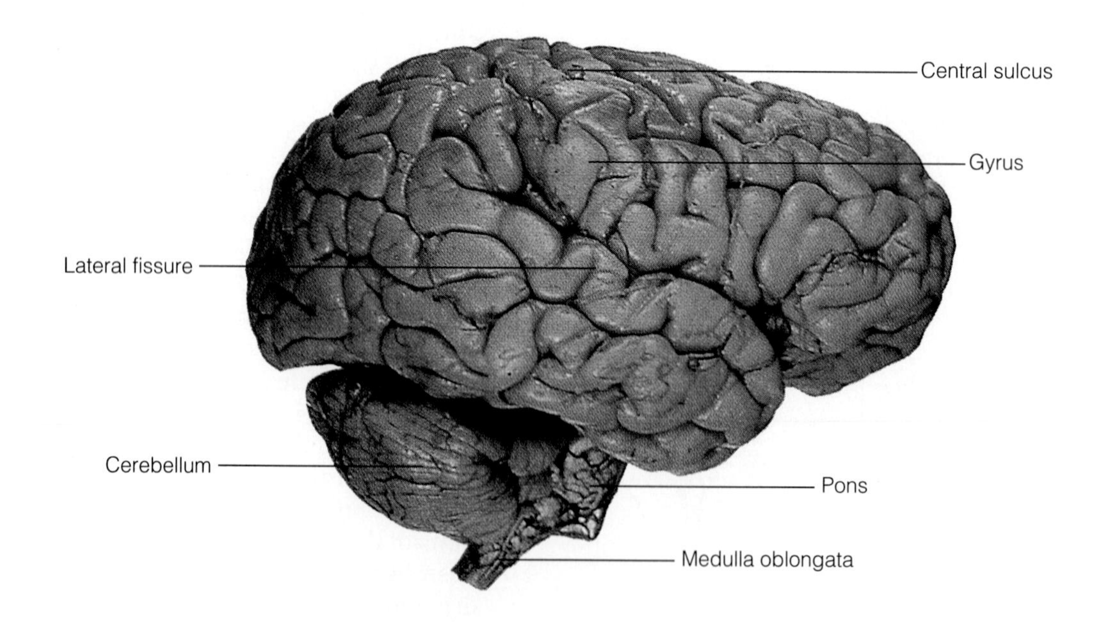

The outermost part of the brain is the cerebral cortex. It is divided into the right and left hemispheres which communicate by a tract within the brain called the corpus callosum. The entire surface of the cerebral hemispheres is marked by convolutions called gyri (sing. gyrus), separated by shallow grooves called sulci (sing. sulcus). Deeper grooves are called fissures.

Biology in Action SEIZURE DISORDERS

Suddenly a person may stare vacantly into space or lose consciousness and fall stiffly to the ground. His or her body may be racked by wild muscle contractions and abrupt jerking movements. After a few minutes the person regains consciousness but is usually very tired. Julius Caesar was afflicted with this "falling sickness" as was St. Paul of the New Testament, who was thought to have had a convulsion on his famous journey to Damascus.

Today we know these are symptoms of one type of **epilepsy,** a neurological disorder caused by abnormal discharges of groups of neurons in the brain. These "electrical storms" of action potentials then spread to different parts of the cerebral cortex. During this uncontrolled activity no other messages can get through to the brain. Research indicates that some types of epilepsy are influenced by genetic factors, but it appears to be more commonly induced by trauma to the brain, such as a blow to the head, or by infections or tumors.

Because of negative past connotations, the term *epilepsy* is not used by most physicians. Instead they use the term **seizure disorder.** Seizure disorders affect about 1% of the U.S. population and are the second most important category of neurological disorders, following stroke. A **seizure** is a period of sudden, excessive activity of the cerebral neurons. Sometimes if the neurons of the motor area of the cerebral cortex are involved, a seizure can cause a **convulsion,** which is wild, uncontrollable activity of the body's muscles. Most seizures, however, do not cause convulsions. There are three basic types of epileptic seizure disorders, as follows.

Petit mal, literally "little bad," is a mild form of epilepsy seen in young children, and it usually disappears by the age of 10 years or so. The child's expression suddenly goes blank, and the child appears to be daydreaming for up to 30 seconds. Slight twitching of facial muscles may occur, but convulsions and unconsciousness do not. **Psychomotor epilepsy,** also called a **Jacksonian seizure,** is accompanied by disorientation, loss of contact with reality, and uncontrolled motor activity of isolated muscle groups, such as those of the hands or lips. Historically these strange manifestations have led to many cases being misdiagnosed as mental illness.

Grand mal, literally "great bad," is the most severe form of seizure disorder. The person loses consciousness and displays intense convulsions. The seizure lasts for a few minutes, and then the muscles relax and the person awakens. Many grand mal sufferers experience a sensory hallucination, such as a taste, a smell, or a flash of light, just before the seizure begins. This characteristic, called an **aura,** is helpful because it gives the person time to lie down and avoid accidents.

Beneath the outer gray matter of the cortex lies the thicker white matter of the cerebral cortex. Here lie most of the myelinated axons of the gray matter's cell bodies. These myelinated nerve fibers allow communication among various portions of the outermost cerebral area and the deeper structures of the brain and the spinal cord.

Functional Divisions of the Cerebral Cortex

Each hemisphere of the cerebral cortex is functionally divided into four major regions or lobes. The major lobes are the **frontal lobe, parietal lobe, occipital lobe,** and **temporal lobe.** The lobes are named for the cranial bones of the skull that overlie them.

There are three major types of areas within the lobes of the cerebral cortex. These are **motor areas,** which control voluntary skeletal muscle contraction; **sensory areas,** which receive sensory information from all over the body and interpret the information as sensations; and **association areas,** which are involved in integrative and interpretative activities and higher order brain functions, such as memory, learning, and language. The specific subregions of the lobes controlling these specific functions are called **cortical areas.** With the exception of language, all other functions are in both cerebral hemispheres (Figure 8-20).

Figure 8-20 Lobes of the cerebrum

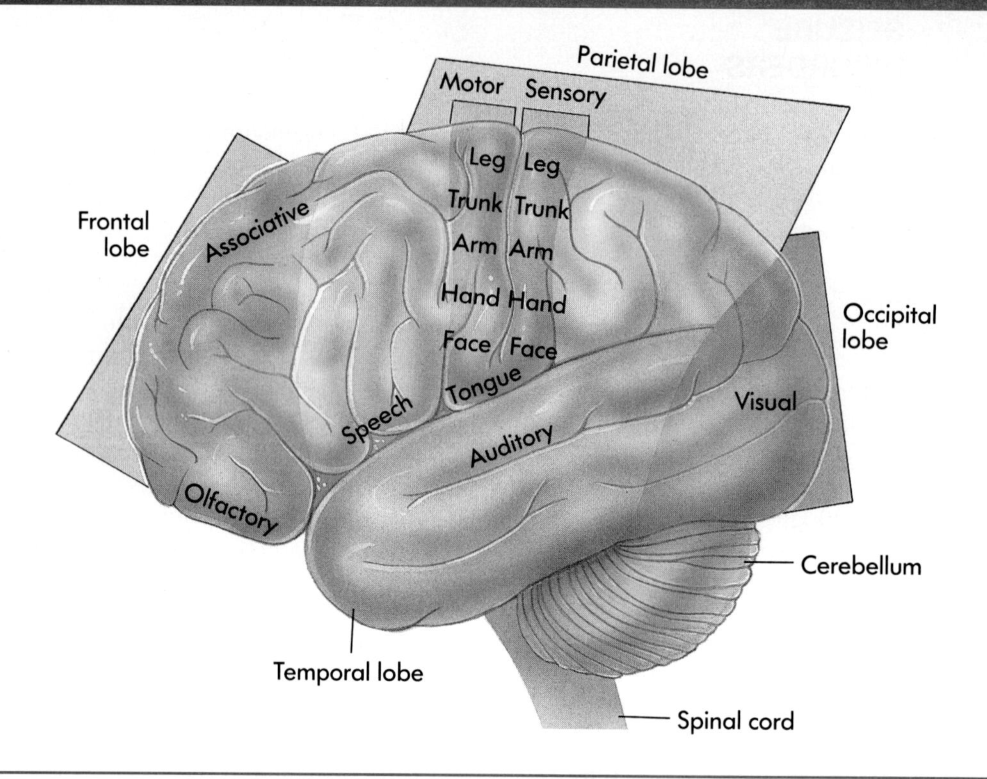

There are specific subregions of the lobes, called cortical areas, that can be grouped into motor, sensory, and association areas. This view of the left hemisphere indicates the four major lobes of the cerebrum and some specific cortical areas.

Motor Areas

Cortical areas controlling motor functions are located in a portion of the frontal lobes called the **primary motor area.** Their neurons allow us to consciously control skeletal muscle movements. Their axons form the major descending voluntary motor tracts that directly connect the cortex and the spinal cord (Figure 8-21, *A*).

The neurons of the motor cortex are grouped according to the body location of the muscles they control. The proportions of motor cortex devoted to various body motor functions are illustrated here in cartoon form. The relatively large amount of motor cortex devoted to the hands and face accounts for the high degree of motor control over these parts, allowing rapid and precise control of the hand and sophisticated speech.

The premotor cortex lies just anterior to the primary motor cortex. The **premotor cortex** controls learned motor skills that are repetitious or patterned, such as playing a musical instrument or typewriting. Neurons in the premotor cortex coordinate several muscle movements simultaneously or sequentially by sending impulses both to the primary motor cortex and to lower brain centers.

Sensory Areas

Unlike the motor areas, which are confined to the frontal lobe cortex, sensory areas are in several lobes of the cortex.

The primary sensory area resides immediately behind the primary motor area. **Primary sensory area** neurons receive information relayed from the sensory receptors in the skin and from those in skeletal muscles, and they identify the body region being stimulated. As with the primary motor area, the body is here represented spatially. The amount of sensory cortex devoted to a particular body region is related to how sensitive that region is and not to its size. In humans the face, lips, tongue, and fingertips are the most sensitive body areas and therefore are the largest parts of the cartoon (Figure 8-21, *B*).

The somesthetic cortex (so′-mis-theh′-tik) is located just posterior to the primary sensory cortex and has many connections with it. The **somesthetic cortex's** major function is to integrate and analyze different somatic sensory inputs, such as touch, pressure, pain, and temperature. For example, the primary sensory cortex tells you that certain impulses are coming from your left hand. Then the somesthetic cortex lets you know, without looking, that you are holding a spool of yarn and not a book. This perception requires the integration of such concepts as size, texture, weight, and the relationship of parts of the object with memories of similar sensory experiences.

The occipital lobe of each cerebral hemisphere contains the primary visual cortex, surrounded by a visual association area. The primary visual cortex receives information that originates in the eyes. The **visual association area** interprets these visual stimuli based on past visual experiences, enabling you to recognize a flower or someone's face.

Each primary auditory cortex is located in the temporal lobe of both cerebral hemispheres. Impulses from the ears are sent to the **primary auditory cortex** where they are interpreted according to pitch, rhythm, and loudness. The surrounding **auditory association area** permits the perception of the sound stimulus, which you "hear" as speech, music, or noise.

Brain and Behavior

Although we use both cerebral hemispheres for almost every activity, there is a division of labor, and each hemisphere possesses unique abilities. One cerebral hemisphere seems to dominate the other. **Cerebral dominance** designates which hemisphere is dominant for language. In about 90% of people the left hemisphere controls language abilities. The left hemisphere is working when we form thoughts into words, memorize a list of numbers, or balance a checkbook. The right hemisphere is more involved in visual and spatial relation skills, music appreciation, and emotion.

Learning and Memory

Learning is the acquisition of knowledge or skills as a consequence of experience, instruction, or both. Although we are far from knowing what specific cellular mechanisms are involved in this process, we do know that in many animals, such as mammals, rewards and punishments are important components of many types of learning. For example, if after responding in a particular way to a stimulus an animal is rewarded, the animal will very likely respond in the same way again to the same stimulus as a consequence of this experience. Conversely, if a particular response is accompanied by punishment, the animal is less likely to repeat the same response to the same stimulus. When behavioral responses giving rise to pleasure are reinforced or those accompanied by punishment are avoided, learning

Figure 8-21 Primary motor and sensory areas of the cerebrum

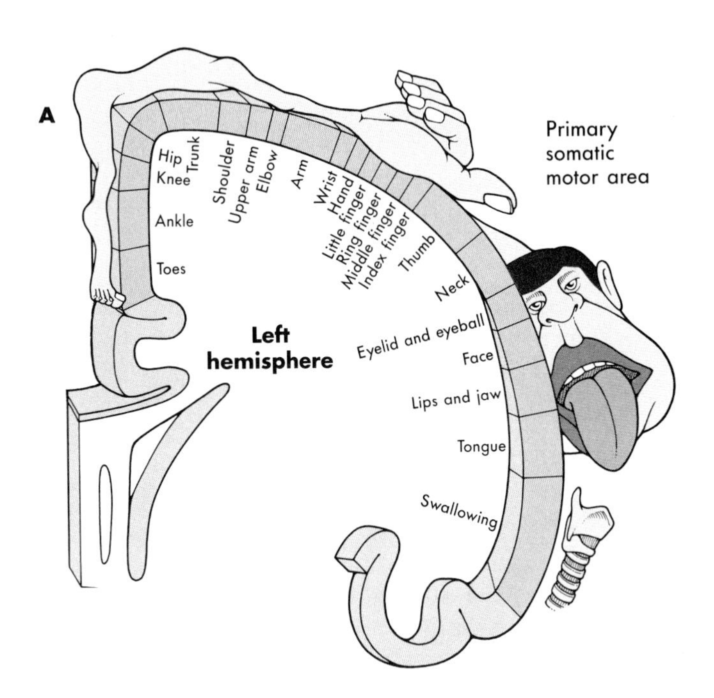

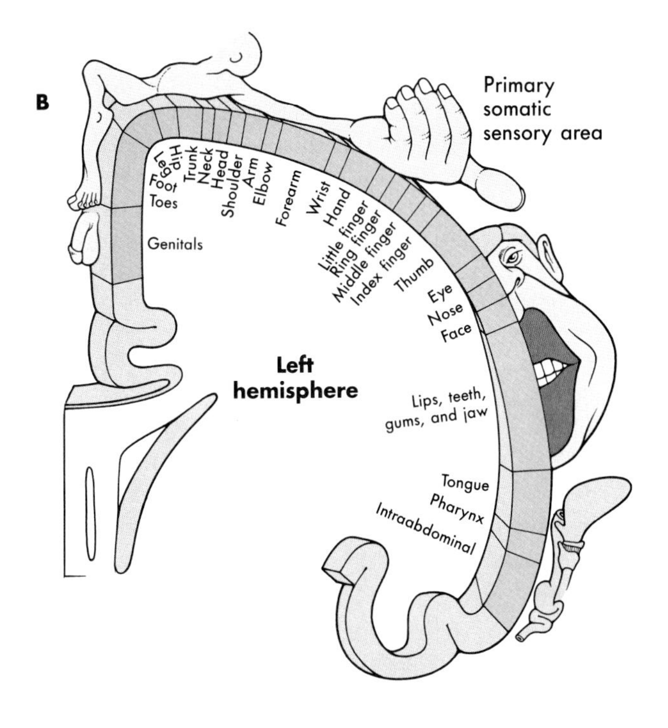

The primary motor area allows for the conscious control of skeletal muscles movements (A). The primary sensory area enables us to perceive sensation from the skin and skeletal muscles, and identify the body region being stimulated (B). The proportional amounts of each primary area devoted to various body parts and regions are illustrated in cartoon form.

Biology in Action SCIENCE AND THE BRAIN

The scientific method and the process of discovery it engenders is often expressed in the example of the controlled experiment. However, some natural phenomena do not lend themselves to the methods of controlled experimentation. How, for example, has science come to understand the structure and function of the human brain? A scientist cannot conduct controlled experiments to discover how the human brain works in the same way someone might to determine how frogs breathe under water. In situations in which controlled experiments are not possible, a different but complementary method of investigation is used. As with any scientific investigation, we begin with observation. Based on observations, **inductive reasoning,** which entails working from observation of particular events to formulating hypotheses, and eventually to general explanations concerning the initial observations is one method used. The science of neuroanatomy, which attempts to understand the structure and function of the human nervous system, offers some dramatic examples of how inductive reasoning is used successfully. Historically, observations of brain-damaged patients have been the foundation for most of our understanding relating human behavior to brain function. Three historical examples can illustrate this inductive reasoning process.

In 1836, Marc Dax, a medical doctor, presented a short paper at a medical society meeting in Montpelier, France. During his long career as a general practitioner, Dax reported he had seen many patients suffering from loss of speech, a condition known as **aphasia,** following damage to the brain. Dax was intrigued by what appeared to be an association between the aphasia and the side of the brain damaged. In over 40 patients with aphasia, Dax recorded damage to the left hemisphere. He did not observe aphasia associated with damage to the right hemisphere. Based on this evidence, Dax concluded that each half of the brain controls different functions and that speech is controlled by the left half.

In 1861, Paul Broca performed autopsies on the brains of patients with aphasia. He found that many had damage to an area in the left hemisphere of the cerebrum between the eyebrow and the temple, an area now known as Broca's area. Broca's area is an association area of the cerebrum that controls the mechanics of human speech.

By 1950, Penfield and Roberts, two Canadian neurosurgeons, developed surgical procedures that relieved a form of seizure disorder that occurs in the left temporal lobe of the cerebrum. Because the brain has no pain receptors, a patient can be given a local anaesthetic and part of the scalp and cranial bone removed to gain access to a particular region of the brain. During the operation, the patient remains awake and alert. Penfield and Roberts took advantage of this. During surgery they applied very small currents of stimulation to Broca's area and the cerebral tissue immediately surrounding it. The patient reported, either verbally or with movements, the specific effects of the electrical stimulation to specific areas of the cerebrum. In this way, the surgeons were able to confirm the initial findings of Dax and Broca and add substantially to our understanding of the structure and function of the human brain.

has taken place. Regardless of how learning occurs, however, a necessary part of the process is the ability to remember previous experiences and encounters with the environment.

Memory is the storage of acquired knowledge for later recall. Memory occurs in two phases. An initial short-term memory is followed by long-term memory. For example, if you look up a number in the telephone book, you will probably remember the number long enough to dial it but forget it soon afterward. This is short-term memory. If you call the number frequently, however, eventually you will remember it more or less permanently. This is long-term memory.

Short-term memory appears to be electrical, involving the repeated activity of a particular neural circuit in the brain. As long as the circuit is active, the memory stays. If the brain is distracted by other thoughts or if the electrical activity is interrupted, the memory disappears and cannot be retrieved no matter how hard you try.

Long-term memory, on the other hand, seems to be structural, involving the formation of new permanent synaptic connections between specific neurons or the strengthening of existing but weak synaptic connections. These new or strengthened synapses appear to last indefinitely. Thus long-term memory persists unless certain brain structures are destroyed. We do not know how short-term memory is converted to long-term memory.

Learning, memory, and retrieval seem to be separate functions, mediated by separate areas of the brain. The hippocampus is involved in learning in some way because intense electrical activity occurs there during learning. Even more interesting is the effect of damage to the hippocampus. Hippocampal destruction does not affect the retention of old memories but results in an inability to learn new things. In a famous medical case one patient with hippocampal damage, while retaining old memories, could read the same magazine article day after day for years, never remembering he had read the article before. Each time the physician came for a visit, the patient needed to be reintroduced.

Retrieval or recall of established long-term memories is localized in the temporal lobes of the cerebral hemispheres. However, the storage site of complex long-term memories is much less clear. Long-term memory appears to be diffused throughout much of the cerebrum. For example, as more cerebral tissue becomes damaged, bits of a memory are lost but not the entire memory. Possibly bits of an individual memory exist in many cerebral cortex areas and do not exist in a single place.

Electrical Activity of the Cerebral Cortex

Consciousness is difficult to define. A good working definition is that it is a subjective awareness of the external world and the self. The self includes the awareness of one's own thoughts, perceptions, feelings, and dreams. The level of conscious awareness resides partly in the cerebral cortex and in deeper brain structures, such as the reticular activating system of the brain stem described earlier in the chapter. Normal brain function involves continuous neuronal electrical activity.

An **electroencephalogram** (ih-lek-tro-en-seh′-fuh-lo-gram), or **EEG,** is a record of the overall electrical activity of the cerebral cortex. An EEG can be made by placing electrodes at various locations on the scalp and then connecting the electrodes to an apparatus that measures the cortex's electrical activity. The resulting patterns of neuronal electrical activity are called **brain waves.** Various states of consciousness are characterized by particular brain-wave patterns.

Two patterns of electrical activity as recorded by the EEG are associated with the normal, alert brain. **Alpha waves** have a frequency of 6 to 13 cycles, or waves, per second. They are highly synchronous, meaning each wave is almost identical to the next. They are also of low amplitude, which means they are not very large. With regard to brain waves, wave amplitude reflects the number of neurons firing together at the same time, not the degree of individual neuronal electrical activity. Alpha waves indicate the cerebral cortex is "idling," that it is in a calm, relaxed state of alertness, usually with the eyes closed (Figure 8-22).

Beta waves have a higher frequency than do alpha waves, about 14 to 25 cycles per second. They

Figure 8-22 The electroencephalogram

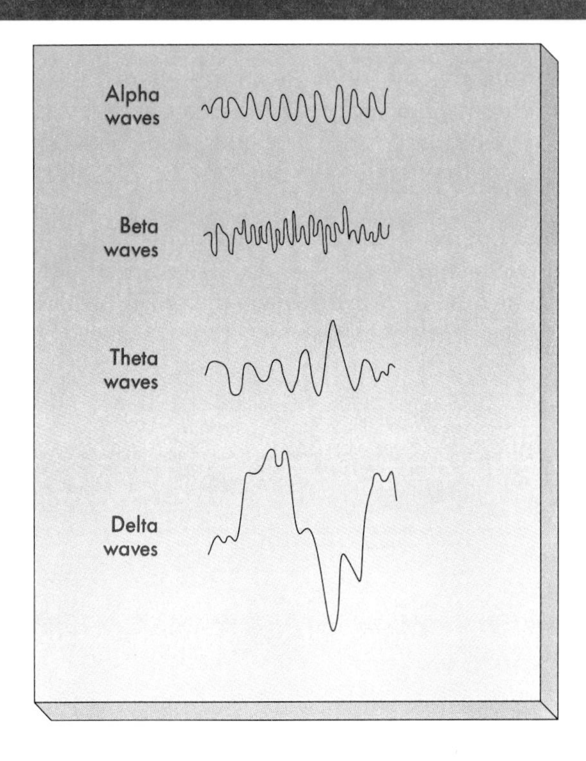

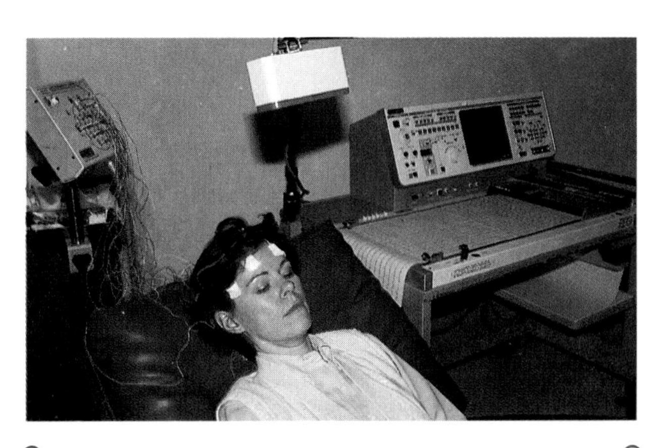

are less synchronous and also have a higher amplitude. Beta waves occur when we are awake and alert or concentrating on some problem with the eyes open.

Sleep

The sleep-wake cycle is the normal cyclical variation in our awareness of surroundings. As you know, when asleep you are not consciously aware of your surroundings, but you may have internally conscious experiences such as dreams. Sleep was once thought to involve a *decrease* in consciousness, but we now know that sleep is an *active* time for the cerebral cortex. We alternate between two stages of sleep during our normal sleep cycle, each characterized by combinations of different brain waves. One is called nonrapid eye movement sleep (NREM), and the other is called rapid eye movement sleep (REM).

During the first 30 to 45 minutes of the sleep cycle, the cerebral cortex is characterized by **theta waves,** which are more irregular than beta waves and have a frequency of about 4 to 7 cycles per second. As we slip deeper into sleep, **delta waves** begin to appear. Delta waves are high-amplitude, 4-cycle-per-second-or-less waves. This stage of sleep, known as **nonrapid eye movement sleep (NREM),** is also called **slow-wave sleep.** About 1 and $\frac{1}{2}$ to 2 hours after sleep begins, an abrupt change occurs in the EEG pattern. Brain waves become very irregular, culminating in the reappearance of alpha waves. **Rapid eye movement sleep,** or **REM,** has begun. This stage of sleep is also called **paradoxical sleep** because of several seemingly contradictory actions displayed by the nervous system and body in general. Although the eyes move rapidly under the lids during REM, most of the body's muscles are inactive. Most dreaming occurs during REM sleep. The sleeper is not easily aroused from REM sleep but is likely to awaken spontaneously and to remember dreams during this time.

During a typical night's sleep, REM occurs about five to seven times and accounts for about 1 to 2 hours of total sleeping time. Successive REM periods each get longer, the first lasting about 5 to 10 minutes and the final one about 45 minutes. They also get closer together over the course of the night, and we therefore tend to awaken more frequently in the early-morning hours.

Many of the complex responses or behaviors controlled by the human nervous system are directly or indirectly in response to external stimuli. However, to appropriately respond to the outside world, the nervous system must first be able to detect or sense it in some way. The ways in which we detect our surroundings are the next chapter's subject.

CHAPTER REVIEW

SUMMARY

1. The human nervous system helps to maintain homeostasis through the regulation and coordination of the body's activities. It accomplishes this by being able to detect, interpret, and respond to changes in both internal and external conditions. The nervous system acts very quickly and is responsible for moment to moment adjustments in maintaining homeostasis.
2. The human nervous system is made up of individual neurons. While there are diverse types of neurons, each consists of a cell body and two types of cytoplasmic extensions covered by a cell membrane. One type of projection is called a dendrite, and the other one is referred to as an axon. There are also neuroglia cells found in the nervous system which are responsible for support and nourishment of the neurons.
3. Neurons produce and conduct electrical nerve impulses called action potentials. Once threshold potential is reached, a neuron generates an action potential that travels down the length of its axon.
4. Neurons generate action potentials because they actively maintain a separation of oppositely charged ions across their cell membrane called a membrane potential. Through the action of sodium-potassium pumps and the permanent retention of negatively charged proteins inside the neuron, the membrane potential is maintained.
5. An action potential usually moves from the cell body and then to the axon which lies in close association with either another neuron, a muscle cell, or a secretory cell of a gland.
6. The gap between one neuron and another neuron is called a synaptic cleft. When an action potential reaches the synaptic cleft, it is chemi-

cally propagated across it by means of neurotransmitter substances. Neurotransmitter substances may depolarize or hyperpolarize the postsynaptic membrane to which they bind.
7. The central nervous system (CNS) is composed of the brain and the spinal cord while the peripheral nervous system (PNS) consists of nerves that carry information between the CNS and other parts of the body.
8. The cerebral cortex is the center for association, integration, learning, and memory in the brain. It also communicates with most other structures of the brain.

FILL-IN-THE-BLANK QUESTIONS

1. Neurons consist of a cell body and two types of cellular projections; usually several ______ and one ______.
2. ______ cells provide nourishment and protection for neurons.
3. A separation of oppositely charged ions across a cell membrane is called a ______ ______.
4. A resting neuron is ______ charged inside and ______ charged outside.
5. Neurotransmitter chemicals are found in ______ vesicles located in ______ ______ which are swellings at the end of axons.
6. An ______ synapse causes the postsynaptic neuron to become hyperpolarized.
7. A simple reflex arc consists of a sensory or ______ neuron and a motor or ______ neuron.
8. The somatic nervous system consists of both ______ and ______ pathways.
9. Both divisions of the autonomic nervous system are composed of nerve pathways containing a preganglionic fiber, a ______, and a ______.
10. The ______ ______ ______ is responsible for the level of alertness and awareness of the cerebral cortex.
11. Sensory neurons enter the spinal cord through the ______ ______ and motor neurons exit from the cord via the ______ ______.
12. The ______ is the center for association, integration, and learning in the human brain.
13. The ______ ______ ______ of the cerebrum controls voluntary control of the skeletal muscles.
14. The ______ ______ controls many vital bodily functions such as swallowing, heart rate, and so on.
15. Short-term memory appears to be ______ in nature, while the long-term memory seems to be ______.

SHORT-ANSWER QUESTIONS

1. How is the neuron's resting potential maintained?
2. What does a stimulus do to a neuron to generate an action potential?
3. Compare and contrast conduction in an unmyelinated and myelinated neuron.
4. Why is there more than one type of neurotransmitter chemical found in the human nervous system?
5. What is the advantage of behavior controlled reflex arcs?
6. Compare and contrast both divisions of the autonomic nervous system.
7. Describe the stages of sleep as indicated by changes in brain wave patterns of the cerebral cortex.

VOCABULARY REVIEW

action potential—*p. 144*
autonomic nervous system (ANS)—*p. 150*
axon—*p. 142*
brain stem—*p. 157*
central nervous system (CNS)—*p. 150*
cerebellum—*p. 158*
cerebrum—*p. 160*
dendrite—*p. 142*
myelin sheath—*p. 143*
neurotransmitter substance—*p. 147*
parasympathetic nervous system—*p. 151*
peripheral nervous system (PNS)—*p. 150*
somatic nervous system—*p. 150*
sympathetic nervous system—*p. 151*
synapse—*p. 147*

Chapter Nine

THE SENSES

For Review

Here are some important terms and concepts that you will encounter in this chapter. If you are not familiar with them, you should review them before proceeding.

Action potential
(page 144)

Primary sensory cortex
(page 162)

OBJECTIVES

After reading this chapter you should be able to:

1. *Define stimulus and differentiate between stimulation and sensation.*
2. *Describe the two ways in which stimulus intensity is conveyed to the brain.*
3. *Define adaptation.*
4. *Describe the function and location of general receptors.*
5. *Name and describe the function of the two major types of encapsulated nerve endings.*
6. *List some special sense receptors and discuss how they differ from general receptors.*
7. *Explain the function of chemoreceptors and describe the two types.*
8. *Describe the structure and function of the eye.*
9. *Explain the differences between rods and cones.*
10. *Define refraction and accommodation and discuss the role each plays in focusing.*
11. *Differentiate between myopia, hyperopia, and presbyopia and define depth perception.*
12. *Discuss the importance of the inner ear in balance.*

See or smell food and your mouth waters. A sudden loud noise causes you to recoil. Place an object next to your skin and you are able to detect whether it is hot or cold, or rough or smooth. These and many other stimuli continually greet you and are interpreted by your nervous system. Although you sense the external environment through the five senses of touch, taste, smell, sound, and sight, other sensory receptors are also continually monitoring your body's internal environment. Through these sensory "gateways" your body is able to respond to a variety of external and internal stimuli, allowing you to interact appropriately with the surrounding world.

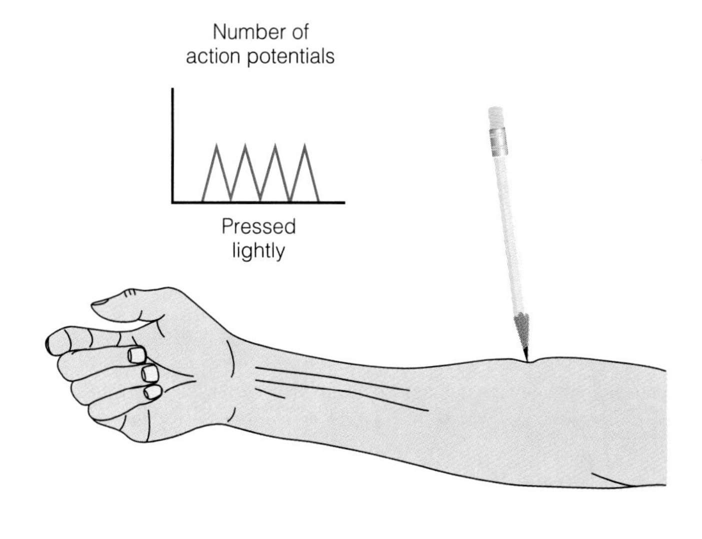

Stimulation and Perception

In this chapter we discuss the means by which the central nervous system (CNS) senses its surroundings. We begin by describing sensation, how particular sensory neurons differ in structure and function, and how the intensity of a stimulus is conveyed to the CNS.

Any environmental change that evokes an action potential in a sensory neuron is called a **stimulus.** Each type of sensory neuron, called a **receptor,** is specialized to respond to one type of stimulus. For example, receptors in the eye are most sensitive to light, and receptors in the ear are most sensitive to sound waves. All input from sensory neurons to the CNS arrives as action potentials. Every arriving action potential is identical to every other one. How then does the brain make sense of this information? How does it distinguish visual sensory input from auditory input? Or the taste of an orange from the fragrance of a rose? The brain makes these interpretations according to where in the brain the action potentials are sent.

In contrast to stimulation, **perception** is the conscious awareness of stimuli received by sensory receptors. How does your brain distinguish a less intense stimulus from a more intense one? How do action potentials from a sensory receptor convey information about variations in stimulus intensities?

Depending on the receptor type stimulus intensity is conveyed to the brain in two ways. The first is through **frequency coding.** When the strength of the stimulus increases, action potentials are fired more frequently by the receptors. For example, take a pencil and press it lightly against your forearm. As you apply greater force the number of action potentials generated by receptors in the skin sensitive to pressure also increases.

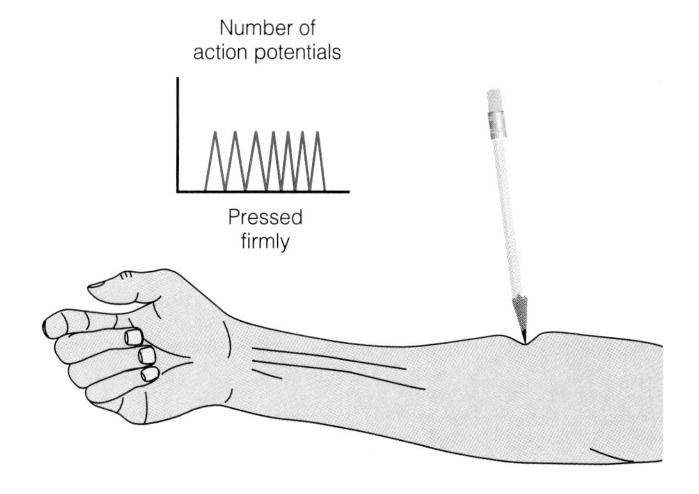

The second way, called **population coding,** occurs when more receptors are stimulated from a larger area of a given tissue as the strength of the stimulus increases. Take your thumb and press it lightly against your forearm. You are activating a small number of pressure receptors, but as you push your thumb more firmly into the same region, more receptors are being stimulated in a larger skin area. The increased stimulation of larger numbers of receptors reflects the increase in stimulus intensity.

Despite continued stimulation at the same strength, some receptors undergo **adaptation. Adaptation** means that these receptors can *decrease* the frequency of action potentials to the brain over time. For example, some sensory receptors in the skin are stimulated by mechanical pressure. Because they adapt rapidly you are not continually aware of wearing shoes. However, when you take your shoes off, you are momentarily aware of their removal because of the *change* in the pressure on the skin and on the skin's sensory receptors.

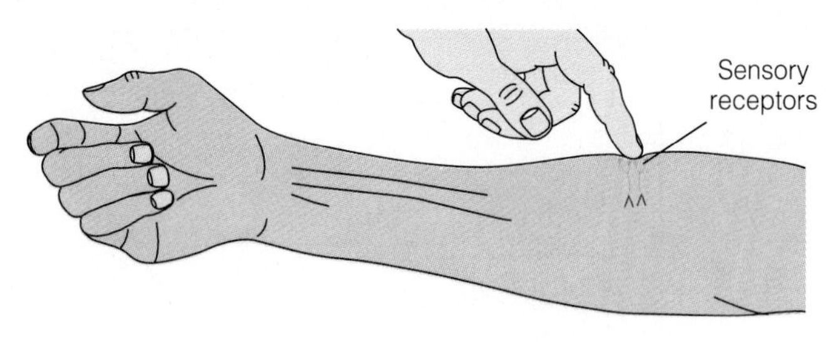

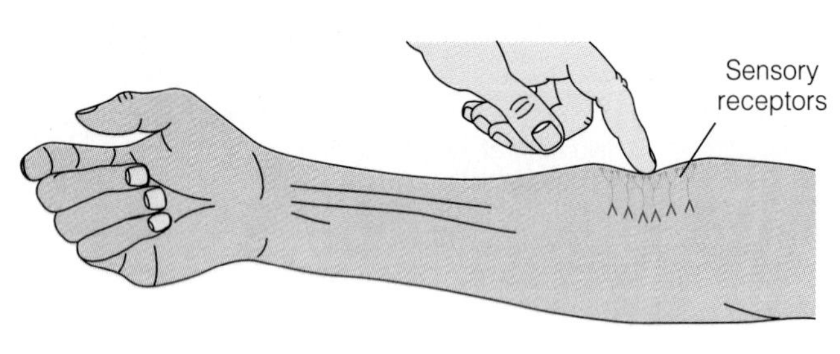

Receptors in the Human Body

Most sensory receptors are modified dendritic sensory neuron endings. In some cases the sensory receptors are actually **sense organs,** which are localized collections of cells, usually of many types, working together to accomplish a particular receptive process. The eye, for example, is a sense organ composed of several different tissues that collectively enhance light-sensitive sensory-receptor function.

Depending on the type of stimulus to which they can respond, human sensory receptors are categorized as follows (Table 9-1):

Photoreceptors are responsive to light.

Mechanoreceptors are sensitive to mechanical energy, such as touch, pressure, and vibration. Examples include the skeletal muscle receptors sensitive to stretch, the receptors in the ear containing fine-hair cells that are bent because of sound waves, and blood pressure—monitoring receptors in the circulatory system.

Thermoreceptors are sensitive to heat and cold.

Chemoreceptors are sensitive to specific chemicals. Some chemoreceptors are the receptors for smell and taste. Those located deeper in the body detect oxygen and carbon dioxide concentrations in the blood.

Nociceptors (no′-sih-sep′-terz), or **pain receptors,** are sensitive to tissue damage, such as crushing or burning, or to tissue distortion. Intense receptor stimulation is also perceived as painful.

Table 9-1 Human Sensory Receptors

RECEPTORS	SENSE	STIMULUS
General		
thermoreceptors	hot-cold	heat flow
mechanoreceptors	touch, pressure	mechanical displacement of tissue
nociceptors	pain	tissue damage
proprioceptors	limb placement	mechanical displacement
Special		
photoreceptors		
eye	sight	light
mechanoreceptors		
ear	hearing balance	sound waves mechanical displacement
chemoreceptors		
taste buds	taste	chemicals
olfactory cells	smell	chemicals

Proprioceptors (pro′-pri-o-sep′-torz) are a special class of sensory neurons monitoring internal conditions and are in only skeletal muscles and tendons. Proprioceptors continually advise the brain of body movements. This information allows you to be aware of the physical position of your body. For example, you are able to walk without watching your feet or to eat without watching the fork on its way to your mouth.

Some sensations you perceive are actually a combination of stimuli from several types of sensory receptors. For example, the sensation of wetness comes from touch, pressure, and thermal receptors. There is no such thing as a "wet receptor."

The body surface is covered with a thin, highly touch-sensitive wrapping called skin. The skin is composed of two layers of tissue. The first is the exposed, thin, outer layer called the **epidermis,** consisting of epithelial cells. The second is the deeper and much thicker layer called the **dermis,** composed of connective tissue. Both layers contain sensory receptors, but the dermis contains a wider variety of nerve endings that represent the major sensory receptors of the skin (Figure 9-1). These sensory neurons are referred to as **general receptors** because they are also located elsewhere in the body, such as in the internal organs, joints, and muscles.

General Receptors

General receptors are sensory neurons located throughout the body. General receptors have specialized dendrites to detect touch, pressure, pain, temperature, proprioception, and internal chemicals. They can be grouped according to their structure as either encapsulated nerve endings or free nerve endings.

Encapsulated nerve endings have one or more terminal dendrites enclosed in a connective tissue capsule. Particular encapsulated nerve endings vary greatly in size, shape, and distribution in the body. In contrast, **free nerve endings** usually have dendrites with small swellings at their ends that respond chiefly to pain, but some of these endings also respond to touch, heat, or cold.

There are two major types of encapsulated nerve endings. The first type, called **Pacinian corpuscles** (pa-che'ni-an kor'-puh-sulz), is located deep in the skin's dermal layer and in the internal organs. In longitudinal section a Pacinian corpuscle resembles a cut onion. Its single dendrite is enclosed by up to 60 layers of Schwann cells. Pacinian corpuscles are mechanoreceptors stimu-

Figure 9-1 Structure of the skin

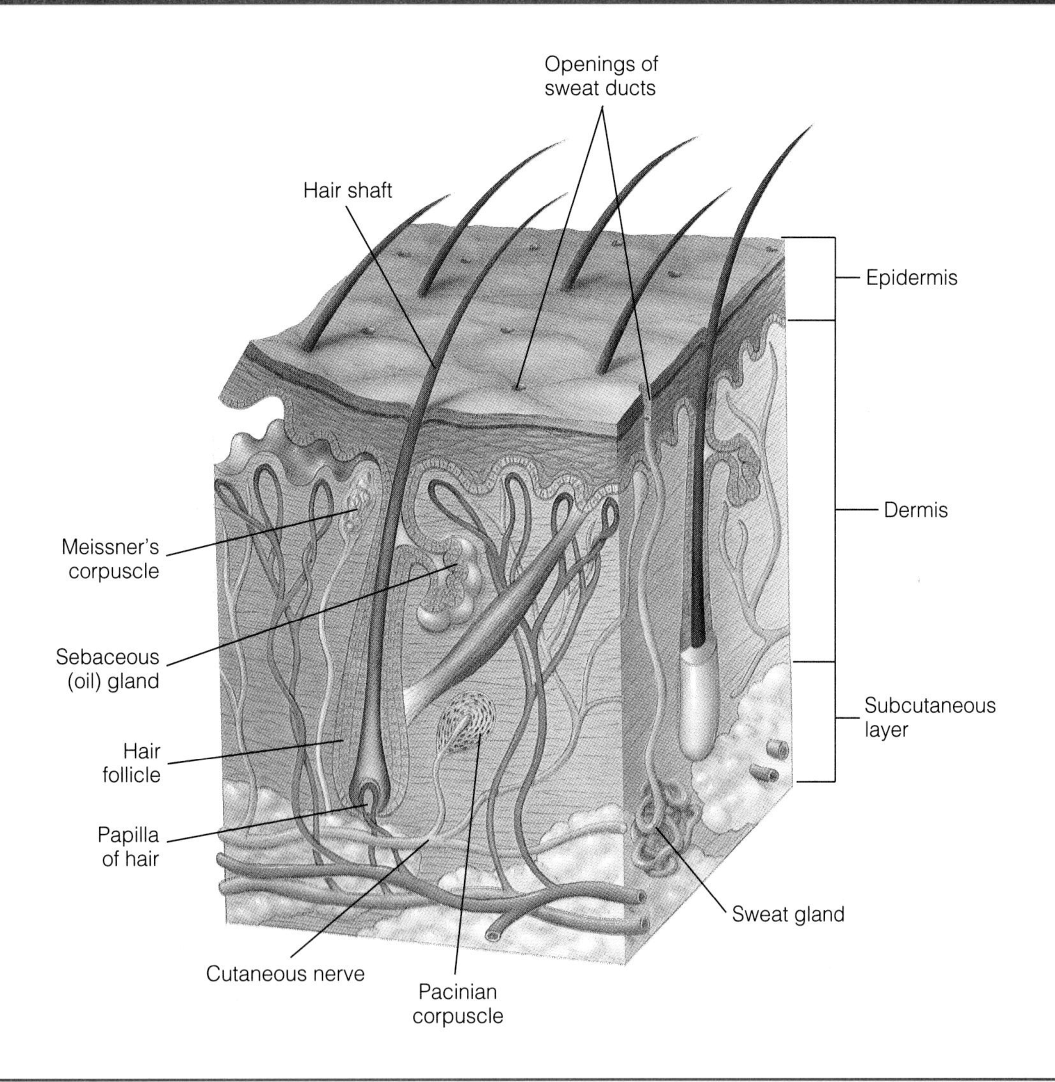

Human skin is composed of two layers of tissue. The outer layer, called the epidermis, consists of epithelial cells. The inner and thicker layer, called the dermis, is composed primarily of connective tissue. The dermis also contains a variety of sensory receptors, collectively known as general receptors because they also occur elsewhere in the body, such as internal organs. The receptors in this illustration are described in the text.

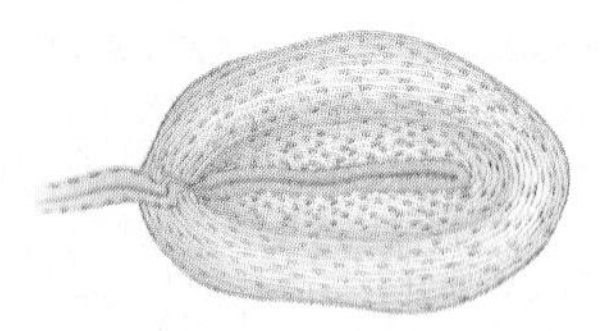

lated by deep pressure and vibration. For example, Pacinian corpuscles allow you to detect pressure changes as you put on and take off your shoes and, within several hundredths of a second, adapt to a new constant stimulation, such as standing on the floor in your bare feet.

The second type of encapsulated nerve ending is called **Meissner's corpuscles** (mis′-nerz kor′-puh-sulz), which are small egg-shaped mechanoreceptors consisting of a few spiraling dendrites surrounded by a thin capsule. Meissner's corpuscles are found at the upper border of the dermis and are especially numerous in the fingertips and lips. Meissner's corpuscles are primarily responsible for your ability to recognize the texture of objects you touch.

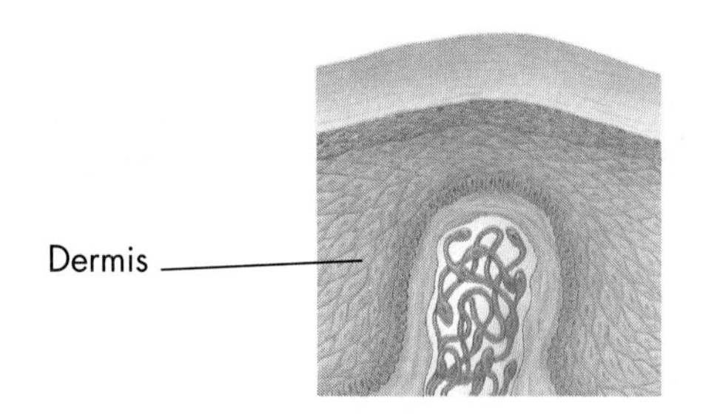

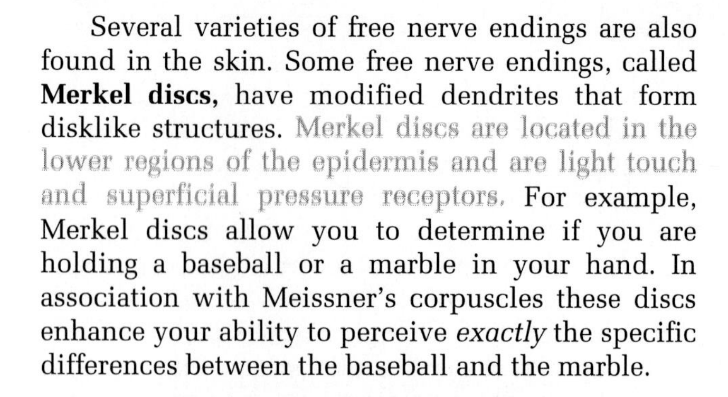

Several varieties of free nerve endings are also found in the skin. Some free nerve endings, called **Merkel discs,** have modified dendrites that form disklike structures. Merkel discs are located in the lower regions of the epidermis and are light touch and superficial pressure receptors. For example, Merkel discs allow you to determine if you are holding a baseball or a marble in your hand. In association with Meissner's corpuscles these discs enhance your ability to perceive *exactly* the specific differences between the baseball and the marble.

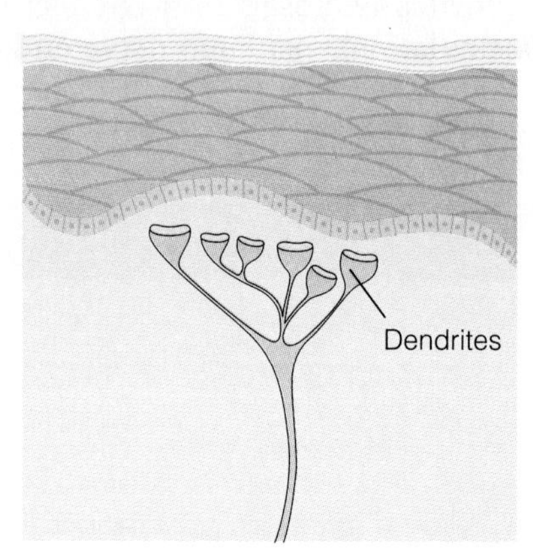

Each hair and its basal nerve fiber, called a **hair end organ,** is also a touch receptor. The slight movement of any hair on the body stimulates a free nerve ending entwining its base.

Figure 9-2 Muscle spindle

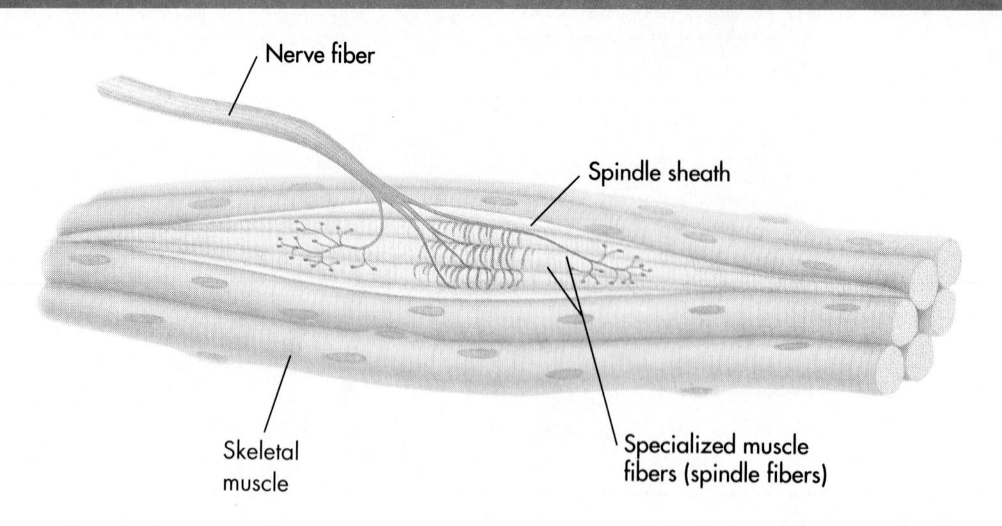

Muscle spindles are encapsulated receptors that are able to detect a skeletal muscle's rate of change or stretch. A sensory nerve synapses with specialized muscle cells in the muscle spindle and when stretched, the sensory nerve is stimulated and an action potential generated.

Proprioception

Earlier we defined proprioceptors as being able to sense body position and movement and to assist in maintaining balance. The particular proprioceptor responsible for detecting this information is the muscle spindle, found throughout all the skeletal muscles. **Muscle spindles** are encapsulated nerve endings that are mechanoreceptors. They are able to detect a muscle's distension, or stretch. The spindle is sensitive to both the muscle's rate of change in length and its final length. If a muscle is stretched, each muscle spindle in the muscle is also stretched. The surrounding sensory neuron fires, sending an action potential to the spinal cord. Once inside the spinal cord the sensory neuron synapses directly on the motor neurons that supply the same muscle, causing the muscle to contract enough to relieve the stretch (Figure 9-2). Muscle spindles are discussed more fully in Chapter 11.

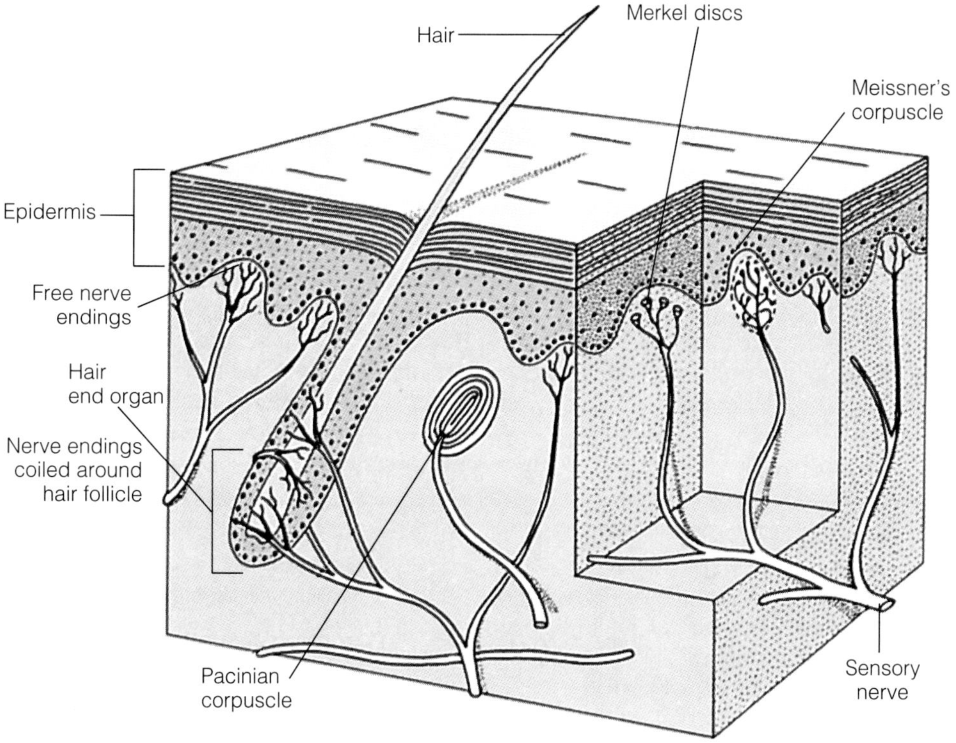

Heat, Cold, and Pain

Many encapsulated nerve endings are thermoreceptors. Those able to detect *decreases* in temperature are called **cold receptors.** Others, called **warmth receptors,** are able to detect *increases* in temperature. The cold and warmth receptors are located immediately under the skin in the lower part of the epidermis. In most parts of the body there are 3 to 10 times as many cold receptors as warmth, and the number in different areas of the body varies. Both cold and warmth receptors respond maximally to *changes* in stimulus. Even after adaptation the receptors continue periodically to send action potentials to the CNS. For example, you experience an initial shock of cold when you jump into a lake or a pool. After a short time the cold feeling subsides. This means that after adaptation of the cold receptors occurs, you are still marginally aware of being in cool or cold water because of slight temperature changes in the water.

Pain is an unusual sense in that it is somewhat nonspecific. For example, whether you cut, burn, or pinch your fingertip, you will feel pain. The common feature of these injuries is that pain is produced by tissue damage, regardless of the cause. It appears that pain perception is actually a special kind of chemical sense. When cells are broken open during tissue damage, their contents flow into the extracellular fluid and blood. The cell contents include enzymes that convert certain blood proteins into a chemical substance called bradykinin (brad′-yki′nin). Pain receptors are free nerve endings whose dendrites have receptor molecules for bradykinin. When bradykinin binds to these receptor molecules, action potentials are generated and sent to the brain, where they are perceived as pain (Figure 9-3). Because each body part has separate pain receptors providing input to particular brain cells, the brain knows where the pain is occurring.

Sensory Area

Two destinations are possible as an afferent impulse reaches the spinal cord. It may become part

Figure 9-3 Pain receptors

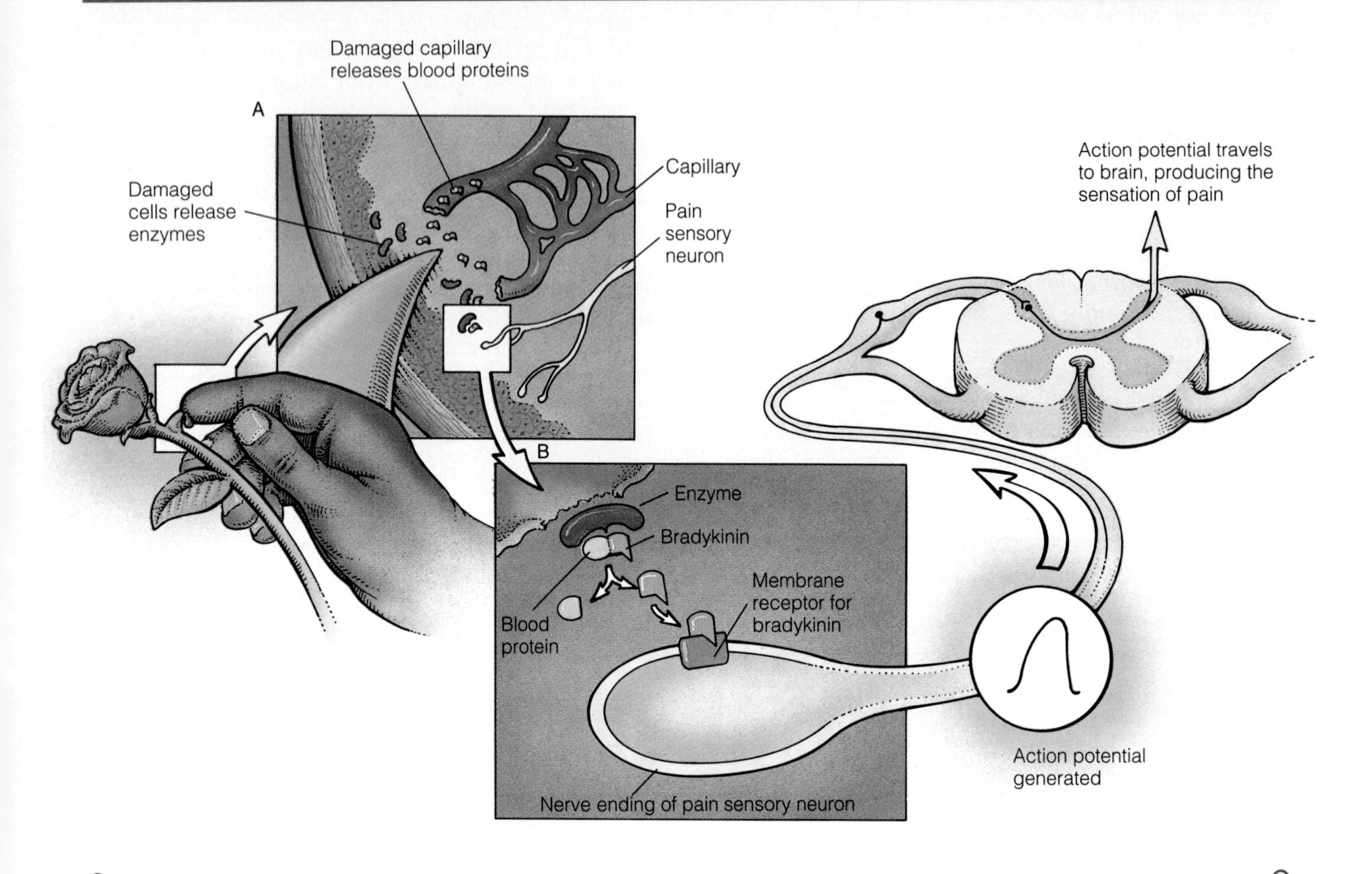

Pain receptors are free nerve endings whose dendrites have receptor molecules for bradykinin. With tissue damage, both capillaries and cells are broken open, releasing their contents into the extracellular space (A). A chemical substance found in the cell's cytoplasm, called bradykinin, is activated by a blood protein enzyme released from damaged capillaries in the injury site. Bradykinin binds to the receptor molecule on the dendrite of a free nerve ending, stimulating it, which in turn generates an action potential. The action potential is sent to the brain where it is perceived as pain (B).

of a reflex arc, stimulating an appropriate motor response, or it may travel up through ascending pathways to be processed by the brain and result in possible conscious awareness. Recall from Chapter 8 that the primary sensory area receives information from sensory receptors in the skin and skeletal muscles. If the information detected by receptors is sent to the primary sensory area, it is considered **sensory information** because it is perceived. Sensory receptors also arise from within the body, such as from internal organs and blood vessels. You are usually unaware of this form of sensory stimuli, and therefore it is not considered to be sensory information.

Special Sense Receptors

Humans are said to possess the five senses of taste, smell, sight, hearing, and touch. As you have just learned, touch is made up of a complex of general receptors. The other four senses of taste, smell, sight, and hearing are called special senses. However, an additional special sense, called equilibrium or balance, is housed in the ear along with the organ of hearing. In contrast to general receptors, the **special sense receptors** are either large, complex sensory organs, such as the ears and eyes, or small, localized clusters of receptors, such as the taste

Figure 9-4 Taste

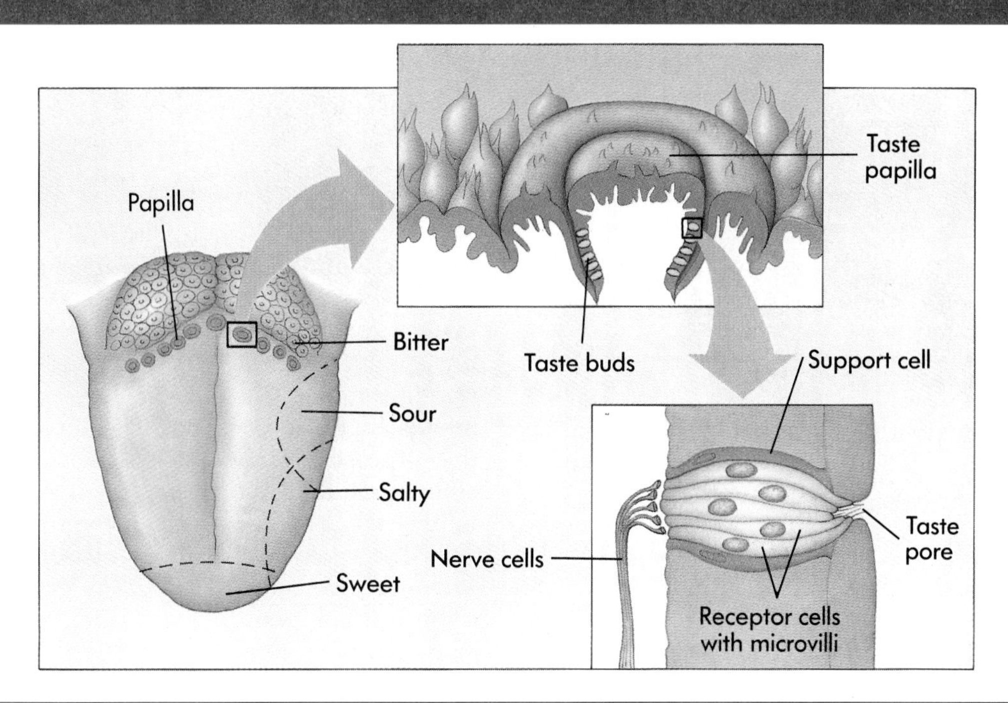

buds or the olfactory cells in the nasal passages. In this section we consider the structures of each of these special senses.

Chemical Senses: Taste and Smell

Chemoreceptors sample the chemical composition of the environment. Humans have two types of special-sense chemoreceptors. One detects chemicals dissolved in an aqueous medium, water, or saliva. The other detects airborne molecules. Both taste and olfactory receptors detect chemicals in solution. Taste is the detection of chemicals dissolved in saliva or water. The mucus-coated, hairlike projections of olfactory receptors are solvents for airborne molecules entering the nasal passages.

The chemoreceptors for taste are packaged in the **taste buds.** Of the 10,000 taste buds in the mouth, most are on the upper surface of the tongue. Taste buds are in projections called papillae (pah-pil′-e) on the tongue's surface. The papillae make the tongue surface feel slightly rough. The taste buds have a small opening, called the taste pore, through which fluids in the mouth come into contact with the tongue's taste receptors. Each taste receptor projects several microvilli into the taste pore. Dissolved chemicals enter the pore and bind to special receptor molecules on the microvilli, producing an action potential interpreted in the brain as a particular taste. (Figure 9-4.)

Although you recognize hundreds of distinct tastes, when taste is tested with pure chemical compounds, there are only four types of taste receptors—sweet, sour, salty, and bitter. Sensitivity to one or another of these basic tastes varies in different tongue regions. The tip of the tongue is most sensitive to sweet tastes, the sides to sour, and the back of the tongue to bitter. The ability to perceive salty taste is more evenly distributed on the tongue, with some emphasis toward the front and side of the tongue.

The great variety of tastes you perceive is produced in two ways. First, a particular substance may stimulate more than one type of taste receptor to different degrees. For example, something might taste "bitter-sour." Second, substances being tasted usually also give off molecules into the air inside the mouth. These molecules diffuse to the olfactory receptors in the nasal passages. Much of your tasting results primarily from your sense of smell. The next time you have blocked nasal passages from a cold, you might notice that even spicy pizza tastes bland.

The sense of smell, or **olfaction** (ol-fak′shun), like taste, also detects chemicals. The receptors for olfaction are sensory neurons with hairlike projections protruding into the roof of each nasal cavity. These hairlike projections are covered by a thin coat of mucus, which is a solvent for the odor molecules entering through the nose. As taste receptors detect chemicals dissolved in saliva, olfactory

Figure 9-5 Sense of smell

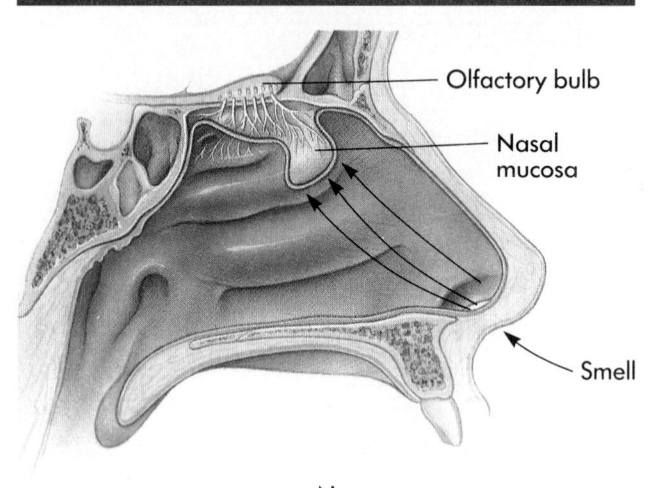

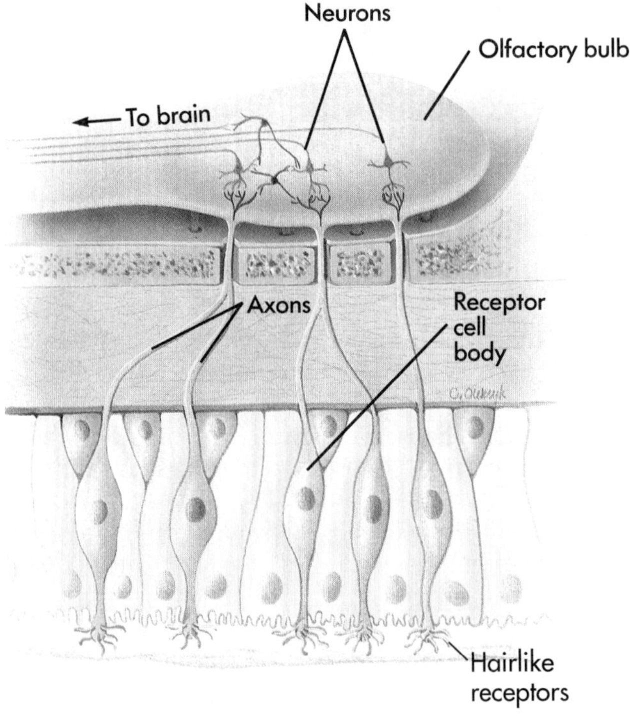

receptors detect airborne molecules dissolved in the thin layer of mucus coating their surfaces (Figure 9-5).

We don't know the exact physiological basis of smell discrimination. Research on smell is difficult because any given odor may consist of many different chemicals. Cigarette smoke, for example, contains more than 100 different molecules, each contributing to its odor. Whatever the mechanism for sorting out and distinguishing different odors, it is extremely sensitive even in humans. For example, methyl mercaptan, or garlic smell, is added to otherwise odorless natural gas, thus enabling people to detect potentially lethal gas leaks in homes and buildings. Humans can detect methyl mercaptan at a concentration of 1 molecule per 500,000,000 molecules of air. As remarkable as this is, other mammals, such as dogs, have an even more developed sense of smell.

Vision

The eye is the sense organ we primarily depend on to perceive our environment. For example, approximately 70% of all sensory receptors are located in the eyes. The sensory receptors of the eye, called photoreceptors, contain molecules, called **photopigments,** that absorb light and chemically change in the process. This chemical change produces an action potential in neurons called **ganglion cells**, to which the photoreceptors synapse. The ganglion cells' axons leave the eye to form the optic nerve. The two optic nerves meet to form the massive **optic tract** of more than 1 million nerve fibers that carry the action potentials from the eyes to the brain. Only the spinal cord tracts controlling the voluntary muscles of the body are larger than the optic tracts. Finally the cerebrum invests the signals from both eyes with meaning, fashioning and making sense of the surrounding world.

The Structure of the Eye

The eye is a hollow sphere approximately 2.5 cm, or 1 inch, in diameter. Its wall is composed of three layers, and its interior is filled with fluids that help maintain its shape.

The outermost layer of the eye, called the **sclera** (skle′-ruh), is composed of dense opaque connective tissue (Figure 9-6). Seen anteriorly it is the "white of the eye," except for the transparent **cornea,** which is the "window," allowing light into the eye. The cornea is well supplied with free nerve endings, most of which are pain receptors. Touching the cornea elicits a reflex blinking response. The cornea contains no blood vessels. If blood moved through the cornea, it would not be transparent.

The second or middle layer of the eyeball is called the **choroid** (ko′-royd), a highly vascular and pigmented membrane. The pigmentation helps absorb light, preventing it from scattering in the eye. Anteriorly the choroid becomes a thickened ring of tissue called the **ciliary body,** which encircles the lens of the eye. The ciliary body is composed of interlacing smooth-muscle fibers called **ciliary muscles,** which control the lens shape. The lens is a convex (thicker in the middle than at either end), transparent, and flexible structure that can change shape to allow precise light focusing on the

Biology in Action PERCEPTION AND IDENTIFICATION

Damage to portions of the brain's visual system has provided investigators with some insights into the anatomy and physiology of visual perception. In particular, damage to the visual association cortex produces a category of deficits known as **visual agnosia.** The word *agnosia* literally means a failure to know. In neuroanatomy *agnosia* is defined as an inability to identify or perceive a stimulus by means of a particular sensory system, even though its details can be detected. People with visual agnosia cannot identify or perceive common objects by sight, even though they have normal visual acuity.

Take a closer look at this definition and note the phrase "cannot identify or perceive." Normally we think of *identify* and *perceive* as being almost synonymous. If we can perceive something, we can also identify it. However, in some types of visual agnosia, normal perception remains intact while the ability to identify what is perceived is impaired. Some recent studies have "opened our eyes" to an intriguing puzzle.

People with **associative visual agnosia** appear to be able to perceive normally but cannot name what they have seen. Even more interesting is that they seem to be *unaware* of these perceptions. In 1982 two neuroanatomists, G. Ratcliff and F. Newcombe of Great Britain, were working with a patient with an unusual set of symptoms. The patient, for example, was able to copy a drawing of an anchor. He could *perceive* the shape of the anchor (Figure A1). However, he could not recognize the sample or the copy he had just drawn. In the next test when asked to draw, not to copy, a picture of an anchor, he was unable to do so (Figure A2). Although the patient could copy an image of an anchor, the word *anchor* failed to produce a mental image of one. On yet another test when asked to define anchor, he was able to do so correctly. He knew what the word meant.

Associative visual agnosia appears to involve a deficit in the ability to transfer information between the visual association cortex and those association areas involved in language. Neuroanatomists believe the syndrome is caused by damage to the white matter underlying the occipital and temporal lobes. The disturbance of these axons disconnects regions of the cerebral cortex mediating visual perception from those controlling verbalization and speech.

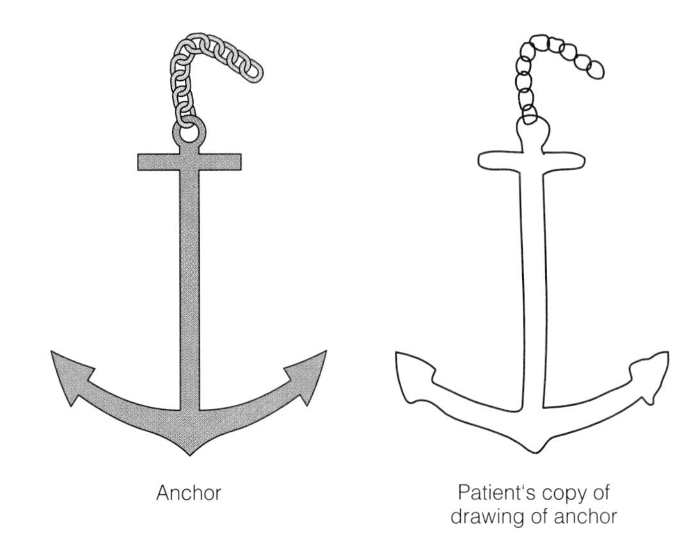

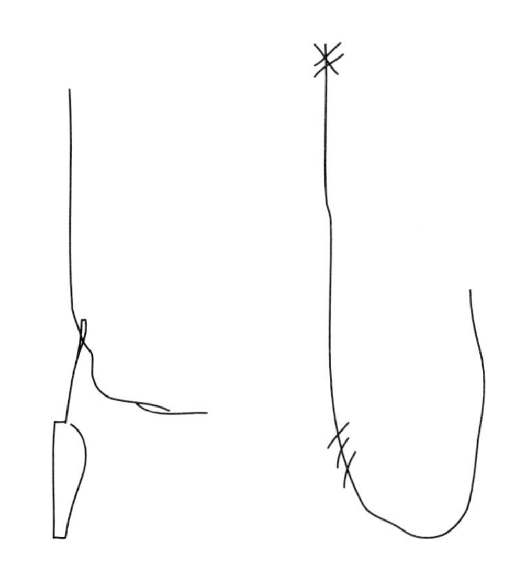

Patient's failed attempts to draw, not copy, a picture of an anchor

Figure 9-6 Structure of the eye

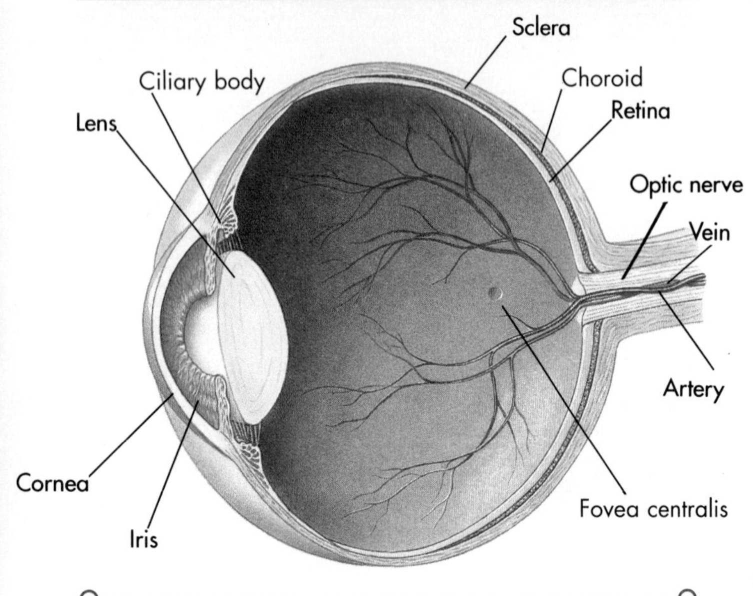

Figure 9-7 The iris

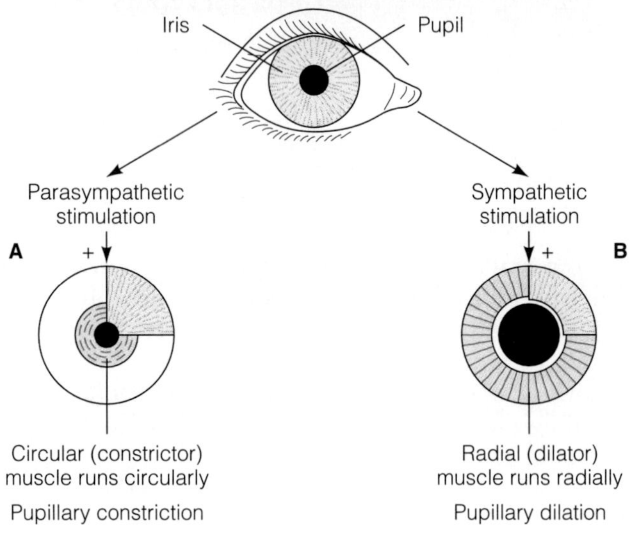

The iris is made up of smooth muscle fibers arranged in two directions. The first direction is circular and is under sympathetic nervous control (A). The second direction is radial and is under parasympathetic nervous control (B).

interior surface of the eye. It is suspended just behind the iris by suspensory ligaments, which are connected to the ciliary body surrounding the lens.

The **iris,** the visible colored part of the eye, is the most anterior portion of the choroid (Figure 9-7). Shaped something like a flattened doughnut, the iris lies between the cornea and the lens. Its round central opening, called the **pupil,** allows light to enter the eye. The iris is made up of smooth-muscle fibers arranged in two directions. The first direction is circular, like a rubber tire on a bicycle rim. The second direction is radially, like the spokes of a bicycle rim. Both are reflexively activated to vary the pupil's diameter. Because muscle fibers shorten when they contract, in close vision or bright light the circular muscles contract, causing the pupil to constrict in diameter. In distant vision or dim light, the radial muscles contract and the pupil dilates or increases in diameter, allowing more light to enter the eye. Pupillary dilation is controlled by sympathetic system stimulation, whereas constriction is under parasympathetic system control. Reflex changes in pupil size also reflect our interest in or emotional reaction to what we see. Dilation often occurs when the subject matter is appealing and during problem solving. On the other hand, boredom or a personally repulsive subject causes pupillary constriction.

The interior of the eye consists of two fluid-filled cavities, separated by the lens, all of which are transparent to permit the passage of light through the eye from the cornea to the innermost layer of the eye. The anterior cavity between the cornea and the lens contains a clear watery fluid, the **aqueous humor,** and the larger posterior cavity between the lens and the innermost layer of the eye contains a semifluid, jellylike substance, the **vitreous humor.** The vitreous humor helps to maintain the spherical shape of the eyeball. The aqueous humor carries nutrients for the cornea and lens, both of which lack a blood supply.

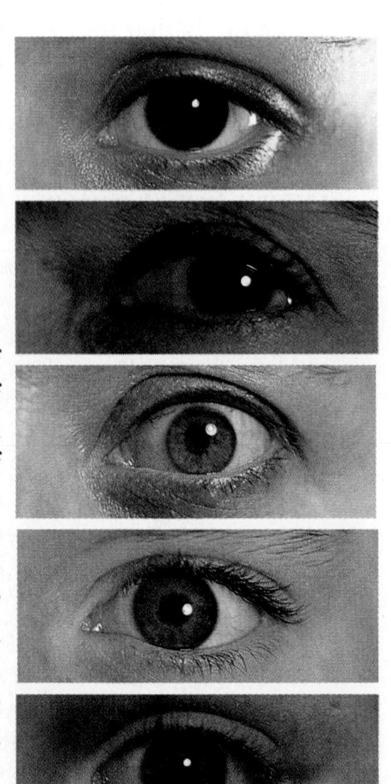

Only one color-producing pigment, called melanin, is present in the iris. Melanin is brown, and the variation in eye color that we perceive is a function of the distribution of this brown pigment in the tissue layers of the iris. If melanin is present in only the deepest layers of the iris, the eyes appear blue because the light waves are bounced back at angles we see as the color blue. A similar effect occurs as sunlight penetrates our atmosphere and

Figure 9-8 The retina

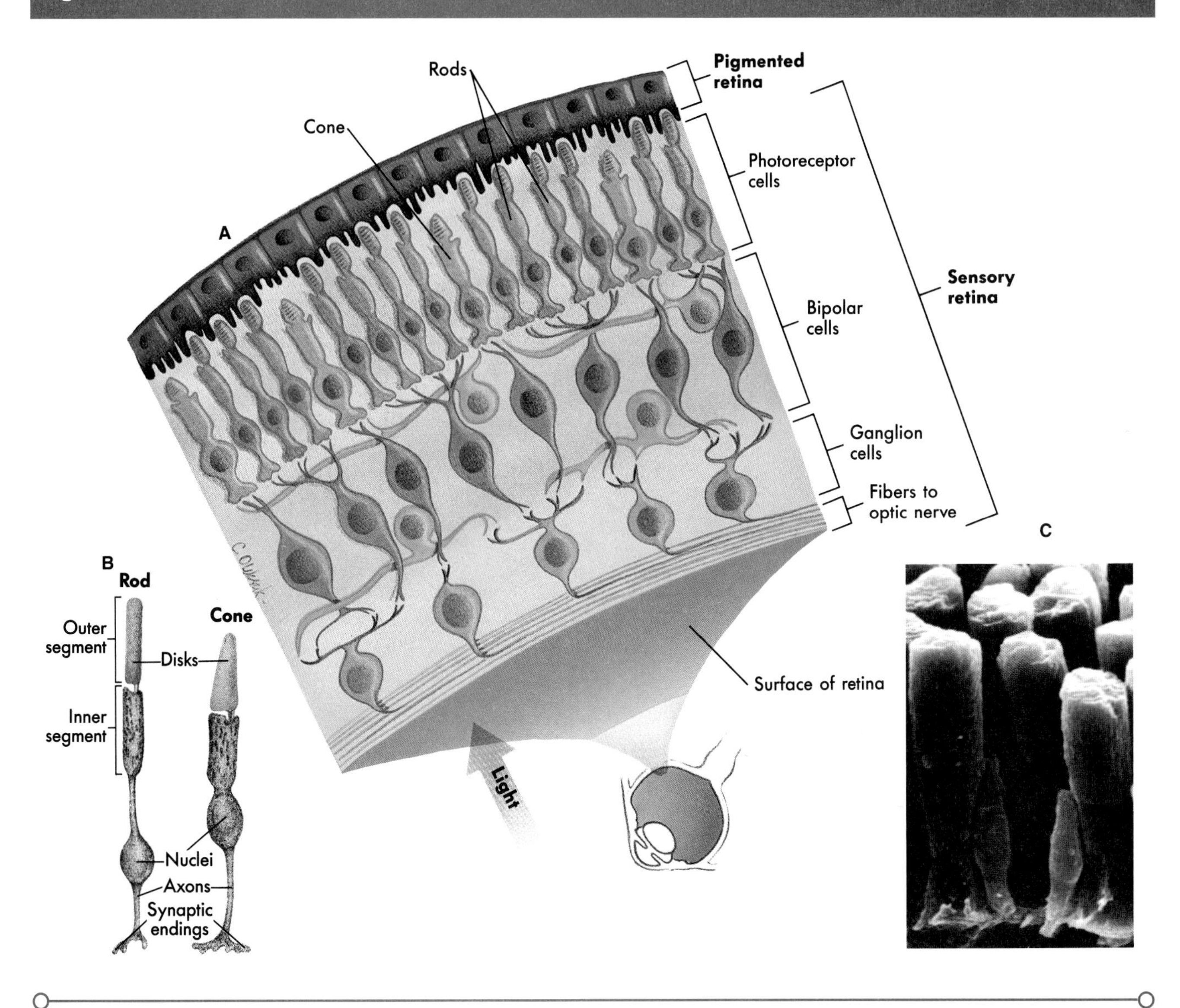

The retina consists of three layers of neurons. The layer closest to the choroid consists of photoreceptor cells called rods and cones. The second layer is made up of bipolar cells and the third layer is composed of ganglion cells (A). Rods and cones get their names from their shapes (B, C). Rods and cones are sensitive to different portions of the light spectrum we perceive.

is scattered by particles suspended in the air, making the sky appear blue. If melanin appears in all layers of the iris, the eye appears brown. Variation in the amount and distribution of melanin in the iris produces the many shades of eye color found in people.

The major function of the eye is to focus light from the environment on the photoreceptor cells of the innermost layer of the eye. We first describe the organization of the innermost layer of the eye and then discuss the nature of light. We then describe the path light takes when it enters the eye, ultimately to stimulate the photoreceptor cells contained in the innermost layer of the eye.

Retina

The innermost layer of the eyeball is the delicate **retina** (reh′-tih-nuh). The retina consists of three layers of neurons. The first layer, closest to the choroid, consists of two types of photoreceptors, called **rods** and **cones**. The second layer of neurons is made up of bipolar neurons. The third and innermost layer, facing the vitreous humor, is composed of ganglion cells (Figure 9-8). This organization of the retina appears to be backward because the light-sensitive photoreceptors are facing *away* from the light. During embryological development the light-sensitive neurons develop first, followed by the other two layers. Consequently light is required to pass through two other layers before it strikes the photoreceptor layer.

The axons of the ganglion cells leave the eye and form the optic nerve, which transmits action potentials to the brain. The area of exit is called the **optic disc,** also known as the **blind spot** because it lacks photoreceptors. Therefore any light focused on it cannot be detected.

A very small pit located in the center of the retina called the **fovea centralis** contains only cones. From the edge of the fovea toward the periphery of the retina, cone density declines gradually. The periphery of the retina contains only rods, which continuously decrease in density from there to the fovea. Just as do general receptors, cones and rods adapt to light stimuli. Therefore your eyes must move rapidly back and forth over an object or scene for you to comprehend it visually.

The **rods** are dim-light visual receptors that allow us to see rather unclearly in gray tones. **Cones** operate in bright light and provide highly accurate color vision. To understand how rods and cones detect light, we need to consider the nature of light.

The Nature of Light

Light is a form of electromagnetic radiation that travels in a wavelike fashion. The distance between two wave peaks is called the **wavelength.** The photoreceptors of the eye are sensitive only to wavelengths between 380 and 760 nm (nm, a nanometer, is 1 billionth of a meter). This **visible light** is only a small portion of the total electromagnetic spectrum. The photoreceptors are maximally stimulated by different wavelengths in the visible spectrum that we perceive as different color sensations. For example, short wavelengths are sensed as violet and blue, whereas longer wavelengths are sensed as orange and red.

Objects have color because they absorb some wavelengths and reflect others. For instance, a red apple reflects mostly red light, and grass reflects more of the green. Things that appear to be white are reflecting all wavelengths of light, and black objects are absorbing them all. Besides having variable wavelengths, light also varies in intensity, or in the amplitude or height of the energy wave. For example, dimming an orange light does not change its color; it simply becomes less intense or bright.

Photoreceptors

Both rods and cones consist of three parts: (1) an outer segment facing the choroid that detects light, (2) an inner segment that contains the cell organelles, and (3) a synaptic terminal that transmits the action potential to the bipolar cells. The outer segment, which is rod-shaped in rods and cone-shaped in cones, is composed of stacked, flattened disks containing photopigment molecules. Photopigment molecules undergo chemical alterations when activated by light. A photopigment consists of a protein called **opsin** and is combined with a substance derived from vitamin A, called **retinal.** There are four different photopigments, one in the rods and one in each of three types of cones. Retinal is identical in all four photopigments, but the opsins vary from one another.

The opsin found in rods, called **rhodopsin,** cannot discriminate between wavelengths in the visible spectrum but detects variation only in their amplitude. Consequently rods provide vision only in shades of gray. The naming of the cones reflects the colors (wavelengths) of light the cones absorb best and that are most efficient at causing the retinal to change shape and detach from its opsin. The blue cones respond best to wavelengths of 455 nm, the green cones to wavelengths of 530 nm, and the red cones to wavelengths of 625 nm (Figure 9-9).

Figure 9-9 Cone types and wavelengths

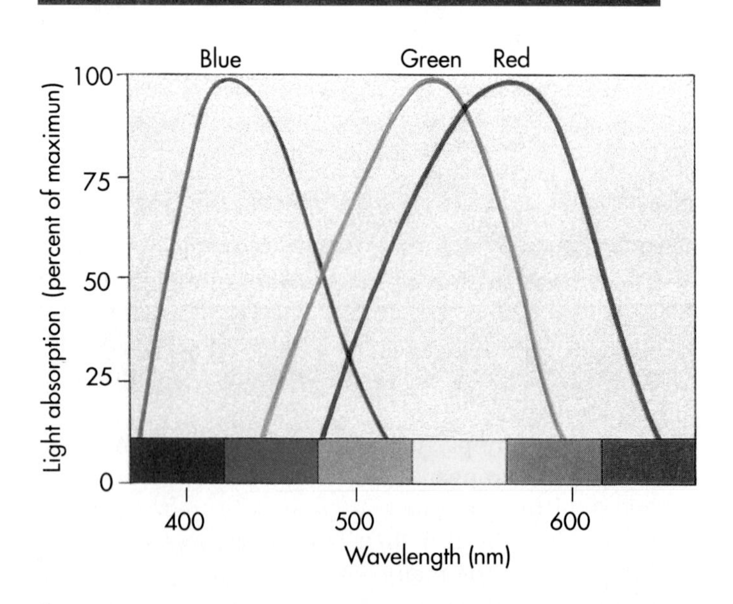

There are three different photopigments found in the cones. The naming of the cones reflects the wavelengths of light that the cones absorb best and are most efficient at causing an action potential to occur. Blue cones respond best to wavelengths of 455 nm, green cones to wavelengths of 530 nm, and the red cones to wavelengths of 625 nm.

Figure 9-10 Convergence

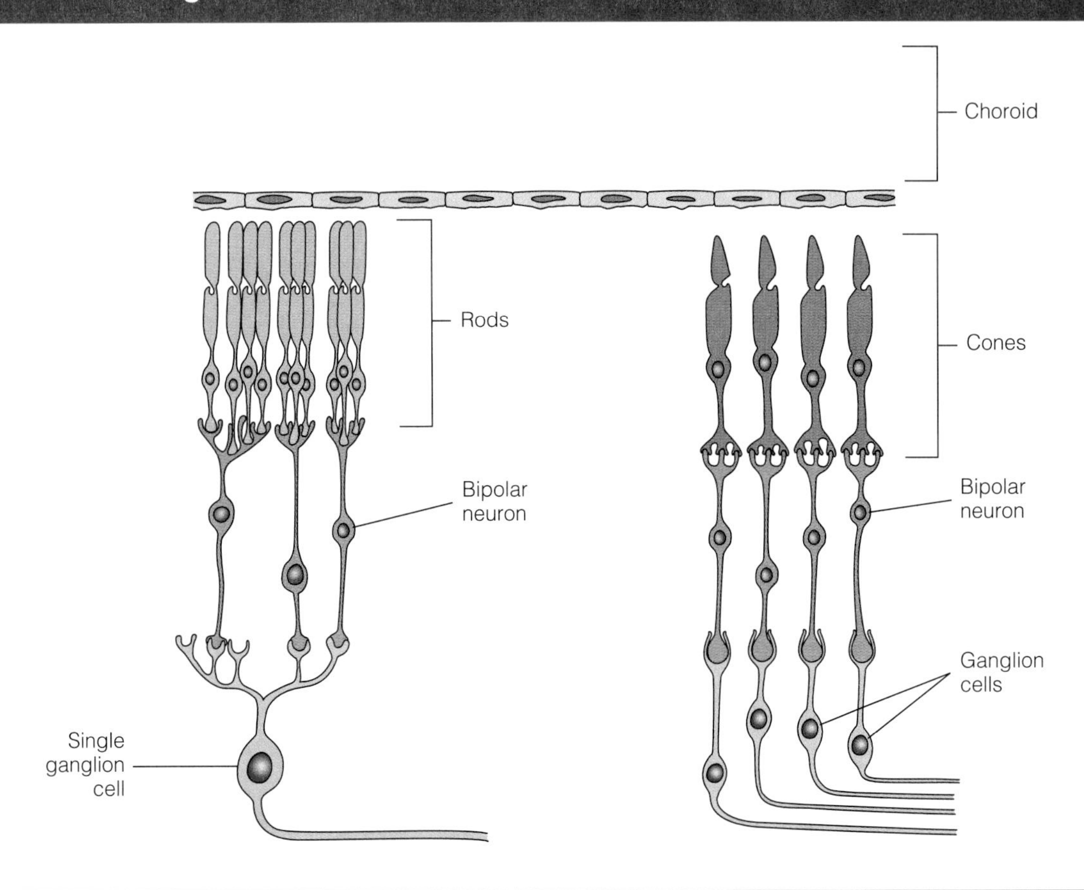

Convergence produces differences in the light sensitivity and acuity of rods and cones. Several rods converge on a single bipolar cell and several bipolar cells converge on a single ganglion cell. Consequently, rods have high sensitivity and respond to dim light but have low acuity, providing rather fuzzy vision. In contrast to rods, a single cone synapses with a single bipolar cell which synapses to a single ganglion cell. This arrangement causes cones to have low sensitivity and respond only in bright light, but to have high acuity, providing sharp, detailed vision.

If there are only three types of cones in the retina, how can you perceive so many different colors? Although the three types of cones respond *best* to certain wavelengths, they also respond in varying degrees to other wavelengths. Our perception of various colors depends on the three cone types' **ratios of stimulation** in response to different wavelengths. For example, the perception of the color yellow arises from the stimulation ratio of 83:83:0. This means the red and green cones are stimulated to 83% of their maximal stimulus while no blue cones are stimulated. In contrast, the ratio for green is 31:67:36. Thus there is 31% of maximal stimulation of the red cones, 67% of the green cones, and 36% of the blue cones. In other words, different combinations give rise to your perception of many different possible colors.

There are approximately 100 million rods and only 3 million cones in the retina of each eye. Remember that cones are stimulated by bright light and provide sharp vision and fine detail. They have low sensitivity to light but high acuity. Rods respond to dim light and provide rather fuzzy vision. They have high sensitivity and low acuity. The differences in the light sensitivity and acuity of rods and cones are caused by differences in the "wiring" between these photoreceptors and the outer layers of neurons of the retina.

This distinction occurs because of convergence. **Convergence** is the relationship between a single neuron and other neurons synapsing on it. For example, a single cone synapses with an individual bipolar neuron, and it in turn synapses with one ganglion cell. This 1:1 ratio of cone to ganglion cells results in high acuity (you are able to distinguish different parts of an image) but decreased light sensitivity. Each cone must be sufficiently stimulated to generate an action potential in the ganglion cell. In contrast, there is a large convergence of the rods on a single ganglion cell. Output from as many as

100 rods may converge via bipolar cells on a single ganglion cell (Figure 9-10). Increasing the number of rods converging on a single bipolar cell and then onto a single ganglion cell allows many weak signals to be added together, generating an action potential in the ganglion cell. The convergence of rods enables you to see in very dim light. That is, it increases the light sensitivity of your eyes. However, the greater the degree of convergence the poorer the rod acuity.

The Path of Light in the Eye

Light travels in straight lines. When traveling in a given medium, the speed of light is constant. However, when it passes from one transparent medium into another with a different density, its speed is altered. Light slows down as it passes into a denser medium. As a consequence of changes in speed, light waves are bent, or **refracted,** whenever they meet the surface of a different medium at an oblique or less than perpendicular angle to its surface. The focusing abilities of a lens are based on the property of refraction.

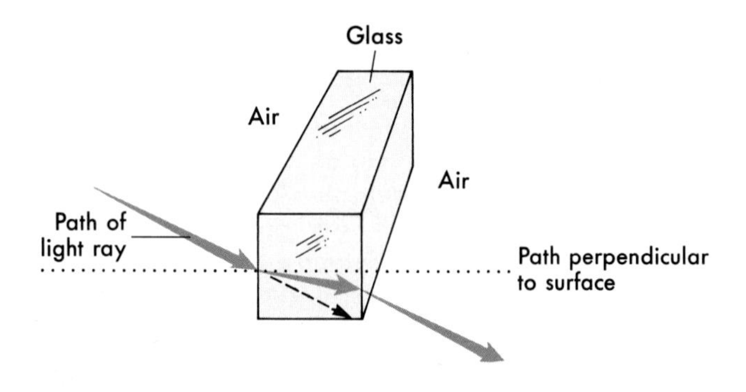

A lens is any piece of transparent material curved on one or both surfaces. Because light hits the curve at an angle and is moving through a medium of different density, it is refracted. For example, if the lens surface is convex (thickest in its center), the light is bent so it converges, or comes together, at a single point called the **focal point.** The more convex the lens, the more the light is bent and the shorter the distance between the lens and the focal point. The image formed by a convex lens is also upside down and reversed from left to right. We can apply these facts to the structure of the eye.

Light that enters the eye first passes through the cornea, continues on through the aqueous humor, and enters the lens by way of the pupil. The rays of light converge as they pass through the lens. The rays pass through the vitreous humor, finally to focus on the retina. The ability of the eye to focus light on the retina is determined by the cornea and the lens. The curved cornea is the first structure light passes through as it enters the eye and contributes significantly to the eye's total refractive ability. This is because the difference in density between the air and the cornea is much greater than the difference in density between the lens and the fluids surrounding it. If the eye is far from an object, at least 20 feet, the cornea alone is needed for focusing. But if the eye is close to the object, additional focusing is required. The lens provides this additional focusing power.

The ability of the eye to adjust the lens so that objects can be focused on the retina is known as **accommodation.** The focusing ability of the lens depends on its shape, which is regulated by the ciliary muscles. The ciliary muscle is a circular ring of smooth muscle attached to the lens by suspensory ligaments. When the ciliary muscle is relaxed, the suspensory ligaments are taut, putting tension on the lens and keeping it flat. However, when observing close objects, the ciliary muscle contract, and the suspensory ligaments become slack. This results in reducing tension on the lens, which then becomes more rounded (Figure 9-11).

Figure 9-11 Accommodation of lens

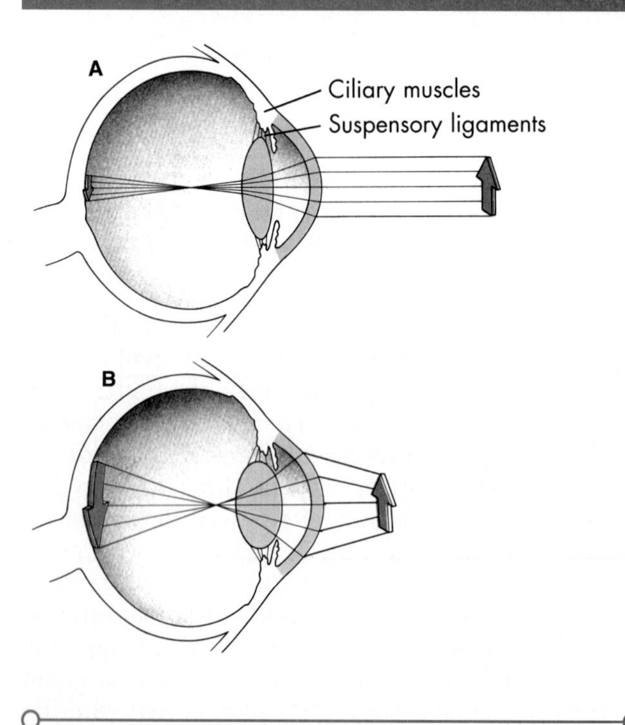

The focusing ability of the lens of the eye depends on its shape. When viewing objects at a distance, the ciliary muscles are relaxed, the suspensory ligaments are taut, and the lens is flat (A). When viewing objects up close, the ciliary muscles are contracted, the suspensory ligaments are slack, and the lens is rounded (B).

Figure 9-12 Correcting visual problems

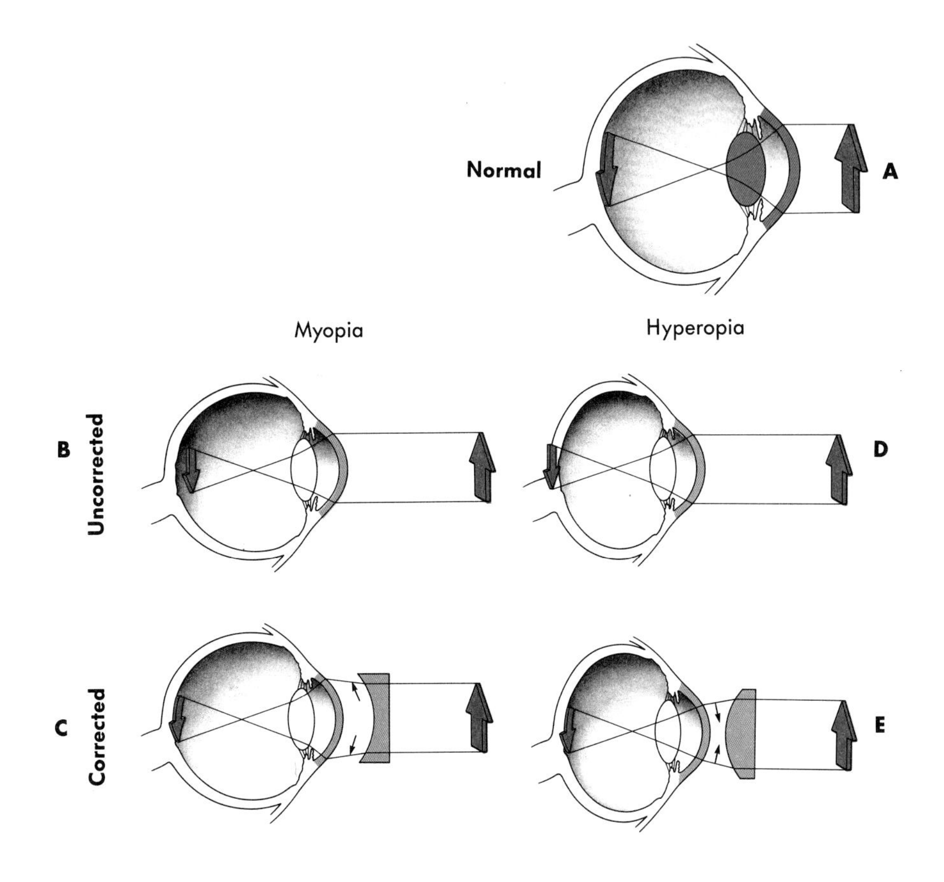

In normal vision the lens focuses visual images directly on the retina (A). In myopia, or nearsightedness, the lens focuses distant objects in front of the retina, but focuses close objects correctly (B). This problem is corrected with a slightly concave lens that diverges the light before it enters the eye (C). In hyperopia, or farsightedness, close objects are focused behind the retina, but distant objects are focused correctly (D). This problem is corrected by a slightly convex lens that converges the light more strongly for close vision (E).

Correcting Visual Problems

When you take an eye examination, if you are able to clearly see a size "20" letter 20 feet away, you have 20/20, or normal, healthy vision.

Myopia (mi-o'-pe-uh) occurs when distant objects are focused in front of the retina. Myopic people can see close objects clearly, but distant objects are blurred. The common name for myopia is **nearsightedness,** which emphasizes the aspect of vision *not affected.* Myopia results from too strong a lens or too long an eyeball. Correction requires a slightly concave lens that diverges the light before it enters the eye.

If a close object is focused behind the retina, the person is said to have **hyperopia** (hi-per-ah′-pe-uh), or **farsightedness.** Hyperopic people see distant objects clearly but cannot focus close objects properly. Hyperopia usually results from either a lens with poor refractory qualities or too short an eyeball (Figure 9-12). Glasses with convex lenses that converge the light more strongly for close vision will correct hyperopia.

As we age so do our eyes. Throughout life only cells at the outer edges of the eye's lens are replaced. Cells making up the center of the lens are doubly at risk as we age. First they are farthest away from the aqueous humor, the lens's source of nutrients. Second these nonrenewable central cells die and become rigid. With the loss of elasticity the lens is no longer able to change to the required shape to accommodate for near vision. This reduction of accommodation for near vision, called **presbyopia** (prez-be-o′pe-a), affects most people between the ages of 45 and 55. The person usually then has to wear a corrective lens for near vision.

Biology in Action WHY SURGICAL GOWNS ARE GREEN

You may have noticed that the most common color for surgical gowns or the two-piece, surgical shirt-and-pants sets is a sort of greenish-blue. When asked why they think the surgical garments are this color, people give responses that range from "it is a peaceful color and relaxes the patients" to "it doesn't show blood as easily as white." There is actually an important technical reason for using this color, based on the neurological workings of our color vision.

The reason surgical gowns are greenish-blue has to do with a visual phenomenon called **negative afterimage.** Afterimage refers to the subjective persistence of an image after the cessation of the visual stimulus. It is negative if the image is seen in complementary colors. How and why does negative afterimage occur?

Recall that the retina contains three types of cones, each type responding to light wavelengths we experience as red, green, and blue. When cones are stimulated by light, a photopigment breaks down, resulting in an action potential. The less photopigment a cone has, the less responsive it is to light until it regenerates more pigment.

When an area of the retina has been strongly stimulated by a colored light, the cones responding to that color will soon become depleted of photopigment. Meanwhile the other two kinds of cones will have been building up their stores of photopigment. For example, if we project a pure blue square on a screen, the blue cones of your eye will become exhausted while the red and green cones remain unaffected. Now we turn off the blue light and immediately project a white square of light on the screen. Remember that light perceived as white contains red, green, and blue wavelengths. But when the light at those three wavelengths strikes the cones in the eye, the red and green cones will be stimulated and the blue cones will hardly be stimulated at all. They are still regenerating their photopigment from the previous encounter with blue light. To the person experiencing this, the square will appear yellow because yellow is what we perceive when red and green cones are stimulated at the same time. This yellow image is the *complementary color* of the original image and is what is meant by a negative afterimage. Now we can return to our operating room example.

We could, for example, put the surgeon and the attending staff in white clothing. The surgeon would be looking down, under very bright white lights, at the incision in a patient. The incision would reflect only the wavelength of the color of blood, bright red. The image of this incision is projected on the surgeon's retina, and it depletes the photopigment in the red cones that outline the shape of the incision. The blue and green cones in that area are still sensitive because they have not been strongly stimulated. Now suppose the surgeon looks up from the operation and across the table to an assistant. Reflected from the assistant's white garment will be white light, that is, light containing red, green, and blue wavelengths. It strikes the cones in the surgeon's retina. However, in the area stimulated previously by the incision, the red cones will be relatively unresponsive while the blue and green cones will be completely responsive. As a result, everywhere the surgeon looks against a background of white, the surgeon will see a greenish-blue negative afterimage of the incision because the blue and green cones are responding but the red ones are not. This illusion disappears after about 1 minute as the blue and green cones use up their photopigments and become less sensitive, and the red cones rebuild their photopigment supply and become more sensitive.

To make life more bearable in the surgical room the solution is simple. Make clothing the color of the negative afterimage, and then one would not have the disorienting experience of perceiving a greenish-blue incision on everyone's clothing no matter where one looked. The greenish-blue surgical garments are then, in a sense, soothing to the surgical staff.

Depth Perception

Normally both eyes are directed and coordinated by the eye muscles toward the same object, and the object is thus focused on corresponding places of the two retinas. However, each eye views the object from a slightly different angle. In humans this angle of difference is about 6 degree. Each eye sends to the brain its own information about the object being viewed. This sensory information is analyzed in two steps. First each half of the brain receives one half of its visual information from each eye. Second the two halves of the primary visual association area of the cerebral cortex communicate, through the corpus callosum, to create a complete three-dimensional interpretation of the object being viewed. This organization gives us **depth perception.** **Depth perception** is the ability to assess the distance between oneself and another object and the relative distances between two objects.

The Ear: Hearing and Balance

The ear is divided into the three major areas of the outer ear, the middle ear, and the inner ear. The outer ear and middle ear structures are involved with hearing, whereas the inner ear functions in both hearing and balance.

Outer Ear

The outer, or external, ear consists of the pinna (pin'ah) and the external auditory canal (Figure 9-13). The **pinna** is what most people call the ear, the shell-shaped projection surrounding the opening of the external auditory canal. The **external auditory canal** is a short, narrow, tubelike chamber about 2.5 cm long. The entire canal is lined with fine hairs and modified sweat glands that secrete a wax-like substance called **cerumen** (ser-oo'-min). This "earwax" helps protect the ear against the entrance of foreign bodies and potential infective agents.

Middle Ear

The middle ear begins at the tympanic membrane, or eardrum. The **tympanic membrane** is a thin membrane that transmits sound vibration to the next section of the ear. The middle ear ends at a bony wall in which are found two small membrane-covered holes, the **oval window** and below it, the **round window**. This bony wall also contains an opening leading to the **eustachian tube** (u-sta'she-an), which runs down at an angle to link the middle ear to the back of the throat, an area called the **nasopharynx** (na'zo-far'ingks). This connection of the middle ear with the nasopharynx ensures there is no difference in air pressure between the middle

Figure 9-13 Hearing

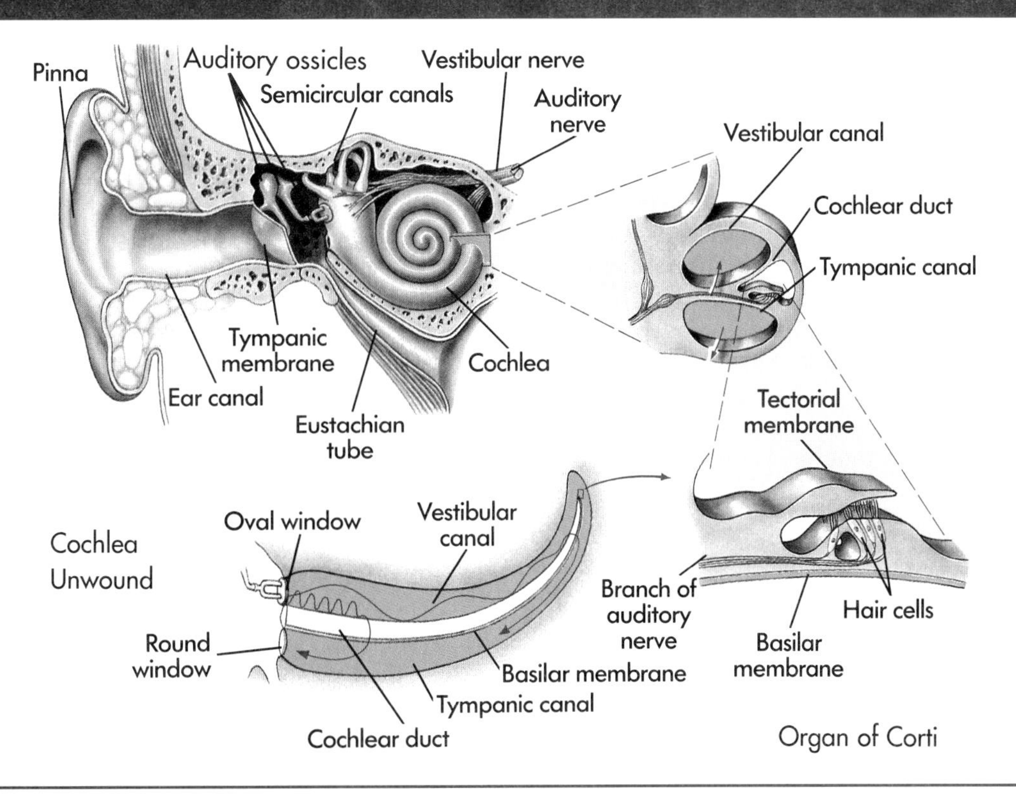

ear and the outer ear. The familiar "ear popping" associated with landing in an airplane or with the rapid descent of an elevator is a result of pressure equalization between the two sides of the eardrum. In children the eustachian tube is horizontal. Therefore the protective fluid in it occasionally does not drain well. As a result, it is often the site of bacterial infections common in children.

Three of the smallest bones in the body are in the middle ear between the tympanic membrane and the oval window, collectively known as the **auditory ossicles** (see Figure 9-13). These bones, named for their shape, are the **malleus** (ma'-le-us), or hammer; the **incus** (in'-kis), or anvil; and the **stapes** (sta-pez), or stirrup. The "handle" of the malleus is fastened to the eardrum, and the stapes fits into the oval window.

Inner Ear

The membrane of the oval window is the door to the inner ear. Whereas the outer ear and the middle ear contain air, the inner ear and its various chambers are filled with fluid. One chamber of the inner ear is shaped like a tightly coiled snail shell and is called the **cochlea,** from the Latin word for snail. Running through its center like a wedge-shaped tunnel is the membrane-covered **cochlear duct.** The cochlear duct divides the cochlea into three separate chambers. These are the **vestibular canal,** connected to the oval window, the **cochlear duct,** and the **tympanic canal,** which terminates at the round window. Although next to each other, the round and oval windows are actually at opposite ends of the cochlea because it is bent back on itself. The cochlear duct houses the **Organ of Corti,** the receptor organ for hearing. The "floor" of the cochlear duct is composed of the **basilar membrane,** which supports the Organ of Corti. The Organ of Corti is composed of supporting cells and mechanoreceptors called **hair cells.** The cilia of the hair cells are embedded in an overlying gel-like "roof" called the **tectorial membrane.** When these extensions are displaced or bent in different directions, the hair cell generates an action potential in a neuron to which it synapses. This action potential is sent to the brain to be analyzed and interpreted.

Hearing

Sound is propagated by the molecules of the medium through which it travels. Just as ripples spread out from the point at which a pebble hits the surface of a pond, sound travels through the air by bumping air molecules into others around them and pushing molecules out from the source of the sound, creating pressure waves, or sound waves. The process of hearing begins when sound waves enter the auditory canal.

The sound waves then hit the tympanic membrane, causing corresponding vibrations to travel through the malleus, incus, and stapes. These bones act together as a lever system, increasing the force of the vibrations. The stapes pushes against the oval window. Because the oval window is smaller than the tympanic membrane, vibration against it produces more force per unit area of membrane. The back-and-forth action of the stapes against the oval window sets up pressure waves in the vestibular canal. Because fluid is incompressible, the pressure must be dissipated as the oval window is pushed inward. As a result of fluid movement in the cochlea, the basilar membrane moves up and down, and the cilia of the hair cells move against the tectorial membrane. This bending of the cilia initiates action potentials that are carried by the auditory nerve to the brain, where they are interpreted as sound.

The basilar membrane is relatively thick and narrow near the oval window. It becomes wider and thinner toward the end of the cochlear duct. Thus different regions of the basilar membrane respond to different sound-wave frequencies. The highest-frequency sounds, such as those of a whistle, vibrate the base of the basilar membrane nearest the oval window. The lowest frequencies, such as those of a fog horn, vibrate the wider and thinner section of the membrane near the tip of the cochlea.

The neurons from each region along the Organ of Corti, which rests on the basilar membrane, eventually arrive at slightly different regions in the auditory association area of the cerebral cortex. The different frequencies of sound you perceive depend on which of these areas is stimulated. On the other hand, loudness is a function of the amplitude of sound waves. Loud noises cause the fluid of the cochlea to move back and forth to a greater degree, which movement causes the basilar membrane to move up and down to a greater extent. The resulting increased stimulation is interpreted by the cortex as loudness.

Balance

The inner ear also functions in the sense of balance called equilibrium. The inner ear contains more than the cochlea. Go back to Figure 9-13 and you will see that off to one side of the oval window and round window are three semicircular canals, one arranged in each of the three dimensions of space. Each semicircular canal is slightly enlarged at its base where it meets the vestibule. In this enlarged space is a very small elevation, called the **crista ampullaris,** containing hair cells with their cilia extending into the enlarged fluid-filled space. (Figure 9-14.)

The **semicircular canals** detect rotational or angular motion, such as spinning, somersaulting, or head turning. Acceleration or deceleration during head rotation in any direction causes fluid movement in at least one of the semicircular canals of each ear because of their three-dimensional arrangement. As the head starts to move, the crista ampullaris and the hair cells also move with the head. However, the fluid does not move immedi-

Figure 9-14 Balance

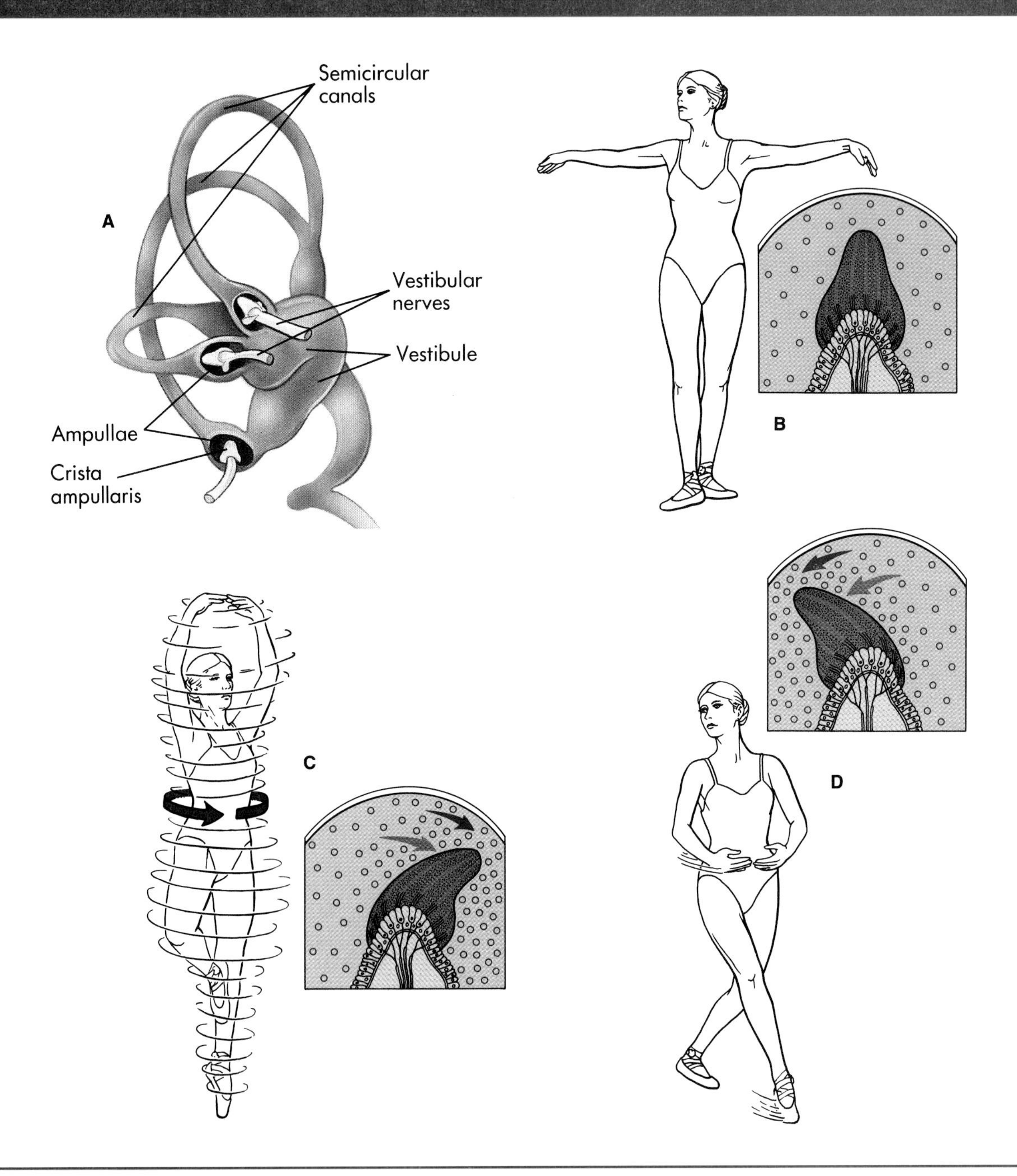

The inner ear houses three semicircular canals, one arranged in each dimension of space (A). At the base of each canal there is a small elevation called the crista ampullaris, which contains hair cells that project their cilia into the fluid-filled space (B). Acceleration or deceleration during head rotation in any direction cause fluid movement in at least one of the semicircular canals. The fluid causes the hair cells to bend and generate an action potential (C, D).

ately but lags behind slightly. As the fluid is "left behind" momentarily while the head starts to rotate, the fluid in the canal in the same plane as the movement of the head bends the hair cells in the opposite direction.

This is similar to tilting your body to the right as the car in which you are riding is suddenly turning to the left. If the head movement continues at the same rate in the same direction, the fluid catches up to the crista ampullaris, so the hair cells return to their upright position. When the head slows down and stops, the reverse action occurs. The fluid briefly continues moving in the direction of the rotation, as the head decelerates to a stop. As a result, the hair cells are bent in the direction of the rotation, which is opposite to the way they were bent during acceleration. For example, after your first ride on a merry-go-round you probably still felt as if you were moving after the ride ended. In this way the semicircular canals detect changes in the rate of the head's rotational movement. They do not respond when the head is motionless or during rotation at a constant speed.

In this chapter you discovered how and in what ways the nervous system senses the external environment and the body's own internal environment to respond appropriately to them. In the next chapter we discuss the body's other primary system for maintaining homeostasis, called the endocrine system.

CHAPTER REVIEW

SUMMARY

1. Any internal or external environmental change that evokes an action potential in a sensory receptor is called a stimulus. Each type of sensory receptor is specialized to respond to one type of stimulus.
2. General receptors are sensory neurons located in the skin and also elsewhere in the body, such as the internal organs. Special-sense receptors are either complex sensory organs, such as the ears and eyes, or small, localized clusters of receptors, such as taste buds. Regardless of receptor type each sends nerve impulses to the brain; it is the brain that interprets the impulses as being "sound," "sight," or "touch."
3. Sensory receptors can be categorized according to type of stimulus: photoreceptors are responsive to light waves, mechanoreceptors to pressure and touch, thermoreceptors to heat and cold, chemoreceptors to specific chemical substances, and pain receptors to tissue damage. Even if activated by a different stimulus, a receptor still produces the sensation usually detected by that receptor type.
4. The eye contains two types of photoreceptors. Rods allow us to see in dim light rather unclearly in gray tones. Cones operate in bright light and provide us with highly accurate color vision. Cones are differentiated into three different types according to the type of photopigment they contain. Each photopigment is maximally sensitive to a particular region of the visible light spectrum.
5. The ear functions in both hearing and the sense of balance, called equilibrium. The sensory cells for both functions, called hair cells because of their hairlike cilia, are found in fluid-filled chambers of the inner ear. Hearing is sensed in the cochlea, while equilibrium is detected in the semicircular canals of the inner ear.

FILL-IN-THE-BLANK QUESTIONS

1. Any environmental change that evokes an action potential in a sensory receptor is called a ______.
2. Despite continued stimulation at the same strength, some receptors undergo ______.
3. Pain receptors are ______ ______ ______ whose dendrites have receptor molecules for ______.
4. Photoreceptors contain receptor molecules called ______.
5. The ______ is used to focus light on the retina when viewing objects at a distance.
6. ______ operate in bright light and provide highly accurate ______ vision.
7. The Organ of Corti is found in the ______ of the inner ear.
8. The hair cells of the ______ and ______ detect static equilibrium.

SHORT-ANSWER QUESTIONS

1. Explain how variations in stimulus intensities are conveyed to the central nervous system from different sensory receptors.
2. Of what value is it to the central nervous system that some sensory receptors undergo adaptation?
3. Describe what occurs from when light first passes through the cornea to when an action potential is sent through the optic nerve of the eye.
4. Trace the path sound takes from when it enters the external ear to when the hair cells are stimulated in the Organ of Corti.

VOCABULARY REVIEW

accommodation—*p. 182*
adaptation—*p. 169*
choroid—*p. 176*
cochlea—*p. 186*
cone—*p. 180*
Meissner's corpuscle—*p. 172*
Organ of Corti—*p. 186*
Pacinian corpuscle—*p. 171*
retina—*p. 179*
rod—*p. 180*
sclera—*p. 176*
semicircular canal—*p. 187*

Chapter Ten

The Endocrine System

For Review

Here are some important terms and concepts that you will encounter in this chapter. If you are not familiar with them, you should review them before proceeding.

Endocrine glands
(page 126)

Homeostasis
(page 123)

Negative-feedback system
(page 124)

Osmosis
(page 40)

Objectives

After reading this chapter you should be able to:

1. *Describe the function of endocrine glands.*
2. *Describe the two mechanisms by which hormones modify cellular activity.*
3. *Explain how negative feedback regulates hormone secretion.*
4. *Explain the hypothalamus–pituitary gland connection and describe the means by which the anterior and the posterior pituitary communicate with the hypothalamus.*
5. *Describe the function of the posterior pituitary gland and the two hormones it releases when stimulated.*
6. *Describe the two functional categories of hormones produced by the anterior pituitary gland and list the hormones in each category.*
7. *Explain how TSH affects the thyroid and describe the importance of the two hormones secreted by the thyroid.*
8. *Describe the structure and function of the adrenal medulla and adrenal cortex.*
9. *Name and discuss the importance of the two types of hormone-producing cells found in the islets of Langerhans.*
10. *Discuss the roles of glucagon and insulin in maintaining blood glucose levels.*
11. *List and describe the function of the most important nonendocrine glands.*

Maintaining homeostasis requires the coordination and integration of all bodily activities. For this to occur communication between cells is necessary. The body relies on two communication systems. The first, called the nervous system, operates through electrochemical impulses that control and regulate the moment-to-moment activities of the body. The second, called the endocrine system, relies on chemical messengers sent through the bloodstream. The endocrine system controls processes that are either prolonged, such as growth and development, or ongoing, such as the production of gametes in adults.

The Endocrine System

In this chapter we discuss the structure and function of the endocrine glands and the substances they secrete. By defining *hormones,* their production and regulation, and how they act on the cells and tissues of the human body, we can gain greater insight into how the human body maintains a dynamic homeostasis through time. You will also begin to gain a perspective fundamental to understanding many essential life processes of the human body, such as reproduction, growth and development, and immunity, and the general role each of the body's organs has in them.

Endocrine glands release their chemical substances, called hormones, directly into the blood. **Exocrine glands,** on the other hand, have ducts through which their nonhormonal products travel to a membrane surface. For example, sweat glands are exocrine glands because they secrete a watery fluid through ducts to the skin's surface.

The endocrine glands include the pituitary, thyroid, parathyroid, the adrenals, and thymus. Several other body organs also contain areas of endocrine tissue that produce hormones. These organs include the pancreas and the gonads, which are the ovaries of females and the testes of males. (Figure 10-1.)

The nervous system either directly or indirectly controls hormone secretion. Conversely, certain hormones also influence nervous system function. The major link between the nervous and endocrine systems is through a region of the brain called the **hypothalamus**, which performs two interrelated functions vital to the body. First it monitors internal organs and influences forms of behavior related to their activities. It is the "control center" of homeostasis. Second the hypothalamus exerts major control over the secretion of several hormones through its influence on the pituitary gland.

Through the hypothalamus–pituitary gland link, integration and coordination of the body's activities to maintain homeostasis are made possible. The major processes controlled and integrated by hormones are reproduction; growth and development; maintenance of water, nutrients, and ion balance in the blood; and regulation of cellular metabolism. Hormones also mobilize the body's defenses against any **stressor,** which is a stimulus that disturbs the body's homeostasis. A stressor could be anything from being hungry, having a viral infection, or childbirth (Table 10-1).

Table 10-1 Hormone Actions

- Growth and development
- Water balance
- Nutrient and ion balance in blood
- Regulation of cellular metabolism
- Mobilization of the body against stress
- Changes in cell membrane permeability

Hormones

A **hormone** is a chemical substance produced and secreted in one part of the body that initiates or regulates the activity of an organ or group of cells in another part of the body. Although the body produces a variety of hormones, nearly all of them can be classified chemically into one of two basic groups of biochemical molecules. Hormones are either amino acid–based structures, or they are derived from lipid molecules and called steroid hormones. Most hormones belong to the first group. Some of these are called peptide hormones because they are comprised of short chains of amino acids. Of the hormones produced by the major endocrine glands, only the gonadal hormones and those hormones secreted by the outer cell layers of the adrenal glands are steroids.

Mechanisms of Hormone Action

Although all major hormones secreted by the endocrine glands circulate through the blood to most tissues, hormones affect only specific cells, which are called **target cells**. The ability of a particular cell to respond to a hormone depends on the presence of specific protein receptors on its cell membrane or in its interior, receptors to which that hormone can chemically bond.

Hormones bring about their characteristic effects on target cells by altering cell activity. They modify normal cellular processes. The particular response

Figure 10-1 The human endocrine system

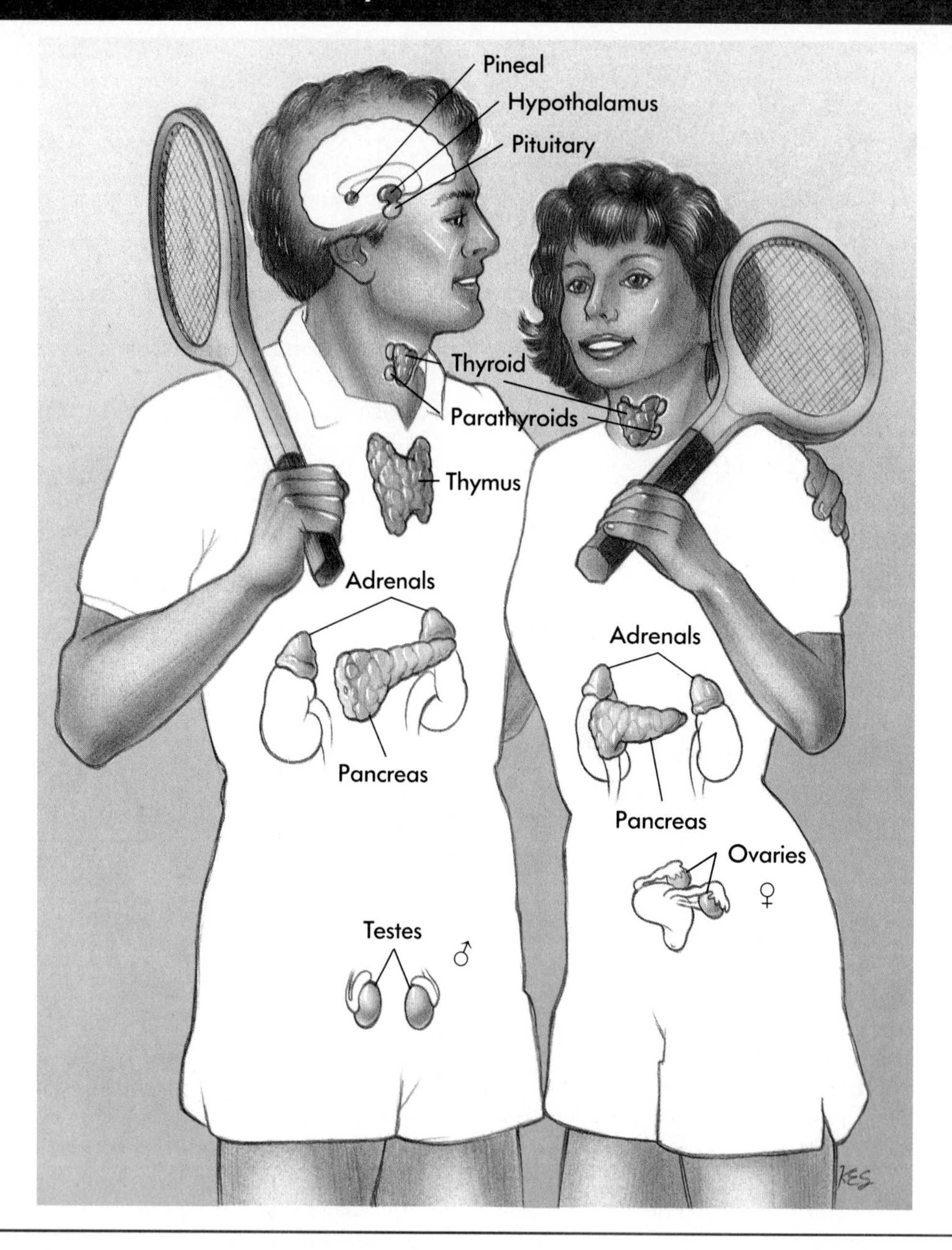

the hormone causes is determined by the particular type of target cell it affects. A hormone affects normal cellular processes in one or more of the following ways: (1) initiating changes in cell membrane permeability, (2) initiating the synthesis of proteins or enzymes in the cell, (3) activating or deactivating enzymes, or (4) inducing secretory activity.

There are two major mechanisms by which hormones modify cellular activity. One involves the formation of one or more intracellular products called **second messengers,** which mediate the target cell's response to the hormone. The other involves direct **gene activation** by the hormone itself. We consider each in some detail.

Amino acid–based hormones do not enter cells. Instead, they cause changes in cellular activity through intracellular second messengers produced as a consequence of the hormone bonding to specific cell-membrane receptors (Figure 10-2). Of the known second messengers, **cyclic AMP** (adenosine monophosphate) is the most common and best understood. A hormone, or the first messenger, first attaches to a protein receptor site on the outer surface of the cell membrane. This protein-hormone complex activates an intracellular enzyme that removes two phosphate groups from an ATP molecule, forming a molecule of cyclic AMP.

Figure 10-2 How amino acid–based hormones work

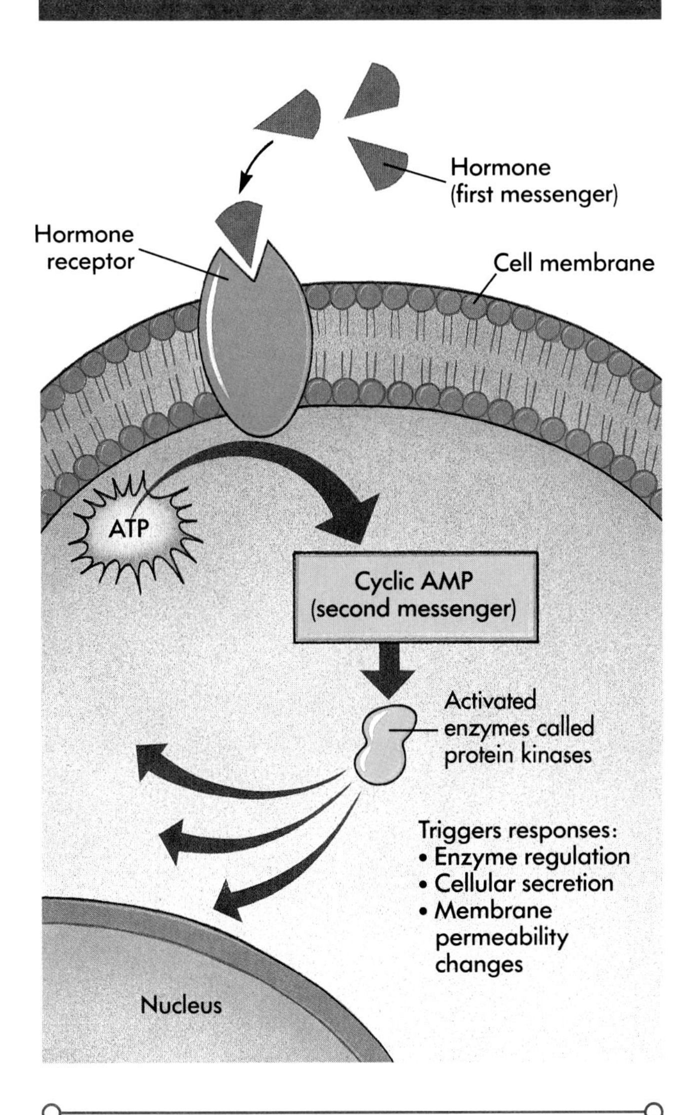

Amino acid–based hormones cannot pass through the cell membrane and therefore must affect cell activity through a second messenger. When the hormone binds to a receptor site, an ATP molecule is converted to cyclic AMP, the second messenger. Cyclic AMP diffuses throughout the cytoplasm to activate a class of enzymes called protein kinases. These in turn influence other enzymes, altering cellular activity.

Acting as a second messenger inside the cell, cyclic AMP can then diffuse throughout the cytoplasm to initiate a cascade of biochemical reactions in which one or more enzymes, collectively called **protein kinases,** are activated. A cell may have several types of protein kinases, each of which interacts with different substrates. Protein kinases usually modify intracellular proteins, many of which are also enzymes. This process eventually activates some enzymes and inhibits others, resulting in a variety of chemical reactions in the target cell.

Steroid hormones modify cellular events through gene activation (Figure 10-3). Because steroid hormones are made from the lipid cholesterol, they are lipid soluble and are able to diffuse easily through the cell membrane of their target cells. Once inside, steroid hormones combine with specific hormone-receptor proteins in the cytoplasm. The chemical bonding of the hormone to the receptor protein changes the shape of the receptor, allowing the hormone-receptor complex to enter the

Figure 10-3 How steroid hormones work

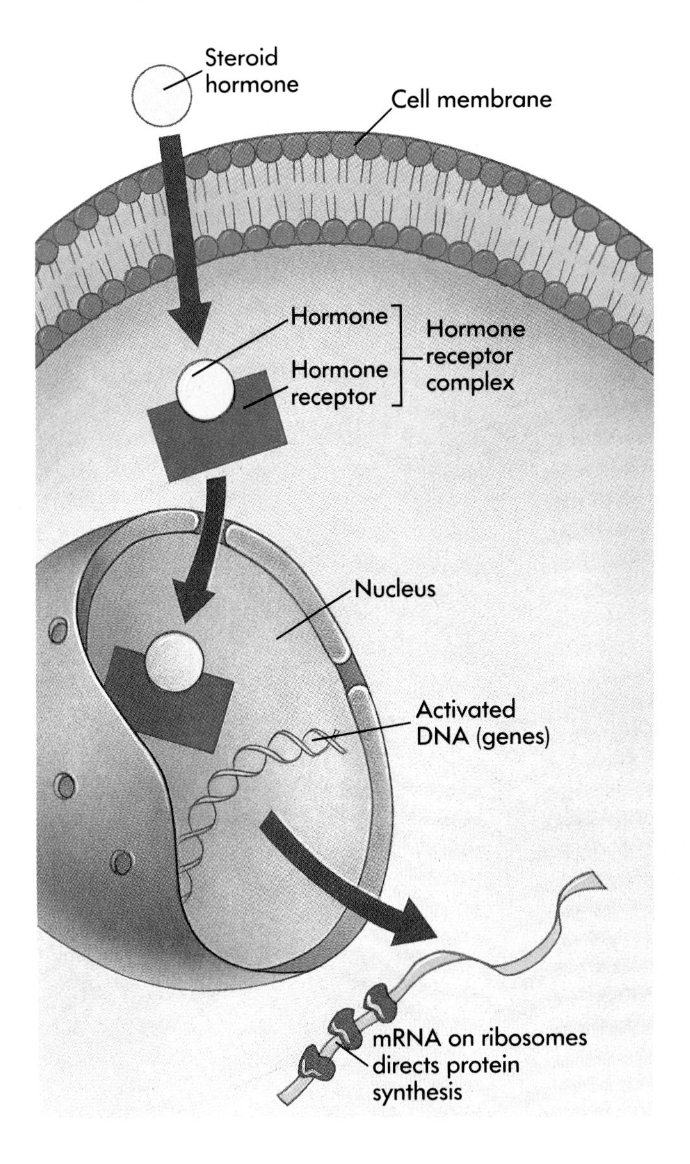

Because steroid hormones are lipid soluble, they can pass through the cell membrane. Once inside the cell the hormone chemically bonds to a hormone receptor, moves into the nucleus, and activates specific genes. Depending on the genes activated, cellular activity is altered.

nucleus and bind to specific regions on the chromosomes. This then triggers the activation of areas of DNA that code for the synthesis of specific proteins, usually other enzymes. Once transcribed and translated, these newly synthesized enzymes initiate specific metabolic functions. Because protein synthesis is the cellular response to steroids, their effects are much slower than those of the amino acid–based hormones that operate by the second-messenger mechanism.

Regulation of Hormone Secretion

Recall from Chapter 7 that negative-feedback mechanisms keep the body in homeostasis. The synthesis and release of most hormones are also regulated by some type of negative-feedback system. Hormone secretion is first triggered by some internal or external stimulus. The rising hormone levels in the blood inhibit further hormone release. As a result blood levels of most hormones vary within only a narrow range. For example, insulin is a hormone that regulates the concentration of glucose in the blood. When blood glucose levels increase after a meal, insulin is secreted. Insulin acts on cells and causes them to take up glucose, resulting in decreasing levels of blood glucose. As the blood glucose level begins to fall, the rate at which insulin is secreted also falls. As insulin levels fall the rate at which glucose is taken up by the target cells decreases, keeping the level of blood glucose from dropping too much.

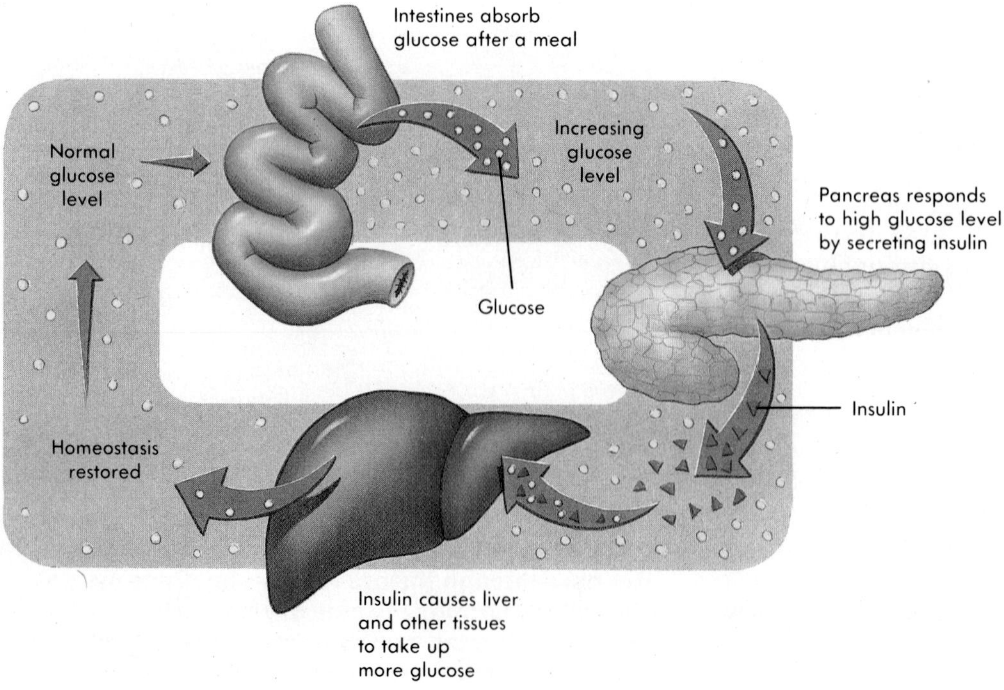

Stimulation of hormone production may be modified by the activity of the nervous system. Without this added control, endocrine-system activity would be strictly mechanical, much like a household thermostat. For example, a thermostat can maintain the house temperature at or around its set value. However, it cannot "know" that your visiting aunt from Florida feels cold at that temperature and then reset itself to another temperature. You must make the adjustment. Similarly the nervous system can override normal endocrine controls to maintain homeostasis. Recall that insulin maintains blood glucose levels within a narrow range. However, when the body is under severe stress, the hypothalamus and other parts of the nervous system are strongly activated and cause blood glucose levels to rise much higher, making greater amounts of energy available to the cells. Let us then look next at where the connection between the endocrine and the nervous systems occurs.

Hypothalamus–Pituitary Gland Connection

The **pituitary gland** (pit-u'i-ter'e), or **hypophysis** (hi-pof'i-sis), is a small endocrine gland about the size of a common garden pea. It is located in a bony cavity at the base of the brain just below the hypothalamus. Try pointing one finger directly between your eyes and another finger horizontally into one of your ears; the imaginary point where these lines would intersect is about the location of your pituitary gland. The pituitary is connected to the hypothalamus by the pituitary stalk, a thin stalk of tissue that contains nerve cells and small blood vessels.

The pituitary gland consists of two anatomically and functionally distinct portions. First is a **posterior lobe,** also called the **posterior pituitary** or neurohypophysis, and the second is an **anterior lobe,** also called the **anterior pituitary** or adenohypophysis (Figure 10-4). The posterior pituitary is derived from an outgrowth of the brain and is composed of nervous tissue. The anterior pituitary consists of cells originally derived from epithelial cells forming

Figure 10-4 Structure of the pituitary gland

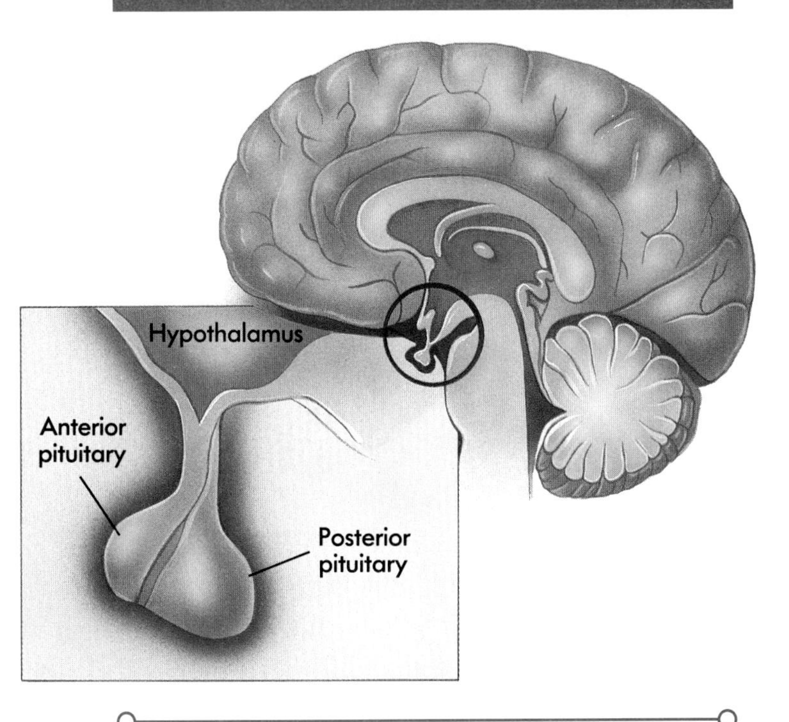

The pituitary gland is connected to and communicates with the hypothalamus by way of the pituitary stalk. The pituitary gland consists of two anatomically and functionally distinct portions, called the anterior and posterior lobes.

the roof of the mouth. The posterior pituitary communicates with the hypothalamus by nerve cells, whereas the anterior pituitary is connected to the hypothalamus by blood vessels.

The hypothalamus responds to information regarding many aspects of the state of the body. This information is received either directly, by hypothalamic cells sensitive to solute concentration of the extracellular fluid that surround them, or indirectly, as signals sent through sensory nerves from other parts of the body and brain. In response to these stimuli the hypothalamus sends out nerve impulses that control hormone release by the pituitary gland.

Hormones of the Posterior Pituitary Gland

The posterior pituitary gland does not actually produce any hormones but simply stores and, on appropriate stimulation, releases hormones into the circulatory system. Passing through the pituitary stalk are approximately 100,000 individual nerve cells, each with its cell body located in the hypothalamus. The cells' axons pass down the stalk and terminate in the posterior lobe of the pituitary gland. Two hormones, **antidiuretic** (an'ti-di-u-re-tik) **hormone (ADH)** and **oxytocin** (ok'se-to'sin), are synthesized by the cell bodies of these neurons (Figure 10-5).

The synthesized hormones are packaged into vesicles that are then transported down the axon and stored in the ends of the axons in the posterior pituitary. Each axon terminal stores either ADH or oxytocin but not both. Depending on whether ADH-secreting or oxytocin-secreting neurons are activated, that particular hormone is released into the capillary blood of the posterior pituitary. The hormones are released in response to action potentials originating in the cell body located in the hypothalamus. Both ADH and oxytocin are peptide hormones, each exerting their influences on their target cells by acting through second-messenger systems.

Antidiuretic Hormone

Diuresis (di'-yer-'e'-sis) means urine production, therefore an *anti*diuretic is a chemical substance that inhibits or prevents urine formation. **Antidiuretic hormone (ADH)** causes the kidney to recover water from urine, returning to the blood water that would otherwise be excreted. The result is the production of a smaller but more-concentrated volume of urine. ADH helps prevent wide swings in water balance that might result in either excessive fluid volumes or dehydration. Specialized sensory nerve cells in the hypothalamus called **osmoreceptors** continually monitor the solute concentration of the blood and the extracellular fluid bathing the hypothalamus. When solutes become too concentrated, for example, as a result of an inadequate fluid intake or excessive perspiration, the osmoreceptor stimulates the hypothalamic neurons responsible for ADH synthesis and release. ADH is then released into the circulatory system by the posterior pituitary gland. ADH binds to particular cells in the kidney, causing them to reabsorb more water during urine formation and return it to the blood. The result is that less water is released by the kidney during urine formation and the blood volume rises to within normal homeostatic levels. In contrast, as solute concentration declines the osmoreceptors are no longer stimulated. This ends ADH release. The result is a greater and more dilute volume of urine formed by the kidney, returning blood volume to within its homeostatic levels.

Figure 10-5 Relationship of posterior pituitary gland to the hypothalamus

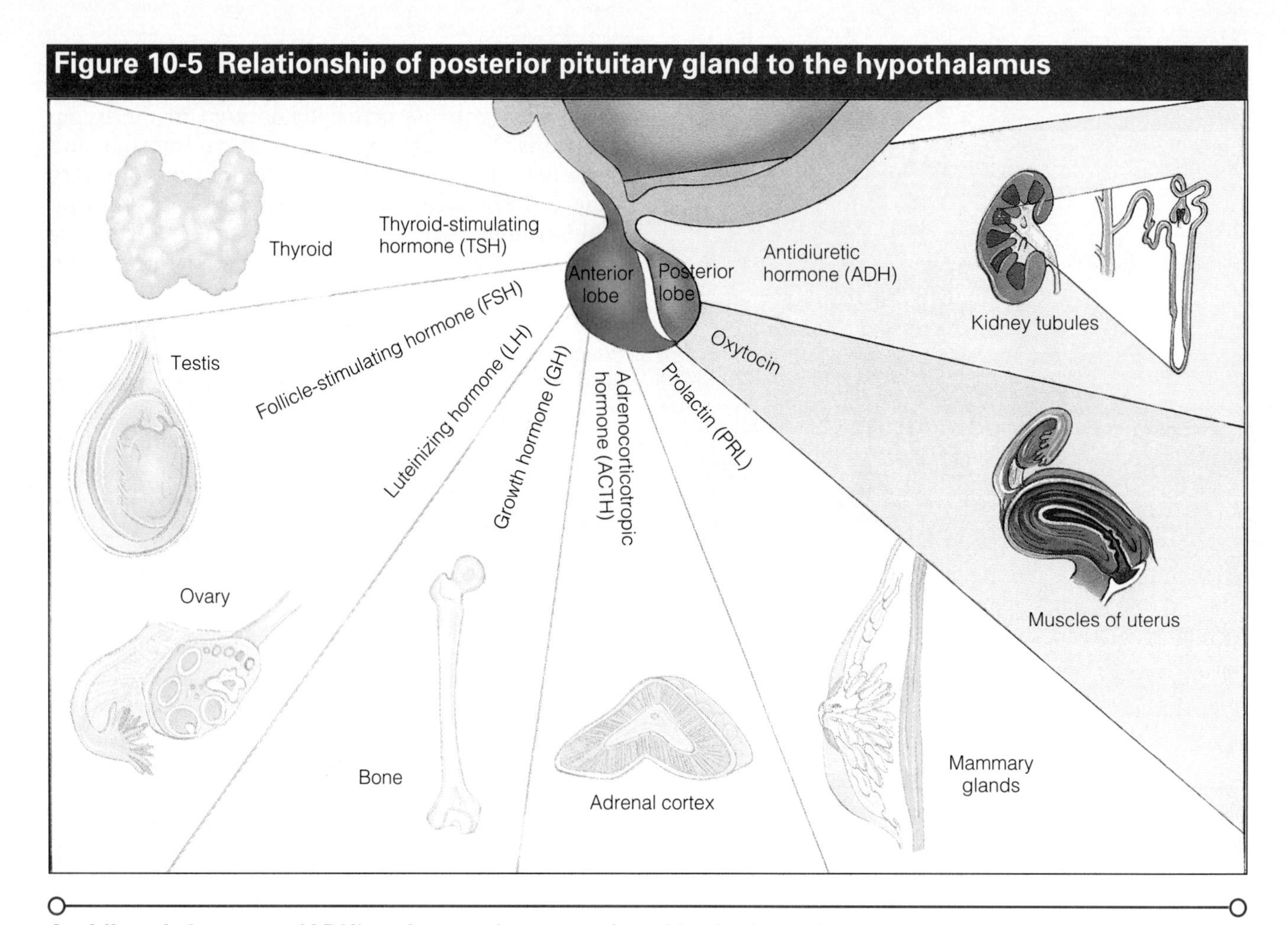

Antidiuretic hormone (ADH) and oxytocin are produced in the hypothalamus. The hormones travel down the pituitary stalk by way of axons whose cell bodies are located in the hypothalamus, and are stored in the posterior pituitary to be secreted when needed.

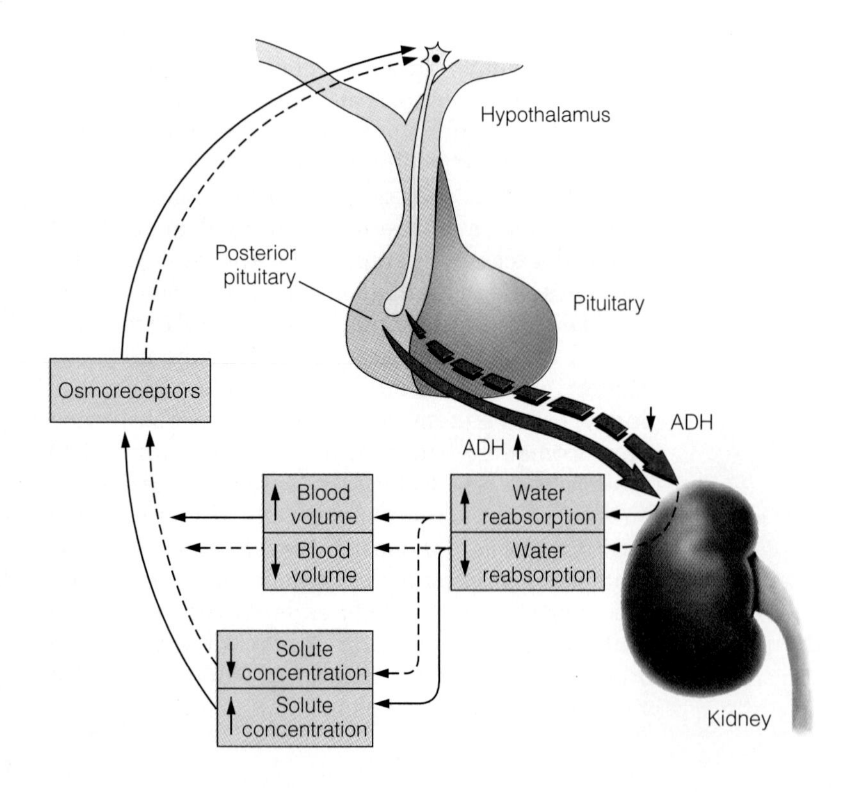

Oxytocin

Oxytocin (ok′-sih-to-sin) is a term derived from Greek meaning swift or sharp birth. **Oxytocin** is a strong stimulant of uterine contraction and is a hormonal trigger for milk secretion or "letdown" in women whose breasts are actively producing milk.

Oxytocin has the unusual property of being controlled by *positive* feedback in the induction of labor during pregnancy. Positive feedback *enhances* the effects of a stimulus, so the response is continued at a heightened rate. As pregnancy progresses the number of oxytocin receptors in the uterus cells increases, making the uterus increasingly sensitive to oxytocin circulating in the blood. With time there are enough receptors for oxytocin to stimulate the smooth muscle of the uterus to contract. During labor the contractions become so frequent and so strong that the fetus is forced down toward the cervix of the uterus. This stretches the cervix, which then causes action potentials to be sent to the hypothalamus. There they stimulate the release of still more oxytocin. The additional oxytocin stimulates the uterus even more, and the contractions become stronger and stronger, until the fetus is forced through the cervix and the baby is born.

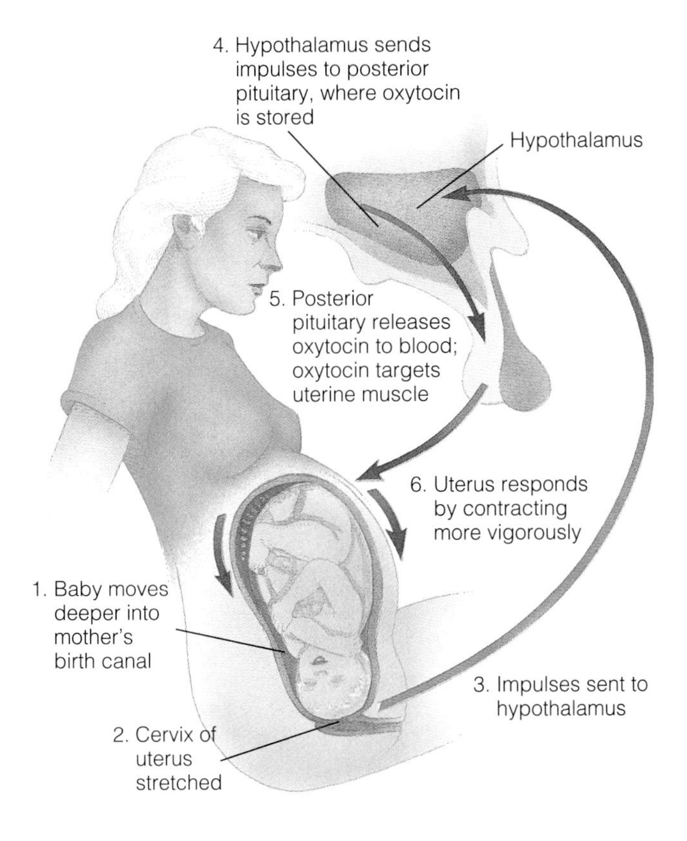

Positive-feedback mechanism also governs milk ejection from the breast. When a baby suckles, the suckling and the visual and auditory stimuli associated with the nursing infant cause action potentials to be sent to the mother's brain. Some of these signals are carried to the hypothalamus, where they induce the release of oxytocin from the posterior pituitary gland. The hormone is then carried by the blood to the mammary glands, where it stimulates the gland's smooth muscles to contract and force out milk that has already been produced and stored by the gland. As milk begins to flow and the baby continues to suckle, additional action potentials are sent from the nipple to the hypothalamus, where they stimulate the release of more hormone. The additional hormone stimulates the gland further and causes the release of even more milk. Thus the more a mother nurses, the more milk she releases.

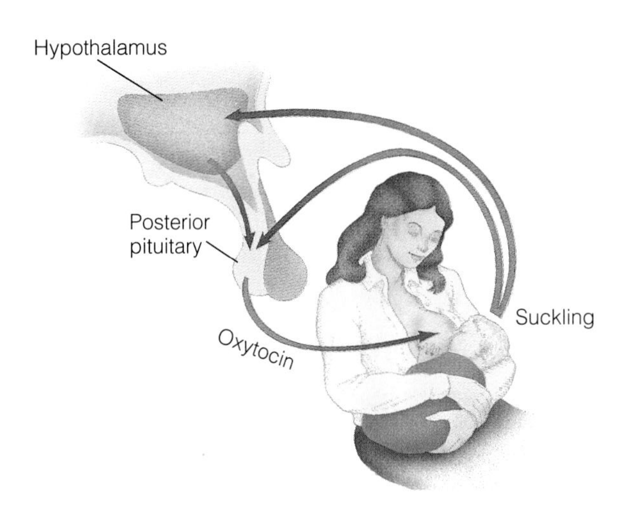

Hormones of the Anterior Pituitary Gland

Unlike the posterior pituitary gland, the anterior pituitary gland synthesizes the hormones it releases into the blood. Different populations of specialized cells in the anterior pituitary gland produce and secrete six different hormones. The six hormones produced by the anterior pituitary gland fall into two functional categories. The first category contains those with a direct effect on nonendocrine target cells. The second are those that stimulate other endocrine glands to produce and secrete hormones. The anterior pituitary gland hormones that stimulate nonendocrine tissues include prolactin (PRL) and growth hormone (GH). The hormones that stimulate endocrine glands are called **tropic hormones** because they activate other endocrine glands. These include thyroid-stimulating hormone (TSH), adrenocorticotropic hormone (a-dre′no-kor′ti-ko-tro′pik) (ACTH), luteinizing hormone (LH), and follicle-stimulating hormone (FSH) (Figure 10-6).

The hypothalamus controls the release of anterior pituitary gland hormones by a different mechanism than the neurological one used to control the release of posterior pituitary gland hormones. Anterior pituitary control is accomplished by releasing, or release-inhibiting, hormones secreted by the hypothalamus. Nerve cells whose cell bodies are located in the hypothalamus synthesize the releasing and release-inhibiting hormones. These hormones enter a capillary bed in the hypothalamus and are transported through portal vessels down the pituitary stalk to a second capillary bed in the anterior pituitary gland. (**Portal vessels** are blood vessels that lie between separate capillary beds or networks.) These hormones cause the anterior pituitary gland either to secrete or to stop secreting a specific hormone (Figure 10-7). We begin by describing

Figure 10-6 Hormones of the anterior pituitary gland

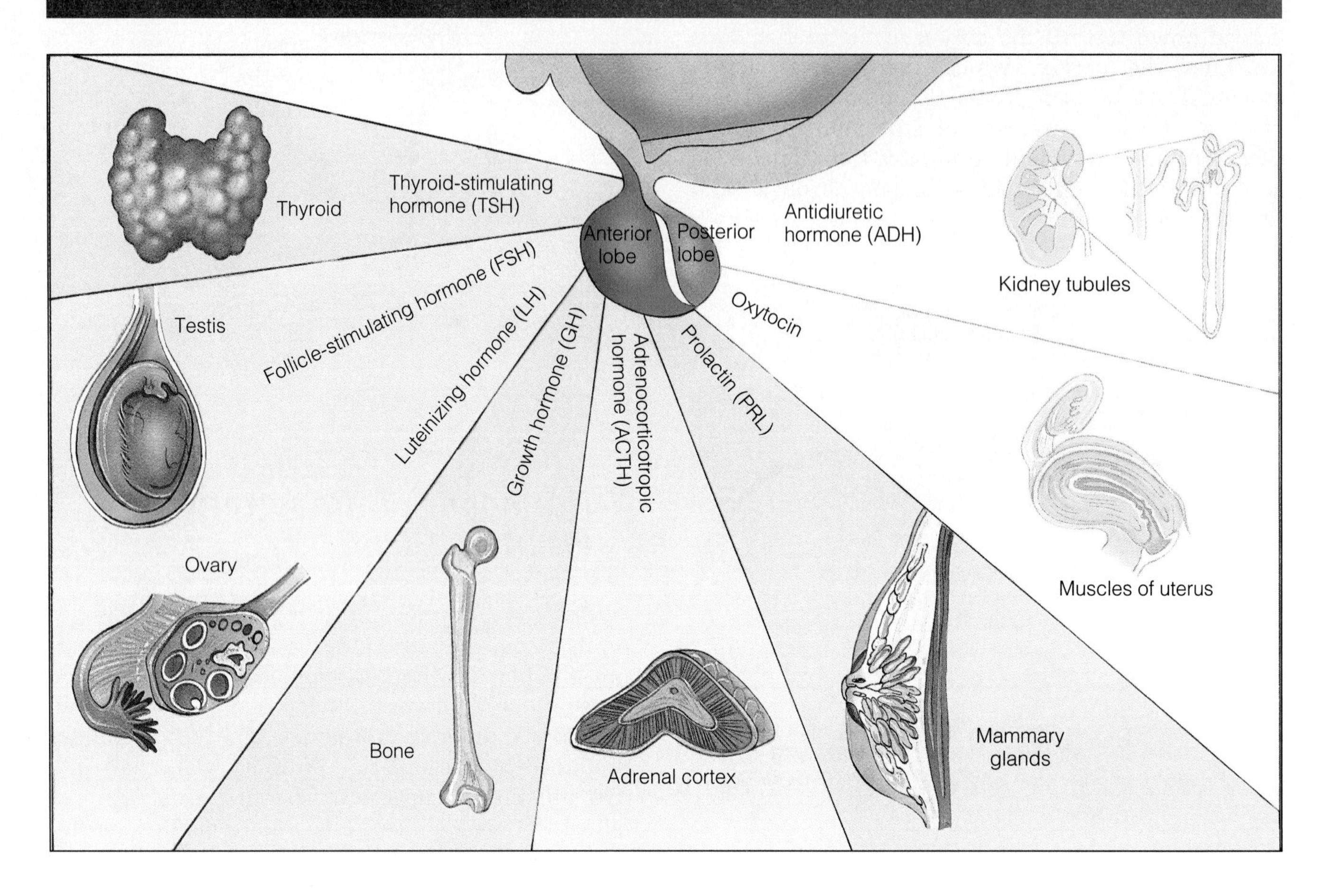

These hormones are produced and secreted by the anterior pituitary lobe of the pituitary gland. Growth hormone and prolactin have direct tissue effects. Thyroid stimulating hormone (TSH), follicle-stimulating hormone (FSH), luteinizing hormone (LH), and adrenocorticotrophic hormone (ACTH) are tropic hormones, meaning that they activate other endocrine glands in the body to secrete hormones.

anterior pituitary gland hormones that stimulate nonendocrine tissue and the primary effects they have on their target tissues.

Prolactin

The protein hormone **prolactin** (PRL) promotes breast development and stimulates milk production in the mammary glands of the female breast. Milk production is distinct from milk secretion, called "letdown," which is controlled by the hormone oxytocin. Because its primary function is milk production, PRL is also called **lactogenic** (lak'to-jen'ik) **hormone.**

PRL levels rise throughout pregnancy, fall after delivery, and rise again in response to suckling. Increased PRL secretion is maintained as long as breast-feeding continues.

Growth Hormone

Growth hormone (GH), also called **somatotrophic hormone,** is the most abundant hormone produced by the anterior pituitary gland, even in adults where growth has ceased. The continued high secretion of GH beyond the growing period indicates this hormone must influence more than body growth.

GH stimulates most body cells to increase in size by promoting the cellular synthesis of proteins. It also increases the rate of cell mitosis. Protein synthesis is increased because GH promotes the uptake of amino acids by cells. GH also stimulates the syn-

Figure 10-7 Relationship of the anterior pituitary gland to the hypothalamus

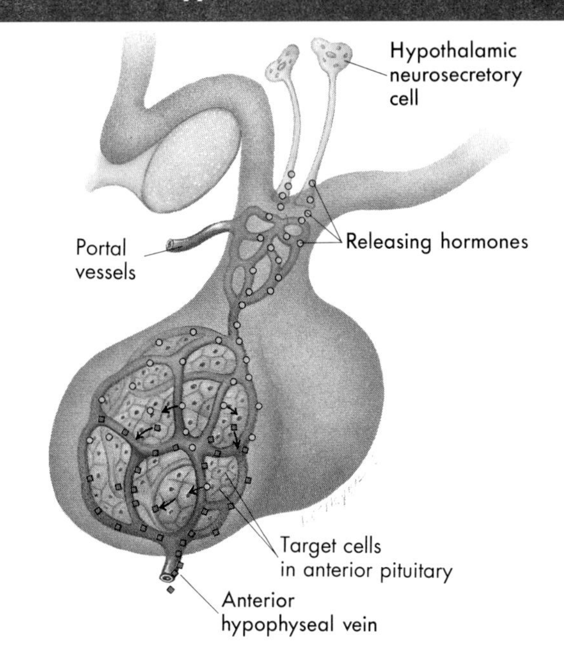

Releasing hormones produced in the hypothalamus travel through portal vessels in the pituitary stalk to a second capillary bed in the anterior lobe of the pituitary gland.

thesis of RNA and ribosomes and increases protein synthesis by ribosomes. It also increases the rate of DNA replication preceding cell division.

The tissues most susceptible to the influence of GH are bones and skeletal muscles. In bone tissue GH increases the activities and numbers of osteoblasts (*osteo* = bone, *blast* = forming) and chondrocytes (kon′dro-sits; *chrondro* = collagen, *cytes* = cells). The osteoblasts secrete mineral salts, composed of calcium ions, into the collagen matrix, giving bone its stiffness, and chondrocytes produce the collagen matrix. Skeletal muscle cells repond to GH by increasing their rate of mitosis and protein synthesis.

Thyroid and Parathyroid Glands

Thyroid-stimulating hormone (TSH) is a tropic hormone secreted by the anterior pituitary gland that stimulates normal development and secretory activity of the thyroid gland. The thyroid gland is a butterfly-shaped tissue mass lying in the front of the neck, just below the larynx, or vocal cords. If you place your finger into the notch at the uppermost border of the breast bone, the tip of your finger will touch the lower edge of the thyroid gland. The thyroid gland, under stimulation from TSH, produces and secretes several important hormones. The two most important hormones are thyroxin (thi-roks′in) and calcitonin (kal′si-to′nin). Thyroxin is made from iodine ingested from various food sources, such as seafood and some vegetables, and the amino acid tyrosine. Iodine is actively transported into the thyroid gland, where it covalently bonds to the amino acid tyrosine, forming thyroxin.

Thyroxin

Thyroxin is responsible for many effects in both the developing and the adult body, but its overall effect is to accelerate the rate of the body's cellular metabolism. Thyroxin does this by stimulating enzymes involved in the cellular metabolism of glucose. With the exception of the brain, spleen, testes, uterus, and thyroid gland itself, thyroxin affects virtually every cell in the adult body.

Falling blood levels of thyroxin trigger the release of TSH from the anterior pituitary gland. TSH causes more thyroxin to be released from the thyroid gland. Rising levels of thyroxin then feed back to inhibit the further release of TSH from the anterior pituitary gland.

Conditions that increase body energy requirements, such as pregnancy or prolonged cold exposure, cause the hypothalamus to increase release of the releasing hormone that controls TSH output from the anterior pituitary gland. In such situations the hypothalamus overcomes the normal negative-feedback controls of the endocrine glands. The result is increased cellular metabolism and body heat production.

Calcitonin

In addition to thyroxin, the thyroid gland also produces and secretes calcitonin (CT). The most important effect of **calcitonin** is to decrease blood calcium levels. About 99% of the body's calcium is deposited in the bones. The remainder is in cells and tissue fluids, where it is important in such processes as blood clotting, nerve conduction, and helping to hold cells together. The level of calcium in the body fluids is controlled and regulated by calcitonin and a hormone secreted by the parathyroid glands. The primary target cells of calcitonin are the osteoblasts in bone. Calcitonin stimulates the uptake of calcium ions from the body fluids by the osteoblasts, where it is then deposited in bony tissue. Calcitonin also increases the excretion of calcium during urine formation by the kidneys.

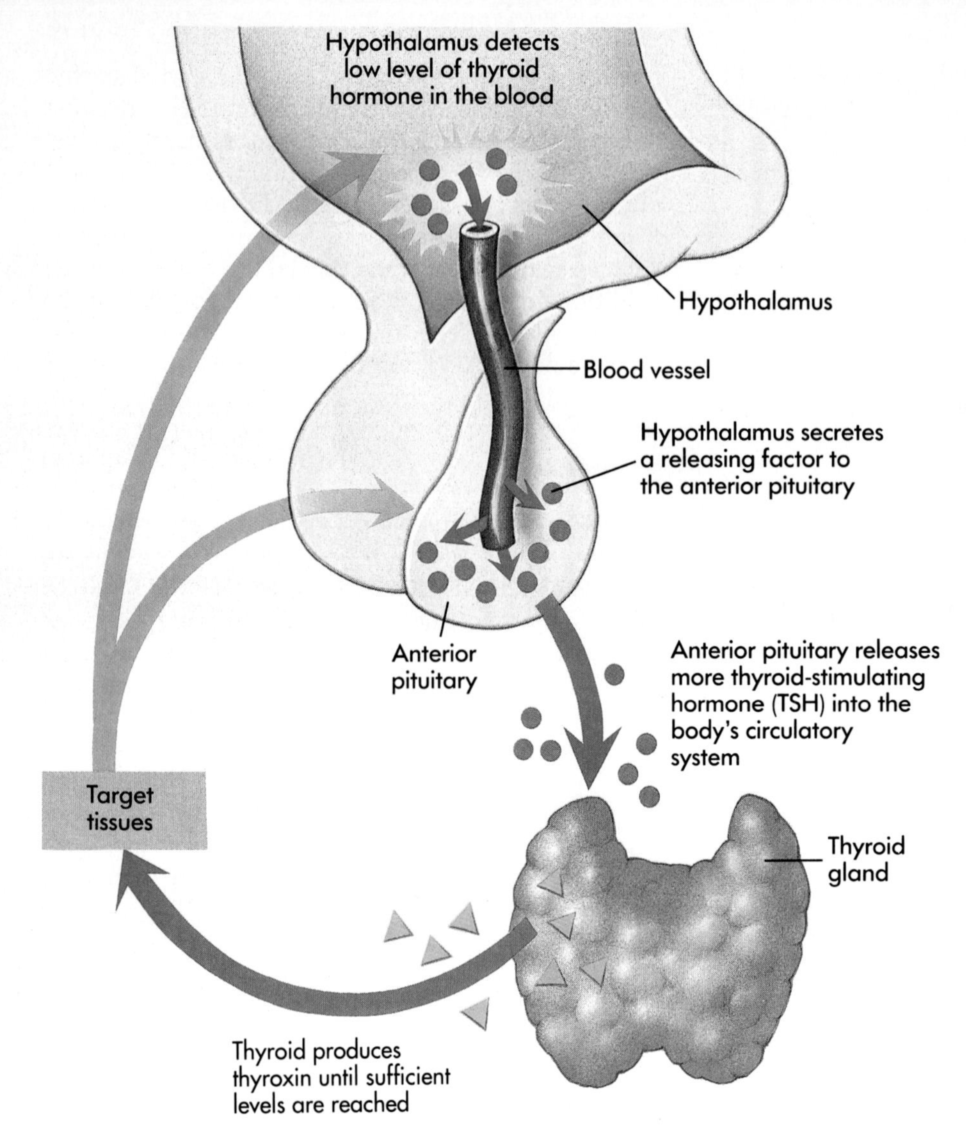

Both calcitonin and parathyroid hormone regulate blood levels of calcium. Decreased calcium blood levels cause parathyroid hormone levels to increase and calcitonin levels to decrease. Increased calcium blood levels cause calcitonin levels to increase and parathyroid hormone secretion to decrease (Figure 10-8).

Parathyroid Hormone

Embedded on the posterior sides of each lobe of the thyroid gland are four small masses of tissue that form the parathyroid glands (*para* = near). The **parathyroid glands** secrete parathyroid hormone (PTH).

Parathyroid hormone is a peptide hormone important in regulating blood calcium-ion homeostasis. Parathyroid hormone secretion is stimulated by decreasing or low levels of blood calcium. The major effect of parathyroid hormone is to increase calcium-ion levels in the blood by stimulating three processes: (1) activation of **osteoclasts** ("bone breakers") in bone tissue, which digest some of the bony matrix and release calcium ions; (2) increased absorption of calcium ions by the kidney; and (3) increased absorption of calcium by the small intestines. Calcium absorption by the intestine is enhanced indirectly. Parathyroid hormone activates Vitamin D, which is required for the absorption of calcium from digested food by the intestines.

The Adrenal Glands

The paired adrenal (a-dre'nal) glands are each about the size of an almond and sit on top of the kidneys. Each adrenal gland is structurally and functionally two endocrine glands in one. The inner **adrenal medulla** (me-dul'ah) is nervous tissue and is functionally part of the nervous system. The outer **adrenal cortex,** which covers the adrenal medulla and forms the bulk of each gland, is derived from epithelial tissue and produces hormones.

Stress, such as intense or prolonged physical activity or low blood glucose levels, stimulates both the adrenal medulla and the adrenal cortex to secrete hormones. The adrenal medulla is stimulated directly by nerves from the sympathetic nervous system. The adrenal cortex is stimulated by the anterior pituitary gland.

Adrenal Medulla

The adrenal medulla is involved with controlling physiological reactions to short-term, immediate stress. The secretion of epinephrine and norepinephrine enhances and prolongs typical actions of the nervous system that occur in so-called fight-or-flight situations. Although this response was discussed in Chapter 8, some general characteristics affected by the adrenal medulla are mentioned here. When the sympathetic division of the autonomic nervous system is stimulated, blood glucose levels rise, blood vessels constrict, the heart beats faster, and blood is diverted from temporarily nonessential organs to the heart and skeletal muscles. At the same time the adrenal medulla is stimulated by nerves from the CNS to release its hormones, which reinforce and prolong the stress response (Figure 10-9). About 80% of the hormones released from the adrenal medulla is epinephrine. Epinephrine and norepinephrine exert the same effects, but epinephrine is the more potent stimulator of the heart and metabolic activities, whereas norepinephrine has the greater influence on constricting the body's peripheral blood vessels.

Adrenal Cortex

Over 20 different steroid hormones, collectively called **corticosteroids** (kor'ti-ko-ster'oids), are synthesized from cholesterol lipid molecules by the adrenal cortex. The corticosteroids fall into two major categories, the glucocorticoids and the mineralocorticoids.

Glucocorticoids influence the metabolism of most body cells and also help to provide resistance to stressors. Glucocorticoids are essential to life. Under normal circumstances the glucocorticoids allow the body to adapt to intermittent food intake by helping to keep blood glucose levels constant, and they maintain blood volume by preventing the shift of water into tissue cells. However, when the body is under severe stress because of infections or some other physical or emotional trauma, there is a marked increase in the output of glucocorticoids, which help the body overcome the stress.

The most important of the glucocorticoids pro-

Figure 10-8 How parathyroid hormone (PTH) and calcitonin (CT) work to maintain normal calcium levels in the blood

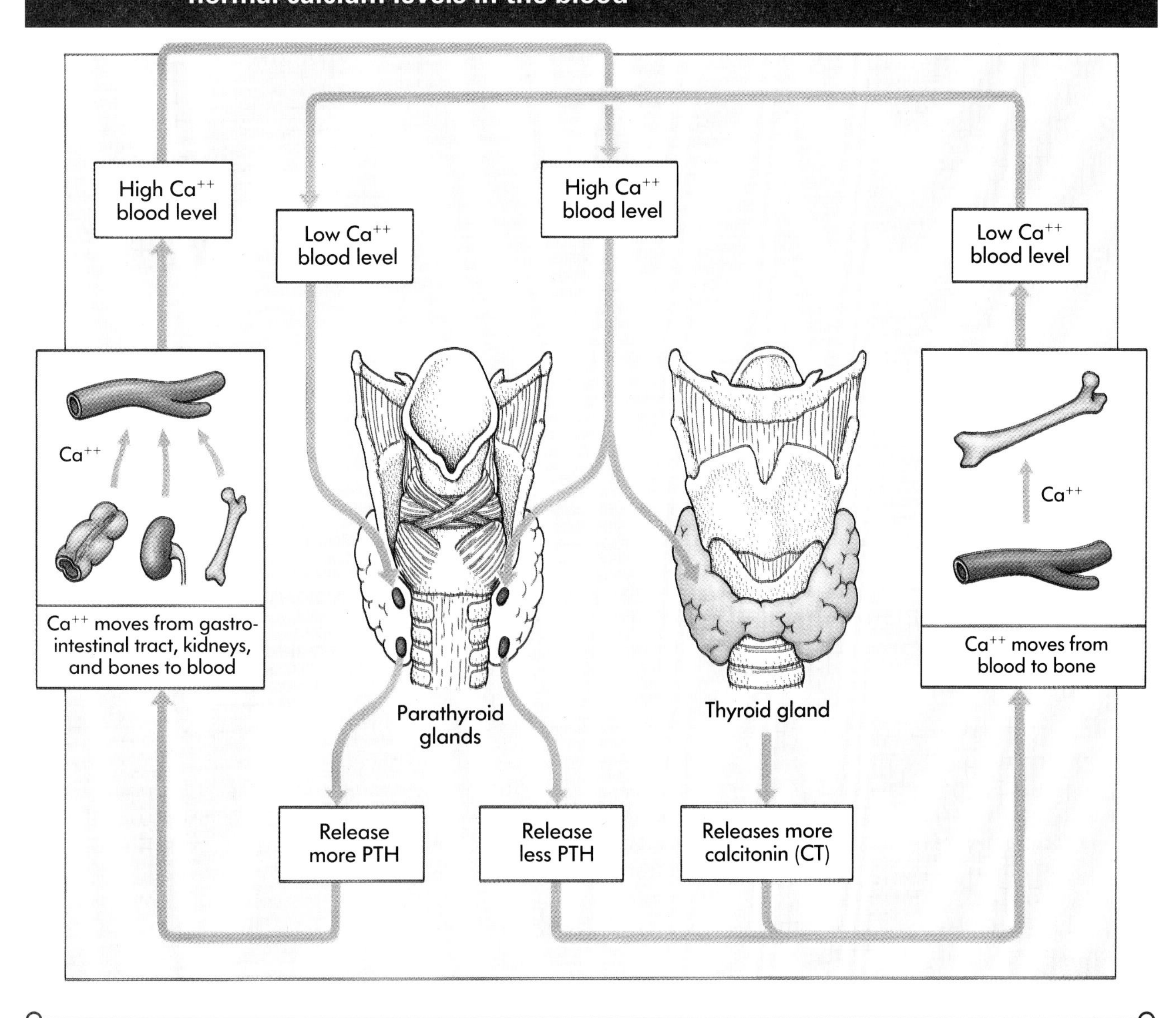

Figure 10-9 Fight-or-flight stress response

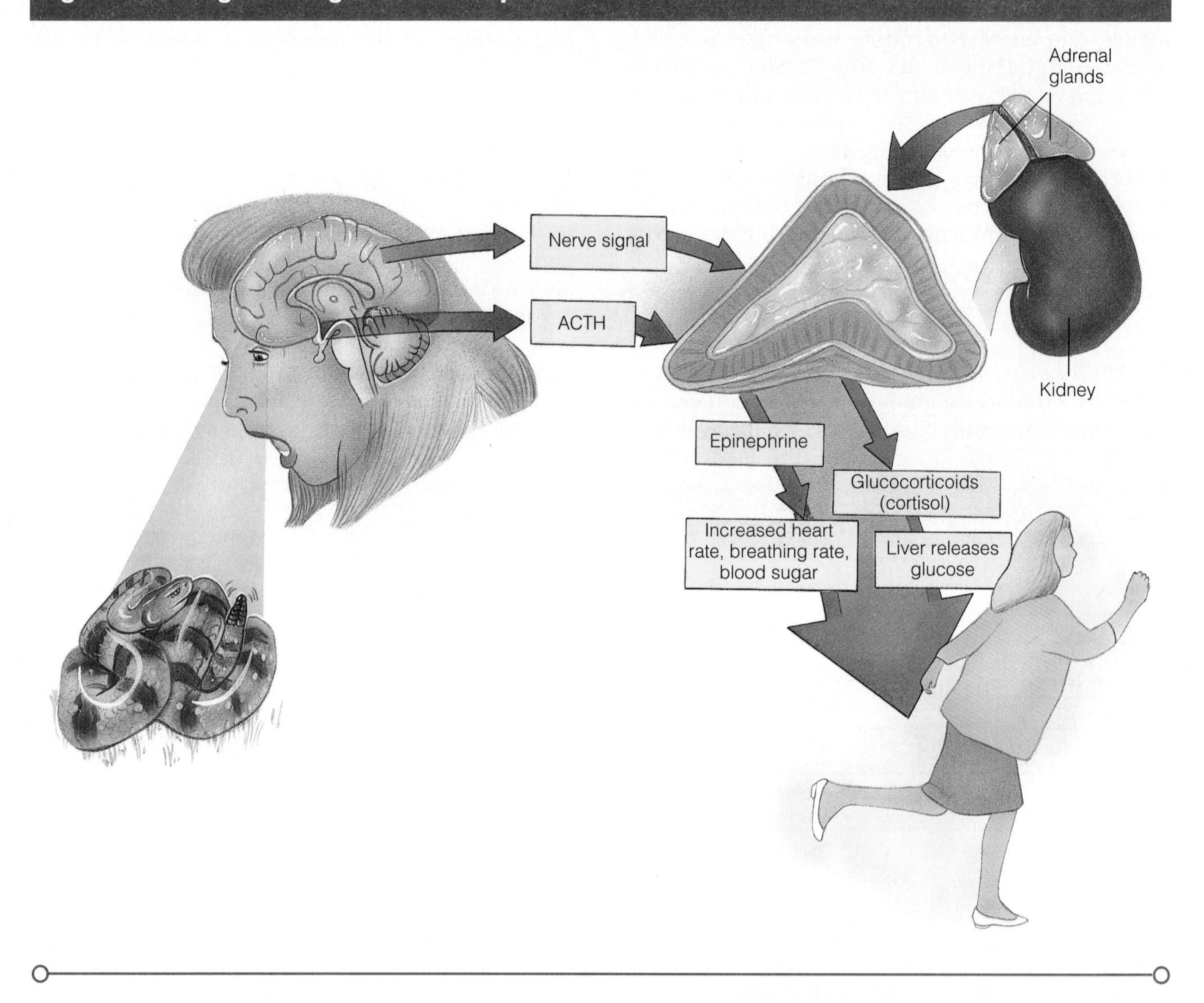

duced by the adrenal cortex is **cortisol** (kor'ti-sol). Control of cortisol secretion is a typical negative-feedback system. Cortisol release is triggered by the release of **adrenocorticotropic hormone (ACTH)** from the anterior pituitary gland. This is triggered in turn by a releasing hormone secreted by the hypothalamus. The liver and the stored adipose fat are the target tissues of cortisol. Cortisol initiates **gluconeogenesis** (glu'ko-ne-o-jen'e-sis), the formation of glucose from noncarbohydrate molecules, such as fats and proteins, by the liver. Cortisol causes the liver to convert amino acids to glucose and also acts on the adipose tissue, causing fat stored in the fat cells to be broken down to fatty acids. The glucose and fatty acids are released into the circulatory system and are taken up by the body's cells and used as an energy source (Figure 10-10).

The **mineralocorticoids,** the second of the hormones secreted from the adrenal cortex, help regulate blood volume and the blood levels of potassium and sodium ions. The most important of the mineralocorticoids is **aldosterone** (al'dah'-ster-on), and it is not under the control of the anterior pituitary gland. The most important function of aldosterone is to regulate ion concentrations of sodium and potassium in extracellular fluids.

The primary target tissue for aldosterone is the kidney, where it stimulates the reabsorption of sodium ions from the forming urine and their return into the bloodstream. The concentration of sodium ions is important to help maintain blood pressure, and it indirectly regulates aldosterone secretion from the adrenal cortex. When the blood volume, and therefore blood pressure, is low the kidneys secrete an enzyme called **renin** (re′-nin). Renin converts a blood protein called **angiotensinogen** (an′-je-o-ten-sin′-o-jin) into its active form, **angiotensin** (an′-je-o-ten′-sin). Angiotensin constricts blood vessels and also stimulates the adrenal cortex to release aldosterone. As aldosterone causes the kidneys to reabsorb sodium, water is also retained, resulting in an increased blood volume. The effect of the **renin-angiotensin system** is to increase blood pressure (Figure 10-11).

Gonads

The male and female gonads also produce hormones. The release of sex hormones is regulated by releasing hormones produced in the hypothalamus that then cause the release of **follicle-stimulating hormone (FSH)** and **luteinizing hormone (LH)** from the anterior pituitary gland. FSH and LH are present in both males and females. In the female the paired ovaries that produce gametes, called ova, also produce the two major steroid hormones estrogen and progesterone (pro-jes′-ter-on). **Estrogen** alone is responsible for the maturation of the female reproductive organs and the appearance of the secondary sex characteristics of females at sexual maturity, such as pubic hair and the fat deposition characteristic of females. **Progesterone** affects the lining of the uterus to prepare it for pregnancy. Acting with progesterone, estrogen also promotes breast development and the cyclical changes in the uterus associated with the menstrual cycle.

The testes of the male produce gametes called sperm as well as the male sex hormone testosterone (tes-tah′-ster-on). **Testosterone** initiates the maturation of the male reproductive organs and the appearance of the male secondary sex characteristics, such as beard growth and body and pubic hair. Testosterone is also responsible for the relatively greater muscle mass of males compared with females.

In Chapter 18 we discuss the roles of each of these hormones, their target tissues, and the delicate balance between them in reproduction and maturation.

Thymus Gland

The **thymus gland** is a two-lobed structure located beneath the sternum, or breastbone, in the chest. The thymus gland produces a group of hormones collectively known as thymosins. **Thymosins** are important for promoting the production and maturation of a class of leukocytes, or white blood cells, in the blood. These leukocytes, known as T-cells, are essential in assisting to protect the body from infection. We discuss the role of the thymus gland in the human immune system in Chapter 14.

Figure 10-10 Structure and function of the adrenal cortex

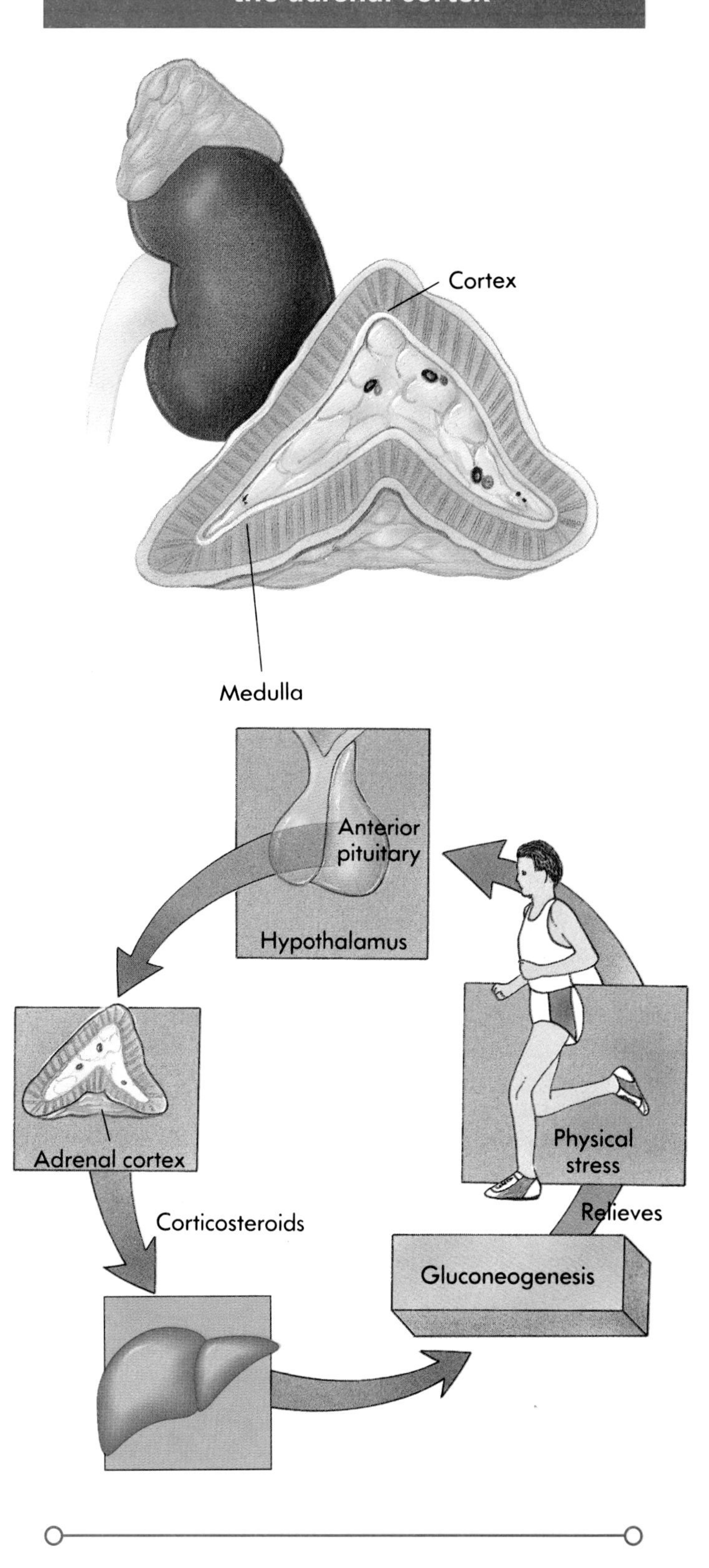

Figure 10-11 Renin-angiotensin system

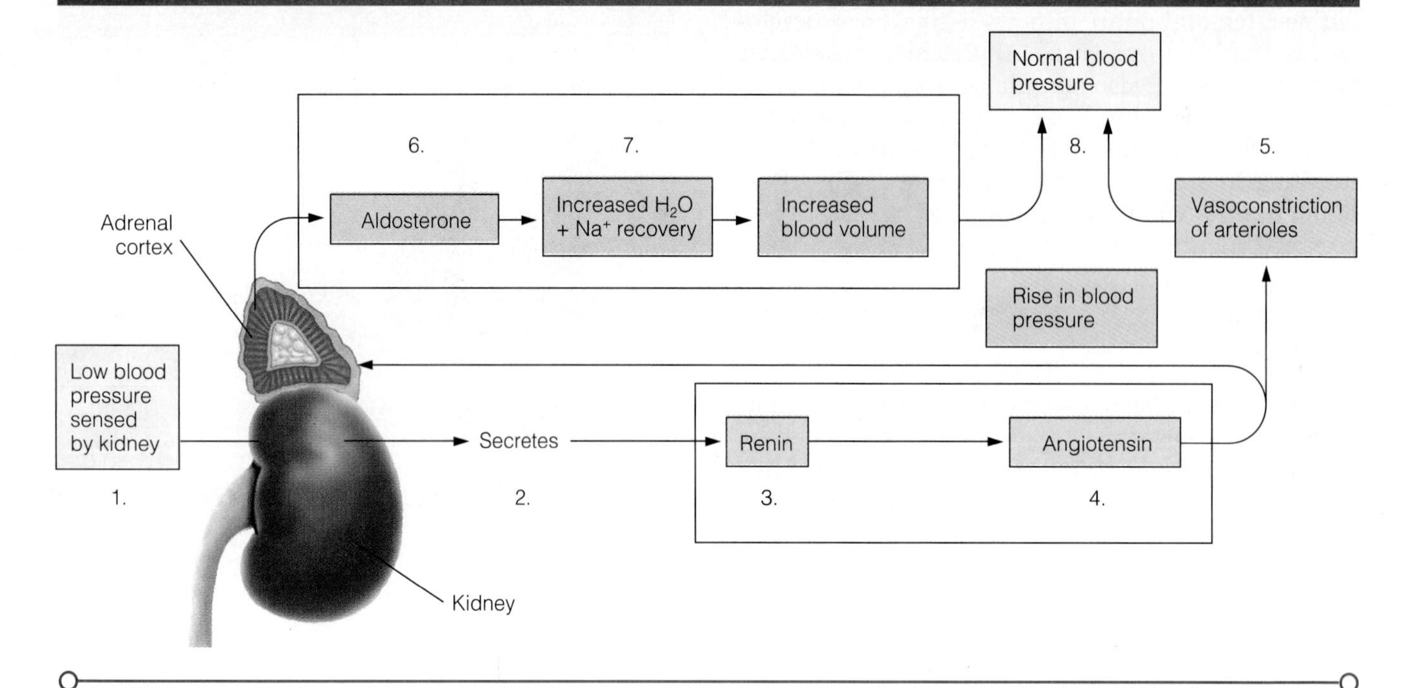

Endocrine Functions of the Pancreas

Located primarily behind the stomach in the abdomen, the pancreas is a compound gland, having both endocrine and exocrine gland cells. Most of the pancreas is composed of **acinar** (as′i-nar, meaning berry or grape) **cells.** These cells produce an enzyme-rich juice that empties into the small intestine during food digestion. We discuss the function of these pancreatic enzymes in Chapter 16.

Scattered among the acinar cells are several million **islets of Langerhans,** clusters of cells that produce pancreatic hormones. The islets of Langerhans contain two types of hormone-producing cells: the **alpha cells,** which synthesize the hormone glucagon (gloo′-kuh-gon), and the more numerous **beta cells,** which are responsible for insulin hormone production.

Glucagon

Glucagon, a peptide hormone, is secreted by the alpha cells of the pancreas when blood glucose levels are low. **Glucagon** increases blood glucose levels by enhancing the conversion of liver glycogen into glucose, which then enters the blood. Glucagon also helps to stimulate gluconeogenesis in the liver. This newly synthesized glucose also enters the circulation.

The secretion of glucagon is regulated by the level of blood glucose. Low blood glucose levels stimulate the alpha cells to secrete glucagon, and high blood glucose levels inhibit glucagon secretion.

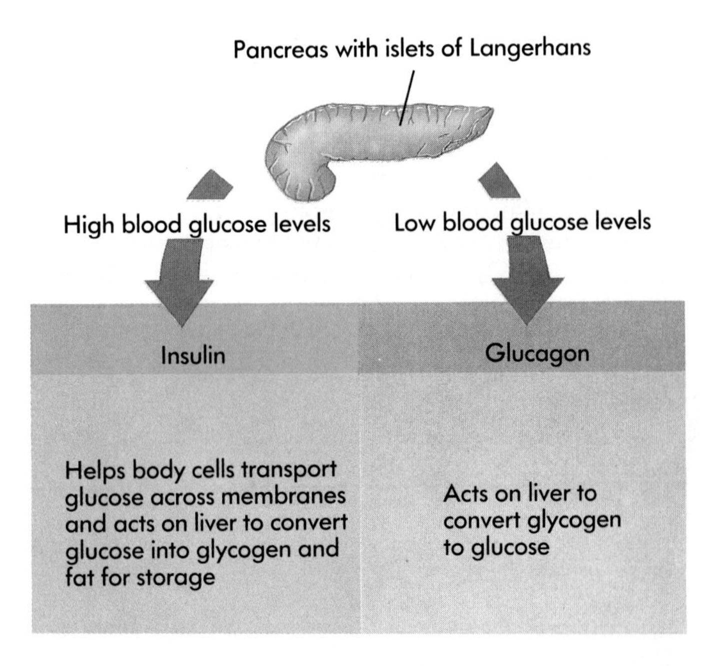

Insulin

Insulin is also a peptide hormone and is secreted by the beta cells of the islets of Langerhans. Insulin is released from the pancreas in response to elevated blood glucose levels or increased blood levels of

Biology in Action: DIABETES MELLITUS

Although there are many problems associated with endocrine gland malfunctions, perhaps the most common is **diabetes mellitus,** which usually results from the hyposecretion of insulin by the pancreas. When insulin activity is absent or deficient, blood glucose levels remain high after a meal because glucose is unable to enter most tissue cells. The two main types of diabetes mellitus are **insulin-dependent diabetes mellitus (IDDM)** and **non-insulin-dependent diabetes mellitus (NIDDM).**

IDDM usually has its onset before the age of 30, often in an adolescent of normal body weight. Approximately 5 people in 1000 are affected. Although heredity does seem to function in its cause, heredity does not appear to have as big a role as it does in NIDDM. IDDM has recently been identified as an **autoimmune disease** because 70% of affected people have circulating antibodies to their own pancreatic islet cells.

Early in IDDM, insulin secretion is decreased, but eventually insulin production may stop completely. The lack of circulating insulin leads to **hyperglycemia**, an increased level of glucose in the blood, because it is not being taken up by tissue cells. Additionally, the liver begins to overproduce glucose through gluconeogenesis. Because of this the synthesis of protein, ATP, DNA, and RNA is impaired.

In contrast, people with NIDDM are usually over age 30 when the disorder appears, are generally overweight, and exercise little. Heredity does seem to be important because there is 95% correlation in identical twins, meaning that if one twin has the disorder, the other will also display it. NIDDM affects between 1% and 5% of the U.S. population. NIDDM is a disorder of insulin reception by tissue cells. Some believe that an insufficient number of insulin receptors exists in the obese person with NIDDM.

Both IDDM and NIDDM cause many secondary disorders. Hyperglycemia damages the Schwann cells that produce and maintain the myelin sheaths around the axons. Hyperglycemia also allows glucose molecules to attach to proteins of the capillary membranes, and this attachment damages these small vessels. This process causes damage in the kidneys and the eyes. In general, kidney failure develops in persons with IDDM, usually causing death. Problems such as heart disease develops in persons with NIDDM, often causing death.

Persons with IDDM are usually treated with a controlled diet including a restricted amount of fats and carbohydrates and are supplied with insulin by injection. Persons with NIDDM are generally required to lose weight and increase exercise and are occasionally given insulin.

amino acids. The major target tissues of insulin are the liver, adipose tissue, and muscles. Circulating insulin lowers blood glucose levels by enhancing the membrane transport of glucose into skeletal muscle cells, fat cells, and liver cells. The glucose is used by these cells as an energy source or is converted to glycogen or fat, and the amino acids are used to synthesize protein.

Nonendocrine Gland Hormones

Many endocrine cells located throughout other body tissues also secrete hormones. The following hormones are among the most important nonendocrine gland hormones found in various body tissues.

Prostaglandins

A class of lipid hormones called **prostaglandins** (pros'ta-glan'dinz) are derived from cell phospholipid molecules and are in most of the body's cells. Unlike other hormones they do not circulate through the blood but have their effects in the tissues where they are produced. Some prostaglandins cause blood vessels to dilate, whereas others cause them to constrict. Prostaglandins produced during childbirth cause the uterus to contract. Some believe that overproduction of prostaglandins by the uterus is responsible for especially painful menstruation, producing such symptoms as severe cramps, nausea, and headaches.

Prostaglandins are also involved in tissue inflammation. They are released by damaged tissues and cause blood vessel dilation, swelling, reddening of the damaged area, and pain.

Neurohormones

We have long known that derivatives from the opium poppy, such as morphine, are able to dramatically reduce human pain perception. Later discoveries show that the human CNS has opiate receptors to which opium derivatives chemically bond and block pain perception. It seems unlikely the human body would have receptors that interact only with substances derived from flowers. Research has indicated the substances in the body that bond with these opiate receptors. In 1974 Swedish researchers isolated several small peptide hormones called **enkephalins** (en-kef′a-linz). Enkephalins are responsible for blocking afferent signals from the body's pain receptors and thereby preventing the brain from perceiving pain. We still do not know exactly how natural pain-suppressing mechanisms are activated. One factor known to modify pain is exercise. The so-called runner's high is thought to be caused by the release of one class of enkephalins called **endorphins.** Other factors also known to cause the release of enkephalins include acupuncture, hypnosis, and stress.

The muscle cells of the heart also secrete a peptide hormone, called **atrial natriuretic hormone** (ANH). Release of ANH occurs when the walls of the heart chambers are stretched because of the entrance of an abnormally large blood volume. The target cells of ANH are the kidney, adrenal cortex, and blood vessels. ANH increases water and sodium ("natriuresis") excretion through the kidney by inhibiting aldosterone secretion from the adrenal cortex and ADH secretion from the posterior pituitary gland. ANH also causes the smaller blood vessels of the circulatory system to dilate. The result of these combined effects is to decrease blood volume and reduce blood pressure. For this reason ANH is being investigated as a treatment for people suffering with chronic high blood pressure.

We can see that the chemical communication occurring between the cells of the body is precise, allowing homeostasis to be maintained under dynamic conditions as diverse as growth and development or reproduction.

CHAPTER REVIEW

SUMMARY

1. Endocrine glands produce and release chemical substances called hormones into the circulatory system, where they are carried throughout the body. Depending upon the specific hormone and its function, hormone release may be intermittent or continuous.
2. Hormones control and integrate such processes as reproduction, regulation of cellular metabolism, growth and development, water and nutrient balance, and ion balance in the blood. Once initiated the metabolic activities are either long-lasting changes or continuous processes through time.
3. Most hormones control these biological processes through negative-feedback mechanisms. The outcome is that the body is able to maintain a dynamic homeostasis. In this way the body maintains homeostasis even as it grows, develops, and becomes reproductively mature.
4. Some hormones, called tropic hormones, stimulate other endocrine glands to release hormones into the circulatory system. The number of target cells varies greatly; some hormones may affect most of the cells of the body while others are specific to only a few cells.
5. A particular hormone influences the activity of only certain cells, referred to as target cells, which have specific protein receptors on their surface membranes or in their interiors to which that hormone can bind. Once the hormone binds to its specific receptor site, it affects cell activities either through a second messenger or by gene activation.
6. Hormones are either amino acid based or steroids derived from cholesterol molecules. Only hormones secreted by the adrenal cortex and gonads are steroid molecules. Amino acid–based hormones can vary from a modified single amino acid to long amino acid chains.

FILL-IN-THE-BLANK QUESTIONS

1. Releasing hormones are secreted by the ______ and affect the ______ pituitary gland.
2. Action potentials originating in the ______ cause hormones to be secreted by the ______ pituitary gland.
3. Hormones bring about their characteristic effects on target cells by altering ______ activity.
4. Steroid hormones are synthesized from ______ and are secreted by the adrenal ______ and also the ______.
5. Insulin is secreted by the ______ of the pancreas and acts to decrease ______ levels in the blood.
6. Cortisol is secreted by the ______ ______ and assists the body in recovering from ______.
7. The secretion of both prolactin and oxytocin is regulated by a ______ feedback system.
8. Growth hormone stimulates most body cells to increase in size by promoting protein ______ and also increasing their rate of ______.
9. Thyroid-stimulating hormone is secreted by the ______ pituitary gland.

SHORT-ANSWER QUESTIONS

1. Select a hormone and describe its function in terms of its negative-feedback mechanism.
2. Compare and contrast the actions of insulin and glucagon.
3. In what ways does the hormone cortisol assist the body in overcoming stress?
4. What effects does growth hormone have on bone tissue and skeletal muscle?
5. Describe how steroid hormones modify the cellular events of their target cells.

VOCABULARY REVIEW

adrenal medulla—*p. 200*
endocrine gland—*p. 191*
enkephalins—*p. 206*
hypothalamus—*p. 191*
islets of Langerhans—*p. 204*
parathyroid glands—*p. 200*
pituitary gland—*p. 194*
prostaglandin—*p. 205*
thymus gland—*p. 203*

Chapter Eleven

THE MUSCULOSKELETAL SYSTEM

For Review

Here are some important terms and concepts that you will encounter in this chapter. If you are not familiar with them, you should review them before proceeding.

Aerobic and anaerobic metabolism
(pages 52, 55)

Bone tissue
(page 131)

Chondrocyte
(page 129)

Muscle spindles
(page 173)

Muscle tissue
(page 131)

OBJECTIVES

After reading this chapter you should be able to:

1. *Name the four functions of the bones.*
2. *Name the two divisions of the human skeleton and list the bones that make up each.*
3. *Describe the function of the pectoral and pelvic girdles of the appendicular skeleton.*
4. *Explain the two functions of joints and describe the three main classifications.*
5. *List and describe the function of the three types of bone cells.*
6. *Differentiate between the two forms of bone growth and development.*
7. *Define extensibility and elasticity.*
8. *Explain the function of tendons and define origin and insertion.*
9. *Describe the events that result in muscle contraction.*

Humans are creatures of almost constant activity. Whether conscious movements, such as walking or talking, or the basic autonomic actions of breathing, swallowing, or the beating of a heart, all of our activity is generated by the same basic mechanism of muscle tissue contraction.

Our muscles use the chemical energy of ATP to supply that force. Your body applies this energy to alter the structural elements in muscle cells, causing the cells to shorten. When many muscle cells shorten all at once, they can exert a great deal of force.

However, if this were all there was to movement, you would not move. Instead, you would lie there pulsating as your muscles contracted and relaxed in countless cycles. For a muscle to produce movement it must direct its force against another object. Some muscles of your body are attached to an internal skeleton of bone, which is both rigid and jointed to be supportive and mobile. The muscles direct their force against the bones to which they are attached.

In this chapter we describe the internal bony framework of the body, called the **skeleton.** We also discuss bone formation, growth, and development. Finally we explore muscle structure and function and the muscles' mutual interaction with the skeleton in coordinating body movement.

The Skeleton

Besides contributing to movement, your bones perform four other related functions. First, bones provide a framework that supports and anchors all the body's organs. For example, the rib cage supports the upper limbs of the body. Second, bones protect organs. The fused bones of the skull, for example, enclose and protect the brain and several sense organs. Third, bones are storage sites for several important substances. Fat is stored in the internal cavities of bone, and bone is a storehouse for minerals, especially calcium. Stored minerals can be mobilized and released into the blood for distribution to other body parts. Mineral "deposits" and "withdrawals" occur in the body almost continuously. Fourth, **hemopoiesis** (he′mo-poy-e′sis), the process of forming and developing the various types of blood cells, occurs in bone.

The human skeleton basically has two divisions. The first division, called the **axial skeleton,** includes the skull; the bones of the spinal column, called **vertebrae;** the ribs; and the breastbone, called the **sternum** (ster′num). The second, called the **appendicular skeleton** (ap′pen-dik′u-lar), consists of the bones of the upper limbs, hands, lower limbs, feet, pelvic girdle, and pectoral girdles (Figure 11-1).

Although the human skeleton has more than 206 bones, each falls into one of four broad categories based on shape. **Long bones** (*A*) are longer than they are wide, and the two ends usually are larger in diameter than the midsection of the bone. For example, the bones of your arms and legs are long bones. **Short bones** (*B*) are nearly the same in length and width, such as in the ankles and wrists. **Flat bones** (*C*) usually consist of a flat plate of bone designed for protection or for numerous muscle attachment sites. Finally, **irregular bones** (*D*) have complex shapes and include such bones as the vertebrae of the spinal column, some facial bones of the skull, and the os coxa, which connects the leg bones to the spinal column.

Functions of Bone

- Framework that supports and anchors all the body's organs
- Protection of organs
- Storage site for important substances, such as calcium
- Process of formation and development of blood cell varieties

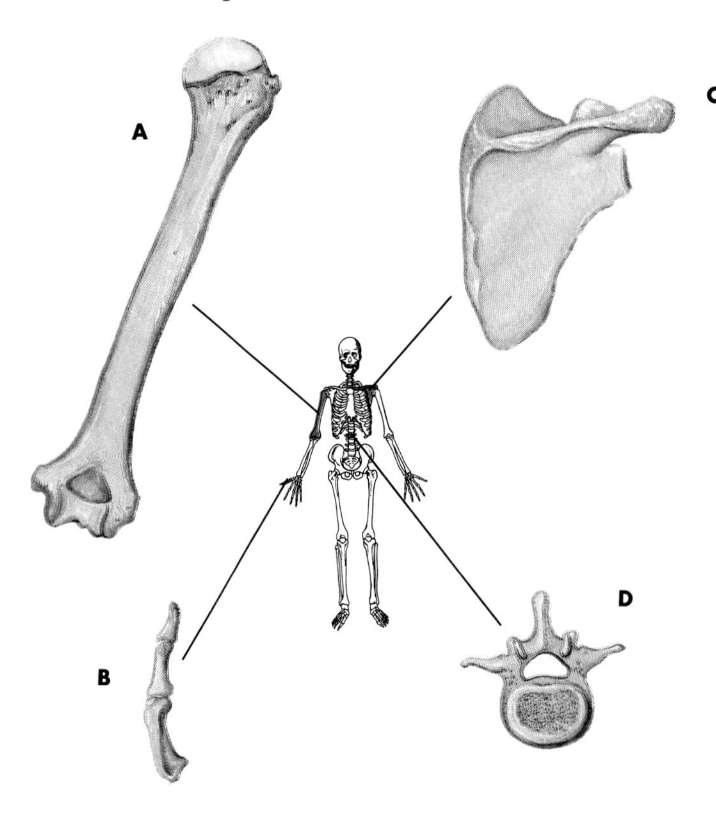

Figure 11-1 Appendicular and axial skeleton

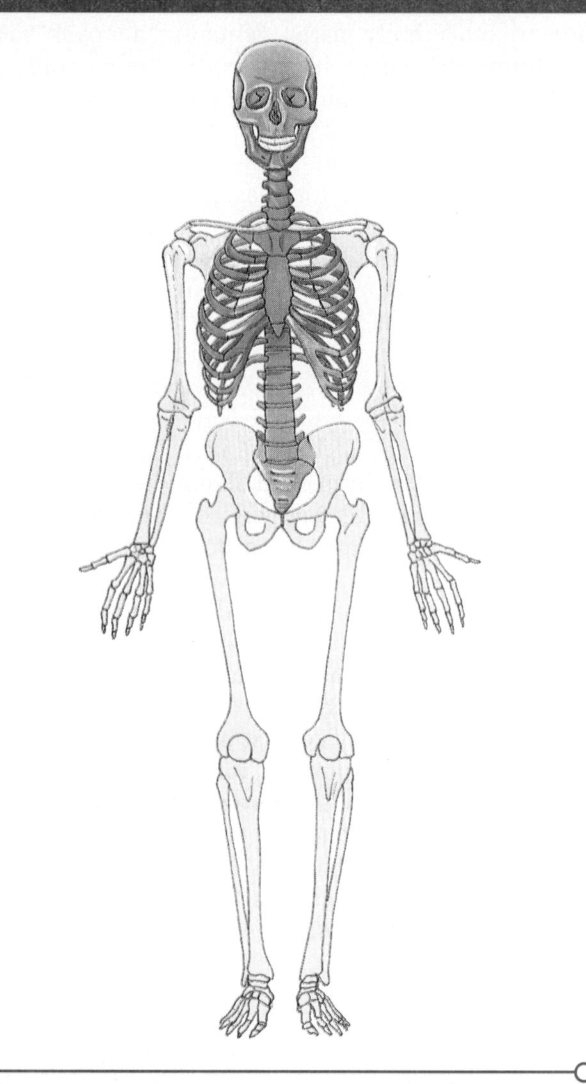

The appendicular skeleton is indicated in yellow and the axial skeleton is indicated in red.

Axial Skeleton

The **skull** is the body's most complex bony structure. It is formed by 22 bones divided into two sets, the 8 **cranial bones** and the 14 **facial bones.** The three tiny bones of the middle ear are sometimes counted as skull bones, but as you recall, because they are involved in hearing, they were considered in Chapter 10. The cranial bones, or **cranium** (kra'-ne-um), enclose and protect the brain and the organs of hearing and equilibrium. The facial bones form the framework of the face, protecting the eyes, providing cavities for the organ of smell, and creating openings for the passage of air and food.

Most skull bones are flat bones. Except for the jawbone, which is connected to the rest of the skull by a freely movable joint, all adult skull bones are firmly connected by immobile interlocking joints called **sutures** (soo'-cherz). In newborns some of the cranial bones are not completely formed, and the spaces between the bones, called **fontanelles** (fon'ta-nelz), are joined by membrane tissues (Figure 11-2). The bones complete their growth and these "soft spots" disappear around 16 months of age. Several bones of the cranium contain **sinuses,** air spaces lined by mucous membranes. Sinuses reduce the weight of the bone without diminishing its strength, much like a steel I beam is as strong as a similarly sized solid steel beam. In both bone and the I beam the reduction of material occurs in areas that do not contribute to the strength of the structure.

The two pairs of matched bones of the cranium each have the name of the lobe of the brain's cerebrum it covers: the parietal and the temporal. There is also a single frontal bone and a single occipital bone. Forming the forehead of the skull is the **frontal bone.** On top of the head and behind the frontal bone are the paired **parietal bones,** which extend to the sides of the skull. Below the parietal bones each **temporal bone** has an opening leading to the middle-ear cavity. The single **sphenoid** (sfe'noyd) **bone** completes the sides of the skull and also contributes to the floors and walls of the eye sockets. Similarly the single **ethmoid bone,** which lies just ahead of the sphenoid bone, is part of both eye sockets and the nasal septum. The **occipital bone** forms the back lower portion of the cranium and the base of the skull. A large opening through this bone is called the **foramen magnum** (fo-ra'men mag'num), through which the spinal cord passes and becomes the brain stem (Figure 11-3).

The facial bones include the **mandible,** or lower jaw bone, which is the largest, strongest bone of the face and is also the only freely movable bone of the entire skull. Tooth sockets are located in the mandible and in the two bones that form the upper jaw, the central portion of the face, and the anterior portion of the hard palate, known as the **maxillae** (mak-sih'le), or **maxillary bones.** The two **palatine bones** form the posterior portion of the hard palate and the floor of the nasal cavity. The irregularly shaped **zygomatic** (zi'go-mat'ik) **bones,** commonly called the cheekbones, also help form the lateral margin of the eye sockets. The two **nasal bones** are thin, rectangular-shaped bones forming the bridge of the nose. The two **lacrimal bones** are very delicate fingernail-shaped bones that help form the medial walls of each eye socket. Finally the **vomer** (vo'-mer) **bone** is located in the nasal cavity, where it forms part of the **nasal septum,** or the wall that separates the nostrils. The entire skull sits atop the vertebral column, which connects the skull to the rest of the skeleton (Figure 11-4).

Figure 11-2 Fontanelles

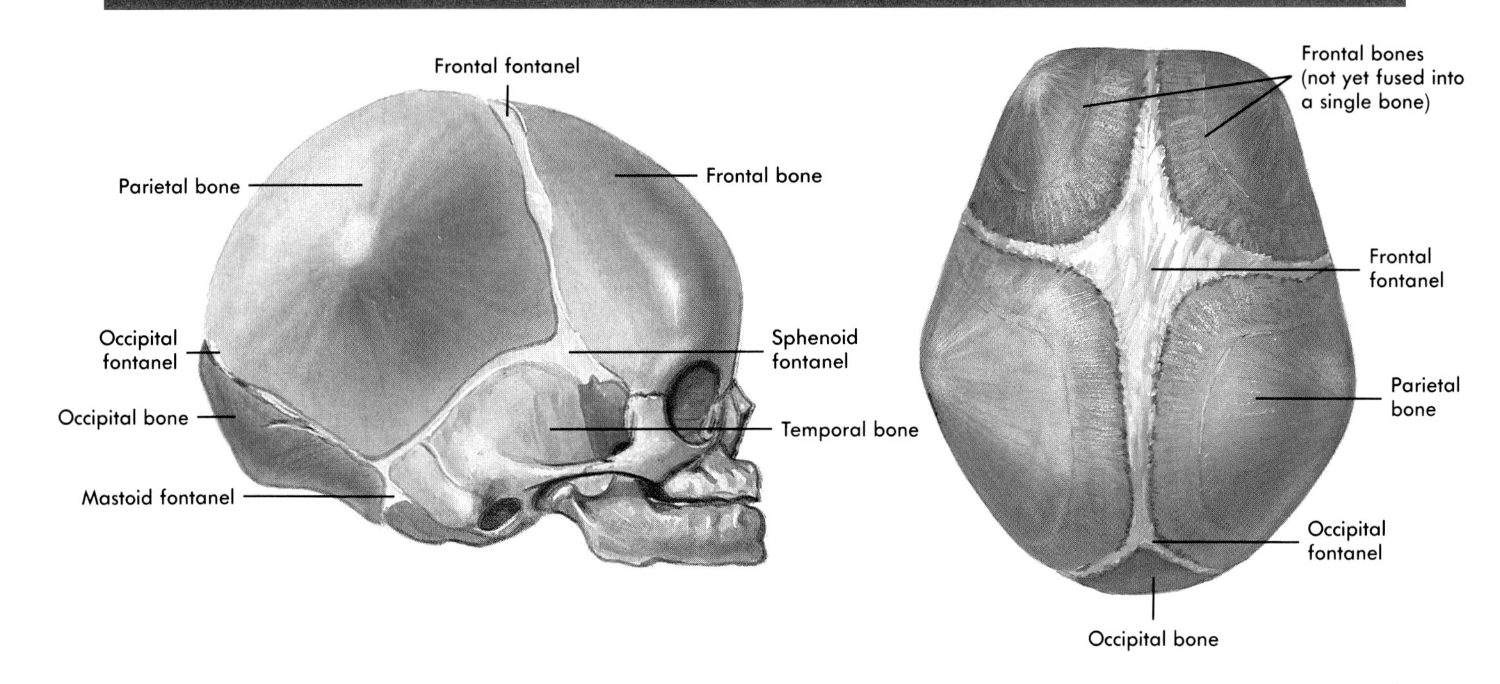

Membranous tissues that cover the spaces between bones and join bones together are called fontanelles. There are several fontanelles in a newborn's skull.

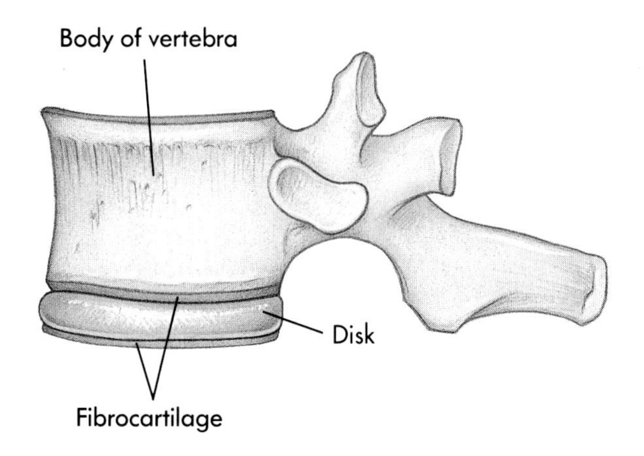

Most people have an image of the **vertebral column,** or **spine,** as a rigid supporting rod. The vertebral column is actually formed from 26 irregular bones, called **vertebrae** (ver'-tuh'bre). As the vertical support of the body trunk, the vertebral column extends from the skull, which it supports, to its anchoring point in the pelvis, where it transmits the weight of the trunk to the legs. As discussed in Chapter 8 the spinal cord runs through the central cavity of the vertebral column. The vertebral column also provides attachment points for the ribs and for the back muscles.

In an average adult the vertebral column is about 70 cm, or 28 inches, long and is divided into five major divisions. The first seven vertebrae of the neck are the **cervical** (ser'-vih-kul) **vertebrae.** All mammals, from the long-necked giraffe to the common mouse, have seven cervical vertebrae. The next 12 vertebrae are the **thoracic** (thor-a'-sik) **vertebrae.** All 12 pairs of ribs join directly to the thoracic vertebrae in the back, and all but 2 pairs connect either directly or indirectly to the **sternum,** or breastbone, in the front of the chest. The five vertebrae supporting the lower back are called the **lumbar** (lum'-bar) **vertebrae.** Below the lumbar vertebrae is the **sacrum** (sa-krum), which is part of the pelvis. The sacrum comprises five fused vertebrae, resulting in a somewhat triangular shape, and helps to strengthen and stabilize the pelvis. At the end of the spine is the tiny **coccyx** (kok'-siks), consisting of four small fused vertebrae (Figure 11-5).

When viewed from the side, the vertebral column has four curvatures that give it an S-like shape. The **cervical** and **lumbar** curvatures are convex *anteriorly,* whereas the **thoracic** and **sacral** curvatures are convex *posteriorly.* These curvatures increase the strength, resilience, and flexibility of the vertebral column. Between the vertebrae are the **intervertebral disks,** which are cushionlike pads composed of an inner gel-like fluid, and a strong outer wall of fibrous cartilage. The disks act as shock absorbers during walking, running, or jumping and also allow the spine to flex forward, backward, and side to side. The disks are thickest in the cervical and lumbar regions, this thickness increasing their flexibility.

Figure 11-3 Bones of the skull

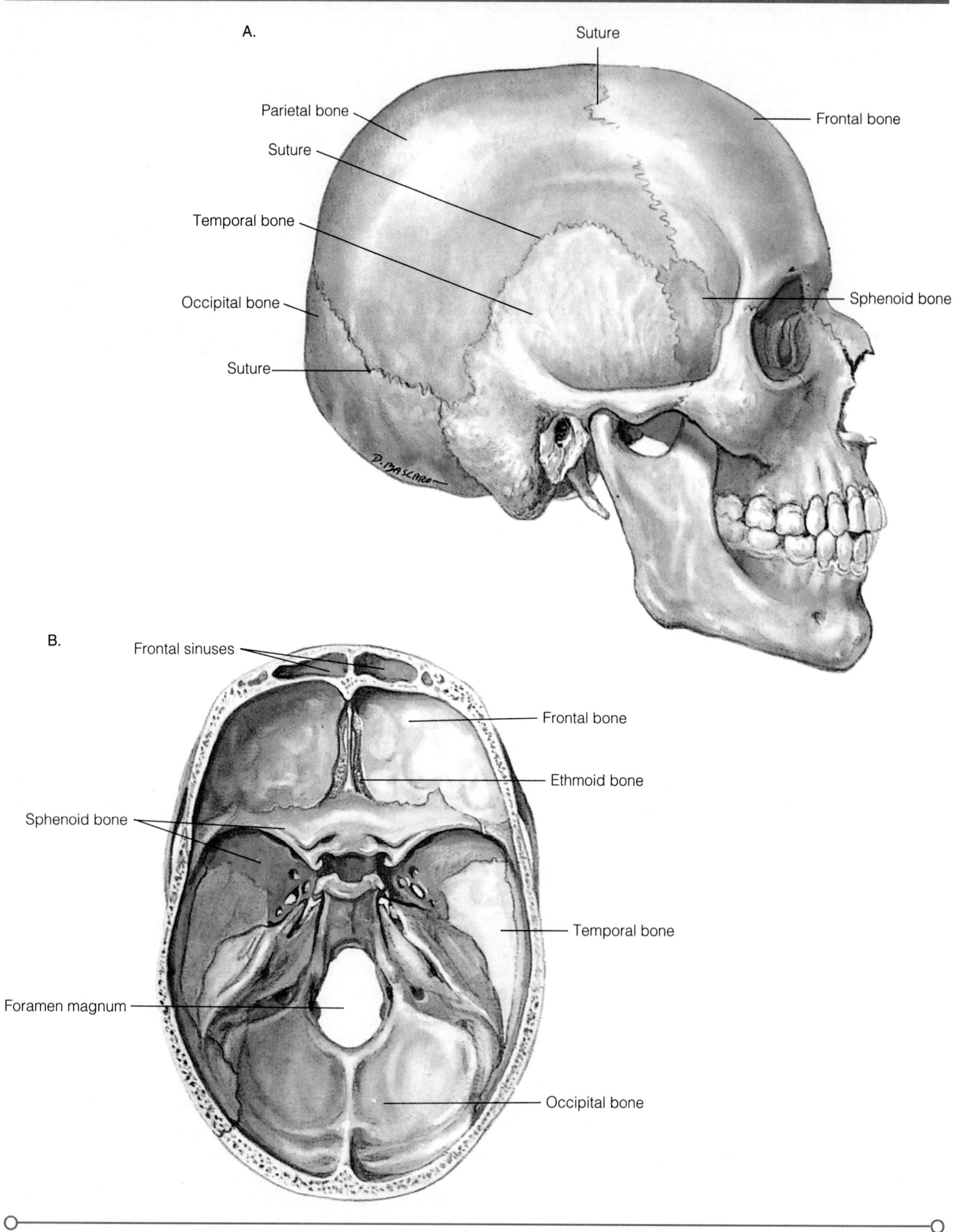

There are eight cranial bones that make up the cranium of the skull. The right lateral view of the skull (A) and the interior view of the skull (B) show the relationships between them.

Figure 11-4 Bones of the face

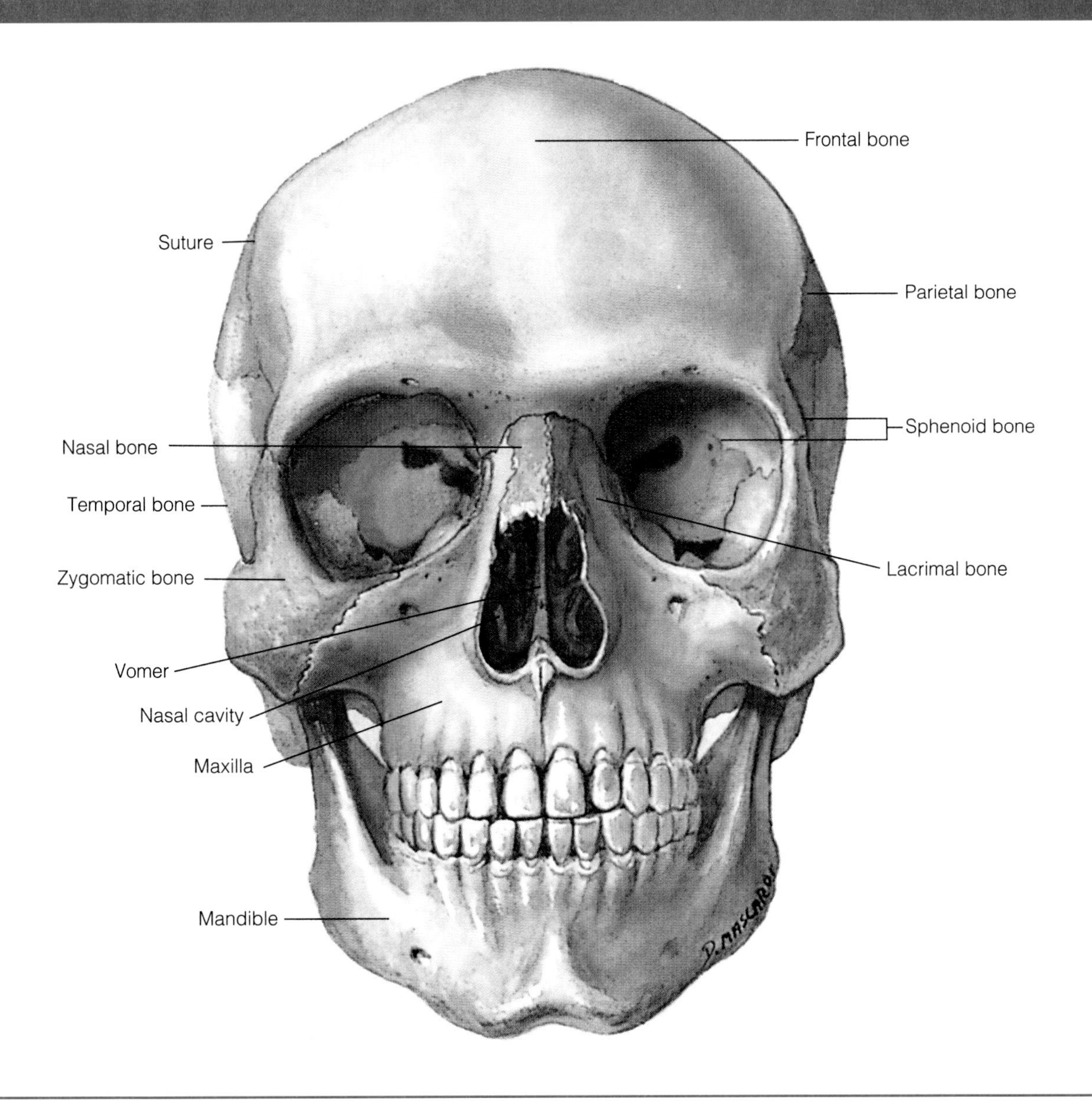

There are 14 bones that make up the face of the skull. Facial bones have complex shapes, as seen in the frontal view of the skull.

Appendicular Skeleton

As the name indicates the bones of the **appendicular skeleton** are "appended" to the vertical axis of the vertebral column by bony girdles anchored to it. The **pectoral** (pek'-ter-ul) **girdle** attaches the upper limbs to the trunk, and the more firmly attached **pelvic girdle** secures the lower limbs to the trunk. Although the bones of the pectoral girdle and the pelvic girdle are different from one another in their functions and mobility, they have the same fundamental structural plan. This means that each limb is composed of three major segments connected by freely movable joints.

Bones of the Upper Limbs

The **pectoral,** or shoulder, **girdle** consists of two bones: the anterior **clavicle** (kla'-vih-kul), or collarbone, and posteriorly the **scapula** (ska'-pyoo-luh), or shoulder blade. The two pectoral girdles, one for each shoulder, and their associated muscles form the shoulders. The medial end of each clavicle joins the sternum, while the distal ends meet the scapula laterally. However, the scapulae are attached to the posterior side of the thorax only by the muscles that cover their surfaces, creating a freely moving scapula and highly mobile arms (Figure 11-6).

Figure 11-5 Vertebral column

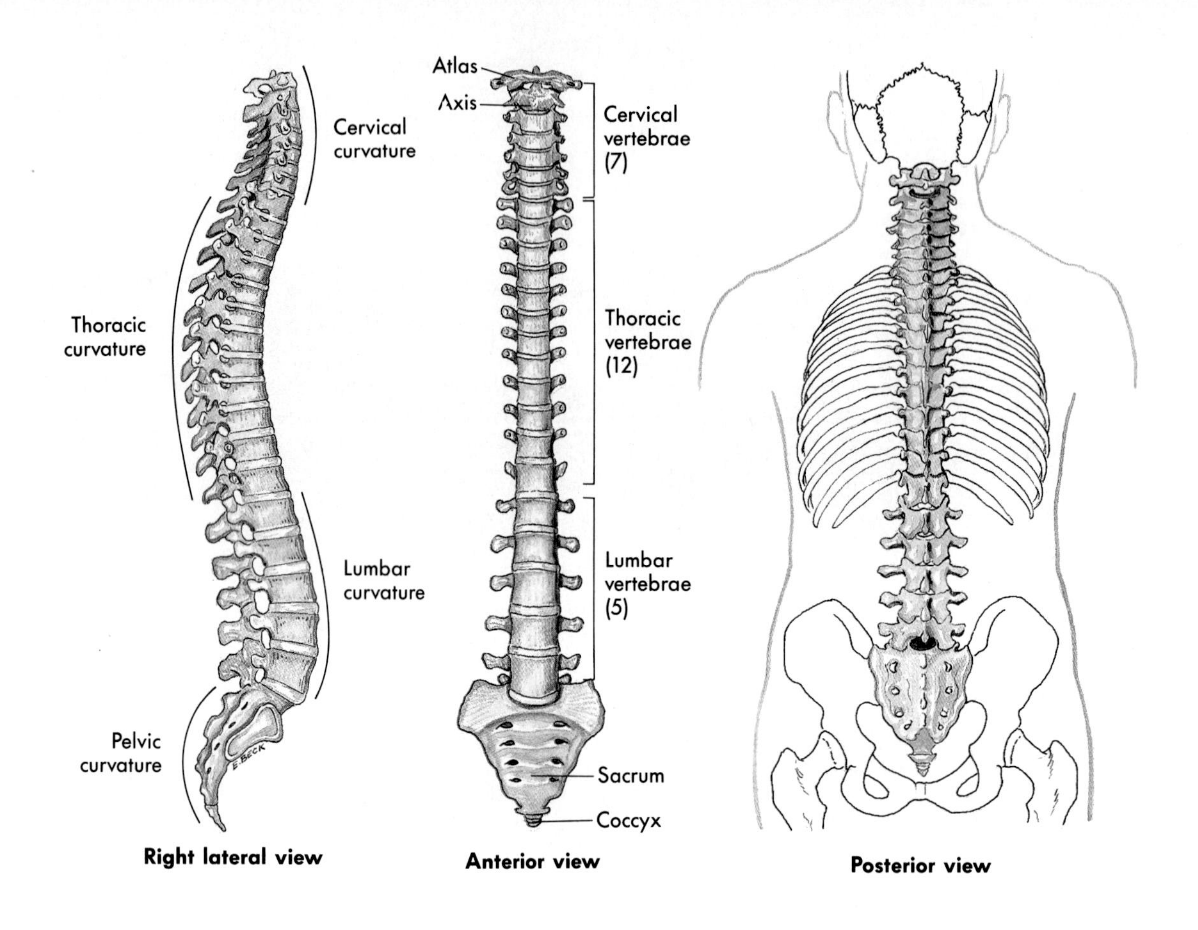

There are four major divisions and four major curvatures of the vertebral column.

The socket of the shoulder joint, called the **glenoid** (glen′-oyd) **cavity,** is small, shallow, and only lightly reinforced by ligaments. Although this arrangement is good for flexibility, it is poor for stability, and as a consequence the shoulder dislocates relatively easily.

Thirty separate bones form each upper limb. Each of these bones may be described as a bone of the arm, forearm, or hand.

The single long bone of the arm is called the **humerus** (hu′merus), the largest and longest bone of the upper limb. Its smoothly rounded head fits into the shallow glenoid cavity of the scapula. The distal end of the humerus joins the two parallel long bones of the forearm, called the **ulna** and the **radius,** at the elbow. When the arm is held to the side of the body with the palm facing the front, the radius lies laterally on the "thumb side" of the arm and the ulna medially on the "body side." As you rotate your hand so the palm faces toward the back, the distal end of the radius crosses over the ulna and the two bones form an X.

The skeletal framework of the hand includes the bones of the **carpus,** or wrist; the **metacarpals,** or palm; and the **phalanges,** or fingers.

The wrist consists of a group of eight marble-sized short bones, collectively known as **carpals** (kar′-pulz), joined by ligaments. These bones glide past one another during movement, giving the wrist its great flexibility. There are five **metacarpals.** The term *meta* is the Greek prefix meaning "after." Therefore these bones come after the carpal bones of the wrist. The metacarpals radiate from the wrist bones like the spokes of a wheel to form the palm of the hand. The proximal ends of the metacarpals join with the carpals of the wrist, whereas the bulbous distal ends meet the finger bones. When you clench your fist, the enlarged distal heads of the

Figure 11-6 The pectoral girdle and arm

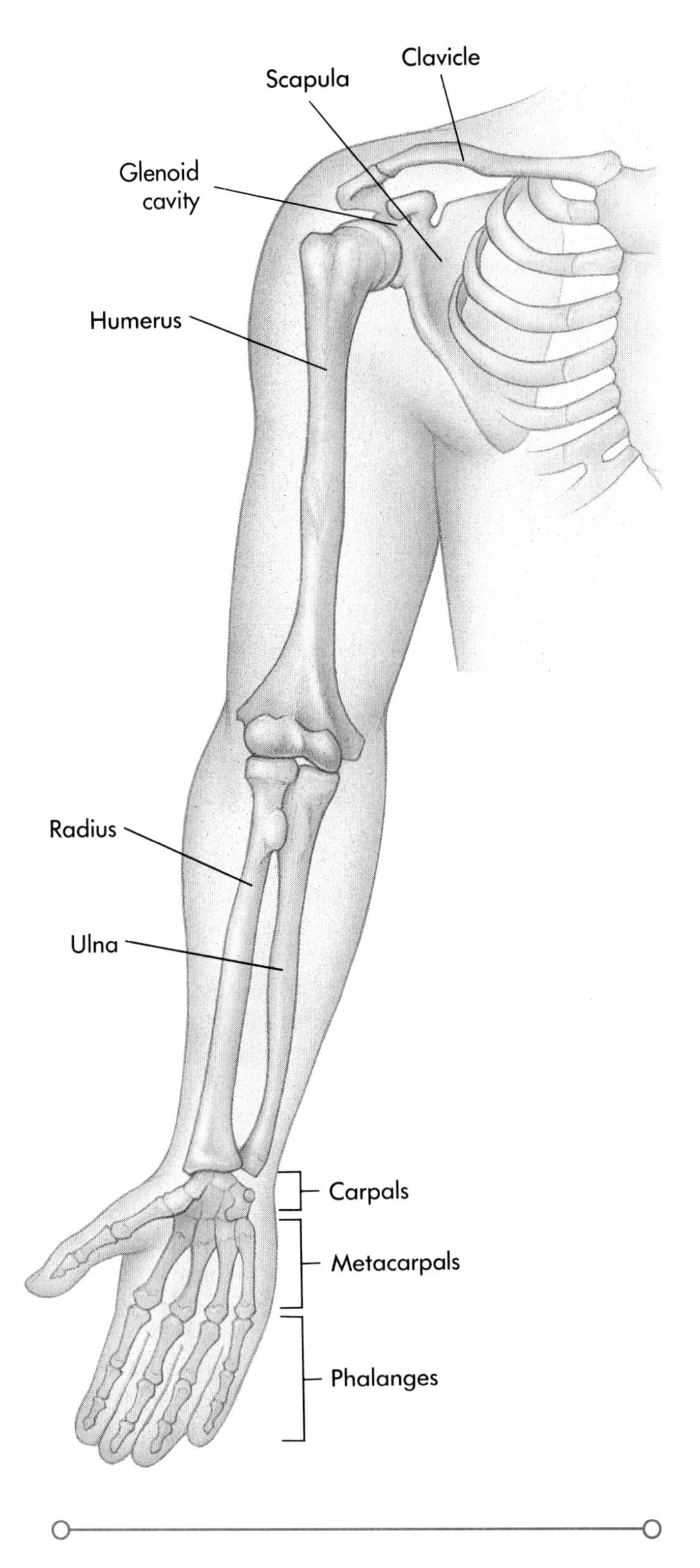

metacarpals become apparent as your knuckles. Metacarpal 1, associated with the thumb, is the shortest and most mobile. The joint between metacarpal 1 and its corresponding carpal bone is unlike the rest and is called a **saddle joint.** This joint gives us our highly mobile thumb, a unique feature that makes the hand a highly efficient instrument for grasping and manipulating objects.

The **digits,** or fingers, are made up of miniature long bones called **phalanges** (fuh-lan′-jez). With the exception of the thumb, which has only two, each finger has three phalanges.

Bones of the Lower Limbs

The pelvic girdle attaches the lower limbs to the axial skeleton and supports the internal organs of the pelvic cavity. In contrast to the relatively freely moving shoulder girdles, the pelvic girdle is fastened to the axial skeleton by some of the strongest ligaments in the body. The thigh bones articulate with the sockets of the pelvis, which are deep, cuplike, and also heavily reinforced with ligaments.

The pelvic girdle is formed by two large bones called the **innominate,** meaning "no name," bones. The innominate bones are anchored to the sacrum of the spinal column posteriorly. Together they form a hollow basinlike structure called the **pelvis.** During childhood each innominate bone consists of three separate bones, the ilium (il′i-um), the ischium (is′ki-um), and the pubis (pu′bis). By the time a person is 23 years of age, these bones are completely fused and indistinguishable, but their names are retained to refer to different pelvic regions. The three bones meet and fuse at the deep socket called the **acetabulum** (a′-sih-ta′-byoo-lum). The acetabulum articulates with the head of the thigh bone.

The **femur** (fe′-mer), the single bone of the thigh, is the largest, longest, and strongest bone of the skeleton. Its length is approximately one fourth of a person's height. The bone's robust anatomy reflects that the stress on the femur during a strenuous jump can reach 2 tons per square inch. Proximally the ball-like head of the femur articulates with the acetabulum of the pelvis. The head is carried on a short **neck** that angles slightly superiorly and medially from the femur shaft. The femur is shaped this way because it articulates with the side of the pelvis. Distally the femur shaft widens into the **medial** and **lateral condyles,** which articulate with the larger of the two lower leg bones (Figure 11-7).

Two bones make up the skeleton of the lower leg and are called the **tibia** and the **fibula.** The bones of the lower leg are less flexible but are stronger and more stable than those of the forearm. The larger tibia, located medially or closest to the midline of the body, articulates proximally with the

Figure 11-7 The pelvic girdle and leg

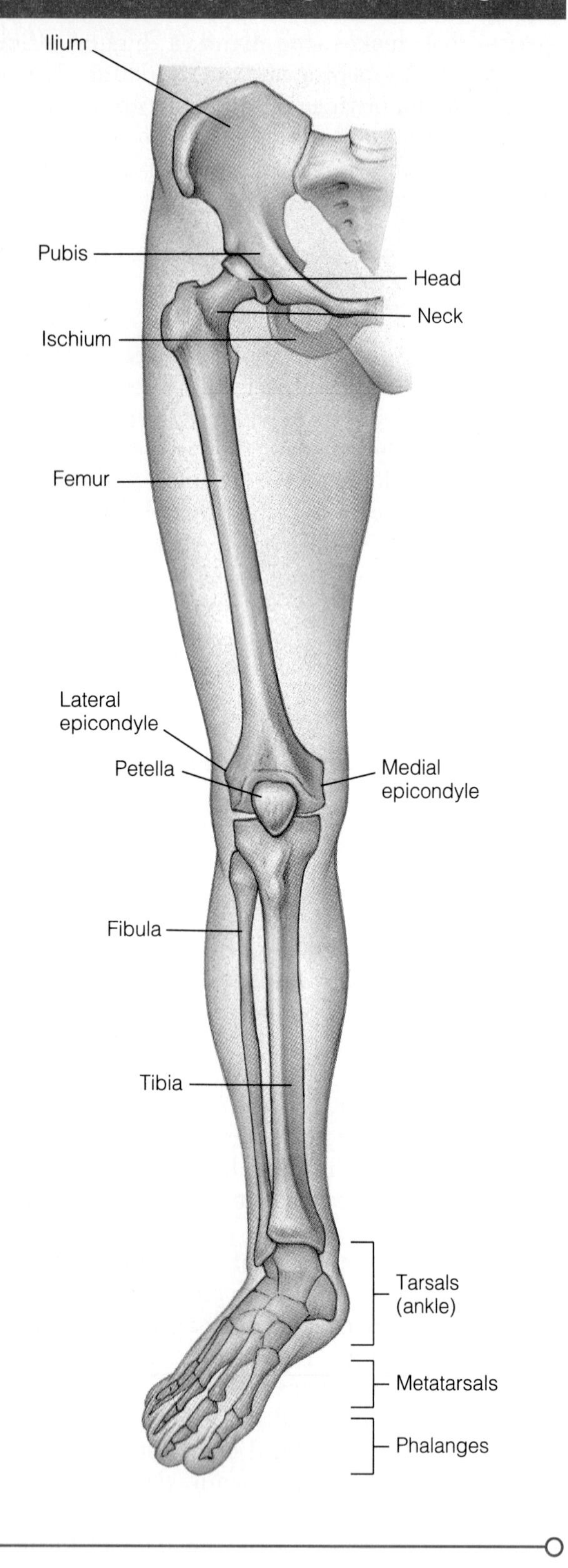

femur to form the hinge joint of the knee and articulates distally with one of the bones of the ankle. The primary role of the fibula, located laterally to the tibia, is to stabilize the ankle joint.

The skeleton of the foot includes the **tarsals** that make up the ankle; the **metatarsals** that comprise the instep, or arch, of the foot; and the **phalanges,** or toe bones. The foot has two important functions. First it supports body weight. Second it is a lever to propel the body forward when walking or running. A single bone could serve both purposes but would be inefficient on uneven surfaces. The segmented construction of the foot skeleton increases its flexibility when encountering uneven surfaces.

Joints

Bones are joined together at **joints** or **articulations.** **Joints** have two functions. They secure the bones together, and they allow the skeleton of rigid bones to be mobile. One way skeletal joints are classified is according to the amount of movement they allow between the articulated bones. Some joints are **immovable,** others **slightly movable,** and still others are classified as **freely movable.** Freely movable joints are generally in the limbs, whereas immovable and slightly movable joints are usually in the axial skeleton, where firm bony attachments and enclosed organ protection are of primary concern. There are also structural differences among immovable, slightly movable, and freely movable joints.

The sutures between the bones of the skull are immovable joints. The wavy edges of the bones either overlap or interlock, and the junction is completely filled by connective tissue fibers. The result is a completely immobile joint (Figure 11-8, *A*). In slightly movable joints the articulating bones are usually held together by cartilage or ligaments. However, the degree of movement between the bones depends in large part on the length of the connecting ligaments. For example, ribs numbered 2 through 10 are joined to the sternum by cartilage, which allows for a slight degree of movement, as when you inhale and exhale deeply. (Figure 11-8, *B*). In contrast, the intervertebral discs between the vertebrae of the vertebral column provide a slightly greater degree of movement, both because of the greater segmentation of the vertebral column and because of the length of the cartilaginous fibers relative to the size of each vertebra. There are several different types of freely movable joints, but what they share in common is there is space between the articulating bones which is usually filled with fluid. We next look at freely movable joints in detail.

Freely moving joints are **synovial** (si-no′ vi-al) **joints.** They consist of a fluid-filled cavity separating the articulating bones. There are several types of synovial joints. However, they each share three related characteristics. First the bone surfaces facing into the joint, called articular surfaces, are covered with a smooth layer of cartilage called **articular cartilage.** Second they have a joint cavity formed by a double-layered **articular capsule.** The external layer is a tough flexible connective tissue, and the internal layer, called the **synovial membrane,** is composed of loose connective tissue whose cells produce the egg white–like substance called **synovial fluid** that fills the joint cavity. Synovial fluid bears most of the weight at the joint surfaces and keeps the articular cartilages from touching each other, thereby reducing friction during joint movement. Third, synovial joints are reinforced and strengthened by ligaments that hold the articulating bones together at the joint (Figure 11-9).

Figure 11-8 Immovable and slightly movable joints

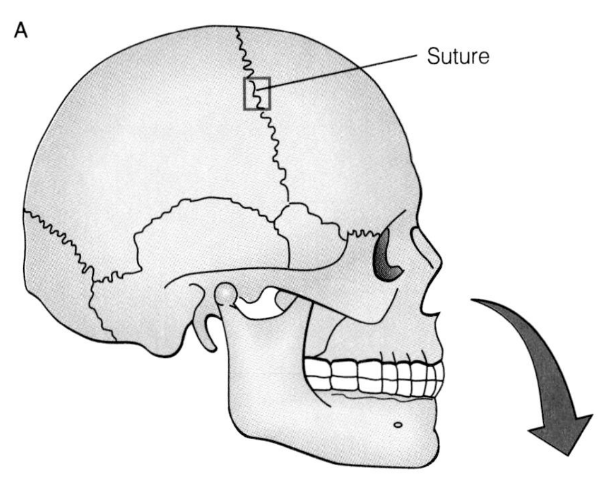

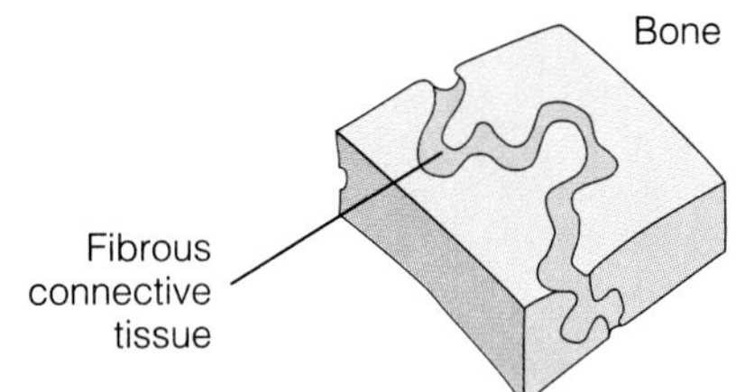

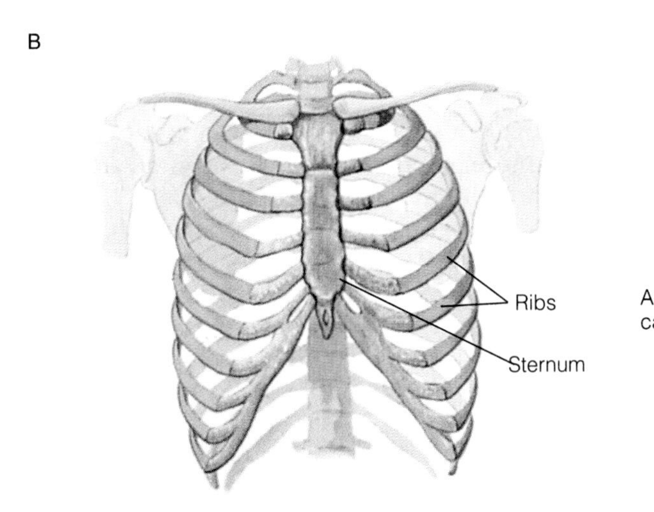

Immovable joints are rigid, such as sutures between cranial bones (A). Slightly movable joints are held together by cartilage or ligaments. Slightly movable joints hold rib number 1 to the sternum (B).

Bone Structure

So far we have described the major types of bone and how they are joined to one another. Now we

Figure 11-9 Detail of a freely movable joint

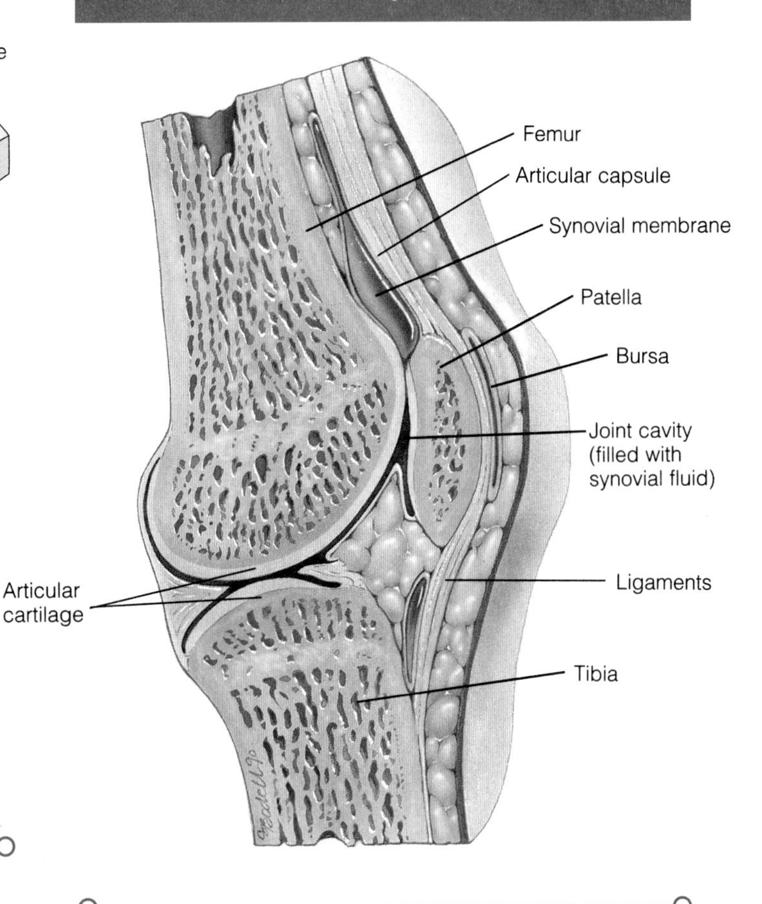

Freely movable joints are synovial joints consisting of a fluid-filled cavity separating the bones of the joint. The knee joint is a synovial joint.

describe what bone is made of and how it grows and develops. We can use a long bone as our standard for describing the overall structure of bone.

The tubular shaft, or **diaphysis** (di-a'-fih-sis), constitutes the long midportion of the bone. It is constructed of a relatively thick wall of compact bone that surrounds a hollow core called the **medullary** (meh'-duh-layr'-e) **cavity.** In adults this cavity contains fat called **yellow marrow.** It also contains a substance called **red marrow**. The cells embedded in the marrow are responsible for **hemopoiesis** (he'mo-poy-e'sis), the process of forming and developing the various types of blood cells. The **epiphyses** are the bone ends. In most cases they are wider than the diaphysis. The epiphyses have a thin outer layer of compact bone, whereas the interior consists of spongy bone. The outer surface of the diaphysis is covered by a shiny, white, double-layered membrane called the **periosteum.** The outer layer of the periosteum is dense connective tissue, and the inner layer, next to the bone surface, consists of bone-forming cells called **osteoblasts** (Figure 11-10).

The periosteum is supplied with nerve fibers and blood vessels that enter the bone through **nutrient canals.** The medullary cavity is lined with a thin membrane called the **endosteum** (en-dos'te-um), which contains both osteoblasts and bone-destroying cells called osteoclasts. Where long bones articulate with one another, the bony surfaces are covered with articular cartilage rather than periosteum.

Figure 11-10 Anatomy of a long bone

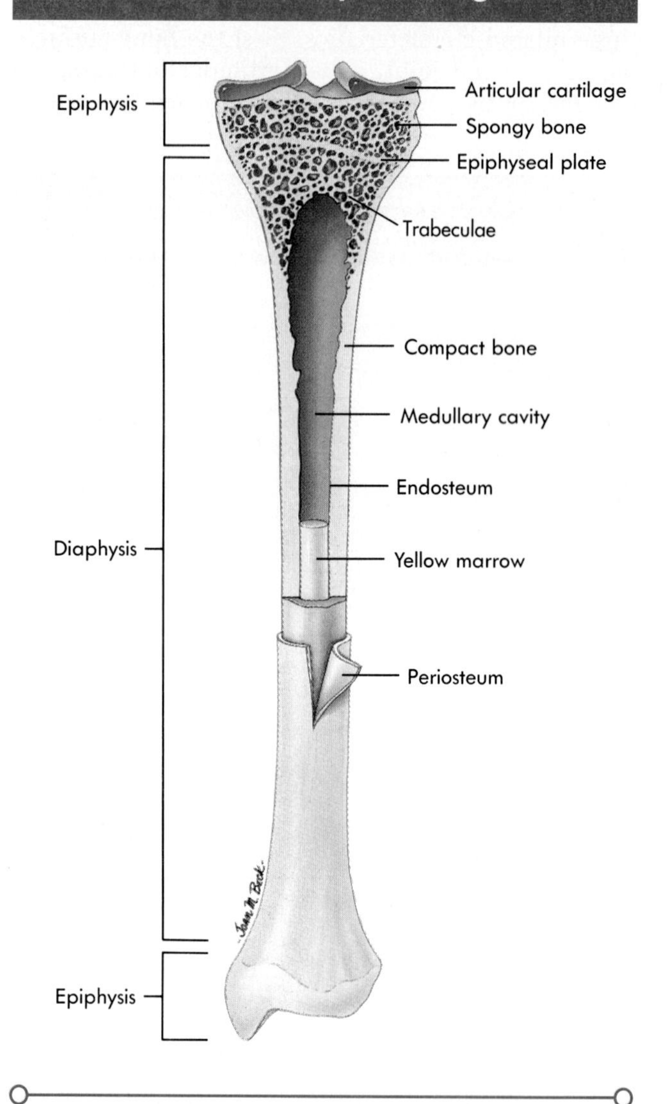

Microscopic Structure of Bone

Compact bone appears very dense. However, when observed under a microscope, it is riddled with canals and passageways through which nerves and blood vessels course. The structural unit of compact bone is called the **osteon,** or **Haversian system.** Each osteon consists mostly of hard inorganic matrix arranged in concentric rings, or **lamellae,** around a central canal, called the **Haversian canal.** The Haversian canals are oriented along the long axis of the bone. Mature bone cells, called **osteocytes** (os'te-o-sitz), lie in small cavities between the lamellae. Hairlike canals connect these cavities to one another and to the central Haversian canal, permitting easy diffusion of nutrients and wastes to and from the blood vessels in the haversian canal (Figure 11-11).

In contrast to compact bone, spongy bone looks poorly organized. However, the arrangement of the trabeculae is not random. They are arranged according to where stress is exerted on the bone. Each trabecula contains irregularly arranged lamellae, with the osteocytes interconnected by the same hairlike canals found in compact bone, but with no osteons. Nutrients reach the osteocytes by diffusion through the hairlike canals from the larger spaces between the trabeculae.

Bone has both organic and inorganic components, and the organic component includes three types of bone cells. The three types of bone cells are **osteocytes,** or mature bone cells that maintain the bone; the **osteoblasts** ("bone-forming"), the bone-creating cells that secrete the hard inorganic component into their surrounding environment; and the **osteoclasts** ("bone-breakers"), which are bone-destroying cells that break down the inorganic component of bone and return it to the bloodstream. The collagen fibers secreted by the osteoblasts also make up one part of the extracellular matrix. Collagen gives bone the ability to resist bending and twisting. The inorganic component that constitutes

Figure 11-11 Microscopic structure of compact bone

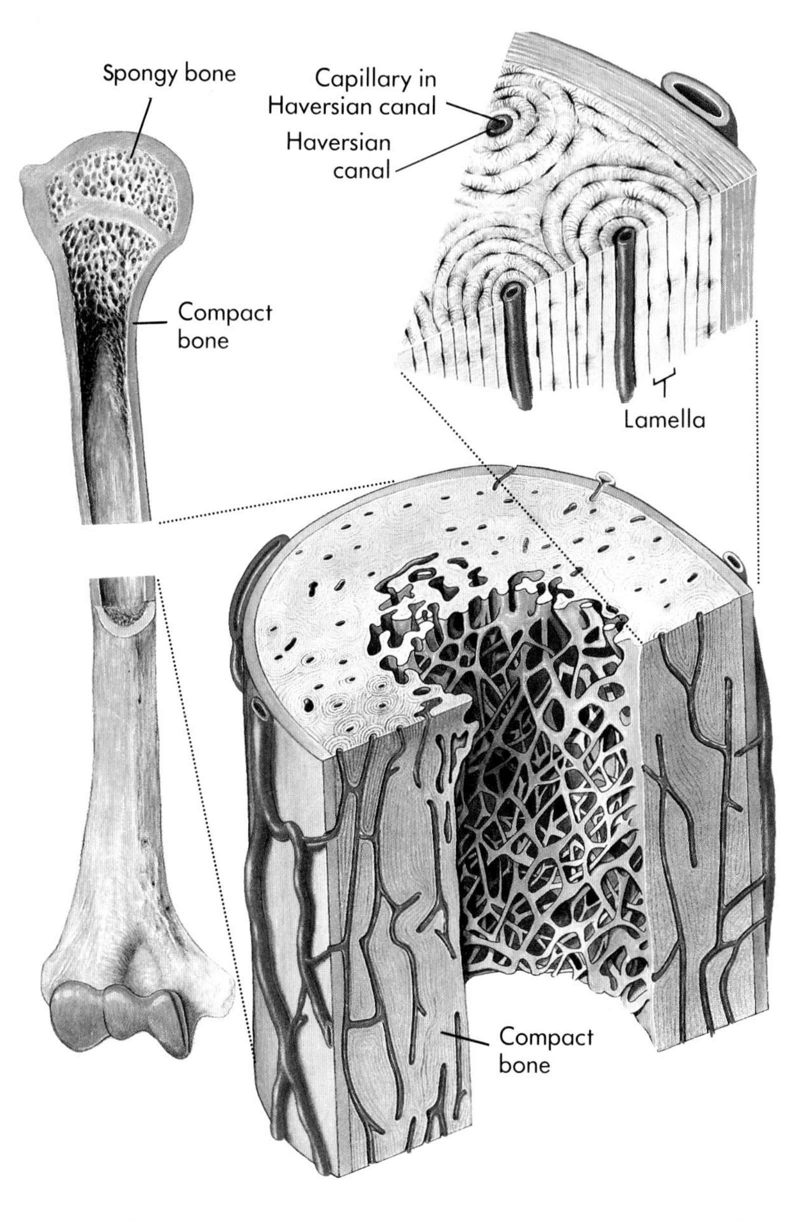

The structural unit of compact bone is the osteon, or Haversian system, consisting of hard inorganic matrix and osteocytes arranged in concentric rings around a canal contain a blood vessel. Haversian canals are oriented along the long axis of the bone.

the other part of the extracellular matrix consists of **inorganic mineral salts,** largely derived from calcium. The mineral salts account for bone's exceptional hardness, which allows it to resist compression, or the crushing forces placed on it.

Bone Growth and Development

Bone growth and development occurs in two slightly different forms. One type of bone growth accounts for the increase in overall size and diameter of the growing bone by the process of **remodeling** the bony material. The old bone is removed by osteoclasts, and the new bone is deposited by osteoblasts. For example, as a long bone increases in length and diameter, the size of the medullary cavity also increases proportionately. The relative thickness of compact bone is maintained by osteoclasts removing bone on the inside and by osteoblasts adding bone to the outside.

The second type, called **osteogenesis** (os'te-o-jen'e-sis), which occurs in developing embryos, forms the bony skeleton from existing cartilage. This process of bone growth goes on until early adulthood as we continue to increase in size. We now describe the process of osteogenesis in some detail.

Osteogenesis

At about 6 weeks after conception, the skeleton of a human embryo consists almost entirely of cartilage. Bone begins to form at this time and eventually replaces most of the cartilage. These cartilage "bones" are models, or templates, for the hard bony tissue that eventually replaces them. A typical long bone, the tibia, is used here to illustrate the process of bone formation, growth, and development.

The change of cartilage to bone, called **ossification** (os'i-fi-ka'shun), involves the death of most of the cartilage cells, or chondrocytes, and the growth of blood vessels and connective tissue into the spaces left behind. First, the bone begins as a cartilage model (Figure 11-12, *A*). The cartilage model develops a periosteum producing a ring, or collar, of bone around it (Figure 11-12, *B*). On the inner surface of the periosteum cartilage cells become osteoblasts, secreting calcium into the collagen matrix. In addition to this, osteoclasts in the center of the bone begin to remove calcium from the collagen matrix, forming a medullary cavity (Figure 11-12, *C*). Increasing numbers of blood vessels penetrate the cartilage model and, because of the change in the quantity of nutrients, more of the cartilage cells become osteoblasts. This process begins in the **primary ossification center** of bone (Figure 11-12, *D*). In this case it is the center of the long bone shaft. This process continues from the primary ossification center out toward the epiphyses of the bone (Figure 11-12, *E*).

At birth most of the long bones have a bony dia-

Figure 11-12 Bone formation

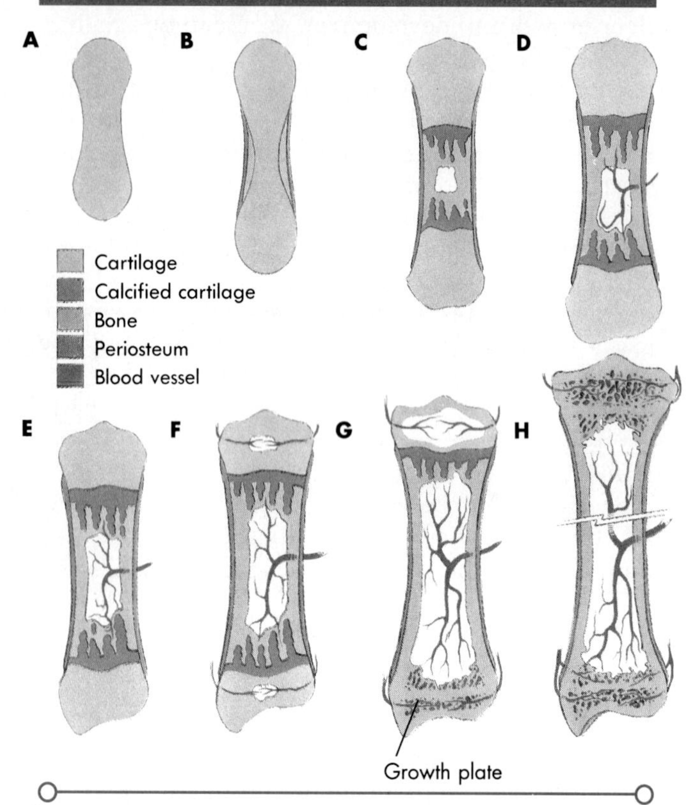

Cartilage model (A). Development of periosteum (B). Formation of the medullary cavity (C). Formation and growth of the primary center of ossification (D, E). Appearance of secondary centers of ossification (F). Bone growth at growth plate (G, H).

physis, a growing medullary cavity, and two epiphyses still consisting of cartilage. Later, bone-forming centers, called **secondary ossification centers,** appear in the epiphyses, reproducing almost exactly the events of primary ossification except that spongy bone in the interior is retained and no medullary cavity is formed (Figure 11-12, *F*). The bone grows in length in a region just below the ends of the bones and above the shaft itself called the **growth plate** (Figure 11-12, *G*). Cartilage cells proliferate in the growth plate and line up in columns parallel to the long axis of the bone, which increases bone length. The cartilage cells then die and are replaced by osteoblasts, which secrete calcium into the extracellular matrix, forming hard bone. Adding bony material at this region causes the bone to lengthen, and the tip of the bone grows farther and farther away from the middle of the shaft. This is an effective means of growth because it does not interfere with the articulation of the end of the bone with the bone next to it (Figure 11-12, *H*).

Muscle Tissue

The remainder of this chapter is concerned with one type of muscle tissue, but as you recall from Chapter 7, there are actually three different types of muscle tissue in the body: skeletal, cardiac, and smooth muscle. Although some skeletal muscles are often activated by reflexes without our conscious control, skeletal muscle is called *voluntary* because it is the *only* type of muscle under our conscious control. In all three types of muscle tissue the individual muscle cell is called a **muscle fiber.** All muscle fibers contain structural elements giving them the ability to contract, or shorten. When muscles contract, they reduce the distance between the parts they connect or the space they surround. For example, contraction of a skeletal muscle that connects two bones brings the attachment points closer together, causing the bone to move. When cardiac muscle contracts, it reduces the space in the heart chambers, pushing the blood in the heart out into the blood vessels. Smooth muscles are usually part of the wall of structures such as blood vessels and digestive organs, so contraction causes the diameter of these tubes to narrow, forcing the contents to move through them. Two other properties of muscle tissue include the ability to be stretched, called **extensibility,** and the ability to return to its original length after having been stretched, called **elasticity.**

However, muscle tissue types differ in their cell structure, where the tissues are in the body, and the means by which they are stimulated to contract. Here we will be describe skeletal muscle tissue.

Muscle Action

Skeletal muscle fibers are packaged into muscles that are attached to and cover the bony skeleton. Muscles are usually attached to bone by fibrous bands of connective tissue called **tendons.** The tendons extend from a muscle to the far side of a joint. When the muscle contracts, one bone remains relatively stationary and the other one moves. The **origin** of the muscle is on the stationary bone, and the **insertion** of the muscle is on the bone that moves. Because muscles can only contract, or shorten, opposite acting muscles are needed at a joint. For example, when the biceps muscle of the anterior side of the arm contracts, the forearm is raised up toward the arm. Another muscle is needed to "undo" this action. The triceps muscle on the posterior of the upper arm contracts and brings the forearm back down and away from the arm. Most skeletal muscles work as **antagonistic pairs,** making movement possible.

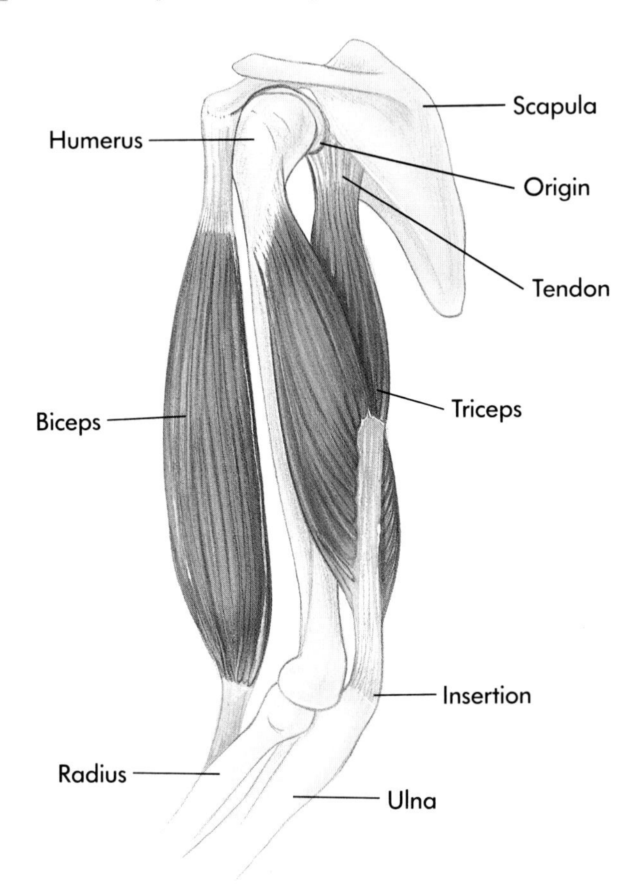

Skeletal muscle has **tone,** meaning there are always some muscle fibers contracted. Muscle tone is important in maintaining reflex responses and posture. Muscle spindles, briefly discussed in Chapter 8, are primarily responsible for controlling muscle tone.

Components of a Skeletal Muscle Fiber

Skeletal muscle tissue consists of long, parallel muscle fibers. Muscle fibers are specially modified to perform their particular function of contraction. Just as it was necessary to learn how an individual neuron produced and conducted action potentials to understand how the nervous system functioned, so it is with skeletal muscle. We need to understand how a muscle contracts by taking a closer look at the structure and function of individual muscle fibers.

Because muscle fibers have particular differences compared with other body cells, some of their cellular components also have modified names. The cell membrane of a muscle fiber is called the **sarcolemma** (sar′ko-lem′ah). Although similar to other cell membranes, it forms tubules that penetrate into the cell's interior. The tubules make up the **transverse tubule system.** These tubules make contact with the fiber's modified smooth endoplasmic reticulum, called the **sarcoplasmic reticulum.** Unlike the smooth endoplasmic reticulum of other cells, the sarcoplasmic reticulum has calcium storage sacs containing calcium ions (Ca^{++}), which are essential for muscle contraction. The last difference is the cytoplasm of the muscle fiber called the **sarcoplasm.** The names of other muscle fiber organelles are unchanged.

When viewed under a light microscope, each muscle fiber contains many rodlike structures called **myofibrils** that run parallel to one another, extending the entire length of the fiber. Depending on the fiber's size, there may be thousands of myofibrils packed into a single muscle fiber. Along the length of each myofibril are a repeating series of alternating dark and light bands. This banding pattern of the myofibrils gives skeletal muscle its striped, or striated, appearance under magnification. The wide dark bands are called **A bands,** and the thinner light bands are called **I bands.** Through the central portion of the dark A band runs a slightly lighter band called the **H zone.** In the middle of each I band is a thin dark line called the **Z line.** The Z line is actually a protein sheet that cuts across the diameter or traverses through the myofibril and is an attachment point for contractile elements in the myofibril (Figure 11-13).

A **sarcomere** (sar′-ko-mer, "muscle part" in Greek) is the region of a myofibril between two successive Z lines and is the smallest contractile unit of a muscle cell. Each myofibril can be thought of as a chain of sarcomere units laid end to end. Imagine a myofibril to be like a train made of a long series of identical boxcars linked together (Figure 11-14).

When viewed under an electron microscope, the banding pattern of a myofibril arises from the orderly arrangement of two types of protein filaments in each sarcomere. Look again at Figure 11-14 and you will notice there are centrally positioned **thick filaments** that extend the entire length of the A band. To either side of the thick filaments are **thin filaments** anchored to the Z lines, extending across the I band and partway into the A band. The H zone of the A band appears less dense than the rest of the A band because the thick filaments are not overlapped by the thin filaments in this region.

Thick filaments are composed of a protein called **myosin.** A myosin molecule has a rodlike tail

Figure 11-13 Muscle fiber

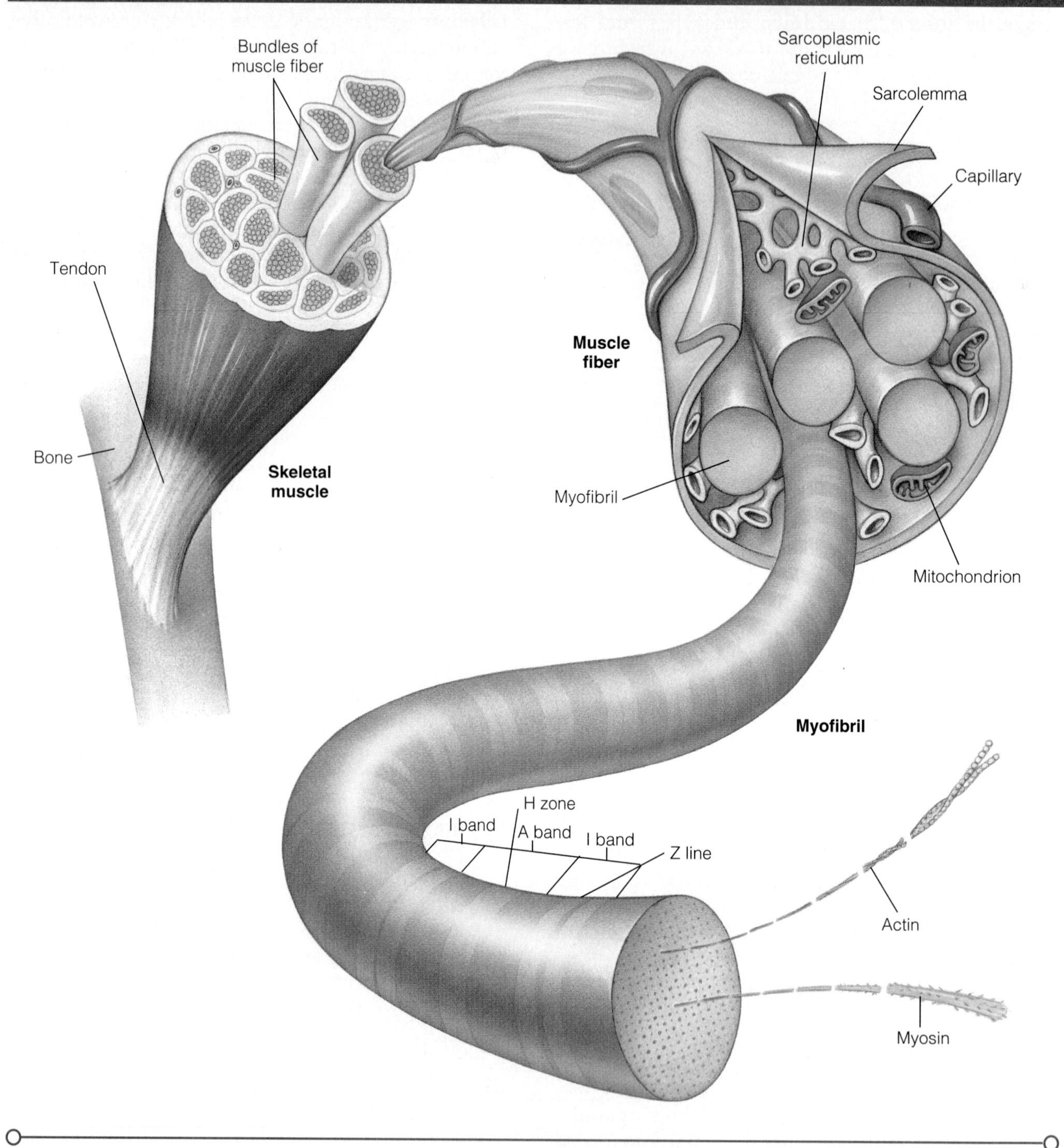

A muscle fiber contains rodlike structures called myofibrils extending the entire length of the fiber. Myofibrils each have a repeating series of stripes, or striations, labeled as the A and I bands. Within the A band there is a lighter stripe called the H zone and in the middle of each I band there is a thin dark line called the Z line.

attached to a globular head. The myosin molecules are bundled together, so the tails form the central portion, or stalk, of the filament and their globular heads face out from it. The thin filaments are composed primarily of the protein **actin.** A thin filament consists of two strands of actin molecules coiled about one another like a twisted double strand of pearls. Next we look at how these basic components of the muscle fiber interact to produce muscle contraction.

Contraction of a Muscle Fiber

Either a single muscle fiber is activated by an action potential and contracts, or it does not. A whole muscle contains many fibers, and the degree of contraction depends on the total number of fibers contracting. Thus the maximum stimulus for a whole muscle is the one at which the maximum number of fibers are stimulated to contract.

As mentioned earlier, contraction occurs at the level of individual sarcomeres. When a muscle fiber contracts, its sarcomeres shorten, reducing the distance between successive Z lines. As the length of their sarcomeres decreases, the myofibrils also shorten, resulting in the whole fiber shortening. However, examining this contraction event reveals that neither the thick nor the thin filaments change in length as the sarcomeres shorten. What then causes muscle fibers to shorten?

Contraction involves sliding the thin filaments past the thick filaments so their overlap increases. In a relaxed muscle fiber the thick and thin filaments overlap only slightly in each sarcomere. However, during contraction the thin filaments move more deeply into the central portion of the A band. The Z lines to which the thin filaments are

Figure 11-14 Sarcomeres

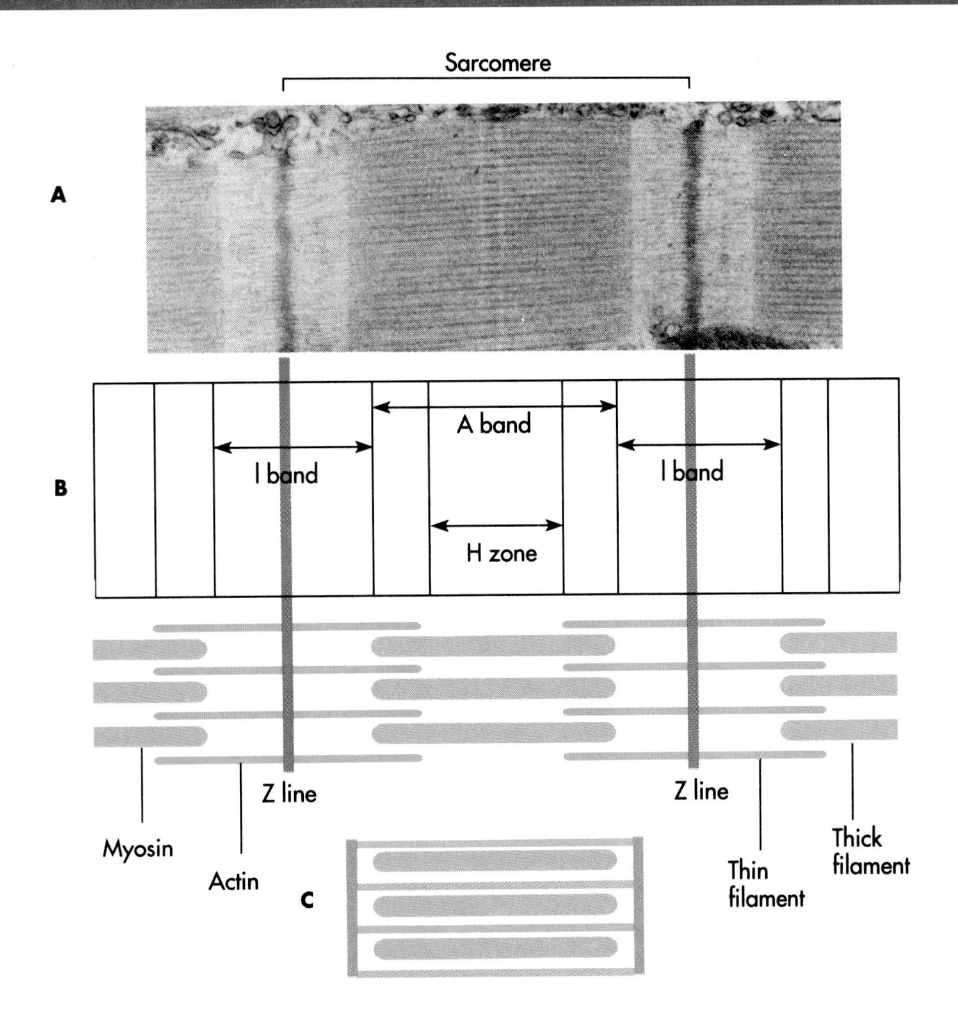

Two successive Z lines define the contractile unit of a myofibril, called a sarcomere. A sarcomere primarily contains two proteins: actin, also called thin filaments, and myosin, referred to as thick filaments. The particular arrangement of these proteins gives the sarcomere of the myofibril its characteristic striation pattern.

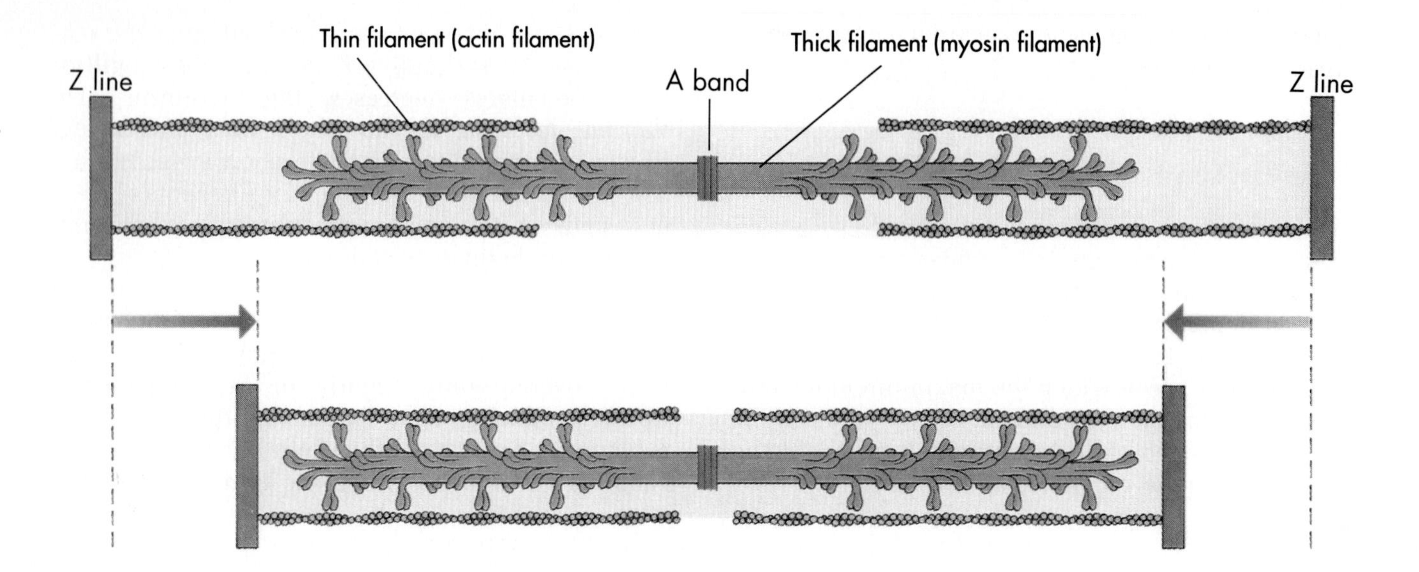

attached are also pulled toward the center of the A band. All of this raises another important question. What causes the filaments to slide past one another?

When muscle fibers are stimulated by the nervous system, the myosin heads attach to particular sites on the thin filaments, forming temporary cross bridges, and the sliding begins. Each cross bridge attaches and detaches up to 100 times during a contraction, acting much like a tiny oar pulling the thin filament toward the center of the sarcomere. Because this event occurs simultaneously in all the sarcomeres throughout the myofibrils of the muscle fiber, the muscle fiber shortens. The attachment of the myosin cross bridges to actin requires calcium ions, and the action potential leading to contraction causes an increase in calcium ions in the muscle fiber.

Stimulation of Muscle Contraction

Skeletal muscle fibers are stimulated by motor neurons of the somatic division of the central nervous system. The axon of each motor neuron branches as it enters a muscle and lies close to a muscle fiber to form a **neuromuscular junction** with it. The end of each branch has vesicles containing the neurotransmitter acetylcholine (ah-se′-tul-ko′-len), or ACh. The neuromuscular junction has the same components as a nerve cell synapse, which are a presynaptic membrane, a synaptic cleft, and a postsynaptic membrane. A neuromuscular junction differs from a synapse in that the postsynaptic membrane is a part of a muscle fiber's cell membrane, or sarcolemma.

In a motor neuron when an action potential reaches the end of the axon, the synaptic vesicles release ACh into the synaptic cleft. The ACh diffuses across the cleft and attaches to ACh receptors on the sarcolemma. Similar to nerve cell membranes, the sarcolemma is polarized. Therefore when ACh binds to it, the result is a depolarization event called a **muscle action potential** that spreads over the sarcolemma and down the transverse tubule system to the sarcoplasmic reticulum. The muscle action potential causes the calcium storage sacs of the sarcoplasmic reticulum to release calcium ions into the sarcoplasm (Figure 11-15). The calcium ions bind to the thin filaments, exposing the myosin binding sites on the filament.

Once the binding sites on actin are exposed, the following events occur in rapid succession:

1. **Cross-bridge attachment.** The activated myosin heads are attracted to the exposed binding sites on the actin, and cross-bridge binding occurs. An activated myosin head means that the hydrolysis of ATP into ADP and one inorganic phosphate molecule (P_i) by the enzyme ATPase, present on the myosin head, has occurred and the energy needed to activate the myosin is present. The ADP and P_i remain attached to the myosin head.
2. **Power stroke.** As a myosin head binds to the actin, it changes from its high-energy shape to its bent, low-energy shape. This causes the head to pull on the thin filament, sliding it toward the center of the sarcomere. At the same time ADP and P_i are released from the myosin head.
3. **Cross-bridge detachment.** As a new ATP molecule binds to the myosin head, the myosin cross bridge is released from the actin.

4. **"Cocking" the myosin head.** Hydrolysis of the newly attached ATP molecule to ADP and P_i provides the energy needed to return the myosin head to its high-energy, or "cocked," position. This returns the myosin head to the proper configuration for the next attachment and power-stroke sequence. We have now returned to the point where we started, and the cycle is repeated. You might think of the myosin cross bridges as "rowing or walking along" the adjacent actin filaments when a sarcomere shortens.

Contraction continues as long as there are calcium ions. When a muscle action potential stops, calcium ions are actively pumped out of the sarcoplasm and back into the storage sacs of the sarcoplasmic reticulum. The muscle fibers now relax (Figure 11-16).

Muscle Fiber Metabolism

As a muscle contracts the energy released by the hydrolysis of ATP is directly linked to cross-bridge movement and detachment. As long as ATP synthesis in the muscle fiber balances ATP use, contraction can continue. However, muscle fibers store limited amounts of ATP. When contraction begins, ATP reserves are soon exhausted, and more ATP must be synthesized for contraction to continue. There are three metabolic pathways by which ATP is produced during muscle activity: (1) by aerobic respiration, (2) by anaerobic respiration, and (3) by the interaction of ADP with a high-energy compound to produce ATP. As you learned in Chapter 5, aerobic and anaerobic respiration are the two primary metabolic pathways by which ATP is produced in all body cells. The third pathway occurs only in skeletal muscle fibers. We describe it here.

As a muscle is exercised vigorously, stored ATP is depleted in about 6 seconds. At this point a supplementary system for regenerating ATP goes into effect until the aerobic metabolic pathway can adjust to the sudden increased demands for ATP. The reaction that fills this temporary "energy gap" couples ADP with a high-energy compound called **creatine** (kre-uh-ten) **phosphate.** The result is the immediate transfer of energy and a phosphate group to ADP to form ATP. Because large amounts of creatine-phosphate are stored in skeletal muscle fibers and because the enzyme catalyzing the reaction is so effective, ATP concentration changes little during the initial periods of vigorous contraction. With reduced energy demands, some of the energy released from the hydrolysis of food molecules is used to resynthesize creatine-phosphate.

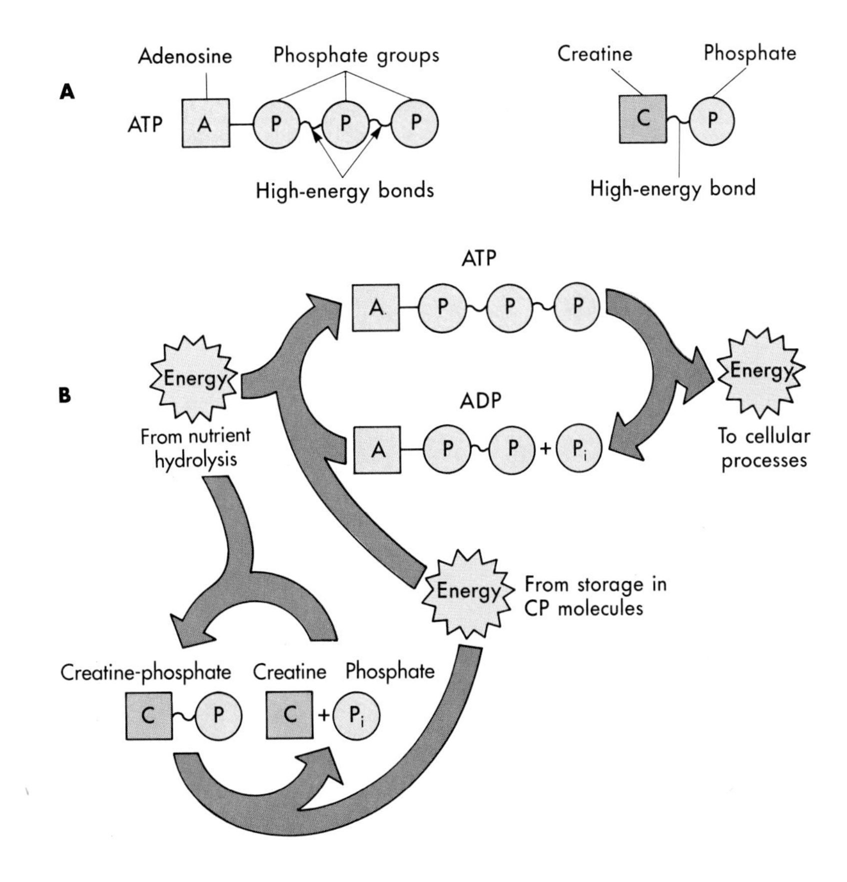

Figure 11-15 Action potential and muscle fiber contraction

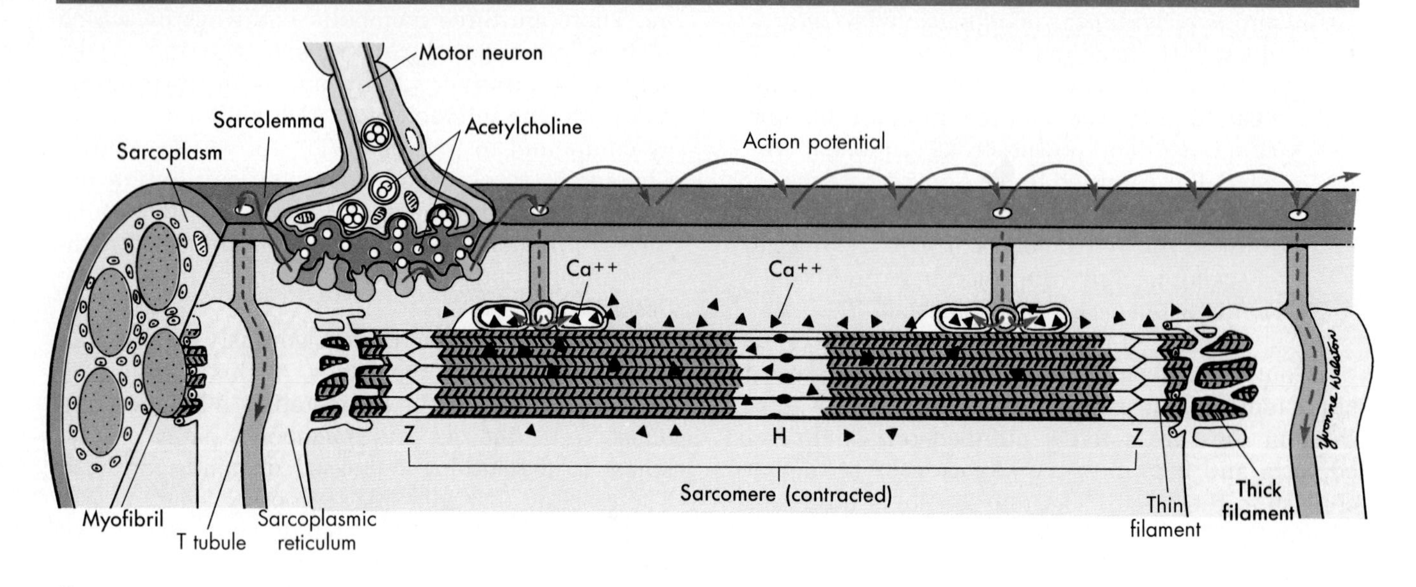

As an action potential reaches a muscle fiber it spreads throughout its sarcolemma and into the interior of the fiber by means of the transverse tubule system. The muscle action potential causes release of calcium ions from the sarcoplasmic reticulum, triggering contraction.

Figure 11-16 Muscle fiber contraction

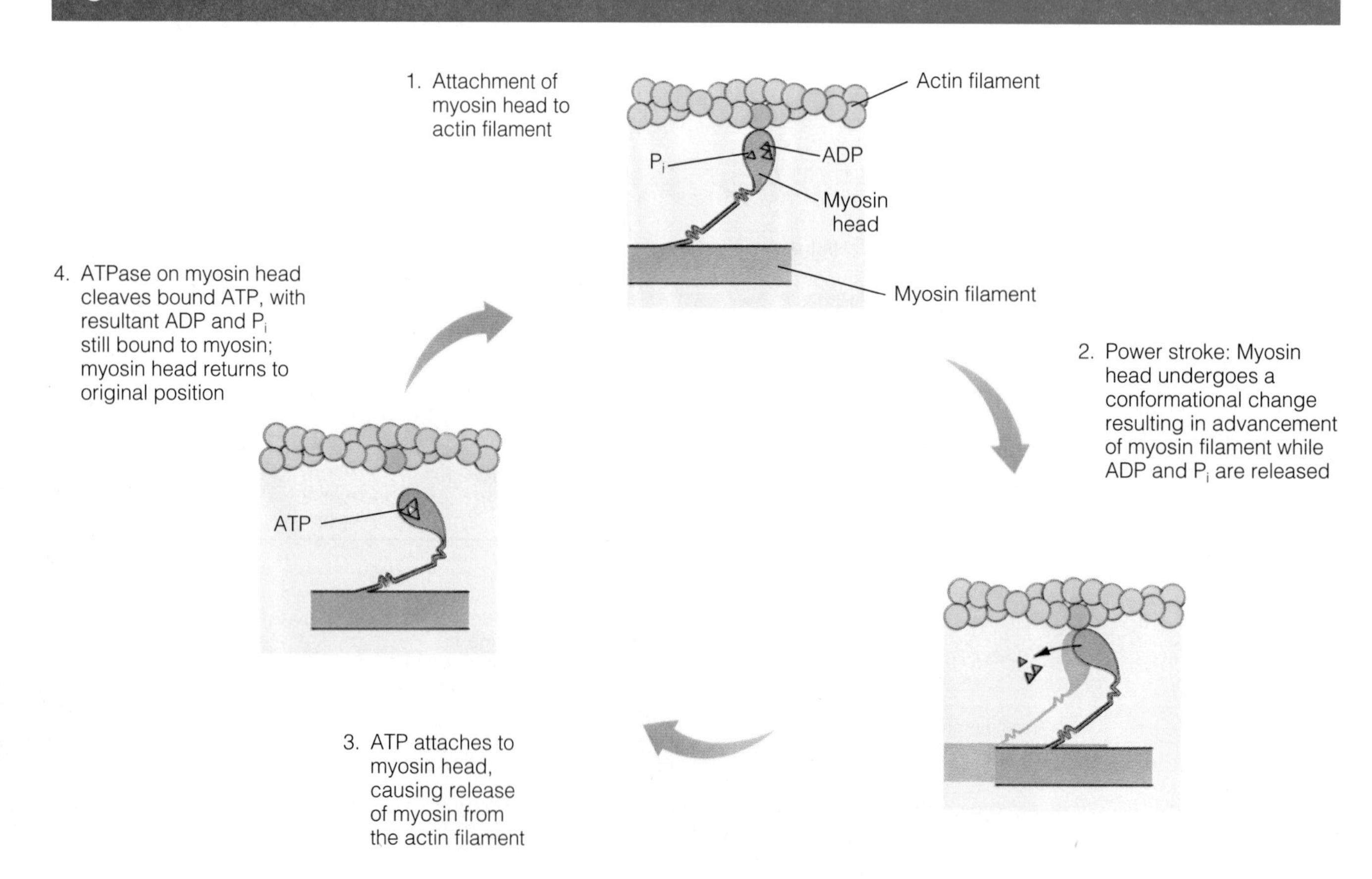

Biology in Action STEROIDS

During the last decade testing athletes for drugs and excluding from competition those testing positive for them or their metabolic by-products has been much publicized. One such group of drugs include the **anabolic androgenic steroids.** *Androgenic* means "male producing." These agents are biochemically related to the male hormone testosterone, which is responsible for, among other things, promoting male secondary sexual characteristics, including the increased muscle mass and size characteristic of males. *Anabolic* refers to their action of causing the increased size of individual muscle fibers.

Before their use was made illegal, anabolic steroids were taken by many athletes who specialized in strength events, such as weight lifting and sprinting, in the hopes of increasing muscle mass. This increase would be translated into increased muscle strength and a competitive advantage. However, in human biology as elsewhere in life, "there's no such thing as a free lunch." Anabolic steroid drugs have adverse effects on the body.

Ironically in males who already have sufficient amounts of their own naturally produced testosterone, sought-after increases in muscle mass and strength are not well demonstrated. Small, variable gains occur only in conjunction with a regular muscle-building exercise program and are not experienced by all who take anabolic steroids. Scientific investigation is limited. Some athletes and bodybuilders take large doses of anabolic steroids, and researchers cannot legally reproduce those same conditions in a controlled laboratory experiment to study their effects directly. However, documentation of the health of individuals who have independently taken large doses of anabolic steroids and experimental research on laboratory test animals has demonstrated the following adverse effects.

Reproductive system. In males testosterone secretion and sperm production by the testes are controlled by hormones released from the anterior pituitary gland. When androgenic steroids are taken, the anterior pituitary gland is inhibited from producing hormones that control testosterone production. Because the testes do not receive their normal input signal from the anterior pituitary gland, sperm production decreases, the testes shrink, and testosterone secretion declines.

The liver. Use of steroids induces liver dysfunction because the liver, which normally inactivates steroid hormones and prepares them for urinary excretion, is overloaded by the excess steroid intake.

Cardiovascular system. Steroid use induces several changes in the cardiovascular system that increase the risk of developing heart attacks and strokes. There is an elevation in blood pressure, and damage to the heart muscle itself has been observed in laboratory animals given steroids.

The point is obvious. The short-term gains that may or may not occur as a consequence of taking steroids will inevitably exact a high long-term cost over the lifetime of the person taking them.

There are limits to the amount of oxygen and glucose that can be delivered to a skeletal muscle. However, this maximum is rarely ever reached. A more immediate limitation lies in the muscle fiber itself. During vigorous exercise, such as during running, even though sufficient oxygen and glucose may be delivered to the contracting muscles, aerobic metabolism may not be able to synthesize ATP rapidly enough to meet the demands of the contracting muscle fibers, especially if the exercise intensity continues to increase. When ATP formation through the aerobic metabolic pathway cannot keep pace with the muscle's demand for it, the muscle fibers rely on glycolysis to generate ATP. Recall from Chapter 3 that this is the first of three metabolic pathways in cellular respiration. Glycolysis has two advantages over the citric acid cycle and the electron transport pathways of cellular respiration. First glycolysis does not require oxygen to proceed because it is anaerobic. Second it can proceed much more quickly than the other two aerobic pathways because glycolysis has far fewer steps. There are, however, two disadvantages in producing ATP by glycolysis. Glycolysis is inefficient, yielding a net of only 2 ATP molecules for each glucose molecule, compared with as many as 36 ATP molecules using the aerobic pathways. Second the end-product of glycolysis, pyruvic acid, when it is unable to enter the aerobic metabolic pathways, is converted to **lactic acid.** Lactic acid interferes with muscle fiber function; its buildup produces muscle soreness and fatigue.

Imagine that you have just finished a strenuous bout of exercise. Even though you are no longer exercising, you continue to breathe deeply and rapidly. This continued intake of large amounts of oxygen is required to complete the metabolism of lactic acid that accumulated in the exercised skeletal muscles. The amount of oxygen necessary to remove this lactic acid is the oxygen debt produced by the strenuous exercise. **Oxygen debt** is the quantity of oxygen taken up by the lungs during recovery from a period of exercise that is in excess of the quantity needed for resting metabolism. During this recovery period the excess oxygen is used in synthesizing new ATP as the lactic acid enters the aerobic respiratory pathways, where it is completely broken down.

CHAPTER REVIEW

SUMMARY

1. The internal bony structure of the body, called the skeleton, has a number of functions, including making possible bodily movement. It also protects vital organs, acts as a mineral reservoir, especially for calcium, and is the primary site of blood cell production.
2. The skeleton consists of two major divisions: the axial skeleton, which includes the skull, vertebral column, and ribs; and the appendicular skeleton, which consists of the bones of the upper limbs, hands, lower limbs, feet, and hip and shoulder girdles. All the bones of the skeleton can also be classified according to shape as flat bones, long bones, short bones, or irregular bones.
3. Skeletal bones are joined, or articulated, at joints. The two primary functions of joints are that they secure the bones together and they allow the rigid bones of the skeleton to be mobile. Joints can be immovable, slightly movable, or freely movable. Freely movable joints have fluid-filled cavities, and are called synovial joints.
4. Bone is made of both organic and inorganic components. The organic component consists of osteoblasts, osteoclasts, osteocytes, and the collagen fibers found in the extracellular matrix. The inorganic component consists of mineral salts largely derived from calcium.
5. Skeletal muscles are attached to the bony skeleton by tendons. Although it is often activated by reflex action, skeletal muscle is called voluntary because it is the only type of muscle under our conscious control. Most skeletal muscles cause movement by working in antagonistic pairs.
6. Each muscle cell or fiber contains a large number of rodlike structures called myofibrils that extend the entire length of the cell. Skeletal muscle tissue is striated, or striped, because myofibrils contain alternating light and dark bands.
7. A myofibril is a long series of attached contractile units called sarcomeres. Each sarcomere is composed of two types of protein filaments: the thin filaments are called actin, and the thick filaments are called myosin. Contraction occurs when the thin filaments slide past the thick filaments, causing the sarcomere to shorten in length.

FILL-IN-THE-BLANK QUESTIONS

1. The two divisions of the human skeleton are known as the ______ skeleton and the appendicular skeleton.
2. The shoulder and hip girdles are also known as the ______ girdle and ______ girdle, respectively.
3. The wrist is composed of a group of marble-sized short bones known as the ______.
4. Immovable joints are also called ______.
5. The structural unit of compact bone is called an ______ or sometimes a __ ____ ______.
6. Skeletal muscles usually work in ______ ______.
7. The thin filaments of a sarcomere are attached to ______ lines that define the borders of this functional contractile unit.
8. The thick filament of a sarcomere is the protein ______, and the thin filament is the protein known as ______.
9. At the level of the muscle fiber, contraction is initiated by the release of ______ ______ from the sarcoplasmic reticulum.
10. ______ ______ fills the energy gap created in muscle cells during the first several seconds of muscle contraction.

SHORT-ANSWER QUESTIONS

1. How does bone remodeling occur?
2. Compare and contrast the structure of compact bone and spongy bone.
3. Describe what happens at the level of the sarcomere when muscle fibers contract.

VOCABULARY REVIEW

actin—*p. 222*
appendicular skeleton—*p. 209*
axial skeleton—*p. 209*
myofibril—*p. 221*
myosin—*p. 221*
ossification—*p. 219*
osteoblast—*p. 218*
osteoclast—*p. 218*
osteogenesis—*p. 219*
remodeling—*p. 219*
sarcomere—*p. 221*
synovial joint—*p. 216*

THE CIRCULATORY SYSTEM

For Review

Here are some important terms and concepts that you will encounter in this chapter. If you are not familiar with them, you should review them before proceeding.

Action potential
(page 144)

Interstitial fluid
(page 123)

Skeletal muscle fiber
(page 220)

OBJECTIVES

After reading this chapter you should be able to:

1. *Name the three major parts of the circulatory system.*
2. *Name the three types of blood vessels found in the circulatory system and describe the function of each.*
3. *Differentiate between vasoconstriction and vasodilation.*
4. *Explain the function of one-way valves.*
5. *List the three layers of the heart wall and describe the structure of each.*
6. *Describe the structure and function of the lymphatic system.*

Whether they constitute a single-celled organism or a multicellular human being, all cells have similar requirements. They incorporate nutrients from the environment and eliminate waste products back into it. How do the many cells of the human body far removed from the environment survive? The answer to this question is through the human circulatory system, which transports nutrients and gases to the extracellular environment and carries wastes and other by-products away from the cell's immediate surroundings. Regardless of their location, all body cells have either direct or indirect contact with the external environment.

The Circulatory System

Imagine for a moment a single-celled organism swimming in water. It is completely surrounded by its environment. Nutrients and gases pass into the cell directly through the cell's membrane, and waste products are directly transported out into the environment. Similarly this is how each cell of your body operates, taking in nutrients from its surroundings and passing waste products back into it. However, humans are multicellular organisms, which means most of our cells are far removed from the external environment. How do these cells avoid starving and stewing in their own wastes? The answer to this question is a system that brings each cell close to a source of food and oxygen and provides a means of carrying wastes and other by-products away from the cell. This circulatory system continually changes the fluid environment immediately surrounding each cell and is the means by which the body maintains the proper environment for each of its cells.

In this chapter we describe the major structural components of the human circulatory system, which is also called the **cardiovascular system.** We also describe a supplementary drainage system called the **lymphatic system,** responsible for returning excess extracellular fluid to the circulatory system.

The human circulatory system consists of three major parts. The first is a fluid called blood that serves as a transportation medium. It carries oxygen, nutrients, and other substances to the body's cells and carries carbon dioxide, wastes, and other products of cellular metabolism away from the body's cells. The composition of blood is described separately in Chapter 13. The second part is a system of vessels to carry the blood throughout the body. Third, the circulatory system uses a pump, called the **heart,** to keep the blood moving through the vessels.

The cardiovascular system is a closed circulatory system where the blood is confined to the pump and a continuous series of connected vessels in the body. We begin by first defining two terms essential to understanding how the circulatory system functions.

Blood pressure is the force per unit area exerted on the wall of a blood vessel by its contained blood. Blood pressure is expressed in terms of millimeters of mercury, abbreviated as mm Hg. For example, a blood pressure of 120 mm Hg means the blood is exerting a pressure against the vessel wall equal to the force able to raise a column of mercury 120 mm high.

Resistance is the opposition to flow and is a measure of the amount of friction the blood encounters as it moves through the vessels. One important source of resistance in the circulatory system is blood vessel diameter. The larger the diameter, the lower the resistance. Blood vessel diameter changes frequently and is an important factor in altering resistance. Fluid close to the walls of a tube or channel is slowed by friction as it passes along the wall. The fluid in the center of the channel therefore flows faster and more freely. For example, if you watch the flow of a river, a leaf placed near the bank will move more slowly than one placed in the center of the stream. So the smaller the channel, or blood vessel, the greater the friction because relatively more of the fluid contacts the walls.

Blood Vessels

All blood vessels are dynamic structures that pulsate, constrict, and even proliferate according to changing body needs. Collectively the blood vessels constitute an approximately 50,000-mile-long system carrying blood on its journey throughout your body.

Three major types of blood vessels make up the circulatory system. First are **arteries,** which carry blood away from the heart. Second are **veins,** which return blood to the heart. Third are **capillaries,** which lie between arteries and veins. It is at the capillary level where the exchange of nutrients, waste products, and other substances between blood and the surrounding cells occurs. We first discuss the structural characteristics common to all blood vessels and then look at each type separately.

The walls of all blood vessels, except for the smallest, have three distinct layers. The innermost layer lines the **lumen,** which is the space within the vessel. It is composed of closely fitting endothelial cells that form a smooth surface, which helps decrease friction as blood moves through the vessel. The middle layer consists of circularly arranged smooth muscle cells and elastic connective tissue fibers. The activity of the smooth muscle is controlled by motor neurons from the autonomic nervous system. Depending on the specific needs of the body at any moment, the motor neurons can cause either **vasoconstriction** or **vasodilation.** **Vasoconstriction** is a reduction in the diameter of the lumen resulting from smooth-muscle *contraction*, and **vasodilation** increases the lumen diameter and results from smooth-muscle *relaxation*. These relatively small changes in blood vessel diameter are able to dramatically affect both blood flow and blood pressure throughout the body. The outermost layer of a blood vessel wall is composed primarily of loosely woven collagen fibers that protect the blood vessel and secure it to surrounding structures. Each of the three major vessel types varies in length, diameter, and relative wall thickness.

Arteries

Arteries carry blood away from the heart. They have thick walls because of an expanded middle layer of elastic fibers and smooth muscle tissue. In the largest arteries of the body there are also additional stretchy sheets of elastic fiber on each side of the middle layer, facing the other two layers. These additional fibers help the large arteries to withstand large pressure fluctuations by allowing the artery to expand when the heart contracts and forces blood into the arteries, and to recoil, or "spring back," as the blood flows forward into circulation during heart relaxation. Because the large arteries can expand and recoil passively to accommodate changes in blood flow, the blood is kept close to a continuous pressure rather than stopping and starting with each heartbeat. The alternating expansion and recoil of arteries during each heartbeat creates a pressure wave, or **pulse**, that is transmitted throughout these vessels. You can feel a pulse in any artery lying close to the skin surface by compressing the artery slightly against firm tissue. This is an easy means of counting heart rate. The point where the radial artery comes close to the skin surface at the wrist is typically used to take a pulse measurement, but there are several other clinically important arterial pulse points. Because these same points can be compressed to stop blood flow into the tissues below them during severe bleeding, they are also called **pressure points** (Figure 12-1). For example, if you cut your hand severely, you can slow the bleeding considerably by compressing the radial artery at the wrist or further up at the brachial artery on the inside surface of your elbow.

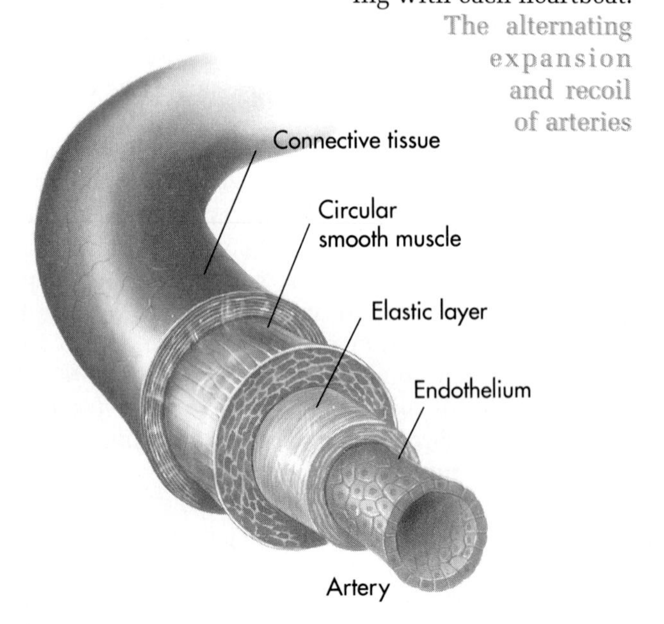

Figure 12-1 Pressure points

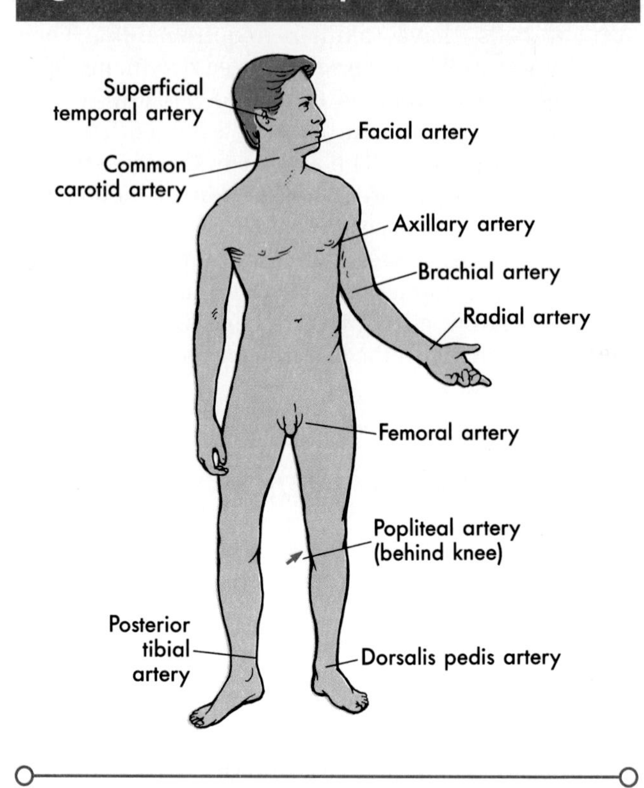

A pulse may be felt at various points on the body where an artery lies close to the skin surface.

Arterioles are the smallest of the arterial vessels. The larger arterioles have all three layers. The middle layer consists of smooth-muscle tissue with only a few scattered elastic fibers. The walls of the smallest arterioles that directly feed into the smallest vessels of the circulatory system, called capillaries, are composed of an inner layer of endothelial cells with smooth-muscle cells periodically coiled about it. The vasoconstriction and vasodilation of arterioles in response to changing stimuli from the autonomic nervous system as well as local conditions of the surrounding tissues are the most important factors determining blood flow and pressure into the capillaries at any given time. When arterioles constrict, the tissues served by the capillaries into which they feed are largely bypassed. When the arterioles dilate, blood flow into the local capillaries increases.

Capillaries

The capillaries are microscopically small vessels that form dense, branching networks throughout nearly all body tissues. The thin walls of the capillaries consist of one layer of endothelial cells, and in some cases just one cell forms the entire circumference of the capillary wall. The average length of a capillary is about 1 mm, and its average diameter is just large enough for red blood cells to move through single file, like a stack of pancakes turned on its side.

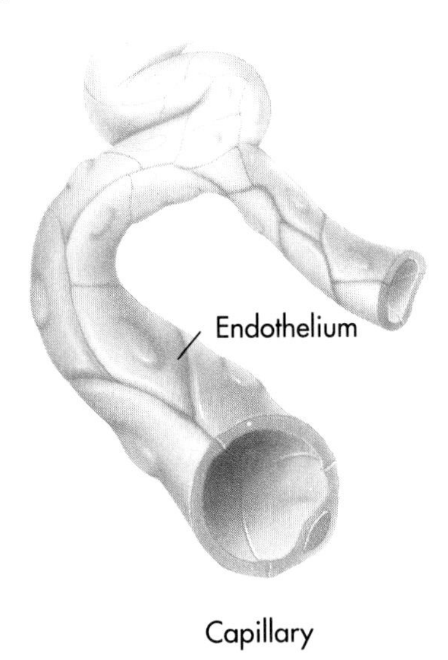

Capillary

Because of their single-cell wall construction, capillaries are well suited to exchange nutrients, waste products, and other materials between the blood and the extracellular fluid surrounding the tissue cells and the capillaries. In most body tissues a network of capillaries, or a **capillary bed,** consists of two capillary types. The first is a **thoroughfare channel,** made of short capillaries that directly connect the arteriole on one side of the bed to the venule, a blood vessel that returns blood to the heart at the opposite end of the bed. The second, the **true capillaries,** vary in number depending on the tissue served, but there are approximately 10 to 100 per bed (Figure 12-2). True capillaries usually branch off near the arteriole side of the thoroughfare channel and return to it near the venule returning the blood to the heart.

Usually a band or cuff of smooth muscle, called a **precapillary sphincter,** surrounds the branching point of the true capillaries from the arteriole, and acts like a "shut-off valve" into those capillaries. When the precapillary sphincters are constricted, blood flows directly from the arteriole through the thoroughfare channel and into the venule. When the precapillary sphincters are dilated, blood flows through the capillary bed before entering the venule. The autonomic nervous system and local conditions determine the path blood may take through the capillary beds of specific body tissues.

Veins

Venules are the first part of the circulatory system responsible for returning blood to the heart. They are formed as the capillaries unite into larger vessels. The smallest venules consist entirely of an inner layer of endothelial cells and a middle layer

Figure 12-2 Capillary bed

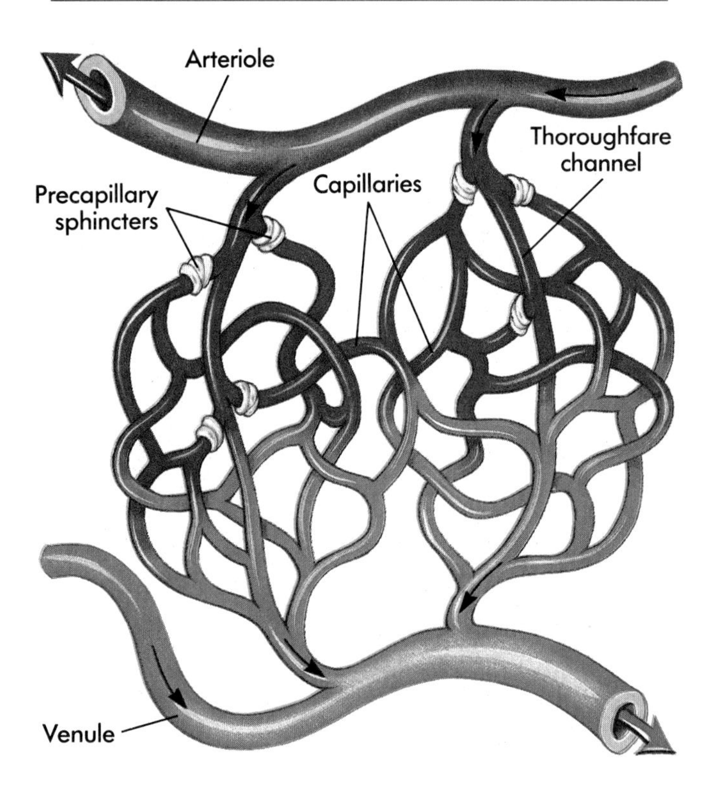

In addition to the true capillaries that make up a capillary bed, there is also a thoroughfare channel directly connecting an arteriole to the venule returning blood to the heart. Depending on the metabolic demands of the tissue, blood can flow through the capillary bed or some blood can flow through the thoroughfare channel.

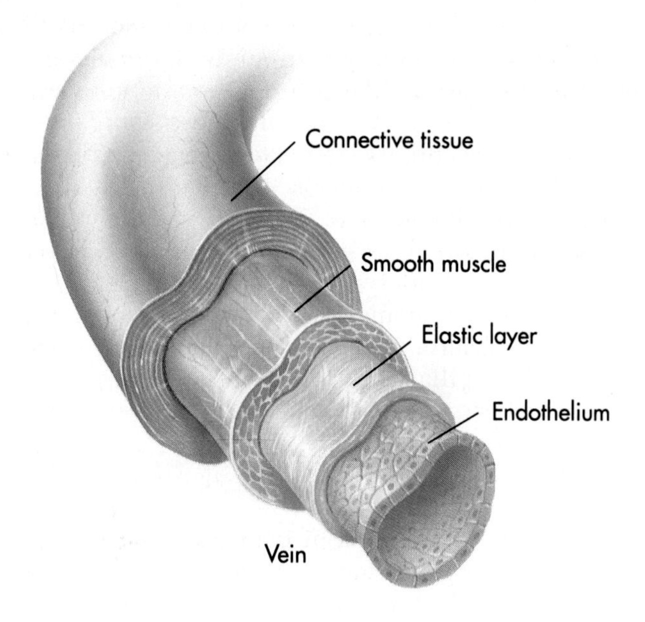

of elastic fibers interspersed with a few smooth-muscle cells.

Venules join to form **veins.** Although veins have three distinct layers, their middle layer contains little smooth-muscle tissue, resulting in thinner walls than those of arteries of the same diameter. Veins also have larger lumens than arteries. Therefore the venous portion of the circulatory system accommodates up to 65% of the body's total blood supply. Venous blood pressure is also lower than arterial blood pressure, primarily because of the distance of veins from the heart. Therefore a structural adaptation is present in veins to assist in returning blood to the heart. One-way valves formed from folds of the inner layer of the vessel walls occur regularly along veins and prevent blood from flowing backward. These valves are common in the veins of the limbs, such as the legs, which must return blood to the heart against gravity.

You can conduct a simple experiment to verify how these valves work. Let one hand hang by your side until the blood vessels on the back side of the hand become distended with blood. Place the first two fingertips of the other hand against one of the distended veins and, pressing firmly, move the index finger toward the wrist and then release it. The vein will remain flattened and collapsed, despite the pull of gravity. Once blood has passed each successive valve, it cannot flow back. A functional adaptation also assists in returning venous blood to the heart. As the skeletal muscles surrounding the veins contract and relax, they "pump" or "milk" the blood toward the heart.

Varicose veins are veins that have become dramatically dilated because the valves are no longer functional. Several factors may contribute to this condition, including heredity, prolonged standing in one position, obesity, or pregnancy. For example, the potbelly of an obese person exerts downward pressure on the vessels of the groin, restricting the return of blood to the heart. Consequently blood tends to pool, or accumulate, in the lower limbs. With time the valves weaken and the venous walls stretch and become greatly distended. Superficial veins close to the skin's surface have less support from surrounding tissue and are therefore especially susceptible.

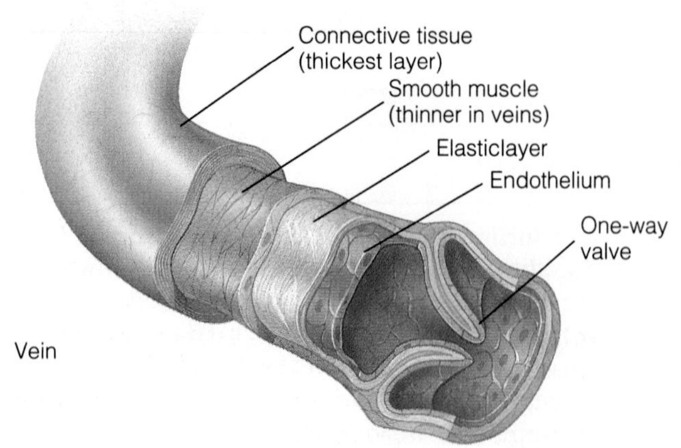

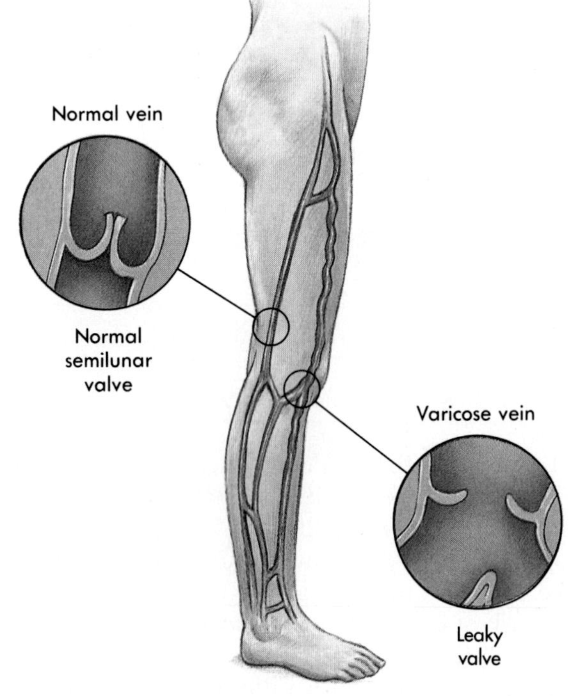

Heart Structure

The human heart is about the size of a fist and weighs less than 1 lb, approximately 350 g. It is a muscular organ in the thoracic cavity between the lungs and behind the sternum. Its somewhat conical shape is oriented with the "point," or **apex,** of the heart pointing down toward the left side of the chest.

The heart wall has three layers. The outermost layer consists of a double-walled membranous sac called the **pericardium** (per-i-kar′di-um). The outer membrane of the pericardium is made of tough fibrous tissue attached to such surrounding structures as the diaphragm, the sternum, and the vessels entering and leaving the heart itself. This attachment anchors the heart so it stays properly positioned in the chest. The thin, smooth, inner membrane of the pericardium produces a small amount of fluid that fills the small space between the inner layer and the outer layer of the pericardium. The fluid allows the layer to move against the surface of the pericardium without creating any friction.

The middle layer of the heart wall is the muscle tissue of the heart, called the **myocardium** (mi′o-kar′di-um). The myocardium, composed primarily of cardiac muscle, forms the bulk of the heart and is the layer that actually contracts. The innermost layer of the heart wall is the **endocardium** (en′do-kar′di-um). It consists of a thin sheet of smooth endothelial cells that line the inner-heart chambers. It serves the same purpose as the endothelium does in blood vessels, reducing friction and resistance to blood flow through the heart.

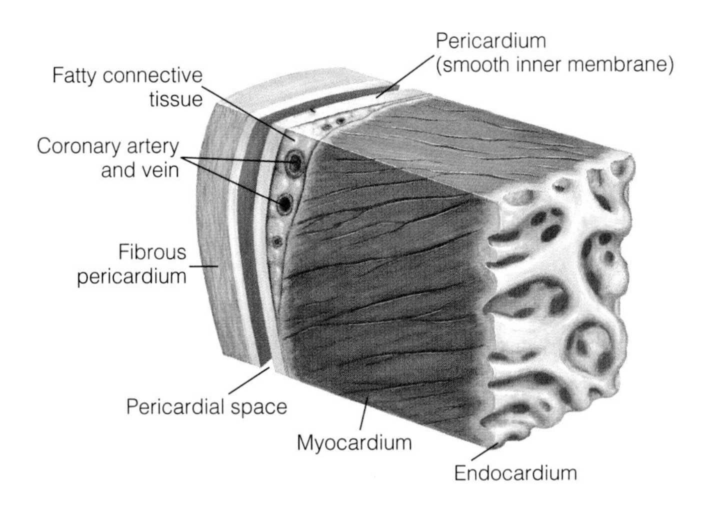

Internally the heart is divided into right and left sides by a wall of tissue called a septum. There is no direct connection between the right and left sides of the heart. Each side has two chambers. The upper chambers are thin walled and called **atria** (sing. *atrium*), and two lower, larger, and thick-walled chambers are called **ventricles.** Valves in the heart direct the blood flow through it in only one direction.

The valves that lie between the atria and the ventricles of each side of the heart are known collectively as the **atrioventricular valves.** They are supported by small collagen cords called **chordae tendineae** (kor′de tendin′ne), a term derived from the Latin meaning "heart strings." These cords are attached to projections on the inside of the ventricular walls. The chordae tendineae support the valves and prevent them from inverting or being pushed back through the opening when the chambers are filled with blood and forcibly contract. The valve between the **left atrium** and the **left ventricle** is called the **mitral valve,** or **bicuspid valve,** because it has two flaps of tissue that resemble the hat, or miter, of a Catholic bishop. The valve separating the **right atrium** and the **right ventricle** is called the **tricuspid valve** because it has three flaps.

A valve separating the right ventricle of the heart from the major blood vessel that carries blood from it to the lungs is called the **pulmonary valve.** A valve between the left ventricle and the **aorta,** the major artery of the body into which it pumps blood, is called the **aortic valve.** Together these valves are called **semilunar valves** because their cusps have a somewhat crescent moon shape (Figure 12-3).

The heart muscle is unable to extract oxygen or nutrients from the blood that circulates in its chambers because the endocardium is a fluid-tight tissue. This means the heart muscle must receive blood through blood vessels like other tissues of the body. Blood to the heart is supplied by the left and right **coronary arteries,** which branch from the aorta as blood leaves the heart and returns via the **coronary veins,** which join to form the **coronary sinus.** The coronary sinus empties into the right atrium.

The Pathway of Blood Through the Heart

The heart pumps blood from the atria to the ventricles and is actually two side-by-side pumps. The right side pumps blood through the lungs, called the **pulmonary circuit,** and the left side pumps blood through the rest of the body, called the **systemic circuit.**

Recall from Chapter 5 that cellular metabolism requires oxygen and makes carbon dioxide as a waste product. Blood returning to the heart is therefore low in oxygen, having delivered it to the cells of the body, and is high in carbon dioxide, having picked it up from the body's cells. This deoxygenated blood is carried back to the right atrium via the **superior** and **inferior venae cavae** (sing. *vena cava*).

Figure 12-3 The four internal chambers of the heart

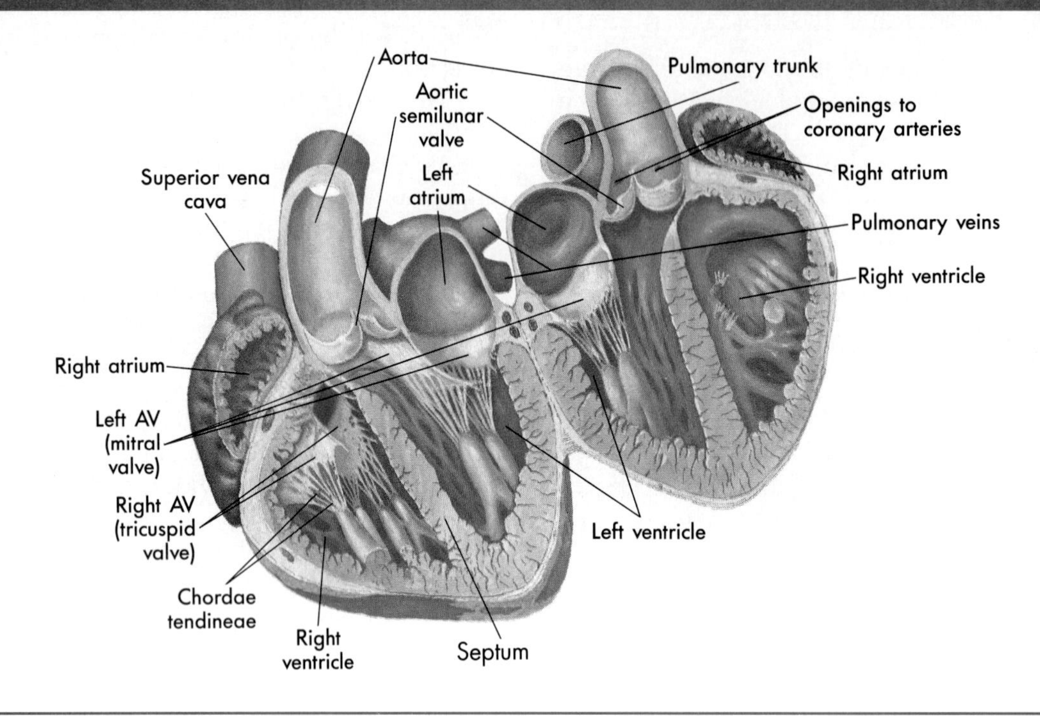

The superior vena cava returns blood from body tissues superior to (above) the heart, such as the brain and upper limbs, whereas the inferior vena cava returns blood from tissues inferior to (below) the heart, such as the lower limbs and the abdominal cavity.

Contraction of the right atrium forces the blood through the tricuspid valve and into the right ventricle. As the right ventricle contracts, the tricuspid valve closes and the blood is then ejected out through the pulmonary valves and into eventually the **pulmonary arteries,** circulating through the lungs where gas exchange occurs. The newly oxygenated blood then moves from the lungs through the **pulmonary veins** and back to the left atrium. The left atrium fills and then contracts, forcing the oxygen-rich blood through the mitral valve and into the left ventricle. After filling, the left ventricle contracts and forces the blood through the aortic valve into the aorta, where it circulates through the rest of the body (Figure 12-4). Because of this, deoxygenated blood never mixes with oxygen-rich blood. Blood must pass through the lungs to reach the left side of the heart.

Both sides of the heart contract and relax in unison. Both atria contract at the same time, followed by simultaneous contraction of the ventricles. To understand how this process works it is first necessary to understand the cardiac muscle itself.

Cardiac Muscle Tissue

Cardiac muscle tissue, like skeletal muscle, is striated, and its contraction occurs by the same interaction between actin and myosin described in Chapter 11. However, there are four major differences between the two. First, cardiac muscle cells are short and branched, in contrast to the long and cylindrical skeletal muscle fibers. Second, cardiac cells interconnect at junctions called **intercalated** (in-ter′ka-la-ted) **disks,** which are areas of the membrane that allow action potentials to spread from one cardiac cell to another. Third, cardiac muscle relies almost completely on aerobic respiration and cannot function more than a few minutes without oxygen. Fourth, it can sustain prolonged contraction.

The Conduction System of the Heart

Nervous control of heart activity is exerted by nerve fibers from the autonomic nervous system. For example, if the body has been accelerated to fight-or-flight mode by sympathetic nervous system stimulation, both the rate and the force of the heartbeat are increased. On the other hand, parasympathetic stimulation causes either a decrease in the heart rate or maintains its typical resting rate.

The ability of cardiac muscle to depolarize and

Figure 12-4 Blood flow through the heart

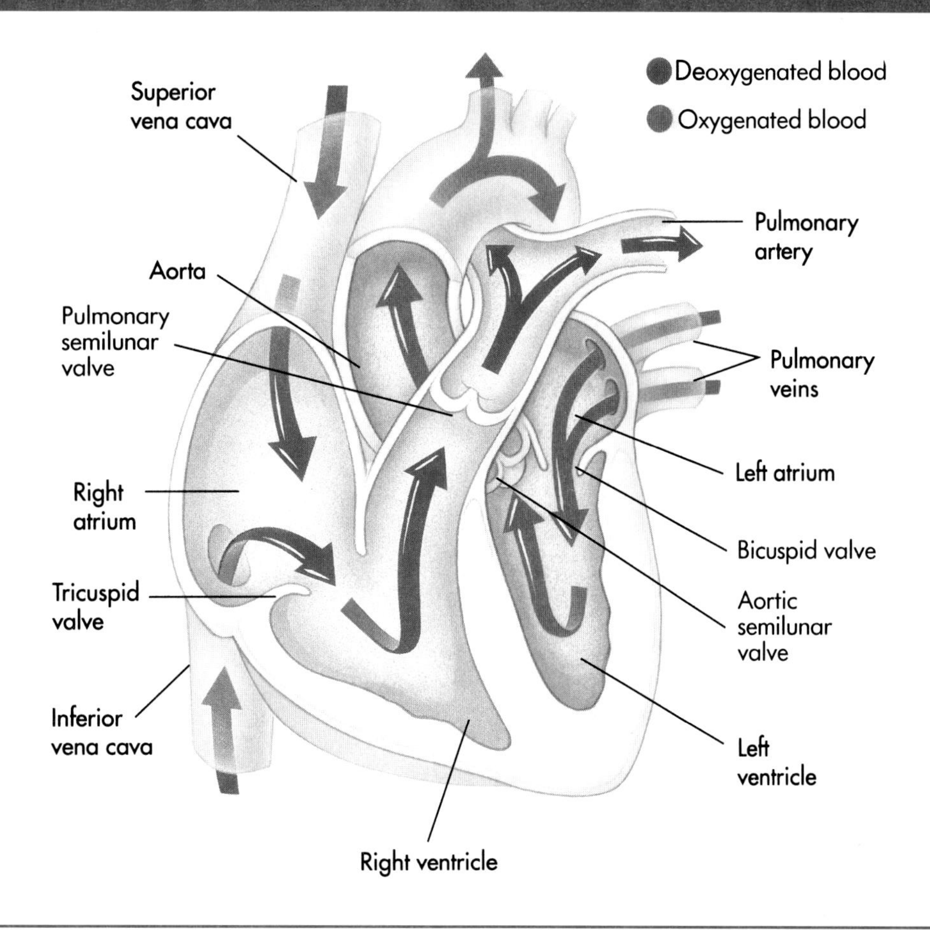

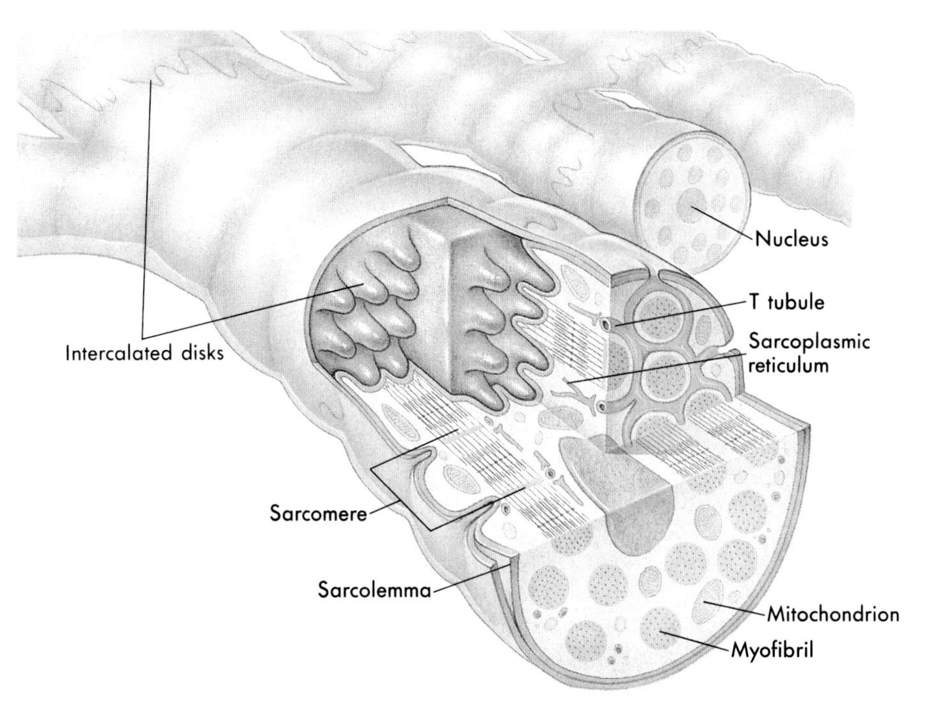

contract is also a property of the heart muscle itself and does not depend on extrinsic nerve impulses. Even if all nerve connections to the heart were cut, the heart would continue to beat rhythmically. The heart's own "built-in" conduction system, called the **cardiac conduction system,** consists of specialized, noncontractile, nervelike cardiac cells that can initiate and distribute impulses throughout the heart, so the myocardium depolarizes and contracts in an orderly, sequential fashion from the atria to the ventricles.

The first component of the cardiac conduction system is the **sinoatrial (SA) node,** a small cell mass located in the right atrial wall, just below the entrance of the superior vena cava. The SA node depolarizes spontaneously at the rate of 70 to 80 times every minute. The SA

Figure 12-5 Cardiac conduction system

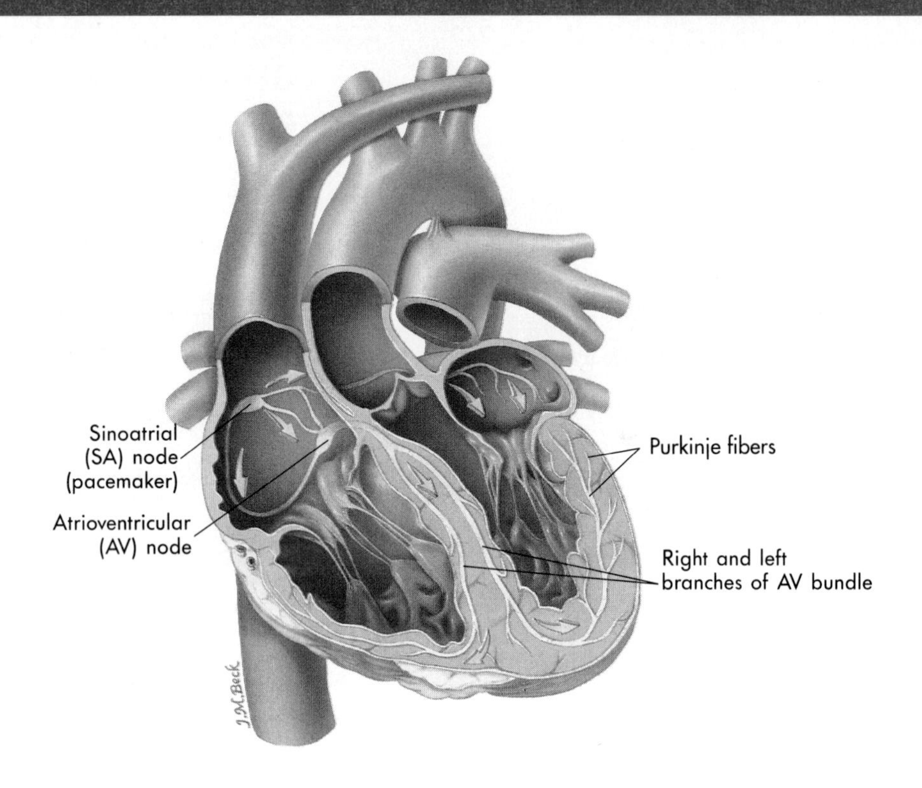

There are specialized heart muscle cells capable of generating and conducting action potentials throughout the heart. The cardiac conduction system coordinates the rhythmic contractions and relaxations of the heart muscle.

node initiates each depolarization wave that travels across the heart and sets the contraction pace for the entire heart. For this reason the SA node is also called the **pacemaker.**

From the SA node the depolarization wave spreads throughout the muscle tissue of both the right and the left atrium. This depolarization wave eventually reaches the **atrioventricular (AV) node** at the base of the right atrium near the septum. The signal then passes rapidly through the **atrioventricular bundle** and **bundle branches** in the septum and finally to the **Purkinje** (Per-kin-ge′) **fibers,** which penetrate into the ventricular myocardium (Figure 12-5).

Ventricular contraction almost immediately follows the ventricular depolarization wave. Because of the organization of the cardiac conduction system, ventricular contraction begins at the apex of the heart and moves *up* toward the atria. It is similar to squeezing toothpaste from a tube. It must be squeezed from the bottom up toward the cap if all the toothpaste is to be extracted. If you squeeze the tube in the middle, the contents below the "contraction" remain in the tube. Similarly for all the blood to be ejected from the ventricles, the ventricles must contract from the apex toward the atria.

Electrocardiograph: Recording the Electrical Activity of the Heart

Because body fluids are good electrical conductors, the action potentials generated and transmitted through the heart can be detected at a distance from the heart, amplified, and recorded with an instrument called an **electrocardiograph.** The graphic recording of the electrical changes during heart activity is called an **electrocardiogram,** or **ECG.**

Three distinct waves are shown on an ECG. The first, called the P wave, represents the excitation of both atria. The QRS complex results from ventricular excitation. The final wave, called the T wave, indicates the repolarization of both ventricles (Figure 12-6).

In a healthy heart the size, duration, and timing of these waves tend to be consistent from heartbeat to heartbeat. Thus any changes in the ECG's pattern or timing may reveal a problem with the heart's

Figure 12-6 Cardiac events and the electrocardiogram

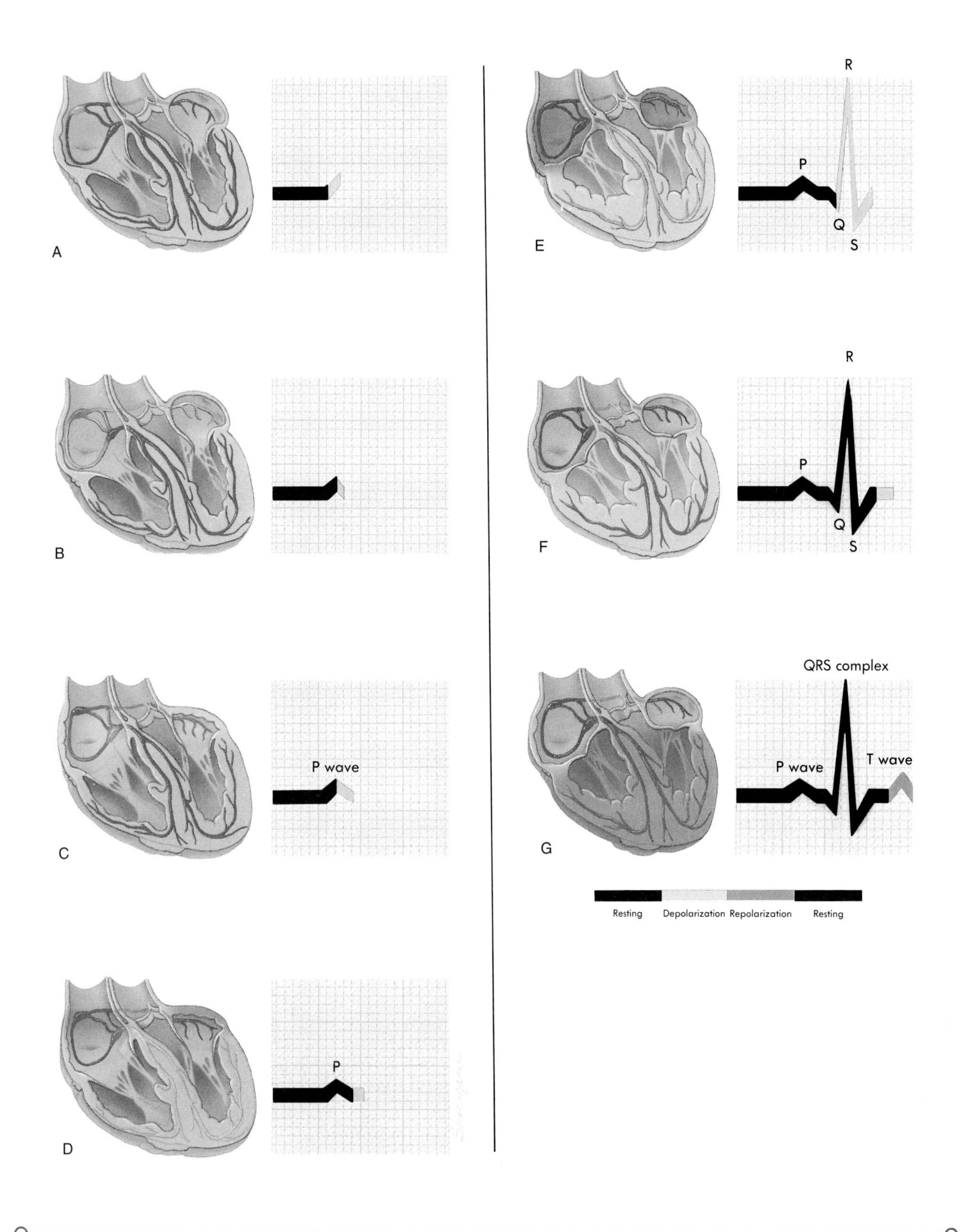

Figure 12-7 Cardiac cycle

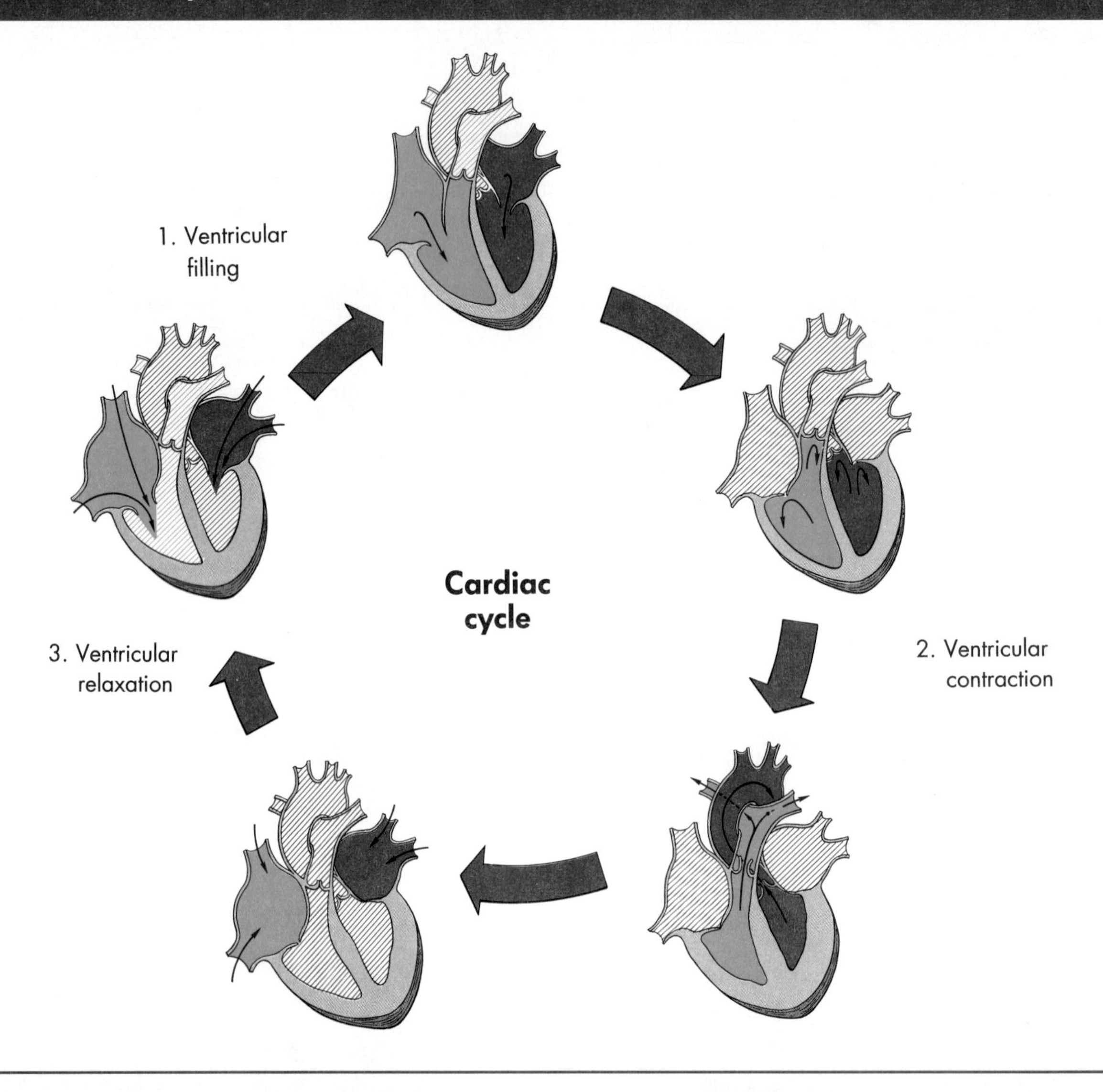

conduction system. For example, if the P wave is missing, it means the SA node is not firing and the AV node has become the heart's pacemaker.

Cardiac Cycle

The heart contracts, or beats, an average of 70 times a minute, or once every .85 seconds. Each heartbeat, known as a **cardiac cycle,** consists of a sequence of pressure and volume changes in the heart and includes all events during atrial contraction and relaxation followed by ventricular contraction and relaxation. We can put together what we have learned about the path of blood flow through the heart and the pattern of the cardiac conduction system that coordinates heart activities to better understand the cardiac cycle. We begin the sequence with the heart in total relaxation, with both the atria and the ventricles quiet (Figure 12-7).

1. Period of ventricular filling

Pressure in the heart is low, and blood returning from both the pulmonary and the systemic circulation is flowing passively into the atria. As the atria begin to fill the blood forces the AV valves open, and the ventricles also begin filling. The pulmonary and aortic valves (semilunar valves) are shut. About 70% of ventricular filling occurs during this period, and the AV valve flaps begin to drift up toward their closed position as ventricular filling proceeds. The stage is now set for atrial contraction. The SA node fires and the atria contract, forcing the remaining blood in them out and into the ventricles. Then the atria relax, and the rest of the cardiac cycle continues.

2. Ventricular contraction

As the atria relax, the ventricles begin to contract. The depolarization wave has reached the AV node and moved down the bundle branches to the Purkinje fibers. The walls of the ventricles close in on the blood in their chambers, and the ventricular pressure rises quickly, closing the AV valves. For a fraction of a second the ventricles are completely closed chambers because the AV valves and the semilunar valves are shut. Blood pressure increases in the ventricles because the ventricles are contracting. As ventricular pressures rise they exceed those in the arteries attached to them, forcing open the semilunar valves. Blood is then expelled from the ventricles into the aorta and the pulmonary arteries. During this **ventricular ejection phase** the pressure in the aorta normally reaches about 120 mm Hg.

3. Ventricular relaxation

During this brief interval the ventricles begin to relax. Because the blood remaining in their chambers is no longer compressed, ventricular pressure drops rapidly and the blood in the aorta and the pulmonary arteries begins to flow back toward the heart, causing the semilunar valves to close. Because blood is just beginning to enter the atria again, the AV valves are still closed and the ventricles are momentarily completely closed chambers. We are now back to the beginning of the cardiac cycle.

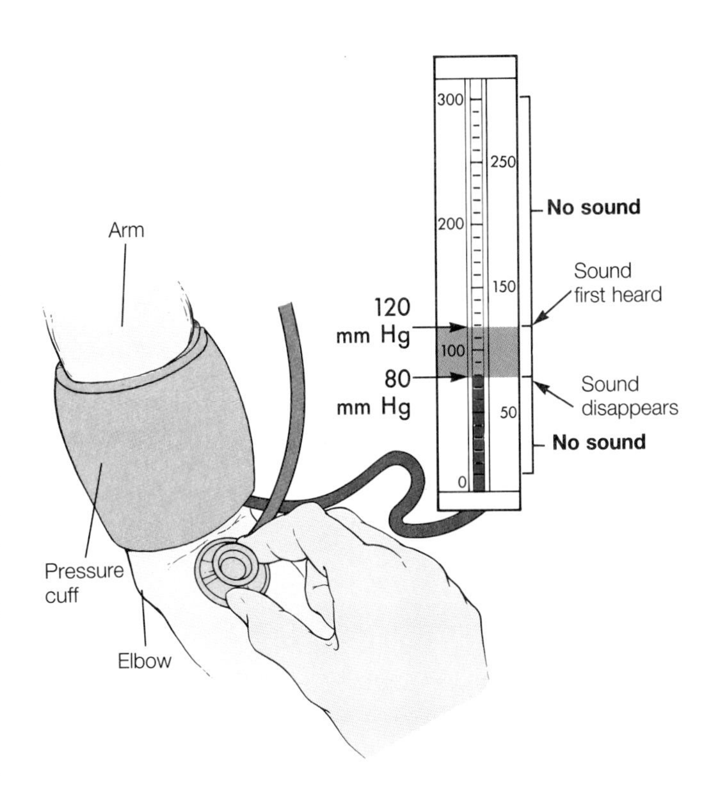

Blood Pressure

The heart undergoes some dramatic movements as it alternately contracts, forcing blood out of its ventricles, and then relaxes, allowing its chambers to refill with blood. The term **systole** (sis′-to-le) refers to these contractions. **Diastole** (di-as′-to-le) refers to the periods of heart muscle relaxation.

One indication of the status of a person's circulatory system is to take pulse rate and blood pressure measurements. Systemic arterial blood pressure is often taken indirectly using a **sphygmomanometer** (sfig′-mo-ma-nom′e-ter), which is an inflatable cuff attached to a pressure gauge. Blood pressure measurement is usually taken at the arm. The cuff pressure can be varied to prevent or permit blood flow in the underlying brachial artery of the arm. The cuff is inflated, usually to greater than 120 mm Hg pressure, which prevents blood flow through the artery. The cuff pressure is slowly reduced until turbulent flow through the artery is first detected using a stethoscope. The pressure at which sound is first detected is the ventricular systolic blood pressure. As the cuff pressure decreases, the turbulent flow diminishes to a smooth flow, which can no longer be detected. The pressure at which sound is no longer detectable is the ventricular diastolic pressure. In healthy adults at rest systolic blood pressure varies between 110 mm Hg and 140 mm Hg, and diastolic pressure varies between 75 mm Hg and 80 mm Hg.

Cardiovascular Disease

Although there has been a 30% decline in death from heart disease since 1963, almost 500,000 Americans will die from heart attacks this year and another 200,000 from related cardiovascular problems. Collectively, cardiovascular disease is the leading cause of death in the United States today. The following describes the major cardiac and circulatory disorders classified under the general term of cardiovascular disease.

Abnormal heart sounds are known as **murmurs.** As long as blood flow is uninterrupted it is quiet. However, if blood strikes obstructions, its flow becomes turbulent and generates sound, just as water "gurgles" when it encounters rocks or a fallen log when flowing down a stream. For example, if an AV heart valve does not close completely, there will be a swishing sound after the characteristic lub sound. Most often in adults murmurs indicate valve problems, caused either by incomplete closure, called **valve incompetence,** or by the inability of a valve to open completely, known as **valve stenosis.**

Hypertension

Periodic elevations in systolic blood pressure occur as normal adaptations to such events as physical exertion or even emotional upset, such as fear or anger. However, **persistent,** or **chronic hypertension,** commonly called high blood pressure, is a

Biology in Action KEEPING THE PIPES CLEAN

Your heart muscle is expected to contract over 2.5 billion times during your lifetime without once stopping to rest. It is also expected to force blood through a series of vessels whose total length is over 50,000 miles. Add to this the possibility that these vessels may become weakened or clogged, and it is easy to understand why the circulatory system is a prime target for malfunction.

Atherosclerosis (ath′er-o-skle-ro′-sis) is a progressive, degenerative arterial disease that causes a narrowing of the vessel lumen, eventually leading to complete blockage of the affected vessels. This disease attacks arteries throughout the body, but for as yet undetermined reasons, the aorta and the coronary arteries are most often affected. What triggers this thickening of the vessel wall and narrowing of its lumen? According to one hypothesis, the initial event is damage to the endothelium caused by blood-borne chemicals, such as the carbon monoxide in cigarette smoke or automobile exhaust, by viruses, or by physical factors, such as a blow or chronic hypertension. Another hypothesis indicates that atherosclerosis begins as **atheromas,** which are cells from the middle muscular layer of the vessel wall that migrate to a position just under the endothelium, where they begin rapidly to undergo mitosis and irregular cell growth.

Regardless of how atherosclerosis begins, once the surface of the endothelium is disrupted, the following events occur. Cholesterol and other lipids begin accumulating in the abnormal cells, producing **plaque.** The plaques bulge into the lumen of the vessel as they continue to enlarge. Eventually the endothelium may tear, causing scar-tissue-forming cells to invade the area. Circulating substances in the bloodstream responsible for clotting begin to adhere to the tear. The clot attached to the damaged vessel is called a **thrombus.** A thrombus that breaks loose from its attachment is called an **embolus.** It may completely plug a smaller vessel as it flows downstream. An embolus may gradually or suddenly block a coronary vessel or possibly an artery in the brain, resulting in myocardial infarctions or strokes, as described in the text.

What can be done when the damage of atherosclerosis occurs to coronary arteries? Aside from coronary bypass operations are some less invasive and traumatic techniques to clear the blocked or narrowed vessel. Blood clots can be dissolved by injecting an enzyme called **streptokinase** into the coronary artery. When done immediately after a heart attack, this procedure can significantly increase the person's chances of survival. A more promising drug, called **tissue plasminogen activator,** or **t-PA,** selectively dissolves clots in vessels without interfering with the body's ability to form clots when needed to prevent bleeding. Another procedure, known as **balloon angioplasty,** involves squashing the plaque deposits flat against the artery walls by inserting a hollow needle with a tiny balloon inside into the obstructed artery. The balloon is inflated, flattening the deposits and restoring greater blood flow.

However, the best way to treat atherosclerosis is to prevent it from occurring. Lifestyle factors, such as smoking, obesity, high-fat diets, and lack of exercise contribute significantly to the initiation of this disease. So why not just advise patients at risk to change their ways of living? As each of us knows, old habits are hard to break; but considering the alternative, more people with diseased arteries are willing to trade lifelong habits for a healthy old age.

widespread and dangerous disease that indicates increased resistance to blood flow. A person is said to have chronic hypertension if the blood pressure is higher than 140 mm Hg over 90 mm Hg. High blood pressure is often called the "silent killer" because, although its symptoms are painless, it often leads to life-threatening problems.

High blood pressure is common in obese people because the total length of their blood vessels is relatively greater than those in thinner people. Additional blood vessels are required to service the extra tissue, and the longer the blood vessels the greater the resistance to blood flow. Because the heart is forced to pump against increased resistance, it must work harder, and eventually the myocardium enlarges. However, when finally strained beyond its capacity to respond, the myocardium weakens and becomes flaccid or flabby. High blood pressure may damage arteries, causing them to thicken, thereby reducing their lumen diameter. It may also damage cerebral blood vessels, leading to a **cerebrovascular accident,** or **stroke.** The kidneys are also affected. Hypertension eventually causes the artery going to the kidney to thicken, reducing blood flow. Reduced blood flow to the kidney leads to water and salt retention, causing increased blood volume and elevated blood pressure. Prolonged hypertension is the most significant factor in cardiovascular disease and is the major cause of heart failure, kidney failure, and stroke.

Heart Attack

A *heart attack* is a general term that refers to the death of a portion of the heart muscle because of a lack of oxygen. For example, a coronary artery or one of its branches may be partially blocked by a blood clot or fat deposits. Under these conditions the myocardial cells die. The death of an area of the myocardium is called an **infarct** (from the Latin meaning "to stuff into"). This is a true heart attack, or **myocardial infarction.** Because cardiac muscle cells do not reproduce, any areas of cell death are repaired with noncontractile scar tissue. Whether a person survives a myocardial infarction depends on the extent of cell death and on the location of the damage.

Angina pectoris, which literally means "choked chest," refers to the chest pain caused by momentary deficient blood delivery to the heart. The myocardial cells are weakened by the temporary lack of oxygen, but they do not die. Angina pectoris is usually a serious symptom of impending cardiac problems.

Sometimes the coronary arteries do not become completely blocked but instead become narrowed. Blood flow to the myocardium may be inadequate, resulting in **ischemia** (from the Latin meaning "to keep back"). This problem can sometimes be overcome through coronary bypass surgery. This involves removing a segment of a healthy blood vessel, usually a vein from the lower leg, and suturing one end of it to the aorta and the other end past the point of the obstruction or damaged coronary artery segment. The newly implanted vessel restores blood flow to that area of the heart formerly supplied by the impaired coronary artery. In the United States as many as 100,000 people a year have coronary bypass surgery.

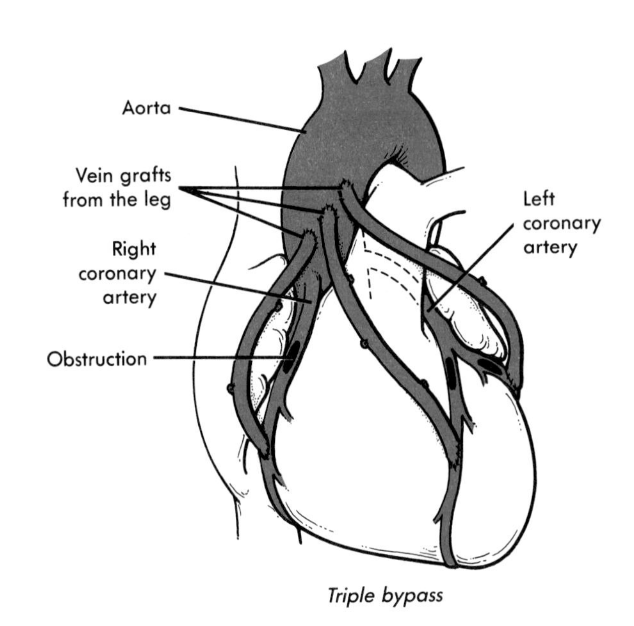

Triple bypass

Stroke

As mentioned earlier a *stroke* occurs when a portion of the brain dies because of a lack of oxygen. A stroke results when an arteriole bursts or becomes blocked. The brain cells fed by the damaged vessel die, and whatever function they controlled is destroyed.

The Lymphatic System

As blood circulates through the body, exchanges of nutrients, wastes, and gases occur between the blood and the surrounding cells through the extracellular fluid. As part of this ongoing process, some of the fluid in the blood also moves into the interstitial space, and most of it is returned to the circulatory system. The fluid that remains behind in the tissue spaces becomes part of the interstitial fluid. However, this also must eventually be returned to the blood if the circulatory system is to have sufficient blood volume to operate efficiently. The **lymphatic system** is a special one-way system of drainage vessels that absorb the excess extracellular fluid and return it to the bloodstream.

Lymphatic Vessels

The lymphatic system begins in microscopic, blind-ended **lymphatic capillaries,** which are between the

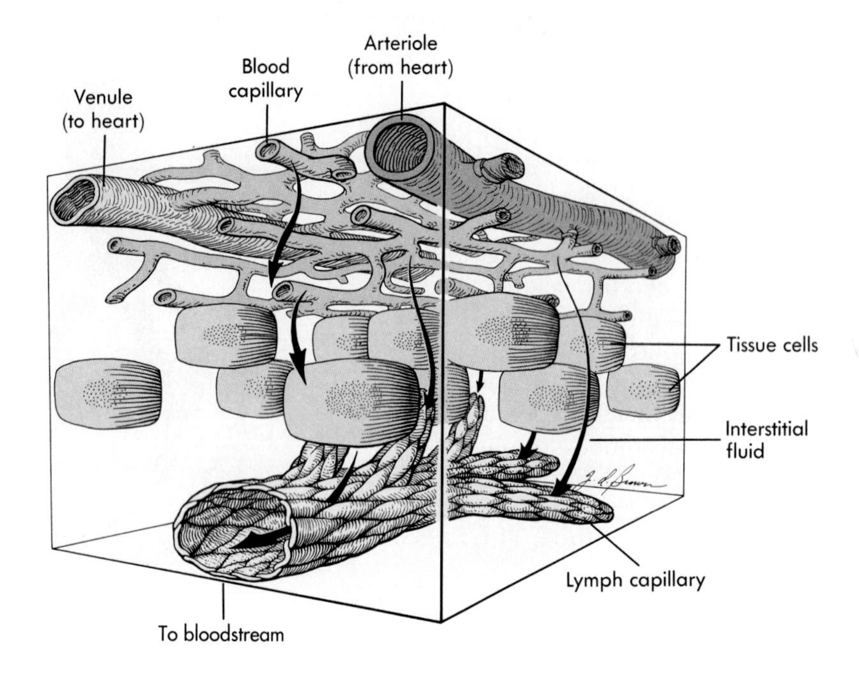

tissue cells and blood capillaries of nearly all body tissues and organs. The only exceptions are the central nervous system, bone, cartilage, teeth, and epidermis. Although similar to blood capillaries, the endothelial cells forming the walls of lymphatic capillaries are not tightly joined. Their edges loosely overlap one another, forming seams into which fluid can pass. The construction is similar to one-way swinging doors. The flaps open when the fluid pressure is greater in the interstitial space compared with the inside of the capillary, allowing fluid to enter. However, when the pressure is greater inside the lymphatic capillaries, the cell flaps press closely together, preventing the fluid from leaking back into the interstitial space. There are also one-way valves in the lymphatic capillaries, preventing backflow of fluid as it moves toward the heart.

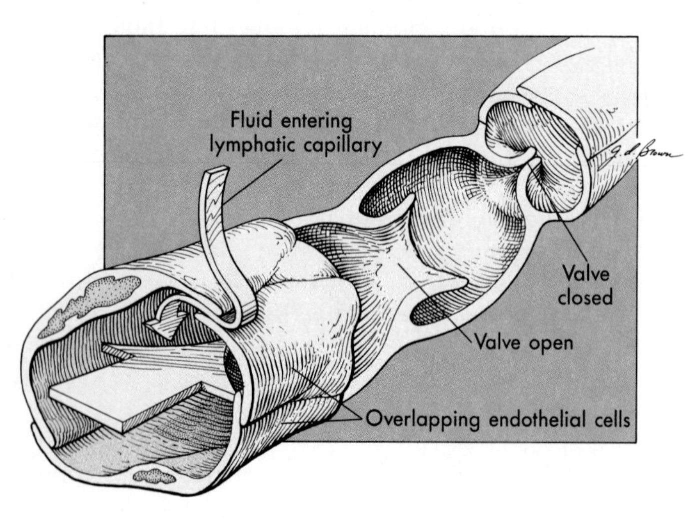

Once extracellular fluid has entered the system, it is called **lymph.** Lymph flows through successively larger and thicker-walled lymphatic vessels eventually to be delivered to one of two large ducts in the thoracic cavity. The **right lymphatic duct** drains lymph from the right arm and the right side of the head and thorax. The much larger **thoracic duct** receives lymph from the rest of the body. These ducts drain the lymph into the left and right subclavian veins, respectively, under the collarbones of the upper chest (Figure 12-8). Lymph is transported to these two ducts by the same mechanisms that move blood through the veins. The milking action of active skeletal muscles and valves prevent backflow.

Lymph Nodes

As lymph is transported, it passes through clusters of specialized tissue called **lymph nodes,** which contain cells that filter out and destroy debris, such as damaged cells and any other particles in the lymph. Lymph nodes also contain lymphocytes that are part of the body's defense system. They destroy microorganisms and other types of foreign matter that enter the lymphatic system. They are discussed in more detail in Chapters 13 and 14.

Figure 12-8 Lymphatic system

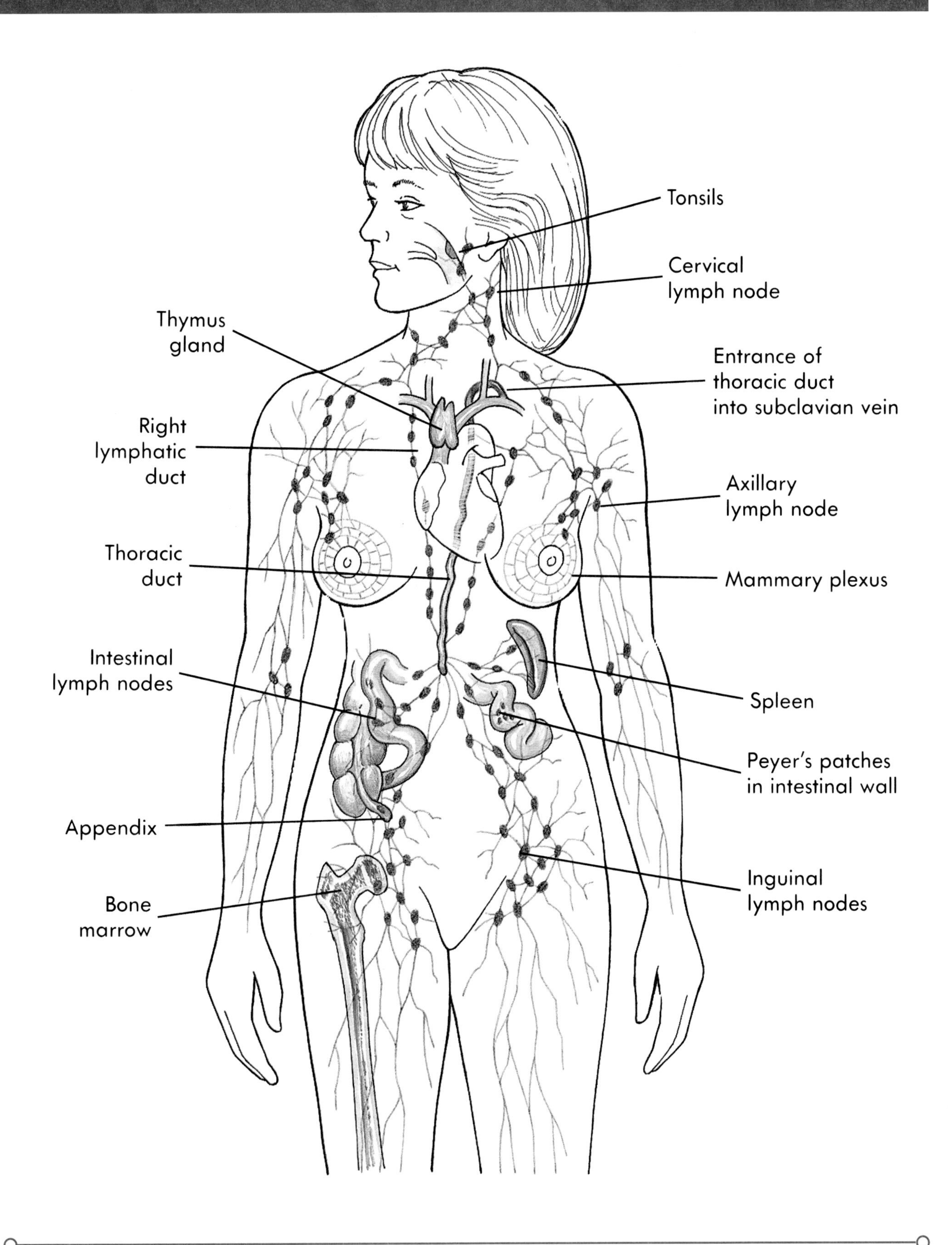

CHAPTER REVIEW

SUMMARY

1. The circulatory system is a continuous series of interconnected vessels through which blood circulates and serves as a medium of transport. A muscular pump, called the heart, keeps the blood moving through the vessels continuously. Blood vessels are dynamic structures capable of vasodilation, vasoconstriction, and even proliferation.
2. The main functions of the circulatory system include bringing each cell of the body into proximity to nutrients and oxygen and providing a means of carrying wastes and other byproducts away from the cell. It is also the medium through which cells communicate and the body defends itself against foreign microorganisms and other harmful elements.
3. The heart consists of four separate chambers. The two upper chambers are called atria, and the two lower and larger chambers are called ventricles. The middle layer of the heart wall, called the myocardium, which contracts and relaxes, produces the force that pumps blood through the circulatory system.
4. The heart functions as two side-by-side pumps, each consisting of an atrium and a ventricle, with no direct connection between them. The right side of the heart pumps blood through the lungs, while the left side pumps blood through the rest of the body. The blood moves through the right side of the heart and lungs first before it enters the left side of the heart to be pumped through the rest of the body.
5. The cardiac conduction system coordinates the relaxation and contraction of the chambers of the heart so that blood moves through both sides of the heart simultaneously. This system consists of specialized noncontractile muscle cells capable of generating and transmitting action potentials to other myocardial muscle cells.
6. Two pairs of valves, the atrioventricular valves and the semilunar valves, allow blood to flow in only one direction through the heart. The mitral valve is located between the left atrium and ventricle, while the tricuspid valve is found between the right atrium and ventricle. The pulmonary and aortic valves separate, respectively, the pulmonary arteries and aorta from the right and left ventricles of the heart.
7. Associated with the circulatory system is a network of vessels and specialized tissue called the lymphatic system that collects excess interstitial fluid from the tissues and returns it to the circulatory system. The lymph nodes contain specialized cells that filter any particulate matter from the lymph before it reenters the circulatory system.

FILL-IN-THE-BLANK QUESTIONS

1. Resistance is the ______ to blood flow through the vessels.
2. Except for capillaries, the walls of all blood vessels are composed of three distinct layers; the innermost consists of ______, the middle layer is made of ______ ______ and ______ connective tissue cells, while the outermost layer consists mostly of ______ fibers.
3. Arteries and arterioles are capable of both increasing and decreasing their diameters, called ______ and ______, respectively.
4. Arteries carry blood ______ ______ the heart, while veins carry blood ______ ______ the heart.
5. Veins have ______ ______ in them to assist blood flow through them.
6. The heart consists of ______ chambers, called ______ and ______.
7. The semilunar valve separating the right ventricle from the blood vessel carrying blood to the lungs is called the ______ ______.
8. Cardiac muscle cells are capable of generating ______ ______ without external nervous stimulation.
9. The first component of the cardiac conduction system is the ______ node and is known as the ______ of the heart.
10. The vessels that return excess interstitial fluid to circulation constitute the ______ system.

SHORT-ANSWER QUESTIONS

1. Describe how blood flow is regulated through capillary beds.
2. What are the mechanisms assisting blood flow through the veins of the circulatory system?
3. Describe the path of blood flow through the heart as it first enters the heart from the superior and inferior venae cavae.
4. Describe the path taken by an action potential initiated by the sinoatrial node as it passes through the cardiac conduction system.
5. List the three major stages of the cardiac cycle.

VOCABULARY REVIEW

aorta—*p. 235*
artery—*p. 231*
atria—*p. 235*
capillary—*p. 231*
diastole—*p. 241*
lymph node—*p. 244*
lymphatic system—*p. 243*
pericardium—*p. 235*
systole—*p. 241*
vein—*p. 231*
vena cava—*p. 235*
ventricle—*p. 235*
venule—*p. 233*

Chapter Thirteen

BLOOD

For Review

Here are some important terms and concepts that you will encounter in this chapter. If you are not familiar with them, you should review them before proceeding.

Osmosis
(page 40)

Phagocytosis
(page 43)

pH scale
(page 24)

Solute
(page 40)

OBJECTIVES

After reading this chapter you should be able to:

1. *List the four major functions of blood.*
2. *Discuss the components of the formed elements.*
3. *Discuss the function of plasma.*
4. *Describe the structure and function of red blood cells.*
5. *Define anemia and list some types of anemia.*
6. *Describe the structure and function of white blood cells.*
7. *Define clotting and describe the major steps in the clotting process.*
8. *Discuss how the direction of flow of plasma and extracellular fluid across capillary walls is affected by capillary and osmotic blood pressure.*

The human body contains between 5 and 6 L of blood in its circulatory system. Recall from Chapter 12 that blood transports many substances in the body through the circulatory system. This essential fluid transports all the materials between the body's cells and the outside world as well as between the cells. Blood also protects the body against foreign invasion from microorganisms, repairs damaged blood vessels, and helps maintain extracellular fluid volume. This chapter describes the blood's structural components and functions.

The Functions of Blood

Blood has four vital functions while maintaining the homeostasis of the body. Blood's first major function is transportation because it carries many substances throughout the body. For example, blood delivers oxygen from the lungs and nutrients from the digestive system to all body cells. It also transports metabolic waste products from cells to elimination sites, such as the lungs to eliminate carbon dioxide and the kidneys to eliminate nitrogenous wastes. Blood also transports hormones from the endocrine glands to their target tissues. Because of its high water content, blood helps maintain body temperature by absorbing and selectively distributing body heat to various body tissues.

Blood's second major function is protecting the circulatory system. When a blood vessel is damaged, several components of the blood initiate clot formation, halting blood loss. Blood's third major function is to assist in protecting the body against infection. Blood contains an assortment of proteins and white blood cells that help defend the body against foreign invaders, such as bacteria, viruses, toxins, and abnormal body cells that cause tumors. Blood's fourth major function is to act as a buffer to prevent excessive or abrupt changes in blood pH, which could disrupt normal cellular activities.

The Composition of Blood

A blood sample drawn from one of your blood vessels appears to be a uniformly red fluid. However, if it is placed into a test tube and then spun at high speed in a centrifuge, it divides into two relatively large and distinct layers separated by a thin third layer.

The reddish layer at the bottom of the tube, as well as the thin layer just above it, contains cells and cellular components collectively called **formed elements.** These cellular components are called formed elements because some are membrane-surrounded particles or fragments and are not all true cells. The heavier red layer consists of **red blood cells,** or **erythrocytes** (eh-rih′-thro-sits), making up 45% of blood. The thin layer just above it, called the **buffy coat,** contains cells called **white blood cells,** or **leukocytes** (lu′ko-sits), which are vital to protecting the body against infections. The buffy coat also contains cell fragments called **platelets** that assist blood clotting. Together they account for less than 1% of blood. Formed elements are produced in the red bone marrow in the sternum, vertebrae, ribs, flat bones of the skull, and ends of the long limb bones. The red bone marrow contains undifferentiated **stem cells** that continuously divide, differentiate, and give rise to formed elements. The clear yellowish fluid at the top of the tube is called **plasma** and accounts for 55% of whole blood. Plasma is about 90% water and contains more than 100 different dissolved solutes, ranging from proteins and nutrients to electrolytes.

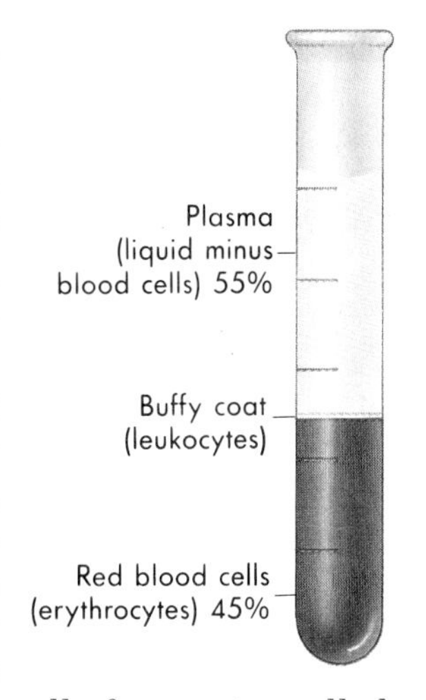

Plasma

Plasma is the fluid portion of the blood and is the medium for carrying materials in the blood. Plasma contains many dissolved substances, including proteins; hormones; nutrients such as glucose, amino acids, and lipids; gases such as carbon dioxide and oxygen; sodium, chloride, calcium, and potassium ions; and nitrogenous waste products. Recall from Chapter 2 that water, in addition to being an excellent solvent, also has a high capacity to hold heat. Plasma is therefore able to absorb and distribute much of the heat generated as a by-product of tissue metabolism with only small changes in the blood's temperature. The heat not needed to maintain normal body temperature is eliminated to the external environment as the blood travels close to the skin's surface.

The most abundant substances dissolved in the plasma are called **plasma proteins.** Although plasma proteins have many functions, they are *not* taken up by the cells to be used as fuel or metabolic nutrients, as are other plasma solutes. They are too large to leave the capillaries and therefore remain in the blood. Because of their presence in the plasma and their absence in the extracellular fluid, plasma proteins establish an osmotic gradient between blood and extracellular fluid. This osmotic pressure is the primary force responsible for preventing excessive loss of plasma from the capillaries into the extracellular fluid.

Plasma proteins fall into three categories, each of which performs a specific task. The first category includes the **albumins,** which are the most abundant and contribute most to maintaining the plasma's osmotic pressure. The second category of plasma proteins includes the **globulins.** Plasma contains three types of globulins. Alpha and beta globulins bind to and transport substances in the plasma, such as hormones, cholesterol, and iron. Both alpha and beta globulins are produced in the liver. Gamma globulins are produced by a special type of white blood cell and are important in fighting infections. The third category of plasma proteins includes **fibrinogen,** a key protein in the blood-clotting process (Table 13-1).

Table 13-1 Plasma Contents

CONSTITUENT	FUNCTIONS
Water	Transport medium; carries heat
Ions	Membrane excitability; osmotic distribution of fluid between extracellular and intracellular fluid; buffering of pH changes
Nutrients, wastes, gases, hormones	No function in blood; merely being transported
Plasma proteins	In general, exert osmotic effect that is important in distribution of extracellular fluid between vascular and interstitial compartments; buffering of pH changes
Albumins	Transport of many substances; greatest contribution to osmotic pressure
Globulins	
Alpha and beta	Transport of many substances; clotting factors
Gamma	Antibodies
Fibrinogen	Inactive precursor for fibrin meshwork of clot

Red Blood Cells

Red blood cells, or **erythrocytes** (e-rith'ro-sitz), function primarily to pick up oxygen in the lungs and transport it to all the body's cells. Red blood cells are the most numerous cells in the blood. The average **red blood cell count** is about 5 million cells per cubic millimeter (mm^3). The female red blood cell count is usually between 4.3 and 5.2 million cells per mm^3. The male red blood cell count is usually between 5.1 and 5.8 million cells per mm^3.

Red blood cells are a wonderful example of what architects describe as "form following function." As red blood cells mature they lose their nucleus and most of their organelles, including mitochondria. As a result, red blood cells generate ATP by *anaerobic* pathways and do not consume any of the oxygen they are transporting. The oxygen-transporting protein contained in the cytoplasm of the red blood cell accounts for about 33% of its total weight. Red blood cells are little more than sacs filled with cytoplasm and oxygen-binding protein. Red blood cells are also very small, approximately 7 micrometers (μm) in diameter, and their shape is unusual. Rather than being spherical, red blood cells are shaped like flattened disks with depressed centers. Their small size and unusual shape provide a large surface area relative to their volume. Thus no point in the cytoplasm is far from the cell membrane surface, and rapid oxygen diffusion can occur through the membrane.

How do red blood cells transport oxygen? All red blood cells contain a substance called **hemoglobin,** consisting of two parts. The **globin portion** is a protein made up of four highly folded amino acid chains. The four protein chains are two identical pairs, designated as **alpha chains** and **beta chains.** Four iron-containing, nonprotein **heme groups** are each bound to one of these amino acid chains. Each iron-containing heme group can combine with one molecule of oxygen. Therefore each hemoglobin molecule can transport four molecules of oxygen

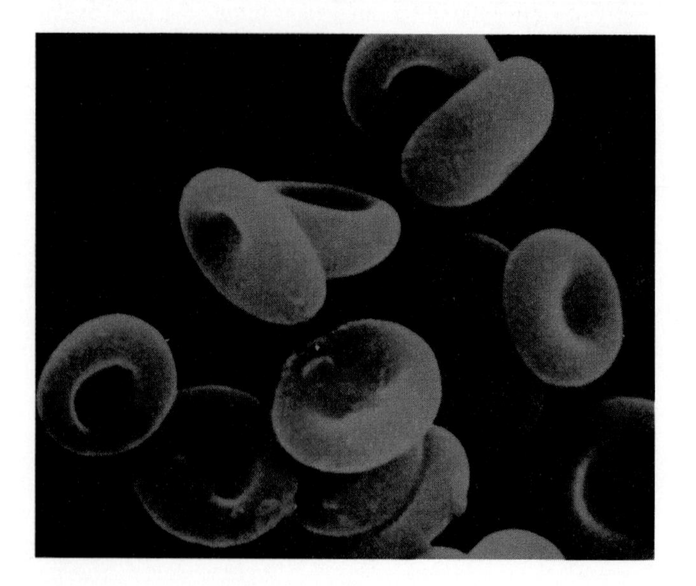

Figure 13-1 Structure of hemoglobin

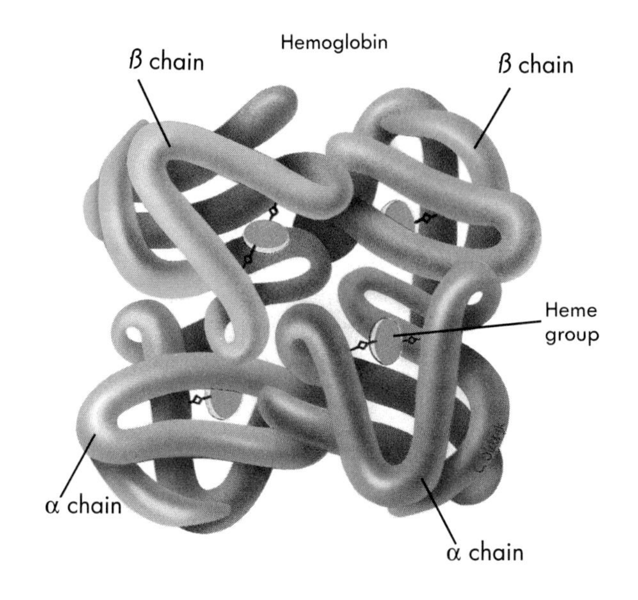

Hemoglobin consists of two pairs of amino acid chains, which are the globin portion of the molecule. One pair is called alpha chains and the other pair is called beta chains. Each amino acid chain also has an iron-containing heme group.

(Figure 13-1). A single red blood cell contains 200 to 300 million hemoglobin molecules, so each of these small cells is capable of transporting about 1 billion molecules of oxygen.

As venous blood moves through the lungs, oxygen molecules diffuse from the air in the lungs through the capillary walls into the blood and then into the red blood cells, where they bind to the heme groups. When oxygen binds to hemoglobin, it assumes a new three-dimensional shape, is now called **oxyhemoglobin,** and becomes bright red. As the red blood cells reach the tissues, the process is reversed. Oxygen dissociates from the iron-containing heme group, causing hemoglobin to resume its former shape, called **deoxyhemoglobin.** The released oxygen diffuses from the blood into the surrounding extracellular fluid and then into the tissue cells.

Carbon dioxide is transported from the body's cells back to the lungs by the blood in three ways. About 7% is directly dissolved in the plasma, and about 20% binds to hemoglobin. The remaining carbon dioxide is converted to bicarbonate ions and is transported in the plasma. If more than 7% or 8% of the carbon dioxide moved directly into the plasma, it would not dissolve but would instead form bubbles, in the same way soft drinks are carbonated. Gas bubbles would disrupt the blood flow and damage the blood vessels. We shall now look more closely at each of the ways carbon dioxide is transported in the blood and plasma.

Unlike oxygen, carbon dioxide binds to the globin protein of hemoglobin rather than to the heme groups. The carbon dioxide combines with the deoxyhemoglobin of the red blood cells, forming **carbamino** (car'-buh-me'-no) **hemoglobin.** Carbamino hemoglobin is formed at the tissues as the carbon dioxide diffuses out of the cells and into the blood, where some of it binds to the deoxyhemoglobin in the red blood cells.

The remaining carbon dioxide is converted to bicarbonate ions (HCO_3) and transported in the plasma. Carbon dioxide moves into the red blood cells and combines with water, forming carbonic acid (H_2CO_3). Most of the carbonic acid molecules immediately break down to bicarbonate ions (HCO_3^-) and hydrogen ions (H^+), as follows:

$$CO_2 + H_2O \underset{\text{lungs}}{\overset{\text{tissues}}{\rightleftarrows}} H_2CO_3 \underset{\text{lungs}}{\overset{\text{tissues}}{\rightleftarrows}} H^+ + HCO_3^-$$

Carbonic acid molecule production occurs thousands of times faster in red blood cells than elsewhere because red blood cells contain an enzyme called **carbonic anhydrase,** which is responsible for the rapid formation of carbonic acid molecules. The bicarbonate ions then diffuse from the red blood cells into the plasma. The hydrogen ions, in contrast, chemically bind to the deoxyhemoglobin in the red blood cells. Once the blood reaches the lungs, the process is reversed. Deoxyhemoglobin releases the hydrogen ions as it combines with oxygen from the air in the lungs. The released hydrogen ions at-tract the bicarbonate ions from the plasma back into the red blood cell. Because of carbonic anhydrase the bicarbonate and hydrogen ions recombine to form carbonic acid. The carbonic acid then breaks down into water and carbon dioxide. The carbon dioxide diffuses out of the red blood cell, through the plasma, out of the capillary, and back into the lungs where it is exhaled.

The Life Span of Red Blood Cells

The price red blood cells pay for their lack of intracellular organelles is a shortened life span. Lacking a nucleus and therefore containing no DNA, red blood cells cannot synthesize proteins for cellular repair, growth, or division. Because of this, red blood cells have a life span of only 120 days. Most aged red blood cells are removed from circulation primarily by the **spleen.** It is situated between the stomach and diaphragm and is considered part of the lymphatic system because it contains lymphoid tissue. The spleen performs various tasks, including blood storage and the destruction of red blood cells. The protein portion of the hemoglobin is broken down into individual amino acids, which are returned to the circulatory system for use as needed elsewhere in the body. The iron from the heme groups is reused in the synthesis of new hemoglobin and red blood cells in the red bone marrow. Iron recycling is not completely efficient, and therefore small amounts of iron must be ingested each day in the diet. The noniron portion of the heme groups is not reused and is eventually eliminated in the feces.

The manufacture of red blood cells from dividing stem cells of the red bone marrow is called **erythropoiesis** (e-rith'ro-poy-e'sis). In a healthy adult erythropoiesis occurs at the astonishing rate of 2 to 3 million cells per second to keep up with the destruction of old red blood cells by the spleen.

The number of circulating red blood cells normally remains within narrow limits because of its importance in transporting the oxygen to tissue cells. The primary stimulus for increased erythropoiesis is reduced oxygen delivery to tissues. The kidneys signal the red marrow to increase red blood cell production. Reduced oxygen to the kidneys stimulates them to secrete a hormone called **erythropoietin** into the blood. On reaching the red bone marrow it stimulates increased erythropoiesis by the stem cells. The increased production of erythrocytes elevates the oxygen-carrying capacity of the blood, which restores normal oxygen delivery to the tissues, including the kidneys.

Blood Types: ABO System

As with all the body's cells, red blood cells have particular recognition markers on the surface of their cell membranes. With red blood cells there is allelic variation in the types of recognition markers attached to the cell membrane. These variations produce the different **blood types.** People with type A blood have one type of recognition marker, and people with type B blood have another type of marker. People with type AB blood have both A and B cell-surface recognition markers, and people with type O have neither A nor B markers.

Red Blood Cell Disorders

Anemia refers to a reduction below normal in the blood's oxygen-carrying capacity and can be caused by inadequate numbers of circulating red blood cells, a deficiency in their hemoglobin content, or both. **Iron-deficiency anemia** occurs when there is not enough iron available for hemoglobin synthesis because of an inadequate dietary intake of iron. The usual number of red blood cells is produced, but their hemoglobin content is reduced. **Pernicious anemia** is caused by an inability to absorb adequate amounts of vitamin B_{12} from the digestive system. Vitamin B_{12} is essential for DNA production, which is critical in controlling stem-cell division in the red bone marrow. Although other cell lines are also affected by inadequate amounts of vitamin B_{12}, because of the high rate of red blood cell production

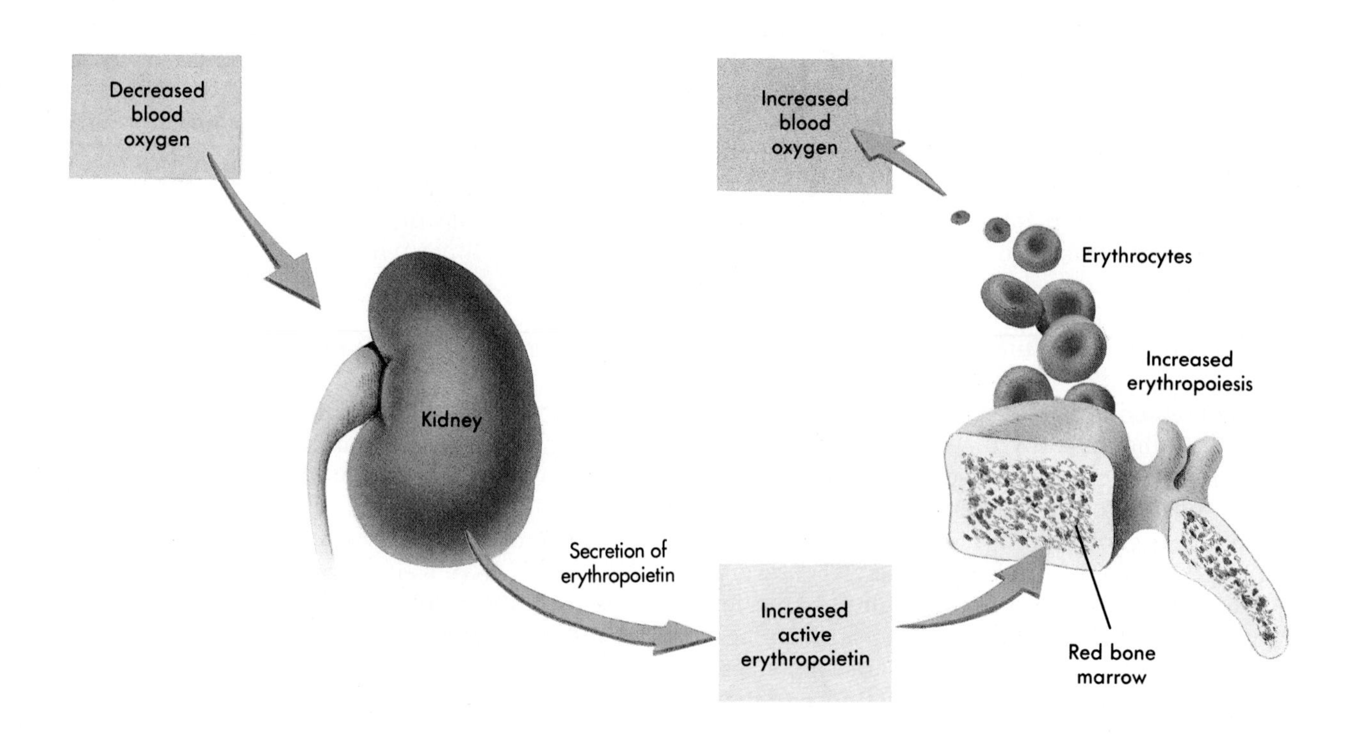

Biology in Action BLOOD CHOLESTEROL

You learned from the topical aside "Keeping the Pipes Clean" in Chapter 12 that several factors contribute to the presence of atherosclerosis, the progressive degenerative disease causing a narrowing of the arteries. One major contributor to atherosclerosis is the appearance of **atheromas,** which originate in the middle layer of the arterial wall and protrude into the lumen of the artery. Atheromas are produced by rapidly dividing cells of the arterial wall. As the cells divide they also begin to fill with lipids, especially cholesterol, and this hastens arterial lumen narrowing. Although some artery narrowing is a function of aging, high levels of blood cholesterol significantly speed up this process.

It may be surprising to learn that the body not only needs cholesterol but actually manufactures it. Unlike other lipids cholesterol is not used as an energy source by cells. Rather it is an important component of cell membranes. It is also essential for the synthesis of steroids, such as sex hormones. Cells can synthesize some of the cholesterol they need from any nutrient molecule ingested as part of the diet, but the rest is manufactured by specialized liver cells.

On average the liver produces between 700 and 800 mg of cholesterol a day. A normal diet usually supplies between 200 and 300 mg a day. The liver is the primary regulator of cholesterol in circulation. If the dietary input drops, the liver increases its cholesterol production. If the dietary input increases, liver output decreases. If it is necessary for our health and is closely regulated, why should cholesterol be such a health concern?

Cholesterol, as well as most other lipids, is insoluble. To travel through the circulatory system, it is attached to carrier proteins. One protein combines with cholesterol to form **low-density lipoprotein,** or **LDL.** LDL carries the cholesterol to most body cells. Another carrier that combines with cholesterol, called **high-density lipoprotein,** or **HDL,** removes cholesterol from cells and transports it, via the circulatory system, back to the liver where it can be used for other purposes. The interplay between LDL and HDL determines the flow of cholesterol between the liver and the body's cells.

The problem begins when blood cholesterol levels exceed the rate at which the liver can absorb, use, and dispose of cholesterol. Atherosclerosis is inversely related to the HDL concentration in the blood. An accurate predictor of developing atherosclerosis is the ratio of blood HDL cholesterol to total cholesterol. The higher the HDL cholesterol concentration in proportion to the total blood cholesterol level, the lower the risk. Compared with HDL concentration, if LDL concentration is high, there is more cholesterol in circulation, including going to cells that cause atheromas in arteries.

Modifying your diet by reducing cholesterol and saturated fat in relation to the total calories consumed can significantly help minimize blood cholesterol levels. Some habits also influence atherosclerosis risk by affecting HDL levels. Cigarette smoking, for example, lowers HDL levels in the blood. In contrast, regular exercise raises HDL levels. We can thus significantly influence, both negatively and positively, the rate of our bodies' changes.

erythropoiesis is where this vitamin deficiency usually first appears. Pernicious anemia is characterized by reduced numbers of circulating red blood cells. It is treated by injecting vitamin B_{12}, which bypasses the defective absorptive mechanism of the digestive system and restores red blood cell production to normal.

Just as there can be a reduction in red blood cell count, there can also be excessive red blood cell production. This condition, known as **polycythemia** (pol′i-si′the′mi-a), is caused by above-normal red bone marrow activity. One cause of polycythemia is a tumor in the red bone marrow causing abnormal numbers of red blood cells to be produced.

Other red blood cell disorders involve hemoglobin. Recall the alpha and beta chains that make up hemoglobin molecules. There are variant forms of these protein chains. One variant form of the alpha chain, known as **hemoglobin S,** has some unusual properties. Red blood cells with this form become crescent or sickle shaped when they unload their oxygen molecules at the tissues. Sickling will also occur when the blood's oxygen content is lower than normal, such as during vigorous exercise. These deformed red blood cells rupture easily, clogging capillaries and disrupting blood flow. Also, because of their shape, they are often removed by the spleen and taken out of circulation. Their life span is significantly shorter compared with that of nonsickling red blood cells, usually 90 days as opposed to 120 days. This type of anemia is called **sickle-cell anemia.**

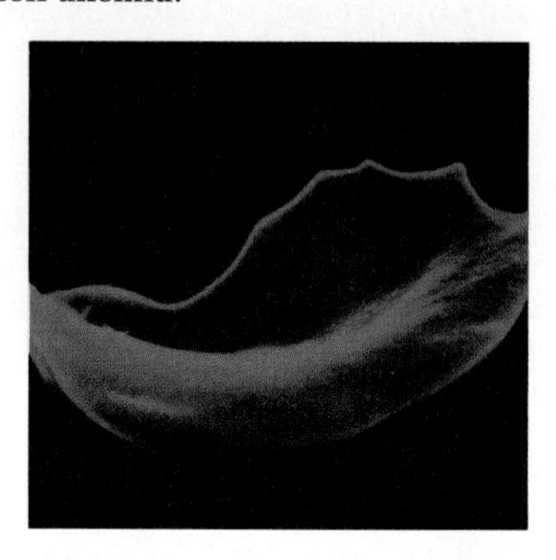

The sickle-cell trait is genetically inherited. A person homozygous for hemoglobin S has two affected alpha amino acid chains and thus has **sickle-cell anemia,** a severe, chronic anemic condition. However, a person heterozygous for hemoglobin S has only one affected alpha amino acid chain and sickling occurs only when the blood's oxygen concentration is extremely low. This person then has the **sickle-cell trait.** About 9% of Americans of West African ancestry have the sickle-cell trait. Approximately 0.2% of them are homozygous for hemoglobin S and have sickle-cell anemia.

White Blood Cells

The body's ability to resist or eliminate potentially harmful foreign materials or abnormal cells is called **immunity. White blood cells,** or **leukocytes** (lu'ko-sitz), are the primary cells of the body's immune system. White blood cells defend against foreign invasion in two different ways: either by engulfing and digesting the foreign substance, called phagocytosis, or by producing substances called **antibodies,** which are proteins that mark, or "tag," invaders for destruction in other ways. White blood cells also destroy abnormal, or cancerous, cells. Some white blood cells also protect the body by cleaning up the body's "litter" by phagocytizing cellular debris, such as dead or injured cells. For example, white blood cells remove aged or damaged red blood cells entering the spleen.

There are several structural differences between white blood cells and red blood cells. White blood cells lack hemoglobin, so they are colorless but appear white when viewed through a microscope. White blood cells are also larger than red blood cells. They range from 9 to 15 μm in diameter and contain nuclei and other organelles. White blood cells are less numerous compared with red blood cells. For every 700 red blood cells in circulation there is 1 white blood cell. The **white blood cell count** ranges from 4000 to 11,000 per mm^3 of blood, with an average white blood cell count of 7000 per mm^3 of blood. All white blood cells come from the same undifferentiated stem cells in the red bone marrow that also give rise to red blood cells and platelets. Unlike red blood cells, which are structurally and functionally identical, white blood cells vary in structure, function, and number.

There are two main categories of white blood cells, granulocytes and agranulocytes, which are determined by the appearance of the nuclei and the presence or absence of granules in the cell's cytoplasm. Five different types are found in the two major categories of white blood cells. Regardless of type, all white blood cells protect the body from foreign invasion or abnormal materials that could disturb homeostasis.

Granulocytes literally means "granule-containing cells." They have nuclei segmented into lobes, and their cytoplasm contains large numbers of membrane-bound granules. Three types of granulocytes can be distinguished on the basis of how they react to particular stains. **Eosinophils** (e'-o-sin-o-filz) are stained by a red dye called eosin. The name of the dye is derived from the Greek word *eos,* meaning "dawn." **Basophils** take up a basic blue dye, and **neutrophils** are neutral, showing no dye preference.

Eosinophils comprise from 1% to 3% of all white blood cells. They specialize in attacking internal parasites by attaching to them and secreting substances that kill them. Eosinophils also contribute to allergic conditions, such as hay fever and asthma, and autoimmune diseases, conditions in which the body's defenses attack their own tissue cells. Basophils, making up less than 1% of all white blood cells, are the least numerous and least understood of the granulocytes. Basophils produce **histamine,** which causes capillaries to dilate and become more permeable. This increases fluid release resulting in swelling of the surrounding tissue. Basophils that take up residence in body tissues are called **mast cells** and also produce histamine. We discuss histamine and mast cells in

Chapter 14. Neutrophils make up from 60% to 70% of all white blood cells circulating in the blood. Neutrophils function as phagocytes and act as the body's first defenders against bacterial invasion. Neutrophils also scavenge any cellular debris found in the body.

Agranulocytes, the second major category of white blood cells, lack cytoplasmic granules. They have nonlobed nuclei and are therefore also called **mononuclear agranulocytes.** The two types of agranulocytes are **monocytes,** which have single nuclei, and **lymphocytes,** which have large spherical nuclei that occupy most of the cell.

Monocytes, like neutrophils, are destined to become phagocytes. Monocytes are released from the bone marrow and circulate for 1 or 2 days before most of them move into body tissues. After their arrival in the tissue, monocytes continue to mature and greatly enlarge to become tissue-bound phagocytes, called **macrophages.** Lymphocytes perform a variety of functions. Some lymphocytes are phagocytic, some produce and secrete substances that stimulate other leukocytes to become active, and others produce antibodies that mark invaders for destruction. Table 13-2 summarizes the cellular and fluid components of the blood.

Table 13-2 Formed Elements of Blood

CONSTITUENT	FUNCTIONS
Erythrocytes	Oxygen and carbon dioxide transport (mainly oxygen)
Leukocytes	
Neutrophils	Phagocytes that engulf bacteria and debris
Eosinophils	Attack parasites; important in allergic reactions
Basophils	Release of histamine, which is important in allergic reactions
Monocytes	Become phagocytic
Lymphocytes	Production of antibodies and immune responses
Platelets	Blood clotting

White Blood Cell Disorders

The term **leukemia** literally means "white blood" and refers to a group of cancerous conditions that involve uncontrolled white blood cell proliferation. However, the body has an inadequate immune response because most of these cells are abnormal or immature and are incapable of performing their normal defense functions. The leukemias are named according to the abnormal cell type involved. For example, lymphocytic leukemia involves abnormal lymphocyte proliferation. Without therapy all leukemias are fatal. Only their course of development differs.

Current treatment involves irradiation and administering drugs that destroy the rapidly dividing cells. These treatments do not cure the disease but are successful in creating symptom-free periods, or **remissions,** lasting from months to several years.

In **infectious mononucleosis,** the number of circulating lymphocytes is significantly elevated and many of them are abnormally structured. A person with infectious mononucleosis, caused by the Epstein-Barr virus, complains of being tired all the time and of having a mild sore throat and sometimes a low-grade fever. Recovery usually occurs within 2 months of the initial onset of symptoms, but it can often take 1 year or more for the person to recover completely.

Blood Clotting

Blood clotting, or **coagulation** (ko-ag-u-la'shun), is the arrest of bleeding from a broken blood vessel. The body's blood-clotting mechanisms are adequate to seal defects and stop blood loss through damaged capillaries, arterioles, and venules. These small vessels are frequently broken by common minor traumas. This is such a routine source of bleeding that we usually are not even aware of any damage. In this section we describe the blood-clotting mechanisms.

Recall that platelets are the third type of formed element in blood. They are small membrane-bound fragments of approximately 2 to 4 μm in diameter that have budded off the outer edges of large **megakaryocytes** (meg'a-kar'i-o-sit). These are derived from the same undifferentiated stem cells that give rise to white blood cells and red blood cells. Because platelets are cell fragments and lack any intracellular organelles, their life span is only 5 to 9 days before they are destroyed in the liver and spleen. Platelets are numerous, ranging from 150,000 to 350,000 per mm^3 of blood. There are also two plasma proteins involved in clotting, called **fibrinogen** (fi-brin'o-jen) and **prothrombin.** Both are manufactured by the liver and released into the circulation.

Figure 13-2 Main events in clot formation

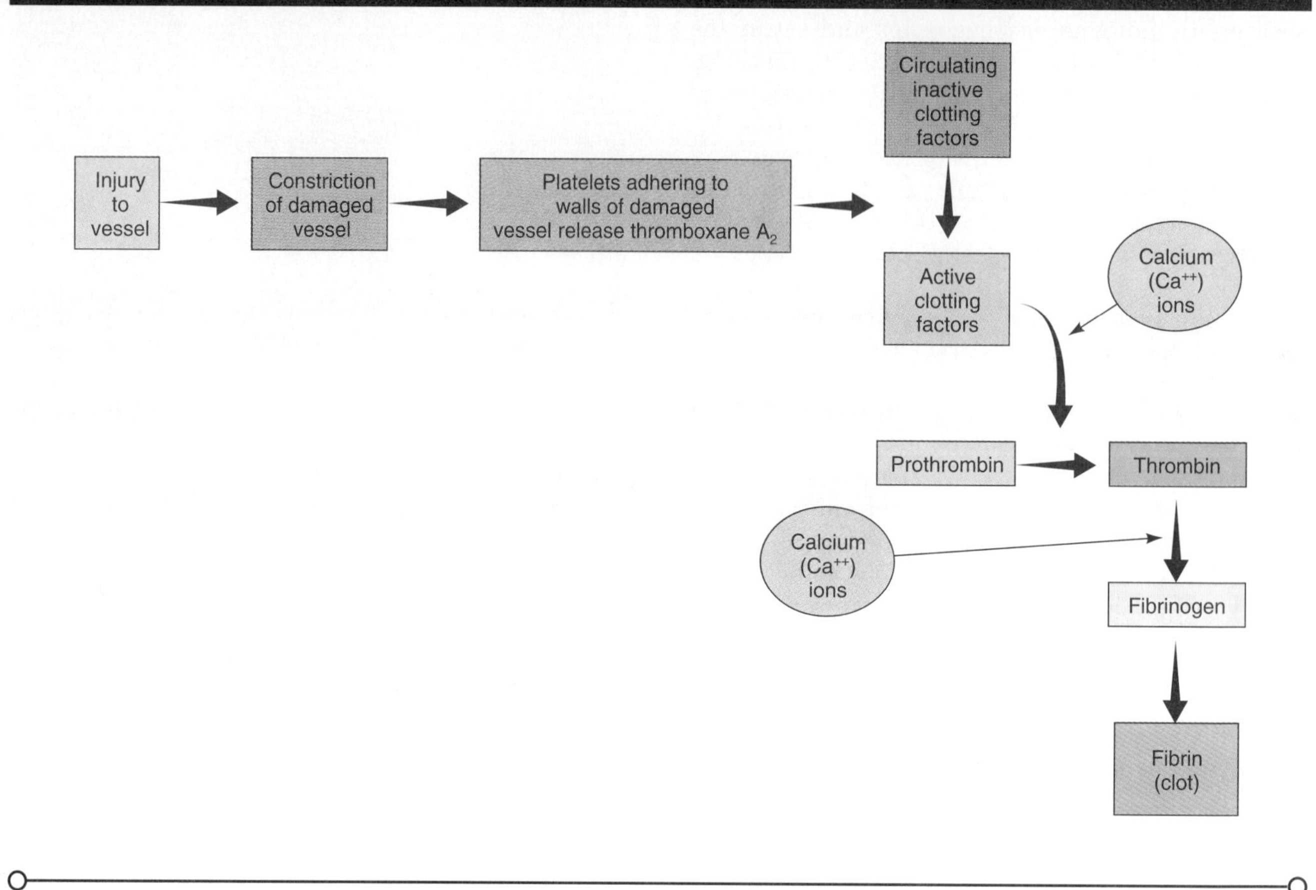

When a blood vessel is cut or torn, the opposing ends of the vessel immediately constrict and adhere to each other, sealing off the damaged vessel ends. This is called **vascular spasm.** It does not completely stop blood loss, but blood flow is significantly reduced through the damaged vessel until other blood-clotting events actually plug the defect. The process is similar to having a water pipe burst and a plumber pinching off the ends with a pair of pliers. Most of the water flow would stop, but some leakage is still possible in the small space between the crimped edges. Thus vascular spasm does not completely shut down the ruptured vessel but significantly restricts its blood loss.

Circulating platelets normally do not adhere to the smooth endothelium of a blood vessel, but when the endothelial surface is damaged or roughened because of vessel injury, platelets attach to exposed collagen fibers found in the wall. As platelets begin to attach to the collagen fibers, they release a substance called **thromboxane A_2,** causing more platelets to stick to the growing platelet plug.

Clotting begins as the injured cells and the clumping platelets release an enzyme that triggers a series of reactions. These result in the conversion of prothrombin to its active form, called **thrombin.** Thrombin is a hydrolytic enzyme that cleaves two short amino acid chains from circulating fibrinogen proteins. In the presence of thrombin the activated fibrinogen proteins begin to join end to end, forming long threads of **fibrin.** The fibrin threads are sticky and form a loose meshwork around and on the platelet plug in the damaged vessel. Red blood cells and other cellular components become trapped in the fibrin network, resulting in clot formation. Once the clot is formed, the platelets trapped in the fibrin network shrink, pulling the damaged vessel's edges closer together and squeezing fluid from the clot.

Calcium ions (Ca^{++}) are needed for all stages of the clotting process to activate prothrombin to thrombin. The synthesis of prothrombin, occurring in the liver, requires vitamin K. If the amount of vitamin K is inadequate, prothrombin synthesis declines and so does the amount in circulation (Figure 13-2).

A clot is not a permanent repair of blood vessel injury. It is a temporary measure to stop bleeding until the vessel can be structurally repaired. During clot formation another circulating plasma protein is also activated. Called **plasmin,** it also becomes trapped among the platelets and fibrin. It is a slow-acting enzyme that gradually dissolves the fibrin network. As plasmin slowly dissolves the clot, new endothelial cells and collagen fibers are produced, eventually repairing the ruptured vessel. Neutrophils and macrophages eventually remove the dissolving clot fragments.

Abnormal Clotting

Despite the specific safeguards the body has to protect itself against inappropriate or abnormal clotting, undesirable clotting sometimes occurs. Persistent roughening of the vessel endothelium, from atherosclerosis or inflammation, might allow the platelets to aggregate at a point in the vessel. A clot that develops in an unbroken blood vessel is called a **thrombus.** If the thrombus is large enough, it may block blood circulation to the cells beyond the blockage and cause tissue death. For example, if the blockage occurs in a coronary artery, the result might be death of the heart muscle and a fatal heart attack. If the thrombus breaks away from the vessel wall and floats freely in the bloodstream, it is called an **embolus.** An embolus is not a problem until it encounters a blood vessel too narrow for it to pass through.

The Physiology of Nutrient and Gas Exchange

After this discussion of plasma proteins and red blood cells, we can now look more closely at capillary exchange.

Extracellular and Plasma Fluid

Exchanges are not made directly between blood and tissue cells. Instead, extracellular fluid acts as an intermediary. Tissue cells exchange materials directly with the interstitial fluid. Movement across the cell's membrane may involve passive diffusion or carrier-mediated transport. Only passive diffusion occurs across the capillary wall between the blood plasma and the extracellular fluid. In diffusion, movement always occurs along a concentration gradient, each substance moving from an area of higher concentration to an area of lower concentration. For example, oxygen and nutrients pass from the blood, where their concentration is relatively high, through the extracellular fluid to the tissue cells. In contrast, carbon dioxide and metabolic wastes leave the cells, where their concentration is relatively high, and diffuse through the extracellu-

Figure 13-3 Nutrient and gas exchange through the capillaries

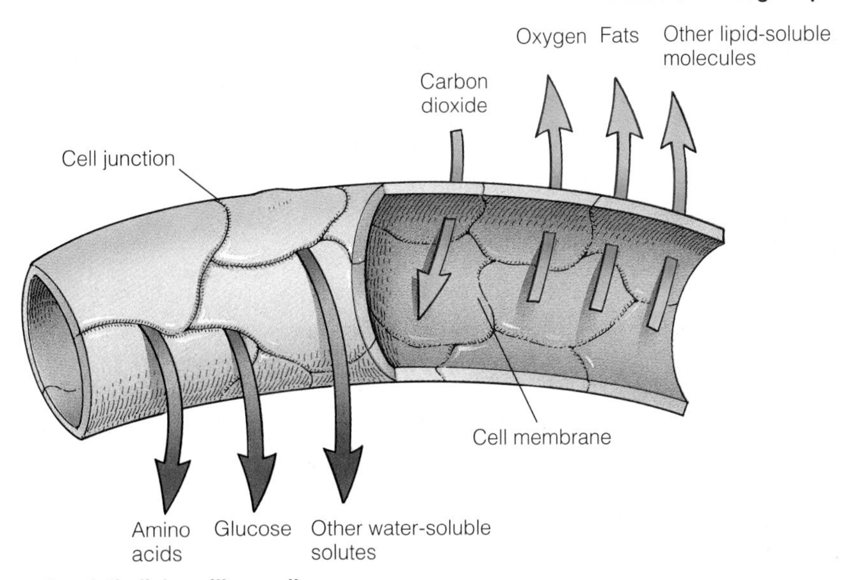

Water-soluble molecules, such as amino acids and glucose, diffuse from the capillaries into the extracellular fluid through junctions between the cells of the capillary wall. Lipid-soluble molecules diffuse through the phospholipid bilayer of the capillary cell wall membrane and into the extracellular fluid.

lar fluid and into the blood. Water-soluble solutes, such as amino acids and glucose, pass between the *junctions* of endothelial capillary cells. Lipid-soluble molecules, such as carbon dioxide, oxygen, and fats, diffuse *through* the phospholipid bilayer of the capillary cell membranes (Figure 13-3).

Extracellular and plasma fluid movement also occurs at the capillary beds. Except for the plasma proteins, the plasma plus its dissolved solutes is forced out of the capillaries through the capillary cell junctions at the arterial end of the bed, but most of it returns to circulation at the venous side. The two opposing forces that determine the direction of flow across capillary walls are capillary blood pressure and capillary osmotic pressure created primarily by plasma proteins in circulation.

First, capillary blood pressure at the arterial end is approximately 40 mm Hg but drops as blood flows through the capillary bed, until at the venous end it is only 10 mm Hg. Plasma proteins develop an osmotic pressure of approximately 25 mm Hg and encourage osmosis whenever the water concentration in the capillaries is lower than it is on the opposite side of the capillary membrane. We can thus see how they affect fluid movement across the capillary membrane.

Because blood pressure is higher at the arterial side of the capillary bed than the osmotic pressure, the plasma fluid, together with dissolved solutes, leaves the capillary. This is called **filtration** because, although the large plasma proteins remain in the capillaries, small solutes and water leave. With the exception of the plasma proteins, the extracellular fluid created by this process is identical to plasma. As blood flows through the length of the capillary, molecules continue to follow their concentration gradients.

At the venous end of the capillary bed, blood pressure has dropped to approximately 10 mm Hg, but the osmotic pressure of the blood remains at 25 mm Hg. Because osmotic pressure is now greater than blood pressure, fluid enters back into the capillaries from the surrounding tissues. As the fluid enters the capillary it is accompanied by cell waste products that are in higher concentration in the interstitial fluid compared with the blood. This method of retrieving fluid by means of osmotic pressure is not completely efficient. Excess fluid enters the lymphatic vessels and ultimately reenters the circulatory system.

One topic covered in this chapter has been the overview of white blood cell structure and function. This is the basis for understanding how the body protects itself against invasion and infection from potential disease-causing agents. This self-defense network of specialized cells, proteins, and enzymes is called the immune system and is the topic of Chapter 14.

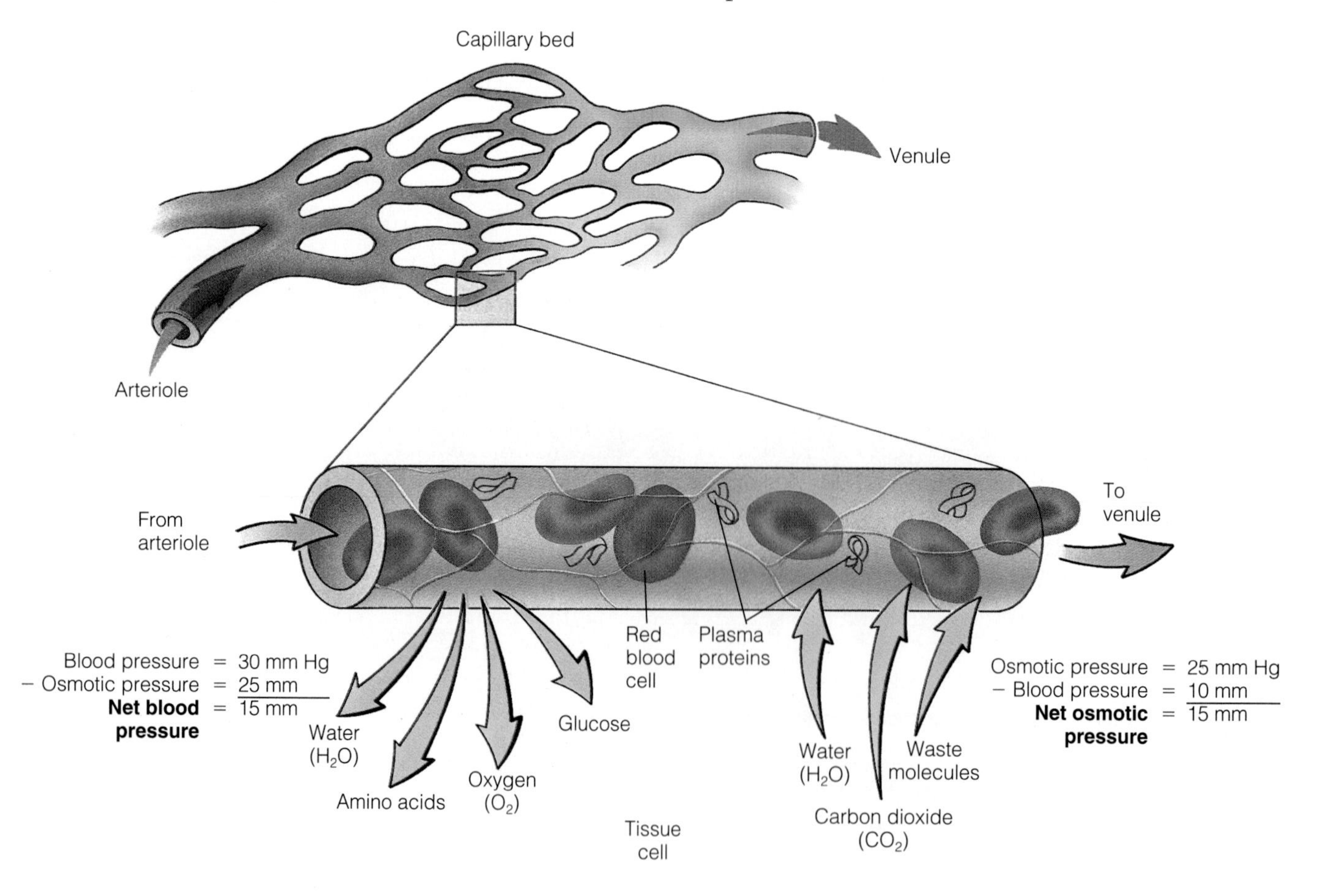

CHAPTER REVIEW

SUMMARY

1. Blood is composed of cells and cellular components, called formed elements, suspended in plasma, which is 90% water. Many other substances are dissolved in blood, including the proteins collectively called plasma proteins. Plasma proteins help maintain osmotic blood pressure.
2. Blood maintains the body's homeostasis by transporting substances between cells and the external environment. It protects the body from infection, blood loss caused by blood vessel damage, and it also buffers the extracellular fluid from abrupt or sudden shifts in pH levels.
3. The exchange of substances transported by the blood and extracellular or interstitial fluid occurs across capillary walls.
4. Red blood cells, called erythrocytes, pick up oxygen in the lungs and transport it to all of the cells of the body. White blood cells, called leukocytes, allow the body to resist or eliminate potentially harmful materials or abnormal cells. Five types of leukocytes are responsible for the body's immunity.
5. The blood-borne substances responsible for stopping blood loss from a broken blood vessel, a process called clotting or coagulation, include the formed elements called platelets and two plasma proteins called fibrinogen and prothrombin. Both plasma proteins are activated under appropriate conditions to initiate blood coagulation.

FILL-IN-THE-BLANK QUESTIONS

1. When blood is spun at high speed in a centrifuge it separates into a fluid layer called ______ and two layers containing ______ ______.
2. Albumin is a(n) ______ ______ that contributes most to maintaining osmotic blood pressure.
3. Red blood cells contain an oxygen-binding substance called ______.
4. Red blood cells lack ______, which significantly shortens their life span.
5. ______ ______ acts as an intermediary between tissue cells and the blood contained in the capillaries.
6. ______ ______ is primarily responsible for forcing fluid through the capillary walls on the arterial side of a capillary bed.
7. ______ are the primary units of the body's immune system.
8. The three types of granulocytes are neutrophils, ______, and ______.
9. Monocytes mature and enlarge to become ______.
10. Fibrinogen is converted into fibrin by an enzyme called ______.

SHORT-ANSWER QUESTIONS

1. What are the structural differences between red blood cells and white blood cells?
2. How do oxygen molecules bind to hemoglobin?
3. Describe the process of capillary exchange.
4. Describe each step in blood clotting, beginning with the damage to the blood vessel and ending with coagulation.

VOCABULARY REVIEW

anemia—*p. 252*
antibodies—*p. 254*
blood clotting—*p. 255*
erythrocyte—*p. 250*
formed elements—*p. 249*
hemoglobin—*p. 250*
leukocyte—*p. 249*
macrophage—*p. 255*
plasma—*p. 249*

Chapter Fourteen

IMMUNITY

For Review

Here are some important terms and concepts that you will encounter in this chapter. If you are not familiar with them, you should review them before proceeding.

Exocytosis
(page 43)

Lysosomes
(page 46)

Transmembrane proteins
(page 39)

White blood cells
(page 249)

OBJECTIVES

After reading this chapter you should be able to:

1. Differentiate between nonspecific immunity and specific immunity.
2. List and describe some nonspecific immune responses.
3. Define the term antigen.
4. Differentiate between antibody-mediated immunity and cell-mediated immunity and describe the cells responsible for each.
5. Describe the structure and function of an antibody.
6. Describe the major functions of each of the four types of cells produced by activated T cells.
7. Discuss the primary and secondary immune responses to an antigen.
8. Describe the action of a vaccine.
9. Define allergy and allergen.
10. Explain two common causes of autoimmune disease.

The human body maintains an internally constant life-sustaining environment for each cell. It is also able to protect this delicate environment from unwanted foreign material and abnormal cells of the body. The immune system recognizes and distinguishes the body's own cells from anything foreign. Any material not recognized as self activates the immune system, which destroys and eliminates the alien substance.

Nonspecific Immune Response

The body's first line of defense against the invasion of disease-causing microorganisms is the skin and the mucous membranes of body cavities exposed to the external environment. As long as the epidermis is unbroken, the skin is an effective barrier to most foreign invaders. As you learned in Chapter 7 the skin has a relatively thin outer covering of epithelial cells called the epidermis. The outermost cells of the epidermal layer are dead. These dead cells are a good mechanical barrier to the underlying skin tissue called the dermis. The skin also protects the body from invasion in other ways. For example, the skin produces a slightly acidic fluid that helps keep it somewhat lubricated and also retards the growth of most bacteria and fungi.

Mucous membranes line body cavities, such as the eyes and mouth and the digestive, respiratory, and reproductive tracts, which open to the external environment. Mucous membranes secrete mucus. Mucus traps particles, microorganisms, and other debris before they can enter the body. The mucous membranes lining the respiratory airways, for example, have cilia protruding from their surfaces. Cilia not only help trap foreign material but also assist in sweeping unwanted material back toward the mouth where it can be eliminated.

However, once the skin or a mucous membrane is broken, the body is open to invasion. When the skin or a mucous membrane is broken and foreign material, such as bacteria or viruses, penetrate into deeper tissues, defense responses that nonselectively defend against any foreign or abnormal material, called **nonspecific immune responses**, defend the body. These responses protect against many threatening factors, including invasions by microorganisms, tissue damage, or both. There are four basic types of nonspecific immune responses the body uses to protect itself. We examine each of these nonspecific immune responses in the following sections.

Inflammatory Response

An **inflammatory response** is a nonspecific response consisting of a sequence of events that occurs when cells are damaged and serves as the body's defense mechanism against trauma. When an injury occurs, capillaries and tissue cells are damaged and usually rupture. A plasma protein released from the ruptured capillaries is activated by an enzyme in the cytoplasm of the tissue cells. This activated protein, called **bradykinin** (brad′-yki′nin), is a chemical that stimulates local free nerve endings, resulting in pain sensation. Bradykinin also stimulates mast cells in the surrounding tissue to release histamine. Recall from Chapter 13 that mast cells are scattered throughout the body's connective tissues, especially near capillaries. They are also concentrated in mucous membranes lining various entry points into the body, such as the lungs and the digestive and reproductive systems. When activated, mast cells release chemicals, such as histamine, that activate several nonspecific immune responses. Histamine causes dilation and increased permeability of the capillaries around the injury site. The resulting increased blood flow around the site and plasma fluid leakage into the injury site cause the localized swelling, redness, and heat characteristic of an inflamed area (Figure 14-1).

Macrophages in and around the isolated tissue begin to phagocytize the foreign invaders. Soon after inflammation begins, the damaged area is invaded by additional neutrophils and monocytes. Neutrophils usually move through the circulatory system randomly, but chemical substances such as bradykinin attract them to the site of injury. This directional movement of cells in response to chemical gradients is called **chemotaxis.** Because of the increased flow of fluid from the capillaries into the injured area, blood flow in the region slows down. The circulating neutrophils squeeze through the enlarged cell junctions of the capillary walls and by means of chemotaxis migrate to the site of injury. Within 1 hour after the inflammatory response starts, neutrophils at the injury site phagocytize bacteria, toxins, and dead tissue cells.

As the attack continues macrophages gradually replace the neutrophils. These new macrophages arise from monocytes entering the area from the circulation. After entering, monocytes undergo a remarkable transformation, swelling to over 10 times their original size, developing a great many lysosomes in the cytoplasm, and becoming active macrophages. You recall from Chapter 3 that lysosomes are organelles filled with hydrolytic enzymes. After a macrophage has ingested the targeted material into its cytoplasm, lysosomes fuse with the

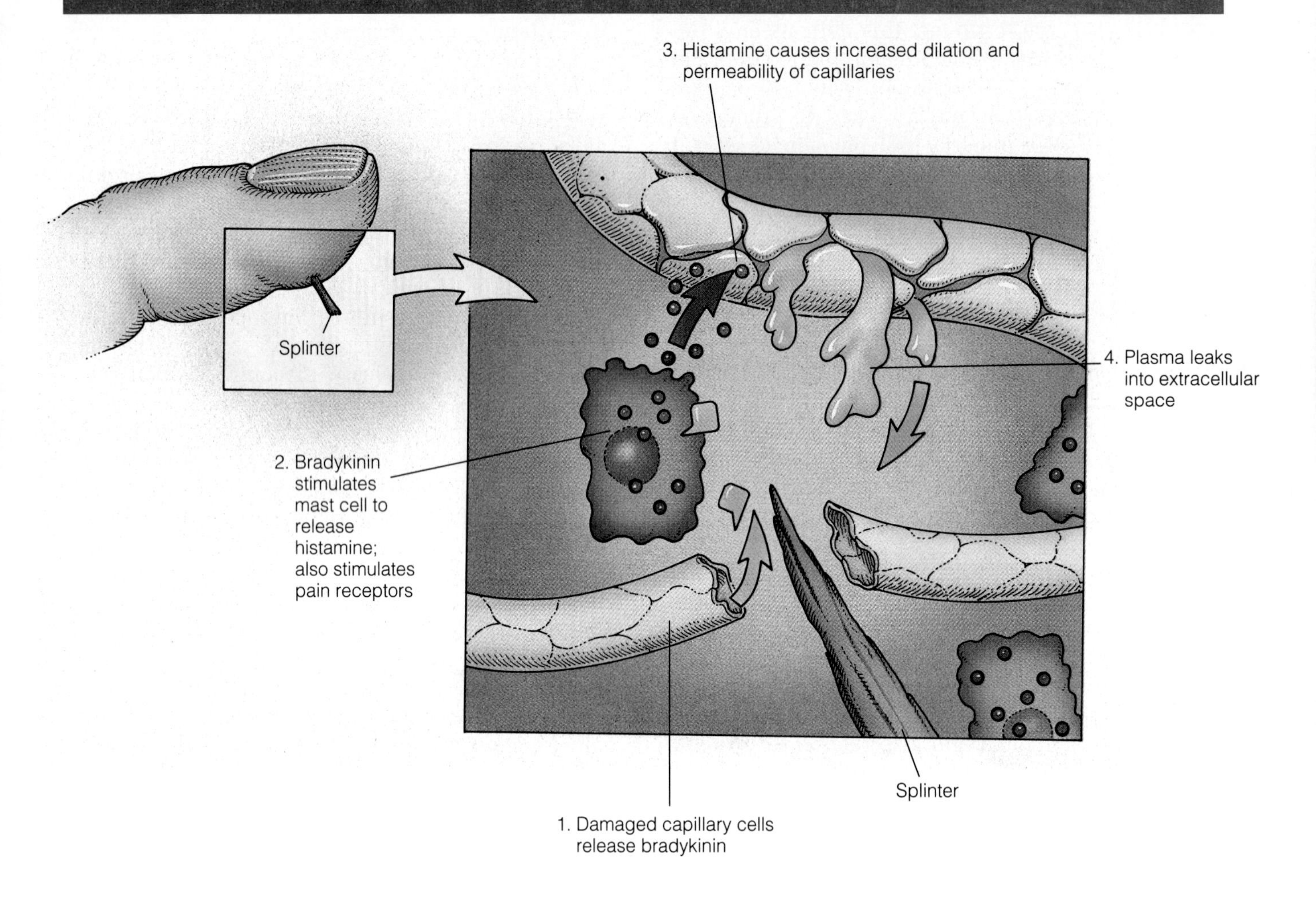

Figure 14-1 Inflammatory response

When an injury occurs, a plasma protein released from ruptured capillaries activates a protein found in the cytoplasm of damaged cells, called bradykinin. In addition to stimulating free nerve endings, bradykinin also stimulates mast cells to release histamine into the extracellular fluid. Histamine causes localized swelling and redness, effectively isolating the damage site from surrounding tissue.

membrane that encloses the engulfed material and release their hydrolytic enzymes into the space of the vesicle. The hydrolytic enzymes break down the entrapped material into its simple organic molecules, such as amino acids and glucose. The macrophage can use these itself, or it can release them into the extracellular fluid for use elsewhere in the body. The phagocytes eventually die, usually because toxic by-products from the degraded foreign invaders accumulate. The creamy, yellowish **pus** that sometimes forms in an infected area is a collection of dead phagocytes, neutrophils, dead tissue cells, and bacteria.

The macrophages are most important in the final disposal of cell debris as the inflammation recedes, hastening the healing process.

Interferon

Viruses survive by invading a cell and taking over its cellular biochemical processes for their own pur-

Figure 14-2 Interferon

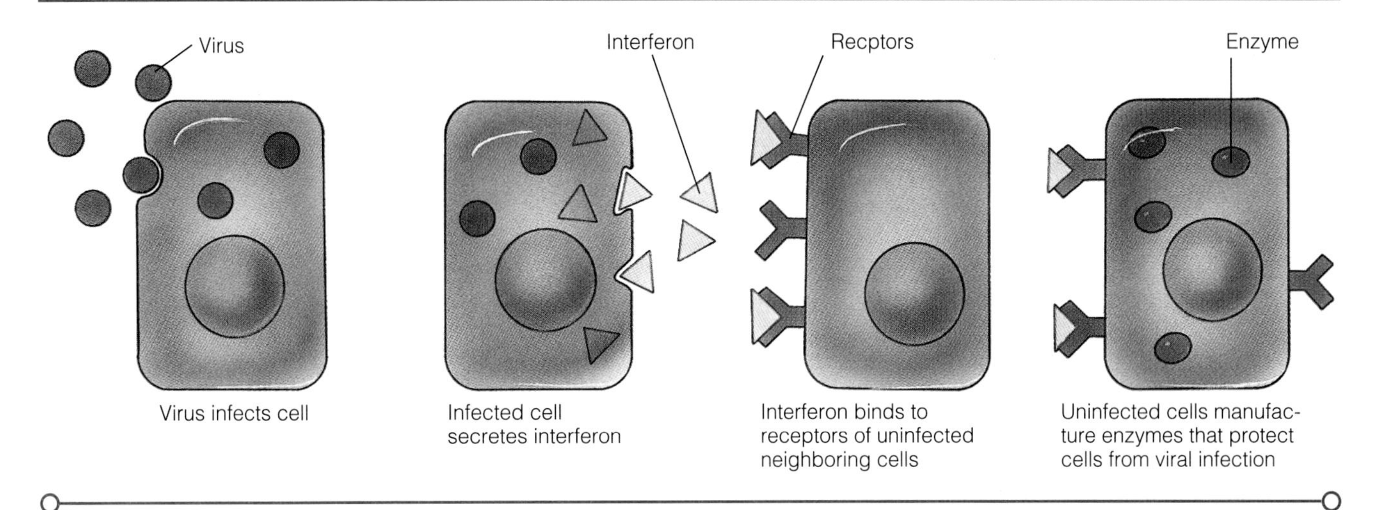
Virus
Interferon
Recptors
Enzyme
Virus infects cell
Infected cell secretes interferon
Interferon binds to receptors of uninfected neighboring cells
Uninfected cells manufacture enzymes that protect cells from viral infection

poses. Because of their ability to invade cells, viruses pose particular problems for the body's immune system because they are *intracellular* disease agents. Because viruses lie inside the body's cells, they are less vulnerable to attack by white blood cells or the body's other immune responses. But the body has a defense system to counteract viruses. When a virus invades a cell, the presence of the viral nucleic acids causes the cell to synthesize chemicals called **interferons** and release them into the extracellular fluid. **Interferons** are proteins that inhibit viral replication. Once released, interferon binds with receptors on the cell membranes of neighboring cells. This interferon-receptor complex stimulates the uninfected cells to produce enzymes that protect them against viral infection. Although the virus is still able to invade these newly warned cells, the enzymes stimulated by the interferon are able to break down the viral hereditary material and inhibit viral protein synthesis (Figure 14-2).

Complement System

The **complement system** consists of at least 11 plasma proteins made by the liver that circulate in the blood in an inactive form. When several of these proteins are activated, a series of biochemical reactions occur. The active proteins of the complement system sequentially attack the membrane of the invading microorganism, such as a bacterium, and produce lesions in it, allowing fluids to leak into it, causing it to burst and die (Figure 14-3). Other proteins of the complement system bind to special receptors on white blood cells, such as macrophages, and enhance phagocytosis by these cells. Finally some of the complement proteins cause mast cells to release histamine and thereby enhance the inflammatory response in the infected tissue.

Fever

When bacteria or viruses invade the body, they often produce **toxins** that are by-products of their

Figure 14-3 Complement system

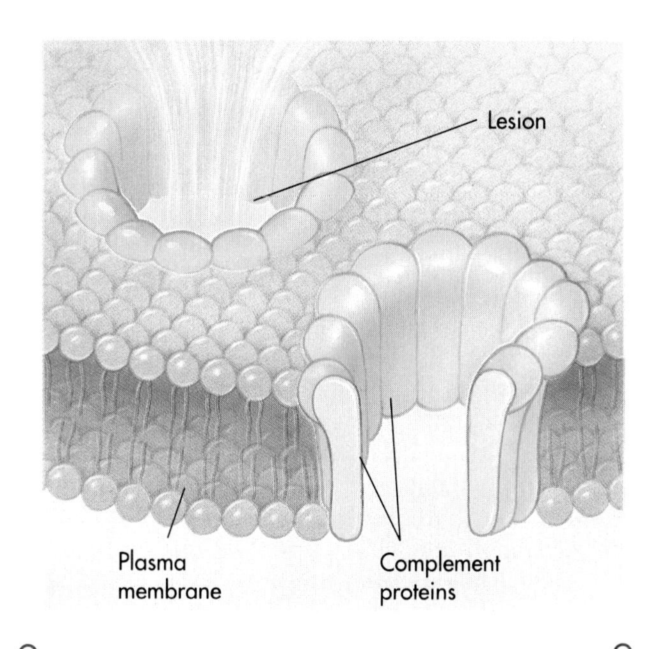
Lesion
Plasma membrane
Complement proteins

The complement system consists of plasma proteins. Once activated, the proteins attack the cell wall of the invading organism and produce lesions in the wall. This allows fluids to leak into the invading organism, causing it to burst and die.

metabolism and are poisonous to the body. The response of the body to toxins is to raise its internal temperature, a condition called **fever.** This response may also be enhanced when the affected cells release chemicals called **pyrogens,** which travel through the circulatory system to the hypothalamus. Pyrogens reset the body's thermostat to a higher temperature. After phagocytizing bacteria, macrophages may also release pyrogens into the circulatory system. Because many bacteria and viruses cannot tolerate temperature fluctuations, the fever response helps kill off the invaders. As the numbers of microorganisms diminish, the concentration of pyrogens in the blood and body temperature both begin to drop, eventually returning to normal.

Specific Immune Responses

Sometimes the body is threatened by an illness or infection that cannot be successfully attacked by nonspecific immune responses alone. In these situations **specific immune responses** are called into action that are selectively targeted against a particular foreign material. The body's specific immune responses are carried out by lymphocytes. Compared with nonspecific immune responses, specific immune responses take longer to develop because they contain mechanisms tailored to particular threats. However, for a nonspecific or specific immune response to work at all, the immune system must be able to recognize foreign substances or abnormal body cell types as different from normal body cells. To understand how this self-defense system operates, we must first describe the self-recognition system the immune system uses to identify and eliminate foreign material and abnormal cells.

Antigens

In Chapter 3 we saw that cells have many different proteins in their membranes. Among the proteins of the body's own cells are recognition markers, or self-markers, that project from the cell membrane surface. You recall from Chapter 5 that these markers are protein molecules specified by genes called the **major histocompatibility complex,** or **MHC.** Consequently the proteins themselves are called **MHC markers.** Bacteria and viruses, cells of transplanted tissues, and abnormal body cells are detected as foreign because the proteins projecting from their cell surface membranes are different from the body's. Consequently the immune system considers them foreign, or nonself, and will respond to them. Any substance capable of stimulating an immune response is called an **antigen.**

Figure 14-4 Antigen

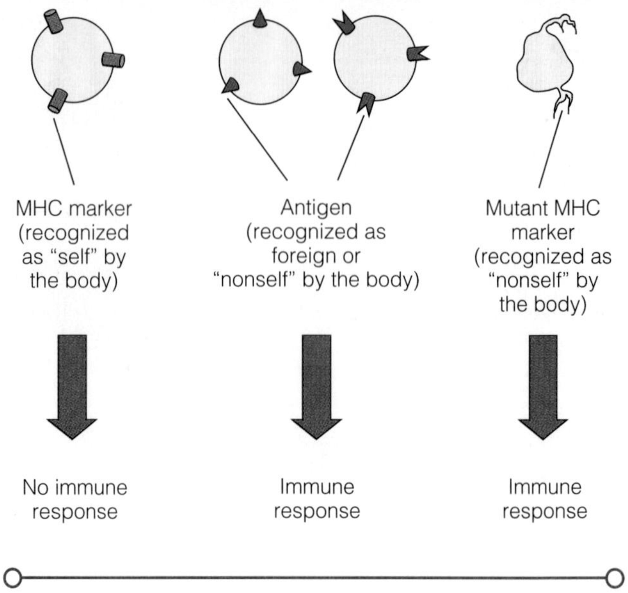

MHC marker (recognized as "self" by the body)
Antigen (recognized as foreign or "nonself" by the body)
Mutant MHC marker (recognized as "nonself" by the body)
No immune response
Immune response
Immune response

The unrecognized foreign antigen triggers a specific immune response against itself when it enters the body. An almost limitless variety of foreign molecules can act as antigens, including practically all foreign proteins, modified proteins such as glycoproteins and lipoproteins, mutant self MHC markers, and many polysaccharides. The immune system recognizes the macromolecules on the membranes of the body's cells as self and all other macromolecules, whether part of a microorganism's membrane or free, such as a bacterial toxin, as foreign, or nonself (Figure 14-4).

As we saw in Chapter 13 lymphocytes and monocyte-derived macrophages are two of the five different types of white blood cells circulating in the blood. From Chapter 12 recall that the lymphatic system is a series of interconnected vessels for returning excess extracellular fluid, called lymph, to the circulatory system. At points along the lymphatic vessels and at their convergences are found lymph nodes, small, round structures filled with macrophages and lymphocytes. As lymph passes through the nodes, foreign materials are filtered from it. For example, during an infection such as a sore throat, the lymph nodes near the infected site may become tender and swollen. This occurs because as the lymph nodes trap and destroy the infective agents, the lymphocytes in them are rapidly dividing and increasing. This slows down the lymph as it passes through the node, causing the lymph node to swell.

Other **lymphoid tissues** in the body include the thymus gland, tonsils, adenoids, spleen, small colonies in the digestive system called Peyer's patches, and the appendix of the small intestine. Some lymphoid tissues intercept foreign microorganisms and abnormal cells before they spread. For example, the lymphocytes in the tonsils and adenoids can respond to microorganisms inhaled through the mouth and nasal passages. Foreign invaders entering the digestive system encounter lymphocytes in Peyer's patches and the appendix of the large intestine. Foreign invaders gaining access to the lymph are filtered through the lymph nodes, where they are exposed to both lymphocytes and macrophages. The spleen, the largest lymphoid organ, performs immune functions on the blood similar to those the lymph nodes perform on the lymph. The lymphocytes and macrophages of the spleen filter the microorganisms and foreign material from the blood and also remove aged red blood cells.

T Lymphocytes

Before birth and into early childhood, some immature lymphocytes, derived from stem cells, are released from the bone marrow, move into the circulation, and travel to the thymus gland. The thymus gland consists of lymphoid tissue and sits just above the heart in the upper chest wall. The lymphocytes that travel to the thymus gland mature and are modified to become **T lymphocytes,** or **T cells.** Some of the T cells leave the thymus to move to other lymphoid tissues, such as the lymph nodes and the spleen while other T cells remain in the circulatory system.

B Lymphocytes

Lymphocytes that remain in the bone marrow and mature become **B lymphocytes,** or **B cells.** After being released into the blood from bone marrow, some mature B cells also move into the lymphoid tissues and others move through the circulatory system.

Although most of these lymphocytes are concentrated in the lymphoid tissues, many are also circulating in the blood, squeezing through the cell junctions of capillary walls and into the body tissues. The tissue cells are under constant surveillance by roaming lymphocytes, each one searching for any foreigner invaders.

The body uses two types of specific immune responses to protect itself. The first specific immune response, called an **antibody-mediated immune response,** relies on the ability of B cells to produce protein molecules, called **antibodies,** that bind specifically to an antigen, thereby tagging them for destruction. The second specific immune response, called a **cell-mediated immune response,** involves the production of activated T cells, which directly attack unrecognized substances and destroy them.

B Lymphocytes: Antibody-Mediated Immunity

Antibodies are produced only by B cells or their descendants. Each mature B cell has a membrane-bound receptor on its surface capable of binding with one particular antigen out of millions of possible antigens. A B cell is capable of binding with only one antigen and producing only one specific antibody to the antigen.

When the membrane-bound receptor binds with a particular antigen, it induces the B cell to differentiate into a **plasma cell,** which produces antibodies able to combine specifically with the particular antigen that stimulated the antibodies' production. This process of activation and differentiation is called **clonal selection** because the B cell with the membrane-bound receptor specific to the foreign antigen it binds to is selected and then undergoes replication and differentiation to produce identical copies, or **clones,** of itself. Not all cells become plasma cells. Some cells do not participate in the immune response. These, called **memory cells,** continue to circulate and become activated only if the body is exposed to the same antigen again. Memory cells enable the immune system to respond much more rapidly when the antigen is encountered again.

During differentiation into a plasma cell, the B lymphocyte enlarges as the rough endoplasmic reticulum in its cytoplasm greatly expands. Remember that the rough endoplasmic reticulum is the synthesis site for proteins to be exported from the cell. Plasma cells become highly active protein antibody factories, producing up to 2000 antibody molecules per second. The plasma cell's protein synthesis machinery is completely committed to antibody production. Therefore the cell is unable to maintain protein synthesis for its own survival and growth. As a result the cell dies after a highly productive 5- to 7-day life span. These antibodies are eventually released into the blood, where they are known as the class of plasma proteins called **gamma globulins,** or **immunoglobulins (Ig).**

Antibody Structure

All antibody proteins share a similar construction. They are composed of four interconnected amino acid chains. Two are long, **heavy chains,** and two are short, **light chains,** both arranged in the shape of the letter Y. The tail portion of the antibody, comprising the two heavy chains and called the **constant fragment (F_c),** contains binding sites to enlist the aid of different cells and substances during the immune response. For example, the F_c of some antibodies enables them to bind with phagocytic cells to enhance their activities. The F_c of other antibodies attaches to mast cells and basophils, triggering their release of histamine.

In contrast to the antibodies' F_c, an antibody protein has two identical antigen-binding sites, one at the tip of each arm. These **antigen-binding fragments (F_{ab})** are unique for each different antibody, so they can bind only to the antigens they specifically match, much like a lock has only one key that fits it. The differences in the amino-acid sequences in the antigen-binding fragments result in the extremely large numbers of unique antibodies in the body.

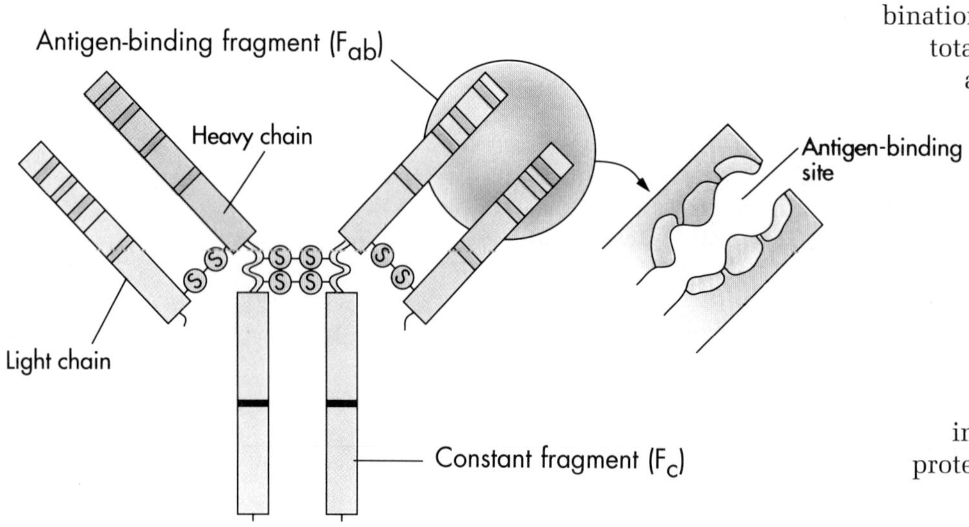

Antibody Diversity

Throughout life we encounter an almost infinite variety of potentially harmful antigens. To protect the body the immune system must be able to respond to each foreign antigen with an appropriate antibody. Antibodies are proteins and so are direct products of genes. Each mature B cell makes only one kind of antibody. Each heavy chain, including its constant and variable fragments, is actually made by a single gene. Similarly the light chains appear to be the products of a single gene. If this is true, how can millions of different antibodies be produced by such a small number of genes?

In the DNA of B cells, antibody genes do not exist as single sequences of nucleotides. Instead, they are first assembled by joining three DNA segments. Each segment, corresponding to a region of the antibody molecule, is encoded at a different site on the chromosome. These sites are composed of a cluster of similar sequences of codons, each varying from the others to a small degree. As each B cell is produced and matures in the bone marrow, its particular membrane-bound receptor is also assembled. One segment is selected at *random* from each chromosomal site, and the DNA sequences selected from these segments are brought together to form a composite gene (Figure 14-5). The process is similar to going to a music store and selecting random types of tapes from different music categories. Few people would come out of the store with the same combination from each category. Because a similar recombination takes place at random in all maturing B cells, the result is the production of many variants of the heavy and light chains. Because a B cell may end up with any heavy-chain gene combination and any light-chain gene combination during its maturation, the total possible combinations are astronomical. Each B cell, however, can synthesize only *one* of the many variants and thus can produce only one type of antibody. The tremendous diversity of antibodies is made possible by reshuffling a small set of genes controlling the antigen-binding fragments of the antibody proteins.

Antibody Function

Antibodies do not directly destroy antigens. Instead, a specific antibody binds to a particular antigen to form an **antigen-antibody complex.** This complex is the biochemical equivalent of a flag or tag that makes the unwanted intruder chemically "visible" to other immune mechanisms that will

Figure 14-5 Composite genes and antibody diversity

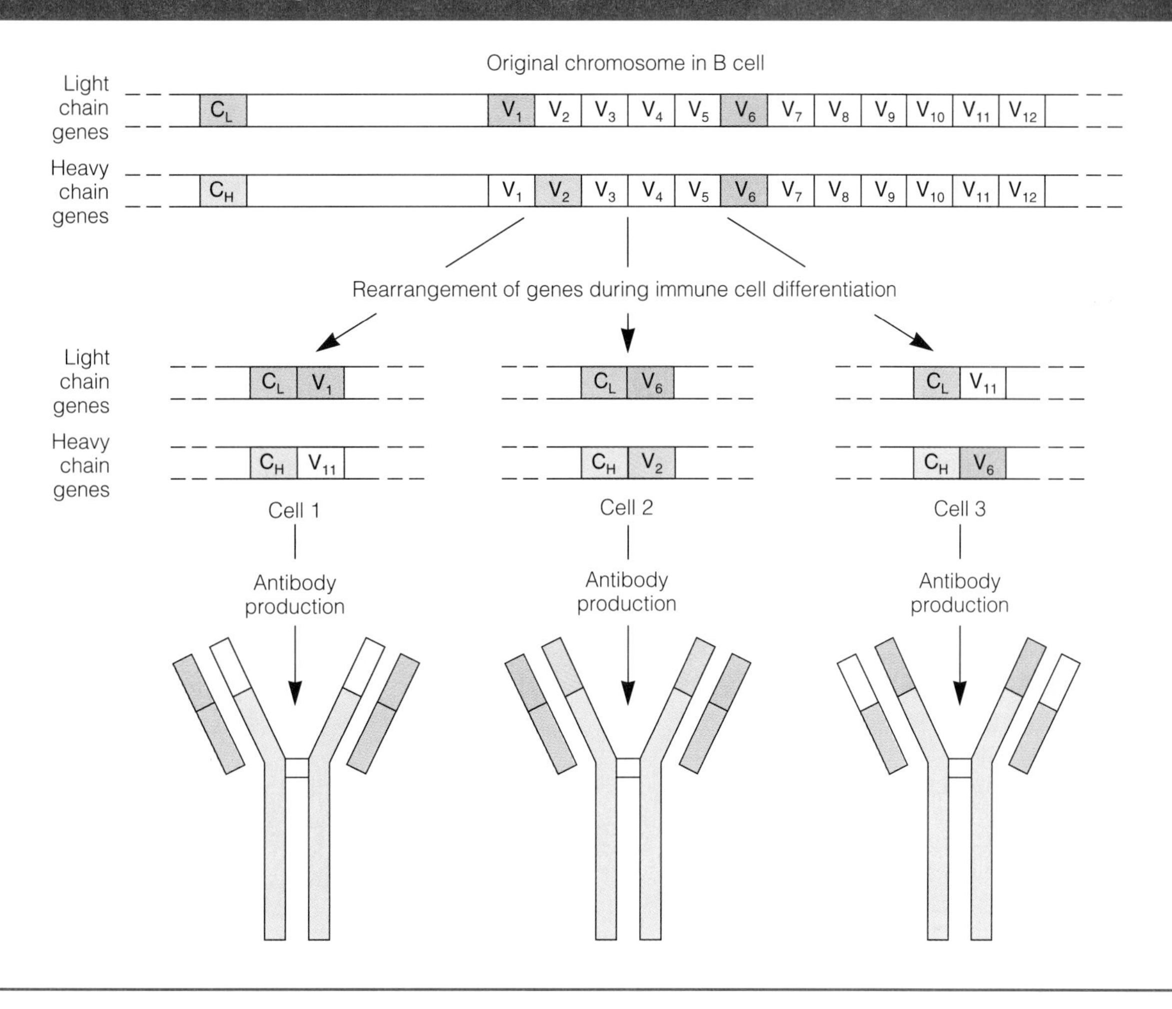

eliminate it from the body. One way the immune system reacts to antigen-antibody complexes is to physically hinder some antigens from destroying or injuring the body's cells. For example, by combining with bacterial toxins, antibodies can prevent these harmful chemicals from interacting with the body's cells. The neutralized toxin is eventually destroyed by phagocytes. Because of its Y-like structure, antibodies have two antigen-binding sites and can bind to more than one antigen at a time. Antigen-antibody complexes can be cross-linked into large clumps. When this cross-linking involves cell-bound antigens, the process causes clumping, or **agglutination** (uh-gloo′-tih-na′-shun), of the foreign cells. These clumps can then be phagocytized.

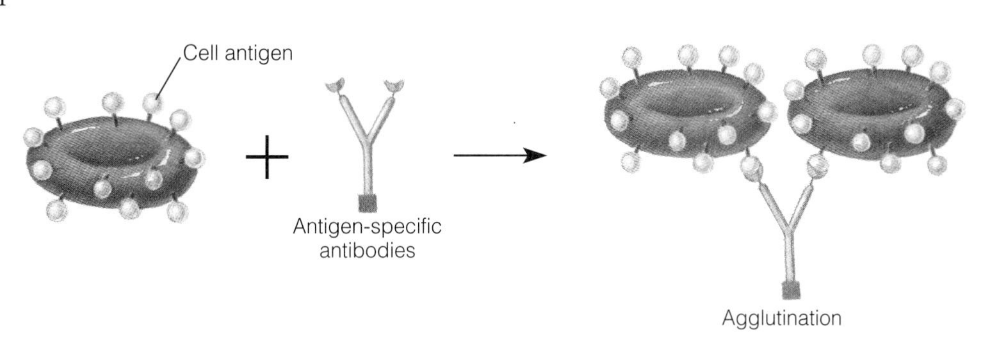

The second way the immune system reacts to antigen-antibody complexes is by enhancing nonspecific immune responses already begun by invaders. Depending on the specific type of antibody that binds to the foreign antigen, one of three nonspecific immune responses may begin.

The first method of destruction begins when an antibody binds with a foreign antigen, activating the complement system, which results in the lysing of the membrane of the invader. The second method of destruction involves enhancing phagocytosis. The constant tail portion of the antibody involved in the

Biology in Action ANTIBIOTICS

Antibiotics are chemical compounds that destroy or interfere with the development of bacteria. Until World War II, infectious diseases caused by bacteria were the leading cause of death in the United States and around the world. The antibiotic penicillin, first used in 1942, inhibits the ability of bacteria to make cell walls during replication. Other antibiotics, such as streptomycin, interfere with protein synthesis in bacteria. Since their discovery in 1928, dozens of antibiotics have been developed to fight various bacterial infections.

By 1969 the medical community was confidently predicting that most infectious diseases caused by bacteria would soon be eliminated. Unfortunately the prediction has proven inaccurate. Today every disease-causing bacterium has versions, or **strains,** that are able to resist at least one of more than 100 antibiotics used by the medical profession. Bacteria may be the most primitive organisms on Earth, but they are wise to Darwin's ideas of natural selection and evolution. Bacteria develop resistance to antibiotics in order to survive and pass the resistance on to future generations.

Bacteria multiply primarily by asexual reproduction, meaning that they produce identical copies, or clones, of themselves. For example, a single bacterium can produce almost 17,000,000 offspring within 24 hours. Occasionally during duplication gene mutations occur. One of these mutations may make the bacterium immune to the antibiotic to which it is exposed. Although the antibiotic kills most of the bacteria, the mutant bacterium survives to produce a next generation, which is now antibiotic resistant.

If a patient has a bacterial infection that is resistant to one antibiotic, it takes the physician additional time to find an antibiotic that will work. In some cases the additional time is sufficient for the infection to overcome the patient. In 1992, 13,000 hospital patients died from bacterial infections that resisted antibiotic treatment. Today 40% of all infections caused by *staphylococcus,* the bacteria responsible for wound infection, "strep throat" and other infections, are resistant to every antibiotic but one, vancomycin.

While bacteria are gaining strength through evolutionary means, doctors and patients seem to have made it easier for the bacteria to adapt to antibiotics. Antibiotics are useless against viral infections like colds and flus, but patients often demand a "quick fix" and insist on antibiotics for any illness. Doctors often feel bound to write prescriptions for anyone who looks sick. Overuse of antibiotics makes it that much easier for resistance to spread. Farmers are also guilty of misusing antibiotics. Farm animals are given antibiotics to treat infections, but the main reason farmers like the drugs is that they make the animals grow faster. Resistant strains emerge in animals just as they do in humans, and some of these strains are passed to humans through undercooked meat. Milk is allowed to contain a concentration of about 80 different antibiotics, and these are passed on to humans with every swallow.

Instead of trying to discover new antibiotics or modify existing ones, recent research efforts have focused on producing drugs that alter bacterial resistance to environmental changes. For example, drugs might make bacteria more vulnerable to temperature changes or pH fluctuations. Other drugs might act as a decoy by binding to the bacterium's enzyme that makes it resistant to a particular antibiotic. This would allow the antibiotic to then attack the bacterium.

Regardless of what new solutions are proposed, the race against disease-causing bacteria is ongoing. The hope is that medical research can stay ahead.

antibody-antigen complex is able to bind with receptors on the phagocyte's surface, ensuring that the foreign invader does not escape from the phagocyte before it can engulf and destroy it. The third method of destruction initiated by antibodies involves activating T lymphocytes, which are cells that are induced to attack an antigen-antibody complex when special receptors on their surface membranes encounter the constant tail portion of antibodies. The cell membrane to which the antigen-antibody complex is attached is lysed, destroying the cell. We discuss the function of T lymphocytes in more detail in the cell-mediated immunity section.

T Lymphocytes: Cell-Mediated Immunity

T cells are responsible for cell-mediated immunity, meaning the T cells themselves specifically bind with antigens and destroy them. T cells, like B cells, are activated by antigens binding to their membrane surface receptors. The tremendous diversity of T-cell surface receptors is thought to occur in the same way as in B-cell surface receptors, except that gene fragments at a different chromosomal site control T-cell surface receptors. There are four different types of T cells: T-cytotoxic or T-killer (T_k) cells, T-helper (T_h) cells, T-memory (T_m) cells, and T-suppressor (T_s) cells. All four types look alike but can be distinguished by their functions.

In contrast to B cells, T cells are unable to directly recognize foreign antigens carried in blood or lymph. They must first be activated. T cells are activated by a foreign antigen only when it is on the surface of a cell that also carries a marker of the individual's own identity. In other words, T-cell activation involves a simultaneous recognition of nonself, the antigen, and self, an MHC protein marker. The cell important in presenting this double-activation signal to T cells is the monocyte-derived macrophage.

The antigens of the foreign material engulfed by macrophages are digested, and a small part of the antigen is displayed on the surface of the macrophage's cell membrane, next to its MHC marker. A T cell whose surface receptor can bind with this particular antigen–MHC marker combination, called a **self-antiself complex,** is activated. The T cell then undergoes mitosis and produces identical copies, or clones, of itself (Figure 14-6). Both T-cytotoxic, or killer (T_k) cells, and T-helper (T_h) cells are activated in this way.

T-cytotoxic, or **T-killer,** cells bind to cells carrying the activating foreign antigen and directly kill them. Once they have destroyed their target, they move onto other cells bearing the same foreign antigen.

An activated T cell may differentiate into **T-helper cells.** The major function of T-helper cells is to chemically stimulate the proliferation of other T cells and B cells that have already become bound to antigen. T-helper cells release an array of substances into the surrounding extracellular fluid called **lymphokines** (lim'fo-kinz). These are chemical substances used to communicate between cells in the immune system. Some lymphokines increase macrophage activities, whereas others stimulate activated B cells to rapidly divide and differentiate. Still other lymphokines stimulate other T cells in the area to divide more rapidly. Without T-helper cells there is no adequate immune response to a foreign antigen. As we discuss later in this chapter, the virus that causes **acquired immune deficiency syndrome (AIDS)** infects and destroys the immune system's T-helper cells. When an activated T-helper cell divides, it produces identical copies of itself. Some clones, however, become T-memory cells and some become T-supressor cells.

An activated T cell may become a **T-memory** cell, which does not participate in the immune response. Instead, like the B-memory cells, T-memory cells remain in circulation to become activated by a subsequent encounter with the same antigen. These will become T-helper cells and T-cytotoxic cells whenever the same antigen enters the body again.

Finally T cells may divide and produce **T-suppressor** cells. T-suppressor cells slow down and then stop the immune response after the foreign antigen that activated the immune response is eliminated.

Primary Immune Response

A **primary immune response** occurs when the immune system encounters an antigen for the first time. The primary immune response is characterized by a lag period that persists for several days after the antigen first appears. For example, this is the time it takes for B-cell activation when the body is first exposed to an antigen. It is also the time required for the B cell specific to that antigen to proliferate and for the differentiation of the plasma cells into antibody-synthesizing factories. The plasma antibody levels slowly rise, reaching peak levels in about 10 days, and then plateau. If the antigen is eliminated, the antibody level slowly falls to barely detectable levels in the blood.

Secondary Immune Response

The production of T-memory and B-memory cells during a specific immune response provides the immune system with an immunological memory. Subsequent challenges by previously encountered antigens, whether for the second or the thirty-second time, evoke a **secondary immune response** that

Figure 14-6 T-cell activation

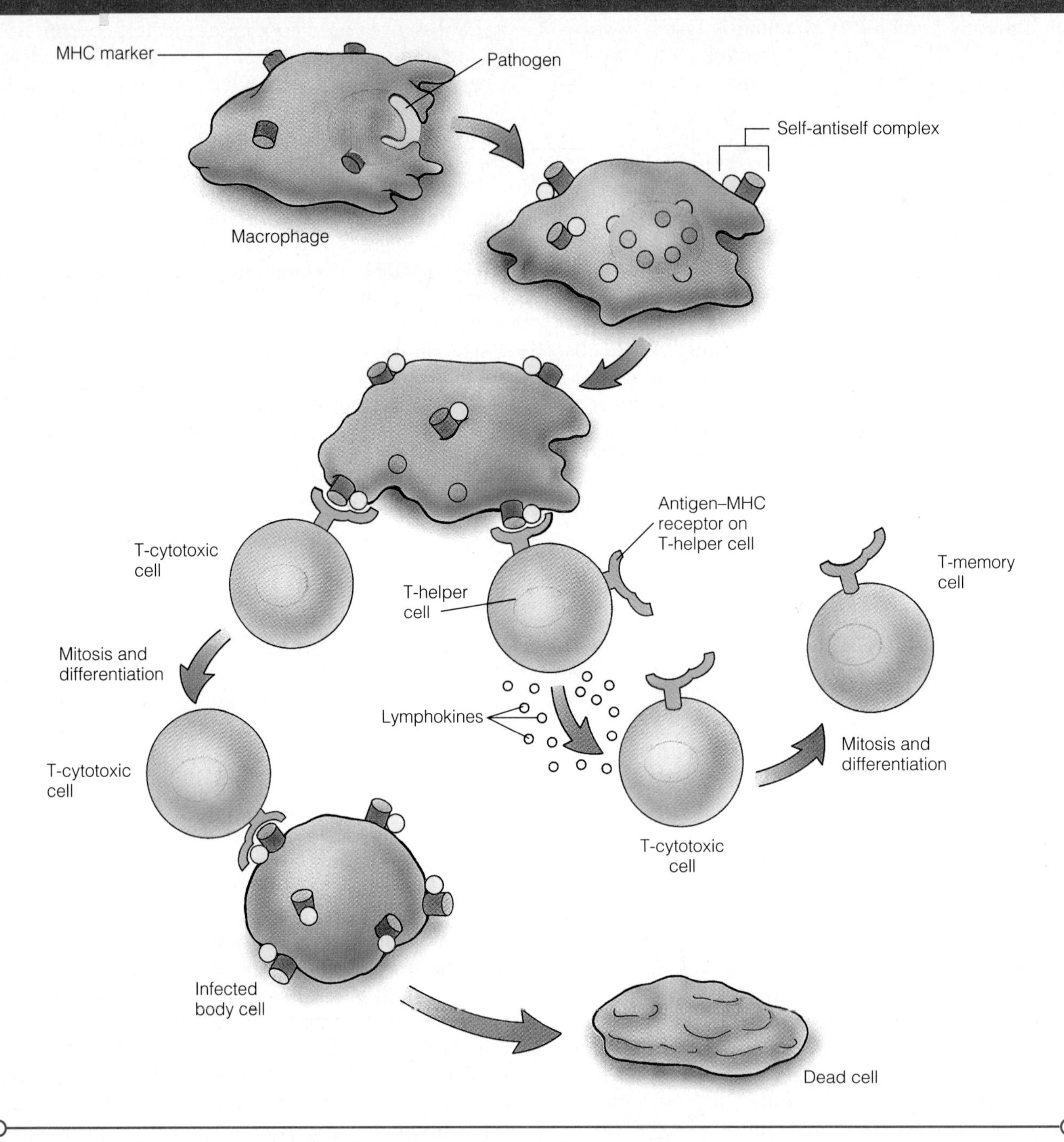

Activated T cells become T-cytotoxic (T-killer) cells or T-helper cells. The T-helper cells secrete lymphokines, which have a variety of effects on other T and B cells.

is faster, more prolonged, and more effective than the first because of the B-memory cells created during the primary immune response. The levels of specific antibody also fall more slowly than after the primary immune response.

This basic immune system behavior makes the development of **vaccines** (vak'senz) possible. **Vaccines** contain either weakened or dead disease-causing microorganisms that cause a primary immune response without causing the actual disease. Vaccines are usually injected into the bloodstream. After the primary immune response several more exposures may be required, sometimes called booster shots because they boost the amount of circulating antibodies to a new level. Antibody levels produced from these secondary exposures may last from months or years to a whole lifetime. Immunological memory is thus firmly established by stimu-

Table 14-1 Immunization Schedule for Children in the United States

AGE	COMBINED INJECTED VACCINES	ORAL VACCINE
1–2 months	Hepatitis B (two injections, one shortly after birth, another before two months old)	
2 months	DPT (diphtheria, whooping cough, tetanus), *Hemophilus influenzae* (pneumonia)	Polio
4 months	DPT, *H. influenzae*	
6 months	DPT, *H. influenzae*	
12–18 months	Measles, mumps, rubella (German measles)	
15 months	Hepatitis B, *H. influenzae*	
18 months	DPT	Polio
4–6 years	DPT	
11–12 years	Measles, mumps, rubella	

lating B-memory cells but sparing the person from most of the signs and symptoms of the disease. Some of the common vaccines include those for measles, pneumonia, smallpox, polio, and diphtheria (Table 14-1).

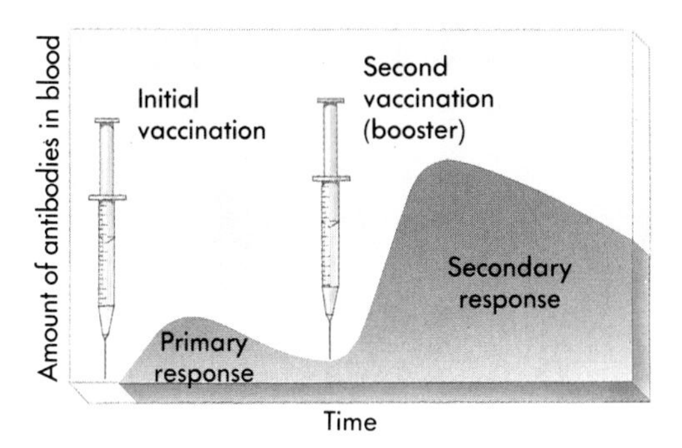

Although described individually, both antibody-mediated and cell-mediated immune responses respond collectively and interact extensively to selectively eliminate foreign antigens. To learn how they work together we describe the action of macrophages, the B and T cells, and their products in response to a first encounter with a foreign antigen.

Suppose you cut your finger, and as a result bacteria enter your body through the wound. At first the bacteria move undetected past many B cells, but eventually they encounter one B cell with a surface receptor capable of binding with the bacterial antigen. When the B cell binds to the antigen, it proliferates and most of the derived cells become plasma cells. These plasma cells begin producing antibodies to the antigen that prompted their production. Some of the clones become memory cells and do not participate in the immune response. Instead, they continue to circulate and become activated if the body is exposed to the same antigen again.

In addition to activating an antibody-mediated response, some of the bacteria may also encounter macrophages. The macrophage engulfs and digests the bacteria, placing part of the bacterial antigen on its membrane surface next to the MHC marker. The macrophage eventually encounters a T cell with a surface receptor that can bind to the macrophage's self-antiself complex. This specific encounter activates the T cell, which begins to proliferate and differentiate. Depending on the surface receptor type, some activated T cells differentiate into T-cytotoxic, or T-killer, cells. These directly attack and lyse cells bearing the antigen that activated their production. Other activated T cells become T-memory cells and do not participate in the immune response. Instead, like the B-memory cells, they remain in circulation to become activated by a subsequent encounter with the same antigen. Other activated T cells differentiate into T-helper cells, whose major function is to chemically stimulate, by releasing lymphokines, the proliferation of other T cells and B cells. Lymphokines also enhance macrophage activity in the invaded area.

Both antibody-mediated and cell-mediated immune responses can be accelerated, decelerated, or stopped. For example, when antigen molecules decrease, fewer B and T cells are activated, and less antibody is produced. There is also another means of regulating the body's immune response. During T-cell activation some T-suppressor cells go into circulation. As the antigenic stimulus for B-cell and T-cell activity decreases, T-suppressor cells begin to secrete a special class of lymphokines that suppress B-cell and T-cell activity. T-suppressor cells moderate and help turn off the normal immune response.

Biology in Action MONOCLONAL ANTIBODIES

Antibodies can now be prepared commercially for use in research, clinical laboratory testing, and the treatment of some forms of cancer. **Monoclonal antibodies,** meaning antibodies produced from a single clone, are prepared in several different ways. One way is to inject an antigen into a laboratory animal, such as a mouse. After a few days the B cells are removed from the mouse. Among the B cells are a few cells that have been stimulated by the antigen. The B cells are mixed with several of a particular kind of cancer cell, called myeloma cells. **Myeloma cells** are cancer cells of the hematopoietic portion of bone marrow. You recall that cancer cells, unlike normal cells, will divide indefinitely. Because of the special environment into which the sensitized B cells and myeloma cells are introduced, they fuse together. The resulting fused cells, called **hybridoma** (hi-brih-do-muh) **cells,** have desirable traits of both parent cells. Like cancer cells, they proliferate indefinitely in the laboratory; like B cells, they produce a single type of antibody. The investigator tests the clones of the hybridomas, one at a time, for reactivity with the original antigen injected into the mouse. The clones that react are then allowed to grow, giving rise to enormous numbers of cells that produce identical antibodies. The antibodies are then extracted from the clone and purified. Monoclonal antibodies can be used to detect antigens that originally prompted their production.

Monoclonal antibodies are used for the diagnosis of such conditions as pregnancy, some sexually transmitted diseases, and hepatitis. Monoclonal antibody tests are also being used for the early diagnosis of colon cancer. However, the most important use of monoclonal antibodies is for cancer therapy. Since 1980 monoclonal antibodies have been used to treat leukemia. When attached to anti-cancer drugs and then injected into a patient's bloodstream, they act as homing devices and seek out the cancerous cells in circulation, delivering the drug only to those abnormal cells.

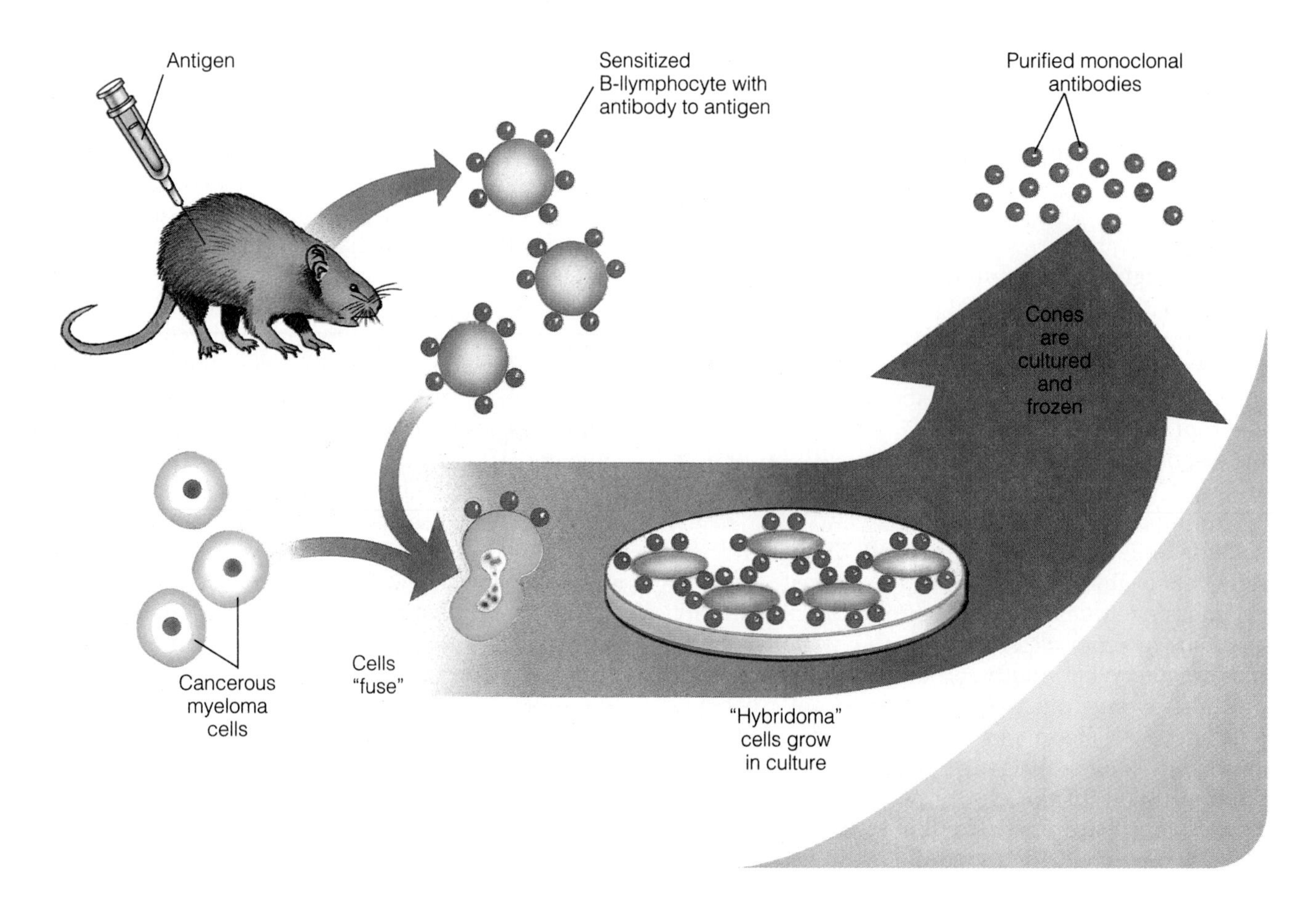

ABO Blood Type and Rh Factor

Besides primary and secondary immunity, certain antibodies occur naturally in the blood, such as the antibodies associated with human blood types. As discussed in Chapter 13 the surface membranes of human red blood cells contain antigens that vary depending on blood type. In the major blood-group system, called the **ABO system,** the red blood cells of persons with type A blood contain A antigens, red blood cells of type B blood contain B antigens, red blood cells of type AB blood contain both A and B antigens, and red blood cells of type O blood do not contain any A or B antigens. Antibodies against red blood cell antigens not on the body's own red blood cells begin to appear in the plasma after about 6 months of age. The plasma of type A blood contains antibodies to type B antigens, type B blood contains antibodies to type A antigens, type AB blood has no antibodies related to the ABO system, and both antibodies to A and B are in type O blood (Table 14-2). High levels of these antibodies are found in all people without ever having been exposed to a different blood type. This is why these antibodies were thought to be naturally occurring antibodies. One line of research suggests that people may be exposed at an early age to small amounts of A-like and B-like antigens associated with common intestinal bacteria. Antibodies produced against these foreign antigens also interact with a nearly identical foreign blood group antigen.

Table 14-2 ABO Blood Types with Corresponding Antibodies

BLOOD TYPE		ANTIBODIES IN CIRCULATION
Genotype	Phenotype	
AA	A	Antibodies to type B
AO		
BB	B	Antibodies to type A
BO		
AB	AB	No antibodies
OO	O	Antibodies to type A and B

Antibody interaction with an antigen bound to a red blood cell may result in agglutination or **hemolysis,** or rupture, of the attacked red blood cells. Agglutination and hemolysis of donor red blood cells by antibodies in the recipient's plasma can cause a serious and sometimes fatal **transfusion reaction.** The agglutinated clumps of the incoming donor cells can plug small blood vessels, impairing circulation and causing tissue damage.

In addition to the ABO system, other red blood cell antigens can cause transfusion reactions. The most important of these is the **Rh factor.** People with the Rh factor are said to have Rh-positive (Rh^+) blood, while those lacking the Rh factor are Rh-negative (Rh^-). In contrast to the ABO system, no naturally occurring antibodies develop against the Rh factor. Rh factor antibodies are produced only by Rh^- people when they are first exposed to the foreign Rh antigen in Rh^+ blood. Rh^+ people never produce antibodies against the Rh factor they possess. Therefore Rh^- people should be given only Rh^- blood, while Rh^+ people can safely receive either Rh^- or Rh^+ blood.

While ABO antibodies cannot cross the placenta, Rh antibodies can. One medical consequence of Rh factor is **erythroblastosis fetalis** (e-rith'ro-blas-to'sis fe-ta'lis), or **hemolytic disease of the newborn** (Figure 14-7). Erythroblastosis fetalis occurs when an Rh^- mother is carrying an Rh^+ fetus. If an Rh^- female becomes pregnant by an Rh^+ male, there is a 50% chance the fetus will also be Rh^+. If fetal blood comes into contact with the maternal circulation, the Rh^+ antigens cause the mother's immune system to attack them. Normally this happens during the birth process, so usually the first pregnancy exposes the mother's immune system to the foreign Rh antigen and the fetus escapes harm. If, however, another pregnancy involves an Rh^+ fetus, the mother's immune system will react much more rapidly against the Rh^+ antigen and will begin to destroy the red blood cells of the fetus. In severe cases of this disorder red blood cells are destroyed, causing anemia, cerebral damage, and death. Fortunately, Rh^- mothers can be treated with an injection of a substance called RhoGAM, which prevents the mother's immune system from forming antibodies to the red blood cells of the fetus.

Immune System Problems

Under certain circumstances the immune system becomes depressed or actually damages the body itself. Such **immunodeficiencies** consist of many genetically determined and acquired abnormalities in the production or function of B cells, T cells, phagocytes, or the complement reaction.

In a congenital disease called **severe combined immunodeficiency disease (SCID),** the affected person is unable to manufacture sufficient numbers of either B or T cells. Therefore the person becomes seriously ill and depleted by infections that begin at about 3 months of age. Some affected children are kept alive and healthy by bone marrow transplants, which provide normal stem cells. You may recall "the boy in the bubble" named David who spent 12

Figure 14-7 Rh factor and maternal fetal incompatibility

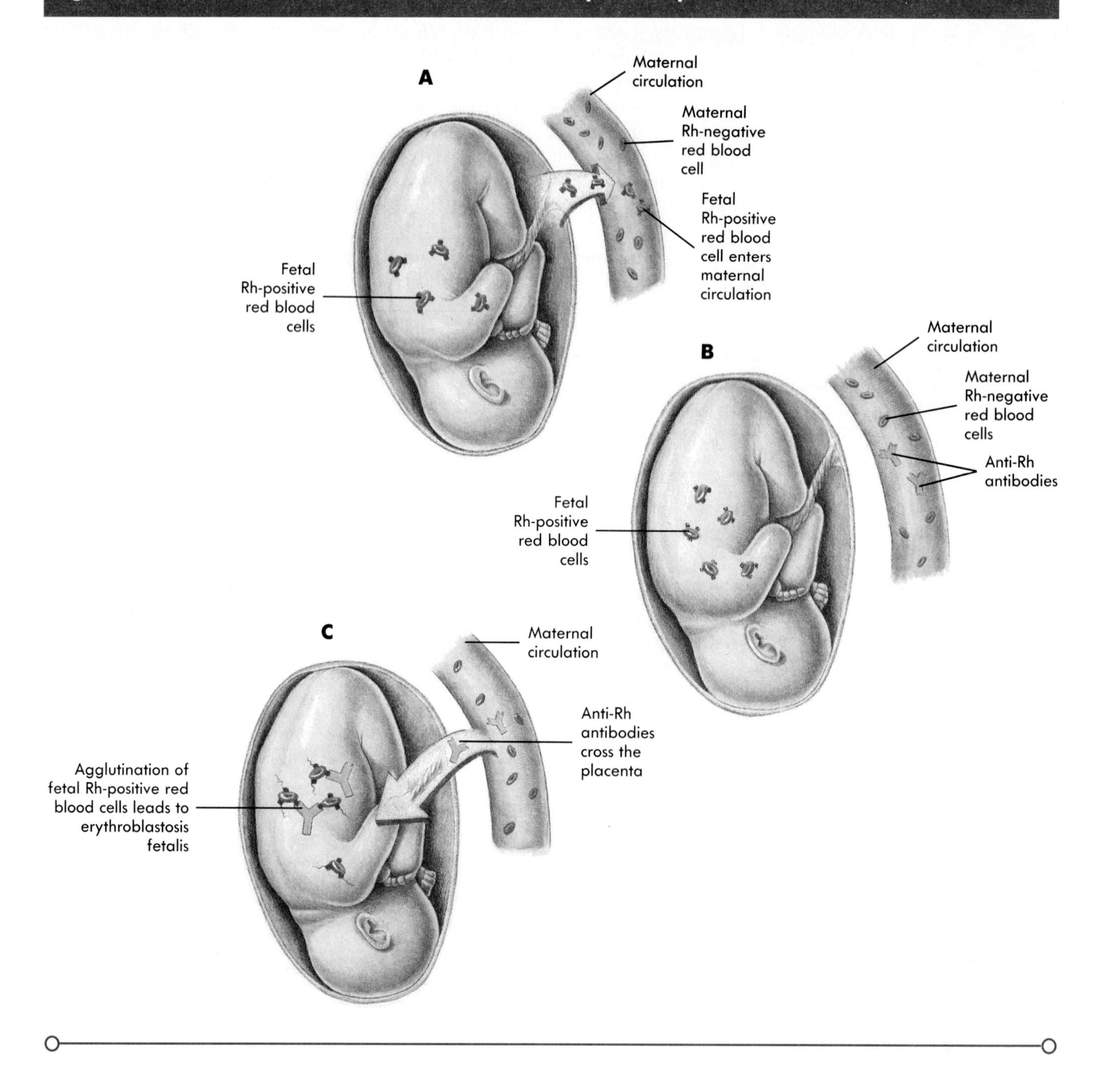

years in a germ-free environment because he had SCID. A treated bone marrow transplant was given by his sister, but it was infected with an Epstein-Barr virus that eventually caused a fatal cancer.

AIDS

The Centers for Disease Control (CDC) located in Atlanta, Georgia, issues a weekly newsletter that monitors the nation's health. In the issue for June 5, 1981, a new listing appeared, a forewarning of what has become one of the deadliest infectious diseases of our time. This new disease was called **acquired immunodeficiency syndrome,** or **AIDS.**

The symptoms of AIDS include weight loss, recurring fevers, repeated viral and fungal infections, and rare forms of pneumonia and cancer. The symptoms are similar to those seen in genetically determined immunodeficiency disease, such as SCID, and this provided the first clue to its possible cause. However, unlike SCID, AIDS is acquired dur-

Special Feature HIV AND AIDS

This feature article was written by Lucy Bradley-Springer, Ph.D., R.N. Co-Director, New Mexico AIDS Education and Training Center University of New Mexico School of Medicine

In 1981 public health officials in the United States recognized a new disease that eventually became known as the acquired immunodeficiency syndrome (AIDS). We now know that AIDS is the final phase of a chronic, immune-function disorder caused by the human immunodeficiency virus (HIV).

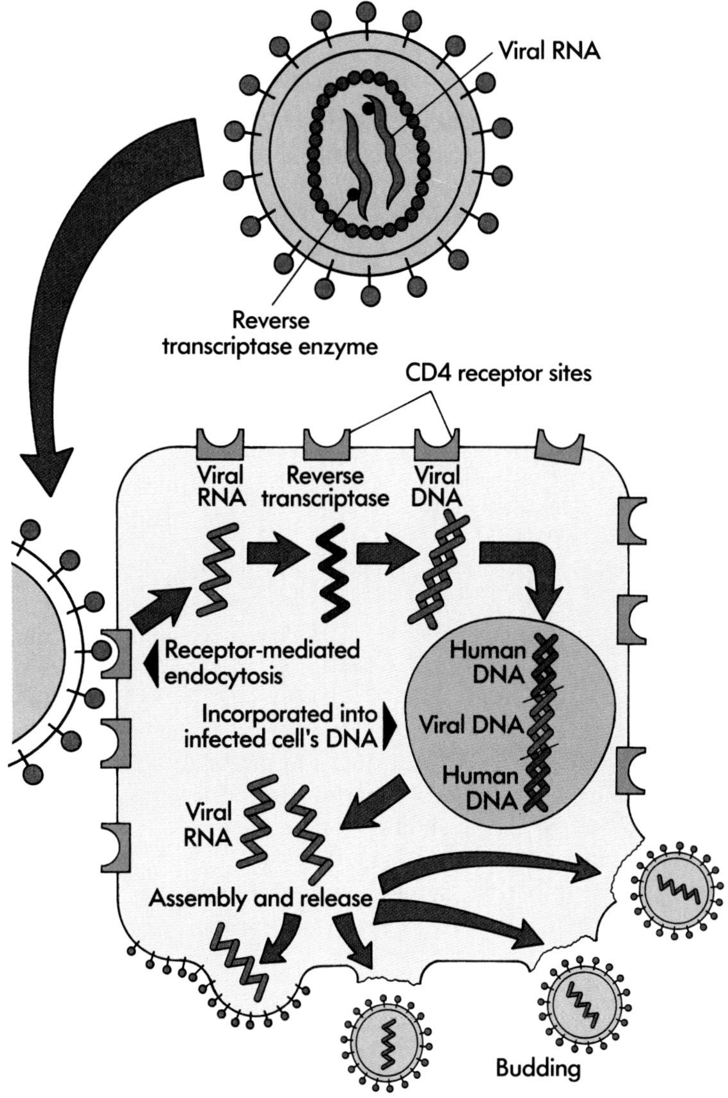

HIV and What It Does

HIV is an RNA virus that was discovered in 1983. RNA viruses are also known as retroviruses. Like all viruses, HIV is an obligate parasite: it cannot survive and replicate unless it is inside a living cell.

HIV can enter a cell by binding to specific (CD4) receptor sites on the cell's surface. Once the virus' RNA enters a cell, it is transcribed into a single strand of viral DNA with the assistance of an enzyme supplied by the virus, called reverse transcriptase. This strand then replicates itself and becomes double-stranded viral DNA. At this point the viral DNA can splice itself into genetic material in the nucleus of the host cell, becoming a permanent part of the cell's genetic structure. Since all the cell's genetic materials are replicated during cellular division, all the cell's daughter cells will also be infected. Also, since the genetic material now contains viral DNA, the cell's genetic codes can direct the cell to make more HIV.

Although HIV can infect several types of human cells, immune dysfunction results predominantly from the destruction of T-helper lymphocytes. These lymphocytes are targeted because they have more CD4 receptor sites on their surfaces than other cells. The T-helper lymphocyte is a kind of white blood cell that plays a pivotal role in the ability of the immune system to recognize and defend against foreign invaders. When HIV replicates within an infected T-helper cell, newly formed viruses bud out from the cell's membrane.

This process not only releases new HIV to infect other cells, it also eventually kills the host cell by punching holes in the cell membrane.

Humans normally have 800 to 1200 T-helper lymphocytes per cubic millimeter (mm^3) of blood. In HIV infection a point is eventually reached where so many T-helper cells are destroyed that there are not enough to direct and coordinate immune responses. This sets the stage for the development of life-threatening illnesses.

The Spectrum of HIV Infection

The typical course of HIV infection follows the pattern shown below.

Early Infection

Initial infection with HIV causes an immune reaction during which HIV-specific antibodies are produced. These antibodies can be measured in a person's blood at a median of 2 months after infection. Some people will have measurable HIV antibody as early as 3 weeks after infection, while others may take as long as 6 months. The most accurate tests that we currently have for HIV detect these antibodies and not the virus itself. This means that some people can be infected (and able to transmit HIV to others) for as long as 6 months before the test would show that they were infected. In some people a flulike syndrome of fever, nausea, and diarrhea accompanies the development of HIV antibodies.

The median time between HIV infection and a diagnosis of AIDS is 10 years. HIV-infected people remain generally healthy during this time, but vague symptoms, including fatigue, headaches, low-grade fevers and night sweats, may occur. Because the symptoms are vague, people may not be aware that they are infected during this asymptomatic phase. During this time infected people are usually able to continue their regular activities, which may include risky sexual and drug-using behaviors. This creates a public health problem because infected people can transmit HIV to others even if they have no symptoms.

Early Symptomatic Disease

Toward the end of the asymptomatic phase and before a diagnosis of AIDS, early sympto-

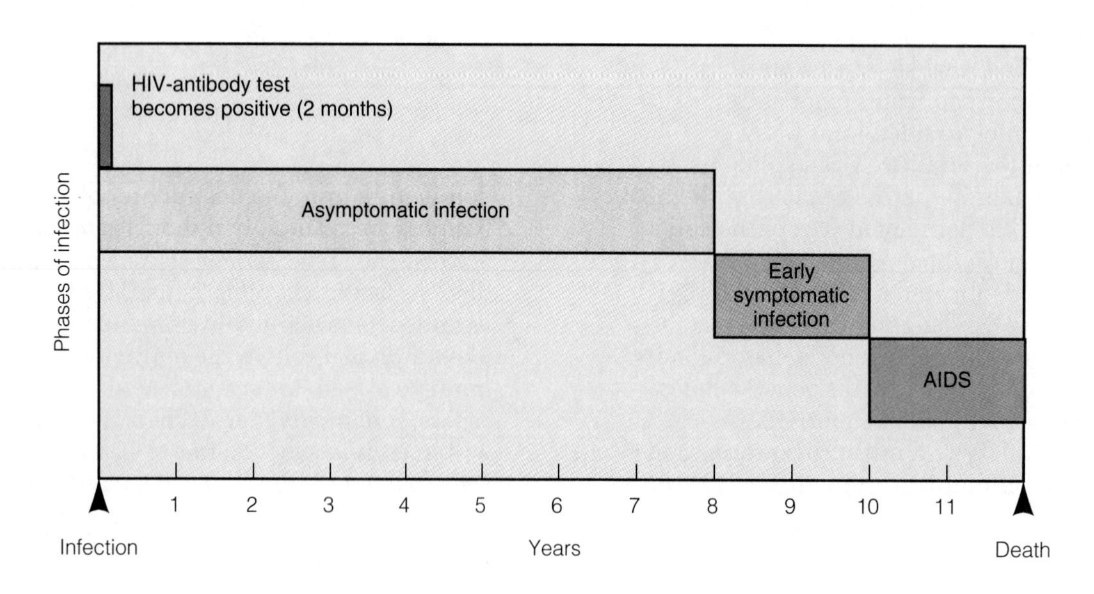

matic disease develops. Early symptoms can include persistent fevers, night sweats, and fatigue. These may be severe enough to interrupt normal routines. Swollen lymph glands and localized infections may also occur. The most common problem at this point is thrush, a fungal infection of the mouth that rarely causes problems in healthy adults. Other infections that can occur at this time include shingles (caused by the *Varicella zoster* virus), persistent vaginal yeast infections, and outbreaks of oral or genital herpes.

AIDS

A diagnosis of AIDS cannot be made until the HIV-infected person meets criteria established by the Centers for Disease Control (CDC). These criteria are more likely to occur when the immune system becomes severely compromised. AIDS is diagnosed when a person with HIV develops at least one of these additional conditions:

1. The T-helper lymphocyte count drops below $500/mm^3$.
2. One of 24 specific opportunistic diseases develops. Opportunistic diseases are diseases, including cancers and infections, that would not occur if the immune system were working properly. The most common HIV-related opportunistic diseases are *Pneumocystis carinii* pneumonia (PCP) and Kaposi's sarcoma (KS), a cancer of the blood vessels.
3. Wasting syndrome occurs. Wasting syndrome is defined as a loss of 10% or more of ideal body mass.
4. Dementia develops. Symptoms of AIDS-related dementia include decreased ability to concentrate, apathy, depression, social withdrawal, and slowed response rates. AIDS dementia can progress to coma.

The median time for survival after a diagnosis of AIDS is 2 years, but this varies greatly. Some people with AIDS live for 6 or more years, while others survive for only a few months. Advances in the treatment and diagnosis of HIV infection have increased survival times, but AIDS fatality rates remain high.

The Impact of HIV

By October 1993 more than 339,000 cases of AIDS and more than 201,000 AIDS-related deaths had been reported in the United States. HIV is estimated to have infected at least one million people in this country. In the United States the fastest growing groups of people with HIV and AIDS are women and adolescents. Of the more than 40,000 cases of AIDS that have been reported in women, 36% were reported between October 1992 and October 1993. During that same time 27% of the more than 12,000 AIDS cases in people aged 13 to 25 were reported.[1] The face of AIDS in the United States is not only changing along the lines of gender and age, it is also becoming an increasing problem for people of color and people who live in poverty. Globally HIV is even more devastating, with a worldwide estimate of more than 10 million infected people.

Transmission

HIV is transmitted from human to human through infected blood, semen, vaginal secretions and breast milk. If these infected fluids are introduced into an uninfected person's body, the potential for transmission occurs. Sexual contact with an HIV-infected partner is the most common method of transmission. Sexual activity provides an opportunity for contact with semen, vaginal secretions, and blood. Although male homosexuals were the initial targets of HIV in this country, heterosexual transmission is becoming more prevalent and is the most common method of infection for women.

Sharing equipment to inject drugs is a major means of transmission to both sexes in many large metropolitan areas and is becoming more common in smaller cities and rural areas. HIV has also been transmitted in blood products. This is now rare in the United States because of a strict program of donor screening, blood testing, and heat or chemical treatment of hemophilia clotting-factor products.

Transmission from HIV-infected mothers to infants can occur during pregnancy, at the time of delivery, or after birth through breast feeding. Studies have found that 9% to 30% of infants born to HIV-infected women will be born infected. This means that 70% to 91% of these infants will *not* be infected. *All* infants born to HIV-infected mothers will be positive on the HIV-antibody test, however, because maternal antibodies cross the placental barrier. It may take 15 months before it is known whether an infant is infected with HIV. Breast feeding is a rare but well-documented HIV-transmission risk. Because of this, HIV-infected mothers in the United States are encouraged not to breast feed their infants.

Prevention

Although advances are being made, there is still no evidence that a cure or a vaccine will be developed soon. The only defense against HIV is prevention through education about risky behaviors and risk-reduction practices. The good news is that behaviors that create a risk for transmission of HIV have been clearly identified and are avoidable.

Drug Use

Risks for HIV infection related to injecting drugs can be *eliminated* by not using drugs or by using only equipment that is never shared. Risk *reduction* during drug use involves cleaning equipment between use with bleach and water. Persons who use drugs of any kind (including alcohol and drugs that are smoked, snorted, or swallowed rather than injected) may put themselves at risk of sexual transmission if they make unwise sexual decisions while under the influence. HIV infection is only one of many health risks associated with drug and alcohol use.

Sexual Activity

Elimination of HIV risks related to sexual activity includes abstinence; sexual activity in a mutually monogamous relationship in which neither partner is infected; or sexual activity in which the penis, vagina, mouth, or rectum does not touch someone else's penis, vagina, mouth, or rectum. Risk *reduction* includes correct and consistent use of condoms, decreasing the number of sex partners, and refraining from unprotected anal intercourse, the most risky form of sexual activity.

Conclusion

By all accounts the global HIV epidemic will continue for the foreseeable future. People who are already infected will be using health-care services more frequently over the next 10 to 15 years. Currently education and prevention of new cases are the only ways to decrease the devastation caused by HIV. Unfortunately these efforts are often hampered by social, cultural, and governmental policies that restrict education about risk-reduction measures. If we cannot overcome these barriers, the HIV epidemic will continue to expand, causing human suffering, premature death, and strains on limited health-care resources.

---

[1]CDC. 1993, October. HIV/AIDS surveillance, third quarter edition. U.S. Department of Health and Human Services.

ing the person's lifetime. In 1983 AIDS was discovered to be caused by a human retrovirus called the **human immunodeficiency virus,** or **HIV.** The AIDS retrovirus selectively infects the immune system's T-helper cells. Like other retroviruses HIV has RNA as its genetic material rather than DNA. After infecting the T-helper cell, the RNA is copied into DNA by a viral enzyme called reverse transcriptase. The newly copied DNA moves from the cytoplasm into the infected cell's nucleus to become part of the cell's chromosomes. At this stage the viral DNA can remain inactive for several years. When the viral genes are activated, they are transcribed by the host T-helper cell, the proteins are synthesized, and new viruses are assembled in the cell's cytoplasm. The replicated viruses move to the cell membrane and bud off, causing perforations in the T-helper cell's membrane resulting in its collapse and death.

Allergies

Hypersensitivities and **allergy,** meaning altered reaction, are synonyms for an antigen-induced state that results in abnormally intense immune responses to an antigen. In an allergic reaction the immune system itself causes tissue damage as it fights off an antigen "threat" that is normally harmless, such as flower pollen, animal hair, or even food. The term **allergen** is often used to distinguish the type of antigen that causes an allergic reaction from foreign antigens that stimulate an essentially protective immune response.

Some allergies are triggered when allergen molecules combine with receptors attached to mast cells. Mast cells are found throughout the body but are primarily in the tissues of the skin, respiratory system, and blood vessels. When stimulated, mast cells release histamine into their surroundings. The body first becomes **sensitized** to the allergen when the plasma B cells produce antibodies against the normally harmless allergen. Then the antibodies attach to the surface membranes of the mast cells. When exposed again, the allergen binds to the antibodies attached to the mast cells, releasing histamine into the surrounding tissue. In addition to increasing capillary-wall permeability, histamine also causes mucus secretion and constricts the smooth-muscle tissue surrounding airway passages in the lungs. The signs and symptoms of local allergic reactions are related to the specific tissues affected. For example, redness, swelling, and hives are typical of skin reactions. Excessive mucus production and itching occur when the nasal passages and other mucous-membrane-lined cavities are affected. Wheezing and labored breathing are typical of constricted air passages caused by lung reactions to an allergen.

Injections of the allergen usually prevent allergic reactions by causing the body to build up high quantities of antibodies. These antibodies combine with allergens received from the environment before they have a chance to reach the antibodies on the mast cells.

Tissue Rejection

Human tissue or organ transplants are becoming increasingly common. A significant obstacle with these transplants is that the recipient's immune system often attempts to reject the transplanted material. There are, however, several different classes of human tissue and organ transplants, each with different rates of tissue rejection. **Autografts** are tissue grafts taken from one part of the body and transplanted to another part of the same body. Skin grafts are usually done in this way. The second class of transplants are **isografts,** which are grafts donated by genetically identical individuals, identical twins. The final and most common class of transplants are **allografts,** which are grafts and transplants involving people who are not genetically identical. However, their degree of biological relatedness does have a marked effect on the likelihood of acceptance or rejection by the graft recipient. The closer the degree of biological relatedness, the more likely the graft will be accepted.

Most graft rejections are caused by cell-mediated immune responses. Therefore, several steps must be taken before attempting an allograft, such as a kidney transplant. The recipient and donor tissues are typed to determine the extent to which their antigens match. Each person has many cell-surface receptors, or markers, including MHC antigens, ABO blood type, and Rh factor. At least 75% of the markers on the recipient's cells must match the donor's cells. Because cell-surface receptors are genetically determined, the probability of matching improves with the degree of biological relatedness between the recipient and the donor. For example, a parent and offspring or two siblings are more likely to match than two unrelated people picked at random.

Following surgery the recipient is treated with **immunosuppressive drugs.** One drug with the fewest number of side effects, called cyclosporine, suppresses T-helper and T-killer cells without disrupting antibody production of B cells to other foreign antigens. Allograft recipients must take immunosuppressive drugs for life. They must therefore be closely monitored for infections and diseases.

Autoimmune Diseases

Occasionally the immune system begins to recognize its own antigens and produces antibodies against its own self-antigens, resulting in **autoimmune diseases.** Although many different factors can trigger this abnormal response, two events are the most common.

First, proteins previously recognized as self may suddenly be recognized as foreign if they are distorted or altered. Antibodies directed against foreign antigens may be damaged or partially degraded by enzymes released during the immune response. As a result the body may not recognize its own antibody molecules. For example, in some forms of rheumatoid arthritis the initial inflammatory response to a viral or bacterial infection in a joint cavity causes a partial breakdown of the antibodies formed against the invaders. The altered antibodies are then attacked by a different class of antibodies, activating the complement reaction and resulting in local tissue damage. The second event occurs as antibodies made against a foreign antigen sometimes cross-react with a self-antigen with a similar structural configuration. Generally the self-antigen is attacked less aggressively because it is not completely identical to the foreign antigen. However, if the invader causes repeated infections, the damage to body tissues is greatly enhanced. Recurring infection by certain types of bacteria may lead to **rheumatic fever.** It is an autoimmune disease that causes swollen joints, rashes, and kidney and heart damage long after the bacterial infection has been reduced.

CHAPTER REVIEW

SUMMARY

1. The specific immune response system is capable of selectively eliminating a particular foreign material from the body. The foreign material is usually destroyed and eliminated by an enhancement of nonspecific immune responses.
2. Two types of lymphocytes derived from stem cells, called B and T lymphocytes, are responsible for the activities of the specific immune response system. B cells control antibody-mediated immunity. T cells control cell-mediated specific immune responses.
3. Both B and T cells recognize the body's own cells by protein self-markers located on the cell membrane surface. The self-recognition markers are specified by genes of the major histocompatibility complex. The ability of B and T cells to respond to specific foreign antigens is a result of where they are prepared and mature in the body.
4. Any substance not recognized as self and that triggers a specific immune response against itself is called an antigen. Antigens can be proteins, modified proteins, or polysaccaharides either attached to cell membrane surfaces or free.
5. B cells produce antibodies to foreign antigens, binding to them to form an antigen-antibody complex. After the antigens are bound by a specific antibody, they can then be eliminated by other components of the immune system.
6. After being presented a foreign antigen by the monocyte-derived macrophages of the immune system, T cells are capable of directly attacking and destroying it. After a macrophage ingests a foreign material, it places a part of the foreign antigen on its cell surface, next to its own MHC marker. This creates a self-antiself complex that the T cell recognizes.

FILL-IN-THE-BLANK QUESTIONS

1. Any substance capable of stimulating an immune response is called a(n) ______.
2. Antibodies are produced only by the ______ cells, or their descendants, of the immune system.
3. Antibodies are released into the blood, where they are known as the plasma proteins called gamma globulins, or ______.
4. All antibodies are composed of ______ interconnected amino acid chains consisting of ______ long, ______ chains, and ______ short, ______ chains.
5. When antibodies bind to antigens they form a(n) ______-______ complex.
6. T cells are activated when a(n) ______ presents a(n) ______-______ complex to the T cells.
7. ______ are released by activated T cells and are generally involved in enhancing ______-mediated immune responses.
8. The activation and proliferation of T and B cells constitute the ______ immune response.
9. Both T and B memory cells provide a(n) ______ ______ for the immune system.
10. ______ are abnormally intense immune responses to a normally harmless antigen.
11. HIV is a (n) ______ virus and also a (n) ______ parosite.
12. HIV primarily infects ______ - ______ cells of the human immune system.

SHORT-ANSWER QUESTIONS

1. Describe how antibody diversity is generated by only a small number of genes in the lymphocytes.
2. Describe the roles of T-helper, T-cytotoxic, and T-suppressor cells in cell-mediated immune responses.
3. Compare and contrast primary immune and secondary immune responses.
4. Describe the nature of the antigens and antibodies of the ABO blood system.
5. What are monoclonal antibodies?
6. What causes hemolytic disease of the newborn?
7. What is an allergy?
8. Describe the usual course of HIV infection.
9. What are the modes of HIV transmission from person to person?

VOCABULARY REVIEW

ABO system—*p. 273*
allergen—*p. 279*
antigen—*p. 264*
B cell—*p. 265*
complement system—*p. 263*
inflammatory response—*p. 261*
interferon—*p. 263*
memory cell—*p. 265*
nonspecific immune response—*p. 261*
plasma cell—*p. 265*
specific immune response—*p. 264*
T-cytotoxic cell—*p. 269*
T-suppressor cell—*p. 269*
vaccine—*p. 270*

Chapter Fifteen

RESPIRATION

For Review

Here are some important terms and concepts that you will encounter in this chapter. If you are not familiar with them, you should review them before proceeding.

Cellular respiration
(page 51)

Diffusion
(page 40)

Hemoglobin
(page 250)

OBJECTIVES

After reading this chapter you should be able to:

1. *Describe the events in the respiration process.*
2. *Name the major organs of the respiratory system and describe the function of each in the process of respiration.*
3. *Differentiate between the respiratory movements that occur during inspiration and expiration.*
4. *Describe the partial pressure gradients for oxygen and carbon dioxide.*
5. *Describe the two ways pulmonary ventilation is controlled.*
6. *Describe how oxygen and carbon dioxide are transported in the blood.*

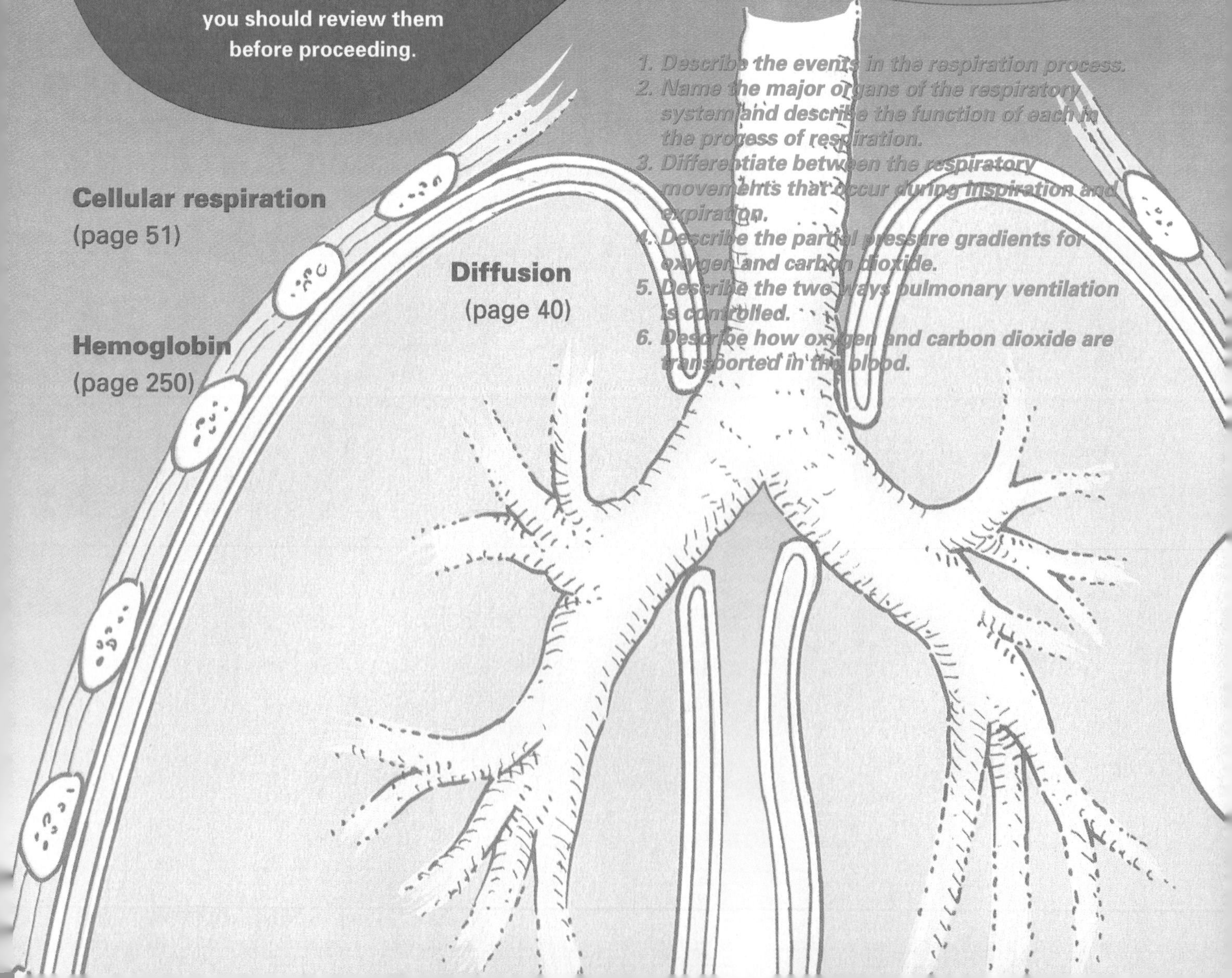

The energy stored in the food we consume becomes available to our bodies through the process of aerobic cellular metabolism, which requires oxygen and produces carbon dioxide as a waste product. The respiratory system acquires oxygen from the atmosphere and eliminates carbon dioxide from the body. The driving force for gas exchange between lungs and blood is simple diffusion. To facilitate gas diffusion from such a large surface area, the pulmonary blood flow each minute equals systemic blood flow to the rest of the body.

Respiration System

The billions of cells that make up the human body require a continuous supply of oxygen to carry out their vital functions. Carbon dioxide is a waste by-product of aerobic cellular respiration. Although it is possible to stop eating food and live for several days, it is not possible to stop breathing for more than several minutes. Most cells can use *anaerobic* ("without oxygen") cellular respiration to produce ATP, but they can do so for only several minutes before lactic acid begins to accumulate in the cell and interfere with other essential cellular functions. Without oxygen, metabolism and energy production eventually cease, ending in cell death.

The respiratory system is a complex of organs and structures that moves air into and out of the body, exchanging oxygen and carbon dioxide between the atmosphere and the blood circulating through the lungs. The process of acquiring oxygen from the external environment and eliminating carbon dioxide from the body is called **respiration.** The primary function of the **respiratory system** is to supply the body with oxygen and dispose of carbon dioxide. Respiration includes the following four events:

1. **Pulmonary ventilation**
   Air must be moved into and out of the lungs for continuous gas exchange. This is also called breathing.
2. **External respiration**
   External respiration is the exchange of oxygen and carbon dioxide between the blood and the air in the lungs.
3. **Internal respiration**
   Internal respiration involves moving oxygen from the blood into the body tissues and moving carbon dioxide, a waste by-product of cellular respiration, from the tissue and back into the blood. The circulatory system accomplishes this using blood as the transporting fluid.
4. **Cellular respiration**
   Cellular respiration releases the energy of the chemical bonds of nutrients, such as glucose, to be used for the cell's metabolic activity.

We discussed cellular respiration in Chapter 3. In this chapter we describe the structure of the respiratory system and the processes of pulmonary ventilation, external respiration, and internal respiration. We also discuss several basic principles governing the physical properties of gases.

Respiratory Organs

The respiratory system consists of the nose, pharynx, larynx, trachea, bronchi, and lungs. During **inspiration,** or inhalation, air is conducted toward the lungs by this series of cavities, tubes, and openings. As air moves along the various air passages toward the lungs, it is filtered, warmed, and humidified (Figure 15-1).

The Nose

The **nose** is the respiratory system's only externally visible part. It has several important functions in respiration. The nose is an airway passage that moistens, warms, and filters the entering air. It is also the site of olfactory, or smell, receptors (see Chapter 9). The nostrils open into two **nasal cavities** separated from one another by the **nasal septum,** a bone and cartilage partition dividing the nostrils. The portions of the nasal cavities just behind the nostrils are lined with skin containing numerous hair follicles. These filter large particles from inspired air. The rest of the nasal cavity is lined with ciliated mucus-secreting cells. Mucus humidifies the air. It also contains an antibacterial enzyme that destroys any bacteria breathed in from the surrounding air. Mucus traps inspired dust, bacteria, and other small particles during inspiration. The cilia of the mucous cells create a gentle wavelike action that moves contaminated mucus to the back of the nasal cavities toward the throat where it can be eliminated. You are usually unaware of this action, but when exposed to cold air, the cilia become sluggish, allowing mucus to accumulate in the nasal cavity. This is largely the reason why you might have a "runny nose" on a crisp, cold day.

The walls of the nasal cavity are somewhat convoluted, enhancing air turbulence in the nasal cavity and also increasing the mucosal surface exposed to the air. The inhaled air swirls and twists, but the

Figure 15-1 Respiratory system

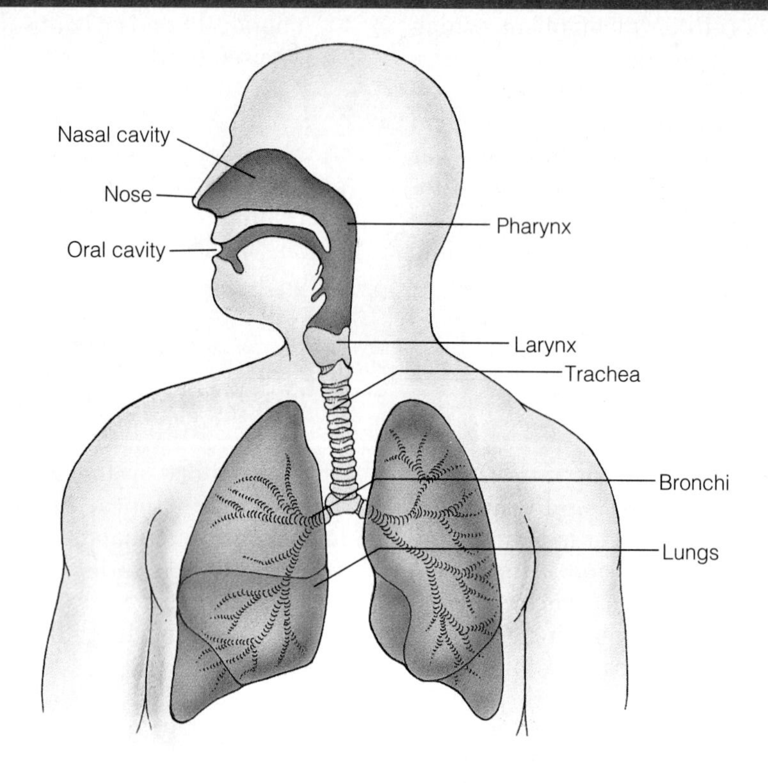

The respiratory system comprises organs and structures that conduct air into and out of the body.

heavier particles tend to strike the mucosa, where they become trapped in the mucus.

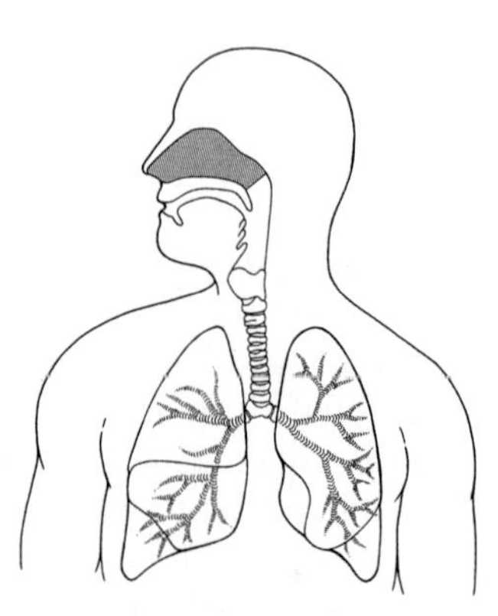

Pharynx

The **pharynx** (fayr′ inks), commonly called the throat, is funnel shaped, about 13 cm, or 5 inches, long and

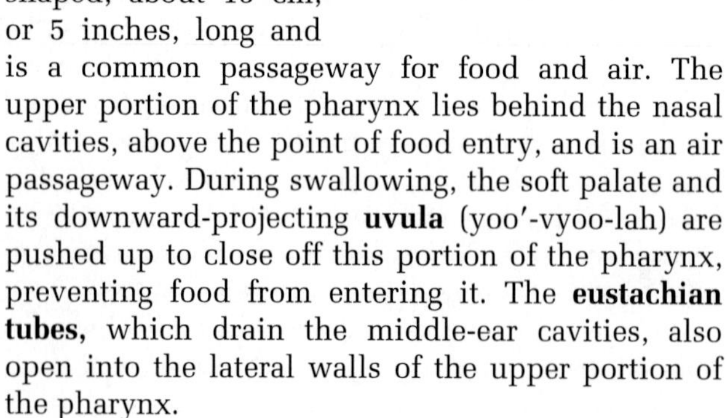

is a common passageway for food and air. The upper portion of the pharynx lies behind the nasal cavities, above the point of food entry, and is an air passageway. During swallowing, the soft palate and its downward-projecting **uvula** (yoo′-vyoo-lah) are pushed up to close off this portion of the pharynx, preventing food from entering it. The **eustachian tubes,** which drain the middle-ear cavities, also open into the lateral walls of the upper portion of the pharynx.

The lower portion of the pharynx diverges into respiratory and digestive pathways. The **esophagus**

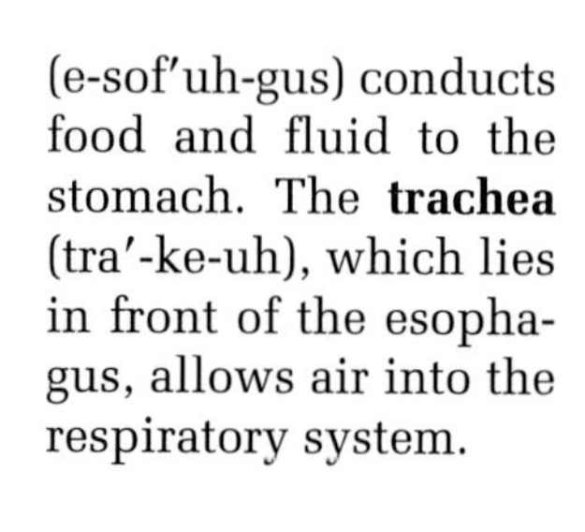

(e-sof′uh-gus) conducts food and fluid to the stomach. The **trachea** (tra′-ke-uh), which lies in front of the esophagus, allows air into the respiratory system.

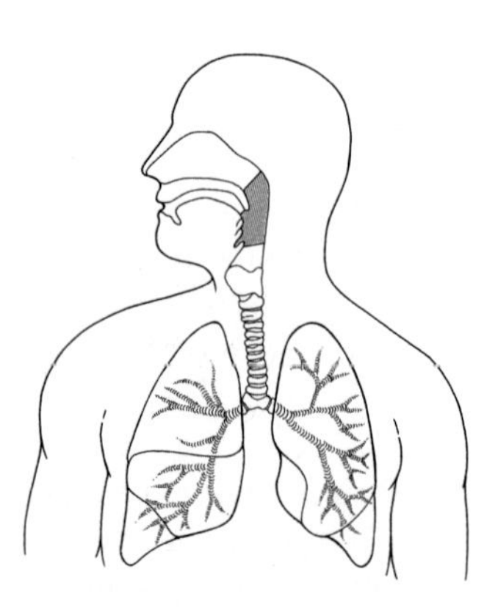

Larynx

The **larynx** is about 6 cm, or 2 inches,

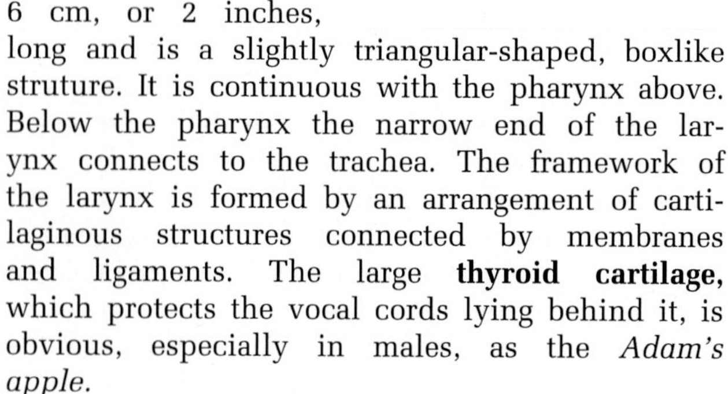

long and is a slightly triangular-shaped, boxlike struture. It is continuous with the pharynx above. Below the pharynx the narrow end of the larynx connects to the trachea. The framework of the larynx is formed by an arrangement of cartilaginous structures connected by membranes and ligaments. The large **thyroid cartilage,** which protects the vocal cords lying behind it, is obvious, especially in males, as the *Adam's apple.*

The larynx is a permanently open airway specialized to perform two functions. Its first function is as a switching station to route air and food into the proper tubes. When food passes through the pharynx, the inlet to the larynx is closed. When only air is flowing through the pharynx, the inlet to the larynx is open, allowing air to pass into the lower regions of the respiratory system. The opening to the larynx is called the **glottis.** When food is being swallowed, a spoon-shaped, tissue-covered piece of cartilage, called the **epiglottis,** covers the glottis so no food or fluid can pass into the larynx. If anything other than air enters the larynx, the cough reflex occurs, which expels the substance from the airway passage.

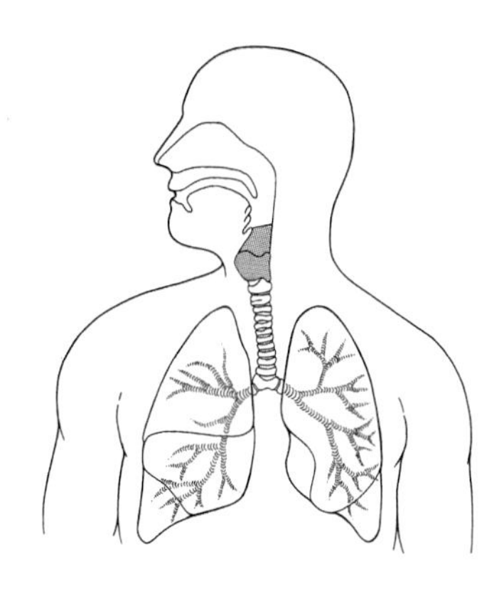

The second function of the larynx is to make speech production possible. Embedded in the mucosal membrane at the sides of the glottis are elastic fibers called **vocal cords.** These cords are attached at the back of the larynx and to the thyroid cartilage at the front. The vocal cords vibrate and produce sound as air rushes up from the lungs. The high or low pitch of the voice depends on several factors, including the thickness, length, and degree of elasticity of the vocal cords and the tension at which they are held. Loudness is controlled by the more forceful expulsion of air through the vocal cords.

Trachea, Bronchi, and Bronchioles

The **trachea,** or *windpipe,* begins at the base of the larynx. It then descends in front of the esophagus into the thoracic cavity, where it branches into two smaller passageways, each connecting to the lungs. The trachea is 12 to 14 cm, or 4 to 5 inches, long and about 2.5 cm, or 1 inch, in diameter. It is held open by 16 to 20 C-shaped rings of cartilage distributed along its length.

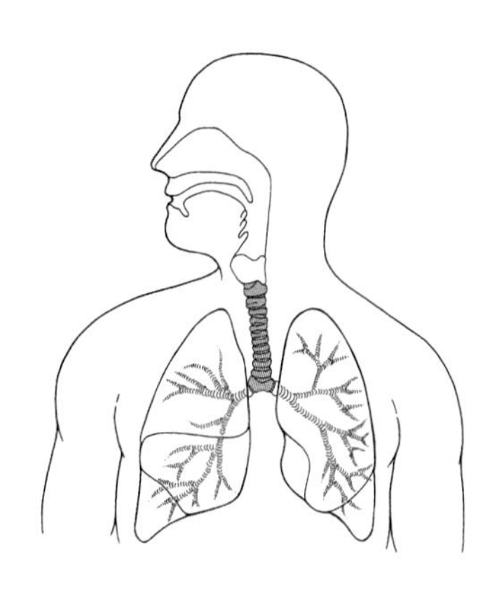

The trachea is lined with a ciliated mucous membrane. The cilia continuously move the mucus and its entrapped debris up to where it can be swallowed or expelled. Smoking diminishes ciliary action and ultimately destroys the cilia. As ciliary function is lost, coughing becomes the only means of preventing mucus from accumulating in the lungs.

Tracheal obstruction is life threatening. An average of eight Americans each day suffocates from choking on a piece of food that suddenly closes off the trachea. The **Heimlich maneuver,** a procdure in which the air in a person's own lungs is used to expel or pop out an obstructing piece of food, has saved many people from becoming victims of what used to be called a "café coronary" (Figure 15-2).

The trachea divides into two passageways, called the **right** and **left primary bronchi** (bron′-ki). Like the trachea both left and right bronchi are kept open by cartilage rings. By the time air reaches the bronchi it is warm, cleansed of most impurities, and humidified. Once inside the lungs each bronchus rapidly divides into smaller **bronchioles** (bron′-ke-ols). The smallest, called **terminal bronchioles,** are less than 0.5 micrometers (0.5 μm) in diameter. The tissue composition of the primary bronchi is the same as in the trachea, but as the conducting tubes become smaller, the tissue type also changes. Toward the ends of the primary bronchi cartilage rings are gradually replaced by irregular plates of cartilage. By the time the bronchioles are reached, all cartilage has disappeared. The mucous membrane changes to nonciliated epithelial cells in the terminal bronchioles. Macrophages remove any debris at this level. The relative amount of smooth muscle in the walls also increases as the passageways become smaller. The bronchioles contain a complete layer of circularly arranged smooth muscle, allowing them to constrict under certain conditions.

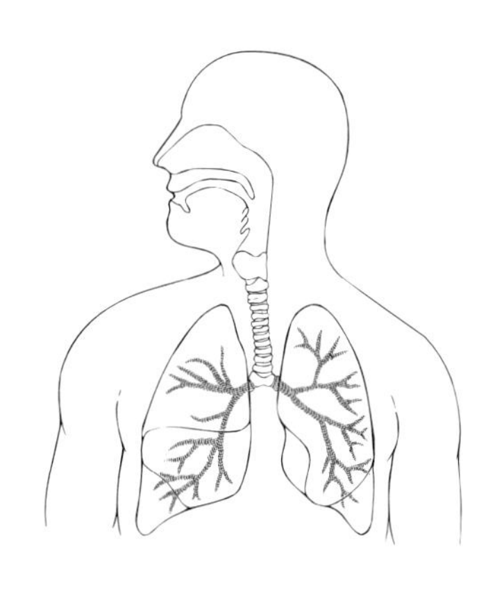

Each terminal bronchiole ends in an elongated space surrounded by clusters of air sacs called **alveoli** (al-ve-o-li). The alveoli resemble bunches of grapes on a vine. Alveoli account for most of each lung's volume and provide a tremendous surface area for gas exchange. There are approximately 300 million alveoli per lung, and their total surface area is 70 to 80 m^2, about 40 times the surface area of the body's skin or approximately the same size as a racquetball court. Being composed primarily of alveoli, lung tissue is spongy and lightweight. A section of lung tissue placed in water will actually float (Figure 15-3).

Figure 15-2 Heimlich maneuver

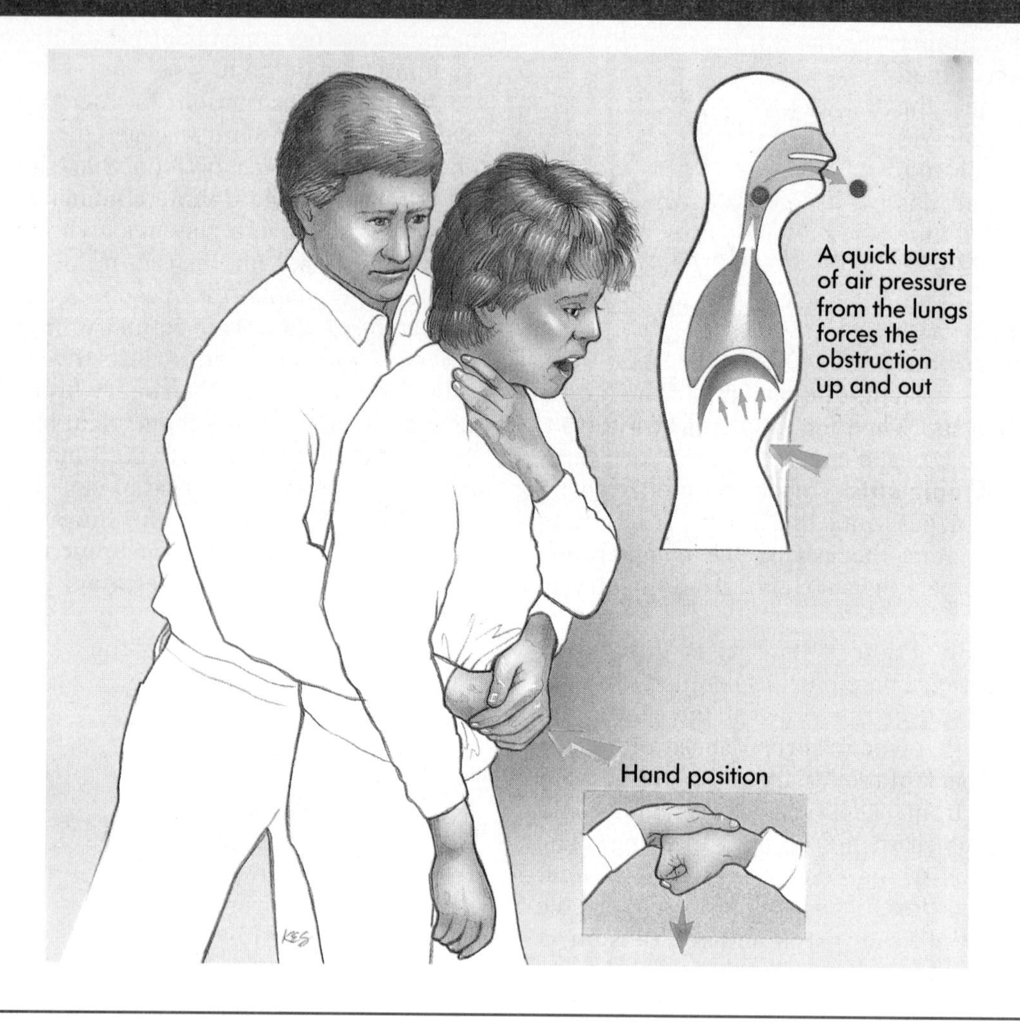

A person stands behind the victim, positions a fist just below the victim's sternum and firmly pulls up and back (repeatedly if necessary) under the ribs. This action forces air out of the lungs, dislodging the obstruction from the airway.

The extremely thin walls of the alveoli are composed of a single layer of cells. The external alveolar surfaces are densely covered with capillaries. Their appearance is similar to a cluster of grapes being covered by a spider's web. Together the alveolar and capillary walls form the respiratory membrane, which has gas on one side and blood flowing past it on the other. Gas exchange across the respiratory membrane occurs by simple diffusion. The oxygen passes from the air in the alveolus into the blood, and carbon dioxide leaves the blood to enter the gas-filled alveolus.

A watery film coats the inside walls of the alveoli. Because water molecules are polar they are more attracted to one another than to the gas. This force of attraction causes the alveoli to reduce to their smallest possible size. If this film were pure water, the alveoli would collapse between each breath. However, scattered among the cells of the alveolar walls are cells that secrete a lipoprotein, called **surfactant** (sur-fak'tant), that breaks up the attraction of polar water molecules. Its action is similar to the way laundry detergent reduces the attraction of water for water, allowing water to interact with and pass through fabrics. Because of surfactant the surface tension of the watery fluid is reduced and less energy is needed to expand the lungs. Surfactant is produced toward the end of fetal development. Prematurely born babies thus sometimes lack enough surfactant to keep the alveoli open. Each time the infant exhales, the alveoli close up, making inhalation difficult. This causes

Figure 15-3 Alveoli

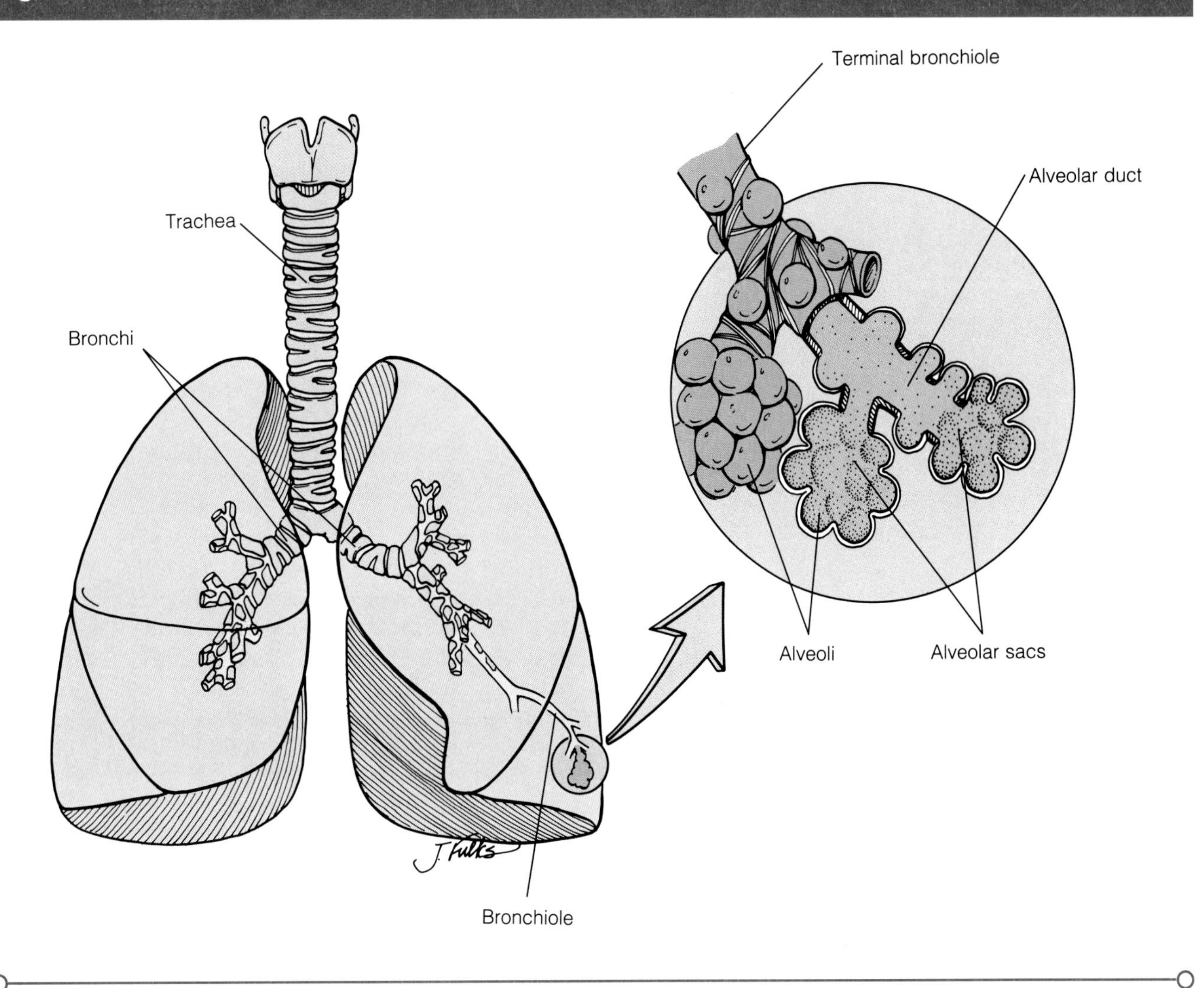

The alveoli are found at the ends of terminal bronchioles. Gas exchange takes place between the air in the alveoli and the surrounding capillaries of the lung.

respiratory distress syndrome of the newborn, which is sometimes fatal.

Lungs

The roughly cone-shaped lungs lie on both sides of the heart, great blood vessels, bronchi, and esophagus in the thoracic cavity. The branches of the pulmonary artery accompany the bronchi and bronchioles to eventually form a network of capillaries surrounding the alveoli. Four pulmonary veins collect blood from the alveolar capillaries and empty into the left atrium of the heart. As you learned in Chapter 13 the newly oxygenated blood moves from the left atrium into the left ventricle, where it is then pumped into the systemic circulation (Figure 15-4).

The lungs are enclosed in a thin, double-walled membrane called the **pleura** (pler′-uh), or **pleural sac.** The outer pleural membrane, called the **parietal pleura,** lines the thoracic cavity walls and covers the pericardium of the heart. The inner membrane, or **visceral pleura,** adheres to the lungs. The membranes secrete a thin film of fluid into the space between the two layers, which lubricates their surfaces as they move past each other during breathing. **Pleurisy,** an inflammation of the pleural membranes, results in sharp pain during each breath because of the friction created between the two membranes as the lung inflates and deflates.

Figure 15-4 Pulmonary circulation

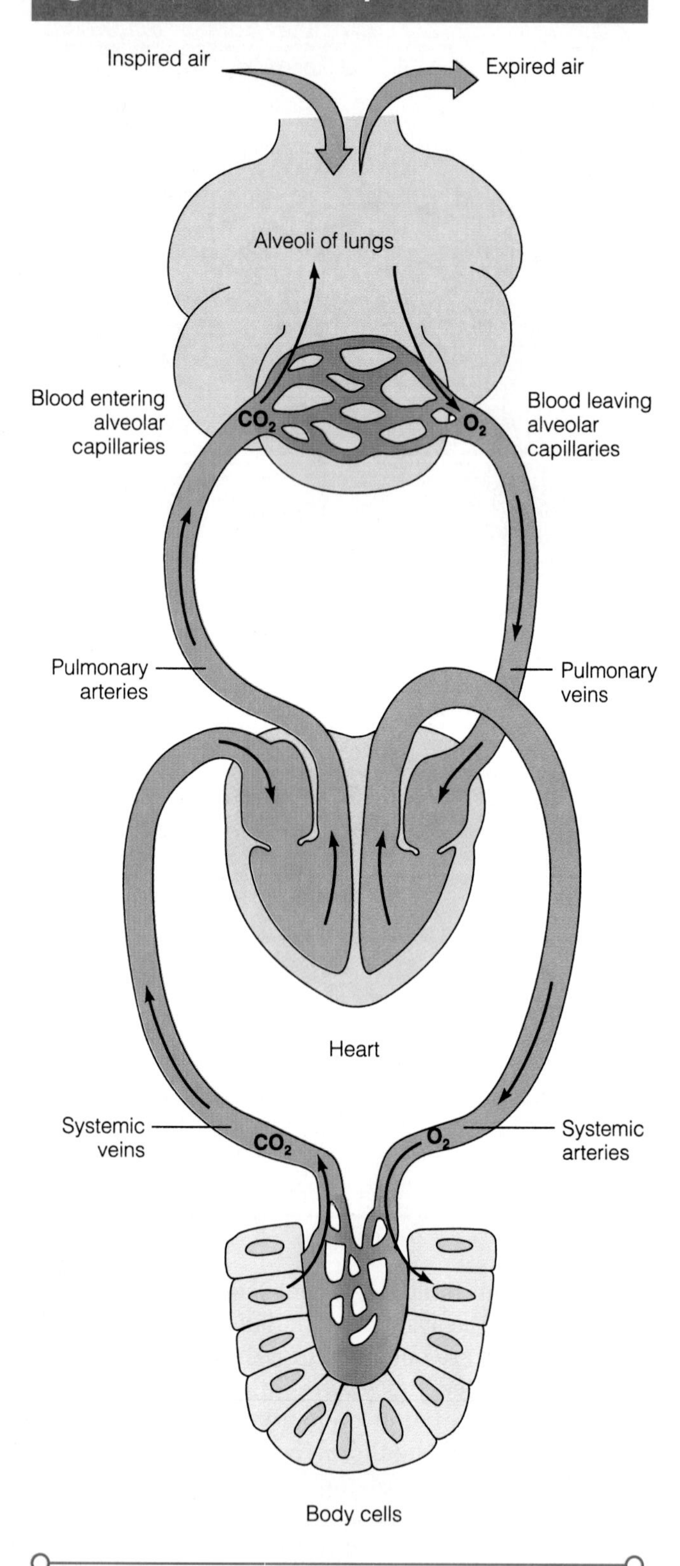

Respiration Process

The lungs lie within a sealed thoracic cavity. The ribs, which are fastened to the vertebral column at the back and to the sternum at the front, along with the intercostal muscles between them, make up the top, sides, and back of the thoracic cavity. The **diaphragm** forms the floor and separates its contents from the abdominal cavity below. The lungs lie in an enclosed space described by the pleural sac.

Pulmonary Ventilation

You learned in Chapter 3 that the diffusion of any substance depends largely on differences in its concentration between two areas. For gases, the differences are expressed in terms of pressure. The greater the number of gas molecules per unit volume of space, the higher the pressure and the greater the force available to drive individual molecules from areas of high concentration toward areas of lower concentration. This is the principle behind using a bellows to fan a fire. As the handles of the bellows are pulled apart, its internal volume increases and the air pressure in the bellows drops. Air moves from the area of high pressure outside the bellows through the opening of the bellows and into the area of lower pressure inside. Then as the handles are pushed together, air pressure in it increases and the volume decreases. As a result, air moves out from the high-pressure area to the low-pressure area outside the bellows. Pulmonary ventilation functions in a similar way.

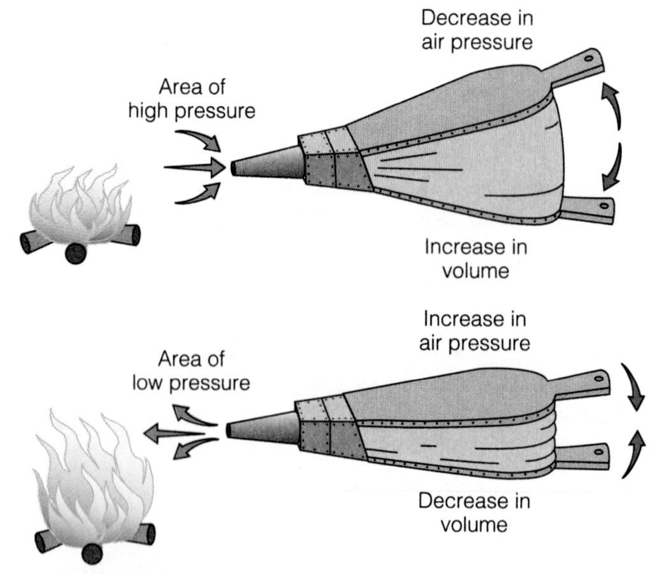

The pressure in the alveoli of the lungs, called **intrapulmonary pressure,** rises and falls with each breath but always equalizes itself with the outside atmospheric pressure. Because the air passages are open, it helps to think of this situation as a continuous column of air that extends from outside the body, through the nasal cavities, all the way to the lungs. During inspiration the thoracic cavity enlarges because the diaphragm contracts and the

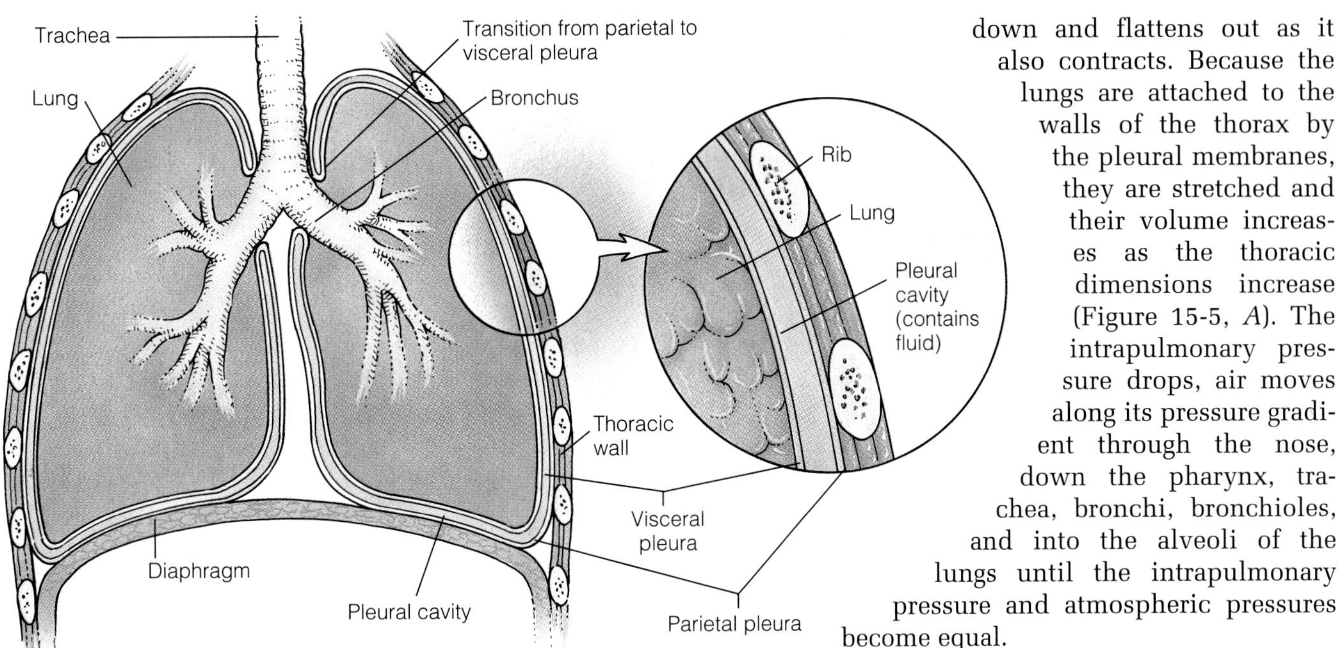

down and flattens out as it also contracts. Because the lungs are attached to the walls of the thorax by the pleural membranes, they are stretched and their volume increases as the thoracic dimensions increase (Figure 15-5, *A*). The intrapulmonary pressure drops, air moves along its pressure gradient through the nose, down the pharynx, trachea, bronchi, bronchioles, and into the alveoli of the lungs until the intrapulmonary pressure and atmospheric pressures become equal.

During deep inspiration, as occurs during exercise, thoracic volume is further increased by the activation of accessory muscles, which raise the ribs even more than occurs during quiet inspiration. Several neck muscles and the pectoral muscles are the primary muscles used.

intercostal muscles between the ribs also contract.

As you breath in, the intercostal muscles contract, causing the rib cage to move up and out. At the same time the dome-shaped diaphragm moves

Figure 15-5 The mechanics of inspiration and expiration

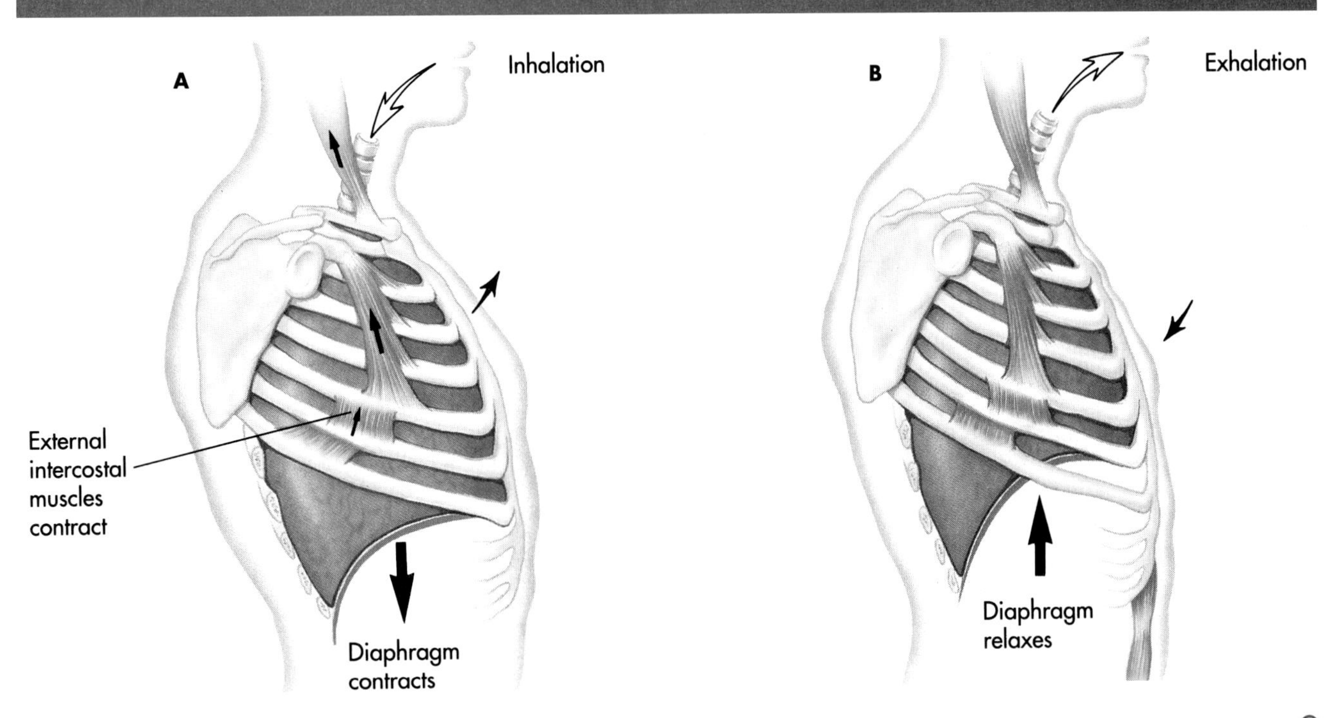

During inspiration the diaphragm contracts, which causes it to flatten, and the intercostal muscles contract, raising the ribs up and slightly out. These actions increase the chest cavity and, as a consequence, lung volume (A). During expiration, the diaphragm relaxes, resuming its domelike shape; the intercostal muscles relax, which lowers the ribs; and the chest cavity and lung volume decrease. The result is air is forced from the lungs and back into the atmosphere (B).

Although the thoracic cavity's size increases by only a few millimeters along each dimension during quiet breathing, its overall volume increases by nearly half a liter. This is the usual volume of air that enters the lungs during normal quiet respiration, such as you are now experiencing as you read these words.

During expiration, or exhalation, the muscles of the ribs and the diaphragm relax, reducing the volume of the thoracic cavity and the lungs. Decreasing volume increases air pressure in the lungs, causing it to flow in the opposite direction, namely, from the alveoli of the lungs, to the bronchioles, bronchi, pharynx, and nose, to the surrounding air (Figure 15-5, *B*).

Quiet expiration is basically a passive process that depends more on lung elasticity than on muscular contraction. **Elasticity** is the ability of a structure to return to its initial size after being stretched. A rubber band, for example, is much more elastic than a strand of copper wire. When stretched and then released, the rubber band returns to its initial length. The copper wire, in contrast, remains distended when stretched because it is not elastic.

The lungs are highly elastic. Because the lungs are attached via the pleural membrane to the thoracic wall, they are always under some elastic tension. This tension increases during inspiration. During expiration the lungs recoil, or snap back, assisting in pushing the air out of the lungs. As the diaphragm and intercostal muscles relax and assume their resting length, the lungs recoil, decreasing the intrapulmonary volume. In contrast to quiet expiration, forced expiration is an active process produced primarily by abdominal wall muscle contraction. For example, a vocalist's ability to hold a note depends on the coordinated activity of several muscles used in forced expiration. Indeed, muscle control of expiration is important during speaking, when precise airflow from the lungs is required. For example, if you put your fingertips on the lower portion of your abdomen while speaking, you can feel the abdominal muscles alternately contracting and relaxing.

Respiratory Control

Pulmonary ventilation is controlled in two related ways. First the control of the basic pattern of breathing is generated by neurons in the brain stem. Several groups of neurons in the medulla oblongata, collectively called the **respiratory center,** fire, sending action potentials to the diaphragm and the intercostal muscles of the rib cage, which then contract. When the lungs expand, sensory neurons scattered throughout the lungs and sensitive to stretch send action potentials back to the respiratory center, causing it to momentarily turn off. Expiration occurs passively as the inspiratory muscles relax and the lungs recoil. This cyclic on-off activity is repeated continuously to produce a quiet respiratory rate of 12 to 15 breaths per minute. Inspiratory phases last about 2 seconds, followed by expiratory phases lasting about 3 seconds.

Both the depth and the rate of breathing can also be modified in response to changing body demands by specialized receptors, called **chemoreceptors,** sensitive to the blood's chemical composition. The **central chemoreceptors** are located in the medulla oblongata of the brain stem, and the **peripheral chemoreceptors** are in the circulatory system.

The most important direct influence on both the central and the peripheral chemoreceptors and therefore on respiration is hydrogen ion concentration or pH. The central chemoreceptors, as well as the entire brain and spinal column, are bathed in cerebrospinal fluid. The pH of the cerebrospinal fluid is determined by the blood's pH. Some carbon dioxide produced by cells dissolves in the blood plasma. Carbon dioxide combining with blood plasma water molecules produces carbonic acid. Carbonic acid molecules dissociate, releasing hydrogen ions into the plasma. When the blood's hydrogen ion concentration increases, it stimulates the chemoreceptors and signals the respiratory center to increase ventilation. With an increased breathing rate, more carbon dioxide is removed

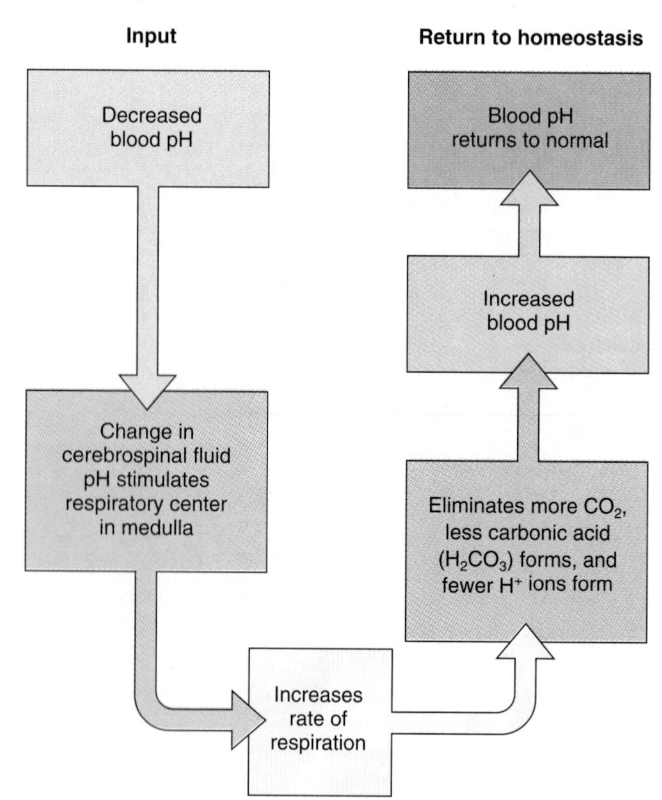

from the blood and the hydrogen ion concentration decreases.

When carbon dioxide levels drop, hydrogen ion numbers drop, the chemoreceptors stop firing, and ventilation decreases. The powerful influence of chemoreceptors on the respiratory center is why you cannot voluntarily hold your breath for much more than 1 minute. When you hold your breath, metabolically produced carbon dioxide begins accumulating in the blood. The respiratory center, stimulated by the chemoreceptors, overrides the voluntary inhibition of respiration, so breathing resumes despite attempts to prevent it.

The respiratory centers can sometimes fail. **Sudden infant death syndrome (SIDS),** or crib death, occurs because the respiratory center stops sending action potentials to the chest muscles and diaphragm, which stops the breathing process. In the United States SIDS is responsible for 10,000 death a year and is the leading cause of death of infants up to 1 year of age. Its cause is not yet known. However, evidence indicates it may be the result of undeveloped central chemoreceptors in the medulla oblongata.

Blood oxygen levels play a relatively minor role in influencing respiration. The peripheral chemoreceptors are stimulated when oxygen levels fall to about 60% of normal values. Because this occurs only in unusual situations, such as with some lung diseases or with exposure to high altitudes, it is not central to regulating normal respiration.

Gas Exchange

Earlier in this chapter we noted that an alveolar wall is only one cell layer thick and the capillary wall also consists of a single cell layer. The space between the wall of an alveolus and a capillary wall is only 0.5 μm. Neither of these walls offers resistance to the passage of gases. Because there is such a tremendous total surface area between the capillaries and the alveoli, diffusion of oxygen and carbon dioxide is rapid. Pulmonary ventilation brings air to this large surface area to facilitate external respiration, the exchange of gases between the air in the alveoli and the blood in the surrounding pulmonary capillaries. Knowing that pressure gradients drives gas diffusion across the alveolar wall, we can look closer at the gas exchange process.

Partial Pressure of Gases

The air that surrounds us, called atmospheric air, is a mixture of gases. Nitrogen makes up approximately 78.6% of air, oxygen about 20.9%, and carbon dioxide only about 0.04%. In addition to these gases, atmospheric air also contains up to 0.5% water vapor. The total pressure exerted by a mixture of gases is the sum of pressures exerted independently by each gas in the mixture. Therefore, the pressure exerted by each gas, its **partial pressure,** is directly proportional to its percentage in the total gas mixture.

The atmosphere at sea level exerts a total pressure great enough to push mercury (Hg) 760 mm, or 29.91 inches, up a vacuum-filled tube (it varies slightly with the weather). Because oxygen represents 20.9% of total atmospheric pressure, the partial pressure of oxygen, written as $\mathbf{P_{O_2}}$, is 20.9×760 mm Hg, or about 159 mm Hg. Atmospheric air contains 0.04% carbon dioxide, so its partial pressure equals $760 \times 0.04 = 0.3$ mm Hg.

Because inhaled atmospheric air mixes with some of the residual oxygen-poor and carbon-dioxide rich air in the alveoli, the partial pressure of oxygen is 105 mm Hg (14%) and the partial pressure of carbon dioxide is 40 mm Hg (5.5%) in the alveoli. The partial pressure gradients of oxygen and carbon dioxide determine the movement of these dissolved gases into and out of cells, extracellular fluids, the bloodstream, and the lungs.

External Respiration

During external respiration the partial pressure of oxygen in the pulmonary capillaries is 40 mm Hg but is 105 mm Hg in the alveoli. Consequently a steep oxygen partial pressure gradient exists between the two regions, and oxygen diffuses rapidly from the alveoli into the pulmonary capillary blood. Carbon dioxide moves in the opposite direction. The partial pressure of carbon dioxide is 45 mm Hg in the tissue cells and 40 mm Hg in the blood, producing a slight partial pressure gradient of about 5 mm Hg. However, carbon dioxide is about 20 times more soluble in water than oxygen is and is thus able to flow from the blood into the alveoli rapidly.

Internal Respiration

Although the partial pressure gradients are reversed, the factors promoting gas exchange between the capillaries and the tissue cells during internal respiration are identical to those acting in the lungs. Because the tissue cells continuously metabolize oxygen, the partial pressure of oxygen in the tissues is always lower than it is in the arterial blood. In contrast, the partial pressure of carbon dioxide in the tissue cells is higher than it is in the capillary blood because it is continuously produced as the waste by-product of cellular respiration. Carbon dioxide moves along its partial pressure gradient from the tissue cells, through the extracellular fluid, and into the blood, whereas oxygen moves in the opposite direction, from the arterial blood into tissue cells. Thus venous blood returning to the heart has a P_{O_2} of 40 mm Hg and a P_{CO_2} of 45 mm Hg.

Gas Transport by Blood

Oxygen carried in the blood is bound to hemoglobin in the red blood cells. As described in Chapter 13 hemoglobin (Hgb) is composed of four polypeptide chains, each bound to an iron-containing heme group. Because the iron groups are oxygen-binding sites, each hemoglobin molecule can combine with four molecules of oxygen. The iron-containing heme group of the hemoglobin molecule has a high affinity for oxygen. When exposed to a P_{O_2} of 105 mm Hg, over 97% of the hemoglobin in the red blood cells combines with oxygen. The remaining 3% dissolves in the blood plasma. In the tissues the reaction reverses because the tissue level of P_{O_2} is lower than the P_{O_2} level in the blood.

To be eliminated from the blood, carbon dioxide must be transported from the tissue cells to the lungs. There are three ways carbon dioxide is transported by the blood to the lungs, where it can be expired.

The first way carbon dioxide is transported is for some of the carbon dioxide to dissolve in the blood plasma. However, this accounts for only 7% to 8% of the total carbon dioxide produced by cells.

Figure 15-6 How gases are transported by blood

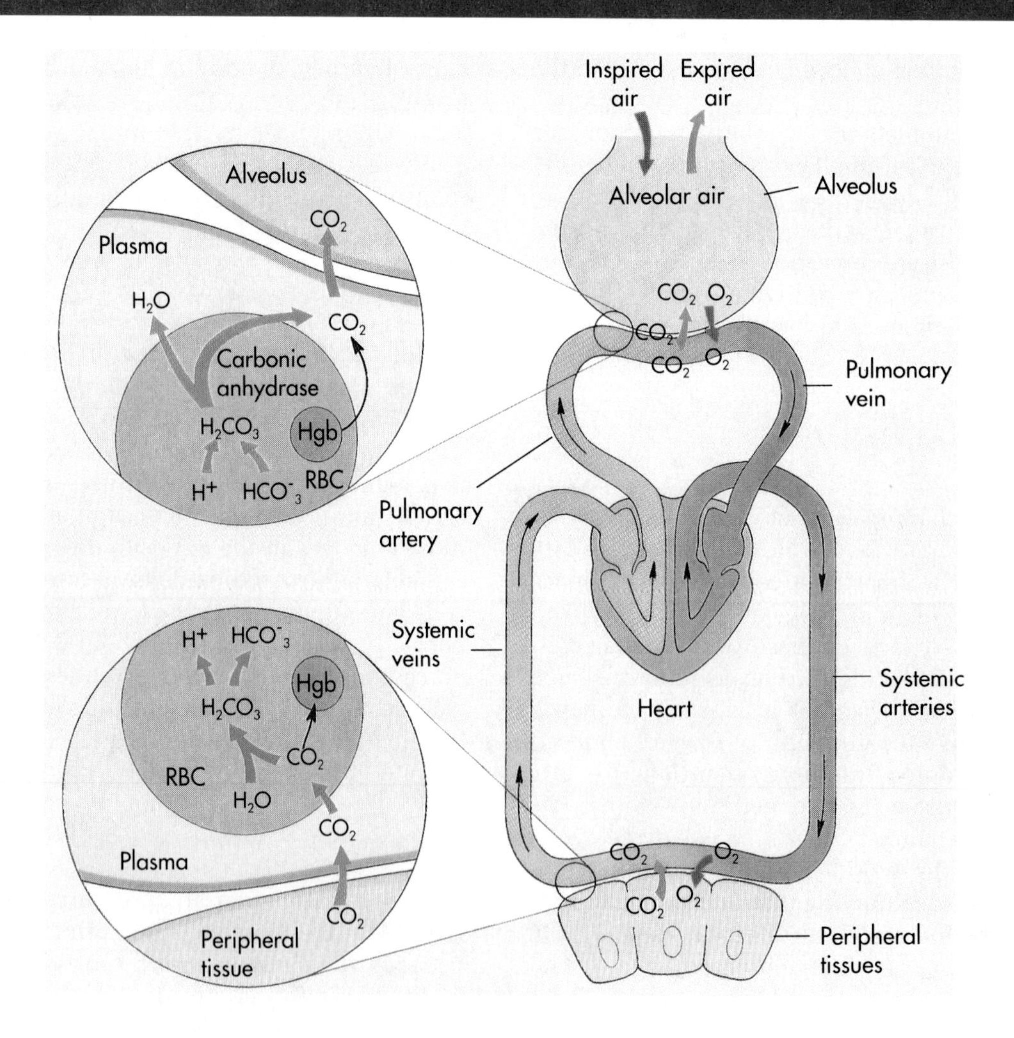

External respiration takes place in the lungs as red blood cells exchange carbon dioxide for oxygen. During internal respiration, oxygenated blood travels to the body's tissues delivering oxygen and picking up carbon dioxide.

Biology in Action HOW MUCH WORK CAN YOU DO?

One of the best single predictors of a person's physical work capacity is to determine the maximum volume of oxygen the person is capable of using per minute to metabolize nutrients for energy production. **Maximum oxygen consumption,** or **max VO_2,** is measured by having a person engage in exercise, usually on a treadmill or stationary bicycle. The workload is gradually increased until the person becomes exhausted. Expired air samples collected during the last minute of exercise, when the person is working the hardest, are analyzed for the percentage of oxygen and carbon dioxide they contain, along with the volume of air expired. The amount of oxygen consumed can be determined on the basis of the percentage of oxygen and carbon dioxide in the inspired air, the total volume of air expired, and the percentage of oxygen and carbon dioxide in the exhaled air.

Maximum oxygen consumption depends on three body systems. First the respiratory system is essential for the exchange of oxygen and carbon dioxide between air and blood in the lungs. Second the circulatory system is needed to deliver oxygen to the working skeletal muscles. Third the skeletal-muscle fibers themselves must have the enzymes available to use oxygen once it has been delivered.

Regular aerobic exercise improves maximum oxygen consumption, and therefore work capacity, by making the heart and respiratory system more efficient in their delivery of oxygen to the skeletal muscles. For example, a well-conditioned heart is able to pump more blood through the pulmonary and systemic circulation with each contraction. The exercised muscles also become better equipped to use oxygen once it is delivered. The number of capillaries serving the muscle tissue increases. So do the number and size of mitochondria in the muscle cells, which contain the enzymes essential for converting nutrient molecules to ATP molecules.

A significant increase in max VO_2 is one of many benefits derived from regular exercise, yet fewer than 20% of Americans exercise regularly. In 1980 the U.S. Public Health Service released the following statement: "The increasing costs associated with health care will compel public policy to emphasize measures such as physical fitness to enhance health." If the benefits of regular exercise, such as increasing the ability to work, are not enough to motivate Americans, perhaps appealing to their wallets will, in the end, induce them to exercise.

If more carbon dioxide moved directly into the plasma, it would not dissolve but would form bubbles, in the same way soft drinks are carbonated. Gas bubbles would disrupt the blood flow and damage the blood vessels. The second way carbon dioxide is transported is for some of it to bind to the amino acid chains of hemoglobin rather than to the heme groups, as does oxygen. About 20% of the carbon dioxide transported in the blood combines with the deoxyhemoglobin of the red blood cells.

The third way carbon dioxide is transported is for the remaining 70% of the carbon dioxide to be converted to bicarbonate ions (HCO_3^-) and transported in the plasma. As carbon dioxide moves from the tissues into the red blood cells, it combines with water molecules in the red blood cell cytoplasm, forming carbonic acid (H_2CO_3). Carbonic acid molecules are unstable and some immediately break down to carbonate ions (HCO_3^-) and hydrogen ions (H^+), as follows:

$$CO_2 + H_2O \rightarrow H_2CO_3 \rightarrow H^+ + HCO_3^-$$

This reaction is driven by the high P_{CO_2} partial pressure in the tissues and an enzyme, called **carbonic anhydrase,** that speeds up the chemical interaction between carbon dioxide and water thousands of times faster in red blood cells than elsewhere in the body. The bicarbonate ions then diffuse out of the cytoplasm into the plasma. Some of the hydrogen ions, however, chemically bond to the deoxyhemoglobin in the red blood cell, and the rest move into the plasma (Figure 15-6).

In the lungs, where P_{CO_2} is lower than in the blood, the process is reversed, as follows:

$$H^+ + HCO_3^- \rightarrow H_2CO_3 \rightarrow CO_2 + H_2O$$

Deoxyhemoglobin releases the hydrogen ions as it begins to pick up oxygen molecules diffusing from the alveoli in the lungs. The released hydrogen ions attract the bicarbonate ions from the plasma back into the red blood cell. The bicarbonate and hydrogen ions recombine because of carbonic anhydrase to form carbonic acid. The carbonic acid then breaks down into water and carbon dioxide. The reformed carbon dioxide then diffuses out of the red blood cell, through the plasma, out of the capillary, and back into the lungs, where it is exhaled.

Homeostatic Imbalances of the Respiratory System

The respiratory system is especially vulnerable to infectious diseases from bacteria, viruses, and airborne irritants because it is open to the surrounding air and all that is contained in it. It is also subject to disorders collectively called **chronic obstructive pulmonary disease (COPD)** and lung cancer. These diseases are not ordinarily caused by infective agents but by airborne irritants. The common link between COPD and lung cancer is that they are usually the consequence of tobacco use. They are probably the best living proof against engaging in this self-destructive behavior. First we look at an example of one respiratory disease caused by infective agents before discussing COPD.

Pneumonia

Pneumonia (nu-mon′ni-ah) is an acute lung inflammation. There are many different kinds of pneumonia, caused by both bacterial and viral infections. Some infective agents affect both lungs. When both lungs are infected, **double pneumonia** occurs. Regardless of the type of microorganism causing pneumonia, the result of the lung tissue inflammation is mucus and fluid production that blocks air passageways, diminishing gas exchange in the alveoli, and reducing the amount of oxygen delivered to the body's tissues.

Biology in Action ASTHMA ATTACK!

One common disease with possibly serious consequences is **bronchial asthma.** A person having an asthma attack has great difficulty breathing and develops a wheezing sound when inhaling and exhaling. What actually happens during one of these episodes?

First there is a blockage of the bronchi. During an attack the smooth muscle in the walls of the bronchi suddenly contracts, narrowing the lumen of the tubes. The spasm continues, making breathing, especially expiration, difficult. Second the mucous membrane lining the bronchi becomes swollen and secretes excess mucus. This reaction also further narrows the bronchial lumen. It also helps to trap air in the lung, further decreasing the amount of air that can be inhaled. The wheezing sound results from air passing through the severely narrowed bronchi. What causes this sudden, spasmodic bronchial contraction?

For many asthma sufferers an asthmatic attack is often brought on as a consequence of an allergic reaction to usually harmless substances, such as dust, pollen, or mold. These people have an excess of a particular antibody circulating in their blood. When usually nonthreatening substances, such as dust, pollen, mold, or even pet hair, are inhaled, they form a complex with the attached antibodies. This antigen-antibody complex causes the mast cells of the mucous membrane to secrete histamine. The histamine, in turn, stimulates smooth-muscle constriction and mucus secretion. Histamine is also a potent vasodilator, which also causes localized swelling of airway passages. An asthma attack is often the result.

There is no cure for asthma. However, attacks usually become less frequent and severe with age. Drug therapy can reduce the incidence and severity of attacks. To control an ongoing attack, substances that dilate bronchi are effective. For example, sprays containing epinephrine are good bronchodilators.

Biology in Action SECONDHAND SMOKE

Today it is common to see "no smoking" or "smoke-free area" signs posted in restaurants, stores, offices, malls and other public places. About 85% of businesses have adopted some form of smoking policy for their employees. In addition, more than 90% of Americans favor restricting or banning smoking in public places. So far 46 states restrict smoking in public places. These laws vary widely from no smoking on school buses to comprehensive clean air laws that ban smoking in all public places. Why is there such concern over keeping indoor air in public spaces free of tobacco smoke?

Environmental tobacco smoke (ETS), or secondhand smoke, is the combination of smoke from a burning cigarette and the smoke exhaled by the smoker. Like the smoke inhaled by the smoker, secondhand smoke contains at least 43 known carcinogens. Concerns that nonsmokers may develop cancer and other illnesses as a result of secondhand smoke have mounted over the last 30 years. In 1986 the Surgeon General and the National Research Council confirmed that ETS causes lung cancer in nonsmokers. Between 1986 and the present, more evidence has accumulated regarding the effects of ETS in nonsmokers. Findings include the fact that each year secondhand smoke kills 3000 adult nonsmokers from lung cancer. Secondhand smoke causes at least 30 times as many lung-cancer deaths as all regulated air pollutants combined. Nonsmokers exposed to cigarette smoke have significant amounts of nicotine and carbon monoxide in their body fluids.

A variety of problems are also seen in children exposed to ETS. Each year between 150,000 to 300,000 children under 18 months of age exposed to cigarette smoke suffer from lower respiratory tract infections, the most common being pneumonia and bronchitis. These infections result in 7500 to 15,000 hospitalizations yearly. Children exposed to ETS also have significantly more middle-ear disease and reduced lung function compared with children not exposed to ETS.

The separation of smokers from nonsmokers within the same space, such as restaurants or offices, does not eliminate the exposure of nonsmokers to ETS. For example, a recent study found that waiters, waitresses, and bartenders are exposed to high levels of secondhand smoke and face a 50% higher risk of lung cancer than does the general population. Levels of tobacco smoke in restaurants are up to twice as high as in offices that allow smoking, and smoke levels in bars are four to six times as high as in offices.

While public debate continues over smoking as a matter of personal freedom of choice, it is clear that it is also very much a matter of public health and concern and must be regarded as such.

Chronic Obstructive Pulmonary Diseases

Two chronic obstructive pulmonary diseases (COPDs), **chronic bronchitis** and **obstructive emphysema** (em-fi-se'mah), are major causes of death and disability in the United States and continue to increase each year. These diseases have several features in common: (1) almost all patients have a history of tobacco use, (2) most patients suffer coughing and frequent pulmonary infections, and (3) most patients ultimately suffer and die from respiratory failure.

Chronic bronchitis is characterized by chronic excessive mucus production in the lower respiratory passageways. Over time tobacco smoke irritates the passageways, causing increased mucus secretion. At the same time it interferes with the ciliary action of the ciliated cells lining the air-conducting tubes. The mucus, including the particles and bacteria trapped in it, begins to accumulate in the bronchi and trachea. Coughing sets in as the body attempts to clear away the mucus. If the irritation persists, the coughing reflex persists. Persistent coughing further irritates the bronchial walls, which become infected and inflamed. Although the cilia diminish, the mucus-secreting cells actually multiply as the body continues working to protect itself against the irritation, infection, and accumulating debris. Eventually the ciliated mucous membrane breaks down and is replaced by fibrous scar tissue that begins to obstruct parts of the damaged respiratory tract.

If the irritation persists, fibrous scar tissue builds up in the respiratory tract and the bronchi become progressively clogged with more mucus. Air becomes trapped in alveoli, causing the alveolar walls to break down. This is called **obstructive emphysema.** As the lung continues to repair itself with scar tissue, it also loses elasticity. As the lungs become less elastic the airways collapse during expiration and further obstruct the outflow of air. People with this disease often have a slightly blue and bloated appearance because of the high levels of carbon dioxide remaining in the circulatory system. Over 1,300,000 people in the United States have emphysema. The emphysema rate is insignificant if one eliminates all smokers from this number.

Lung Cancer

Lung cancer appears to closely follow the pattern of oncogene-activating events that were described in Chapter 5. Ordinarily, mucus and the ciliary action of the cells that line the air-conducting tubes effectively protect the lungs from irritants. When a person smokes, these mechanisms eventually break down and become nonfunctional. However, the 43 or more carcinogens in tobacco smoke result in lung cancer, eventually causing mucus cells to divide uncontrollably.

In the United States lung cancer accounts for approximately 15% of all cancer cases. Lung cancer accounts for about 30% of all cancer deaths. More than 90% of lung-cancer patients have a history of tobacco use. Lung cancer is difficult to detect in its early stages. Symptoms such as chest pain and mucus streaked with blood do not usually appear until the cancer is well advanced. Although treatment may include surgery combined with radiation and chemotherapy, the 5-year survival rate is only between 7% and 13%.

CHAPTER REVIEW

SUMMARY

1. The process of acquiring oxygen from the external environment and eliminating carbon dioxide from the body is called respiration. The respiratory system fulfills these demands by supplying the body with oxygen and disposing of carbon dioxide.
2. Gas exchange between the lungs and the blood occurs as a result of the passive diffusion of oxygen and carbon dioxide along their respective partial pressure gradients. Although the partial pressure gradient for carbon dioxide is only 5 mm Hg, carbon dioxide is highly soluble and is therefore able to diffuse from tissue cells to the blood rapidly.
3. The conducting airway passages of the lungs terminate in clusters of microscopic air sacs called alveoli, which are surrounded by capillaries. Together the alveolar and capillary walls form the respiratory membrane, which has gas on one side and blood flowing past it on the other.
4. Alveoli account for most of each lung's volume and provide a tremendous surface area for gas exchange between the gases they contain and blood in the surrounding capillaries. As a result, the combined total surface area of the lungs is about 70 to 80 m^2.
5. Air movement into and out of the lungs is a consequence of changing dimensions of the thoracic cavity, caused by contractions of the diaphragm and the intercostal muscles between the ribs. Lung movement occurs passively because of the elasticity of the lungs.
6. Oxygen binds to hemoglobin contained in red blood cells and is transported through the blood to tissue cells. In contrast, most carbon dioxide is transported from the tissue cells to the lungs in the blood modified as bicarbonate ions (HCO_3^-) and hydrogen ions (H^+). The transformation of most carbon dioxide into bicarbonate ions occurs in the red blood cell before diffusing into the blood plasma.
7. Almost all people with the most disabling of respiratory disorders, called chronic obstructive pulmonary disease (COPD), have a history of tobacco use. If tobacco smoking were eliminated, COPD would become an insignificant health problem in the US.
8. Over 90% of lung cancer patients were also tobacco users. Most forms of lung cancer closely follow the pattern of oncogene activation with at least 1 of the 43 known carcinogens in tobacco smoke being a causative agent.

FILL-IN-THE-BLANK QUESTIONS

1. Breathing is also called ______.
2. The pressure exerted by each gas in the atmosphere is referred to as its ______ ______.
3. At the level of the larynx, the structure that routes air and food into the proper passageways is called the ______.
4. The trachea branches into two air-conducting tubes called the ______ and ______ ______ ______.
5. Microscopic air sacs found at the end of each terminal bronchiole are called ______.
6. Gas exchange occurs by simple ______.
7. Both the ______ and ______ chemoreceptors are sensitive to the hydrogen ion in blood and cerebrospinal fluid.
8. Most carbon dioxide is transported in the blood as ______ ions.
9. Oxygen is transported in the blood bound to ______ in the red blood cells, which forms ______.
10. Most chronic obstructive pulmonary disease patients have a history of ______ ______ in common.

SHORT-ANSWER QUESTIONS

1. What are the four major events of respiration?
2. What occurs to the thoracic cavity and the lungs during ventilation?
3. Compare and contrast gas exchange during external and internal respiration.
4. What events occur during the exchange and transport of oxygen and carbon dioxide by the blood?

VOCABULARY REVIEW

alveoli—*p. 285*
bronchi—*p. 285*
bronchiole—*p. 285*
diaphragm—*p. 288*
external respiration—*p. 283*
glottis—*p. 285*
internal respiration—*p. 283*
larynx—*p. 284*
pharynx—*p. 284*
pleura—*p. 287*
pulmonary ventilation—*p. 283*
trachea—*p. 285*

Chapter Sixteen

NUTRITION AND DIGESTION

For Review

Here are some important terms and concepts that you will encounter in this chapter. If you are not familiar with them, you should review them before proceeding.

Cellular respiration
(page 51)

Calorie
(page 48)

Synthesis
(page 26)

Hydrolysis
(page 26)

Coenzyme
(page 51)

OBJECTIVES

After reading this chapter you should be able to:

1. *Define nutrients and essential nutrients.*
2. *List the six basic nutrients essential for good health.*
3. *Differentiate between complete and incomplete proteins and give examples of each.*
4. *Differentiate between and give examples of fat- and water-soluble vitamins.*
5. *Define calorie, metabolic rate, and basal metabolic rate.*
6. *Define digestion and describe the three major processes of digestion.*
7. *List the major organs of the digestive system and tell whether each is part of the GI tract or is an accessory digestive organ.*
8. *Describe the four layers of the GI tract.*
9. *List the major functions of the liver and pancreas and name the enzymes produced by each.*

The act of eating does not automatically make the macromolecules of food available to your body's cells. The food must first be broken down into simple organic molecules. This is the digestive system's function. You might think of the human digestive system as a disassembly line in which food is mechanically and chemically broken down as it moves through the system until it is reduced to simple molecules, which can then be absorbed into the bloodstream and delivered to the body's cells.

How Energy Is Stored in the Food We Eat

A common characteristic of all life-forms is the ability to acquire energy and materials from their environments, convert them into new forms, and use these forms to grow and maintain themselves. To understand how life-forms acquire energy and materials from their environment, we can review some concepts introduced in earlier chapters.

In Chapter 2 you learned that the creation of chemical bonds between atoms and molecules requires energy and that these bonds represent stored, or potential, energy. When chemical bonds are broken, energy is released. We use some of the complex organic molecules we ingest as food, along with the oxygen we breathe, to release this stored energy. Carbon dioxide and water are by-products of this process. This energy-acquiring process is summarized as follows:

$$\underset{\text{(in food)}}{\text{organic molecules}} + O_2 \xrightarrow[\text{respiration}]{\text{human cellular}} \underset{\text{(ATP)}}{\text{energy}} + CO_2 + H_2O$$

This energy, in the form of ATP, is used in synthesis pathways to construct complex macromolecules from the simple organic molecules necessary to grow, develop, and maintain cells. Cells are provided with simple organic molecules to carry out cellular metabolism by the food we eat.

However, eating food does not automatically make either the energy of the complex macromolecules or the simple organic molecules from which they are made available to the body cells. The food must first be broken down into simple organic molecules so they can be absorbed into the circulatory system for distribution to the cells. The simple organic molecules are then taken up by the body's cells. There they can be used as an energy source and as the basic materials for making the macromolecules the cells need. The digestive system is responsible for disassembling the food we eat. Food is mechanically and chemically broken down as it moves through the digestive system until it is reduced to simple molecules the body uses as both an energy source and a basis for growth and maintenance.

We begin here by discussing the food we eat. We then describe the basic structure of the human digestive system. Finally we discuss the specific digestive processes for different regions of the digestive system.

Nutrients

The human body grows and is able to maintain itself when it is kept supplied with energy obtained from the environment. In particular our bodies require minimum amounts of certain substances found in the varieties of food we eat. **Nutrients** are substances in food used by the body to promote normal growth, maintenance, and repair. The nutrients in food can be conveniently organized into six groups: carbohydrates, lipids, proteins, vitamins, minerals, and water. The body's cells, especially those of the liver, are also able to convert one type of nutrient molecule into another. These **interconversions** allow the body to use many substances from different foods and to adjust to the consumption of varying amounts of food. But as you might suspect, there are limits to this ability to construct new molecules from old. **Essential nutrients** cannot be made by such interconversions and must be provided by the diet. However, as long as all the essential nutrients are ingested, the body is able to synthesize the hundreds of additional molecules required to grow and maintain good health. Here we look at each of the basic types of nutrients essential for good health.

Carbohydrates

Most of the body's main sources of energy come from carbohydrates. With the exception of lactose (the milk sugar in milk) and small amounts of glycogen in meats, all the carbohydrates we eat are derived from plants. Sugars, both monosaccharides and disaccharides, come from fruits, sugar cane and sugar beets (from which granulated sugar is produced), honey, and milk. Starches, or polysaccharides, come from grains and leafy and root vegetables. We do not digest the polysaccharide cellulose found in most vegetables and fruits. However, it is important in our diet because it provides roughage, or fiber, which increases the bulk of feces and facilitates defecation.

Lipids

The most common types of lipids in the diet are neutral fats, or triglycerides. In Chapter 2 we discussed two general types of fatty acids. Fatty acids whose carbon atoms are each bonded to two hydrogen atoms are said to be saturated because they contain the maximum number of hydrogen atoms possible. Other fatty acids contain double covalent bonds between one or more pairs of carbon atoms. Because the double bonds replace some of the hydrogen atoms, these fatty acids are called unsaturated. Saturated fatty acids come from meat and dairy products and a few plant products, such as coconuts. Unsaturated fatty acids come from plant products, such as seeds, nuts, and most vegetable oils. The liver can convert one fatty acid into another, but it cannot synthesize linoleic (lin-o-le-ik) acid. Linoleic acid is thus an essential fatty acid. It is required to manufacture phospholipids. Fortunately, it is readily available in most vegetable oils and is also the most common unsaturated fatty acid in our food.

Lipids are important for several reasons. Lipids are a concentrated energy source, they are essential for building cell membranes and constructing myelin sheaths around nerve fibers, and they are necessary for steroid hormone synthesis. However, over 40% of the stored energy in American diets is currently derived from fats. The American Heart Association recommends that fats represent no more than 30% of the total stored energy of the food we consume. A diet high in saturated fats contributes to cardiovascular disease (see Chapter 12, Biology in Action: "Clean Pipes"). Additionally, studies have shown that women with a high fat diet are more likely to develop breast cancer.

Protein

Proteins, the most diverse type of macromolecules in the body, provide structure to organs and cells, transport substances throughout the body, serve as enzymes in chemical reactions, and make up hormones. Animal products, as well as some vegetables, nuts, and grains, are rich sources of protein. Protein formation in the body requires all 20 types of amino acids. Through interconversions cells can assemble many amino acids. They can also transform one type into another type, which may be required in protein formation. However, nine amino acids, called **essential amino acids,** must be directly acquired from the diet because they cannot be manufactured (Table 16-1). The body's cells cannot produce their carbon backbone, cannot put an amine group on the carbon backbone, or simply cannot do the whole process fast enough to keep up with demand.

Table 16-1 Amino Acids

Alanine	Leucine*
Arginine	Lysine*
Asparagine	Methionine*
Aspartic acid	Phenylalanine*
Cysteine	Proline
Glutamic acid	Serine
Glutamine	Threonine*
Glycine	Tryptophan*
Histidine*	Tyrosine
Isoleucine*	Valine*

*An essential amino acid

Animal proteins are called **complete proteins** because they provide all the essential amino acids we cannot manufacture. Foods such as beans, peas, nuts, and cereals are also rich in protein. These proteins are **incomplete** because they are either low in or lack one or more of the essential amino acids. If you eat foods that do not contain a complete balance of essential amino acids, you will not have all the amino acids needed for protein synthesis. Protein synthesis stops when the supply of any one essential amino acid is depleted. The amino acid in shortest supply is called the **limiting amino acid.** Most of us, however, eat such a varied assortment of proteins that the combined amino acid contributions yield sufficient quantities of all nine essential amino acids.

What about diets that consist either of too little or no animal protein? It is possible to combine nonanimal protein sources to acquire all the essential amino acids. For example, when eaten together, cereal grains and beans provide all the essential amino acids. A combination of beans, rice, and flour or corn tortillas can provide the body with a well-balanced combination of all the essential amino acids required for protein synthesis.

Vitamins

Vitamin (*vita* means "life giving") is the name given to a group of organic compounds needed for important human metabolic reactions. Most plant cells can synthesize the vitamins they need, but animals, including humans, cannot. We must therefore obtain vitamins from food. Unlike other nutrients we consume, vitamins are not used for energy or as building blocks. However, without vitamins all the carbohydrates, proteins, and fat we eat would be useless. Vitamins are needed in almost all body cells, so a deficiency in one vitamin can have wide-ranging effects.

Humans require at least 14 different vitamins. Most vitamins act as coenzymes, meaning they act with an enzyme to accomplish a particular metabolic reaction. However, we need vitamins only in small amounts. On average we need about 1 ounce for every 150 pounds of food we eat.

Vitamins are classed as either **water-soluble** or **fat-soluble.** The water-soluble vitamins, which include the B vitamins and vitamin C, are absorbed with water from the digestive system. The body does not store water-soluble vitamins. Therefore we need to obtain them from our diet. Fat-soluble vitamins bind to ingested lipids and are absorbed with them from the digestive system. Unlike water-soluble vitamins, fat-soluble vitamins can be stored in various body tissues, such as the liver and body fat, to be used as needed. Fat-soluble vitamins include vitamins A, D, E, and K.

Vitamins come from all major food groups, but no one food contains all the required vitamins. As a result, a balanced diet is the best way to ensure obtaining a full range of vitamins. However, as we discuss later in this chapter, vitamins are not wonder drugs. The notion that taking huge doses of vitamin supplements will prevent illness or stop the aging process is useless at best and at worst may cause certain health problems, especially with fat-soluble vitamins.

Table 16-2 is a brief overview of some of the most important fat- and water-soluble vitamins and their functions.

Minerals

In addition to the various organic nutrients just described, the body also needs inorganic nutrients. Called **minerals,** these are involved in a wide array of body functions. Some, such as calcium and phosphorus, are vital bone components. Other minerals, such as potassium, sodium, and chloride, are necessary constituents of bodily fluids. Still others are important for protein and enzyme functioning. For example, iron is needed for the functioning of hemoglobin in red blood cells, and without the mineral cobalt, vitamin B_{12} will not properly function as a coenzyme in cellular respiration.

Minerals are categorized as **major** and **trace** minerals on the basis of the amount humans need per day. Major minerals are required in relatively large quantities. For example, 800 to 1200 mg of calcium is the recommended daily amount required for adults. In contrast, trace minerals, also essential to maintain physiological processes and health, are found in such minute quantities in the body that analysis yields a presence of nearly zero amounts. In general the best sources of minerals come from animals, especially seafood. This is because minerals are more concentrated in animal tissues than they are in plant tissues. As an animal eats plant material, the minerals concentrate in the animal's tissues. Sea animals concentrate the minerals in sea water (Table 16-3).

Water

Although the amount of water in the human body varies somewhat with age, between 50% and 70% of our body weight consists of water. Water has three major bodily functions: as a solvent, as an essential part of temperature regulation, and as a part of the body's waste removal system.

As you discovered in Chapter 13, the plasma portion of blood is about 90% water. Nutrients dissolved in it travel from the digestive system to the other body cells. Hormones that regulate the body's activities are also carried in the plasma. Waste products are carried by it from the cells to the kidneys and the lungs, where they are eliminated. Enzymes dissolved in the watery cytoplasm of the cells control the chemical reactions essential to the cells' survival. Water is thus perhaps our most important nutrient because without it life is not possible.

Water is essential for regulating body temperature in two ways. First, water acts as a heat sink, absorbing heat as it is produced as a by-product of metabolism. Second, water helps regulate body temperature by producing perspiration. When the internal temperature of the body increases from its normal temperature, the body secretes a watery substance, perspiration, onto the skin's surface. Evaporating perspiration requires heat energy, so as perspiration evaporates, heat energy is taken from the skin, cooling it in the process. Each liter of perspiration evaporated from the skin requires 600 Kilocalories of heat energy. The net result of this process is to lower internal body temperature back to its normal homeostasis.

As you will learn in Chapter 17, water-soluble waste products are removed by the kidneys and excreted from the body as a watery substance called urine.

Metabolic Rates and Food Intake Requirements

Recall from Chapter 2 that the unit of measurement

of energy contained in the food we eat is the **calorie,** which is the heat energy needed to raise the temperature of 1 g of water 1° C. This is a minute unit of measurement. It is useful, for example, in determining how much energy a mosquito derives from the blood it removes from your arm, but it is too small to be useful when calculating the amount of energy we derive from food. Consequently we use the **kilocalorie,** or **Kcal,** unit instead. A kilocalorie equals 1000 calories. *Kilocalories* is also abbreviated, using a capital **C.** To avoid confusion, here we use the abbreviation "Kcal" when counting food calories. For example, the caloric content of 1 g of carbohydrate metabolized by the body yields

Table 16-2 Major Vitamins

VITAMIN	MAJOR FUNCTIONS	FOOD SOURCES	MINIMUM REQUIREMENT (mg)
Water-Soluble Vitamins			
Vitamin B_1 (thiamine)	Coenzyme that converts pyruvate molecules to acetyl-CoA in cellular respiration.	Meat, grain, legumes, milk.	1.1–1.5
Vitamin B_2 (riboflavin)	Part of coenzyme FAD and FMN, which are important in metabolism, maintenance of red blood cells and skin cells.	Liver, green leafy vegetables, milk, eggs, wheat germ.	1.3–1.7
Vitamin B_3 (niacin)	Part of coenzyme NAD and NADP, needed in cellular respiration.	Meat, enriched grain and cereal, nuts.	15–19
Pantothenic acid	Part of coenzyme A, needed in cellular respiration.	Liver, eggs, milk, green vegetables, grains.	4–7
Biotin	Coenzyme in carbohydrate and fat metabolism and synthesis of fatty acids and nucleotides.	Liver, egg yolk.	0.03–0.10
Vitamin B_6 (pyridoxine)	Protein metabolism.	Vegetables, fish, liver, wheat, cereal.	1.6–2.0
Vitamin B_{12} (cobalamin)	Coenzyme in amino acid production and nucleic acid metabolism.	Liver, meat, milk, eggs, cheese.	0.002
Vitamin C	Collagen synthesis, formation of connective tissue, especially bone and blood vessels, wound healing.	Fruit, green leafy vegetables.	60
Fat-Soluble Vitamins			
Vitamin A	Formation of rhodopsin and other light-receptor pigments, formation and maintenance of epithelial cells.	Green and yellow vegetables and fruits, liver, milk products.	0.8–1.0
Vitamin D	Absorption of calcium and phosphorus, bone and teeth formation.	Vitamin D fortified milk, fish liver oil.	0.005–0.010
Vitamin E	Antioxidant that helps prevent cell destruction.	Vegetable oils, wheat germ.	8–10
Vitamin K	Synthesis of blood clotting factors.	Green leafy vegetables, liver, egg yolk.	0.06–0.08

Table 16-3 Important Minerals

MINERAL	MAJOR FUNCTIONS	FOOD SOURCES	REQUIREMENT (mg)
Calcium (Ca)	Formation of bone and teeth, blood clotting, nerve transmission, muscle action.	Dairy products, egg yolk, green leafy vegetables.	800–1200
Chlorine (Cl)	Acid-base balance, HCl synthesis, water balance, nerve transmission.	Salt, processed food.	750
Fluorine (F)	Increases hardness of bones and teeth.	Drinking water, fish, tea, foods cooked in fluoridated water.	1.5–4.0
Iodine (I)	Thyroid hormone production, which regulates cell metabolism, basal metabolic rate (BMR).	Iodized salt, seafood.	0.15
Iron (Fe)	Hemoglobin synthesis, oxygen transport.	Liver, meats, enriched bread and cereal.	10–15
Magnesium (Mg)	Bone and teeth formation, muscle and nerve function.	Meat, seafood, milk, cheese, whole grains.	280–350
Phosphorus (P)	Bone and teeth formation, ATP production, component of nucleic acids.	Dairy products, meat, whole grains, nuts, legumes.	800–1200
Potassium (K)	Muscle and nerve function, osmotic pressure.	Fruits and vegetables.	2000
Sodium (Na)	Muscle and nerve function, osmotic pressure regulation.	Salt, seafood, processed foods.	500
Zinc (Zn)	Component of several enzymes, carbon dioxide transport.	Seafood, liver, meat, whole grain.	12–15

4.1 Kcals; 1 g of protein, 4.2 Kcals; and 1 g of fat, 9.3 Kcals.

Energy Requirements

How much energy does your body require to maintain itself over time? To answer this question we begin with a definition. The **metabolic rate** is the sum of the heat produced by all the chemical reactions and mechanical work of the body. Because many factors influence the metabolic rate, it is usually measured under standardized conditions. A person who has not eaten for at least 12 hours reclines and physically and mentally relaxes. The temperature of the room is maintained between 20° to 25° C, or 68° to 77° F. The measurement calculated under these conditions, called the **basal metabolic rate, or BMR,** is the energy the body requires to perform only the most essential activities of breathing, blood circulation, and cellular metabolism. Basal metabolic rate amounts to about 1 Kcal per kg of body weight per hour in men and 0.9 Kcals per kg of body weight per hour in women. For example, a 70-Kg, or 154-lb, adult male has a BMR of approximately 70 Kcal/hr. If the BMR is extended over 24 hours, the rate would be 1680 Kcals. In contrast, an adult female weighing 50 kg, or 110 lb, would have a BMR of 45 Kcals, or 1080 Kcals, over 24 hours.

The BMR depends primarily on **lean body mass,** that part of the body free of fat. Lean body mass includes muscle, bone, organs, some types of connective tissue, and skin. Because males have on average greater lean body mass, especially muscle, compared with females, their BMR is usually slightly higher.

Most of us do considerably more than rest 24 hours a day. We go to school, work, exercise, shop, walk, or drive. This **voluntary work** also requires energy. On average during a moderate daily routine an adult male will burn about 4 Kcals per kg of body weight per hour. The total energy needed for 24 hours is simply the sum of the BMR added to the energy cost of voluntary work.

Dieting

When energy intake and energy expenditure are in balance, body weight remains stable. When they are unbalanced, weight is either gained or lost. Body weight in most people is surprisingly stable, which has led many researchers to look for the physiological mechanisms that control food intake, energy expenditure such as metabolic rate, or both. However, the nature of such control mechanisms remains elusive.

If the amount of food consumed exceeds the energy needs of the body, the excess ATP produced cannot be stored for significant lengths of time. Instead, this excess energy is used to synthesize fats which are then stored in adipose tissue. For example, when large amounts of glucose and amino acids are left over from a large meal, most of their carbon atoms are used to synthesize lipids. This process requires ATP. These lipids are then transported to fat storage cells of the body, called **adipose** cells. The result is weight gain. How much weight you gain is directly related to the amount of excess food you consume.

What regulates the body's metabolism and weight? Some have theorized there is a set point in the hypothalamus that determines the amount of body fat a person will have, plus or minus about 10%. One nutritionist has described the set point as a coiled spring. The further you move from your usual weight, the harder the force acts to pull you back to it. For example, a 70-kg male would be able to lose about 7 kg, or 15 lb relatively easily, but beyond that it would become increasingly difficult to lose more weight. There is much controversy over the set point. Indeed, many researchers even question its existence. However, whether a set point exists or not, there are relatively simple ways of maintaining a comfortable body weight.

Each pound of fat stored in the body's adipose tissues represents about 3500 Kcal of stored energy. If you wish to lose weight, you must induce the body to use this stored energy. There are two ways to do this. First you could maintain the same activity throughout the day and reduce your food intake. Second you could increase your activity level while also reducing your food intake. This second means is more effective than the first. Some argue that exercise combined with reduced food intake is more effective than reducing food intake alone because the body somehow gets around the limitations of weight loss imposed by the body's set point if exercise is included in dieting.

Also, the BMR is roughly the same in most people, about 25% to 30% of total energy use. The energy needed by the body to digest food plus the BMR accounts for about 60% to 75% of total energy use. The big difference in energy use among people results from different activity levels. Some people are active, whereas others are sedentary.

Food Requirements as a Function of Age and Sex

The BMR decreases about 2% every 10 years after age 30. As we discuss in Chapter 20 this decrease in BMR is mainly caused by a decrease in the total number of metabolically active body cells. On the basis of this change, although it is important to decrease the total number of Kcals consumed each day, it is still important to ingest the same amount of nutrients. For example, your need for calcium is as great at age 70 as it is at age 30. However, for pregnant women caloric and nutrient needs change markedly. We discuss the specific changes in Chapter 19, on human reproduction.

With an understanding of nutrients and the body's energy requirements, we can now turn our attention to the digestive system that prepares nutrients for absorption into body cells.

Gastrointestinal Tract

The mechanical and chemical breakdown of food and its absorption into the circulatory system for distribution to tissue cells is called **digestion.** As you can tell from the definition, digestion involves three major processes. First, **mechanical digestion** involves the physical breakdown of the food into smaller fragments. Mechanical digestion includes such processes as chewing, mixing these small fragments with a fluid secreted by accessory glands into the mouth, and once swallowed, churning and mixing food in a region of the digestive system called the stomach. Second a hydrolytic process called **chemical digestion** involves breaking food macromolecules down to simple molecules. Chemical digestion occurs as enzymes are secreted into the

Biology in Action BODY FAT: HOW MUCH IS TOO MUCH?

Most of us know that being overweight has many well-documented health risks, such as a higher incidence of arteriosclerosis, high blood pressure, coronary artery disease, and diabetes. But how much weight is too much? The five basic components of weight are the skeletal bones, muscles, organs, water, and fat. Biologists refer to **lean body mass** as the proportion of total body weight made up of bones, muscle, water, and organ tissues. Body fat, represented as a percentage of total body weight, comprises all nonessential adipose tissue. Bone and muscle weigh more per unit than does fat. Knowing this means that the bathroom scale is probabily not an accurate indicator of your body weight because it reveals little about body composition. For example, a skilled dancer or long-distance runner with well-developed muscles and dense bones may have considerably less body fat than an inactive person of the same size and weight. The important component, then, is not total body weight alone but what percentage of total body weight is fat.

Generally people are considered to be overweight or obese when they are at least 20% heavier than their ideal weight as published in standard insurance company tables. These tables simply list for any height the weight associated with maximum life span. In other words, the table does not tell what weight will make you healthiest while alive. It states only the average weight most often associated with longevity. A 260-lb muscular football player may be overweight according to height and weight standards yet actually have much less body fat for his height than average. Similarly a 45-year-old woman may weigh exactly the same as when she was in high school, yet now have a considerably different body composition.

Other methods are available to assess body composition. Perhaps the most accurate and least expensive assessment of body fat is the skin-fold caliper test. Using a set of calipers, a technician grasps a fold of skin at several different spots on one side of the body, just hard enough to pinch. A dial on the calipers indicates the thickness of the fat just beneath the skin at each site. These measurements are put into an equation that calculates body-fat percentages. The equation was derived from measurements of thousands of people of all shapes and sizes to develop correlations between the thickness of subcutaneous fat at different places on the body and the percentage of total body fat. In other words, "this much fat right here means that much all over."

So how much fat is too much? It depends. According to the American College of Sports Medicine, if you are in good general health and want to be a bit more athletic, a healthy range would be about 11% to 18% for men and 16% to 23% for women. Other researchers indicate the healthy range for women, for example, may run as high as 32%. However, all agree that above 35% for women and 25% for men *is* overweight and a significant health risk.

digestive system. All digestive enzymes break down food molecules by hydrolysis, which involves adding a water molecule to each molecular bond broken, or *lysed.* Third, **absorption** involves transporting simple molecules from the digestive system to the circulatory system for distribution to the rest of the body.

The organs of the digestive system can be separated into two major categories. These are the **alimentary canal,** also called the **gastrointestinal (GI) tract,** and the **accessory digestive organs.** The alimentary canal is a continuous, hollow, coiled tube that winds through the body cavity and is open to the outside environment at both ends. The organs of the alimentary canal include the mouth, pharynx, esophagus, stomach, small intestine, and large intestine. The large intestine includes the terminal ending, called the anus. The alimentary canal is approximately 7 m, or 22 feet, long. The most important accessory organs include the salivary glands, the liver, the gall bladder, and the pancreas. Each of these structures lies outside the alimentary canal (Figure 16-1).

The walls of the GI tract have four basic layers with regional modifications. A **mucous membrane,** or **mucosa,** lines the lumen, or the space within the

Figure 16-1 Human digestive system

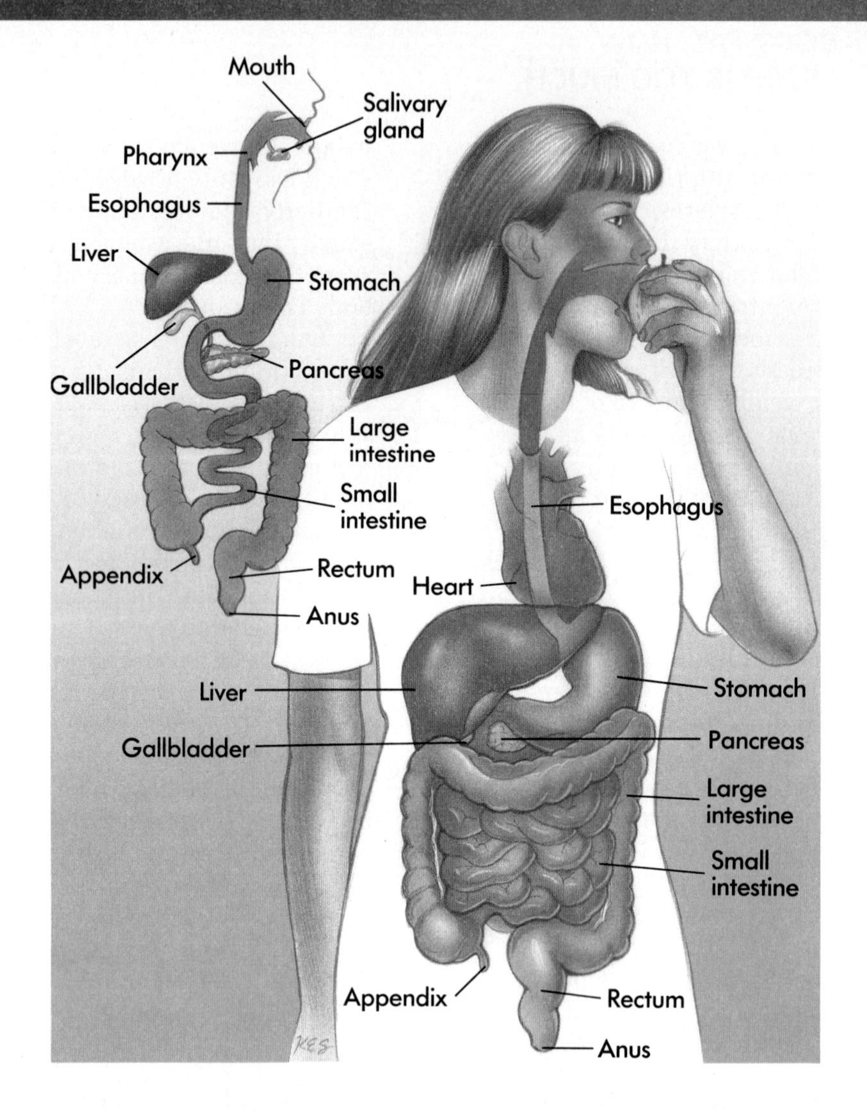

The digestive system consists of the gastrointestinal tract and the accessory digestive organs.

tube, of the alimentary canal. The mucous membrane is composed of epithelial cells, an underlying layer of connective tissue, and capillaries. Its major functions, depending on the region of the alimentary canal, are secreting mucus and digestive enzymes, absorbing the end products of digestion into the blood, and protecting against invasion by foreign microorganisms. The mucus produced by this membrane also protects the digestive system from being digested itself by the enzymes working on the food as it passes through the GI tract.

The second layer of the wall of the GI tract is the **submucosal layer,** which consists of nerves, blood vessels, lymphatic vessels, and connective tissue. Its rich capillary network supplies the alimentary canal tissues with blood and also carries away molecules and other substances absorbed from the alimentary canal. Its nerve supply is part of the autonomic nervous system. The third layer of the GI tract is the **muscularis layer,** which consists of both circular and longitudinal smooth-muscle fibers. The protective outermost layer consists of tough connective tissue and is called the **serosa.**

Mouth and Associated Organs

The **mouth,** or **oral cavity,** is the site of food entry, or **ingestion,** into the alimentary canal. A primary reason you probably enjoy eating food is because of the combined sensation of olfaction and taste in the mouth. As you recall, the olfactory receptors, in the

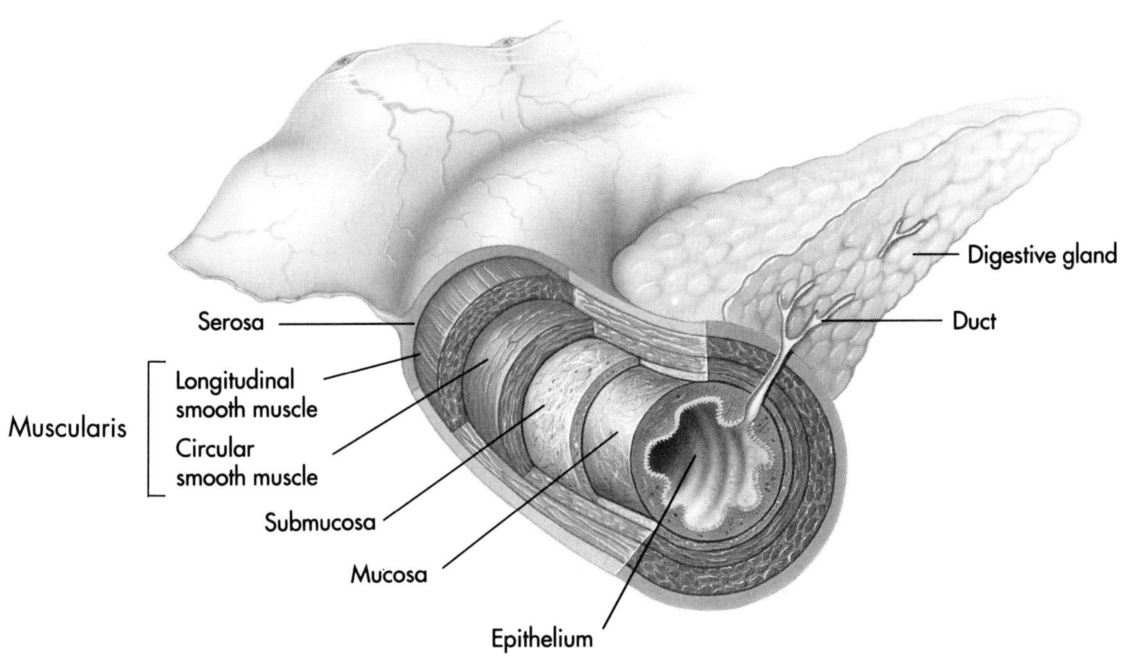

roof of the nasal cavities, are responsible for odor, or smell, detection. As food is brought toward the mouth, its odor is detected as air is inhaled through the nostrils. Taste itself is a function of the taste buds, located on the surface of the tongue.

The **palate,** forming the roof of the mouth, has two distinct parts. The **hard palate,** toward the front of the mouth, consists of tissue-covered bone and forms a rigid surface against which the tongue can force food during chewing. The **soft palate** extends back from the hard palate into the pharynx. Along with the small cone-shaped uvula suspended from it, the soft palate automatically rises to close off the nasopharynx region when we swallow food.

The **tongue** occupies the floor of the mouth. It is composed primarily of skeletal muscle covered by mucous membrane. The anterior two thirds of the tongue are in the oral cavity, whereas the remaining one third is actually in the pharynx. Besides making movements for speech, the tongue mixes food with saliva during chewing, forms food into a compact mass called a **bolus** (bo′-lus), and initiates swallowing. The thick mucous membrane over the tongue is covered with tiny projections called **papillae** (puh-pih′-lie). Papillae give the tongue its roughness and provide friction for manipulating food.

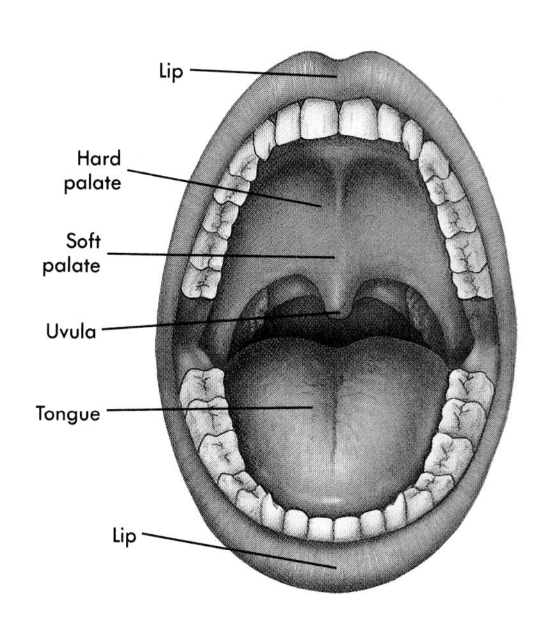

Glands outside the oral cavity produce and secrete a watery substance called **saliva,** which has three important functions in the digestion process. First, saliva cleanses the oral cavity and dissolves chemicals so they can be tasted. Second, saliva moistens food and helps compact it into a bolus. Third, saliva contains **salivary amylase,** an enzyme that begins the breakdown of starchy foods. Saliva is about 97% to 99.5% water. In addition to salivary amylase, saliva also contains bicarbonate ions, which are buffers keeping the salivary pH between 6.5 and 7.5 even when acidic foods are in the mouth. Another component of saliva is a glycoprotein called **mucin** (myoo′-sin), which gives saliva its stickiness, or viscosity, and helps to bind bits of food together into a bolus.

Saliva is produced by three pairs of exocrine glands called **salivary glands,** which lie outside the oral cavity. Each salivary gland secretes its contents into the oral cavity via a duct. The combined output of these glands is from 1 to 2 L of saliva every 24 hours. The large **parotid** (puh-rah′-tid) **glands** lie just below and in front of the ears. The parotid duct opens next to the second molars of the upper jaw. Mumps, a common childhood disease, is an inflammation of the parotid glands.

The **submandibular glands** lie along the lower jaw, or mandible, toward the inside of the oral cavity. Their ducts open alongside the membrane that fastens the tongue to the floor of the mouth. The small **sublingual glands** lie in front of the submandibular glands under the tongue. Their ducts open into the floor of the mouth (Figure 16-2).

Figure 16-2 Salivary glands associated with the oral cavity

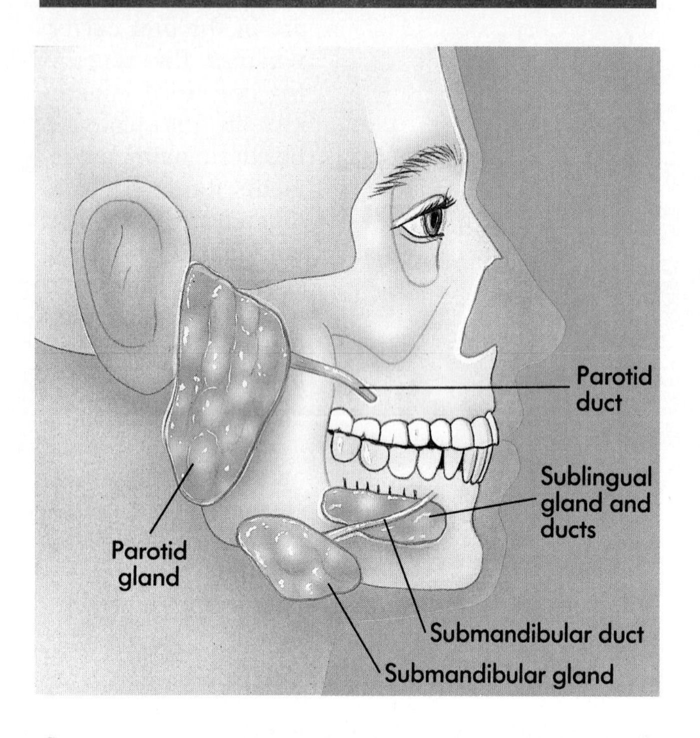

Teeth

The role of the teeth in food processing is to mechanically break down food so it can be swallowed. We chew, or **masticate** (mas'ti-kat), by opening and closing the jaws and moving them from side to side. At the same time, food is moving between the teeth and the tongue and mixing with saliva. During this process the teeth tear and grind the food, breaking it down mechanically into smaller fragments.

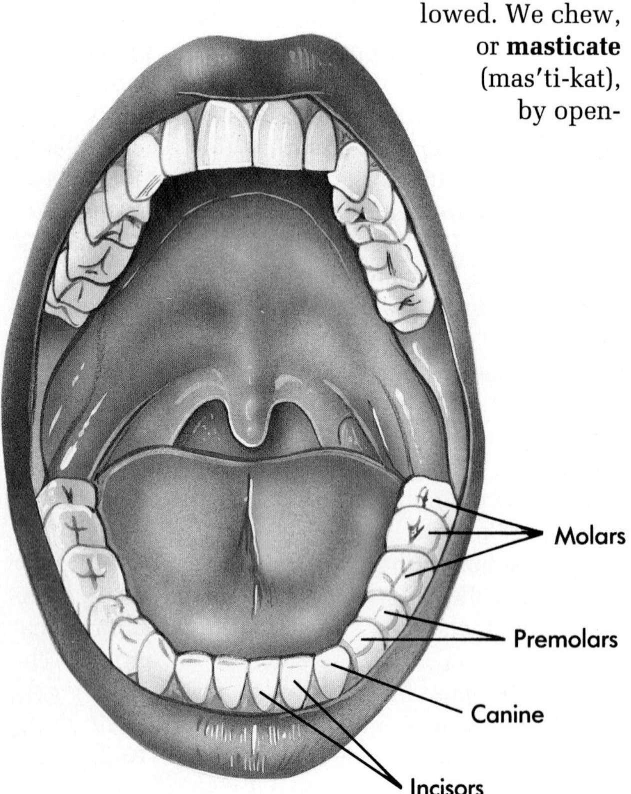

On average by age 21 two sets of teeth have formed. The first set consists of the **deciduous** (de-sid-u-us, *deciduus* = falling off) **teeth,** sometimes called **milk** or **baby teeth.** Deciduous teeth first begin to appear at about age 6 months. An additional pair of teeth erupts at 1- to 2-month intervals until all 20 milk teeth are through the gums by age 20 to 24 months.

As the second set, or **permanent teeth,** develop and enlarge in the jaw, the deciduous teeth loosen and eventually fall out, sometime between the ages of 6 and 12. Generally all the permanent teeth, with the exception of the third molars, have erupted by the end of puberty. The third molars, sometimes called **wisdom teeth,** emerge between the ages of 17 and 25. There are a total of 32 permanent teeth, but the wisdom teeth often fail to erupt or are completely absent in many people.

Pharynx

As described in Chapter 15, the pharynx is a short, muscular common passageway for food, fluids, and air. When food and fluids are swallowed, they must be propelled from the back of the mouth through the pharynx into the tube leading to the stomach. However, food must be prevented from entering the larynx and the trachea. The swallowing process ensures that food usually goes into the GI tract.

The swallowing process consists of both voluntary and involuntary, or reflex, phases (Figure 16-3). In the mouth once the tongue has formed the food into a bolus, the voluntary phase of swallowing begins. During this phase the tip of the tongue is placed against the hard palate. As the tongue muscles contract, the bolus is forced back and into the upper part of the pharynx, where it stimulates tactile receptors and passes out of voluntary control. The involuntary phase of swallowing is controlled by the autonomic nervous system from the swallowing center in the medulla oblongata. Once the tactile receptors in the pharynx send their impulses to the swallowing center, subsequent events occur automatically.

Once food enters the pharynx, all routes it might take, except for the one into the GI tract, are blocked off. The tongue blocks off the mouth, the soft palate rises to close off the upper region of the pharynx leading to the nasal cavities, and the larynx rises so its opening into the respiratory airway,

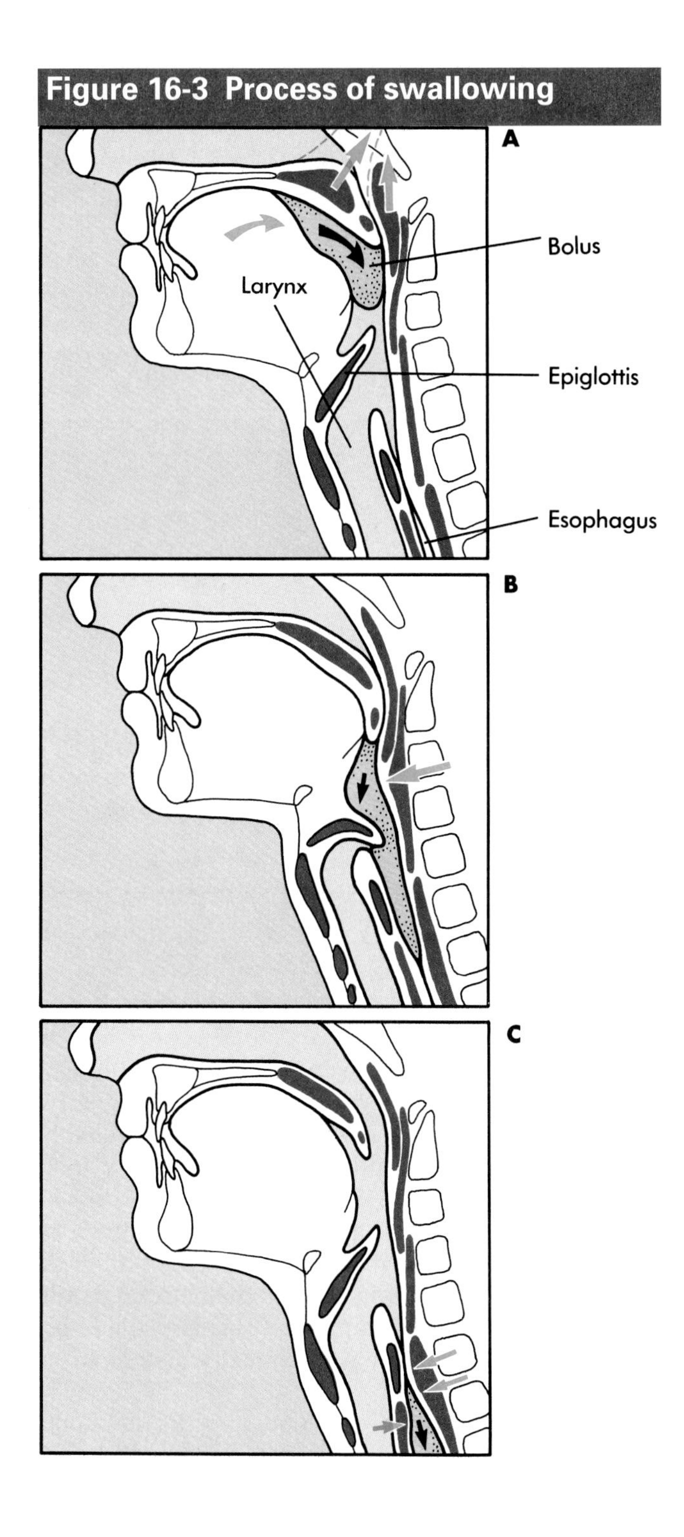

During the voluntary phase of swallowing, the tongue contracts and forces the bolus back into the upper part of the pharynx. Tactile receptors are stimulated, and the soft palate rises to close off the upper regions of the pharynx (A). The larynx reflexively rises so that the glottis is covered by the flaplike epiglottis (B). The bolus enters the esophagus where wavelike contractions continue to move the bolus downward (C).

called the glottis, is covered by the flaplike epiglottis. This particular movement is easy to observe. When food is eaten, the up-and-down movement of the *Adam's apple* indicates the movement of the larynx up to the epiglottis.

Esophagus

The **esophagus** (e-sof′-uh-gus) is a tube, approximately 25 cm, or 10 inches long, that conducts food from the pharynx to the stomach. Unlike the trachea in front of it, the esophagus is normally relaxed and collapsed, except when food or fluids are passing through it. As the bolus passes from the pharynx into the esophagus, an involuntary, wavelike muscle contraction begins. This moves the bolus down the esophagus and is also generally responsible for moving food through the entire GI tract. **Peristalsis** (payr′-ih-stal′-sis) involves alternate waves of esophageal muscle wall contraction and relaxation. The effect is to squeeze food down the esophagus toward the stomach. In the esophagus peristaltic waves are so powerful that food and fluids will reach your stomach even if you stand on your head. Peristalsis also occurs in the stomach and intestines, assisting in pushing food through the digestive tract (Figure 16-4).

The entrance of the esophagus into the stomach is marked by a slight thickening of the circular smooth muscle called the **gastroesophageal constrictor.** This thickened ring of muscle is normally constricted, closing off the stomach from the esophagus. However, when food is swallowed, the peristaltic wave causes the constrictor to relax temporarily, allowing the bolus to pass through into the stomach. After the bolus has passed through the gastroesophageal constrictor, it closes, preventing the stomach contents from moving back up into the esophagus. Occasionally after an unusually large meal, the stomach contents do flow back into the lower regions of the esophagus, despite the closure of the gastroesophageal constrictor. The acidity of these contents cause a burning sensation we call heartburn.

Vomiting is the forceful expulsion of the stomach contents out through the mouth. Vomiting is not a reversal of peristalsis, nor is it caused by a sudden stomach contraction. Surprisingly, the major force for expelling the contents of the stomach comes from the contraction of the diaphragm and the abdominal muscles. The vomiting reflex begins with a deep inhalation and a closure of the glottis. The contracting diaphragm moves down forcefully on the stomach. At the same time, the abdominal muscles contract, compressing the abdominal cavity. The result is that the stomach is

Figure 16-4 Peristalsis

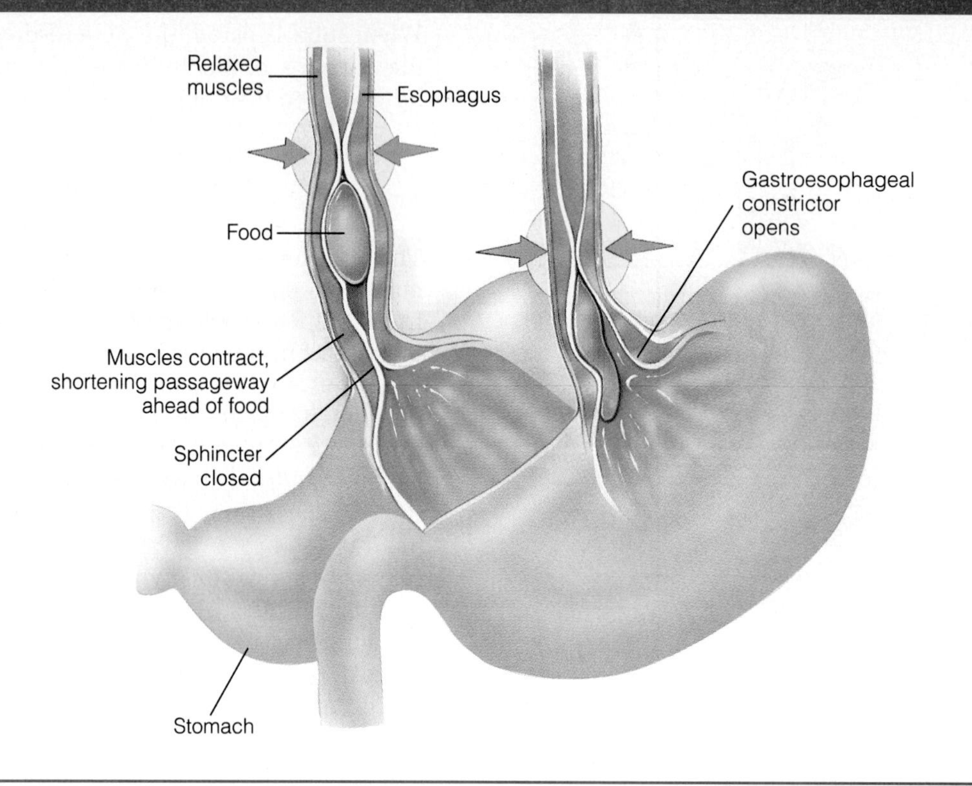

squeezed from both sets of muscles, forcing its contents up through the gastroesphogeal constrictor into the esophagus and out of the mouth.

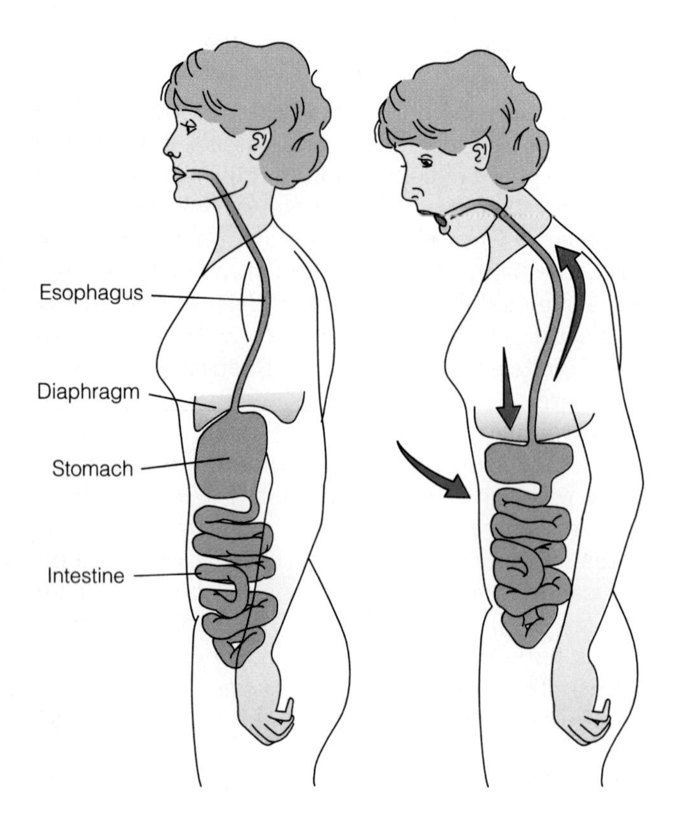

Stomach

Below the esophagus tucked under the diaphragm in the upper left section of the abdominal cavity, the GI tract greatly expands to form the **stomach.** The stomach is a temporary storage area for food and an important site for the mechanical breakdown of most food and the chemical breakdown of some types of food. The stomach is approximately 25 cm, or 10 inches long, but its diameter depends on how much food it contains. It usually holds up to 1.5 L of food, but when it is greatly distended, it can hold as much as 4 L.

During the 2 to 6 hours food stays in the stomach, it turns into a semiliquid called **acid chyme** (kim), with a pH of between 2.0 and 3.0.

The stomach wall is made of the four layers typical of the alimentary canal generally, but two of its layers are modified for special stomach functions. Besides the usual circular and longitudinal layers of muscle, the stomach also contains another layer of smooth muscle that runs obliquely to the other two. This layer allows the stomach not only to move food along the tract but also to churn and mix the food and physically break it down into smaller fragments. It also contributes to the stomach's ability to distend greatly. Stomach contractions, accomplished by the three layers of smooth-muscle tissue in the stomach wall, compress, knead, and continually mix the food with gastric juice, enhancing mechanical digestion. Because the food is moved around in several directions at once, like clothes in a washing machine, this unique version of peristalsis is called **mixing action.**

A second modification occurs in the mucous membrane lining the stomach. This membrane is dotted with millions of microscopically small glands called **gastric glands,** which collectively produce a secretion called **gastric juice.** Passage of food from the stomach into the next segment of the alimentary canal is controlled by a thickened ring of smooth muscle encircling the exit, called the **pyloric** (pi-lor′-ik) **sphincter** (Figure 16-5).

Small Intestine

The **small intestine** is the major digestive segment of the GI tract and is where usable food is finally prepared for its journey into the body cells. Chemical digestion is completed with the assistance of secretions produced by the liver, the pancreas, and the small intestine.

The small intestine is a convoluted tube that begins at the pyloric sphincter of the stomach. It is the longest section of the alimentary canal, with an average length of 6 meters, or 20 feet, but its diameter is only about 2.5 cm, or 1 inch. It is called the small intestine because of its small diameter, not because of its length. The first 25 cm, or 10 to 12 inches, of the small intestine is called the **duodenum** (doo′-o-de′-num). The ducts delivering enzymes and other substances from the liver and the pancreas enter at the duodenum. The next 2.5 meters, or 8 feet, following the duodenum is called the **jejunum** (je-ju′num). The final 3.5 meters, or 12 feet, is called the **ileum.** The small intestine ends at the **ileocecal valve,** where it joins the large intestine (Figure 16-6).

Figure 16-5 Stomach

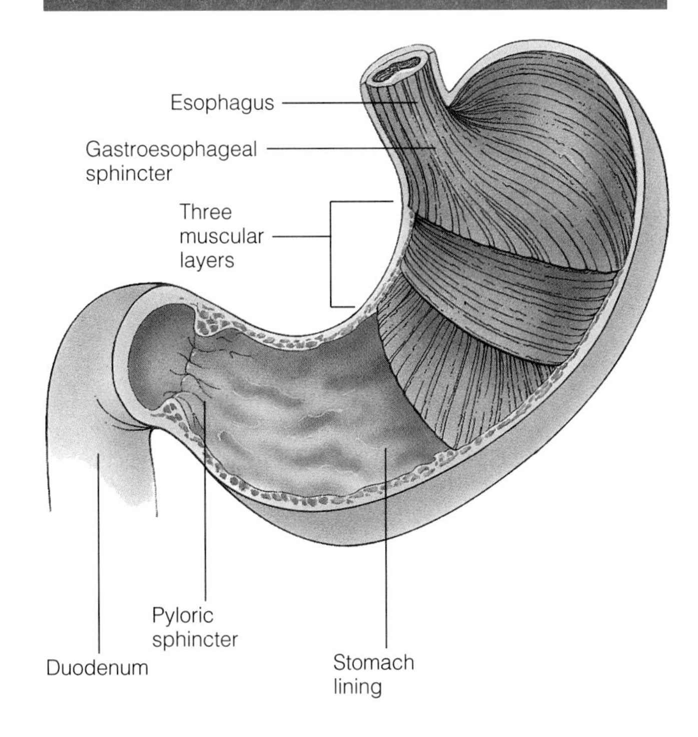

The wall of the small intestine has the typical four structural layers common to the alimentary canal, but because it is specialized for absorption, the small intestine differs in several ways from other portions of the GI tract. The mucosal layer has three structural modifications that increase the absorption surface tremendously. First there are deep, permanent folds in the mucosal and submucosal layers of the entire small intestine that increase its surface area. Most of these folds extend around the circumference of the small intestine but are also spread further apart down the length of the small intestine.

Second, fingerlike projections, called **villi** (sing. *villus*), extend from the mucous membrane, giving it a slightly fuzzy appearance. Villi are about 1 mm long, but they decrease both in number and length in the later segments of the small intestine. The villi increase the surface area of the small intestine about 600 times over what it would be if it were a smooth surface. Third there are tiny projections of the cell membrane of the absorptive cells covering the villi, called **microvilla** (sing. *microvillus*) or the **brush border.** Beneath this single layer of absorptive cells covering each villus is a dense capillary bed and a modified lymphatic capillary called a **lacteal** (lak′-te-ul). Not only do villi increase surface area but their structure also brings blood and lymph vessels close to the absorptive surface of the small intestine to enhance molecular absorption into the circulatory system (Figure 16-7).

Accessory Glands Associated with the Small Intestine

The **liver** and the **pancreas** (pan′kre-as) are two glands that lie outside the small intestine but are connected to it by ducts, through which their secretions move into the duodenum.

The liver is the largest gland in the body, weighing almost 1.5 kg, or about 3 lb, in the average adult. The liver is made up of cells called **hepatocytes** (heh-pa′-to-sitz), which have many functions.

Figure 16-6 Small intestine

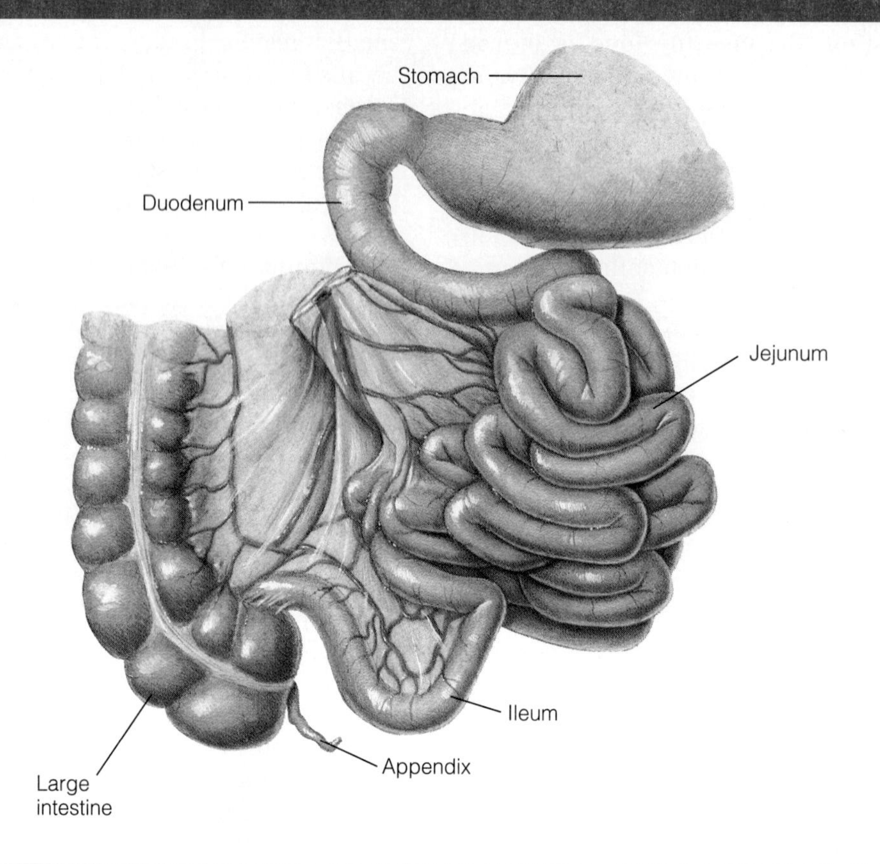

The small intestine is the longest segment of the gastrointestinal tract and is where usable food is finally prepared for its journey into the body's cells.

They have diverse functions because of the specialized organelles in them. The liver is located under the diaphragm, extending to the right side of the abdominal cavity.

The hepatocytes are in direct contact with blood from two sources: venous blood from the GI tract and arterial blood from the aorta. The materials absorbed from the GI tract do not go directly into circulation to be distributed to body cells. Instead, they are sent through the **hepatic portal vein** to the liver. The hepatic portal vein then branches into specialized capillaries in the liver. The **hepatic artery,** a major branch of the aorta, carries oxygenated blood to the liver. The arterial blood nourishes the hepatocytes. Blood from these two sources mixes in the liver and then leaves the liver through the **hepatic veins,** which empty into the inferior vena cava.

The hepatocytes produce **bile,** a mixture of substances primarily responsible for assisting in the breakdown of fats in the small intestine. Bile also contains lecithin, a lipid called cholesterol, and a substance called **bilirubin.** Bilirubin is a by-product of the heme portion of hemoglobin found in red blood cells that have been destroyed by the liver. Bilirubin gives bile its greenish color and gives feces its characteristic brownish color. Bile contains toxic chemicals that have been converted to a harmless form by the liver. Bile also contains sodium bicarbonate.

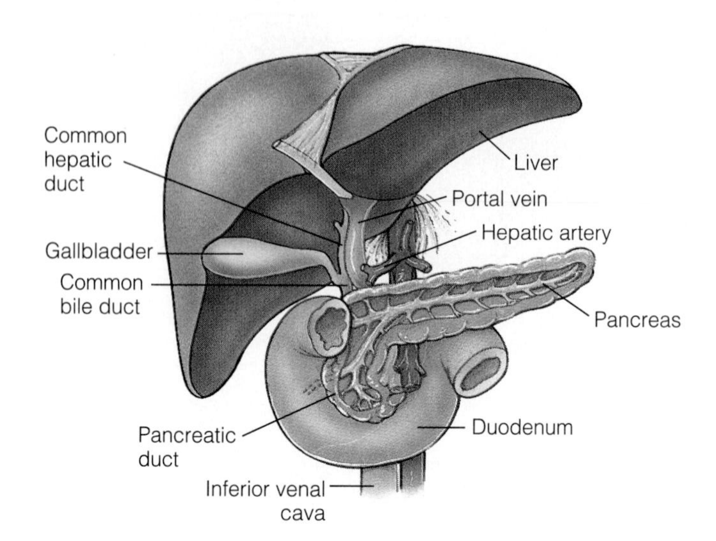

Figure 16-7 Villi

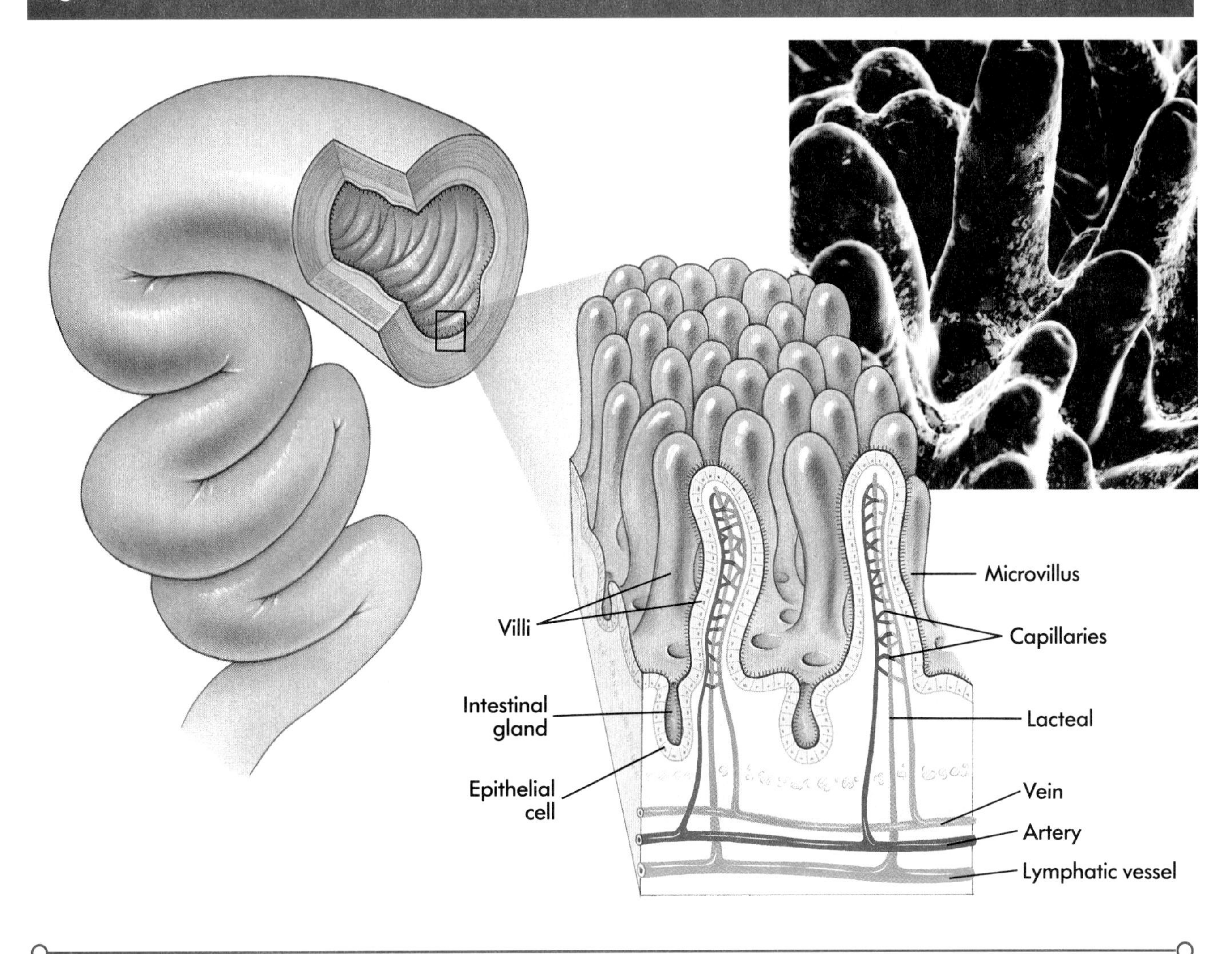

Villi are fingerlike projections of the small intestine's mucous membrane. Within each villus is a capillary bed and a modified lymphatic capillary called a lacteal. Molecules derived from digested food are absorbed across the villi and eventually into the circulatory system.

Bile is continuously being secreted by the hepatocytes, leaving the liver through the **hepatic duct** and traveling toward the duodenum. However, when digestion is not occurring, the bile is temporarily stored in the gallbladder. The **gallbladder** is a pear-shaped sac which serves as a reservoir for bile. During digestion of fats, the gallbladder contracts, ejecting bile through the **common bile duct** into the duodenum.

Although an important liver function is to produce bile, the liver also has many other important functions, making it essential for life. Blood glucose levels are always about 0.1% of the total blood volume even though we eat only intermittently. For example, after a meal any excess glucose in circulation is removed and converted by the liver to glycogen or fat for storage under the influence of the hormone insulin. In contrast, between meals if the glucose supply in the blood is low, the liver, under the influence of the hormone glucagon, breaks down glycogen to glucose and releases it to the blood.

The liver can also convert amino acids to glucose molecules. Recall that all amino acids contain nitrogen atoms, and glucose contains only carbon, hydrogen, and oxygen atoms. Consequently, before amino acids can be converted to glucose molecules, nitrogen atoms must be removed. The liver, through a metabolic process called **deamination,** eventually converts these nitrogen-bearing groups to a substance called **urea.** Isolated nitrogen groups released into the blood are toxic to cells. Converting them to urea makes them safe to transport in the blood to the kidneys for removal and excretion in urine. Also recall from Chapter 13 that blood-borne proteins, such as the blood-clotting proteins prothrombin and fibrinogen, are manufactured by the liver and iron and several fat-soluble vitamins necessary for proper cellular metabolism are stored by the liver.

The pancreas is a pinkish-colored gland weighing approximately 100 g, or about 4 oz. It has both endocrine and exocrine functions. As you learned in Chapter 10, the pancreas contains clusters of endocrine cells that secrete two hormones, insulin and glucagon, into the circulatory system. The rest of the pancreatic tissue cells produce and secrete a substance called **pancreatic juice,** which contains enzymes that break down proteins, lipids, nucleotides, and starches; it is also alkaline, helping neutralize the acidic liquid entering the small intestine from the stomach. Pancreatic juice is secreted into a network of ducts in the pancreas. These ducts feed into a centrally located **pancreatic duct,** which leaves the pancreas and merges with the **common bile duct** from the gallbladder that empties into the duodenum.

Large Intestine

The **large intestine** is the final section of the GI tract. It extends from the end of the small intestine to the exterior of the body and is approximately 1.5 meters or 5 feet long and about 7 cm, or 2.5 inches, in diameter. Its position relative to the small intestine is like that of a picture frame, with the large intestine surrounding the coiled small intestine.

The large intestine begins in the lower right side of the abdominal cavity. The small intestine joins the large intestine, so there is a "dead end" to one side of the large intestine. This small sac, called the **cecum** (se′kum), has a small projection, called the **appendix,** about the size of your little finger. Although the appendix contains lymphoid tissues that produce white blood cells, it is also an ideal place for bacteria to accumulate. This can sometimes cause the appendix to become infected and inflamed, a condition called **appendicitis.** Symptoms include pain in the lower right section of the abdomen, a high fever, and diarrhea. If the inflamed appendix ruptures, its bacteria-laden contents are released into the abdominal cavity, leading to a generalized infection that is often fatal. To prevent this situation, the inflamed appendix is surgically removed and the remaining infection treated with antibiotics. As mentioned earlier in the chapter, the ileocecal valve separates the end of the small intestine from the cecum of the large intestine. It is a one-way valve comprising two flaps of tissue that allow the material to flow in only one direction, that is, from the small to the large intestine.

The first segment of the large intestine is called the **colon.** The ascending colon begins at the cecum at the end of the small intestine and goes up the right side of the abdominal cavity to about the level of the liver. The transverse colon crosses the abdomen just below the liver, and the descending colon passes down the left side of the abdominal cavity. The last segment, which is about 15 cm, or 6 inches long, is called the **rectum.** The opening of the rectum to the exterior is called the **anus,** which has two sphincters that normally keep it closed, an internal involuntary (smooth muscle) and an outer voluntary (skeletal muscle) sphincter.

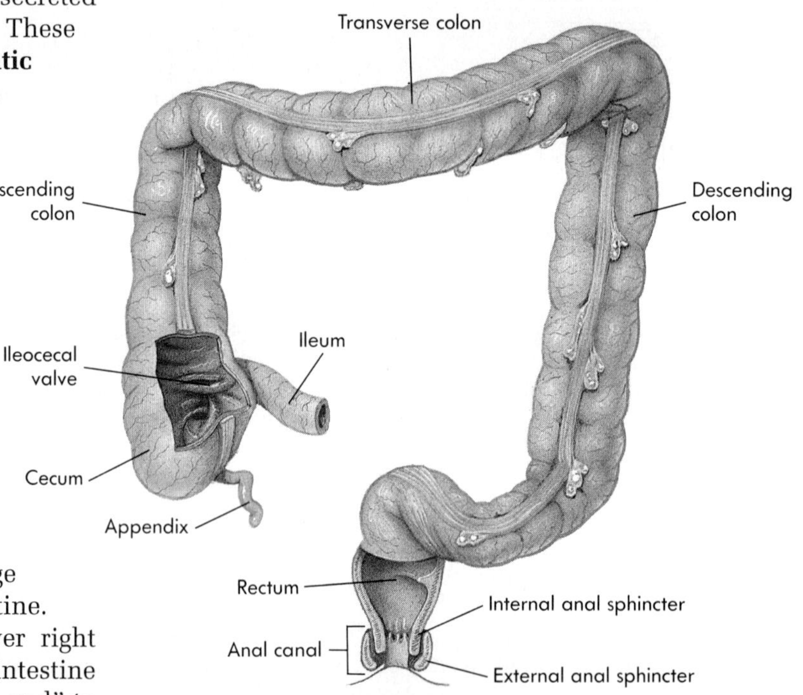

Digestion in the Gastrointestinal Tract

Both mechanical digestion and limited chemical digestion occur in the mouth. Salivary amylase, a hydrolytic enzyme, splits complex carbohydrates, or starches, into smaller fragments of linked glucose molecules.

Besides being a holding area for ingested food, the stomach continues the breakdown process begun in the oral cavity by the teeth and salivary amylase, further degrading food both physically and chemically. As food reaches the stomach the bolus is exposed to gastric juice secreted by the stomach's inner lining. Gastric juice contains hydrochloric acid and a hydrolytic enzyme called **pepsinogen** (pep-sin′o-jen). When both are secreted into the lumen, the pepsinogen becomes activated by hydrochloric acid and converts to **pepsin.** The hydrochloric acid in gastric juice assists in further breaking down large food particles into smaller fragments, but they are still not small enough yet to be absorbed.

Pepsin begins the process of breaking down protein by splitting the peptide bonds linking the amino acids. This is the only chemical digestion occurring in the stomach. The only commonly ingested substances known to be absorbed by the walls of the stomach and into circulation are lipid-soluble substances, such as alcohol and aspirin, ibuprofen, and acetaminophen (a-set-a-me′no-fen). Although alcohol is water-soluble, it is also somewhat lipid-soluble, so it can diffuse through the phospholipid membranes of the cells lining the stomach and enter the capillaries of the mucosa. Aspirin and aspirinlike substances are weak acids. When they enter the acidic stomach environment they do not dissociate but remain intact. They are lipid-soluble and can also pass through the stomach wall.

The small intestine completes the chemical digestion of the chyme that enters from the stomach and absorbs the final products into the blood and lymph. Entering the duodenum, pancreatic juice and bile, produced in the liver, as well as many enzymes secreted from the walls of the small intestine, are responsible for the final chemical digestion. One important substance in pancreatic juice is sodium bicarbonate ($NaHCO_3$), which neutralizes the highly acidic chyme as it enters the duodenum. It is so effective in neutralizing acid chyme that the pH of the fluid chyme is raised to 8.5. Alkaline conditions are necessary for the hydrolytic enzymes to chemically digest food. Here we will describe the completion of the chemical digestion of carbohydrates that enter the small intestine.

Carbohydrates

Carbohydrates not broken down by salivary amylase are chemically digested by **pancreatic amylase** in the small intestine. Within 10 minutes after entering the small intestine, starch is broken down into disaccharides. Maltose and other disaccharides, such as sucrose and lactose, are broken down into monosacchrides by specific enzymes secreted by microscopic intestinal glands embedded in the brush border of the small intestine. For example, the enzyme **maltase** splits the disaccharide maltose into two separate glucose molecules. The enzyme **lactase** splits lactose, or milk sugar, into glucose and galactose. These monosaccharides are then absorbed across the intestinal wall into the bloodstream.

Protein

Protein fragments that enter the small intestine from the stomach are split into individual amino acids by enzymes produced by the brush border of the small intestine. One group of enzymes, called **aminopeptidases,** split off one amino acid at a time from the end of the peptide chains bearing the amine group. Another group of enzymes, called **carboxylpeptidases,** split amino acids from the end of the peptide chains bearing the carboxyl group. The result is free amino acid molecules ready for absorption.

Lipids

Neutral fats, or triglycerides, are the most abundant type of fat in the diet. Because of the mixing action of the stomach, relatively large fat globules are broken up into small fat droplets. The mechanical breakup of fat globules into smaller fat droplets is called **emulsification.** The fat droplets then enter the duodenum where they are quickly coated by bile salts secreted by the gallbladder. The bile-coated fat droplets are called **micelles. Lipase,** an enzyme in the pancreatic juice, hydrolyzes the lipids in the micelles into free fatty acids and a glycerol molecule with one attached fatty acid chain, called a **monoglyceride.** These simple organic molecules are now ready for absorption by the small intestine.

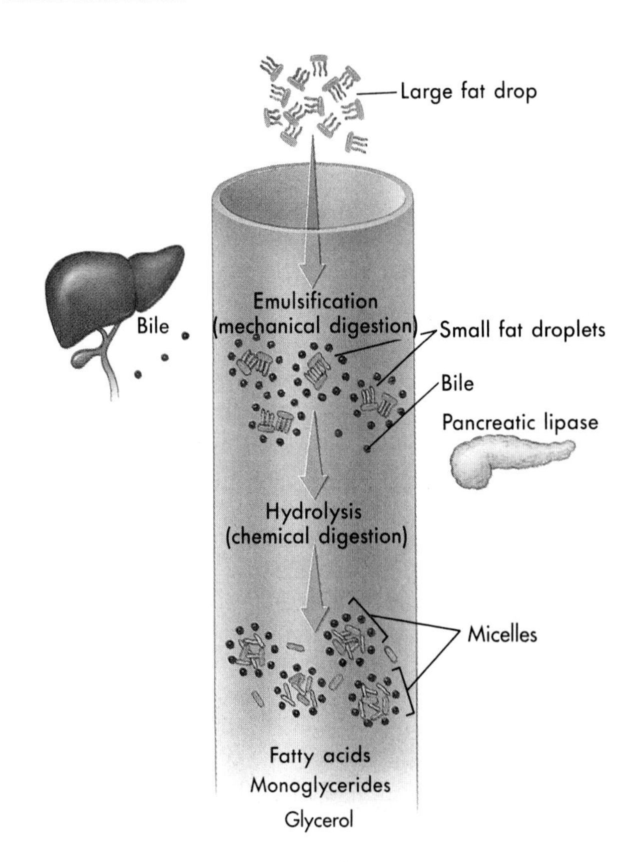

Biology in Action DIET AND CHOLESTEROL

As you read in previous chapters, high levels of cholesterol in the blood have a number of negative consequences for continued good health and longevity. Cholesterol levels are dramatically affected by what we eat and how active we are. The good news is we can control our dietary cholesterol intake.

The American Heart Association recommends that dietary fat should be no more than 30% of our total daily caloric intake. Today dietary fat contributes about 40% of the total caloric intake of the average American diet. The American Heart Association also recommends that dietary cholesterol be limited to less than 300 mg per day. More than any other type of fat, the *amount of saturated fat* in the diet affects blood cholesterol levels. Because of this, saturated fats should provide no more than 10% of total kcals consumed each day. There is strong evidence to support this view. For example, only 25% of people lower their blood cholesterol level when they eat less cholesterol. However, almost 90% of people who lower saturated fat intake lower an elevated cholesterol level up to 20%. How are these percentages translated into the types and amounts of food we should eat each day?

Contrary to popular belief, virtually all meats, whether red or white, contain about the same amount of cholesterol at the same serving size. What varies between different meats is the saturated fat content. For example, marbled beef contains streaks of fat. Therefore, it is important to eat only lean, nonmarbled beef. Trimming away any excess fat and removing the skin from meat, whether beef, poultry, or pork, also removes most of the saturated fat.

Drinking low-fat or skim milk and using margarine instead of butter are also relatively easy ways to reduce saturated fat intake without sacrificing nutrients.

Probably the hardest of obstacles to overcome in lowering your saturated fat intake is that the rich-tasting and highly flavored foods you probably enjoy most are the ones you must cut down on. This is because triglycerides and other lipids in foods provide texture and carry flavors in many foods. Many flavors are fat-soluble. They dissolve in fat, and the fat carries them to the sensory cells in the taste buds. Consequently we quickly learn to associate flavorful foods with fatty foods. A healthier diet, however, need not be bland. Spices and nonsalt seasonings can significantly enhance foods without adding to saturated fat intake.

Absorption in the Small Intestine

The small intestine is responsible for as much as 90% of the water absorbed by the digestive tract. It absorbs about 95% of the kcal it receives in the form of acid chyme from the stomach. Nutrient absorption takes place across the villi in the small intestine. Monosaccharides produced by carbohydrate digestion and amino acids produced by protein digestion move across the mucosal cells lining the small intestines by active transport. The energy of ATP powers a pump similar to the sodium potassium pump in neurons, transporting monosaccharides and amino acids across the cell membrane. After crossing the mucosal wall, the molecules move into the capillaries of the villi by simple diffusion.

Fat absorption in the small intestine differs markedly from monosaccharide and amino acid absorption because fat is insoluble in water. Fat undergoes a sequence of changes to transport it through the watery cytoplasm of the cells of the villi. Bile-coated micelles adhere to the brush border of the villi where the monoglycerides and free fatty acid chains enter the mucosal cells by simple diffusion. Once inside the mucosal cells, they enter the cell's smooth endoplasmic reticulum, where the monoglycerides and free fatty acids are resynthesized into triglycerides and are packaged into tiny protein-covered droplets in the Golgi apparatus called **chylomicrons** (ki-lo-mi'kronz). The chylomicrons leave the cells and enter the lacteals. Fat enters the lymphatic system and the lymph, traveling to the thoracic duct and emptying into the venous blood. Once in the blood, the lipids in the chylomicrons can then pass through the capillary walls to be used by tissue cells or stored as fat occurring in the small intestine in adipose tissue (Figure 16-8). Table 16-4 summarizes the digestive process of the three major macromolecule groups.

Figure 16-8 Fat absorption

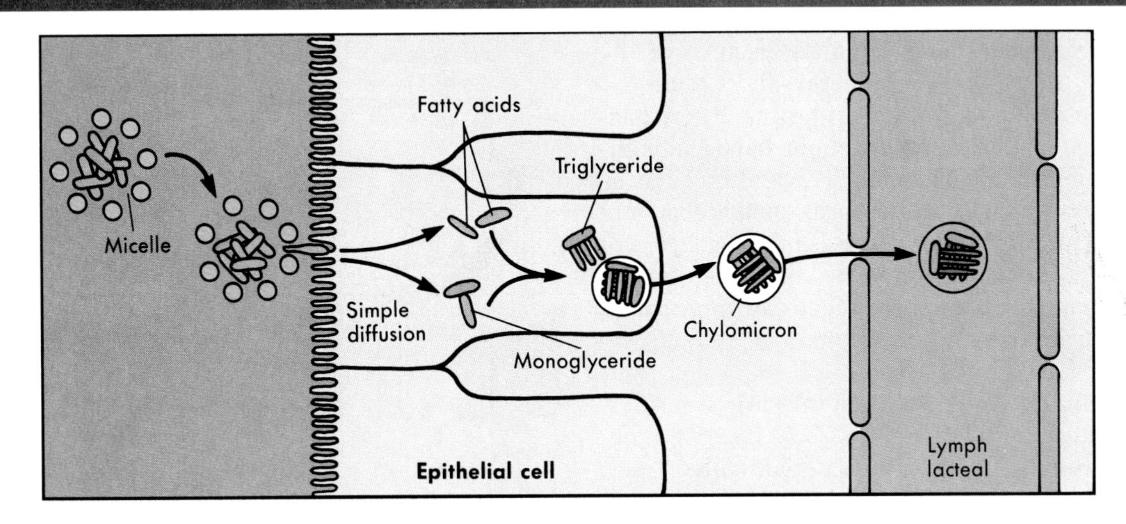

Bile-coated micelles adhere to the brush border of the villi, and their contents diffuse into the mucosal cells. Fatty acids and monoglycerides are resynthesized into triglycerides and packaged into protein-covered droplets called chylomicrons. The chylomicrons then enter the lacteal where they are ultimately carried to the circulatory system.

Table 16-4 Digestion of the Three Major Macromolecule Groups

CATEGORY OF FOODSTUFFS	INTERMEDIATE BREAKDOWN PRODUCTS	END PRODUCTS OF DIGESTION: ABSORBABLE UNITS
Carbohydrates		
Polysaccharides (starch and glycogen) — Pancreatic amylase	Maltose — Maltase	Monosaccharides (glucose, fructose, galactose)
Disaccharides: Sucrose — Sucrase		
Disaccharides: Lactose — Lactase		
Proteins		
Carboxypeptidase	Peptide fragments — Aminopeptidase	Amino acids
Fats		
Triglyceride — Lipase		Monoglyceride; Free fatty acids

= Glucose = Fructose = Galactose = Glycerol = Fatty acid = Amino acids = Enzyme

Absorption in the Large Intestine

The large intestine's main function is to absorb electrolytes, largely sodium and chloride; most of the remaining water; and some minerals and vitamins.

Bacteria reside in the large intestine. These bacteria, collectively called **endogenous bacteria,** manufacture vitamin K and some B vitamins, such as biotin. Endogenous bacteria also metabolize any remaining simple sugars and amino acids not absorbed in the small intestine, releasing carbon dioxide, methane, and hydrogen sulfide gases, which contribute to fecal odor.

The result of the absorption of water and remaining nutrients by the large intestine is a semisolid product called **feces** (fe′-sez). Feces largely contain cellulose from plant material, called fiber; some connective tissues from animal products, such as meat; a great deal of bacteria; and some bile salts and bilirubin.

The feces move into the last 15 cm, or 6 inches, of the large intestine, the rectum. When the rectum fills, its walls are stretched, initiating the **defecation reflex.** This reflex has both involuntary and voluntary components. The involuntary spinal-cord-mediated reflex causes the rectal walls to contract and the internal anal sphincter to relax. As the feces are forced down, messages reach the brain allowing us to decide whether the external anal sphincter, under voluntary control, should remain open or be constricted to stop passage of the feces. If the decision is to delay defecation, the reflex contractions end within a few seconds. With the next movement in the rectum, the defecation reflex is initiated again (Figure 16-9).

Figure 16-9 Defecation reflex

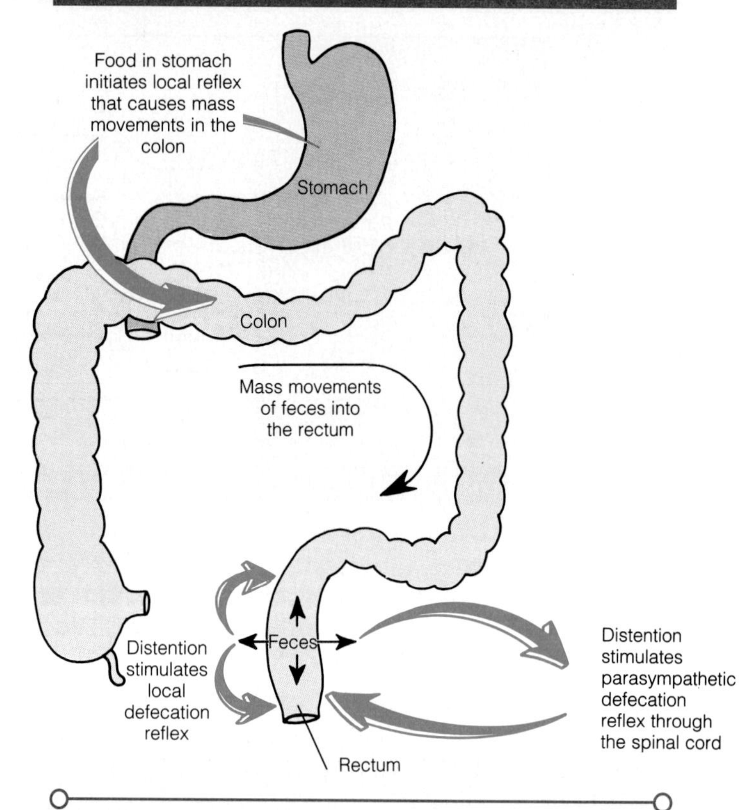

The defecation reflex is under both autonomic and voluntary control. As the rectum fills, its walls distend, signaling parasympathetic nerves of the autonomic nervous system to contract the rectum and relax the internal anal sphincter. When these messages reach the brain, the person decides whether the external anal sphincter should remain open or be constricted.

CHAPTER REVIEW

SUMMARY

1. Food contains substances, called nutrients, used by the body to promote growth, maintenance, and repair. Essential nutrients are substances that cannot be made by the body's cells and therefore must be provided directly by the diet. Essential nutrients are found in all major categories of nutrients humans eat.
2. The food we eat must first be broken down into small, simple organic molecules so they can be absorbed into the circulatory system for distribution to the cells of the body. The human digestive system disassembles ingested macromolecules into simple organic molecules.
3. The digestive system, called the gastrointestinal tract, is responsible for the physical and chemical breakdown of food and its absorption into the circulatory system. Most mechanical digestion occurs in the mouth and stomach, whereas most chemical digestion occurs in the small intestine.
4. The gastrointestinal tract is a continuous, hollow, coiled tube with several regional specializations that assist in digesting food. The primary areas of specialization are the mouth, esophagus, stomach, small intestine, large intestine, and rectum.
5. Hydrolytic enzymes are responsible for the chemical breakdown of the macromolecules in food to simple organic molecules that can be

absorbed into the circulatory system. The primary macromolecules are complex carbohydrates, proteins, and neutral fats.

6. The liver and the pancreas produce hydrolytic enzymes and other substances responsible for most of the chemical digestion occurring in the gastrointestinal tract. Most chemical digestion occurs in the first 12 inches of the small intestine, called the duodenum.
7. Absorption of simple molecules occurs through fingerlike projections of the inner layer of the small intestine, called villi. Villi contain capillaries, including a lymphatic capillary called a lacteal. Glucose and amino acids are actively transported into the villi, and fatty acids and monoglycerides passively diffuse into the villi.

FILL-IN-THE-BLANK QUESTIONS

1. The human digestive system takes ______ that make up food and breaks them down into ______ ______ molecules.
2. The digestive system consists of a hollow, muscular tube called the ______ canal and several accessory digestive ______.
3. Swallowing consists of both a voluntary and ______ phase.
4. Alternate waves of contraction and relaxation of the muscles in the walls of the alimentary canal are called ______.
5. Gastric glands secrete ______ ______ and ______ into the lumen of the stomach.
6. Both the liver and the pancreas secrete their substances into the section of the small intestine called the ______.
7. Absorption in the small intestine occurs via the ______ of the mucosa.
8. ______ ______ cannot be made by cellular interconversions and therefore must be obtained directly from the ______.

SHORT-ANSWER QUESTIONS

1. Describe the main phases of the swallowing reflex.
2. What occurs to food while in the stomach?
3. Describe how fat is digested and absorbed from the small intestine.
4. What is the main function of the large intestine?
5. List the specific substances found in pancreatic juice.
6. What are the major categories of nutrients found in the food humans eat?

VOCABULARY REVIEW

absorption—*p. 305*
basal metabolic rate—*p. 303*
bile—*p. 312*
calorie—*p. 302*
complete protein—*p. 300*
digestion—*p. 304*
esophagus—*p. 309*
hard palate—*p. 307*
nutrients—*p. 299*
peristalsis—*p. 309*
salivary glands—*p. 307*
villi—*p. 311*

Chapter Seventeen

URINARY EXCRETION

For Review

Here are some important terms and concepts that you will encounter in this chapter. If you are not familiar with them, you should review them before proceeding.

Aldosterone
(page 202)

Antidiuretic hormone
(page 195)

Osmosis and tonicity
(pages 40, 41)

pH scale
(page 24)

Positive feedback
(page 196)

Urea
(page 313)

OBJECTIVES

After reading this chapter you should be able to:

1. *List the structures that make up the urinary system.*
2. *Describe the main functions of the kidneys.*
3. *Describe the structure and function of the nephron.*
4. *Discuss the three processes of urine formation.*
5. *Discuss the importance of urine concentration and describe the process involved in the concentration of urine.*
6. *Explain the ways in which the kidneys help control homeostasis in the body.*

Each of the body's cells is bathed in extracellular fluid through which it communicates and interacts directly with other cells and indirectly with the outside world. The kidneys are primarily responsible for maintaining the extracellular fluid composition. They are able to filter waste products of cellular metabolism from body fluid by first separating most of the plasma from the blood's formed elements and plasma proteins. Most of the water, nutrients, and ions are then returned to the blood, leaving behind waste products and some water in the urine, which is then excreted.

Urinary System

Cellular metabolism produces many waste products that diffuse from the cells into the extracellular fluid and from there into the circulatory system. Homeostatic maintenance of the blood is the result of the coordinated activities of the nervous, endocrine, and circulatory systems. These systems are in turn aided by organs that exchange materials directly with the external environment, such as the lungs and the digestive organs. For example, waste products such as carbon dioxide are removed from the blood by the lungs. Other waste products of cellular metabolism also enter the extracellular fluid and blood. These waste products include the nitrogen-containing compound called urea, which results from the breakdown of amino acids, and hydrogen and bicarbonate ions. The cells of the digestive system, through which most substances enter the body, are relatively unselective. Any molecule that can move into the body through the intestinal lining does so, including an excess of water, nutrients, salts, and minerals, and some drugs. With all these solutes entering the blood, how does the content of the extracellular fluid and the blood plasma of the body remain constant?

Maintaining the proper balance of bodily fluid composition is the primary function of a pair of organs called the **kidneys.** The kidneys are part of the **urinary system,** which also consists of the **ureters, bladder,** and **urethra** (Figure 17-1). Whereas the kidneys filter the blood and produce **urine,** a fluid consisting primarily of waste substances and some water filtered from the blood, the rest of the system transports, stores, and excretes the urine. We begin this chapter by first describing the major structures of the urinary system, tracing the pathway of waste products through it. We then describe the structure and function of the kidneys and their role as homeostatic organs. We then describe the types of waste products and other substances removed from the fluid portion of the blood by the kidneys. We conclude by discussing some kidney-function problems.

Kidney Structure

Kidneys are paired, reddish-brown organs located on the dorsal wall of the abdominal cavity on either side of the vertebral column just above the waist. Each kidney weighs about 150 g, or 5 oz, and is approximately 13 cm, or 5 inches, long, 8 cm, or 3 inches wide, and about 2.5 cm, or 1 inch thick. The kidneys are each covered by a layer of connective tissue called a capsule. The kidneys have two major, related functions: to excrete cellular waste products from the body and to maintain body fluid composition. They perform both functions simultaneously by filtering the blood. The kidneys first separate and collect the fluid portion of the blood. From this fluid most of the water and important nutrients are reabsorbed back into the blood, while waste prod-

Figure 17-1 Urinary system

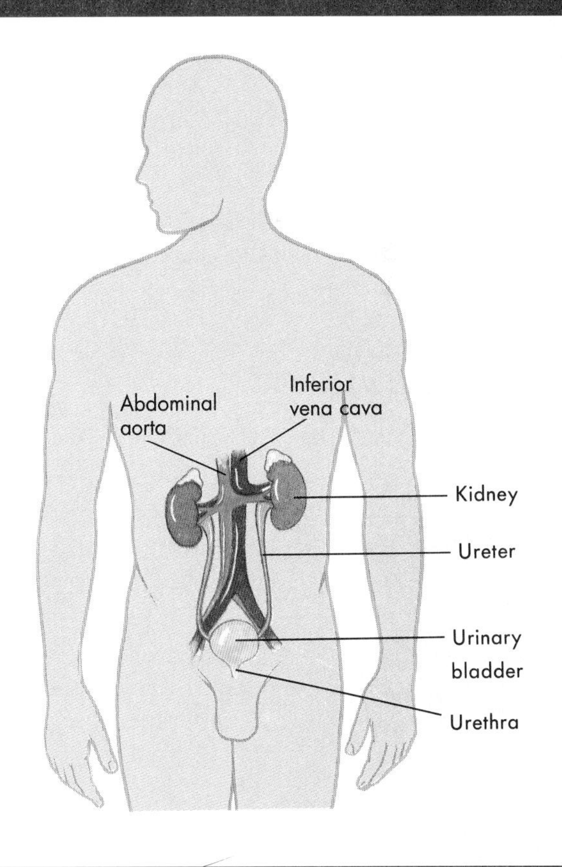

The urinary system consists of the paired kidneys, ureters, bladder, and urethra. Urine produced by the kidneys moves through the ureters to the bladder where it is temporarily stored. When released from the bladder, urine leaves the body via the urethra.

ucts, toxic substances, and some water are left behind to be eliminated from the body as urine.

Blood carrying cellular waste products enters each kidney through a **renal artery.** Each renal artery branches directly off the abdominal aorta. After being filtered the blood leaves the kidney by way of the **renal vein,** which connects directly to the inferior vena cava. Urine leaves each kidney through a duct called the **ureter** (u-re'ter). The ureters transport urine by peristaltic contractions to the **bladder,** a hollow chamber that collects and temporarily stores urine. Urine completes its journey to the external environment through a single narrow tube called the **urethra** (u-re'thrah). The urethra is about 4 cm, or 1.5 inches, long in females and about 20 cm, or 8 inches, long in males.

The bladder walls are composed primarily of smooth-muscle tissue and are capable of considerable distension. Urine is held in the bladder by two sphincter muscles called the **internal sphincter** and the **external sphincter,** located at the base of the bladder just above its connection to the urethra. As urine collects, the bladder walls stretch and sensory receptors in the walls send signals to the spinal cord, initiating reflexive contractions of the internal sphincter muscle. The lower external sphincter, composed of skeletal muscle tissue, is under voluntary control, so the release of urine can be suppressed by signals from the brain until bladder distension becomes acute. Although the average adult bladder can hold 700 to 800 ml, about $2\frac{1}{2}$ cups, of urine, the urge to urinate is initiated by accumulations of between 200 to 400 ml, about 6 to 12 oz, of urine.

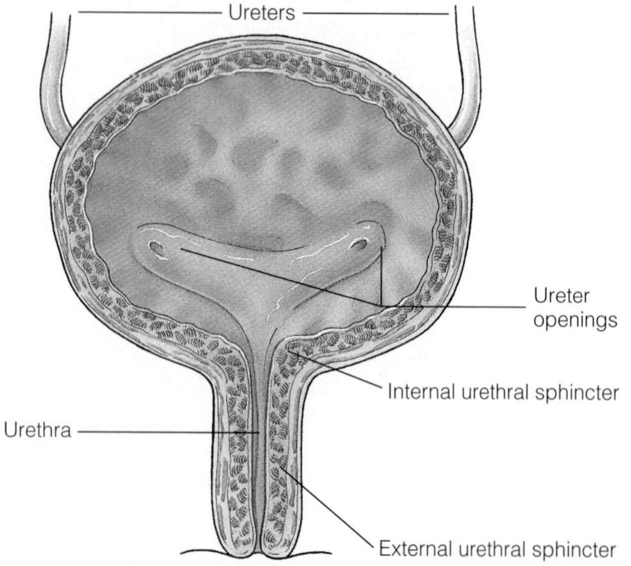

Viewed in longitudinal section the kidney consists of a solid outer layer and a hollow inner chamber called the **renal pelvis.** The renal pelvis is a branched collecting chamber that directs urine into the ureter. The solid outer layer of kidney tissue is divided into two zones. The outer zone, called the **cortex,** is slightly granular, and the inner zone, called the **medulla,** consists of tissue with a striated cone-shaped pattern called **renal pyramids.** When viewed microscopically these zones contain small structures called **nephrons.** The upper portion of each nephron is in the cortex, while its lower portion is in the medulla. This structural arrangement gives the renal pyramid its striped appearance (Figure 17-2).

Nephron Structure

Each microscopic nephron is a filtering device that releases its urine into the central chamber of the renal pelvis and serves as the basic functional unit of the kidney. Each kidney contains about 1 million

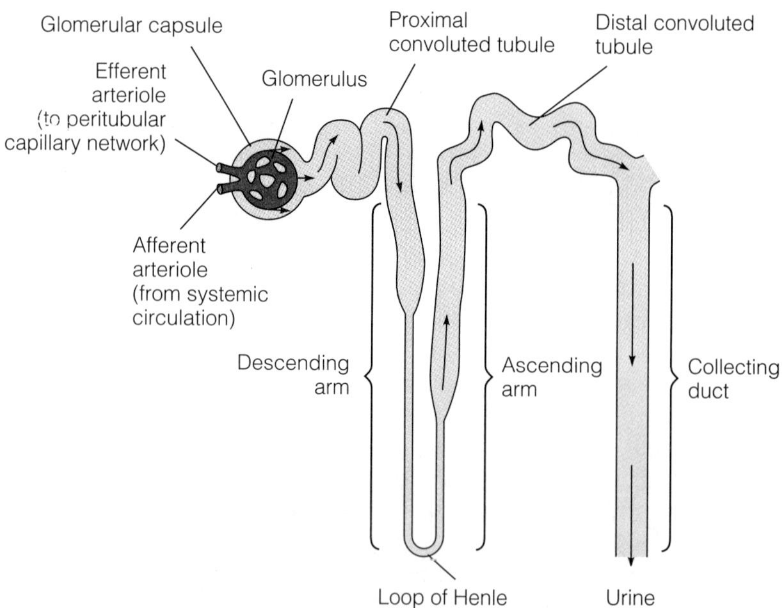

nephrons. The nephron consists of a series of branching capillaries, collectively called the **peritubular capillary network,** surrounding a hollow tubular structure called the **renal tubule.** The renal tubule consists of four major regions. The epithelial cells that make up the walls of each section of the tubule are slightly different and are specialized to perform different functions. Consequently modifications are made to the fluid, altering its composition as it moves through the tubule. We discuss the specific function of each region in the next section.

The first region of the renal tubule is spherically shaped and pushed in on itself to form a cuplike structure called the **glomerular** (glo-mer'yooler) **capsule**. Its shape is similar to that of a balloon after a fist has pushed into one side of it. The sec-

Figure 17-2 Kidney structure

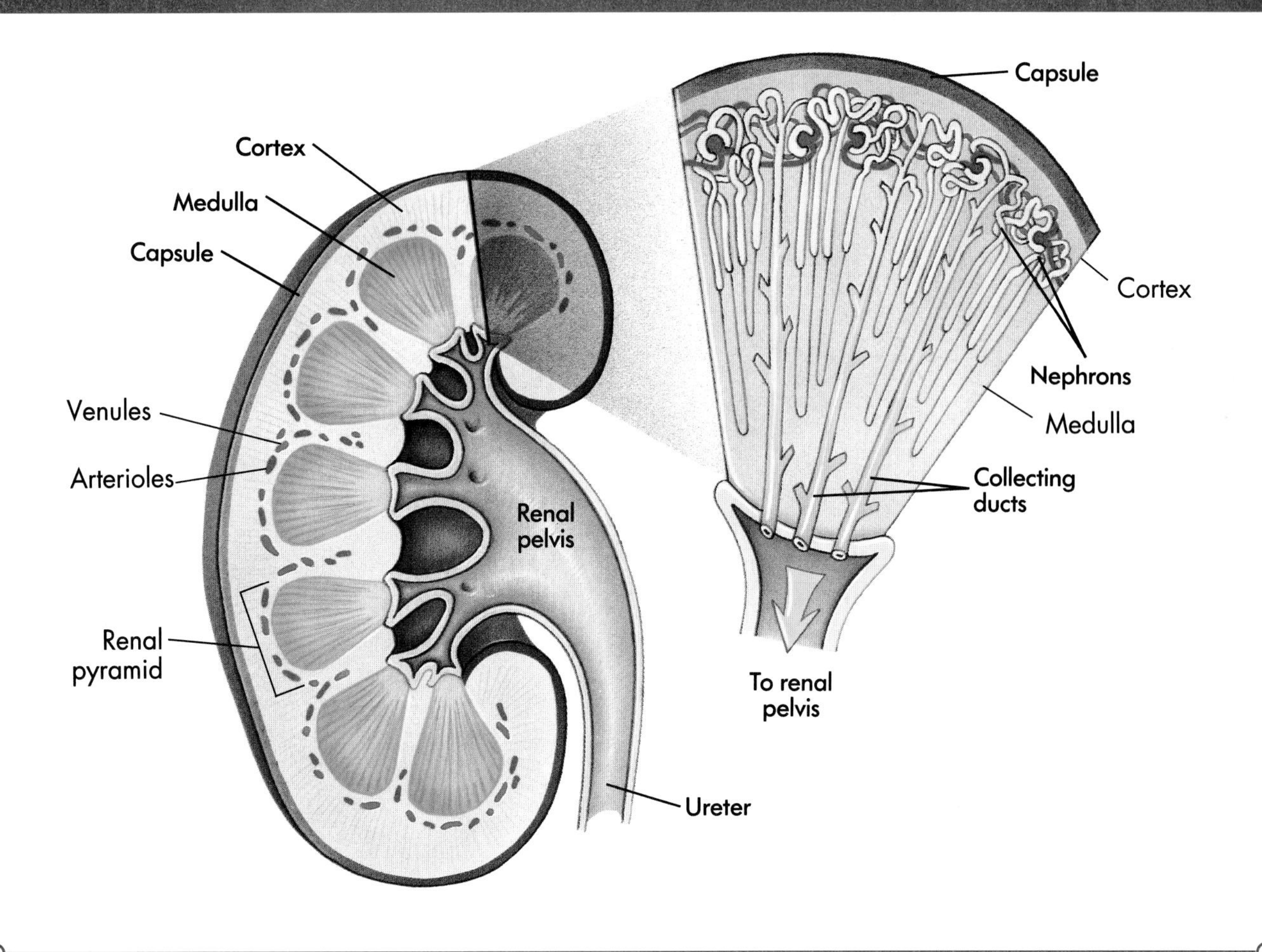

Each kidney is covered by a fibrous outer capsule. Inside, the kidney is composed of a cortex, containing primarily arterioles and venules; and a striated inner region, called the medulla, which consists of renal pyramids. The renal pyramids are made up of microscopically sized structures called nephrons and collecting ducts.

ond region is called the **proximal convoluted tubule.** The convoluted tubule is proximal, or closer to, the glomerular capsule. The proximal convoluted tubule is followed by the third region, a long U-shaped area called the **loop of Henle.** The fourth region is called the **distal** (*distal* means "away from") **convoluted tubule.** After leaving the distal convoluted tubule, the urine empties into the **collecting tubule** that also receives urine from several other nephrons. The collecting tubules empty into the renal pelvis and then into the ureters.

Blood is conducted to each nephron by an arteriole called the **afferent arteriole** (*afferent* means "leading to"). The afferent arteriole immediately subdivides into numerous small capillaries that form an intertwined mass or tuft, called the **glomerulus** (glo-mer'u-lus). The glomerulus sits in the glomerular capsule. Past the glomerulus, the capillaries immediately reunite to form an arteriole called the **efferent arteriole** (*efferent* means "leading from"). The efferent arteriole then branches into smaller, highly porous capillaries, called the **peritubular capillaries** (*peri* means "around"), which form an extensive network of vessels surrounding the renal tubule. The peritubular capillaries merge to form venules that unite with venules from other nephrons to eventually form the renal vein, which carries blood away from the kidney and back to the heart (Figure 17-3).

Urine Formation

Urine is formed in the nephrons by the three processes of filtration, reabsorption, and secretion. Each process is defined in the context of where it primarily occurs in the nephron.

Blood flows through the afferent arteriole and into the glomerulus of each nephron. Because the diameter of the efferent arteriole leaving the

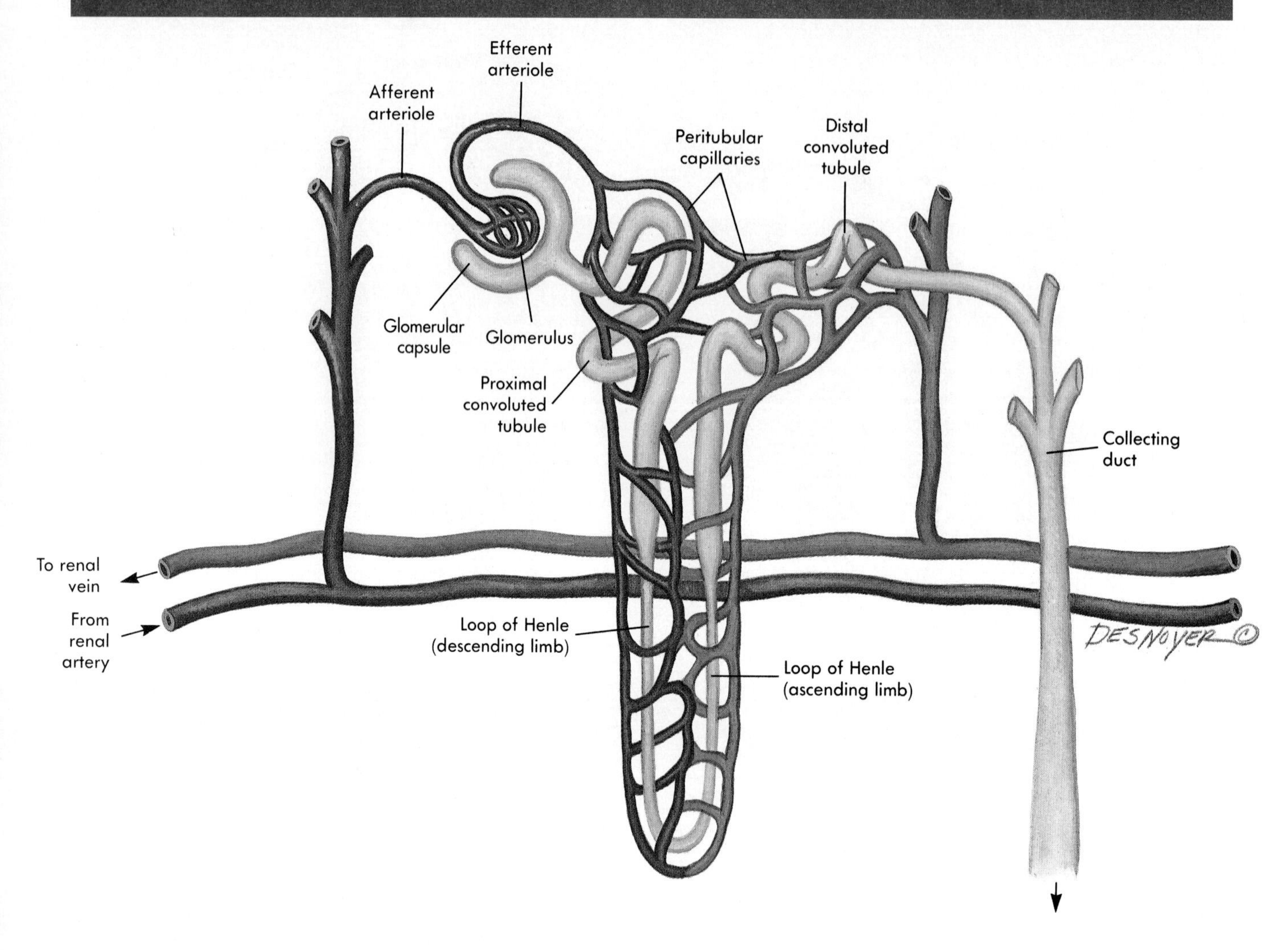

The afferent arteriole leads to the nephron where it immediately subdivides into a tuft of capillaries, called the glomerulus. The glomerulus sits inside the blind end of the renal tubule, called the glomerular capsule. The efferent arteriole leads away from the glomerulus and then branches into numerous capillaries, forming a meshwork around the rest of the renal tubule, called the peritubular capillaries. The peritubular capillaries merge with a venule, eventually uniting with other venules to form the renal vein.

glomerulus is smaller than the diameter of the afferent arteriole entering it, the glomerulus has a relatively high blood pressure of about 60 mm Hg. Glomerular capillaries are also porous. As a result, glomerular capillaries are almost 400 times more permeable to plasma and dissolved solutes than are the skeletal muscle capillaries. Because of the high glomerular blood pressure and the porous glomerular capillaries, most of the plasma, which is primarily water, and small dissolved solutes from the blood are driven through the capillary walls of the glomerulus and into the glomerular capsule. This process is called **glomerular filtration.** The blood remaining in the glomerular capillaries contains formed elements, such as blood cells and platelets, plasma proteins, and a small amount of plasma. **Glomerular filtrate** formation is the first step in urine formation. Filtration is a continuous process in which the blood plasma is filtered through the kidneys about 60 times a day. About 20% of the body's total blood supply passes through the kidneys at any given time.

With the filtrate removed, the blood leaving the efferent arteriole from the glomerulus is concentrated. Remember that during filtration *any molecule* in the blood plasma small enough to move through these pores does so. Therefore most of the filtered plasma and molecules, such as glucose, amino acids, other nutrients, water-soluble vitamins, and many essential ions, must be recaptured. The function of the rest of the renal tubule is to return the nutrients, ions, and most of the water to the blood while retaining and concentrating waste products for elimination. A simple analogy may help you to visualize the process. It is similar to cleaning a room by first taking everything out of it, cleaning the room thoroughly, and then putting back only those items you want to keep in the room.

As the glomerular filtrate passes into the proximal convoluted tubule, most of the solutes and water are moved from the tubule and put back into the blood circulating around it in the peritubular capillary network. This process is called **reabsorption.** Reabsorption begins in the proximal convoluted tubule and continues throughout the renal tubule and the collecting tubule. Reabsorption of ions and nutrients, such as amino acids, glucose molecules, and vitamins, occurs by active transport through the tubule's wall into the extracellular fluid and then into the peritubular capillaries.

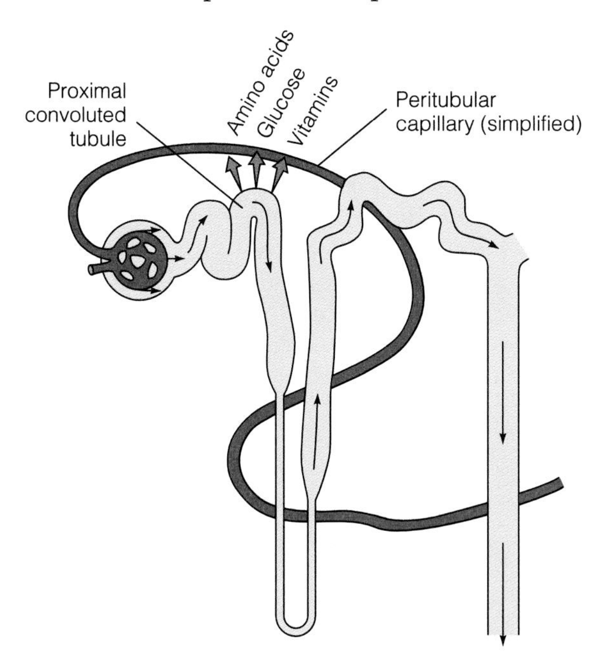

Different solutes are reabsorbed from the proximal convoluted tubule at different rates and therefore to different degrees. The concentration of transported molecules that saturate the carriers and achieve maximal transport rate is called the **transport maximum.** For example, the transport maximum for the active transport of glucose is approximately 180 mg per 100 ml of blood plasma passing through the kidney. However, if a particularly large quantity of glucose is consumed during a meal, this upper threshold may be exceeded and glucose will spill over and appear in the urine. This can be illustrated with a simple example. Assume there are equal numbers of red and green apples moving on a conveyor through a spray washer. At the other side of the washer are two people. Each is responsible for picking either the red or the green apples off the conveyor. Under normal conditions all the apples are picked off the line. However, periodically the number of green apples on the conveyor is doubled. If the person is still picking them up at the previous rate, the added apples are lost down the conveyor and not picked up. Similarly under normal conditions most actively transported solutes, such as glucose, are completely reabsorbed from the glomerular filtrate, but once the transport maximum is exceeded, the excess molecules will be eliminated from the body via the urine.

One of the symptoms of diabetes mellitus is the appearance of large amounts of glucose in the urine. This occurs because the person produces insufficient insulin and therefore glucose uptake by body cells is diminished. Consequently a great deal of glucose is in circulation, which, because of the glucose transport maximum, is excreted by the kidneys into the urine.

Some substances are not removed from the blood during glomerular filtration and are therefore left in the peritubular capillary network. As mentioned earlier a third way exists to move a substance from the blood and into the renal tubules. This mechanism, called **secretion,** is a process by which some substances remaining in the peritubular capillaries are actively transported back *into* the renal tubule. These substances include hydrogen and potassium ions and some drugs, including antibiotics such as penicillin. Secretion of hydrogen ions from the blood back into the renal tubule assists in maintaining the proper pH of the body's fluids.

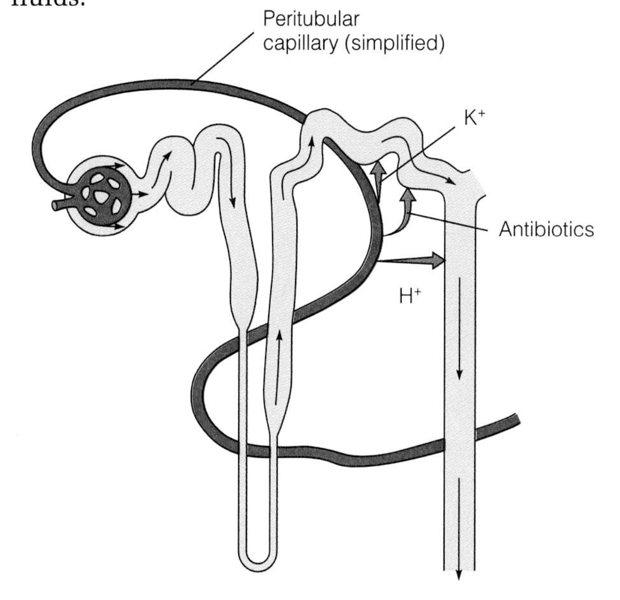

Urine Concentration

If the kidney was unable to concentrate solutes during urine formation, our bodies would produce about 25 L of dilute urine every 24 hours instead of 1 to 2 L of very concentrated urine. Recall from Chapter 3 that water moves down its concentration gradient in response to the solute concentration. We can apply this important concept to kidney function and urine concentration.

After the filtrate leaves the glomerular capsule, it enters the proximal convoluted tubule. At this point the filtrate in the tubule is relatively *isotonic* to the plasma in the surrounding peritubular capillaries. The filtrate is isotonic because, even though much of the plasma is in the filtrate, the *solute concentration* in both the filtrate and the blood found in the peritubular capillaries is equal. Water cannot be recovered from the filtrate by osmosis unless the solute concentrations of the filtrate and the surrounding extracellular fluid are changed. The solute concentration is changed by the *active transport* of sodium ions (Na^+) from the filtrate to the extracellular fluid surrounding the tubule. Negatively charged chloride ions (Cl^-) also passively follow the positively charged sodium ions out of the tubule. As a result of the active transport of sodium chloride (NaCl) out of the proximal convoluted tubule and into the extracellular fluid, an osmotic gradient is generated. About 65% of the water in the filtrate moves out of the proximal convoluted tubule back into the extracellular fluid. The salt and the water then diffuse into the peritubular capillaries.

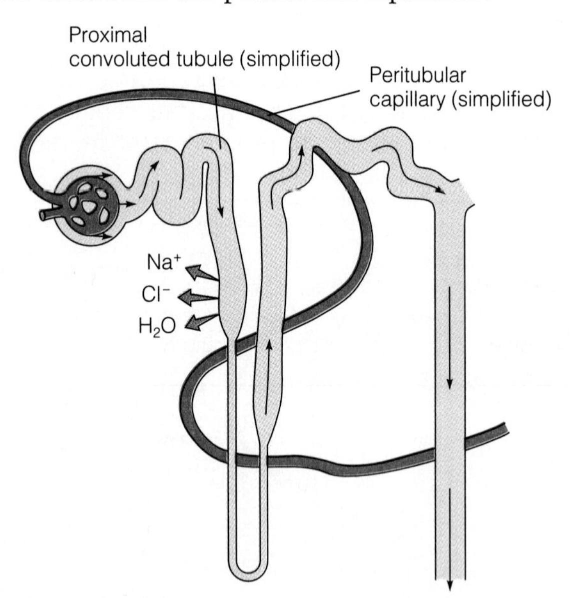

Waste products remain in the tubule and become concentrated as water leaves. Although there is a relatively greater concentration of urea and other waste products in the filtrate compared with the blood, *overall* there is approximately the same concentration of solutes per unit volume of filtrate compared with the blood. If this *ratio* of solute to solvent remained in the filtrate, the kidneys would produce about 25 L of very dilute urine every 24 hours. The loop of Henle, the next region of the renal tubule, allows the kidney to produce a very *hypertonic* urine with about four times the concentration of waste products compared with blood plasma concentrations. This results in the excretion of about 1 to 2 L of urine every 24 hours.

Three facts must be kept in mind to understand how the loop of Henle concentrates urine. First the loop of Henle and the collecting tubule are positioned closely together and are roughly parallel to each other in the medulla of the kidney. Second the loop of Henle maintains a solute concentration gradient in the *extracellular fluid* surrounding the loop and the collecting tubule, with the highest concentration at the bottom of the loop. Third, fluid is always moving through the tubule and blood through the glomerulus and peritubular capillaries. This is a dynamic system in which several events are occurring simultaneously. You may think of this concentration process as the mental equivalent of juggling several balls in the air at the same time. Understanding each part of the process depends on a quick glance at the other parts at the same time.

The urine leaves the proximal convoluted tubule and passes down the **descending limb** of the loop of Henle. The walls of the descending limb are impermeable to solute but are freely permeable to water. Because the surrounding tissue fluid has a relatively high concentration of solute, primarily salt, compared with the urine, water flows out the descending limb and back into the extracellular fluid by osmosis. This results in a more concentrated urine in the tubule as the urine moves deeper into the medulla. At the bottom of the loop, the

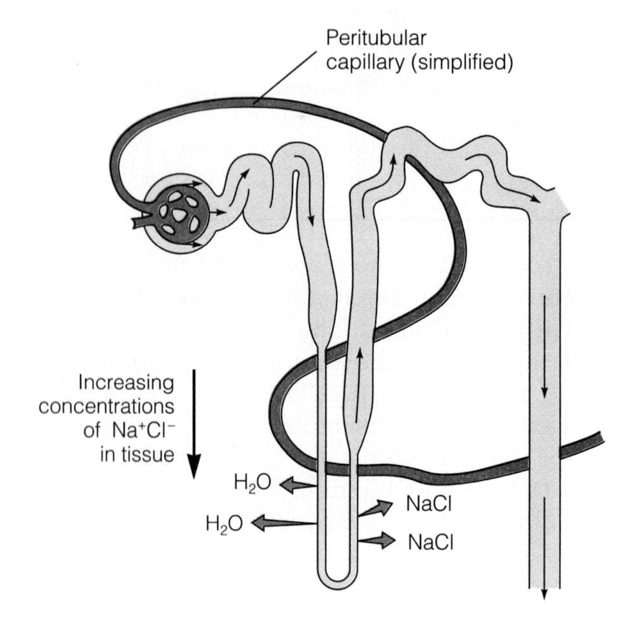

walls of the tubule become permeable to salt but impermeable to water. Although the greatest *solute concentration* exists in the tissue surrounding the bottom of the loop, *salt* still diffuses out of the tubule because the surrounding tissue does not contain as much *salt* as does the urine. The **ascending limb** of the loop of Henle is responsible for an increasing solute concentration gradient of the extracellular tissue fluid surrounding the loop of Henle and the collecting tubule.

As the urine moves up the **ascending limb** of the loop, the wall's cells actively transport salt out of the urine and into the surrounding tissue fluid. Water does not follow because the ascending limb is impermeable to water. As the urine moves up the ascending limb, there is less salt to actively transport into the surrounding tissue fluid. It is as if, when walking up a flight of stairs, you began to empty your pockets. As you move up each step you would have less to discard. The action of the ascending limb maintains the solute concentration gradient in the medulla. Because of the solute concentration gradient in the surrounding tissue fluid, water leaves the descending limb of the loop all along its length.

The urine flows from the ascending limb of the loop of Henle into the distal convoluted tubule. Although low in salt, more salt is actively transported out of the urine by distal convoluted tubule cells. Water continues to move down the osmotic gradient produced because of salt movement.

By the time the urine reaches the collecting tubule, little salt is left in it and about 99% of the water that entered the tubule originally has been reabsorbed and returned to the peritubular capillary network. The collecting tubule passes the urine down through the increasingly concentrated extracellular fluid gradient of the medulla created by the nearby loop of Henle. Under certain conditions, the entire collecting tubule is permeable to water, but the lower portion of the collecting duct is also semipermeable to urea. The movement of a small amount of urea into the surrounding extracellular space around the bottom of the loop of Henle gives this area the greatest overall solute concentration while still allowing salt to move down its own concentration gradient from the urine to the surrounding extracellular fluid.

Urine Composition

Normally urine is about 95% water and 5% solutes. Urea makes up most of the solutes in urine. This is largely because we consume more amino acids than are required to synthesize new proteins. Remember from Chapter 16 that blood flows from the digestive system via the hepatic-portal circulation directly to the liver. This blood contains the small organic molecules absorbed from the digestive system. Because the excess amino acids in circulation cannot be stored, they are broken down in the liver through a metabolic process called deamination, either to provide energy or to eventually be converted to storable fats. In deamination the amino group ($-NH_2$) is removed from each amino acid. The initial waste product formed from the amino group is ammonia (NH_3). Ammonia, however, is highly toxic to cells and therefore must be converted to a less toxic substance. This occurs in the liver in a complex series of biochemical reactions called the **urea cycle,** in which a carrier molecule picks up carbon dioxide and two molecules of ammonia to produce a molecule of urea. Another nitrogen-containing compound, called **uric acid,** is also produced in the liver when nucleotides are broken down. Occasionally people whose diets are rich in meat and milk products produce excess amounts of uric acid that precipitate out of the plasma and accumulate in the fluid surrounding the skeletal joints. These crystals of uric acid cause a painful ailment called **gout.**

Other solutes normally in urine include, in order of decreasing amounts, sodium, potassium, phosphate, and sulfate ions. Much smaller but highly variable amounts of calcium, magnesium, and bicarbonate ions are also found. The color of urine normally varies from clear and pale to deep yellow. The yellow color is due to the presence of a by-product of hemoglobin breakdown that occurs in the liver. Generally the more concentrated the urine, the deeper the yellow color.

Control of Kidney Function and Homeostasis

One of the most important kidney functions is to regulate water balance in the body. By changing the volume of urine excreted to match changes in the volume of fluid taken into the body, the kidney is able to maintain homeostasis of total body water volume. The kidney does this by regulating the amount of extracellular fluid surrounding the tissues and the plasma portion of the blood.

Water Balance

We can put kidney function in the larger context of water balance in the body. The three primary sources of fluid intake are the liquids we drink, the water in the foods we eat, and the water formed by the hydrolysis of foods and other substances in the body, called metabolic water. We also lose water to our surroundings in many ways. The kidneys, lungs, skin, and large intestine each secrete water from the body. The fluid output that changes the most is from the kidneys. The body maintains homeostasis of body water volume mainly by changing the volume of urine excreted to match changes in the volume of fluid taken into the body.

How much water is reabsorbed into the blood is controlled by the presence of antidiuretic hormone, or ADH, circulating in the blood. One way to understand the function of ADH is to understand what its name means. *Diuresis* is derived from the Greek word meaning "increase urine." Therefore *anti*diuresis means to decrease or suppress urinary excretion. Antidiuretic hormone causes the distal convoluted tubule and the collecting tubule to become permeable to water. This allows more water to be reabsorbed from the urine and returned to the blood. The release of ADH is regulated by receptor cells in the hypothalamus that monitor the osmotic concentration of the blood and by receptors in the heart that monitor blood volume. For example, you have just completed a strenuous run of several miles on a warm summer day. Because of profuse sweating and increased breathing during the run, your body is dehydrated. Thus the osmotic concentration of your blood rises and blood volume falls, triggering the release of ADH. This increases water reabsorption, primarily from the collecting tubules of the kidney, and produces urine more concentrated than the blood. If too much water is in the blood, ADH release is inhibited, causing the distal convoluted tubule to become less permeable to water. The result is a more watery and dilute urine.

Drinking alcohol causes diuresis, or *increased* urinary excretion, because it inhibits ADH secretion. The resulting dehydration contributes to the hangover symptoms of the next morning. Drinking coffee, tea, or carbonated beverages that contain caffeine also causes water loss. The caffeine also inhibits ADH secretion, resulting in a more dilute urine output.

Blood Pressure and Blood Volume

Recall from Chapter 3 that the maintenance of blood pressure is essential for nutrient and gas exchange between capillaries and surrounding tissues. Blood pressure is also vital to kidney function because without adequate blood pressure, glomerular filtration is inhibited. Recall that the blood pressure in the glomerulus drives the filtration process. Embedded in the walls of the afferent arterioles are specialized cells called **juxtaglomerular** (juks'ta-glo-mer'u-lar) **cells,** or **JG cells** (Figure 17-4). These cells are part of the juxtaglomerular apparatus located at the connection between the afferent arteriole and the distal convoluted tubule. The prefix *juxta* means "near" or "next to," indicating this region is near the glomerulus. The JG cells are sensitive to changes in blood pressure. If their normal stretch is reduced, by lowered blood pressure, for example, they release an enzyme called **renin.** When released into circulation, renin converts a large circulating plasma protein, called **angiotensinogen** (an-je-o-ten'sin), into its active form, called **angiotensin.** Angiotensin elevates blood pressure in two ways. First, with the exception of the afferent arterioles in the kidney, it contracts all the body's arterioles, causing an increase in systemic blood pressure. The rise in blood pressure increases glomerular filtration in the nephrons. Second angiotensin stimulates the cortex of the adrenal gland to secrete the hormone **aldosterone.** Recall from Chapter 10 that aldosterone causes the distal convoluted tubule to increase sodium reabsorption into the extracellular fluid. This in turn increases the reabsorption of water by osmosis, increasing the blood volume in the circulatory system, and this increase elevates the blood pressure.

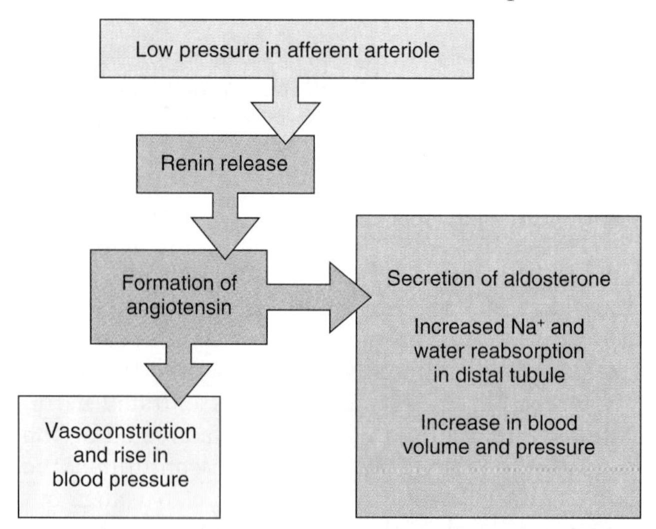

Acid-Base Balance

Recall from Chapter 2 that chemical reactions occurring in cells in the extracellular fluid require a stable pH. Great fluctuations in pH would disrupt many essential chemical reactions. Through the kidney the body is able to maintain the proper acid-base balance, or appropriate pH, of the blood and extracellular fluid. Normally the pH of the blood and extracellular fluid is between 7.3 and 7.4. To maintain the pH within that narrow range, various acidic and basic substances entering the blood from both the digestive system and normal cellular metabolism must be either neutralized or eliminated. Normally most of the excess acids in extracellular fluid and blood plasma come from cellular

Biology in Action WATER LOSS AND DEHYDRATION

In addition to the water we lose through the process of urine formation in the kidney, there are two other routes by which our bodies lose water. Each breath we take exposes the moist inner surfaces of the respiratory passages to air. Consequently, with each exhalation we lose a small amount of water to the air. During a cold day the water vapor condenses, and we can "see" our breath. This type of water loss is called insensible perspiration.

We also lose water when it is secreted to the surface of the skin as perspiration or sweat, which is a solution consisting of water, sodium, and urea. As the body heats up, the sweat glands excrete perspiration. The body is cooled as the perspiration evaporates.

Although sweating maintains body temperature, excessive water loss from sweating can actually adversely affect cardiovascular function. After a few hours of intense exercise in the heat, **dehydration** can occur. For example, marathon runners may lose more than 5 L of fluid during a race. When a person becomes dehydrated, blood plasma volume drops, causing an increased heart rate. Sweating is greatly reduced, making thermoregulation more difficult.

When water loss reaches about 4% of body weight (1 fluid pt equals approximately 1 lb of body weight), physical work capacity is greatly impaired. For example, walking endurance was reduced by 48% among a group of college students dehydrated by 4.3% of body weight. However, even moderate dehydration will affect performance. For example, among a group of distance runners with an average dehydration of only 1.9% of body weight, athletic performance decreased by 22%.

The thirst mechanism is an imprecise guide to the body's need for water. If rehydration were left to the thirst sensation, it would take several days to reestablish fluid balance. How, then, do we prevent dehydration?

Drinking water before exercising provides some protection against dehydration. The extra water increases sweat rate during exercise, which helps reduce internal body temperature. For example, drinking 400 to 600 ml (13 to 20 oz) of water before beginning moderate exercise in the heat, such as aerobics or jogging for 25 minutes, is recommended. However, this doesn't replace the need for continual fluid replacement. Table 17-A represents the recommended amount of fluid intake for a person engaged in a 90-minute workout. For example, if an individual lost 5 lb of body weight because of water loss during exercise, he or she would need to drink 10 to 12 oz of fluid every 15 minutes during exercise.

Table 17-A Recommended Amount of Fluid Intake for 90-Minute Workout

WEIGHT LOSS		MINUTES BETWEEN WATER BREAKS	FLUID PER BREAK	
LB	(KG)		OZ	(ML)
5	(2.3)	15	10–11	(311)
$4\frac{1}{2}$	(2.1)	15	9–10	(281)
4	(1.8)	15	8–9	(251)
$3\frac{1}{2}$	(1.6)	20	10–11	(311)
3	(1.4)	20	9–10	(281)
$2\frac{1}{2}$	(1.1)	20	7–8	(222)
2	(0.9)	30	8	(237)
$1\frac{1}{2}$	(0.7)	30	6	(177)
1	(0.5)	45	6	(177)
$\frac{1}{2}$	(0.2)	60	6	(177)

Whatever type of physical exercise you enjoy, fluid replacement is essential for continued activity. It is very difficult to drink too much fluid, but drinking insufficient amounts has a significant impact on activity levels and health.

metabolism. Recall from Chapter 2 that once in solution, acidic substances dissociate to release hydrogen ions (H^+), which lowers pH. In the body, lowering pH is minimized when excess hydrogen ions react with buffers, such as bicarbonate ions (HCO^-_3), that can accept hydrogen ions and neutralize them. The kidneys are also able to eliminate hydrogen ions into the urine. All the hydrogen ions that leave the body in urine are *secreted* into the urine.

The distal convoluted tubule cells respond directly to the pH of the extracellular fluid and are

Figure 17-4 Juxtaglomerular cells

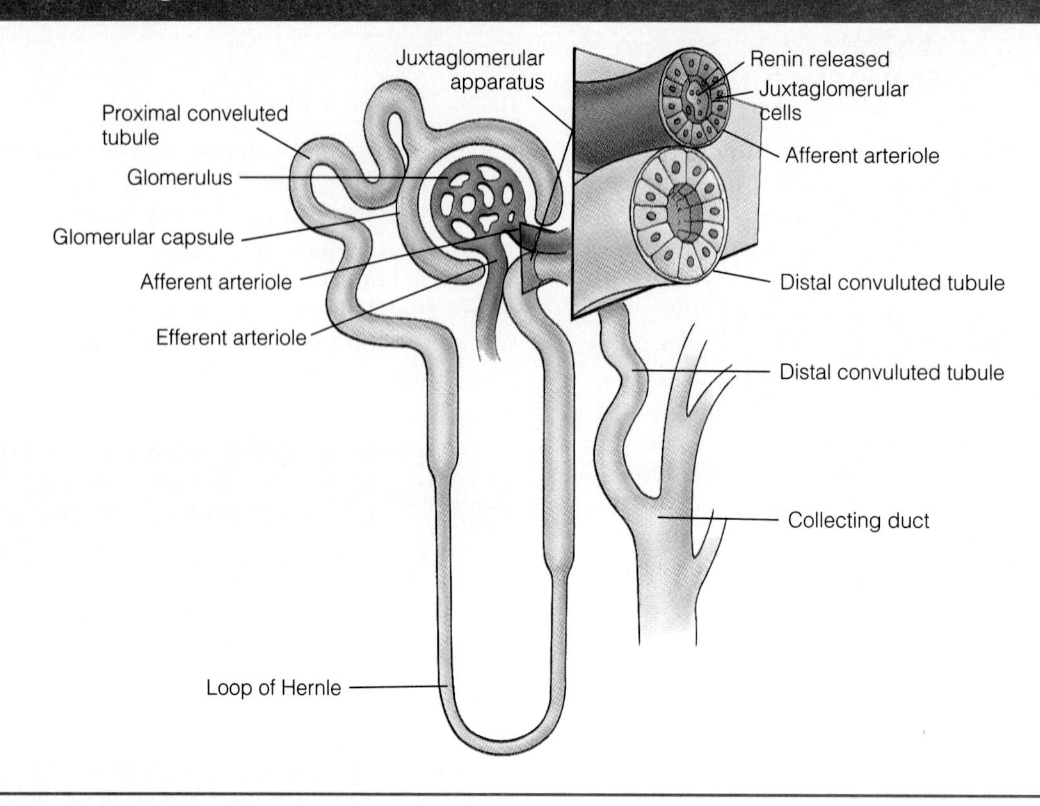

JG cells are specialized cells located in the wall of the afferent arteriole where it lies next to the distal convoluted tubule. If the normal stretch of the JG cells is reduced, they release an enzyme into the blood called renin. Renin activates angiotensin, which in turn elevates blood pressure by contracting the body's arterioles. Angiotensin also stimulates the release of aldosterone from the adrenal gland, which increases sodium reabsorption into the extracellular fluid.

able to alter their rate of hydrogen ion secretion accordingly. The distal convoluted tubule cells use a biochemical reaction already familiar to you from Chapter 13. It is the same biochemical reaction found in the blood plasma when carbon dioxide enters from the tissues to be transported back to the lungs, as follows:

$$CO_2 + H_2O \xrightarrow[\text{carbonic anhydrase}]{} \underset{\text{carbonic acid}}{H_2CO_3} \longrightarrow H^+ + HCO_3^-$$

When the blood pH is low, carbon dioxide diffuses from it into the cells of the distal convoluted tubule walls. Once inside the cells, carbon dioxide and water form carbonic acid, which dissociates into bicarbonate ions and hydrogen ions. The cells exchange the hydrogen ions for sodium ions in the urine in the distal convoluted tubule. In other words, they move hydrogen ions out of the cells and into the urine while drawing sodium out of the urine back into the body. The hydrogen ions are actively transported into the urine, while the bicarbonate ions diffuse into the extracellular fluid and then into the peritubular capillaries of the nephron.

Some of the secreted hydrogen ions combine with bicarbonate ions in the urine to form water molecules and carbon dioxide. Carbon dioxide is returned to the blood and excreted by the lungs. However, this process eliminates only a small fraction of the hydrogen ions secreted into the urine. If the remaining hydrogen ions were not buffered in some way, the pH of the urine would actually be about 1.4. The distal convoluted tubular cells are also capable of producing ammonia (NH_3) by metabolizing amino acids. However, instead of being further processed metabolically, the ammonia molecules immediately diffuse from the distal convoluted tubule cells into the urine. There they combine with the hydrogen ions to form ammonium ions (NH_4^+), which can be safely excreted in the urine.

If the blood is acidic, hydrogen ions are secreted into the urine and bicarbonate ions are produced and put into circulation via the peritubular capillaries until the proper pH is restored. If the blood is alkaline, fewer hydrogen ions are secreted into the urine and fewer bicarbonate ions are produced to be put into the peritubular capillaries.

Nervous Control

Kidney function is also affected by stimulation of the sympathetic division of the autonomic nervous system. When the sympathetic division innervating the kidneys is stimulated, the arterioles of the nephrons constrict dramatically, lowering the glomerular filtration rate and decreasing kidney function. Because the kidneys normally contain approximately 20% of the circulating blood at any given time, sympathetic stimulation, during stressful situations requiring the mobilization of various body resources, allows a large amount of blood to be shunted to other body organs, such as the skeletal muscles.

Urinary System Disorders

As we have just discussed, the body maintains much of its homeostasis through the kidneys. Therefore any disruption of renal function can be life threatening.

Infection

Infections of the urethra are relatively common. This is especially true for women, for two reasons. The urethra is much shorter in women than in men, and the urethral opening is not far from the anus, which is a potential source of bacterial contamination. Therefore after urination if a woman wipes from the urethral opening toward the anus, instead of the other direction, she greatly reduces the possibility of bacterial infection. If the infection is confined to the urethra, it is called **urethritis** (u're-thri-tis). If the infection moves into the bladder, it is called **cystitis** (sis-ti'tis). In both men and women the primary symptoms of cystitis are a frequent urge to urinate accompanied by a burning or stinging sensation as the urine passes through the urethra. Treatment usually involves antibiotic therapy and increased fluid intake to help flush the bacteria from the infected portion. A more serious infection, called **nephritis,** refers to a bacterial infection that has spread to the kidneys.

Kidney Failure

Physical damage, disease, or certain drugs can cause the kidneys to cease functioning, called **kidney failure.** In **acute kidney failure** the kidney stops functioning abruptly. For example, a rapid drop in blood pressure caused by sudden blood loss, or the blockage of urine through the ureters may cause the kidneys to stop functioning. In **chronic kidney failure** repeated inflammation or injury to the kidneys damages the tissues, gradually reducing their efficiency because of a build-up of scar tissue. For example, hypertension slowly but inevitably damages the kidney nephrons. As the damage progresses, waste products and other solutes build up in the blood and abnormal amounts of water are excreted in the urine.

Dialysis

If kidney damage is great, a person's blood may have to be filtered through an external filtration system. An artificial kidney is capable of removing waste products, excess ions, and other substances from the blood by means of simple diffusion, or **dialysis** (di-al'i-sis). The artificial kidney is a relatively simple device. Blood is pumped from one of the patient's arteries through plastic tubing, which is connected to a series of cellophane coils placed in a bath solution, and is then returned to one of the patient's veins.

The cellophane coils are immersed in a fluid with the same chemical composition and solute concentration as normally filtered extracellular tissue fluid. The cellophane separating the patient's blood from the chemical solution has almost the same permeability characteristics as do capillary walls. It therefore allows ions and other small molecules to move down their concentration gradients and diffuse through it, but is impermeable to formed elements and plasma proteins. To maintain the concentration gradient necessary for diffusion to continue, fresh fluid must also be continually circulated through the chamber holding the cellophane coils (Figure 17-5).

Kidney dialysis does have complications. It is time consuming and uncomfortable, and patients are subject to infections at the site of connection to the dialysis machine. Because dialysis is not as thorough in removing wastes and other solutes from the blood as normal kidney functioning, specific control of the patient's homeostasis is difficult to maintain over time.

Kidney Transplantation

People with chronic kidney failure may sometimes have kidney transplant operations, in which they receive a functioning kidney from a donor. The donor's health will not be impaired because, although we have two kidneys, only one is necessary for survival and health. The first successful

Figure 17-5 Dialysis

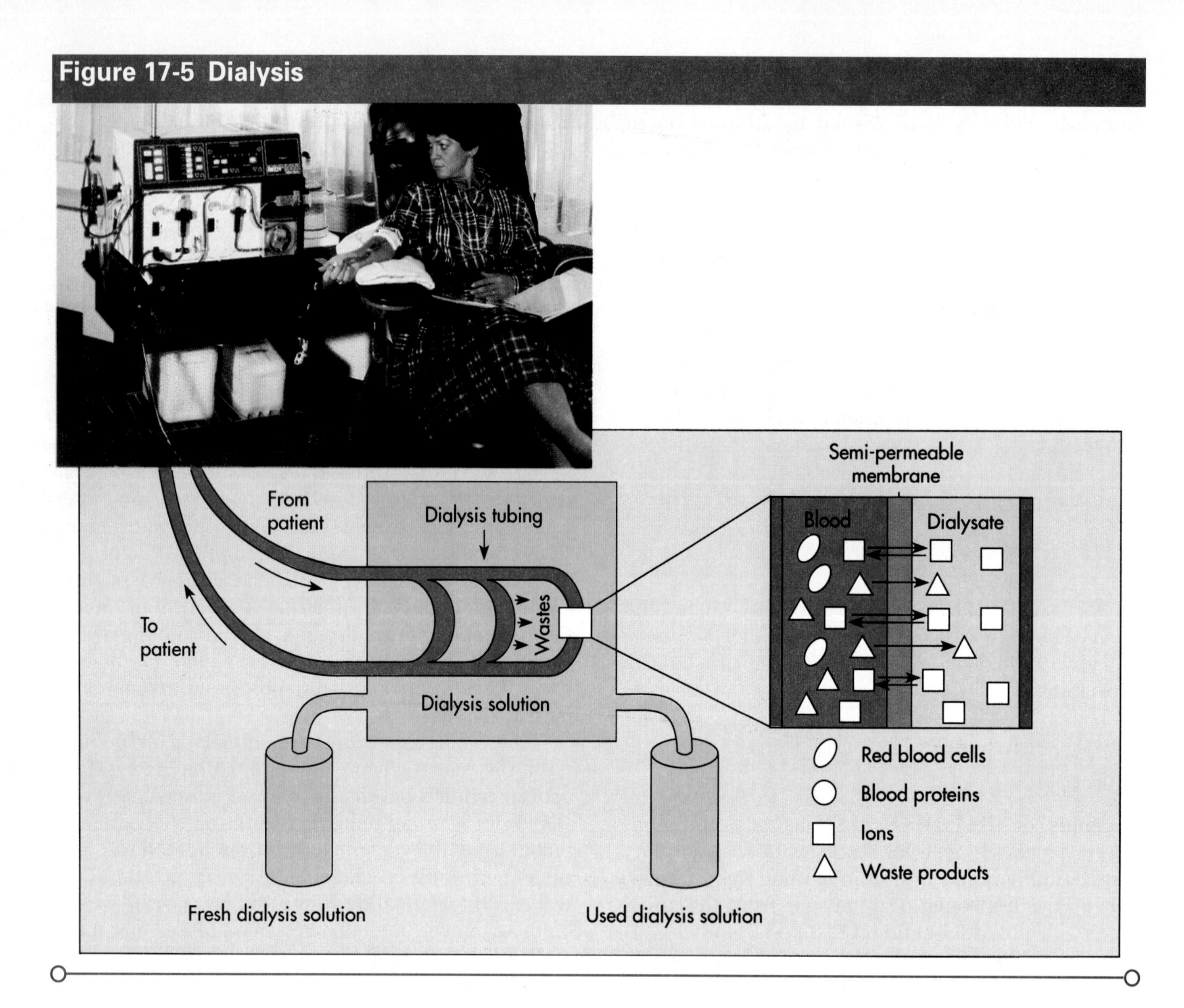

kidney transplantation was performed in 1954 on identical twins. Since then more than 87,000 successful transplants have been performed in the United States. The major problem is the immunological response that leads to rejection of the transplanted kidney. Recall from Chapter 14 that the body's immune system is able to distinguish self from nonself due to the protein markers made by the major histocompatibility complex (MHC) genes. As with other organ transplants, receiving a kidney from a close biological relative decreases the probability of rejection by the recipient's immune system because it is more likely that two biologically related people share a greater number of MHC protein markers than do two people picked at random. It is therefore highly likely that siblings will have similar MHC protein markers because they share the same parents. The 5-year survival rate of transplant recipients receiving kidneys from a sibling is approximately 90%. In contrast, it is 65% if the donor is biologically unrelated to the recipient.

Related Kidney Problems

Other organ system problems may also affect kidney function. Normally renin secretion from the juxtaglomerular kidney cells is controlled by negative feedback. As the blood pressure in the afferent arteriole rises to appropriate levels, renin secretion is inhibited. However, the system may function incorrectly because of disease in other body tissues. For example, if atherosclerosis reduces the lumen diameter of the renal arteries, blood flow and pressure to the kidneys is reduced. This reduction stimulates the JG cells to continuously secrete renin, causing an abnormal elevation in blood pressure called **renal hypertension.** Another related problem, called **congestive heart failure,** occurs when a weakened heart is unable to maintain an adequate blood flow through the kidney. This releases angiotensin, which increases aldosterone secretion by the adrenal cortex. Because the net effect of aldosterone is sodium and water reabsorption, blood volume greatly increases. The increased blood volume elevates the blood pressure in all the capillaries, dramatically increasing extracellular tissue fluid, which results in tissue swelling, or **edema** (e-de'mah).

CHAPTER REVIEW

SUMMARY

1. The kidneys are responsible for maintaining the correct composition of internal body fluids. They remove many waste products of cellular metabolism and other substances, such as excess ions and water absorbed from the digestive system. The primary functions of the kidneys include maintaining homeostasis of water content, blood pressure, blood volume, and the correct pH of the body's intercellular environment.
2. The kidneys remove waste products and other substances from body fluid by filtering the blood through microscopic structures called nephrons. The nephrons span both the cortex and the medulla of the kidney. Each kidney contains approximately 1 million nephrons.
3. Each nephron consists of a series of branching capillaries that surround a hollow tubular structure called the renal tubule. The capillaries consist of two regions, the glomerulus and the peritubular capillary network. The four regions of the renal tubule are the glomerular capsule, the proximal convoluted tubule, the loop of Henle, and the distal convoluted tubule.
4. As blood flows into each nephron, it enters a tuft of capillaries, called the glomerulus, that resides in the blind end of the renal tubule. Blood pressure forces out most of the blood plasma without its proteins, called filtrate, into the renal tubule to be further processed. The filtrate consists of small solutes, which include waste products, nutrients, and ions.
5. The rest of the renal tubule is responsible for reabsorbing nutrients and most of the water of the filtrate, and for concentrating the remaining waste products for excretion. The cells comprising the walls of each region of the renal tubule are slightly different and specialized to perform different functions as the filtrate passes through.

FILL-IN-THE-BLANK QUESTIONS

1. Blood carrying cellular waste products enters each kidney through the ______ ______.
2. The kidney consists of an outer layer of solid tissue and a hollow inner chamber called the ______ ______.
3. The renal tubule consists of ______ major regions.
4. The glomerulus sits within ______ ______ of the renal tubule.
5. The walls of the descending limb of the loop of Henle are freely permeable to ______ and impermeable to ______.
6. The target tissue of ADH is primarily the ______ ______ ______ of the renal tubule.
7. Most of the solute in urine is ______.
8. One way the kidney controls blood pressure is through the ______ apparatus, located at the connection between the afferent arteriole and the distal convoluted tubule.
9. Cessation of kidney function is called ____ ____.

SHORT-ANSWER QUESTIONS

1. Describe the means by which the loop of Henle maintains a solute concentration in the extracellular fluid surrounding the loop, with the highest concentration at the bottom of the loop.
2. Compare and contrast tubular reabsorption and tubular secretion.
3. Describe how the kidneys maintain blood pressure and blood volume.
4. Describe how the kidneys maintain the pH of the extracellular environment.
5. How does an artificial kidney machine filter blood?

VOCABULARY REVIEW

aldosterone—*p. 328*
angiotensin—*p. 328*
bladder—*p. 322*
collecting tubule—*p. 323*
distal convoluted tubule—*p. 323*
glomerulus—*p. 323*
juxtaglomerular cells—*p. 328*
kidney—*p. 321*
loop of Henle—*p. 323*
nephron—*p. 322*
peritubular capillary—*p. 323*
proximal convoluted tubule—*p. 323*
ureters—*p. 322*
urethra—*p. 322*

Chapter Eighteen

HUMAN REPRODUCTION

For Review

Here are some important terms and concepts that you will encounter in this chapter. If you are not familiar with them, you should review them before proceeding.

Haploid (page 70)

HIV (page 275)

Hormones (page 191)

Meiosis (page 70)

OBJECTIVES

After reading this chapter you should be able to:

1. *List the primary organs and accessory glands of the male reproductive system.*
2. *Describe the process of gamete production in the male.*
3. *Describe the hormones that affect the male reproductive system and describe the function of each.*
4. *List and describe the primary organs and accessory structures of the female reproductive system.*
5. *Describe the process of gamete production in the female.*
6. *Describe the steps of the ovarian and menstrual cycles.*
7. *List the hormones associated with the female reproductive system and describe the hormonal regulation of the ovarian and uterine cycles.*

A species is sustained over time through reproduction. In sexually reproducing species, such as ourselves, both males and females make a genetic contribution to the offspring they produce. However, the female reproductive system not only produces gametes but also must prepare an environment in which the fertilized egg can be protected and nourished during its period of development. This dual role makes the female reproductive system physiologically and hormonally complex compared with that of the male.

Reproduction System

Depending on your state of mind, the term *reproduction* may bring up a number of images involving courtship, exotic honeymoons, or cute babies. However, as George Gaylord Simpson, an evolutionary biologist, was once reported to have said, romance and courtship are the frills that have evolved only because they enhance the real purpose of reproduction, to pass one's genes to another generation. Although this statement may offend our romantic sensibilities, from an evolutionary perspective it rings true. A species is "successful" as long as its members are able to reproduce another generation. Furthermore, any characteristic that increases the likelihood of producing offspring is selectively advantageous and will increase in frequency in subsequent generations.

From the more narrow perspective of observing human behavior, there are probably few other physical activities more pleasurable to humans than sexual activity. Although the biological drive to reproduce is powerful in all animals, cultural, social, and sometimes emotional factors can enhance or restrain its expression in humans. However, it must also be a strong influence on human behavior because regardless of culture or social group, it is always bounded by many rules and customs. In virtually all modern industrialized cultures, television, newspapers, and magazines promote and sell products by using advertisements that play on our natural curiosity about and interest in sex. However, few other subjects are more private and also of such general interest as sexual behavior.

Although male and female reproductive organs are different, they share a common purpose. This purpose is to produce offspring. The male's role in this reproductive scheme is to manufacture **gametes** called **spermatozoa** (sper'ma-tozo'ah) or **sperm** and to deliver them to the female reproductive system for fertilization. Similarly, the female's role is to produce gametes called **ova** (sing. *ovum*). When events are appropriately timed, a sperm and an ovum fuse to form a fertilized egg, or **zygote** (zi'got), the first cell of a new individual. After fertilization the female reproductive system must also provide the protective environment in which the zygote develops until birth.

In this chapter we describe the structure, function, and hormonal regulation of the male and female human reproductive systems. Within this context we also discuss birth control and the most prevalent sexually transmitted diseases.

The Structure of the Male Reproductive System

Both the male and the female reproductive systems consist of a pair of primary reproductive organs, called **gonads** (go'-nadz), and accessory reproductive structures. The male gonads are the **testes,** which produce spermatozoa, or sperm. The testes also produce male sex hormones, collectively called **androgens.** The most important male sex hormone is testosterone.

Primary Reproductive Organs

The paired, oval-shaped testes (sing. *testis*) of the male lie suspended in pouches, called the **scrotum** (skro'-tum), outside and below the pelvic region. The scrotum consists of a thin outer layer of skin loosely connected to an underlying layer of smooth-muscle tissue. Internally the smooth-muscle layer forms a partition that divides the scrotum into two separate pouches, one for each testis (Figure 18-1).

Each testis is about 4 cm, or 1.5 inches, long and 2.5 cm, or 1 inch, in diameter. It is surrounded by two layers of connective tissue. Each testis is divided internally into 250 to 300 wedge-shaped lobules. Each lobule usually contains several tightly coiled **seminiferous** (seh-mih-nih'-fer-us) **tubules** in which sperm are produced. Although each testis is relatively small, about 125 m, or nearly 140 yards, of seminiferous tubules are packed into each one. Lying in the connective tissue surrounding the tubules are the **interstitial** (in'-ter-stih'-shul) **cells** that produce the hormone **testosterone.**

The **epididymis** (ep-i-did'i-mis) is a coiled tube about 6 m, or 20 feet, long that lies just outside each testis. The immature sperm are produced in the testes but are stored temporarily in the epididymis. During their movement through the epididymis the sperm mature and become mobile. Most of the sperm are stored in the region of the epididymis closest to the next section of the duct system, called the **vas deferens** (vas deh'-fer-enz).

Figure 18-1 Male reproductive system

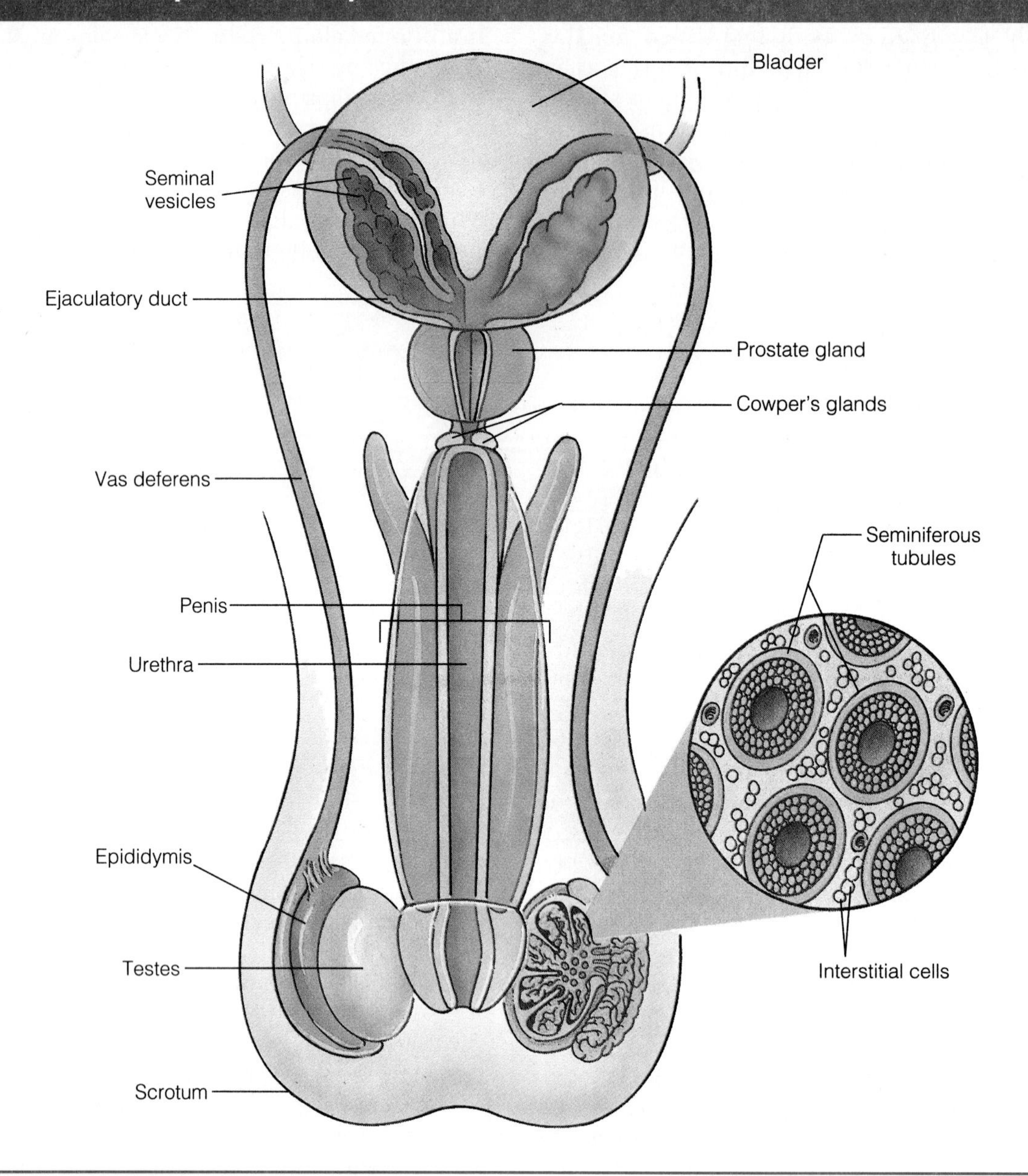

The vas deferens, leading from the epididymis of each testis, is about 45 cm, or 18 inches, long. It runs upward from the epididymis through a space called the **inguinal** (ing′gwi-nal) **canal.** It then enters the abdominal cavity, where it curves around the bladder and connects to the urethra. The vas deferens's main function is to propel sperm from their storage sites, the epididymis and a small part of the vas deferens adjacent to it, into the urethra. Under appropriate conditions the smooth-muscle tissues in the walls of the vas deferens create peristaltic waves that rapidly squeeze the sperm forward in an action called **ejaculation.** The urethra serves both the urinary system and the reproductive system in males, which is not the case in females (See Figure 18-1).

Accessory Glands

As the sperm move through the male reproductive system's ducts, they mix with fluids secreted from three different accessory glands (See Figure 18-1), producing a sperm-bearing mixture of fluids called **semen** (se′-min). The first accessory glands are the paired **seminal** (seh′-mih-nul) **vesicles,** located near the base of the bladder and connected by a short duct to the vas deferens. The seminal vesicles produce about 60% of the semen's fluid volume. The fluid they secrete contains primarily fructose and prostaglandins (see Chapter 10) and also leukocytes. The **prostate** (prah′-stat) **gland** is a single gland about the size of a walnut that encircles the upper part of the urethra, just below the bladder. The secretions from this gland, accounting for the rest of the semen volume, form a milky, alkaline fluid. These fluids enter the urethra through several small ducts as the smooth-muscle tissues of the prostate contract during ejaculation. The final accessory glands are called **Cowper's glands.** They are two pea-sized glands just below the prostate gland. They produce a thick, clear mucus, which drains into the urethra. This secretion appears during sexual excitement before ejaculation and serves primarily as a lubricant during sexual intercourse.

Semen provides the transport medium, nutrients, and chemical substances that protect the sperm and facilitate their movement through the female reproductive tract. Mature sperm cells are basically packages of DNA and contain little cytoplasm or stored nutrients. The fructose in the seminal vesicle secretions provides most of their energy fuel. Once the semen enters the female reproductive tract, the prostaglandins in the semen cause rhythmic contractions of the tract walls, which assist in moving the sperm through it. The alkalinity of the fluid secreted by the prostate gland, with a pH of about 7.5, helps neutralize the acidic environment of the female reproductive tract, protecting the sperm until they reach the ovum. Sperm are most active in an environment with a pH of between 7.1 to 7.6, so the prostate gland secretions also enhance sperm movement. Although the amount of semen propelled out of the penis is relatively small, only 2 to 6 ml, there are between 50 to 100 million sperm in each milliliter of fluid.

The **penis** delivers sperm into the female reproductive tract. The penis and scrotum make up the external reproductive structures, or **external genitalia** (jeh-nih-ta′-le-uh), of the male. The penis is a long shaft with an enlarged tip called the **glans penis.** Skin covering the penis is loose and forms a cuff, called the **prepuce** (pre′-pyoos) or **foreskin,** around the glans penis. The foreskin is frequently surgically removed shortly after birth, a procedure called **circumcision** (sur-kum-sizh′un).

Internally the penis contains the urethra and three columns, or cylinders, of spongy connective tissue collectively called **erectile tissue.** Erectile tissue also consists of smooth-muscle cells and small vascular spaces. **Erection** of the penis occurs when the vascular spaces fill with blood, causing the penis to enlarge and become rigid. Normally when a man is not sexually aroused, the arterioles supplying the erectile tissue are constricted and the penis is flaccid or limp. However, during sexual excitement erection of the penis is initiated by the parasympathetic nerves, causing these arterioles to dilate. As a result, the vascular spaces of the erectile tissues fill with blood, enlarging and erecting the penis. Erectile tissue expansion also compresses its drainage veins, slowing the outflow of blood and helping maintain the erection (Figure 18-2).

Male Gamete Production

Spermatogenesis (sper′-mah-to-jeh′-nih-sis) is the development of sperm cells within the coiled seminiferous tubules of the testes (Figure 18-3). Spermatogenesis begins during puberty, about 14 years of age, and continues throughout life. Both spermatogenesis and its counterpart in the female, called **oogenesis** (o-o-jen′ih-sis), involve meiosis. As you recall from Chapter 4, meiosis consists of two consecutive nuclear divisions without a second replication of the chromosomes between divisions. This produces four cells. Each cell has *one half* as many chromosomes as the original cell. Here we will follow the major stages of meiosis in sperm production.

Sperm are produced in the seminiferous tubules of the testes. Inside, forming part of the inner lining of the seminiferous tubule, are rapidly dividing, diploid cells called **spermatogonia** (sper′-mah-togo-ne-uh). Other cells located in this inner cell layer,

Figure 18-2 Accessory glands of the male reproductive system

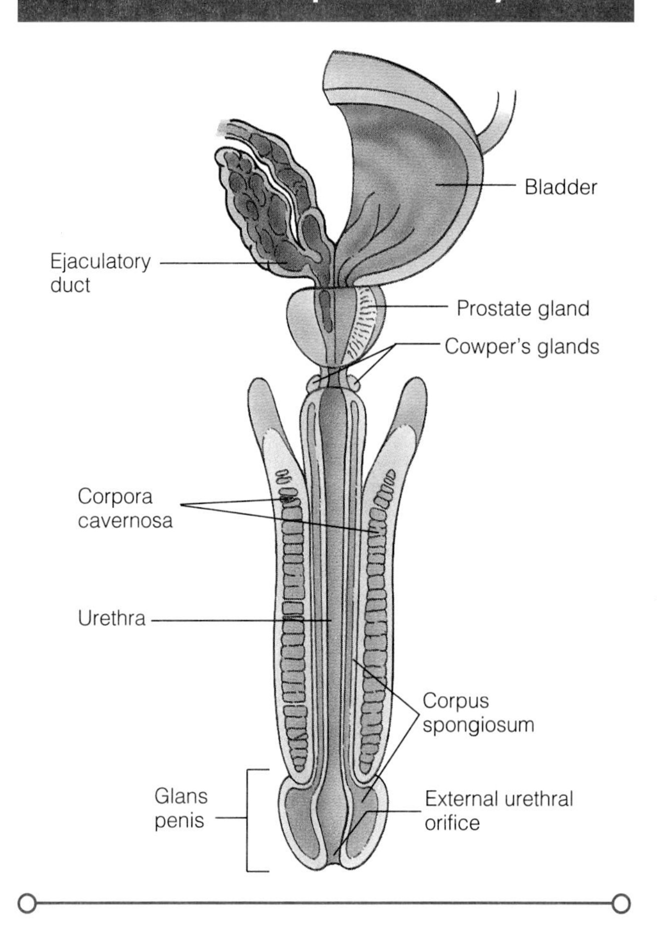

Figure 18-3 Spermatogenesis

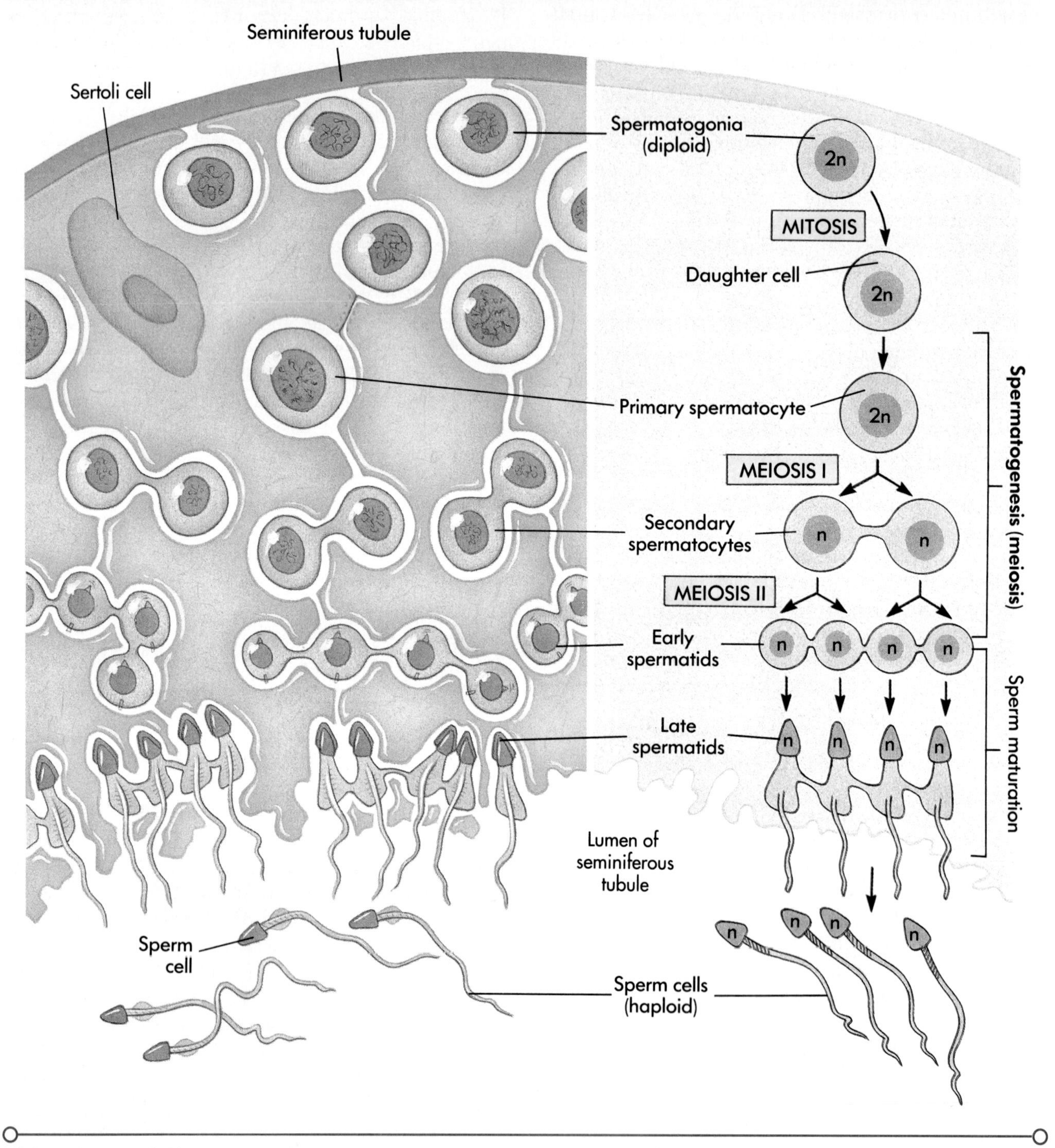

Lining the seminiferous tubules are diploid (2n) spermatogonia cells that undergo meiosis, giving rise to haploid (n) sperm cells.

called **Sertoli cells,** regulate spermatogenesis and support and nourish developing sperm cells. Spermatogonia divide in one of two ways. Some undergo meiosis, eventually giving rise to sperm cells. Other spermatogonia undergo *mitosis* to produce more identical spermatogonia, thereby maintaining a continual supply of these cells.

The spermatogonia that undergo meiosis first enlarge and are forced away from the tubule wall to become **primary spermatocytes,** which are still diploid. Each primary spermatocyte then undergoes meiosis I, producing two **secondary spermatocytes.** Recall from Chapter 4 that although each secondary spermatocyte is *haploid,* each chromosome is still in a duplicated state. It consists of two *sister chromatids.* Almost immediately after their formation, each secondary spermatocyte undergoes meiosis II. This means the sister chromatids of each chromosome are separated from each other and move into newly forming cells. The resulting haploid cells are called **spermatids** (sper′-mah-tidz). They then begin to lose most of their cytoplasm and develop tails, gradually developing into mature sperm. The spermatids are released into the lumen of the seminiferous tubule. (See Figure 18-3.)

From formation of the primary spermatocyte to the release of the spermatids into the tubule, spermatogenesis takes from 64 to 72 days. However, spermatids are still not mobile, nor are they capable of fertilizing an ovum. They are moved through the tubule into the epididymis by peristalsis and by the developing sperm entering the tubule behind them. There they mature further, increasing in mobility and in the ability to fertilize an ovum.

A mature sperm consists of a head, midpiece, and tail, which correspond to genetic, metabolic, and locomotor functions, respectively. The head contains the nucleus, which is packed with DNA. A caplike structure called an **acrosome** (ak′ro-som) covers most of the head and contains hydrolytic enzymes that will enable the sperm to penetrate the layers of cells surrounding the ovum. The midpiece consists of mitochondria, which supply the energy necessary for sperm mobility. The tail contains microtubules typical of cilia in other body cells. The tail moves in a whiplike fashion to propel the sperm at a speed of 1 to 4 mm per min.

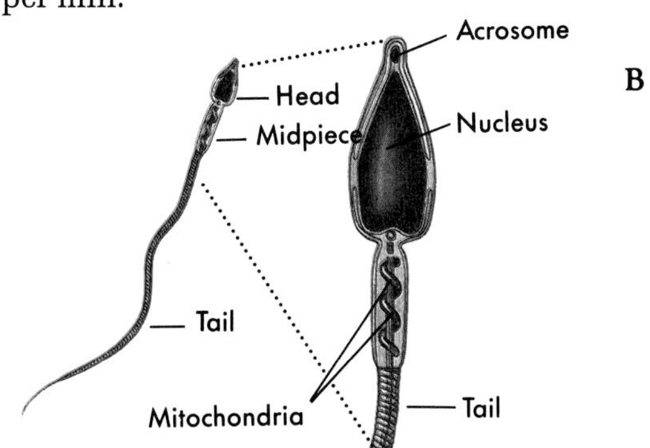

Hormonal Regulation in the Male

Hormonal regulation of spermatogenesis and the testes's production of androgens involves interactions among the hypothalamus, the anterior pituitary gland, and the testes. Beginning around 14 years of age, there is increased production and release of **gonadotropin-releasing hormone (GnRH)** from the hypothalamus. This releasing hormone is carried via the portal system within the pituitary gland's stalk to the anterior pituitary, where it stimulates the secretion of two gonadotropic hormones. These hormones, **follicle-stimulating hormone (FSH)** and **luteinizing hormone (LH),** are secreted in males and females but are named according to their activities in females.

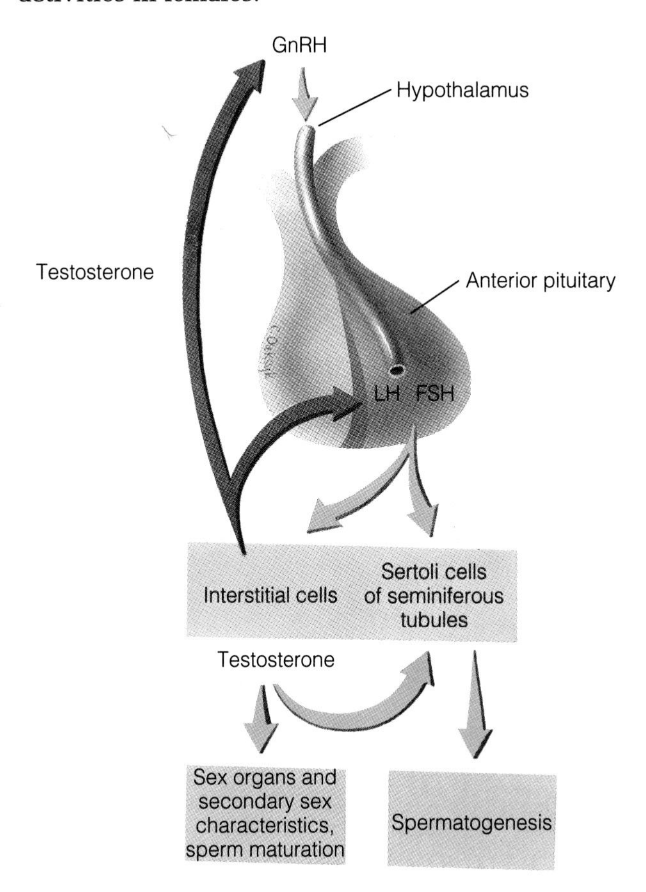

Both LH and FSH act on the testes. LH stimulates the interstitial cells to produce testosterone. Testosterone has at least four related functions in the male. First, testosterone assists in stimulating spermatogenesis. Second, it is necessary for the initial growth and then maintenance of the whole male reproductive system. Third, testosterone initiates the growth of **secondary sex characteristics.** These are the physical characteristics associated with males, such as increased muscle size and tone; body size; facial, body, and pubic hair; and voice changes. Fourth, testosterone promotes the growth of most body tissues, especially bone and skeletal muscle. It is the primary reason why males are on average larger than females.

The specific target tissues of FSH are the seminiferous tubules, where it stimulates spermatogenesis. FSH stimulates spermatogenesis by acting on Sertoli cells embedded in the seminiferous tubules's walls. Testosterone causes the Sertoli cells to release a substance called **androgen-binding protein (ABP).** ABP binds to the spermatogonial cells, which can then be stimulated by testosterone to undergo meiosis and produce sperm cells.

Figure 18-4 Female reproductive system

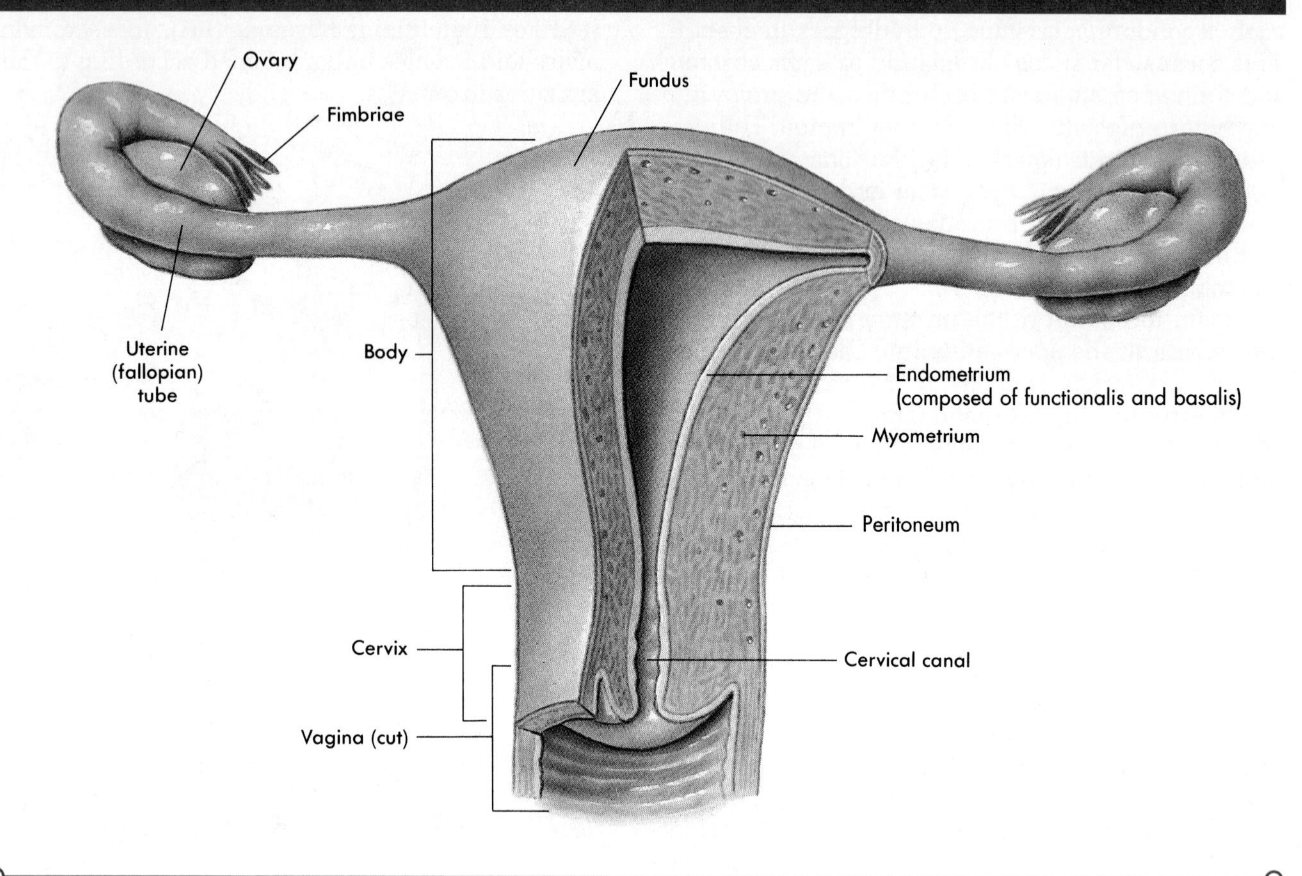

Structure of the Female Reproductive System

The female's reproductive role is more complicated than the male's reproductive role. Recall that not only must she produce gametes, but her body must prepare itself to nurture a developing zygote for approximately 280 days.

Primary Reproductive Organs and Accessory Structures

The female reproductive system consists of paired structures called **ovaries** (o-ver-ez), which are the primary reproductive structures, and accessory structures that accept sperm, conduct it to the ovum, and nourish the developing zygote. Like the male testes, the ovaries also have the two major functions of producing gametes and sex hormones.

The ovaries are located on either side of the upper pelvic cavity and are about the size and shape of almonds (Figure 18-4). When viewed in cross section under a microscope, an ovary is seen to be composed of an outer cortex and an inner medullary region. Within the cortex are many saclike structures called **ovarian follicles** (*follicle* is derived from the L. for "little bag"). Each follicle contains a cell called a **primary oocyte** (o′-o-sit), which is surrounded by one or more layers of cells.

Adjacent to each ovary but not connected directly to it is a **uterine** (yoo′-ter-in), or **fallopian, tube** (Figure 18-4). This provides a fertilization site and also conducts ova to the next region of the female reproductive system. Each uterine tube is about 10 cm, or 4 inches, long and 1 cm, or approximately $\frac{1}{2}$ inch, in diameter. Fingerlike projections called **fimbriae** (fim′-bre-e) extend from the enlarged distal portion of the uterine tube and drape over the ovary. When an ovum is released from the ovary, it must make its way through this open space to enter the uterine tube. The fimbriae assist its journey by "sweeping" the ovum into the tube. Each uterine tube's walls are lined with ciliated mucous membrane. Ciliary movement and smooth-muscle tissue contraction in the uterine tube's walls propel the ovum through the lumen toward the next reproductive tract section.

The uterine tubes connect to the **uterus** (L. for "womb"), located in the pelvic cavity in front of the rectum and just above and behind the bladder (See

Figure 18-4). It is a hollow, thick-walled structure about the size and shape of a pear with the narrow portion pointed down. After childbirth it is usually slightly larger. The main portion is called the **body,** and the upper rounded region, including where the uterine tubes enter it, is called the **fundus.** The narrowed "neck" and outlet of the uterus is called the **cervix.**

The uterine wall is composed of three layers. The outer layer, called the **perimetrium,** consists of fibrous and elastic connective tissue. The **myometrium** (mi'-o-me'-tre-um) is the thick middle layer composed of smooth-muscle tissue. The inner mucous membrane lining the uterus is called the **endometrium.** The endometrium in turn is composed of two layers. The innermost layer, called the **functionalis** (funk'-shuh-na'-lis), undergoes cyclic changes in response to changing blood levels of hormones secreted by the ovaries and is shed approximately every 28 days. The deeper **basalis** (buh-sa'-lis) layer does not respond to hormones. It generates a new functionalis after the previous one is shed. **Endometriosis** (en'-do-me-tri-o'sis), a painful sensation accompanied by severe cramping and discomfort, is caused by the presence of endometrial tissues in reproductive tract regions other than the inner uterine layer.

The cervix projects slightly into the structure immediately below it, called the **vagina** (Figure 18-4). The vagina is a thin-walled muscular tube approximately 8 to 10 cm, or 3 to 4 inches, long. It lies between the bladder and the rectum and extends from the cervix to the body's exterior. The vagina's primary function is to receive the penis and semen during sexual intercourse. It also provides a passageway from the uterus for the delivery of an infant. Thus it is also called the **birth canal.**

The structures located at the body surface and surrounding the vaginal opening are called the external genitalia (Figure 18-5). The **mons pubis** (monz pyoo'-bis), anterior to the pubic symphysis, is a rounded area with underlying fatty tissue. At puberty the mons pubis becomes covered with hair. Running slightly down and back from the mons pubis are a pair of fatty skin folds covered with hair on their external surface, but with a smooth internal surface, called the **labia majora** (la'-be-uh mah-jor'-uh). Within the cleft formed by the labia majora are two thin folds called the **labia minora** (mih-nor'-uh). The labia minora enclose the **vestibule,** or opening to the vagina, which contains several structures. Within the vestibule and closest to the mons pubis is the **clitoris** (klit'-o-ris). It is a small protruding structure composed of erectile tissue and developmentally derived from the same tissue as the male penis. The clitoris contains sensory receptors that make it a sexual structure extremely sensitive to touch. Below the clitoris is the external opening of the urethra, followed by the vaginal entrance. The vagina may be partially closed by an incomplete ring of tissue called the **hymen** (hi'-min). Because the hymen is highly vascular, it tends to bleed during the first sexual intercourse. Its durability varies among individuals, and in some cases it may be ruptured during physical activities, a routine pelvic examination, or tampon insertion.

Female Gamete Production

As you recall, male sperm production begins at puberty and usually continues throughout life. The situation is different in females. The total supply of gametes a female produces is determined by the time of her birth. The time during which they are released extends from puberty, beginning between the ages of 11 to 14 years, to about the age of 50 years. We will examine female gamete formation here in some detail.

Meiosis occurs in the ovaries to produce oocytes, a process called **oogenesis,** meaning "the beginning of an egg." The process of oogenesis begins during female fetal development. During the first 3 months of fetal development, cells called **oogonia** proliferate in the ovary by mitosis. The oogonia then enlarge, storing nutrients in their cytoplasm, to become **primary oocytes.** The primary oocytes also become surrounded by a single layer of cells called **follicular cells.** Primary oocytes and their follicular cells are called **primordial follicles.** The DNA replicates, and the primary oocytes begin meiosis I but do not complete it.

Figure 18-5 External genitalia of the female

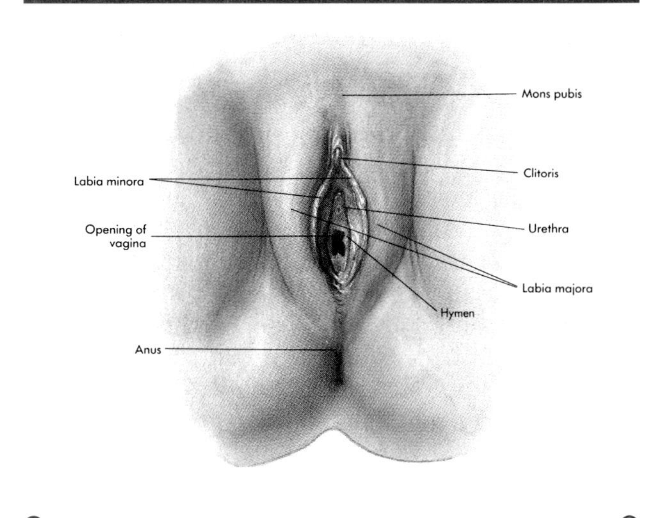

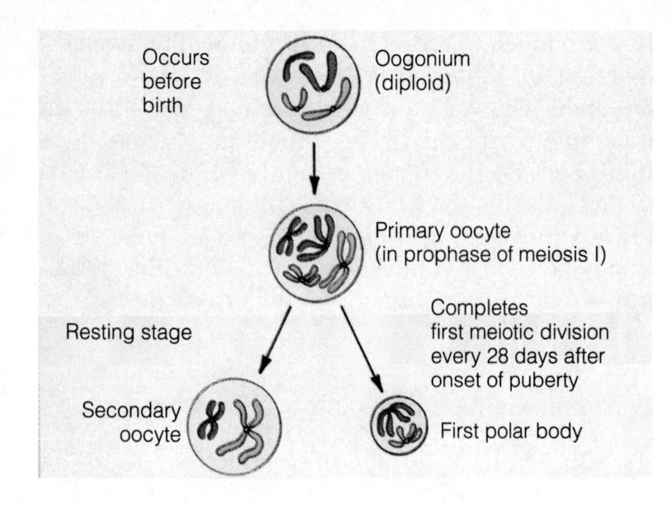

Instead, they remain in this arrested state until the onset of puberty, at least 10 to 14 years later.

Beginning at puberty one primary oocyte completes its first meiotic division approximately every 28 days. This monthly cycle continues until about 50 years of age. The events occurring to the primary oocyte, to the follicle that contains it, and to the uterus are called the **ovarian cycle.**

Before the primary oocyte completes meiosis I, the follicle also undergoes a series of changes. The follicular cells begin to multiply to form several layers of cells around the oocyte. These cell layers, now called **granulosa cells,** begin to secrete a substance that forms a thick transparent membrane, called the **zona pellucida** (peh-loo′-sih-duh), around the oocyte. Fluid also begins to accumulate between some of the granulosa cells, eventually forming a fluid-filled space called the **antrum** that makes up much of the follicle. The primary oocyte, surrounded by the zona pellucida and several layers of granulosa cells, is now almost completely surrounded by fluid accumulating in the antrum.

The two haploid cells produced as the primary oocyte completes meiosis I are different in size. One of them, called the **first polar body,** has 23 replicated chromosomes but is small because it contains almost no cytoplasm. In contrast, the other cell, called the **secondary oocyte,** is large because it contains nearly all the cytoplasm, nutrients, and enzymes. Because it lacks cytoplasm and nutrients, the first polar body eventually disintegrates.

About 36 hours before **ovulation,** or the release of the oocyte from the follicle, the mature follicle, now called the **graafian follicle,** begins to balloon out near the ovarian surface. At ovulation the ovarian wall ruptures and releases the secondary oocyte, now called an **ovum,** into the abdominal cavity near the opening of the nearby uterine tube. The ovum is still surrounded by its granulosa cells, which are now called the **corona radiata** (from L. for "radiating crown"). The tiny polar body temporarily lies between the cell membrane of the ovum and the zona pellucida until it gradually disintegrates (Figure 18-6).

The second meiotic division occurs only if the ovum is fertilized by a sperm. The haploid nuclei of the sperm and ovum fuse, producing a zygote, the first cell of a new individual. After the second meiotic division is completed the **second polar body** is produced and dies. Fertilization is described more fully in Chapter 19.

After the graafian follicle ruptures, releasing its ovum and antrum fluid, it undergoes further changes. The remaining granulosa cells in the follicle grow larger. They secrete lipids into the antrum and transform into a **corpus luteum** (L. for "yellow body"), which begins to secrete the hormones progesterone and estrogen. The corpus luteum continues to develop for about 10 days following ovulation. If fertilization does not occur, the corpus luteum degenerates and a new round of follicular maturation begins. However, if fertilization occurs and pregnancy results, the corpus luteum is maintained until the fifth or sixth month of the pregnancy.

The cyclic changes that occur in the ovary during oogenesis are also closely coordinated with changes in uterine condition. Recall from the beginning of this chapter that in addition to producing gametes, the female reproductive tract must also prepare an environment able to support a fertilized egg. All the changes in the uterine endometrium are collectively called the **uterine,** or **menstrual** (men′-stroo-ul), **cycle.** On average both the ovarian and the uterine cycles occur over a 28-day period, with ovulation occurring on day 14. These events are coordinated by hormones secreted from the hypothalamus, the anterior pituitary gland, and the ovaries. We will next look more closely at this hormonally controlled process.

Hormonal Regulation of the Ovarian and Uterine Cycles

Both the ovarian cycle and the uterine cycle are regulated and coordinated by the same hormones from the hypothalamus and the anterior pituitary gland that also control the male reproductive system. The difference is in the hormones produced by the ovaries. Two important hormones produced by the ovaries, called **progesterone** (pro-jes′ter-on) and **estrogen,** replace the hormones secreted by the testes. Progesterone primarily causes the uterine endometrium to periodically grow and become thicker. The ovaries secrete at least six chemically related *estrogens.* Estrogen is necessary for the development and maintenance of the ovaries,acces-

Figure 18-6 Follicular maturation

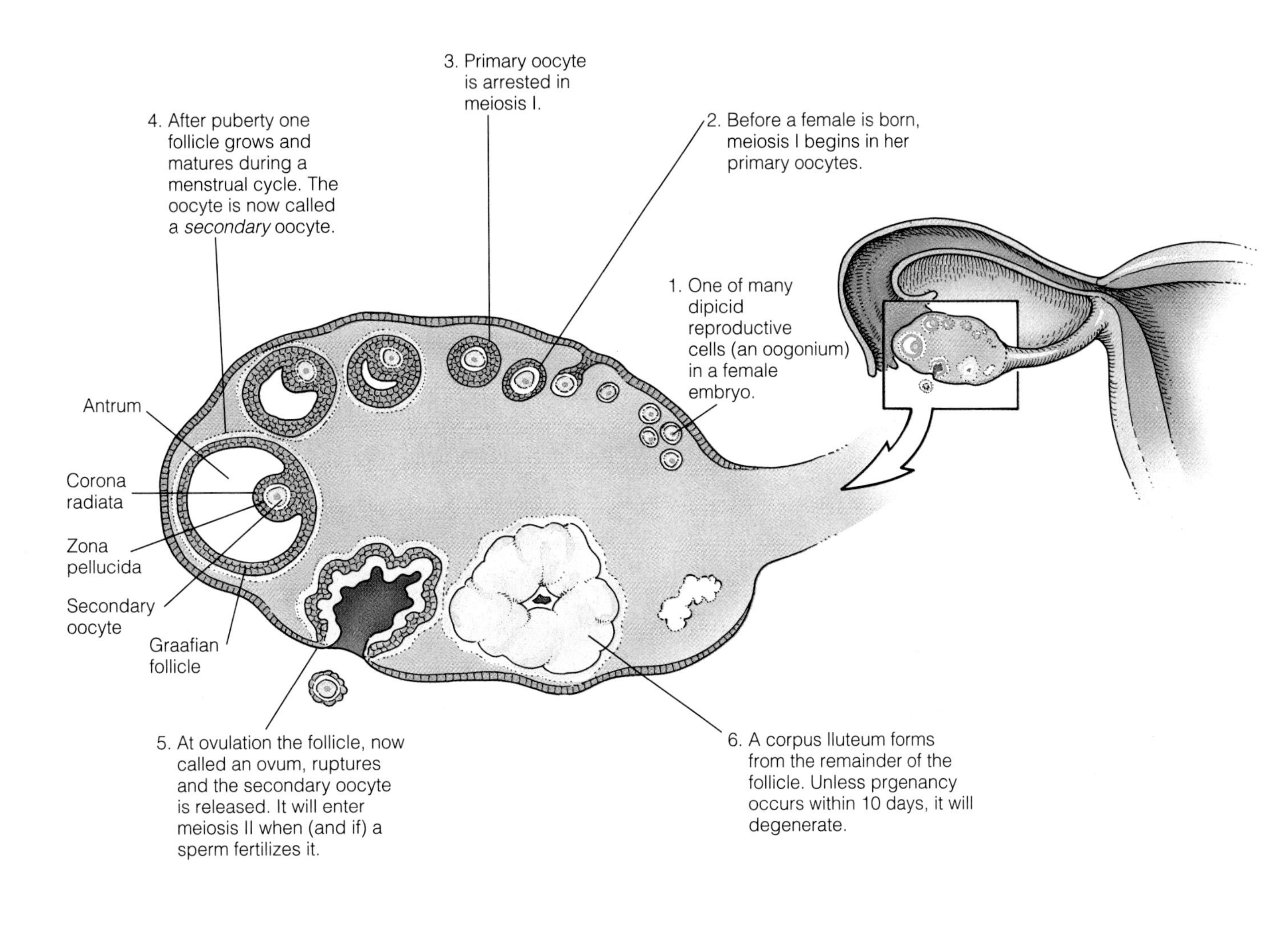

sory reproductive structures, and female secondary sexual characteristics.

Estrogen levels rising during puberty cause the growth of female reproductive structures. The uterine tubes, uterus, and vagina grow larger and become functional. Estrogen also initiates the rapid growth spurt at puberty that makes girls grow more quickly than boys at this time. However, this growth is relatively short because increased estrogen levels also cause earlier closure of the epiphyses of the long bones. Consequently, females usually reach their full adult height between 15 and 17 years of age as compared to males, who may continue growing until 20 years of age. Secondary sexual characteristic development in females includes breast growth and development; increased subcutaneous fat deposits, especially in the hips and breasts; pelvic widening; body and pubic hair growth; and several metabolic changes. These metabolic changes include low blood cholesterol levels and increased calcium uptake, which helps maintain bone density.

Ovarian Cycle

The ovarian cycle is divided into two major phases (Figure 18-7). The **follicular phase** is the first 14 days of the ovarian cycle and ends at ovulation. This is the period of follicle growth. Days 15 through 28 are marked by corpus luteum activity and are called the **luteal phase** of the ovarian cycle.

The beginning of the follicular phase, starting on the first day of menstrual flow, initiates follicle growth and oocyte maturation by the interaction of FSH, LH, and estrogen.

On day 1, gonadotropin-releasing hormone (GnRH) is released from the hypothalamus, stimulating the anterior pituitary gland to release both FSH and LH into the bloodstream. Both hormones stimulate ovarian follicular growth. As the follicle enlarges, the granulosa cells it contains begin secreting estrogen.

Blood estrogen levels, slowly rising during the first 7 to 10 days of the cycle, begin to exert negative feedback on the anterior pituitary gland and the hypothalamus. This feedback causes the pituitary output of FSH and LH to fall slightly.

However, in the ovary the increased amount of estrogen *intensifies* the effect of FSH on follicular growth and oocyte maturation. This *increases* estrogen output by the follicle. The result is a dramatic rise in the blood concentration of estrogen in the last few days before ovulation.

Although the initial rise in estrogen blood levels has a negative-feedback effect on the hypothalamus and the anterior pituitary gland, high concentrations of estrogen produced by the follicle have the opposite effect. They exert a *positive-feedback* effect on the hypothalamus and the anterior pituitary gland. Recall that positive feedback means that the conditions responsible for generating a particular effect increase as the effect itself increases or is enhanced.

By day 13, high levels of estrogen begin a series of events. First is a sudden increase in LH and to a lesser extent, FSH, by the anterior pituitary gland. This LH increase causes meiosis I to resume in the primary oocyte of the graafian follicle. The primary oocyte completes meiosis I, producing the secondary oocyte and the first polar body. LH also causes the ovarian wall to rupture, resulting in ovulation on day 14. The role of FSH in this process remains unknown.

Shortly after ovulation, estrogen levels decline, possibly because of the damage done to the estrogen-secreting follicle during ovulation.

The sudden surge of LH also transforms the ruptured follicle into a corpus luteum. LH stimulates the corpus luteum to begin producing progesterone and smaller amounts of estrogen.

The rising blood levels of progesterone exert a negative-feedback effect on LH and FSH secretion from the anterior pituitary gland. Declining LH and FSH levels inhibit new follicular development during the luteal phase of the ovarian cycle.

As LH levels slowly decline, corpus luteum activity also decreases, and the corpus luteum begins to degenerate. As the corpus luteum degenerates, ovarian hormonal secretion also drops sharply.

The lowered levels of progesterone and estrogen trigger the GnRH release from the hypothalamus, initiating a new cycle.

The progesterone and estrogen produced by the developing ovarian follicle markedly affect the uterus. Estrogen initiates the growth of the functional layer of the endometrium. Progesterone stimulates changes in the growing endometrium that enable it to support a fertilized egg.

Uterine Cycle

There are three sequential phases of the uterine cycle (See Figure 18-7). The **menstrual phase** begins with menstruation. As mentioned earlier the functional layer breaks down and is discharged through the vagina. This process is initiated by corpus luteum degeneration. The sharp drop in progesterone causes the arterioles that supply blood to the functional layer of the endometrium to constrict. Without an adequate blood supply the functional layer's cells begin to die. At the same time, the constricted arteriolar walls also begin to deteriorate, releasing blood into the uterus. This blood, along with the tissues of the endometrium's functional layer, makes up the menstrual flow. As menstruation begins a growing ovarian follicle starts producing estrogen.

The **proliferative phase,** lasting from approximately day 6 to day 14, is marked by rising levels of estrogen. Under the influence of estrogen the endometrium's functional layer cells begin to grow or proliferate. At the end of menstruation the functional layer is approximately 1 mm thick. By the end of the proliferative phase it is about 3 mm thick. Estrogen also stimulates the synthesis of progesterone receptors in the functional layer cells, preparing them for interaction with progesterone. The proliferative phase ends with ovulation.

The **secretory phase,** from day 15 to day 28, begins immediately after ovulation. It is marked by rising levels of progesterone produced by the corpus luteum. Progesterone acts on the estrogen-primed endometrium. This causes the arterioles of the functional layer to grow and spread through this tissue layer. If fertilization has not occurred, the corpus luteum begins to degenerate toward the end of the secretory phase as LH levels drop. Progesterone levels fall, depriving the endometrium of a sufficient blood supply. The endometrial cells begin to die, and the constricted arterioles begin to leak blood into the uterus. The cycle starts again with the first day of menstrual flow.

Sometime between 45 to 55 years of age (the average is 47 years of age), the menstrual cycle begins to become irregular. The most likely reason is that over time the ovaries become less and less

Figure 18-7 Ovarian and uterine cycles

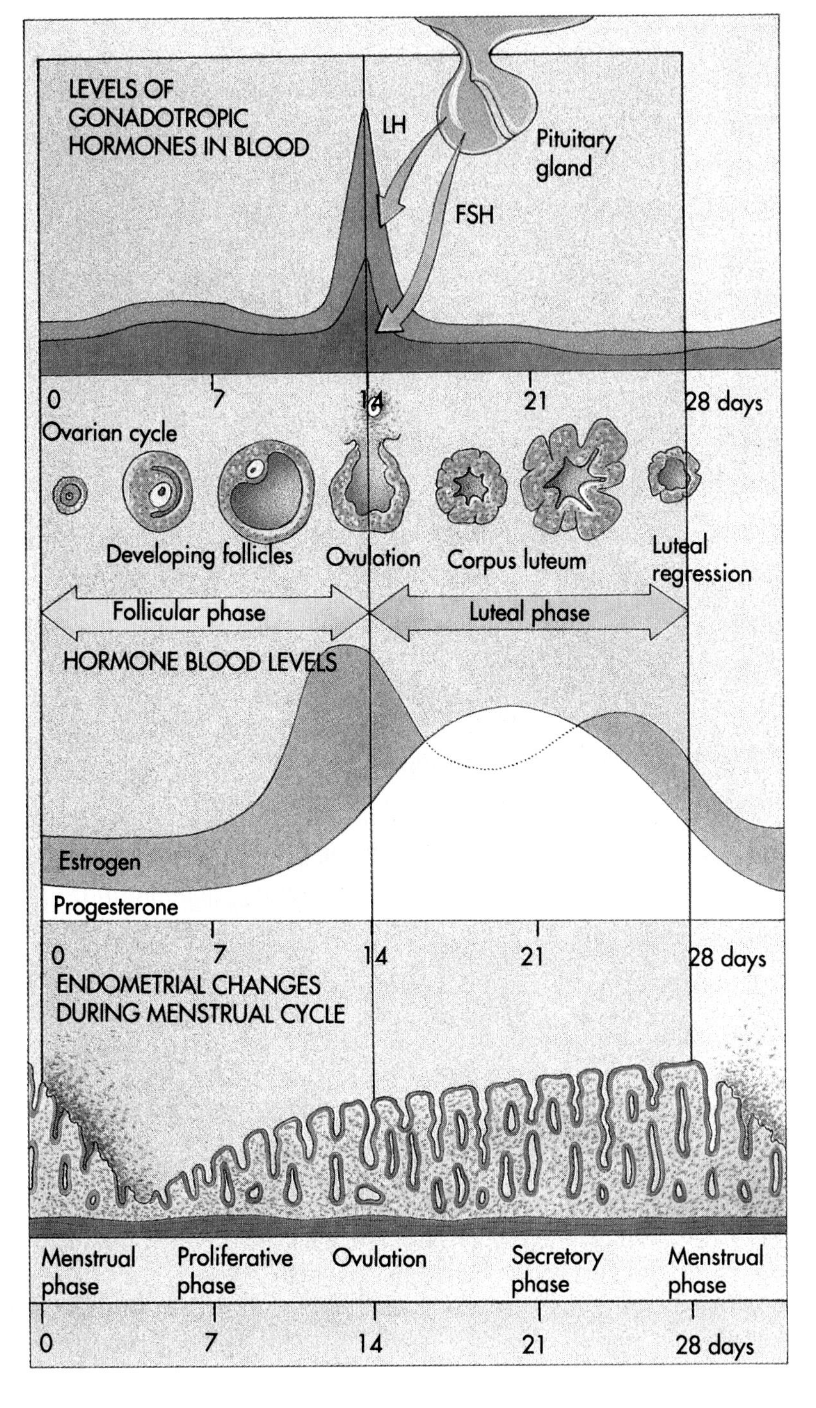

The development of the oocyte and its follicle is called the ovarian cycle. Also occurring at the same time are changes to the functionalis layer of the endometrium, collectively called the uterine or menstrual cycle.

responsive to GnRH signals. As estrogen production declines many ovarian cycles become anovulatory. **Anovulatory** means they do not produce a secondary oocyte, and menstruation becomes erratic and shorter. Eventually ovulation and menstruation cease. When 1 year has passed without menstruation, **menopause** has occurred.

With lower amounts of estrogen, the reproductive structures and breasts begin to change. Vaginal mucus secretions decrease. Because of a lack of this protective fluid, vaginal infections may become common. Other consequences of a decline in estrogen sometimes include periodically strong vasodilation of the skin's blood vessels, which causes uncomfortable "hot flashes" and night sweats; loss of bone mass; and slowly rising blood cholesterol levels, which increase the likelihood of the woman developing cardiovascular disorders.

Birth Control

Although currently many birth control, or contraceptive (*contra* = against, *cept* = taking), methods are available, they can be grouped into four major categories. These are the rhythm method, barrier methods, surgical methods, and chemical methods. The effectiveness of contraceptive methods varies considerably. Effectiveness is determined by the number of women out of 100 who become pregnant using a particular method. This is called the **failure rate,** expressed as a percentage. Failure rates for several birth control methods are listed in Table 18-1. In the final analysis, however, the only 100%-effective means of birth control is *total abstinence.*

The **rhythm method,** or **fertility awareness method,** is one means of preventing the gamete union and is often described as being a natural method. This method is based on recognizing the ovulation, or fertility, period and avoiding intercourse during this time. This may be accomplished by two means. First, daily basal body temperatures can be recorded. The body temperature drops slightly, 0.2° to 0.6° F, immediately before ovulation and then rises slightly, 0.2° to 0.6° F, after ovulation. Second, changes in the consistency of vaginal mucus are recorded because the mucus first becomes sticky and then clear and stringy during the fertile period. Both of these rhythm techniques require accurate recordkeeping

Table 18-1 Failure Rates for Reversible Birth Control Methods*

METHOD	FAILURE RATES (%)
Rhythm method	13–21
Withdrawal	9–25
Condoms	3–15
Diaphragm	4–25
Cervical cap	4–25
Sponge	15–30
Spermicides (foams, creams, jellies, vaginal suppositories)	10–25
Oral contraceptives	1–5
Norplant	0.3
Intrauterine device	1–5

*Approximate effectiveness of reversible methods is measured in pregnancies per 100 actual users per year.

for several cycles before they can be used with confidence. However, for women with regular 28-day cycles, fertility awareness methods have a high success rate.

Surgical methods, such as **tubal ligation** or **vasectomy**—cutting the uterine tubes or vas deferens, respectively—are nearly foolproof. They are the choice of approximately 33% of couples of childbearing age in the United States. However, these techniques are usually permanent, making them unpopular with people who still plan to have children but want to select the time (Figure 18-8).

The most widely used **chemical** method in the United States is the female oral contraceptive, or **birth control pill.** It is a preparation containing minute amounts of synthetic female hormones, usually estrogens and progesterones. Because this maintains relatively constant blood levels of ovarian hormones, the hypothalamus and the anterior pituitary glands respond as though ovulation has already occurred. Consequently, the hypothalamus does not secrete GnRH, nor does the anterior pituitary gland secrete FSH or LH. Ovarian follicles do not develop, and ovulation ceases. The major disadvantage of birth control pills is that they provide no protection against sexually transmitted diseases.

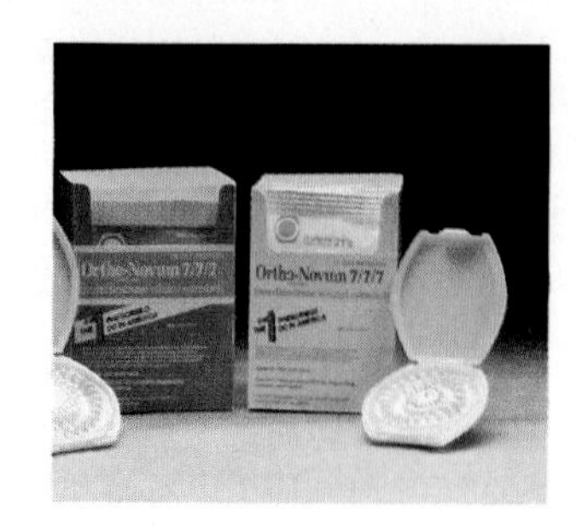

Another chemical method of birth control, called by the manufacturer's name of **Norplant,** was first introduced in 1990. The Norplant system consists of 6 matchstick-sized capsules (each about $1\frac{1}{2}$ inches long) containing a synthetic progesteronelike hormone. It is implanted under the skin of the upper arm in a 15-minute surgical procedure. After implantation the hormone is continuously released in small amounts through the capsule walls into the bloodstream. The hormone prevents pregnancy primarily by inhibiting ovulation. Norplant provides contraception for up to 5 years. The capsules can be removed at any time, returning the woman to her previous fertility level. If after 5 years she wishes to continue using Norplant, the old capsules are removed and a new set is implanted.

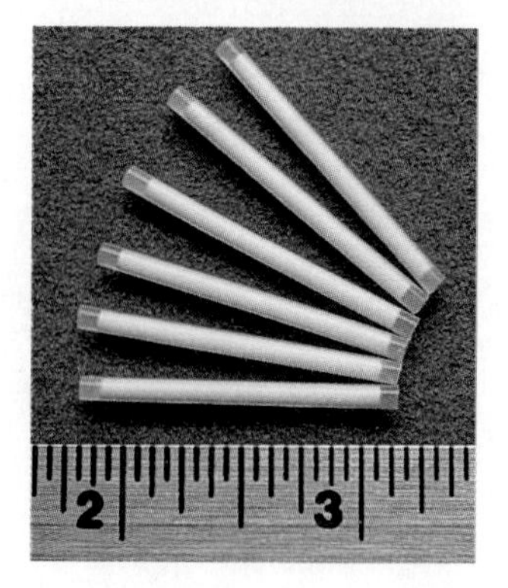

A high dose of synthetic estrogen called *diethylstilbestrol,* (di-eth'il-stil-bes'-trol), or DES, was once available to be used as a "morning after pill," or a postcoital contraceptive. Its precise mechanism of action is unclear, but it is thought to prevent implantation of the developing embryo. However, after being on the market a few years, it was linked to vaginal cancer in the daughters of women who used it. Additional side effects, such as severe nausea and cramping, also developed. Today it is given only to victims of rape or incest.

A recent drug available in some European countries is called **RU 486** and works by blocking the normal action of progesterone. Progesterone is necessary to maintain the blood-rich thickened endometrium. RU 486 binds to the progesterone receptors of the endometrium, preventing progesterone from maintaining the endometrium. If implantation has occurred and RU 486 is administered, along with an injection of prostaglandins, the endometrium and the developing embryo are sloughed off and discharged through the vagina. This drug is highly effective. Approximately 96% of women receiving RU 486 and prostaglandins in the first 9 weeks after conception abort the embryo within 1 day. Minor pain, cramps, and nausea are the reported side effects, but patients claim these effects are indistinguishable from those associated with heavy menstruation. Primarily because of the

Figure 18-8 Surgical methods of contraception

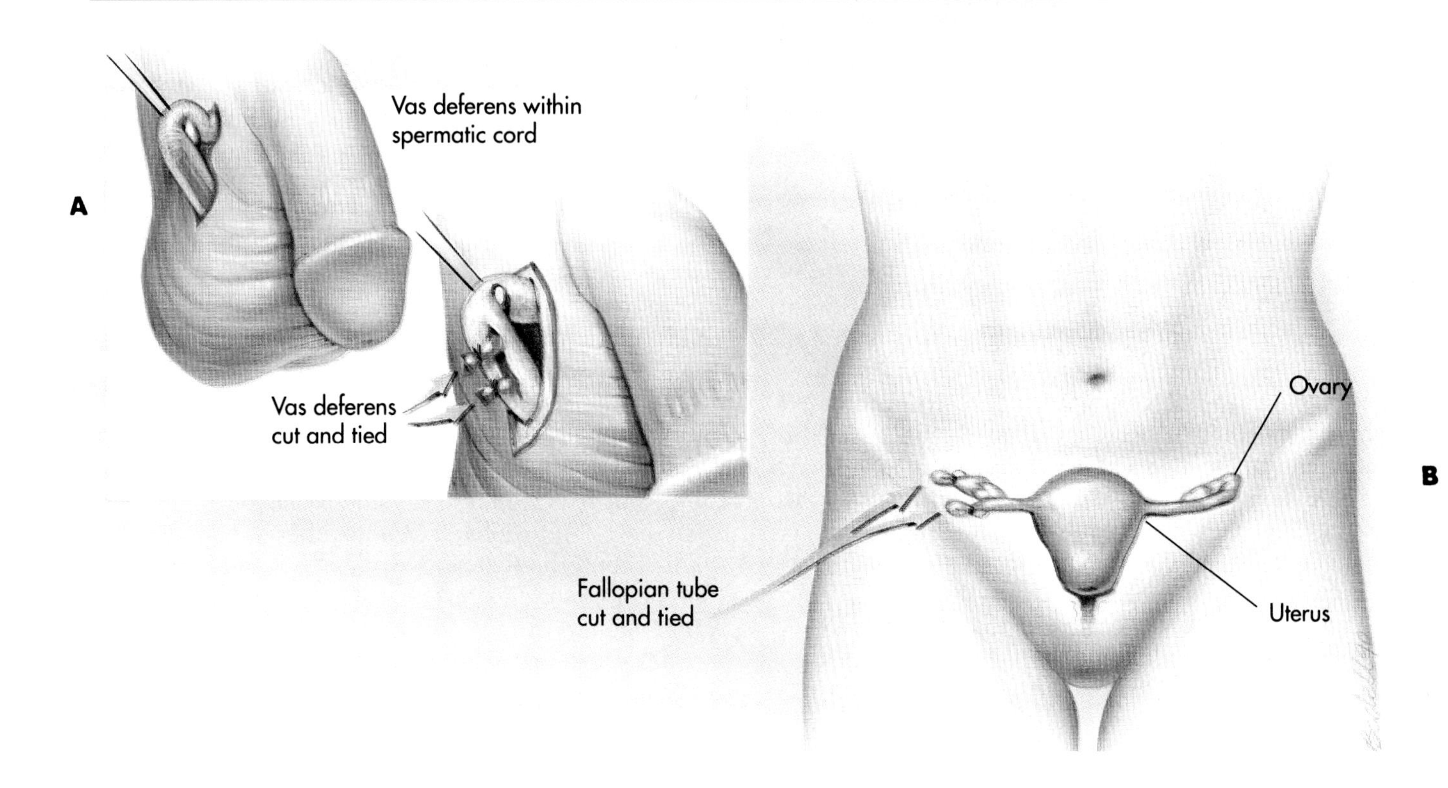

Vasectomy involves the cutting of the vas deferens, which prevents sperm from leaving the testes (A). Tubal ligation involves cutting the fallopian tubes, which prevents sperm from reaching an ovum released from the ovary.

current intensely politicized debate concerning this new "abortion pill," RU 486 has not been approved for distribution in the United States.

Barrier methods prevent or interfere with sperm movement toward the ovum. A common barrier method involves using a **condom,** a latex or natural skin sheath that fits over the male's erect penis, trapping semen before it can enter the female reproductive tract. Although the failure rate is about 12%, condom use has significantly increased over the past few years because it is an effective barrier against the transmission of HIV and the herpes virus. Both of these are discussed in the following section.

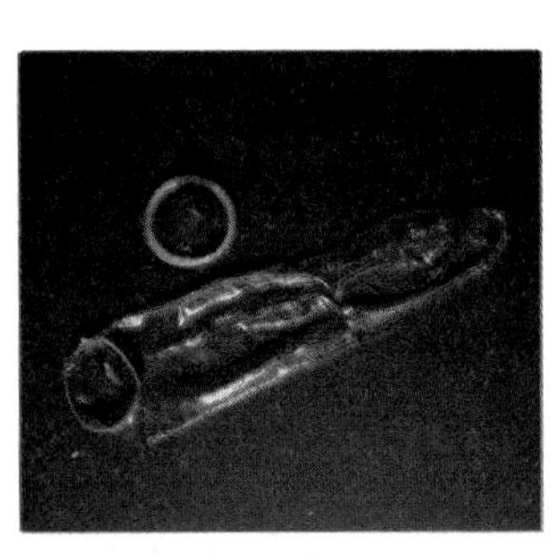

A female condom was introduced in 1993, offering another birth control option for responsible couples. The female condom is a prelubricated, polyurethane sheath with a flexible ring at each end. The condom is inserted like a tampon and is designed to line the vaginal walls. The lower portion of the condom extends outside the vagina and covers the labia. The condom provides contraceptive protection and prevents the transmission of sexually transmitted diseases by preventing the penis and sperm from making physical contact with the vaginal walls.

Sperm-killing chemicals, called **spermicides,** can be placed in the vagina before sexual intercourse. Spermicides come in the form of foams, creams, jellies, and vaginal suppositories and destroy sperm before they can enter the uterus. Creams and jellies are less concentrated than foams or suppositories, and all

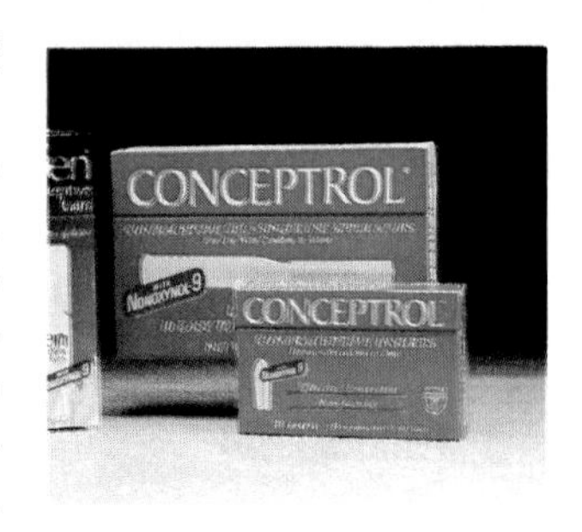

spermicides are recommended for use in conjunction with other contraceptive methods. Spermicides alone provide no protection against sexually transmitted diseases.

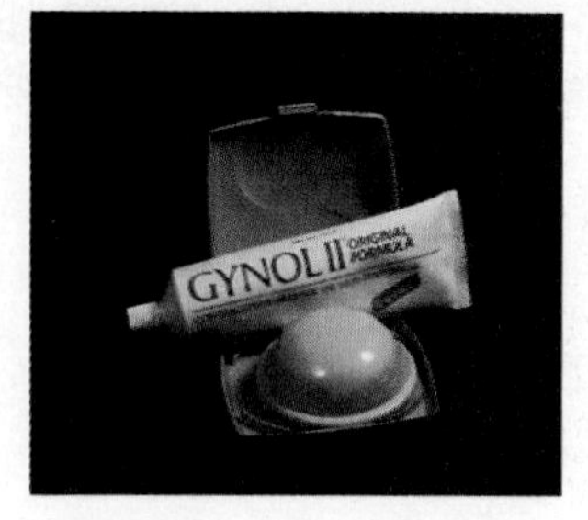

Another barrier method involves using a **diaphragm** (di'uh-fram), a circular dome-shaped structure made of latex rubber. The diaphragm is inserted into the vagina at the cervical opening just prior to intercourse and prevents sperm from entering the uterus. The effectiveness of the diaphragm is greatly increased when a spermicidal jelly or cream is placed in the dome before insertion. A smaller and similar barrier is the **cervical cap,** which fits closely over the cervix. As with the diaphragm, a spermicide is often used with the cervical cap to increase its effectiveness in preventing the passage of sperm to the uterus. The cervical cap can be inserted up to 6 hours before intercourse and can be left in place at least 6 to 8 hours after intercourse. To ensure that the diaphragm and the cervical cap completely cover the cervix and because the diameter of the cervix varies among women, both should be prescribed and fitted by a physician.

The **contraceptive sponge** contains spermicides and is also placed at the cervical opening. The device blocks sperm from entering the uterus, and the spermicides kill the sperm before they are able to enter the uterus.

For several years the second most used contraceptive method in the United States was the **intrauterine device,** or **IUD,** a plastic or metal device inserted into the uterus that prevented the developing embryo from being implanted in the endometrium. The failure rate of the IUD was nearly as low as that of the birth control pill. However, with the exception of two products, IUDs have been taken off the market because of such complications as uterine perforation, severe infections, cramping, and severe menstrual bleeding.

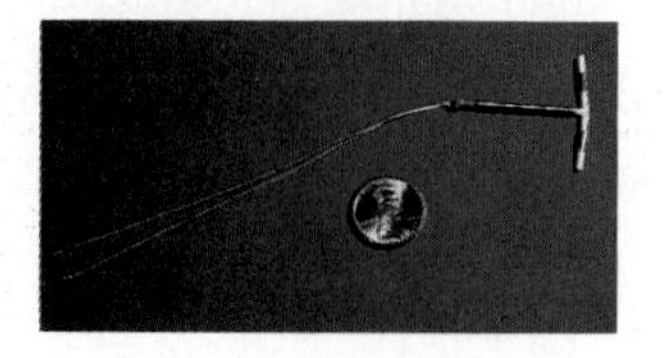

Sexually Transmitted Diseases

Sexually transmitted diseases, also called *venereal* (ve-ne're-al) *diseases,* are infectious diseases spread from one person to another by some form of sexual contact. Until recently the most common sexually transmitted diseases were caused by bacterial infections, but viral infections have begun to spread during the past decade. AIDS, caused by the human immunodeficiency virus (HIV), was discussed in Chapter 14. In this chapter we discuss its transmission more fully. First we will describe the three most common types of bacterial infections and their treatment, followed by two types of viral sexually transmitted diseases.

The infective agent causing **gonorrhea** (gah'-nor-e'uh) is a kidney-bean-shaped bacterium called *Neisseria gonorrhoea.* It invades the mucous linings of the reproductive and urinary system. In males the most common symptom of gonorrhea is urethral infection and inflammation accompanied by painful urination and a penile pus discharge. In females the symptoms vary, ranging from none (about 20% of women) to lower abdominal discomfort, vaginal discharge, or urethral symptoms similar to male symptoms.

If untreated, gonorrhea can infect and inflame all the reproductive system ducts in both males and females, leading to sterility. Since the 1950s treatment has involved using large doses of antibiotics, usually penicillin. However, strains resistant to penicillin and other antibiotics are becoming more common, making the disease now more difficult to cure.

Syphilis (sih'-fih-lis) is caused by a corkscrew-shaped bacterium called *Treponema pallidum.* It is usually spread by sexual contact. However, it can be contracted by a fetus from an infected mother. The bacterium commonly gains entry into an uninfected person by penetrating the mucous membranes of the vagina, the urethra of the penis, or the mouth. From there it enters the lymphatic system and bloodstream, and within a few hours of exposure the infection is in progress. The incubation period, during which there are no symptoms, is usually 2 to 3 weeks, but may be as long as 60 days. The first symptom to appear is a painful sore, called a **chancre** (shank'-er), that appears at the invasion site. In males this is usually on the penis, but in females the chancre is usually in the vagina and often goes undetected. The chancre persists for 1 or more weeks then heals and disappears.

If syphilis goes untreated, secondary symptoms appear within 1 month. A pink skin rash appears all over the body, usually accompanied by fever, joint

Biology in Action THE FUTURE OF MALE BIRTH CONTROL

In the past 25 years many new contraceptive methods have become available, especially methods geared for use by females. While researchers work to improve the safety, convenience, and reliability of current methods, their efforts in the development of new methods have expanded to also include males. Currently, male birth control options are limited to vasectomy and condoms. However, attempts at both direct and indirect inhibition of sperm production by chemical means are being investigated.

Several years ago, *gossypol,* a chemical extracted from cottonseed oil, was shown to prevent sperm production by binding to the FSH receptor sites of the Sertoli cells of the seminiferous tubules. Because the receptor sites are blocked, sperm production ceases. A major drawback discovered during experimentation with gossypol is that it caused sterility in 20% of the test population.

Another study was conducted in seven countries throughout Europe and Asia in which men were given weekly injections of testosterone. Increased testosterone levels in the blood inhibit the hypothalamus from signaling the anterior pituitary gland to release FSH and LH. Among the group of men tested the average time of sperm production cessation was 4 months. Sperm production stopped in only 70% of those tested. After the injections were stopped, full sperm production resumed in about 6.5 months.

In both men and women FSH and LH production by the anterior pituitary gland is controlled by the hypothalamus through a releasing factor called **gonadotropin-releasing hormone (GnRH).** In men FSH stimulates sperm production and LH regulates the release of testosterone by the testes. Perhaps the most promising study to date involves giving men a synthetic compound, Nal-Glu, that binds to GnRH. Nal-Glu combines with GnRH, preventing it from stimulating the anterior pituitary gland to secret FSH and LH.

In the study men were given daily injections of Nal-Glu as well as a weekly injection of testosterone for 20 weeks. The testosterone injections were necessary because without LH, the interstitial cells of the testes do not produce testosterone. Without testosterone, several secondary sexual characteristics are affected, such as a reduction in skeletal muscle mass, and there is also a marked reduction or loss of sexual energy, or libido. As a result of the Nal-Glu injections, sperm production stopped completely in 75% of the men tested. In the remaining 25% of those tested, sperm production only *decreased.* However, when injections containing larger doses of Nal-Glu were given to those men, sperm production also ceased. Within 12 to 14 weeks after the injections were stopped, sperm production was normal among all the men tested.

A primary problem in developing a chemical method of male contraception is in determining the correct dosage. Whereas FSH and LH production in females varies on a 28-day cycle, production of these two hormones in males is continuous but variable between individuals. Variation in FSH and LH production may reflect individual variation in GnRH production. Consequently, a given dose of a so-called male birth control pill, such as Nal-Glu, may stop sperm production in some men and only reduce it in others. Factors such as body weight and health also influence how these chemicals affect sperm production. Despite these problems, research continues in the hope that one day men too will have a form of birth control pill available to them.

pain, and a general feeling of illness. These symptoms usually disappear in 3 to 6 weeks. Then the disease enters a latent period, which may last a person's lifetime. However, the bacteria may be killed by the immune system or may spread, leading to a tertiary (ter'shi-ar'i) stage. This stage is characterized by destructive lesions, called **gummas** (guh'-muz), of the CNS, blood vessels, bones, and skin. Penicillin, which interferes with the ability of dividing bacteria to synthesize new cell walls, is usually the treatment for all stages of syphilis.

Chlamydia (kluh-midh'-de-uh) is caused by

Chlamydia trachomatous, a bacterium with a virus-like dependence on host cells—the epithelium of the urethra in both males and females and the cervix in females. Symptoms, which often go unrecognized, include a penile or vaginal discharge and a burning sensation during urination. In males this infection not only can cause widespread urinary system infection, but also can cause skeletal joint damage. In females its most severe consequence is sterility. Newborns infected in the birth canal develop eye inflammations or respiratory infections such as pneumonia. The disease can be diagnosed by special cell-culture techniques and is treated with the antibiotic tetracycline.

There are many types of human **herpes** (her'-pez) viruses, including the Epstein-Barr virus responsible for mononucleosis and one type responsible for chickenpox. The **herpes simplex virus** consists of two types. Type I causes cold sores and fever blisters. Type II causes **genital herpes.** Genital herpes is transmitted through infected secretions. Immediately after infection the virus begins to multiply rapidly. Blisters appear at the infected site within 20 days of infection. Once the blisters break, the lesions take from 5 days to as long as 3 weeks to heal.

After the lesions heal the infection remains dormant, and blisters recur. There is no cure for herpes. A drug called acyclovir, taken orally, only prevents the outbreaks of blisters. Infection of the newborn occurs if it comes in contact with a lesion in the birth canal. From 1 to 3 weeks after birth the infant may show signs of neurological disorders ranging from blindness to brain damage.

Acquired immune deficiency syndrome, or **AIDS,** is caused by **human immunodeficiency virus (HIV).** Its transmission is primarily by the semen or vaginal secretions during sexual intercourse or by blood or blood products, such as extracted clotting agents from whole blood. All these fluids contain some leukocytes that will contain the HIV if the person is infected. Generally the likelihood of infection is dramatically increased if tissue membranes are broken during exposure to the infected fluid. For example, microscopic tears, or lesions, may occur in the mucosal lining of the vagina or rectum during sexual or anal intercourse. These expose the circulatory system of the uninfected person to the virus in the semen of the infected person. The route is obvious and also more direct in the case of blood transfusions or in sharing infected needles, a common practice among intravenous drug users.

As you learned from Chapter 14, HIV infects and destroys T-helper cells, which are essential components of active immune-system responses to foreign invasion and infection. Once the immune system is immobilized, diseases and infections take their toll and eventually prove fatal. Initial infection by HIV may be followed by a short illness with flu-like symptoms. When these symptoms disappear after a few weeks, the person typically has antibodies to HIV and appears to be healthy. The virus then enters a latent stage. However, as current available information indicates, in most individuals the virus becomes active, meaning that it replicates and spreads to other T-helper cells. With the immune system thus severely weakened, **opportunistic infections** result. These are infections a healthy immune system routinely eliminates before they can spread through the body.

Because HIV is transmitted only through sexual intercourse, anal intercourse, or in blood and blood products, it should be relatively easy to protect against infection. If used properly, latex condoms should prevent HIV transmission during sexual contact. Intravenous drug users should avoid sharing hypodermic needles with others. Viral transmission in transfused blood and blood products has been reduced to extremely low levels now that they are carefully screened for HIV antibodies. However, there are rare cases in which infected people have not produced HIV antibodies, and they, as well as their blood products, have gone undetected. Finally, avoiding behavior that increases the possibility of exposure should also be considered a real and vital option in preventing infection.

CHAPTER REVIEW

SUMMARY

1. Males and females produce gametes through meiosis. Male gametes, called spermatozoa or sperm, are produced in the testes. Female gametes, called ova, are produced in the ovaries. After puberty, male gamete production throughout life is continuous, but females produce gametes until about the age of 55 years.
2. In addition to producing gametes, the female reproductive system must also prepare an environment able to protect and nourish an embryo for several months. This is the reason why the hormonal regulation of the female reproductive system is much more complex

than it is in males.

3. Meiosis in the male results in four haploid sperm cells being produced from each original cell. In the female, meiosis results in the production of only one ovum from the original cell, with the other cells being discarded during cell division. In the female, the discarded products of meiosis I and II are called polar bodies.
4. In addition to producing gametes, both the testes and ovaries produce sex hormones. The testes produce testosterone, and the ovaries produce estrogen and progesterone.
5. Hormonal regulation of the male and female reproductive systems is controlled by the same hormones. Gonadotropin-releasing hormone is secreted by the hypothalamus. It induces the anterior pituitary to secrete follicle-stimulating hormone and luteinizing hormone. Control of these hormonal secretions, however, differs between males and females.
6. The concentration of male sex hormones in the blood is relatively constant, but the concentration of female sex hormones varies on a 28-day cycle. During each cycle, the ovaries produce one ovum and prepare the endometrium for implantation of a fertilized egg.

FILL-IN-THE-BLANK QUESTIONS

1. Male gametes are called ______ and female gametes are called either ______ or ______.
2. The primary male sex organs are called ______ and the primary female sex organs are called ______.
3. Spermatozoa are produced in the ______ ______, located in the ______ of the male.
4. Oocytes are produced in ______ ______, located in the ______ of the female.
5. The ______ cycle produces gametes, whereas the ______ cycle prepares the ______ for implantation of a(n) ______.
6. Ovulation usually occurs on day ______ of the 28-day cycle.
7. The uterine cycle is marked by the beginning of ______.
8. Generally, in a woman in her late 40s, when 1 year has passed without menstruation, ______ has occurred.
9. Generally, the two main categories of sexually transmitted diseases are those caused by ______ and those transmitted by ______.

SHORT-ANSWER QUESTIONS

1. Compare and contrast oogenesis and spermatogenesis.
2. What influence does the corpus luteum have on the endometrium during the secretory phase of the uterine cycle?
3. What factors cause ovulation?
4. Describe the three stages of syphilis.
5. What effects do FSH and LH have on the male reproductive system?
6. Describe one type of birth control method from each of the four major categories.
7. List the secondary sexual characteristics of males and females.
8. Describe the events that occur during each phase of the ovarian cycle.

VOCABULARY REVIEW

acrosome—*p. 339*
endometrium—*p. 341*
epididymis—*p. 335*
estrogen—*p. 342*
gametes—*p. 335*
oogenesis—*p. 337*
ovarian cycle—*p. 342*
ovaries—*p. 340*
progesterone—*p. 342*
seminiferous tubules—*p. 335*
spermatogenesis—*p. 337*
testes—*p. 335*
uterine cycle—*p. 342*
zygote—*p. 335*

Chapter Nineteen

HUMAN DEVELOPMENT

For Review

Here are some important terms and concepts that you will encounter in this chapter. If you are not familiar with them, you should review them before proceeding.

Acrosome (page 339)
Circulatory system (page 231)
Endometrium (page 341)
Gametes (page 335)
Meiosis (page 70)

OBJECTIVES

After reading this chapter you should be able to:

1. ***Describe the events of fertilization.***
2. ***Describe the events of cleavage.***
3. ***Describe the blastocyst and the process of implantation.***
4. ***List the three germ layers and describe their formation.***
5. ***Define organogenesis and describe the formation of the major organ systems.***
6. ***Define and give examples of teratogens.***
7. ***Describe the major stages of labor and the events that occur during lactation.***
8. ***Differentiate between growth and development.***
9. ***List the three stages of growth and development preceding adulthood and describe the major events associated with each stage.***

Human development is a series of biological events that begins at the moment a sperm and ovum fuse and encompasses an individual's entire life. This ongoing series is divided into several developmental stages. Prenatal development begins with the union of two gametes and ends with the birth of the child. The human postnatal developmental stages of infancy, childhood, and adolescence make up the structural and functional transformation of a newborn into a reproductively mature adult.

Fertilization to Birth: An Introduction

Although the birth of a child is a familiar event, we sometimes lose sight of what an amazing biological accomplishment it is. It begins with a single fertilized cell, the zygote, and ends at birth with an extremely complex human being consisting of billions of cells. The entire process takes place **in utero** (Latin for "in the uterus") of the female. Significant maternal changes must therefore also occur in response to the demands of the developing infant.

In Chapter 18 you learned about what happens in the female reproductive system if fertilization *does not* occur. In this chapter we describe the general sequence of events that occurs when fertilization *does* occur. We begin by first describing the fertilization process. We then discuss the birth event itself and pregnancy's effects on the female. Next are the major developmental stages of human growth and development *in utero.* We conclude with a description of growth and development after the birth of the fetus to reproductive maturity, the beginning of adulthood.

The term **pregnancy** refers to the events that occur in the female from fertilization until the baby is born. The time during which this development occurs is called the **gestation** (jes-ta'-shun) **period** and is approximately 266 days from fertilization. However, it is conventionally calculated from the onset of the last menstruation, making the gestation period approximately 280 days. **Parturition** (par'-tu-rish'-un), or giving birth to the baby, usually occurs within 15 days of the calculated due date.

Although development is continuous, gestation is usually divided into three successive stages. For the first two weeks following fertilization, called the **preembryonic period,** the rapidly dividing mass of cells, derived from the zygote, is called a preembryo. From the third through the eighth weeks after fertilization, called the **embryonic period,** the former preembryo is now called an **embryo** (em'-bri-o). In the final stage, extending from the ninth week through birth, called the **fetal period,** the developing human being is called a **fetus.**

Fertilization

Fertilization is the fusion of the sperm from the male with the ovum from the female. Fertilization results in a **zygote,** meaning fertilized egg, which has a full complement of 23 pairs of chromosomes, one half of each pair from each parent. For fertilization to occur the sperm must reach and fuse with the ovum, which normally occurs in the upper third of the uterine tube.

Why does fertilization occur so far away from where the sperm initially enter the female reproductive system? Several factors are involved. The smaller sperm are self-propelled and capable of moving much more rapidly than the large ovum, which must rely on the rhythmic beating of the cilia lining the uterine tube. Consequently, fertilization normally occurs in the upper third of the uterine tube.

The passage of sperm from the vagina through the uterus and into the uterine tube is difficult. Of the millions of sperm ejaculated into the vagina, perhaps only several thousand will reach the upper region of the uterine tube. The acidic conditions of the vagina kill millions of sperm initially. The remaining sperm then swim toward the cervix. Once inside the uterus, many thousands are destroyed by the female's immune system. Wandering phagocytic leukocytes (see Chapter 14) attack the sperm cells because they are recognized as foreign. This is why a male's sperm count must be at least 20 million sperm per milliliter of semen to be capable of fertilizing an oocyte.

After these difficulties are surmounted, sperm are still incapable of fusing with an ovum. Recall from Chapter 18 that an ovulated ovum is surrounded by an outer layer of cells called the corona radiata and encapsulated by an inner layer of clear, mucuslike secretions called the zona pellucida. Both of these protective layers must be crossed before the sperm can make contact with the oocyte's membrane. Penetration of these protective layers and the subsequent fertilization of the ovum by a sperm involves several steps. The head of the sperm is covered by a membranelike structure called an acrosome containing hydrolytic enzymes. First hundreds of acrosomes must rupture, releasing their hydrolytic enzymes, as they come into contact with the corona radiata of the ovum. The enzymes break down the intercellular material holding the cells of the corona radiata together. Once a path has been cleared, the enzymes continue to digest their way through the zona pellucida. The sperm taken in by the ovum is usually one that comes along later, after other sperm have used enzymes stored in their acrosomes to clear a path. Once a single sperm contacts

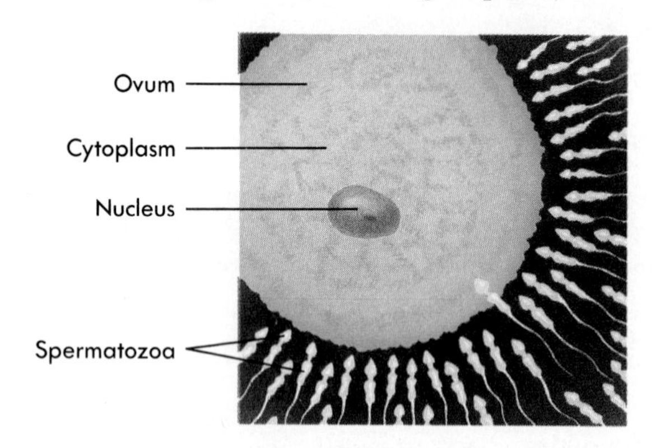

the ovum cell membrane, its nucleus is quickly engulfed by the endocytosis process into the ovum. After the sperm's nucleus enters the ovum, the haploid nuclei of the sperm and ovum are called **pronuclei** (pro-noo′-kle-i). Entry of the sperm triggers the ovum's second meiotic division, and the second polar body is cast off and degenerates. The pronuclei then fuse, producing a single diploid cell called a zygote. Fertilization is now complete.

Soon after the zygote is formed, it begins to divide, with the first mitotic division occurring about 30 hours after fertilization (Figure 19-1).

Figure 19-1 Fertilization

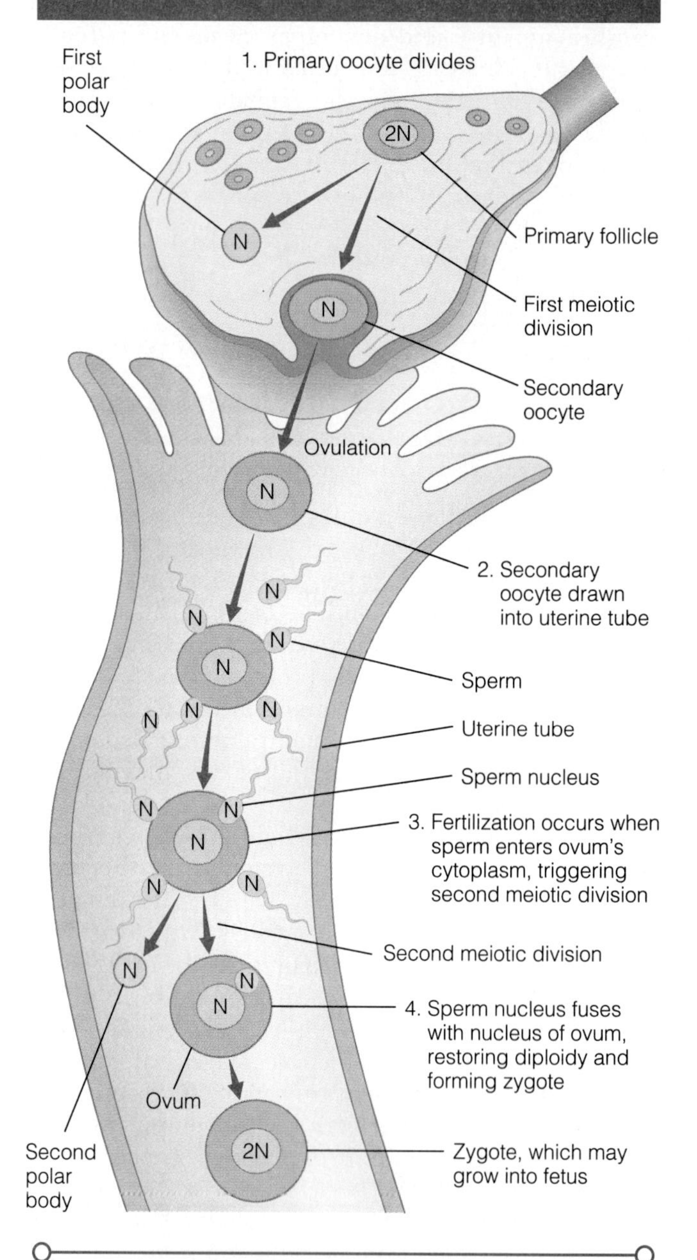

The sperm and ovum meet in the upper third of the uterine tube. Fertilization occurs when the nucleus of a single sperm enters the ovum's cytoplasm. The ovum then undergoes a second meiotic division and discards the second polar body. The sperm and ovum pronuclei fuse, forming a single diploid cell called a zygote.

Preembryonic Development

Preembryonic development begins with fertilization, continuing as the zygote travels down the uterine tube, floats free in the uterus, and finally embeds itself in the uterine wall's endometrial lining. We discuss the major changes that occur during this period in the following sections.

Cleavage

The dividing cell mass takes about 3 days to travel down the uterine tube to reach the uterus. During its journey through the uterine tube the zygote undergoes a series of fairly rapid mitotic divisions called **cleavage.** Cleavage produces a small group of cells that become the basic building blocks for constructing a human being (Figure 19-2). Imagine the difficulty of constructing a house from a single large pine tree. If you think about how much easier construction would be if the tree were first cut into hundreds of boards, you can appreciate the importance of cleavage.

Because there is no time for cell growth between divisions, the resulting cells become smaller and smaller. This results in a mass of cells the same size as the original zygote from which they came. By the third day, cleavage results in a mass of about 100 cells, now called a **morula** (mor′-uh-luh, Latin for "little mulberry"), floating freely in the upper reaches of the uterus.

Figure 19-2 Cleavage

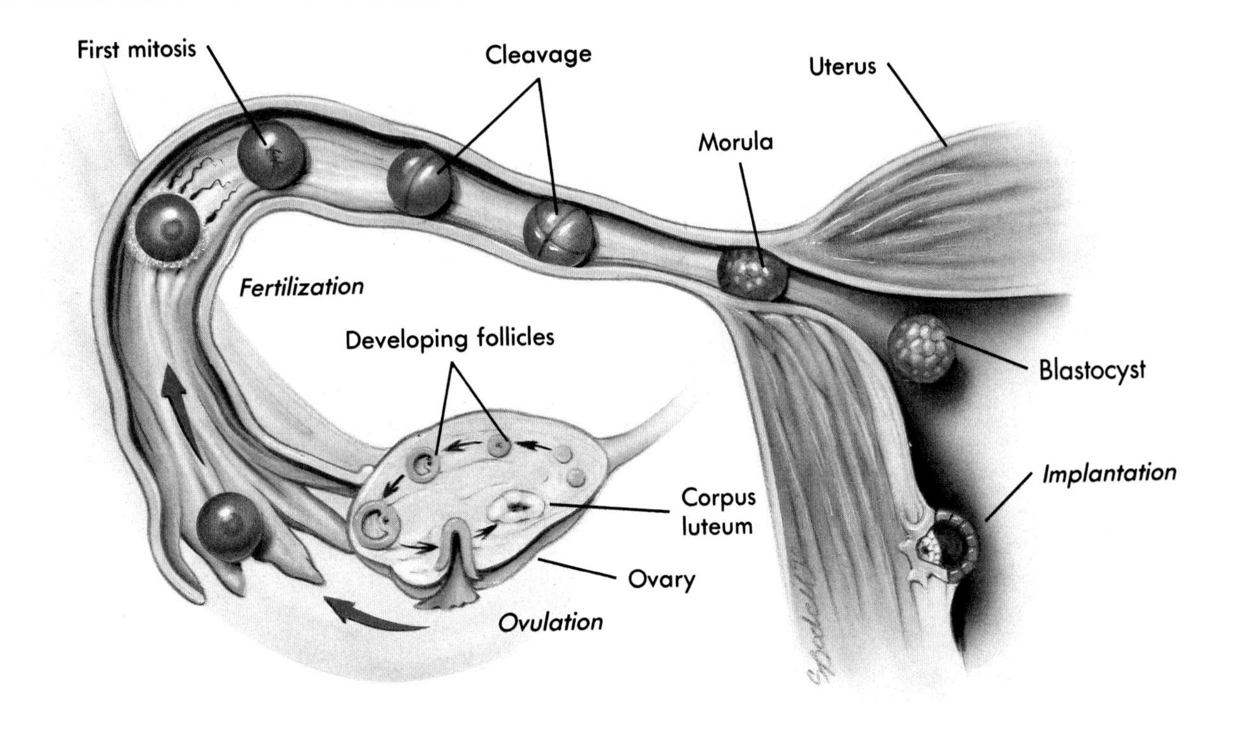

The zygote undergoes mitosis as it travels down the uterine tube toward the uterus. By the time it reaches the uterus it has become a solid mass of cells, called a morula.

The morula floats free in the uterus for 1 or 2 additional days, undergoing changes so it can begin to implant in the uterine wall. First the zona pellucida surrounding the morula begins to break down as fluid accumulates between the morula's cells. This results in the formation of a **blastocyst** (bla′-sto-sist). A **blastocyst** is a hollow sphere composed of a single layer of large flattened cells, called **trophoblast** (tro-fo-blast) **cells** (*troph* = nourish; *blast* = sprout or bud) and a small cluster of rounded cells, called the **inner cell mass**, attached to one side. The inner cell mass will eventually become the fetus. The trophoblast eventually becomes part of the tissue that nourishes the developing fetus.

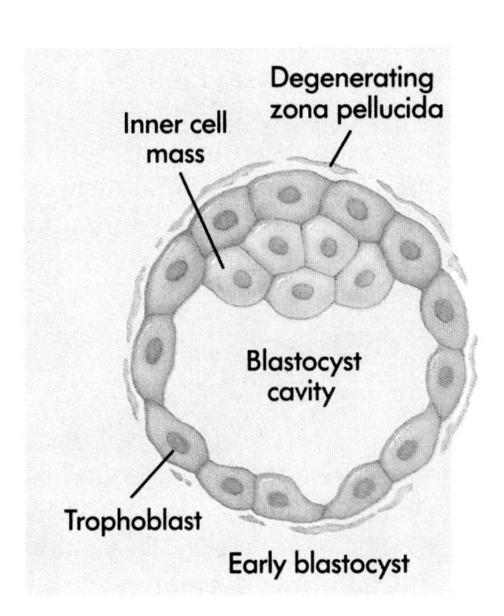

Twinning

Each cell of the inner cell mass has the genetic potential of becoming a complete individual. Occasionally the inner cell mass splits, and two embryos start developing rather than one. These two embryos develop into **monozygotic,** or **identical, twins** and possess identical chromosomes. In contrast to monozygotic twins, **dizygotic,** or **fraternal, twins** arise from two ova released during the same menstrual cycle, each being fertilized by a different sperm (Figure 19-3). Because they arise from two different zygotes, they do not share identical chromosomes. For example, you may know fraternal twins, one male and the other female. Fraternal twins have the same degree of genetic similarity as other siblings born at different times, but their birth dates are identical.

In the United States, twins occur about 1 out of every 90 pregnancies, and about 66% of these are dizygotic twins.

Figure 19-3 Twinning

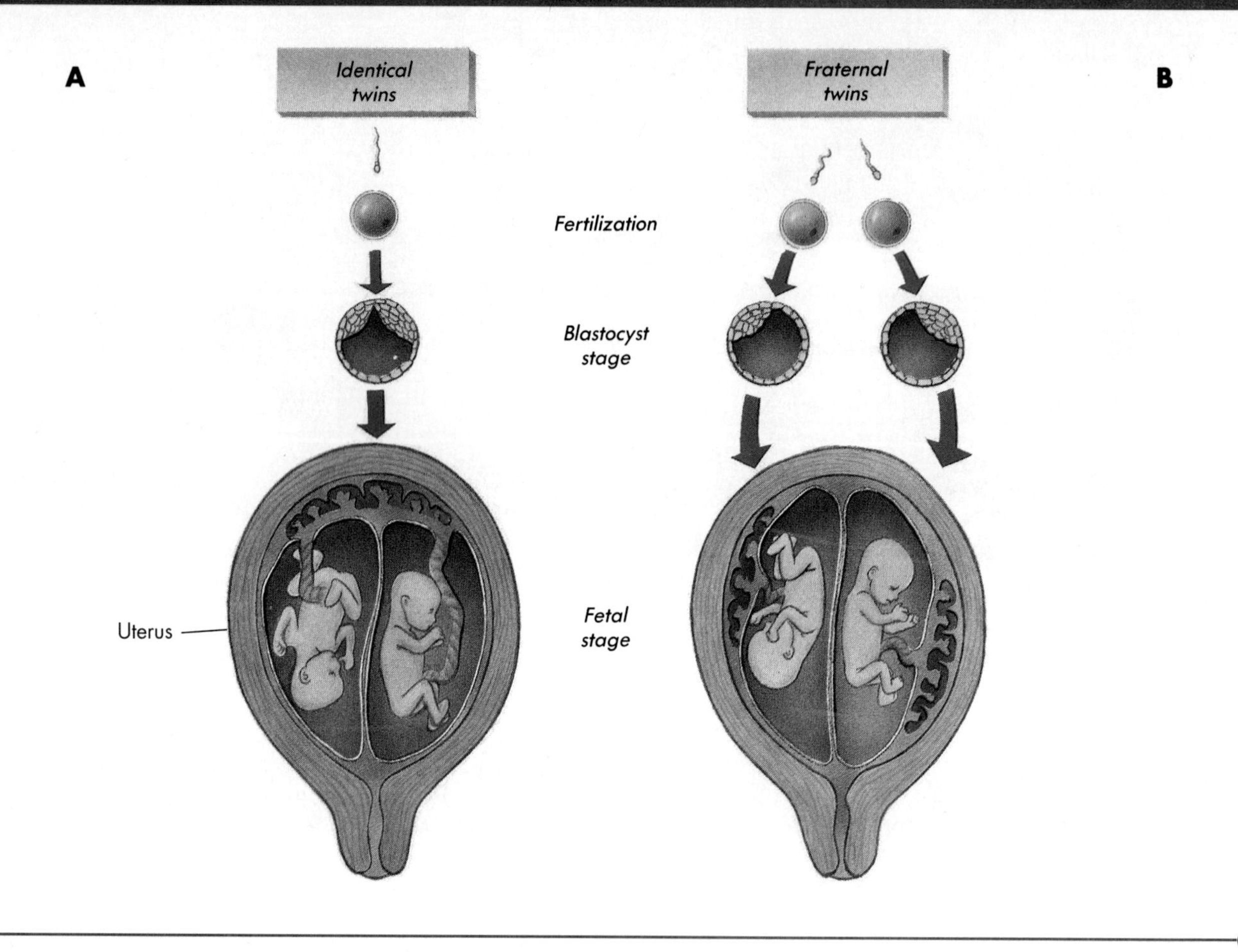

Monozygotic twins, whose chromosomes are identical, occur when the inner cell mass splits and two embryos develop. Dizygotic twins are the result of two ova having been released during the same menstrual cycle, each being fertilized by a different sperm. Dizygotic twins have the same degree of genetic similarity as other siblings born at different times.

Implantation

Implantation is the process of the blastocyst attaching to the endometrium and occurs about 6 or 7 days after fertilization. The blastocyst usually attaches in the upper third of the uterus, the trophoblast cells adhering to the endometrium (Figure 19-4). At the attachment site the trophoblast cells secrete digestive enzymes that begin to erode the endometrial surface. In addition to allowing implantation, these digestive enzymes also create the metabolic fuel and raw materials necessary for the continued growth of the inner cell mass's dividing cells. The trophoblast cells also begin multiplying, first at the attachment site and then over the entire blastocyst. The dividing trophoblast cells at the attachment site also extend fingerlike projections into the eroding endometrium. As the endometrium is progressively digested, the blastocyst becomes embedded in the tissue and surrounded by a pool of blood leaked from the surrounding blood vessels of the endometrium. The endometrium's epithelial cells begin to grow over the embedded blastocyst. Implantation is usually completed by the fourteenth day after ovulation. To maintain the endometrium the trophoblast cells secrete **human chorionic gonadotropin hormone (HCG),** which causes the corpus luteum to continue secreting progesterone and estrogen, thereby maintaining the endometrium.

Placenta

The digested endometrial cells are capable of providing nourishment for the preembryo only during its first days of development. To sustain it for the rest of its intrauterine development, a temporary

Figure 19-4 Implantation

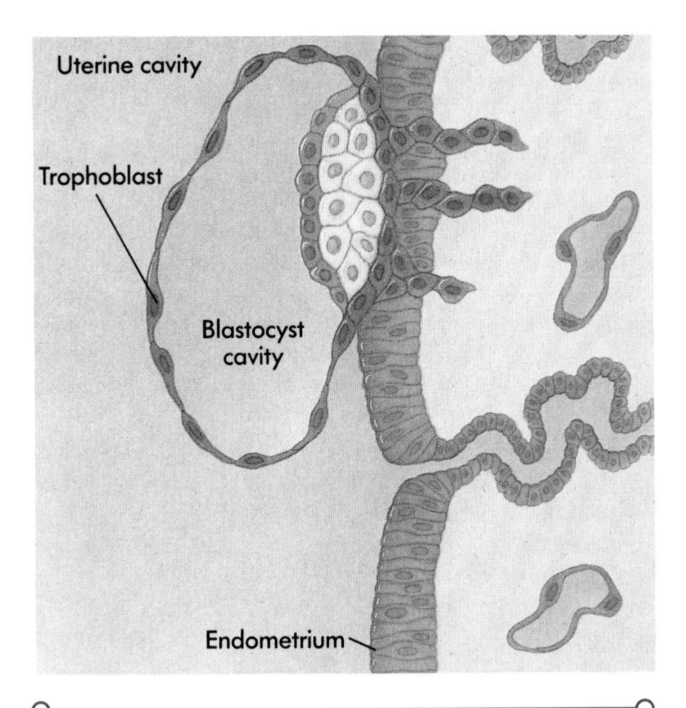

After the blastocyst attaches itself to the endometrium of the uterus, the outermost layer of cells of the blastocyst, called the trophoblast, begins secreting digestive enzymes that erode away endometrial cells. In this way, the blastocyst burrows into the endometrium and other endometrial cells eventually grow over it. Implantation is complete by the fourteenth day after ovulation.

specialized organ of exchange between the maternal circulation and the circulatory system of the embryo, called the **placenta** (pluh-sen′-tuh), rapidly develops (Figure 19-5). The embryo is attached to the placenta by the **umbilical cord.**

The placenta originates from both trophoblastic and endometrial tissues. Approximately 14 days after fertilization the trophoblast is two cell layers thick and is now called the **chorion** (kor′ri-on). As the chorion cells continue to proliferate and secrete enzymes, they create open spaces or cavities in the endometrium. These fill with blood leaking from the surrounding endometrial blood vessels. The chorion eventually sends fingerlike projections of tissue into these blood-filled spaces. The embryo soon sends newly formed blood vessels into the finger-like projections, transforming them into **chorionic villi.** Each chorionic villus consists of embryonic capillaries surrounded by a thin layer of chorionic tissue. This will eventually separate the embryonic blood supply from the maternal blood

Biology in Action TAKING A LOOK INSIDE

In recent years many techniques have been developed for prenatal diagnosis of the fetus, usually when there is a known risk of a genetic disorder or birth defect. The most common type of testing is **amniocentesis** (am′-ne-o-sen-te′-sis). This is a relatively simple procedure in which a hollow needle is inserted through the mother's abdominal wall and into the amniotic sac, where about 10 ml of amniotic fluid is withdrawn. Because there is a chance of injuring the fetus before sufficient amounts of fluid are present, this procedure is not normally done before the fourteenth week of pregnancy. With the development of **ultrasound imaging,** in which high frequency sound waves are bounced off of the fetus, and picked up by a receiver, and create a visual image of the fetal position, the risks associated with amniocentesis have been greatly reduced.

Biochemical tests can be done on the fluid itself to determine the presence of chemicals, such as enzymes or metabolic by-products, that are indicators for specific diseases. However, most tests are done on the sloughed-off fetal cells in the fluid. These cells are isolated, grown in laboratory dishes over several days, and then the chromosomes are examined for genetic markers for abnormalities. The cells are also karyotyped to check for chromosomal abnormalities, such as Down syndrome.

A more recently developed procedure, called **chorionic** (kor-e-ah′-nik) **villi sampling,** involves removing small amounts of the chorionic villi from the placenta. A small tube is inserted through the vagina and cervix and guided by ultrasound imaging to an area where a sample of the placental tissue can be safely removed. This procedure allows earlier testing, at about 8 weeks gestation, and karyotyping can be done almost immediately on the rapidly dividing cells of the chorionic villi sample. However, there is a slightly increased risk of spontaneous abortion of the embryo.

Because these procedures are invasive, there is a risk to both the fetus and the mother. Although they are routinely done for pregnant women over 35 years of age (due to a higher probability of Down syndrome), they are done for younger women only when the probability of finding fetal defects is greater than the probability of doing harm to the fetus.

Figure 19-5 Placenta

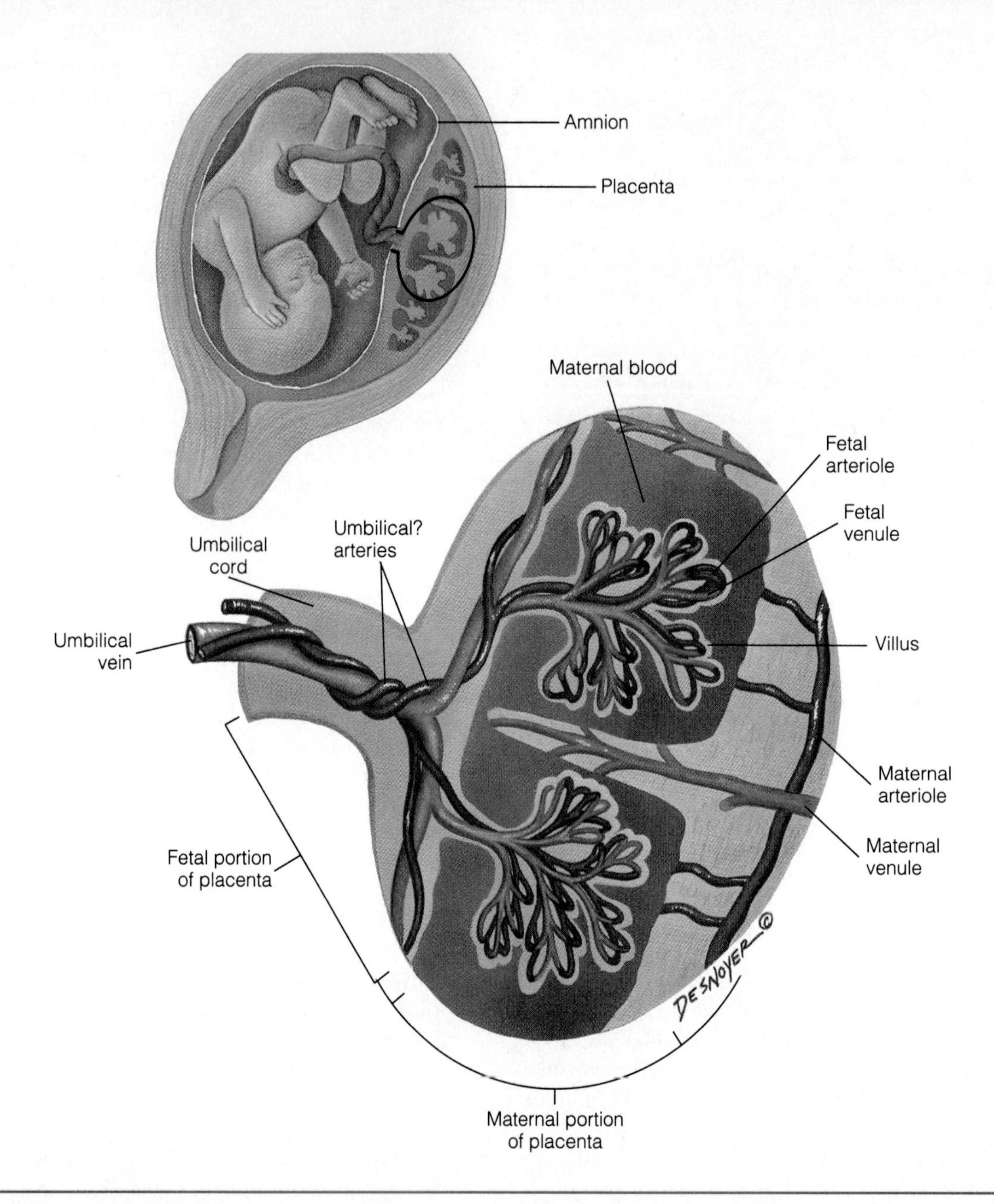

The placenta originates from both trophoblastic and endometrial tissues. Chorionic villi containing embryonic capillaries extend into blood-filled spaces of the endometrium. Embryonic blood and maternal blood are separated by the chorionic membrane.

pooled in the spaces. There is no actual mixing of the embryonic blood with maternal blood, but the chorionic membrane between them is thin.

It may help to visualize this relationship within the placenta by thinking of your hands (the capillaries of the embryo) in rubber gloves (the chorionic tissues) immersed in water (the pool of blood in the endometrial spaces). Only the rubber glove separates your hand from the water. In the placenta, across the thin layer of chorionic membrane all materials are exchanged between the two bloodstreams.

The portion of the endometrium that lies between the chorionic villi and the underlying basal layer of the endometrium and the embryonic chorionic villi make up the placenta. As the embryo develops, the size and thickness of the placenta also increase. The placenta is well established and functional about 4 weeks after implantation. By this time the rudimentary embryo heart is beating and blood is circulating through the chorionic villi, exchanging materials with the mother's blood in which the villi are bathed.

Formation of Primary Germ Layers

The transition from preembryo to embryo occurs with the appearance of three layers of cells, collectively called the primary **germ layers,** which eventually give rise to all the body's specialized tissues and organs (Table 19-1). Development is a continuous process, and even during implantation the blastocyst is changing. By the ninth day after fertilization, during implantation, the inner cell mass begins to differentiate into two layers. The layer of cells facing the fluid-filled cavity of the blastocyst is called the **endoderm,** meaning "inner skin." The other layer of cells, facing toward the trophoblast layer, is called the **ectoderm,** which means "outer skin." The ectoderm is separated from the trophoblast cells by a fluid-filled space called the **amniotic** (am'-ne-ot-ik) **cavity.** Together the endoderm and ectoderm are called the **embryonic disk.** By 12 days after fertilization, the embryonic disk is formed (Figure 19-6, *A,B,C*).

By the beginning of the third week, the embryonic disk begins to elongate and broaden at one end, becoming a pear-shaped plate of cells. A depression forms down the center of the plate, called the **primitive streak.** By day 17, ectodermal cells move toward and into this groove. The entering cells then push laterally between the upper ectodermal and lower endodermal layers. The cells sandwiched between the two layers are called the **mesoderm,** meaning "middle skin" (Figure 19-6, *D*).

The fluid-filled amniotic cavity and the cells that come to line the cavity to form a transparent membranous sac are together called the **amnion** (am'-ne-on). This sac will eventually extend all the way around the embryo, protecting it from physical trauma.

Embryonic Development

With the formation of the three germ layers during the embryonic period, the stage is set for **organogenesis** (or'-gah-no-jeh'-nih-sis), body organ and organ system formation. During organogenesis the

Table 19-1 Structures Produced from the Primary Germ Layers of the Embryo

ECTODERM	MESODERM	ENDODERM
All nervous tissue	Skeletal, smooth, and cardiac muscle	Epithelium of digestive tract (except that of oral and anal cavities)
Epidermis of skin and epidermal derivatives (hairs, hair follicles, sebaceous and sweat glands, nails)	Cartilage, bone, and other connective tissues	Glandular derivatives of digestive tract (liver, pancreas)
Cornea and lens of eye	Blood, bone marrow, and lymphoid tissues	Epithelium of respiratory tract, auditory tube, and tonsils
Epithelium of oral and nasal cavities, of paranasal sinuses, and of anal canal	Endothelium of blood vessels and lymphatics	Thyroid, parathyroid, and thymus glands
Tooth enamel	Serosae of ventral body cavity	Epithelium of reproductive ducts and glands
Epithelium of pineal and pituitary glands and adrenal medulla	Synovial membranes of joint cavities Organs of urogenital system (ureters, kidneys, gonads, and reproductive ducts)	Epithelium of urethra and bladder

Figure 19-6 Formation of the primary germ layers

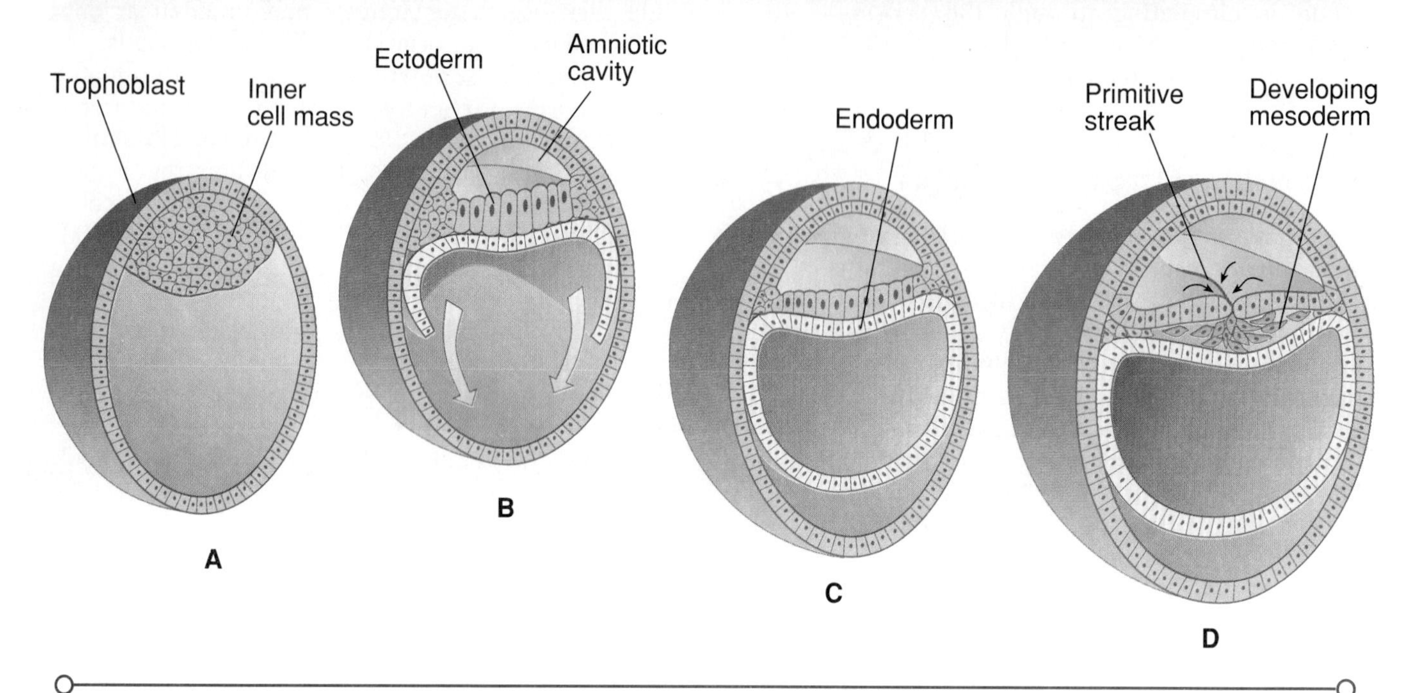

The inner cell inside the blastocyst begins as an aggregation of undifferentiated cells (A). Formation of the primary germ layers begins with the appearance of the amniotic cavity, separating the inner cell mass into two layers (B). The developing cell layer facing toward the large fluid-filled cavity of the blastocyst is the endoderm, and the cell layer facing toward the amniotic cavity is called the ectoderm (C). As the primitive streak forms, cells move into it. These entering cells push between the ectodermal and endodermal layers to form the last of the primary germ layers, called the mesoderm (D).

cells of the three germ layers rearrange themselves, forming clusters, rods, or membranes that differentiate into distinct organs and tissues (See Table 19-1). By the end of the embryonic period the embryo is 8 weeks old, about 22 mm, or 1 inch long, and all the organ systems are recognizable. To illustrate this process we will briefly describe the general development of several major organ systems.

Organogenesis

The nervous system is the first organ system to form. By the seventeenth day of development, the mesodermal cells beneath the primitive streak aggregate, forming a solid rod of mesodermal cells called the **notochord** (no-to-kord). Chemical signals from the notochord stimulate the ectoderm overlying it to thicken and form the **neural plate.** By day 21 the raised edges of the neural plate form two neural folds (Figure 19-7). Two days later the raised edges of the neural plate come together and fuse, forming a **neural tube,** which detaches from the ectodermal layer of cells to lie beneath it along the embryo's midline. The anterior end of the neural tube develops into the brain and associated sensory organs, such as the eyes and ears, and the posterior portion becomes the spinal cord. Other cells along the ridges of the neural folds migrate throughout the embryo, eventually giving rise to the neurons of the peripheral nervous system. By day 28 the major parts of the brain have begun to form. The nervous system continues to develop and increase in complexity throughout most of the ensuing fetal period.

Heart development begins in the third week. Like other muscles of the body (see Table 19-1) the heart is derived from the mesoderm. It starts out as two simple tubes that fuse to form a single chamber pumping blood by the fourth week after fertilization. During the next 3 weeks, this enlarged chamber undergoes structural changes that convert it into a four-chambered organ capable of functioning as a double pump. After the seventh week few changes other than growth occur in the heart until birth.

At the fifth week of development, limb buds

Figure 19-7 Formation of the neural tube

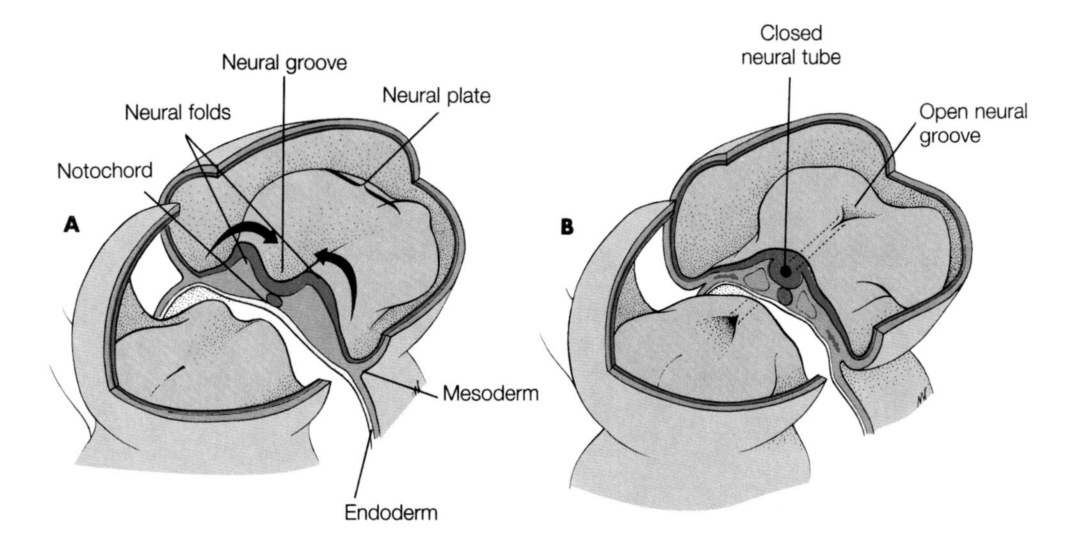

Chemical signals from the notochord stimulate the ectoderm to thicken and form the neural plate. The edges of the neural plate raise up, come together, and fuse, forming the neural tube (A). The neural tube detaches from the ectodermal layer to lie beneath it and above the notochord along the midline of the embryo (B). The neural tube becomes the central nervous system, and the notochord becomes the vertebral column that encases the spinal cord.

appear and will develop into arms and legs. Each limb bud has a flattened paddlelike end from which hands and fingers, or feet and toes will develop. Fibrous membranes and cartilage also begin to form the embryonic skeleton. These structures are then systematically replaced by bone cells.

The last of the major organ systems to form is the reproductive system. The gonads begin to differentiate in the sixth week and complete the process in males by the eighth week. In females gonad formation is not complete until the twelfth week.

We now can describe the mechanisms that control the precise timing and organization of what begins as undifferentiated cells into specifically different tissues and organs characteristic of the human body.

Control of Tissue and Organ Development

Organogenesis depends on localized tissue growth, which contributes to the changes in the size, shape, and proportions of various body parts in the developing embryo and fetus. Each of these processes is governed in some way by differential genetic expression of the DNA contained in the cells' nuclei. At least two primary mechanisms are known to govern human development. The first is cell-to-cell interactions, and the second is controlled cell death.

One type of cell-to-cell interaction that occurs during both preembryonic and embryonic development is **active cell migration.** This involves cells moving to particular destinations and establishing specific relationships with the cells already there. For example, the precursor cells of the peripheral nervous system migrate to precise spots in the embryo, ultimately giving rise to all the neurons outside the CNS. Cells move in response to chemical gradients (chemotaxis). The gradients are created when specific substances are released from target cells or tissues.

Another important form of cell-to-cell interaction during development is **induction.** Again recall the notochord formation process. After the mesoderm is formed, the tissue just at the midline of the embryo forms the notochord. The cells of the notochord then induce, or stimulate through the release of particular chemical substances, the overlying ectodermal cells to form a neural plate, the beginnings of the CNS.

Organ formation and growth also depend on the process of **controlled cell death.** This involves eliminating cells and tissues that are used for only short periods in the embryo or fetus. For example, cell death helps transform the paddlelike appendages in human embryos into hands and feet. Cells between lobes of the paddlelike portion of the limb bud die on cue, leaving separate fingers and toes.

Abnormal Development

Although the placenta is an effective filter against many harmful substances reaching the developing embryo, many drugs, other chemical substances, and some viruses and bacteria can still cross the placenta and enter the developing embryo or fetus. Agents causing the death of the embryo or birth defects of varying degrees are called **teratogens** (ter′-ah-to-jinz). The embryo, however, is not equally susceptible to a teratogen at all developmental stages. If exposed to a teratogen during preembryonic development, the preembryo may either be completely unaffected by the teratogen or abort spontaneously. However, in general the earlier the exposure to a teratogen during *embryonic* development, the more severe will be the defect. The probability of a teratogen causing a birth defect decreases during the fetal period. About 18 weeks after fertilization, most teratogens have little if any developmental impact on the fetus.

One exception to this general rule is syphilis. The microorganism causing syphilis is able to move across the placenta after the twentieth week of gestation and cause many fetal malformations. If the female contracts syphilis *after* becoming pregnant, the syphilis nearly always causes malformations in the embryo. The tissues most often affected are the bones, the lungs, the liver, and the spleen.

Following are brief descriptions of the teratogenic effects of several infectious diseases, some prescription drugs, alcohol, and tobacco use on the developing embryo (Figure 19-8).

Infectious diseases such as syphilis, gonorrhea, and rubella (ru-bel′ah, commonly called German measles) are teratogens. If left untreated in the mother, gonorrhea and syphilis will be transmitted to the fetus and may be fatal. The rubella virus may cross the placenta and cause eye, ear, and heart malformations. However, the nature of the defect depends on when the fetus is exposed to the virus. For example, heart defects occur from infection during the fifth week of gestation, blindness and other eye problems during the sixth week, and hearing abnormalities occur from exposure during the ninth week.

During the first 3 months, or **first trimester,** the developing embryo is highly sensitive to many drugs. In Europe during the early 1960s a tranquilizer called **thalidomide** (thuh-lih′-duh-mid) was routinely prescribed to alleviate morning sickness. When taken during the first 2 months of pregnancy, thalidomide affected limb bud formation, resulting in infants born with severely deformed or missing limbs. Other drugs, such as streptomycin, cause hearing loss. Tetracycline, a commonly prescribed antibiotic, causes stained and yellowed teeth to form in the infant after birth.

Some drugs have detrimental effects only much later in the postnatal life of individuals exposed to them while still *in utero.* For example, **diethylstilbestrol** (di-eth′il-stil-bes′trol), or **DES,** a synthetic form of estrogen, was taken by pregnant women in the United States between 1945 and 1970 to prevent miscarriage. The harmful effects were not noticeable until the offspring reached reproductive maturity. There was a significantly higher incidence of cervical cancer and problem pregnancies in females and deformities in the reproductive organs of males for those exposed to DES prenatally, compared with fetuses not exposed to the drug.

Cigarette smoking has an adverse effect on fetal growth and development. Although it is unclear whether the observed effects are caused by nicotine or by the other potentially harmful substances in tobacco smoke, the major known effect of smoking on the fetus is oxygen deprivation. This occurs in two ways. First, nicotine causes blood vessels to constrict, reducing blood flow and therefore the flow of oxygen from the mother to fetus. Second, the increased levels of carbon monoxide from cigarette smoke reduce the capacity of the fetal blood to carry oxygen to body cells. These combined effects, whose magnitude appears to increase directly with the number of cigarettes the mother smokes, result in a higher incidence of prematurity, low birth weight, and spontaneous abortion for those infants whose mothers smoked during their pregnancy than for those infants whose mothers did not smoke.

Alcohol passes freely across the placenta and has the same kind of physiological effects on the fetus as on the woman who drinks it. In addition, alcohol is teratogenic, producing a cluster of four basic problems in the fetus called **fetal alcohol syndrome,** or **FAS.** First are defects of the CNS, including mental retardation, a small head, and hyperactivity and irritability in infancy. Second, low birth weight and overall poor height and weight growth postnatally occur. Third a distinctive facial appearance, including a narrow forehead, small nose and midface, and a wide space between the upper lip and nose results. Fourth many other defects, including cardiac abnormalities, eye and ear malformations, and skeletal joint problems are common.

It is unclear how alcohol produces the characteristics of FAS. Evidence derived from experiments with mice suggests that the facial abnormalities are

Figure 19-8 Effects of teratogens on embryonic and fetal development

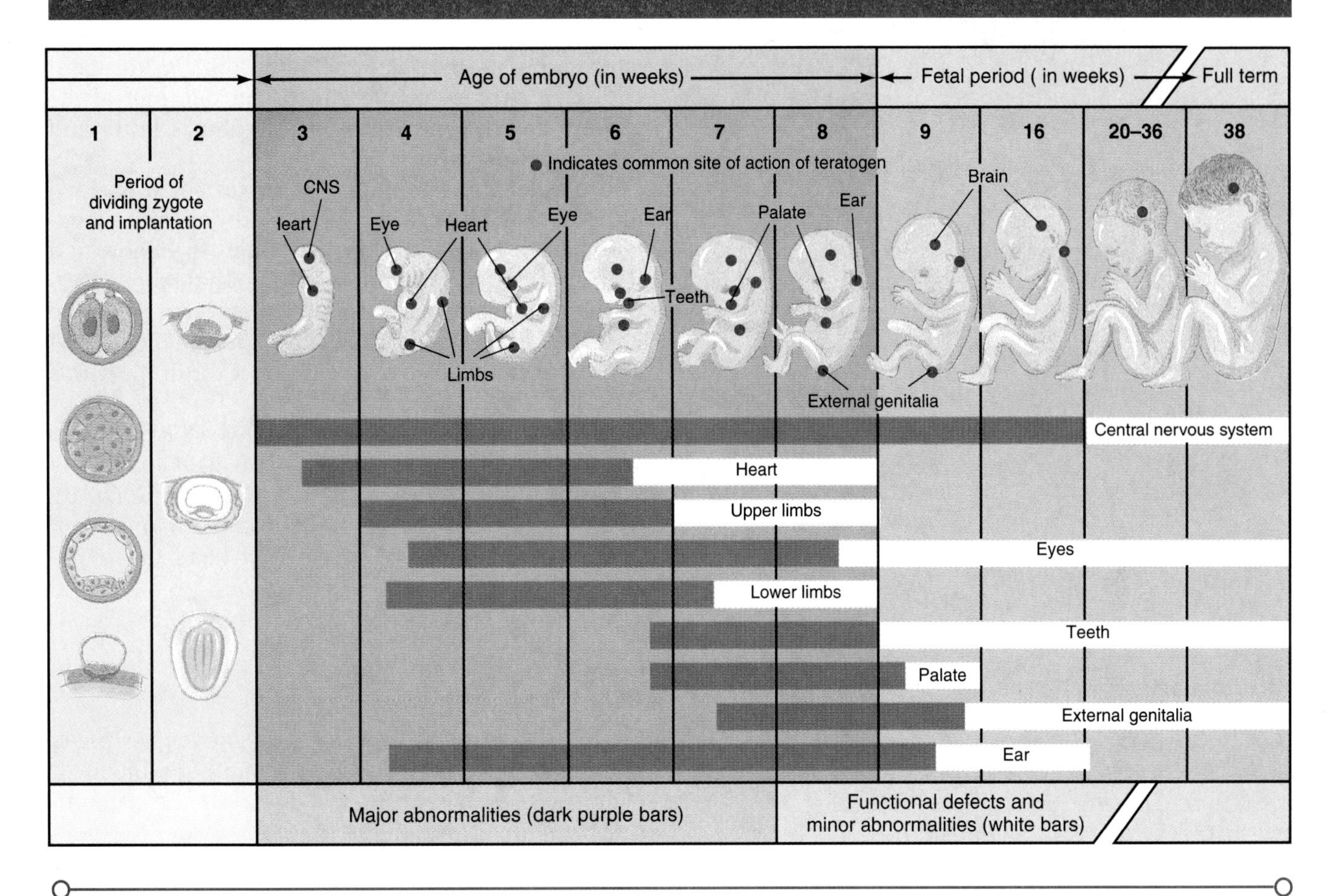

caused by early exposure to alcohol, corresponding to the third week of a human pregnancy. Exposure to alcohol late in pregnancy probably also has harmful effects on the fetus. For example, because the fetal brain is growing, it is still susceptible to alcohol in the third trimester.

It is also still unclear if there are safe amounts of alcohol to consume during pregnancy. Some researchers suspect that even two drinks a day during pregnancy may be harmful to the fetus. At this level of consumption, low-birth-weight babies are common. Because of this uncertainty, physicians are urging total or near alcohol abstinence during pregnancy.

Cocaine affects both the mother and the fetus in several ways. Because cocaine is a stimulant it accelerates the maternal heart rate and elevates her blood pressure. It also constricts blood vessels and reduces blood flow to the uterus. It depresses appetite and can lead to malnourishment of both the mother and the fetus. Cocaine also impairs the fetal nervous system. Vision and muscular coordination are affected, and often mental retardation and other mental function problems, such as difficulties with memory and problem-solving abilities, can result.

Fetal Development

The **fetal stage** of development begins as bone cells begin to replace the embryo's cartilage skeleton at the eighth week of gestation. Although the 8-week-old fetus weighs about 5 g and is about 3 cm long, growth now accelerates.

Growth and System Development

The most prominent growth during the first trimester is in the head and brain. During the fourth month the beginning of the second trimester, rapid body growth begins to catch up with head growth.

By the third month, blood is being formed in the bone marrow and the kidneys begin to function.

Figure 19-9 Fetal alcohol syndrome

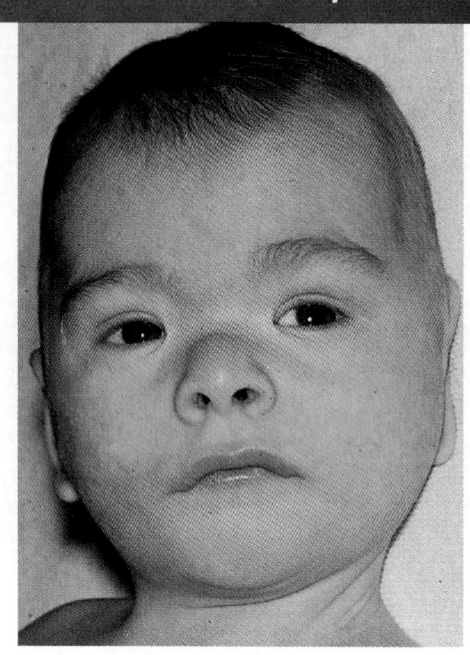

Alcohol consumed by the mother produces four types of defects in the fetus: CNS defects, including mental retardation, hyperactivity, and irritability in infancy; low birth weight and slow growth and slow weight gain postnatally; distinctive facial features, including a low forehead and wide spaces between the upper lip and nose and also the eyes; and often, cardiac abnormalities and skeletal joint problems.

Bones continue to form from the cartilaginous skeleton, and the long bones ossify independently of one another. During the fourth month the lungs are formed but are collapsed and do not function.

At 5 months the fetus weighs approximately 225 g, or about $\frac{1}{2}$ lb, and is more than 7 inches long. Movements become more pronounced, and the fetus becomes sensitive to many external stimuli. The eyes are sensitive to light, but there is still no hearing. The skin is well formed but cannot adjust to temperature changes. Because the skeleton begins to develop rapidly during the second trimester, by the end of the fifth month the mother must now supply large quantities of calcium for the ossifying fetal skeleton.

During the sixth month the fetus clearly looks like a miniature human being. Scalp hair, eyebrows, and eye lashes are apparent, and the eyelids have separated.

At the seventh month, the fetus now weighs approximately 1 to 1.5 kg, or 2 to 3 lb, and is more than 30 cm, or 12 inches, long. The brain is growing, and the cerebral surface becomes grooved. By this time about 85% of the calcium the mother consumes goes to the fetal skeleton, along with about the same percentage of iron. About one half of the protein she eats now goes to the fetus's brain and other nervous tissues.

In the eighth month, growth rates begin to slow, although the fetus gains weight and deposits subcutaneous fat under its skin. Although the lungs and digestive systems are not yet functional, about 75% of the babies born at the eighth month survive. The fetus has filled the available space in the uterus and is no longer able to move freely as it did in previous months.

During the last month the fetus weighs somewhere between 2.5 to 3.5 kg, or 6 to 8 lb, and is between 48 to 53 cm, or 19 to 21, inches, long. By this time the fetus has usually changed its position, dropping lower in the pelvis in a head down position.

Birth

During the last few weeks of pregnancy, estrogen reaches its highest levels in the mother's blood. It causes oxytocin receptors to form on the smooth-muscle cells of the uterus. As birth nears, oxytocin is released from the posterior pituitary gland, stimulating the placenta to release a type of **prostaglandin** that sensitizes uterine cells to oxytocin. Both of these hormones stimulate the uterus to begin contractions (Figure 19-10). The increased physical stress of uterine contractions stimulates the posterior pituitary gland via the hypothalamus to greatly elevate its oxytocin secretion. This sets a positive-feedback loop in motion. Greater contractions cause the release of more oxytocin, which causes greater contractions, and so on. The series of events that accompany birth are collectively called **labor,** which is separated into three stages.

The first stage of labor, called **dilation,** involves dilation of the cervix to allow the fetus to pass from the uterus through the vagina and to the outside. Stage two is **expulsion,** or delivery of the baby. The third stage, called the **placental stage,** is the delivery of the placenta.

Either at the outset or sometime during dilation, the **amnion** ruptures and the fluid in it escapes out of the vagina. During this stage the cervix begins to dilate, usually to a maximum of 10 cm, or approximately 4 inches. This is the longest and most variable stage of labor, lasting from 8 to 24 hours, with an average of 12 hours for a woman's first pregnancy and from approximately 3 to 12 hours for subsequent pregnancies. The head of the fetus acts as a

Figure 19-10 Onset of labor

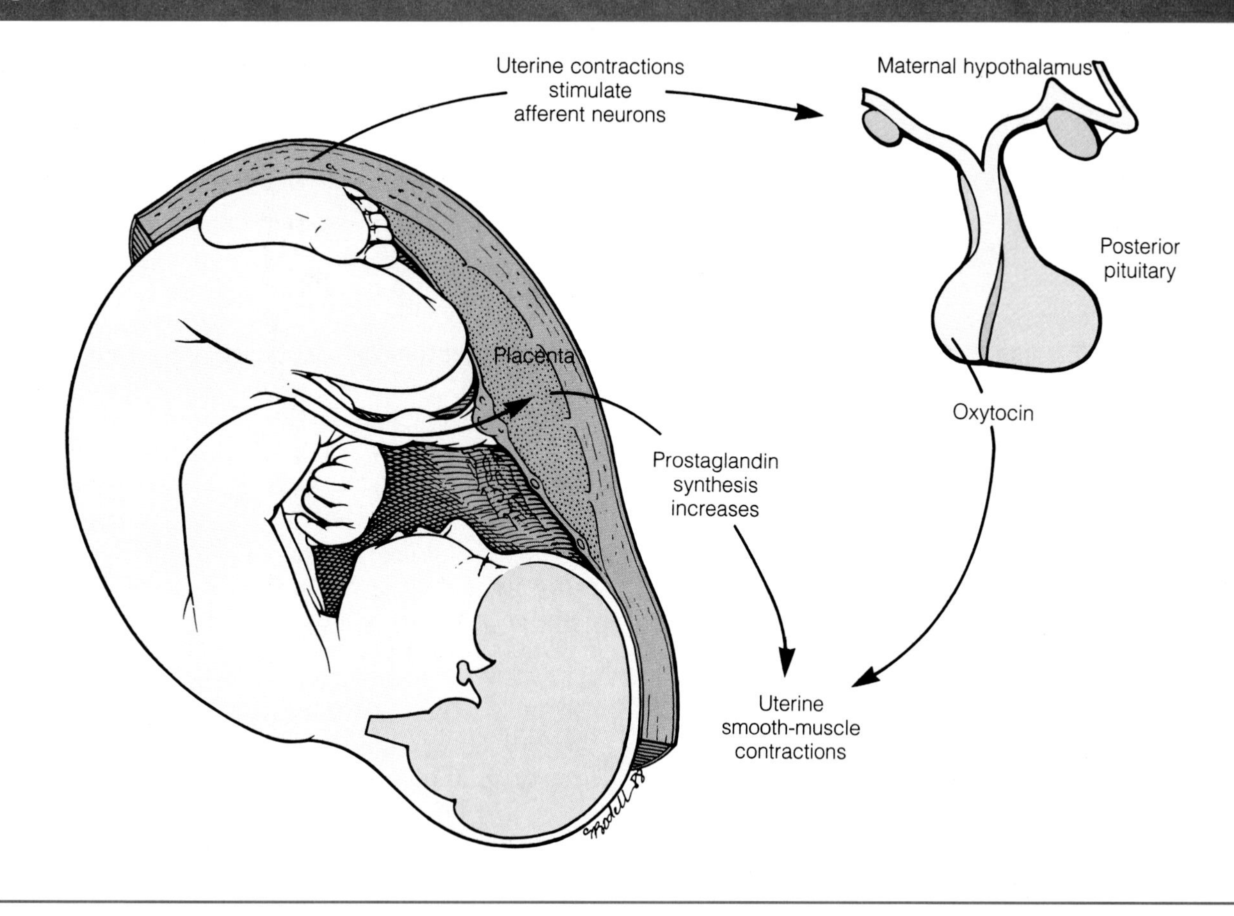

Both oxytocin and a type of prostaglandin stimulate the uterus to begin contractions. Uterine contractions, via stretch receptors, stimulate the posterior pituitary gland to secrete more oxytocin, setting a positive feedback loop in motion that continues until after the birth of the baby.

wedge and assists in cervical dilation. If another part of the baby's body instead of the head is against the cervix, the result is called a **breech birth** and the cervix is less effectively dilated. Because the head has the largest diameter of the infant's body, if the baby approaches the cervix feet first, for example, the cervix may not dilate enough to permit the head's passage. If a breech birth or some other complication arises, the baby may be delivered by **caesarian section,** which involves cutting directly into the abdomen and uterus to remove the baby.

Expulsion, the second stage, begins once cervical dilation is complete. By the time the cervix is fully dilated, strong contractions are occurring every 2 to 3 minutes and last about 1 minute. Although this phase can last up to 2 hours, the average time spent in delivering the baby is 50 minutes in a first birth and about 20 minutes in later births. Within a few seconds of the birth the umbilical cord that has connected the fetus to its life-sustaining placenta is tied off and cut.

The placental stage, the third stage, is complete within 15 minutes after delivery. Forceful contractions continue after birth and compress the uterine blood vessels. This prevents bleeding and causes the placenta to detach from the uterine walls. The placenta and its attached fetal membranes, collectively called the **afterbirth,** are then expelled and delivered.

After delivery the uterus shrinks to its prepregnant size. This process is called **involution** and usually takes 4 to 6 weeks. During involution the remaining endometrial tissue that was not expelled gradually disintegrates, producing a vaginal discharge called **lochia** (lo'ki-ah).

About 1 minute after birth the newborn is given a test called the **APGAR Screening Test,** which is a quick assessment of the newborn's status (Table 19-2). The APGAR score, ranging from 0 to 10, is based on the five signs of appearance, or color; heart rate; reflexes as determined by stimulation of the soles of the feet; muscle tone; and respiratory rate. Each of these signs can be given a score of 0, 1, or 2, with 2 being the best condition. Generally

Table 19-2 APGAR Screening Test

Score	A Appearance (color)	P Pulse (heart rate)	G Grimace (reflex irritability)	A Activity (muscle tone)	R Respiration (respiratory effort)
0	blue, pale	absent	no response	limp	absent
1	body pink, extremities blue	slow (below 100)	grimace	some flexion of extremities	slow, irregular
2	completely pink	rapid (over 100)	cry	active motion	good, strong cry

newborns with low APGAR scores, from 0 to 3, must be closely monitored, whereas APGAR scores of 7 to 10 indicate a healthy baby.

Effects of Pregnancy on the Female

Pregnancy causes anatomical, metabolic, and physiological changes in the mother. As pregnancy progresses the female reproductive organs become increasingly vascularized and engorged with blood. For example, the vagina develops a purplish hue that is used to diagnose pregnancy. Another frequent early clinical sign of pregnancy is **morning sickness,** a daily bout of nausea and vomiting that usually occurs in the morning but can actually occur at any time. Morning sickness usually appears shortly after implantation and coincides with peak progesterone and estrogen secretion. This may be because this hormone triggers the symptoms, possibly by acting on the chemoreceptors of the vomit reflex in the brainstem. The most obvious change during pregnancy is uterine enlargement. The uterus expands and increases in weight more than 20 times, exclusive of either the fetus or the placenta. As birth nears, the uterus reaches the level just beneath the sternum of the rib cage and occupies most of the abdominal cavity.

The female breast contains from 15 to 25 lobules, called **mammary glands,** each with its own duct leading outside to the nipple. The lobules consist of sacs lined with milk-producing cells. **Lactation** is milk production by the hormone-prepared mammary glands, which normally occurs after the infant's birth.

Toward the end of pregnancy the anterior pituitary gland begins to secrete the hormone prolactin. After birth, prolactin release begins to diminish to prebirth levels. Continued milk production depends on mechanically stimulating the mother's nipples, normally provided by the sucking infant. Nipple receptors send nerve impulses to the hypothalamus, resulting in a pulselike release of prolactin from the anterior pituitary gland. Prolactin stimulates milk production for the next feeding. The same impulse also causes the hypothalamus to release oxytocin from the posterior pituitary gland. Oxytocin causes the milk **let-down reflex,** the actual ejection of milk from the nipple. Oxytocin also helps the recently stretched uterus to continue contracting to its prepregnant size. For this reason, as well as because of the advantages of breast milk for the infant, many obstetricians recommend that mothers nurse their babies. Nursing mothers also have a reduced incidence of breast cancer compared with nonnursing mothers.

A normal pregnancy causes considerable weight gain. Because some women are overweight or underweight before pregnancy begins, it is difficult to determine an ideal weight gain. However, adding up the weight gain resulting from fetal and placental growth, maternal reproductive organ growth, breast growth, a weight gain of about 12 kg, or 24 to 28 lb, is standard.

Body systems other than the reproductive system must also adjust. For example, blood volume increases by 30% to accommodate the growing placenta. Blood pressure and pulse rate typically rise and increase by 20% to 40% at various stages of pregnancy. Respiratory activity increases by about 20% to handle the additional fetal requirements of oxygen uptake and carbon dioxide removal. Urinary output increases as the mother's kidneys excrete the additional wastes produced by the fetus.

Good nutrition is necessary all through pregnancy if the developing fetus is to have all the building materials, especially proteins, calcium, and iron, necessary to form tissues and organs. However, it is not simply a matter of the mother having to now eat "for two." A pregnant woman needs only about 300 additional calories daily to sustain normal fetal growth. The emphasis should be on eating food rich in vitamins and minerals, not just on eating more food (Table 19-3).

It is particularly necessary to increase vitamin intake, especially vitamins D and B_6, and folate (see Chapter 16). Vitamin D helps to increase calcium absorption and distribution. The developing fetal bones require that the mother's intake of vitamin D double to about 10 mcg per day. That is equivalent to one quart, or four cups, of milk. Folate assists in DNA synthesis and red blood cell formation. Because both DNA synthesis and red blood cell formation occur in the developing embryo and fetus, the mother should double her intake of this vitamin, from 200 mcg to 400 mcg per day.

Minerals, such as calcium and iron, should also be increased because they are necessary for fetal bone formation and red blood cell formation.

First Breaths

The **neonatal period** is the first 4 weeks of the infant's life immediately after birth. Our interest in this section is confined to the first few hours after birth.

Cutting the umbilical cord disrupts the only source of oxygen the infant has known. Without maternal respiratory exchange via the placenta, carbon dioxide quickly accumulates in the baby's blood. This carbon dioxide buildup triggers the respiratory center in the brainstem, which in turn sends nerve impulses to the diaphragm. The contraction of the diaphragm causes the baby to gasp for breath. The exhaling cry means the baby is breathing on its own. Inflating and keeping the lungs expanded is much more difficult for premature infants. Sometime during the last 2 months of prenatal development, the alveoli of the lungs begin secreting a lipoprotein called **surfactant,** which assists in preventing the alveoli from collapsing when the first breaths are taken after birth.

After considering all the changes involved in human development, from fertilization until birth, it is easy to forget this is just the beginning. The drama of prenatal development and birth sets the stage for many other equally important changes in postnatal development.

Dynamics of Growth

One major characteristic of human biology is body growth and development. What are the processes that determine human growth and development?

Table 19-3 Vitamin and Mineral Requirements During Pregnancy

VITAMIN OR MINERAL	% ABOVE RECOMMENDATIONS FOR NONPREGNANT WOMEN (25–50 Years of Age)
Vitamin A	20
Vitamin D	100
Vitamin E	25
Vitamin C	15
Thiamin	35
Riboflavin	25
Niacin	15
Vitamin B_6	38
Folate	120
Calcium	45
Phosphorus	45
Magnesium	15
Iron	95
Zinc	18
Iodine	15

Growth and Development

To understand the growth process we must first add to what we have learned about the cell, as first discussed in Chapter 3. **Growth** involves either the production of new cells or an increase in the size of existing cells. Cell growth has three phases. The first phase, called **hyperplasia** (hi′per-pla′zi-ah) or the **proliferative phase,** is the increase in the number of cells by means of mitosis. Next is a **transitional phase,** where the rate of mitosis begins to slow down and the newly produced cells simultaneously undergo enlargement as the cytoplasm and organelle content increase in amount and number. In the third phase, called **hypertrophy,** cell size alone increases and mitosis either ceases completely or continues only to maintain necessary cell numbers.

The second component of our definition of growth involves the increase in the size of existing cells. In fact, postnatal growth is for many tissues a period of developing and enlarging the cells produced during prenatal growth. For example, all the neurons of the brain are present at birth. Because

Table 19-4 Developmental Stages of Human Growth

Neonate	First 2 weeks after birth. By 2 weeks neonate has regained original birth weight.
Infancy	From 2 weeks to about 1 year. Birth weight has tripled, length increased by one half. Increased brain growth, teeth begin to erupt. Learning period has begun.
Childhood	One to about 12 years. Baby teeth erupted. Brain growth nearly complete. By about 9 years the child's body proportions foreshadow those of the adult.
Puberty	Secondary sexual traits appear. For girls this begins between 10 to 15 years, and for boys it begins between 12 to 16 years.
Adolescence	After puberty. More behavioral than biological changes occur. Emotional maturation begins.
Adulthood	Body growth is finished.

neurons do not reproduce, they are the only neurons an individual will ever possess. The neurons, however, enlarge and establish more dendrites and dramatically increase their number of synapses as the brain becomes neurologically more integrated during postnatal development. Much of the postnatal increase in brain size is a consequence of an increase in the number of the supporting neuroglial brain cells and not of the neurons themselves.

Development refers to the appearance of new structures or new body functions. Table 19-4 lists the developmental stages of growth. For example, during puberty the increase in the female breast size results from the growth of mammary glands and the increase in subcutaneous fat surrounding them.

Development also extends to functions of the whole organism, which we call behavior. Although we have defined growth and development as individual processes, they most often simply describe different aspects of the dynamic interrelated quality of any organism. Infants, for example, usually begin to walk alone by 12 to 15 months of age. The development of this new behavior is an outcome of the growth of skeletal muscles, bones, motor neurons, and their integration. The growth or development of any individual body part alone does not produce this behavior.

Specific body tissues have individual patterns of growth. Growth curves are usually represented as the size attained at a given time as a percentage of total adult size (Figure 19-11). As mentioned earlier the brain and head develop earliest, most of the development taking place prenatally and in early infancy. At birth the brain is already 25% of its adult size. By the age of 5 it is approximately 90%, and by age 10 it is about 95% complete. The curve for general growth, which includes factors such as body dimensions, musculature, and internal organs, is different from that of the brain and head. General growth is relatively rapid during the first 4 years. It then slows down but accelerates again during puberty, before leveling off at adulthood. In contrast, lymphoid tissue, which includes the appendix, the spleen, and the thymus gland, has a completely different growth curve. It reaches its maximum size, about *double* its adult size, before puberty. Then under the influence of sex hormones, it *declines* to its adult size.

Figure 19-11 Growth curves of different parts and tissues of the body

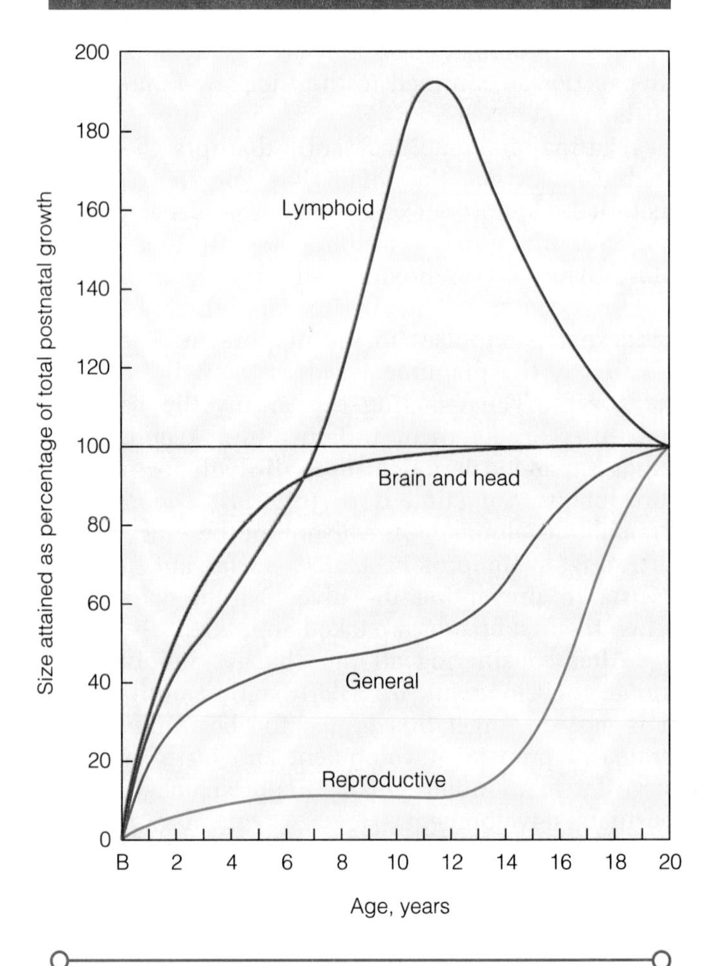

Specific tissues have individual patterns of growth. Growth curves show the size of different parts and tissues of a child's body relative to the final adult size.

Figure 19-12 Changing proportions of the body during prenatal and postnatal growth

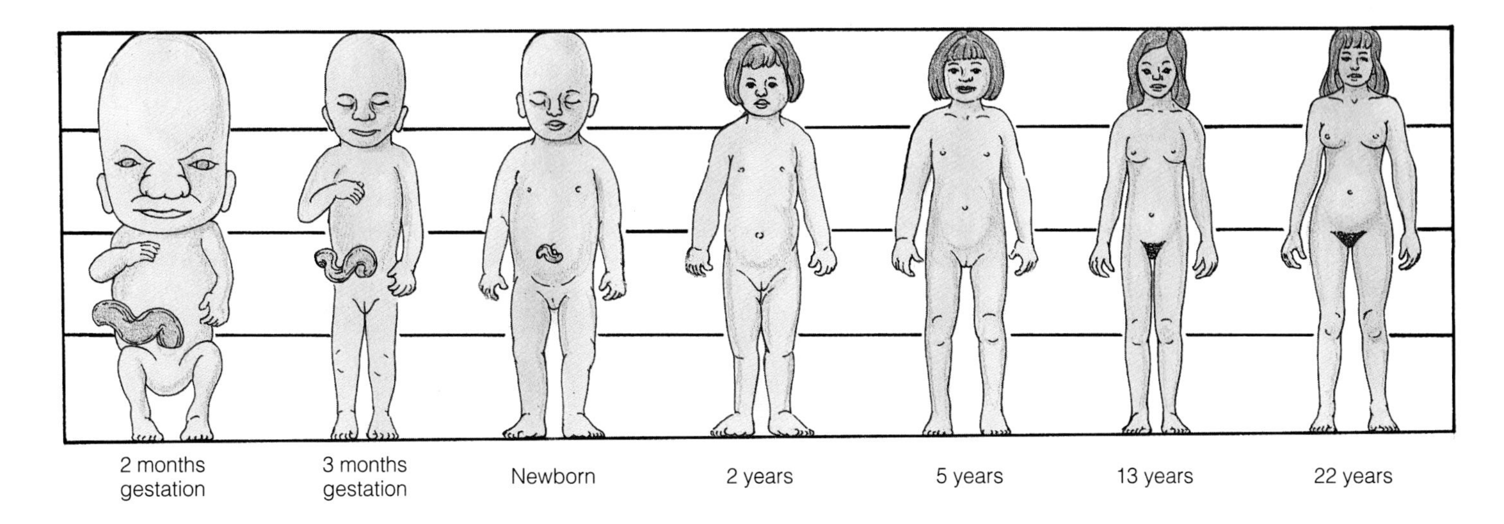

The shape of the embryo and fetus prenatally and the change in shape from infancy to adulthood in postnatal development occur because different body tissues and organs grow at different rates.

Because different body tissues and organs grow at different rates, they are responsible for the appearance of the characteristic shape of the embryo and fetus prenatally and for the change in shape from an infant to an adult in postnatal development. These different growth rates produce clusters of characteristics that define the several recognizable stages of human development (Figure 19-12).

Hormones

As you recall from Chapter 10, hormones also affect growth. Growth hormone, secreted by the anterior pituitary gland, controls overall body size and acts on all the body's cells by increasing protein synthesis and carbohydrate and lipid metabolism. In particular, growth hormone causes bone growth by stimulating increased activity primarily of the osteoblasts of the epiphyses of the long bones. It also dramatically increases the rate of mitosis and the uptake of amino acids in skeletal muscle cells.

The thyroid hormone called thyroxine is important in regulating the body's metabolic rate and in developing the brain and the nervous system. Fetal production of thyroxine occurs between the fourth and fifth month of gestation and remains relatively high in the fetus and young infant until the age of 2, when it begins to decline. This is the time of the most rapid development of the brain and nervous system. As the brain approaches maturity, thyroxine levels decline.

Finally, sex hormones, particularly testosterone in the male and estrogen in the female, have important functions in the normal patterns of growth and development. Testosterone is responsible for increased height and weight and the development of male secondary sex characteristics at puberty. Estrogen is responsible for characteristic female changes.

Stages of Growth

Keeping these concepts in mind, we can now discuss the four general stages of human growth and development preceding adulthood and some of the primary changes that occur during each stage.

The first stage, **infancy,** includes the first 2 years of life after birth. **Childhood** includes the period after infancy and extends from the age of 2 years of age to approximately 10 years of age. The third stage, which extends to approximately 18 years of age and is marked by the onset of **puberty** and reproductive maturity, is called **adolescence.**

Infancy

When infants are born, they are already equipped with many reflexes. Some of these reflexes, such as sucking, are clearly related to initial survival, but other reflexes are not. Reflexes actually develop in utero. For example, the sucking reflex develops between the second and third month of gestation.

Biology in Action SOONER TO GROW, SOONER TO MATURE

A female born in the United States at this time probably will have her first menstrual period when she is between the ages of 12.5 and 13 years. A female born in America in 1900 first menstruated at about age 14. This change is not isolated. Historical data from many European and Scandinavian countries and from Japan have indicated that girls in the nineteenth century did not begin to menstruate until they were 15.5 to 17 years of age. A similar historical pattern has also been observed in the onset of puberty among boys worldwide. Not only do children mature more rapidly than they did 150 years ago, but each generation has also tended to be taller and heavier on average than the generation preceding it. For example, between 1880 and 1950, height gain during adolescence increased about 1 inch per decade, and weight gain increased about 15 lb per decade. Biologists call this earlier maturation and increase in average body size the **secular trend.**

Why is the secular trend occurring? As we attempt to answer this question, we must keep two observations in mind. First the secular trend has not been occurring throughout recorded history. As far as can be determined from skeletal remains, children from the Middle Ages until the nineteenth century appear to have matured at the same rate and reached the same average adult stature as children do today. Puberty appears to have been delayed primarily during the nineteenth century. Second the secular trend has been slowing in recent generations, coming to a complete stop in some parts of the world. In other words, we are not getting indefinitely taller and reaching puberty indefinitely earlier.

Many factors have been cited as either causing or contributing to the secular trend. Improved nutrition and generally improved environmental conditions, extending from better health care to smaller family size, have probably each been important. However, the secular trend has been noted among most socioeconomic classes during the last century, which makes the impact of environmental changes less clear. Recent evidence suggests that genetic factors may help to explain observed secular trends. For example, genes for height appear to be dominant over genes for shortness. When a person from a tall family marries a person from a short one, the children tend slightly toward the taller side rather than simply being halfway in between the two. The dramatic increase in the migration of people and the subsequent gene flow between formerly isolated, inbreeding villages and regions has spread and increased in frequency the genes for tallness.

As with any other multivariate problem of human biology, observation and simple association between variables is only the beginning in attempting to determine the cause or causes for observed changes in human populations through time. As we come to learn more about ourselves perhaps more specific answers to these questions will be forthcoming.

Many of these reflexes do not persist throughout infancy, and usually disappear within the first 6 months of postnatal life. They are then replaced by more complex behavioral patterns. Grasping, for example, is a reflex response to an object being placed in the hand and is replaced by deliberate grasping at about 4 months of age.

Infant reflexes are important for at least three reasons. First, as indicated in the APGAR score, reflexes reflect the maturation and development of the infant's nervous system. Second, reflexes clearly indicate that human babies are not "empty" but have a number of "built-in" responses to the environment. Third, reflexes are significant because they form the foundation of more complex, learned behaviors that occur later in life.

Childhood and Adolescence

For the most part, growth is relatively steady during childhood. For example, most children lose the round appearance and "baby fat" of infancy, and the internal organs steadily increase in size and become functionally stable. Adolescence, however, is the time of developmental transition from childhood to adulthood and involves physical changes so profound that virtually every body tissue is affected. **Puberty** (pu'ber'ti) refers specifically to the period of sexual development that culminates in sexual maturity. The chronological timing of the beginning, progression, and completion of the unique changes associated with adolescence and puberty is considerably variable among individuals and between the sexes. We will outline the patterns of adolescent and pubertal growth and development here, but because of the often dramatic individual variations that occur in the appearance of these changes, we must also look at the range of variability for each characteristic.

Puberty

The primary events associated with puberty include growth and weight-gain acceleration and the development of sexual characteristics. These changes in growth and development are the result of hormonal changes. The specific actions of the hormones involved in reproductive maturity and function were described in Chapter 18. Our goal in this chapter is to look at their effects on adolescent growth and development.

The accelerated growth rate during puberty is called the **adolescent growth spurt.** The typical growth spurt is marked by a sharp acceleration in the growth rate that results in a rapid height increase. For example, for boys at the time of their peak growth rate, 3 to 5 inches may be added to their height in 1 year. For girls, 2.5 to 4.5 inches is typical. Girls begin their adolescent growth spurt and reach their peak-growth rate earlier than boys, and their peak-height rate is a bit shorter than that of boys.

As you recall from Chapter 11, ossification begins *in utero* and continues throughout childhood. There is a slight increase in the ossification rate at puberty. During adolescence the process is largely completed. During childhood there are minor differences in skeletal structure between males and females, but at puberty testosterone causes the cartilage in the bones of the male upper thorax and shoulder area to grow rapidly, whereas estrogen has a similar effect on the female pelvis. The result is that during adolescence a boy's shoulders expand noticeably compared with other trunk areas. A girl's hips develop more rapidly than other areas, producing a trunk with wide hips compared with shoulders.

Another pubertal change involves body composition. The increase in muscle and fat during puberty is not simply a consequence of the adolescent growth spurt. The accumulation of muscle and fat occurs at different rates, and each contributes to total body weight in different proportions during growth. Although muscle growth increases steadily in both sexes before the onset of puberty, this muscle growth during puberty is more dramatic in males because of increases in testosterone levels. In contrast, because of the influence of estrogen, fat accumulation increases after the attainment of peak height velocity in females but not in males. Given these hormonal differences, males actually have a relative *decrease* in body fat accumulation as a percentage of total body weight because of the disproportionate gain in muscle weight before reaching the peak-height growth rate.

Perhaps the most dramatic changes of puberty involve the development of the reproductive organs and the appearance of secondary sex characteristics. The internal reproductive organs enlarge and change shape for both boys and girls during puberty. In girls the appearance of secondary sex characteristics ranges from breast enlargement to axillary (underarm) hair development. In boys changes include the appearance of facial hair and a deepening voice (Table 19-5).

Differences in Growth Rate During Adolescence

As you might recall from your own adolescence, nowhere is the difference in the rate and timing of adolescence more evident than in a junior high school classroom. A group of 12- to 13-year-olds will consist of some children who have reached puberty and are adultlike in appearance and some children who are still physically childlike. This variation occurs because the age of onset of puberty varies from one child to the next.

In boys after the initial growth of the testes and penis, the *sequence* of events is much less variable than the *age* at which they take place. A boy's testes may begin to enlarge as early as 9.5 years of age or as late as 13.5 years of age. Once genital growth occurs, other changes consistently follow. The height spurt is normally next. It may begin as early as 10.5 years of age or as late as 16 years of age. It may end by 13.5 years of age, or it may not end until 17.5 years of age. A particular boy of 14 years of age may still resemble a child; or his testes may have begun to develop, but he may show no other signs of puberty; or he may have completed his height spurt, and his genitals may resemble those of an adult male's.

As among boys, among girls there is a variation in the time at which the growth spurt begins. However, girls on average reach puberty 2 years earlier than boys. Breast development is usually the first sign of puberty, with the uterus and vagina developing almost simultaneously. **Menarche** (me-nar′ke), the first menstruation, almost always occurs *after* the peak-height rate has passed. However, the onset and completion of these events varies. A girl's breasts may begin to develop as early as 8 years of age or as late as 13 years of age. A girl can have her first period as early as 10.5 years of age or as late as

Table 19-5 Development of Secondary Sexual Characteristics and Other Changes that Occur During Puberty

IN GIRLS

Characteristic	Age of First Appearance	Hormonal Stimulation
Appearance of breast bud	8–13	Estrogen, progesterone, growth hormone
Pubic hair	8–14	Estrogen
Menarche (first menstrual flow)	10–16	Estrogen and progesterone
Axillary (underarm) hair	About two years after the appearance of pubic hair	Estrogen
Sweat glands and sebaceous glands acne (from blocked sebaceous glands)	About the same time as axillary hair growth	Estrogen

IN BOYS

Characteristic	Age of First Appearance	Hormonal Stimulation
Growth of testes	10–14	Testosterone, FSH, growth hormone
Pubic hair	10–15	Testosterone
Body growth	11–16	Testosterone, growth hormone
Growth of penis	11–15	Testosterone
Growth of larynx (voice lowers)	Same time as growth of penis	Testosterone
Facial and axillary (underarm hair)	About two years after the appearance of pubic hair	Testosterone
Sweat glands and sebaceous glands; acne (from blocked sebaceous glands)	About the same time as facial and axillary hair growth	Testosterone

15.5 years of age. Again as is the case with boys, there is a time in the development of girls when some may not have reached puberty and others have completed the process.

From infancy to reproductive maturity, postnatal growth and development encompass dramatic structural and functional changes and transformations. However, adulthood, the next stage of development, is equally dynamic and is the topic of Chapter 20.

CHAPTER REVIEW

SUMMARY

1. Pregnancy refers to the events that occur in the female from fertilization until the baby is born. The time during which development occurs is called the gestational period and is approximately 266 days.
2. The fusion of sperm and ovum, called fertilization or conception, usually occurs in the upper third of the uterine tube. Fertilization in the uterine tube occurs because of several factors, such as differences in male and female gamete mobility and longevity.
3. The prenatal developmental process is divided into three stages, called the preembryonic period, the embryonic period, and the fetal period. Fertilization marks the beginning of the first period. Formation of the primary germ layers marks the transition between the preembryonic and embryonic periods. Ossification of the cartilaginous skeleton denotes the onset of the fetal period.
4. The developing embryo gains oxygen and nutrients from the mother by way of a temporary organ, called the placenta, that develops in the uterus during pregnancy. The placenta

consists of both maternal and embryonic tissue. The umbilical cord connects the embryo to the placenta.
5. During embryonic development, tissue specialization and organ formation occur in a process called organogenesis. If exposed to a teratogen, the developing embryo will be affected. Depending on when it is exposed, the developing embryo will respond differently to different teratogens. Fetal development is primarily organ growth and elaboration.
6. Labor, the series of events that accompany birth, is divided into three stages: dilation, expulsion, and placental.
7. The first four weeks of an infant's life is called the neonatal period. Physical growth ultimately involves producing new cells and increasing the size of existing cells.
8. Each tissue has its particular rate of growth called a growth curve. The differences in the growth curves of particular body tissues produce clusters of characteristics that define the major stages of human growth and development.
9. Hormones and external environmental factors significantly influence growth and development. The impact of environmental factors on growth depends on when developing tissues are first exposed to them.
10. The transition from childhood to adulthood, called adolescence, is marked by the onset of puberty, a highly regular pattern of sequential changes culminating in reproductive maturity. The onset of the adolescent growth spurt is governed by the increase in sex hormone levels of testosterone in boys and estrogen in girls.
11. Individual differences during adolescence result from the variation of the age of the onset of puberty. The differential onset of puberty often results in marked temporary physical differences among children of the same age.

FILL-IN-THE-BLANK QUESTIONS

1. Once penetration of the ovum by a sperm has occurred, the ovum's ______ ______ prevents other sperm from entering.
2. Cleavage results in a mass of cells called a(n) ______.
3. Occasionally, ______ twins result when the inner cell mass splits into two equal halves during preembryonic development.
4. Initially, to maintain the endometrium, the ______ cells of the blastocyst secrete ______ ______ ______ ______, causing the corpus luteum to continue secreting ______ and ______ hormones.
5. The ______ membrane of the placenta prevents direct contact between maternal and fetal blood.
6. The ______, ______, and ______ are the primary germ layers which eventually give rise to all of the body's specialized tissues and organs through a process called ______. The embryonic disk is made up of the ______ and ______ layers.
7. Agents causing the death of the embryo or birth defects of varying degrees are called ______.
8. As birth nears, ______ is released from the ______ pituitary gland, stimulating the placenta to release ______.
9. An early clinical sign of pregnancy is ______ ______.
10. The first stage of labor is complete when the cervix is completely ______ to about ______ cm in diameter.

SHORT-ANSWER QUESTIONS

1. Describe the structural characteristics of the placenta.
2. Describe the effects of two known teratogens on the development of the embryo or fetus.
3. What are the hormones controlling labor?
4. In both boys and girls, what is initiated by rising levels of sex hormones?
5. List the sequences of events that occur during puberty in males and females.

VOCABULARY REVIEW

amnion—*p. 359*
blastocyst—*p. 355*
chorion—*p. 357*
cleavage—*p. 354*
ectoderm—*p. 359*
endoderm—*p. 359*
fertilization—*p. 353*
implantation—*p. 356*
lactation—*p. 366*
menarche—*p. 371*
mesoderm—*p. 359*
teratogen—*p. 362*
trophoblast cell—*p. 355*

Chapter Twenty

ADULTHOOD AND AGING

For Review

Here are some important terms and concepts that you will encounter in this chapter. If you are not familiar with them, you should review them before proceeding.

Blood vessels (page 231)
Basal metabolic rate (page 303)
Collagen (page 30)

OBJECTIVES

After reading this chapter you should be able to:

1. *Define gerontology and aging.*
2. *Describe the general effects basic to the human aging process.*
3. *Describe the major aging changes on each of the organ systems.*
4. *Name the two general groups of hypotheses that explain aging and list theories from each group.*

Around age 30, humans are at their peak physiologically. After that, more body cells began to die than are reproduced. This results in a decrease in functional capacity and a lowered ability to maintain homeostasis. Although several changes associated with aging are inevitable, the rate of aging is highly variable. By maintaining good health practices throughout life, we can hope to prolong health and longevity.

The Aging Process

Each of us recognizes, both in ourselves and in others, many signs of the aging process. These include changes in appearance, overall strength, and stamina that distinguish an older person from a younger one (Figure 20-1). We also have images of what people should look and act like at certain chronological ages. The study of the aging process is called **gerontology** (jer-on-tol'o-je). Why do we grow old? What does "growing old" mean in terms of the changes at cellular and tissue levels? What does it mean to say that grandfather "died of old age"? Often we compare older adults to today's younger people. But is that an accurate way to assess age? Older individuals grew up in a world in which nutrition, medicine, and other environmental factors were different from those today. Are many of the medical problems we associate with older adults an inevitable consequence of age or a possible consequence of "harder times"?

It is sometimes difficult to know what is "normal." Often, it is defined as what we find to be most common. Osteoporosis is a loss in bone density caused primarily by calcium mineral loss. This loss eventually weakens bones and increases the probability of fracture. Is osteoporosis a "normal" consequence of aging, however, or is it what we currently find to be most common? In women over age 65 about 65% show evidence of osteoporosis. Of these women about 30% have one fracture resulting from osteoporosis. We are now finding that many factors, including diet and exercise, influence the likelihood that osteoporosis will develop as one ages. These behavioral and nutritional practices, not just aging, strongly influence the incidence of osteoporosis among older women. We may find in the future, as a new generation of women reach age 65, that osteoporosis is not necessarily a common occurrence in aging.

In this chapter we address three fundamental questions concerning what aging is, how we age, and why we age. The first question involves defining some basic processes and perspectives concerning several biological processes common to the human life span. We then describe the primary structural and functional changes that occur to body organs and tissues during the aging process. We also discuss some of the major hypotheses proposed to explain why we age. In Chapter 19 we focused on the cell's role in the growth and development process. Here we return to this basic biological building block to focus on why cells deteriorate and die, ultimately resulting in the death of the body.

Figure 20-1 Aging

Changes in appearance are usually the most obvious indicators of the aging process.

Aging

Aging is defined as the sum total of all the changes that lead to functional impairment and an increased probability of death. How old we get before this functional impairment begins to occur depends on many factors. Cultural and technological changes, specifically in nutrition and medical science, have greatly increased the average age to which people now live. For example, between 1900 and 1970 the proportion of the population in the United States over age 65 increased from 4.1% to 10%. This is projected to approach 13% by the year 2000 (Figure 20-2).

Not only has the number of older people increased, but a higher percentage of people are living to be over age 80. The fastest-growing segment of the U.S. population is the group over age 85. However, the maximum age of the longest lived individuals in a population, called the **life span,** has not increased significantly over time. Because the longest lived humans die at about 100 to 120 years of age, there appears to be a biological limit on the human life span.

Figure 20-2 Projected elderly percentage of population

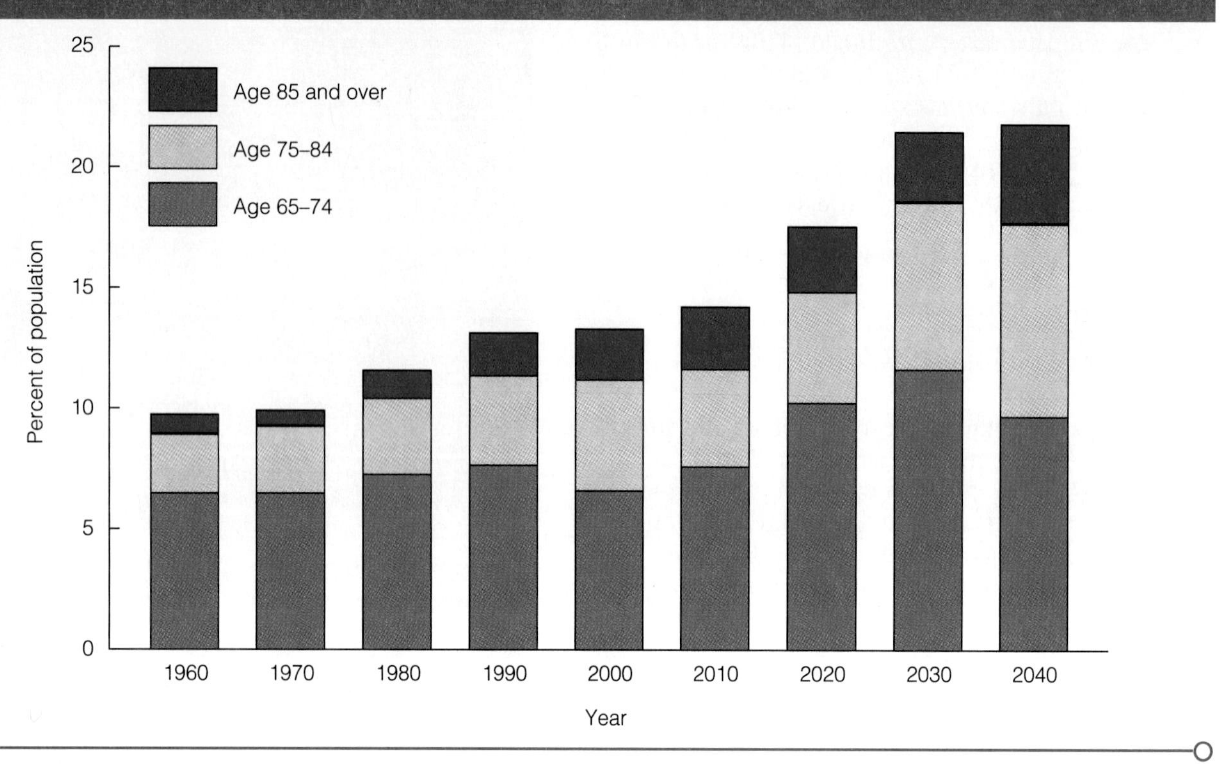

General Changes in Aging

Aging is a continuing process beginning at conception and ending with death. All of us are in various phases of aging as we progress through life. The effects of aging, however, tend to become more apparent after about age 40. Three related general effects basic to the human aging process are reduced cell division, decreased functional capacity, and altered connective tissue.

Remember that some body cell types are nonrenewable. Although they actively divide during embryonic development, they no longer continue to divide after birth. These cells are sometimes called **postmitotics.** For example, nerve cells and skeletal muscle fibers are postmitotic cells. Their cell number is thus fixed, and damaged or destroyed cells are not replaced. Tissues composed of postmitotic cells show different aging characteristics than tissues whose cells are still able to undergo mitosis. In postmitotic tissues, cell numbers and tissue mass eventually decline with aging. In tissues whose cells can divide, age-related changes are less well defined. Cell division thus seems to gradually slow down, but other effects of aging are more variable.

The second general effect of aging is a lowered or decreased capacity at the cell and organ levels to function properly. Much of this lowered functional capacity results from deteriorating body components. Because of lowered function, the aging body is not able to respond to internal or external stimuli as effectively as it once did. The reduced capacity to respond to a changing environment makes it increasingly difficult for the body to maintain homeostasis. As homeostasis wanes, various dysfunctions become more common, increasing the likelihood of death.

Finally, significant changes occur in the collagen and elastin protein fibers in the connective tissues. Collagen fibers account for almost 30% of total body protein. With aging many structural changes occur. The number of cells that produce collagen is reduced, as is the number of collagen fibers in tissues. The remaining collagen fibers

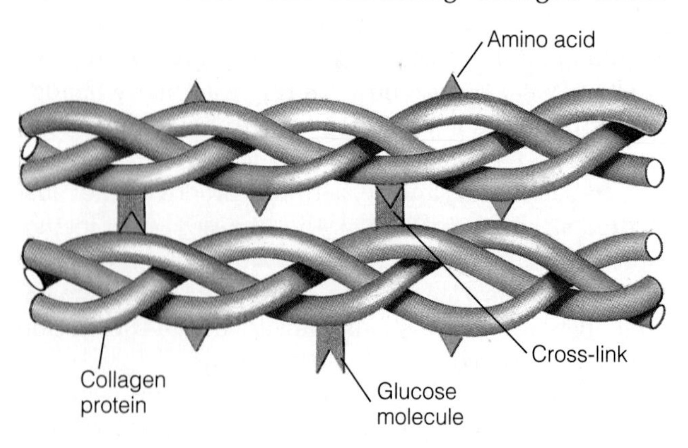

become larger, reducing the amount of space between fibers, decreasing the amount of intercellular substances such as fat and water. During aging, collagen and elastin fibers also become less resilient. Glucose molecules bind to amino acids in the collagen protein fibers, eventually forming cross-links between the fibers. Elastin tends to undergo slight calcification and becomes less flexible. Both of these processes occur throughout much of the body wherever collagen and elastin are located.

Changes in Organ Systems

Just as variability exists among individuals in a population for almost all biological characteristics, variability also exists for the growth, development, function, and aging of those characteristics. All organs do not age at the same rate in an individual, nor does any specific organ age at the same rate in different people. However, although aging rates vary, average changes do occur. We describe in the next sections some of the primary structural and functional changes that occur with aging.

Epidermis

The skin, one of the most easily observed organs of the body, shows obvious aging changes. The changes are influenced by a variety of factors, including heredity, nutrition, and levels of various hormones in the body. External factors, such as the amount of sun and wind exposure, can also significantly affect the rate of aging of the skin.

Most skin changes that result from aging are not life threatening, nor do they greatly affect physical health. The outer skin layer is generally thin. Unlike most body tissues the epidermis has no blood vessels or nerves. The primary result of aging on the epidermis is that it becomes thinner because of a declining cellular division rate. The number of melanocytes in the epidermis also decreases. However, the remaining melanocytes tend to grow larger and group together, forming visible darkly pigmented areas called **aging spots** typical of older people (Figure 20-3).

Most of the cells in the underlying dermis are fibroblasts. These form the collagen fibers. The number of fibroblasts and collagen fibers decreases with aging. As with the epidermis, the dermis thins and the remaining collagen fibers grow larger. This reduces the amount of space between fibers, causing the dermis's fat and water content to decrease with age. Elastin fibers also decrease in number. The ultimate outcome is that the skin is less resilient and unable to smooth out. Consequently, skin wrinkles and sags as we age (Figure 20-4).

Figure 20-3 Aging spots

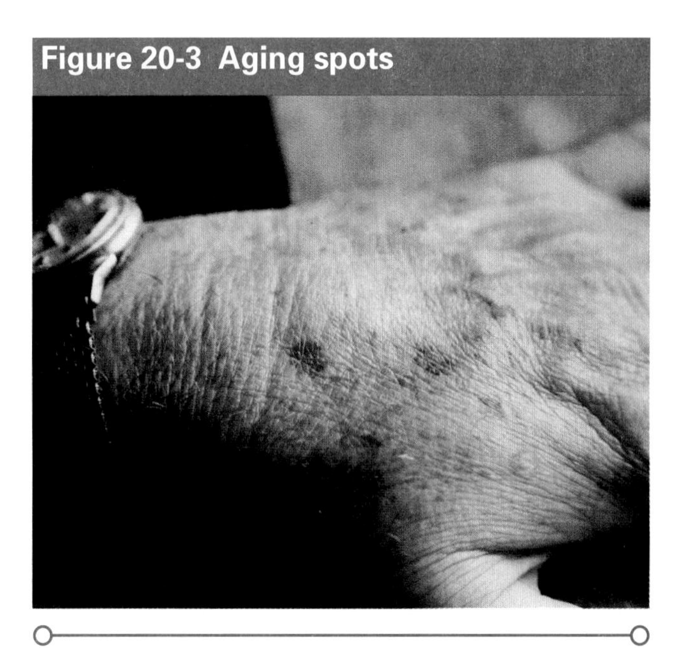

Melanin-producing melanocytes in the lower layers of the epidermis change as we age. While decreasing in number, melanocytes tend to grow larger and group together, forming visible, darkly pigmented areas, sometimes called aging spots.

Musculoskeletal System

The major age-related skeletal-system change, especially in women after menopause, is calcium loss from bone. This loss generally begins at about age 40. Because some degree of osteoporosis occurs in 65% of women over age 65 but in only about 21% of men, a decrease in estrogen levels associated with menopause is one factor in its development.

Inactivity also contributes to osteoporosis. Weight-bearing activities, such as walking briskly or jogging, produce a long-term bone calcium increase. For example, 30 to 45 minutes of moderate walking three times a week greatly slows the calcium loss in older women. If continued for 1 year, this activity can begin to reverse the demineralization process.

The most apparent aging change in the muscular system is a reduced muscle tissue mass. Muscle fibers are postmitotic. Therefore, new fibers cannot be produced to replace those lost. Associated with muscle mass loss is reduced muscle strength. The strength and endurance of skeletal, cardiac, and smooth muscle all gradually decrease with age. However, in most people this is small, usually only 10% to 20% of muscle strength up to age 70. In skeletal muscle the ability to coordinate muscle-group activity also diminishes. This is probably caused by CNS changes. These changes also apply to other muscle cell types. Mass and strength reduction occur in muscles less frequently used. Regard-

Figure 20-4 The aging skin

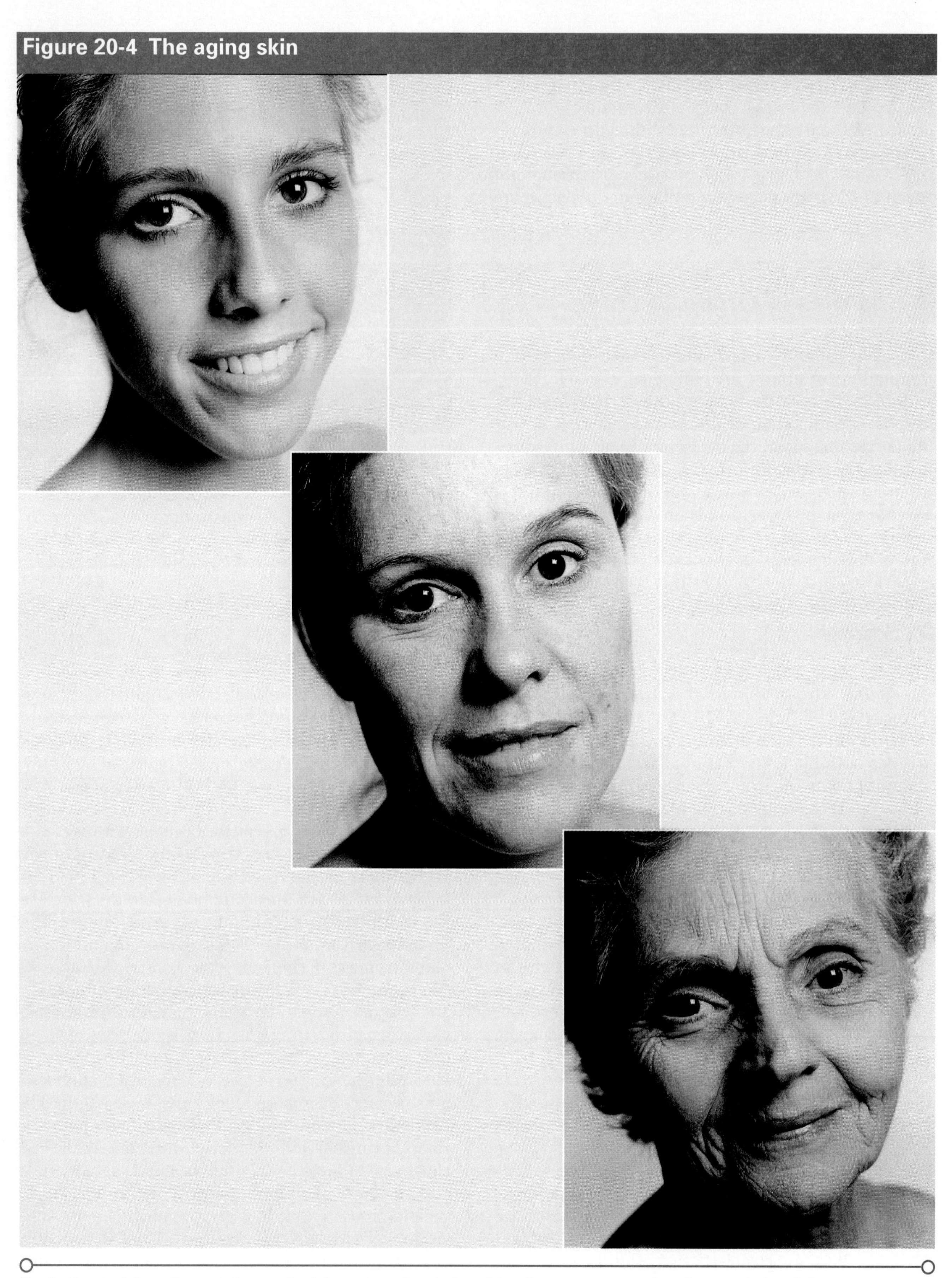

Both the epidermis and the underlying dermis of the skin change as we age. As a result of these changes, skin becomes less resilient, wrinkling and sagging with the passage of time.

less of age, regular moderate exercise not only slows the rate of reduction but can actually increase both skeletal and cardiac muscle strength in people over age 70.

Nervous System

The primary aging changes in the nervous system are a decrease in neuron conduction rate, an increase in the time it takes for neurotransmitter substances to cross a synapse, and neuron loss.

The myelin sheath that surrounds the neuron becomes thinner, resulting in a decreased conduction rate. The increased time required for neurotransmitter substances to cross a synapse results from decreases in the amount of neurotransmitter substances the presynaptic neurons can produce and a decreased number of receptors on the postsynaptic membranes.

Many nervous tissue cells die each day in the normal aging process. However, nerve cells are also postmitotic, so any neurons lost are not replaced. There are thus fewer axons in the peripheral nerves and fewer neurons in the CNS itself. How these changes affect nervous system functions varies among individuals and depends on where the decrease occurs. For example, the brain may ultimately lose as much as 10% of its maximum weight by age 90. The cerebellum alone loses about 25% of the cells responsible for coordination of movement, whereas other areas of the brain (such as speech-association areas of the cerebrum) remain essentially constant throughout life. Neuron loss in the cerebellum is probably responsible for the decreased ability to coordinate skeletal muscle groups as we age.

Another change associated with reduced nervous system function involves learning new skills and information. However, the decline is in the *speed* of response and the ability to integrate what is learned and not in verbal ability or memory. This means that although people learn more slowly as they age, they can acquire new material and remember it as well as when younger.

Cardiovascular System

Both the heart and the blood vessels of the cardiovascular system are affected by aging. A general reduction in the size of cardiac muscle cells results in a progressive loss of cardiac muscle strength and a decrease in the volume of blood pumped with each contraction. The decrease in heart size, often thought to be a direct consequence of aging, is largely caused by a reduction of the muscle mass of

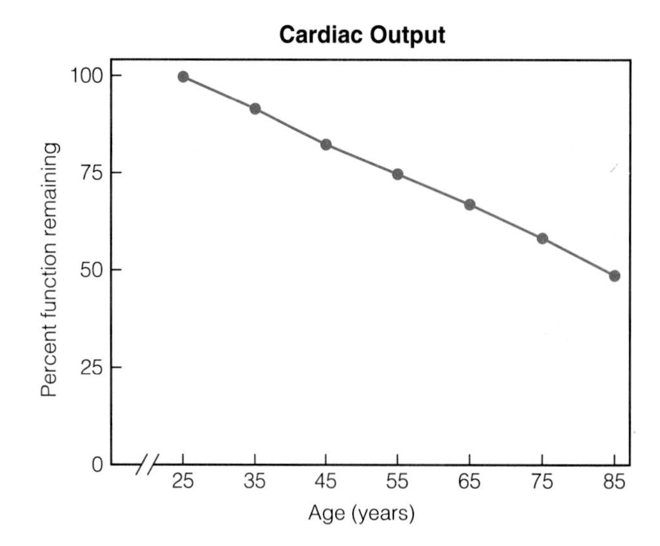

the left ventricle. This is because of diminished demands placed on it by the lower levels of physical activity common with aging.

One of the most common age changes in blood vessels is a decreased elasticity of their walls. One cause is the structural change that occurs to both elastin and collagen fibers. A common functional effect of these structural changes is the wall's contribution to increased systolic blood pressure. Recall from Chapter 12 that blood pressure is determined by both the amount of blood pumped out of the heart and the peripheral resistance blood encounters in the vessels. Elasticity determines the vessels' capacity to return to the original diameter in response to blood flow alterations as the heart contracts and relaxes. The decrease in elasticity of the arteries gradually increases peripheral resistance and therefore blood pressure. For example, in men and women age 30 a systolic pressure of 120 mm Hg and a diastolic pressure of 80 mm Hg are considered normal. In people age 65 and older a systolic pressure of 160 mm Hg and a diastolic pressure of 90 mm Hg is considered normal.

Respiratory System

A major aging change in the respiratory system is a decreased ability of the respiratory organs to acquire and deliver oxygen to the circulatory system. Aging involves a gradual deterioration of the walls separating adjacent alveoli. This reduces the surface area across which gas exchange can occur. Also, because of cross-linkages occurring to the collagen fibers in the alveolar walls, the remaining alveoli do not expand as much during inspiration. As a result, ventilation decreases. This further reduces the volume of inspired air and the amount of oxygen delivered to the body tissues.

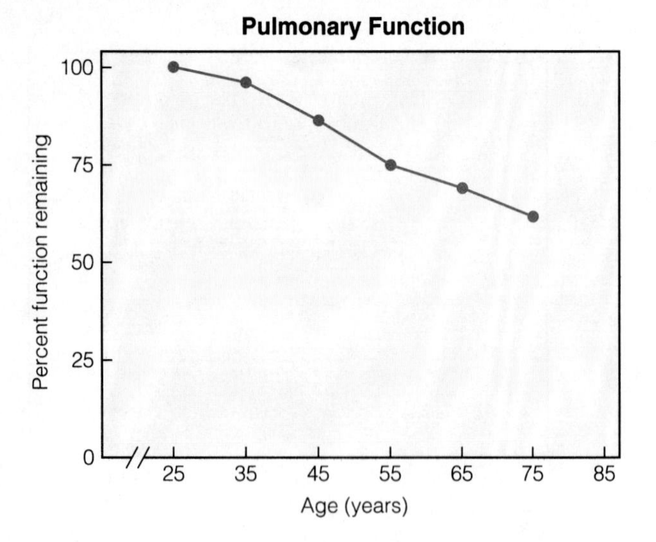

Digestive System and Related Dietary Changes

As we age, digestion takes longer because digestive gland secretions tend to diminish, and both peristalsis and the rate of nutrient absorption slow. Diminishing amounts of digestive enzymes decrease the rate of chemical breakdown of food. Peristalsis is reduced because the smooth muscle tissue in the muscularis layer of the alimentary canal walls generally weakens, slowing down food moving through the digestive tract. Throughout the digestive system the mucous membrane thins as the rate of mitosis of the epithelium decreases. Age-related changes also occur in various digestive tract regions. For example, because of decreases in the mucosal lining of the small intestine, the shape of the villi alters, reducing the surface area across which nutrients can be absorbed.

No major changes in nutritional needs result from aging. The old, like the young, require a diet of carbohydrates, fats, proteins, vitamins, minerals, and adequate amounts of water. The amount of total calories needed changes, however. Energy requirements gradually decline because of both loss of cell numbers and marked physical inactivity. Fewer calories are needed when energy requirements are less. Recall from Chapter 16 that the basal metabolic rate (BMR) is the energy the body requires to perform only the most essential activities, such as breathing, blood circulation, and cellular metabolism. The BMR declines approximately 5% every ten years between the ages of 55 and 75. The reduction in basal metabolic rate is mostly a result of a decrease in body cell numbers.

The amino acids contained in proteins are necessary for tissue growth and maintenance. Although these activities decline with age, protein intake should remain fairly constant because amino acids are also necessary to form essential enzymes. Finally because fats provide about twice as many calories as do carbohydrates, a good way to reduce caloric intake is simply to reduce the amount of dietary fat to no more than 25% of the total nutrients consumed daily.

The aging body is unable to handle glucose as efficiently as it used to because beta cells in the islets of Langerhans producing insulin and the secretory cells producing other digestive enzymes diminish. Older people are thus more likely to have significant blood sugar fluctuations. Noninsulin-dependent diabetes mellitus, or type II diabetes, occurs mostly after age 40. Although most people with type II diabetes produce insulin (see Chapter 10), either the amount is inadequate or there is a problem with the cell's insulin receptors. Most people with type II diabetes are overweight. Excess weight actually accounts for over 90% of cases of type II diabetes. One important method used to control blood sugar levels is to increase the amount of complex carbohydrates and reduce the amount of simple carbohydrates in the diet.

Senses

The senses that change the most in the aging process are vision and hearing.

Many changes often occur with age that cause most older adults some visual problems. The iris usually hardens with aging because of collagen

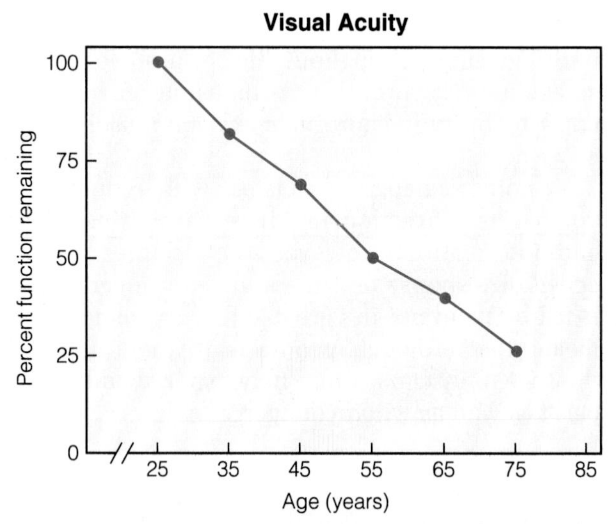

cross-linking. The muscles in the iris responsible for pupil size begin to diminish. Consequently the pupil is no longer able to dilate as quickly or completely. Thus the amount of light that reaches the photoreceptors of the retina by age 70 may be only one third the amount at age 30.

The lens also begins to become less transparent with age. The lens is one of the few body tissues that actually *increases* cellular growth with age.

Biology in Action WHAT IS OLD?

What life will be like when you are older depends largely on what you think it will be like. Of course your physical health and financial success make a difference, but your own expectations also contribute to your future.

Most people have a largely negative unconscious stereotype of what it is like to be old, and then they become that way. Many psychological surveys and profiles of individual attitudes toward aging have found that negative attitudes develop early. One survey found, for example, that on a given evening of television programs, an average of 15 positive age-related expressions were used, but 75 negative ones. In many portrayals older adults are demeaned either by exclusion from social life or by being depicted as confused, helpless, bossy, or socially inept. Interestingly enough, a Harris poll conducted in 1981 showed that *expectations* of problems of the elderly and the problems *actually* reported by the elderly are significantly different.

In reality, older people are an incredibly diverse group. For the most part they are self-sufficient, socially sophisticated, mentally lucid, fully participating members of society who consider themselves happy and healthy. People over age 70 have often greatly contributed to society. Age, of course, does not automatically make one brilliant or otherwise outstanding. But to discount a person simply because of chronological age wastes valuable human experience and wisdom.

To discriminate against older adults is to be prejudiced against everyone, including your own future self. The next time you find yourself thinking an older person is about to expire and turn to dust, just keep in mind Betty Davis's statement on her eighty-first birthday that "growing old ain't for sissies, it's tough work."

Harris Poll Regarding Problems of the Elderly

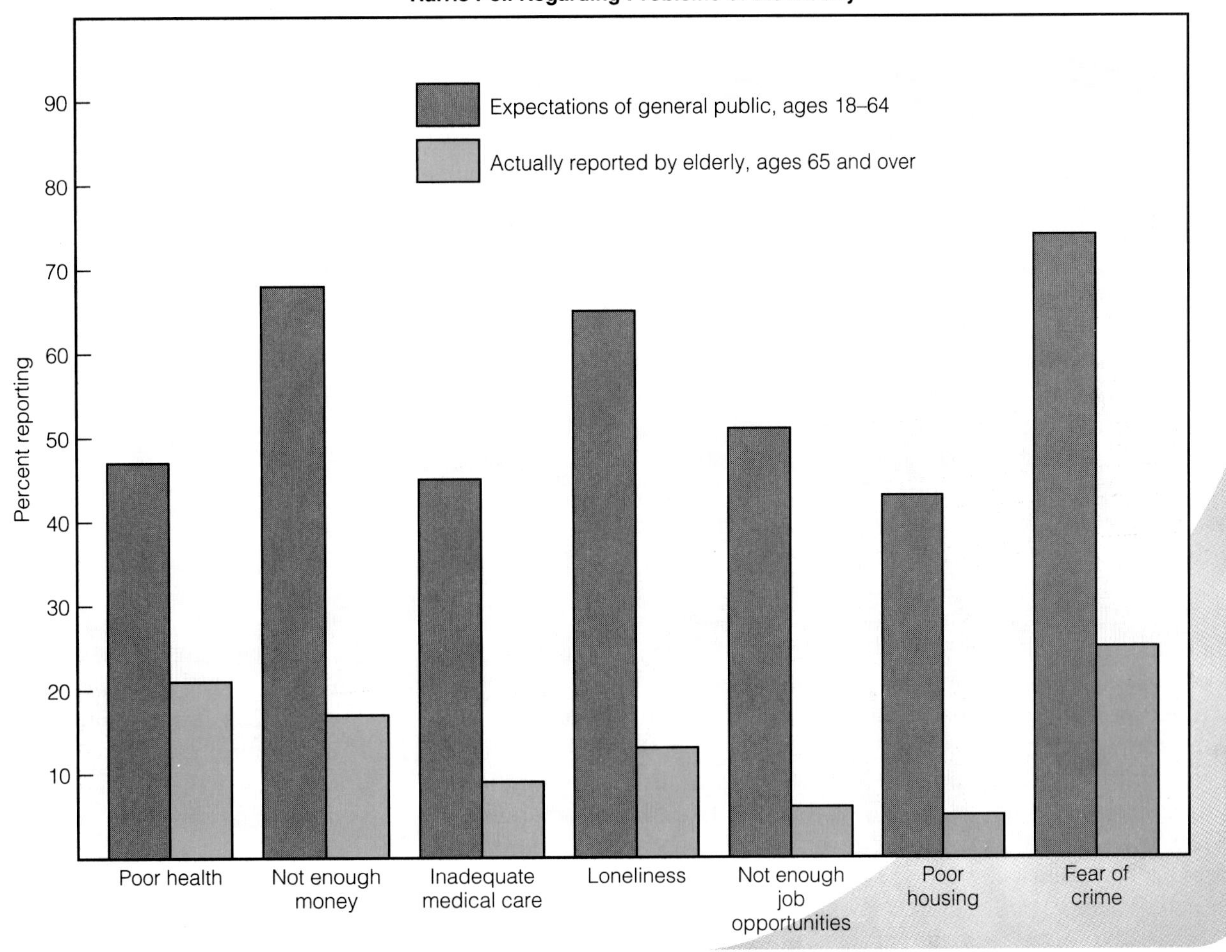

This causes most of the aging changes of the lens. For example, the new cells are laid down on the outside surface of the lens, and the older cells are pushed and squeezed toward the center of the lens. Because of this cellular growth the lens thickens and flattens, contributing to farsightedness, called **presbyopia.**

The gradual hearing loss that often begins by age 40 results from changes in inner ear structures. The sensory receptors in the cochlea begin to degenerate by age 50 primarily because the capillary walls supplying them thicken. **Presbycusis** (prez-bi-koo′sis) is the term used to indicate the loss of the ability to hear high-frequency sounds that typically occurs with aging. For example, a young child can hear high-pitched sounds with frequencies of about 20,000 cycles per second. The upper range for the average 65-year-old is about 8000 cycles per second.

Theories of Aging

Embryonic growth and development involve at least two mechanisms. First some cells die during development unless they receive a "survival signal" from somewhere. For example, far more spinal column motor neurons are produced during embryonic development than are produced postnatally. Embryonic motor neurons are programmed to die *unless* they innervate a skeletal muscle, which releases a chemical preventing the death of its own motor neuron. Second, for other cells the situation is just the opposite. These cells live unless they receive a "death signal" from other cells. Recall from Chapter 19 that embryonic structures such as the webbing between the toes and fingers disappear in the developing human fetus.

Our body cells also appear to have a definite life span. During the 1960s researchers discovered that when fetal cells were cultivated in the laboratory, they underwent about 50 duplications and then died. The same experiment was done using adult cells, and they duplicated only 20 times. In later studies, when one half of a group of aged cells was put into "young" plasma, they replicated more times than the other half of the aged cells put into "old" plasma. Whatever the final outcome, results such as these indicate that cells are capable of only a limited number of divisions.

In contrast to this apparent limitation of a normal cell's ability to replicate, cancer cells will divide indefinitely, acting as if the "off" switch were broken. Thus it appears that cell death in some way is a part of each cell's genetic program.

The two general groups of hypotheses that explain why aging occurs are those that view aging as genetically programmed into every body cell and those that view aging as resulting from some form of damage or wearing out, eventually causing death.

Genetic Theories

The **gene theory of aging** states that aging is the result of normal processes controlling earlier development and differentiation. For example, during growth and development, for connective tissue to function properly, some cross-linkages between collagen fibers are necessary. However, as we age these cross-linkages continue to be made. These result in rigid connective tissue and the structural and functional impairment of many body tissues.

During the normal life cycle, mutations occur in the cell genes composing most of the body's tissues. The **gene mutation theory** proposes that as an individual ages, mutations become more numerous. Some of these mutations "switch on" gene-controlled processes important during earlier cell growth and development. Oncogenes and cancer, discussed in Chapter 5, are two such examples. More malfunctions will occur and will ultimately increase the probability of death.

The **hormonal theory** of aging states that the aging process may be a by-product of reduced levels of various hormones. Recall from Chapter 10 that human growth hormone declines steadily after the adolescent growth spurt. A decline in hormone levels is caused by decreased numbers of hormone-producing cells. Recent studies involving men over age 70 who were given a small amount of growth hormone showed significant improvement in bone and muscle mass and increased skin thickness. As mentioned earlier, decreased estrogen levels in women are one factor involved in the increased incidence of osteoporosis. Estrogen supplements can slow down or even reverse osteoporosis in postmenopausal women.

Wear-and-Tear Theories

The wear-and-tear theories propose that organ systems simply wear out as we age, primarily because of decreased cell numbers. As one system fails, others also become dysfunctional. It is similar to one component of an automobile engine wearing out. Eventually the worn out part will affect the efficient operation of the entire engine and result in its failure.

The **cellular garbage theory** of aging is a wear-and-tear theory that focuses on waste accumulation in the cell. Recent research in cellular biology has shown that as cells age they accumulate increased amounts of various substances in their cytoplasm. This is particularly common in postmitotic cell lines. These substances, some of which are reactive and some of which are inert, are by-products of normal cellular metabolism. Particular metabolic by-products, called **lipofuscins** (lip-o-fus′inz), consist of chemically inert and strongly linked molecules that cannot be broken down by normal enzyme action. Lipofuscins accumulate in lysosomelike structures in the cytoplasm for the life of the cell and appear, under microscopic examination, as brown granules. Lipofuscins are thought to be cellular garbage. The cellular garbage theory of aging states that the gradual accumulation of lipofuscins interferes with normal cellular functioning by physically displacing other cell organelles. They also interfere with biochemical processes in the cytoplasm. Eventually cellular dysfunction and death occur.

The **autoimmune theory of aging** holds that the immune system declines with advancing age, giving rise to three related problems. First, the immune system becomes less likely to detect and eliminate abnormal body cells, which are the forerunners of cancer. Second, the immune system takes longer to respond to foreign invasion, increasing the likelihood of physiological dysfunction. Finally, it is often increasingly common in advancing age for the immune system to mistakenly attack a person's own body cells and tissues, leading to a variety of degenerative diseases. The accumulated damage from declining immune system function is seen as being responsible for many of the degenerative changes associated with the aging process.

It is becoming increasingly common to view aging as having many internally and externally induced causes. For example, several hypotheses state that aging may be initiated by the accumulation of mutations or cellular garbage, both in the cell and in the connective tissues. The rate, however, at which damage accumulates might be dependent on genetic factors such as the cell's susceptibility to various kinds of damage and how well it is able to repair that damage. It might also be dependent on external factors, such as diet, stress, or exposure to environmental hazards.

What Is Inevitable

Although the only sure things in life may be death and taxes, we might also add that so is aging. No matter what you do to prevent it, your body will age (Figure 20-5). You can, however, slow the process of aging and prolong your health as you age. Good health in old age begins with choices you make every day that either promote or harm your health. These choices and their effects are compounded by time and dramatically influence the physiological quality of the aging process. For example, a study of 7000 adults focused on health habits and identified six factors that had a maximum impact on physiological age. These are regular and adequate sleep, regular meals, regular physical exercise, abstinence from smoking tobacco, abstinence from or moderation in alcohol use, and weight control. These effects are cumulative. Those who followed all six practices were in better health, even if older in chronological age, than those who did not. The physical health of those who reported all positive health practices was consistently about the same as people *30 years younger* who followed none of

Figure 20-5 Health and aging

Good health habits throughout life slow the aging process and improve the "health quality" of life.

these practices. Perhaps one of the most important things to keep in mind is that these practices significantly improve one's "health span," one's general health and physical status, even though one's overall life span may not change.

Many of the negative functional changes associated with aging seem to be balanced by many positive ones. For example, although the ability of the immune system to ward off infection declines with age, older people get infections significantly less often than younger people because their immune systems have developed a large immunological memory over time. Another important factor contributing to prolonged health and longevity is attitude. Although older people are less able to recover physically from physiological stress, such as illness or injury, they are usually more resilient psychologically. They have seen more of life and have learned to adjust to the unexpected. They have lived through the fears and restlessness of youth and at least one major life tragedy. It seems that as we grow older we also grow into our happiness. We often learn how to balance spontaneity with planning, so that life is neither too rigid nor too aimless. Thus it appears that how well we care for ourselves during our youth and earlier adulthood years results in the level of quality of our lives in our later years.

CHAPTER REVIEW

SUMMARY

1. Aging is the total of all of the changes that eventually lead to functional impairment and the increased probability of death. All tissues in the body do not age at the same rate, nor do people of the same chronological age physiologically age at the same rate.
2. General trends of aging include slowed cellular division, decreased functional capacity, and significant structural changes in the connective tissues. Changes in the connective tissues significantly affect whole body functioning because they are found everywhere in the body.
3. Although there is a decline in speed of response and ability to integrate new information resulting primarily from changes in the CNS, the ability to learn and remember new material does not significantly change with age. This decline in speed and ability is because the conduction rate along neurons and between neurons declines with age.
4. The most significant factor contributing to decreased physical activity among older adults is the decline in the respiratory system's ability to acquire and deliver oxygen to the tissues of the body. The walls separating adjacent alveoli gradually deteriorate, resulting in decreased lung surface area and increased rigidity of the lung walls during aging.
5. Theories about why we age fall into two categories. First is the view that aging is genetically programmed into every cell. Second is the view that aging results from some form of damage or wearing out that eventually results in death. However, there are probably multiple causes of aging rather than a single factor governing the process.
6. Prolonged health and longevity is, to a marked extent, the result of lifelong health habits. If they are positive practices, the probability of healthy aging is high. Psychological health also seems to be important in the quality of the aging process.

FILL-IN-THE-BLANK QUESTIONS

1. The maximum age of the longest lived individuals in a population is called the ______ ______.
2. Cells that no longer divide after birth are called ______ cells.
3. With age, glucose molecules bind to amino acids of ______ fibers eventually forming ______ ______ between the fibers.
4. ______ is the progressive loss of calcium from skeletal bones.
5. Reduction of heart size occurs primarily in the ______ ______ because of decreased physical activity.
6. The lens of the eye changes with age primarily because of continued cellular ______ of the lens.

SHORT-ANSWER QUESTIONS

1. List the reasons why relatively little is known about the aging process.
2. Describe one primary change that occurs with aging in each of the major organ systems.
3. Describe the changes in dietary requirements that should accompany aging.
4. List the factors that appear to promote prolonged health and longevity.

VOCABULARY REVIEW

gerontology—*p. 375*
lipofuscin—*p. 383*
postmitotic cell—*p. 376*
presbycusis—*p. 382*

Chapter Twenty-One

THE HUMAN ENVIRONMENT

For Review

Here are some important terms and concepts that you will encounter in this chapter. If you are not familiar with them, you should review them before proceeding.

Photosynthesis
(page 5)

Biological species
(page 112)

Breeding populations
(page 109)

OBJECTIVES

After reading this chapter you should be able to:

1. *Define ecology.*
2. *Describe the properties common to all ecosystems.*
3. *List the four trophic levels, giving examples of organisms found in each.*
4. *Define pollutant and discuss the three ways in which a substance can act as a pollutant.*
5. *Define sustainable development and describe the ways it can be achieved.*

All biological organisms share several common characteristics. Fundamental to these shared characteristics, however, is that biological species do not live in isolation from one another but instead make up a complex network of interrelationships among other species and their environment. This is called an ecosystem. Disrupting one part of the network destabilizes the whole system in ways we are only now beginning to understand.

Ecology and the Fabric of Life

Perspective, or having a point of view about one's surroundings, is a slippery idea. Imagine it is a warm spring day and you have decided to take a break from studying human biology and go outside to sit under a shady tree. As you sit there you notice may different insects moving on the ground. Ants are busy burrowing into the ground, beetles and grasshoppers are climbing about, stopping to eat a blade of grass now and then, and a spider is diligently weaving a web among the vegetation to ensnare and eat some insects. As you look at the scene more closely, what you first thought to be simply "grass" is actually several different species of grasses. Some varieties appear to consist of tall and thin blades, others are coarse and short, and some form a dense mat of entangled vinelike blades. Again you change your perspective. As you take in more of your surroundings, you see several species of birds flying overhead, some swooping down close to the lawn to catch insects on the fly, while others momentarily alight and then busily burrow into the grass, pulling worms or grubs from the soil and flying away, and some roost in the tree over your head to eat its seeds. Each time you enlarge your view of the environment you notice more life-forms and the interactions between them and their environment. So it is with all life on this planet. Almost any aspect of it viewed up close seems to stand alone and isolated. But as you enlarge your view it becomes apparent that each life-form is one part of a larger, more intricate system.

If we can enlarge our view of things, we also sense that human beings are also bound up, in some fashion, in this web of life. Many questions we have about biology probably stem from our intuitive notion of this system. Why, for example, does the field across the street never seem to need fertilizer or water, but the front lawn does? Why did the barn owls disappear from the farm even though there were plenty of mice for them to eat? Why do our eyes burn when we are stuck in traffic? **Ecology** is the science concerned with understanding the interrelationships among living things and their environments.

This chapter has two related goals. First, we describe the basic organizing principles and factors that shape the structure of all biological populations and the nature of the interrelationships among them and their environments. Second, we apply these principles to human populations to determine the nature of our interactions with the environment and the other organisms that inhabit it. This application is called **human ecology.**

From Populations to Ecosystems

As you learned in Chapters 2 and 3, only a few requirements are necessary for life on this planet. These include organic and inorganic molecules to construct cells and tissues, energy to power them, liquid water to act as a medium or solvent for metabolic reactions, and a moderate temperature in which to carry out these processes. If any of these basic elements is significantly altered, so is life.

Wherever an organism lives, it interacts with its environment, which consists of both living and nonliving elements. The living component of any environment, or **biotic** portion, includes all life-forms in the environment. The **abiotic** portion consists of such things as water, temperature, and soil. Each organism in an environment belongs to a local population, which, as you recall from Chapter 6, is a group of organisms of a single species that interbreed. A **community** includes all the local populations of different species in a particular area. The specific locale of a population in the environment is its **habitat.** A habitat can be described in several ways. For example, ants can be said to occupy a prairie habitat, or more specifically, the sandy topsoil of that prairie. All the biotic and abiotic habitat's components related to an organism's activities are called its **niche** (nitch). The niche occupied by the ants includes such factors as the sandy soil in which they live, the predators, such as spiders, that prey on them, the leaf litter that ants may rely on for food, and the blades of grass that help anchor the soil and modify the climate's effects. Ants also influence their environment in other ways. By digging they help aerate and loosen the soil so plants can establish themselves and water can drain easily through the soil. Ants also remove and break down leaf and other plant litter. They may also be somebody's next meal.

The complex, interrelated network of living organisms and their nonliving surroundings is called an **ecosystem.** An ecosystem might be a pond, a field, the aquarium sitting in the family

CO_2 CO_2 CO_2

room, or the Sahara desert. A group of ecosystems that share a similar climate and support the same types of plants is called a **biome.** Because biomes are determined to a great extent by climate, they cover the earth's land masses as broad bands or stripes. For example, the tropical rain forest biome falls across the equator wherever there is sufficient rainfall. To the north and south lie deserts, which get less than 10 inches of rainfall a year. Grasslands are farther north and south. To the far north and south lie taiga (tie-gah), characterized by great forests of pine, spruce, fir, and hemlock trees. At the poles is the polar biome, marked by continuous cold temperatures, ice, six-month-long nights, and food chains almost exclusively confined to the surrounding sea. Finally the earth's surface and all living organisms on it constitute the **biosphere.**

The interactions of populations within communities making up an ecosystem can be quite complex. However, ecosystems are ordered systems with definable relationships among all of their components. The energy that powers an ecosystem enters through the process of photosynthesis, and the basic chemicals supporting the biotic component of the ecosystem cycle through it to be used over and over again. These two fundamental properties are common to all ecosystems.

Energy Flow

Recall from Chapter 5 that the sun's energy is captured by organisms capable of photosynthesis. In photosynthesis the solar energy is used to combine carbon dioxide and water into simple sugar molecules, which store energy in chemical bonds. Some of this stored energy is then released through the metabolic pathways of cellular respiration to power other chemical reactions, creating other molecules necessary for maintaining the organism.

Photosynthetic organisms are called **autotrophs** (from the Greek words meaning "self-feeders") because they produce food for themselves. In other words, autotrophic organisms are capable of producing their own food from inorganic molecules and an environmental energy source, such as sunlight. The organisms relying on the high energy molecules made by autotrophs are collectively called consumers, or **heterotrophs** (meaning "other-feeders"). Heterotrophic organisms can store energy by synthesizing organic molecules and macromolecules, but they must obtain that energy by eating energy-rich molecules contained in other organisms. Ultimately, that chemical energy comes from autotrophs.

Energy flows through communities in one direction, from the photosynthetic autotrophs through several layers of consumers. Each consumer level is called a **trophic level,** or **feeding level.** Living organisms are classified according to their role in the energy flow through the ecosystem. The energy captured by photosynthetic organisms and made available to other members of the ecosystem is influenced by many environmental factors. These include the quantity of chemicals available to them, the amount of sunlight and water, and the temperature. For example, a desert lacks water. As a consequence, plant production is limited. In contrast, in a large lake, light and chemicals are limited and so is plant life. However, where these resources are abundant, as in a tropical rain forest, plant life is also abundant.

The autotrophs, whether pine trees, algae, or plants, are called **producers** and are the first trophic level of an ecosystem. Some organisms feed directly and exclusively on producers. These organisms, called **herbivores,** ranging from grasshoppers to rabbits to elephants, are **primary consumers** and form the second trophic level. **Carnivores,** meaning "flesh eaters," such as spiders, mice, and cheetahs, usually feed on herbivores. They are **secondary consumers,** making up the third trophic level. Some carnivores, such as hawks, are called **tertiary consumers** and form the fourth trophic level because they also eat other carnivores.

An important component in an ecosystem is the biological equivalent of "sanitary engineers." **Decomposers** are primarily bacteria and fungi that break down organic molecules of dead organisms and their by-products and contribute to recycling both organic and inorganic molecules back to the environment. Ultimately all living matter is reduced to simple molecules and atoms again. For example, carbon dioxide and water return to the atmosphere, and minerals return to the soil (Figure 21-1).

To illustrate the feeding relationships in an ecosystem, it is common to identify a representative of each trophic level that eats the representative of the trophic level below it. This linear feeding relationship between trophic levels is called a **food chain.** However, communities rarely contain well-defined groups of primary, secondary, and tertiary consumers. For example, some animals, such as raccoons, rats, bears, and humans, are **omnivorous** ("all eating"), which means they are able to live on both animal and plant foods. Therefore, depending on circumstances, omnivores may be primary, secondary, and occasionally tertiary consumers. An owl, for example, is a secondary consumer when it eats a mouse but a tertiary consumer when it eats a shrew, which feeds on insects. What trophic level do you occupy when eating a hamburger, fries, and a milk shake? A **food web** describes the *actual* feeding relationships in a given community (Figure 21-2).

Figure 21-1 Trophic levels

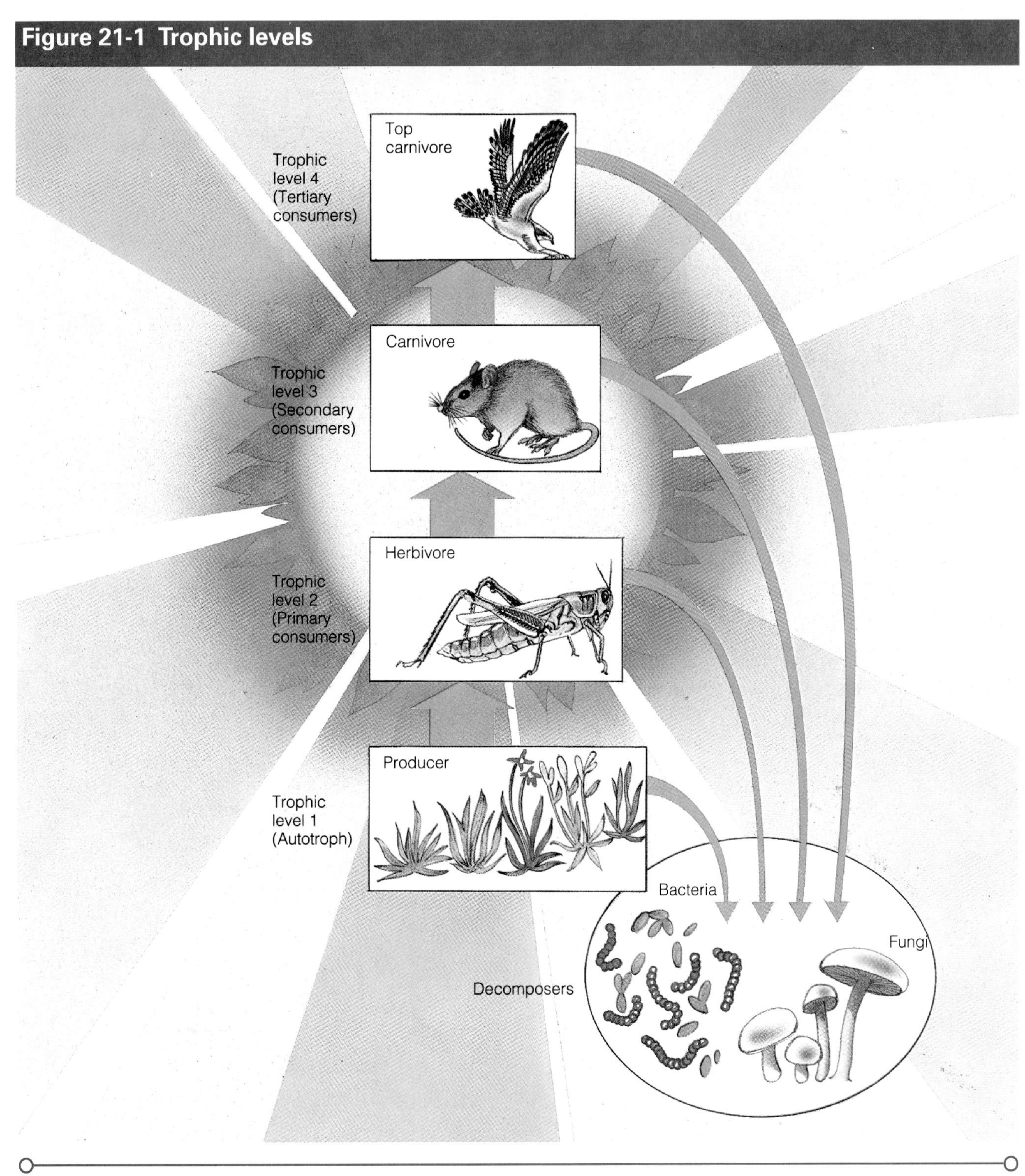

Autotrophs, such as plants, obtain their energy from the sun. Organisms that eat plants occupy trophic level 2. Organisms that eat primary consumers occupy trophic level 3, and organisms that eat secondary consumers are in trophic level 4. Decomposers are found at all trophic levels.

All organisms take in food to survive and maintain themselves. As discussed in Chapters 3 and 16, the chemical bonds of complex molecules that organisms ingest are hydrolyzed, and the energy released is used by the organism's cells. Also, in Chapter 3 you learned that according to the second law of thermodynamics, energy use is never completely efficient. For example, as an automobile engine converts the energy stored in the chemical bonds of gasoline molecules to the mechanical ener-

Figure 21-2 Food webs

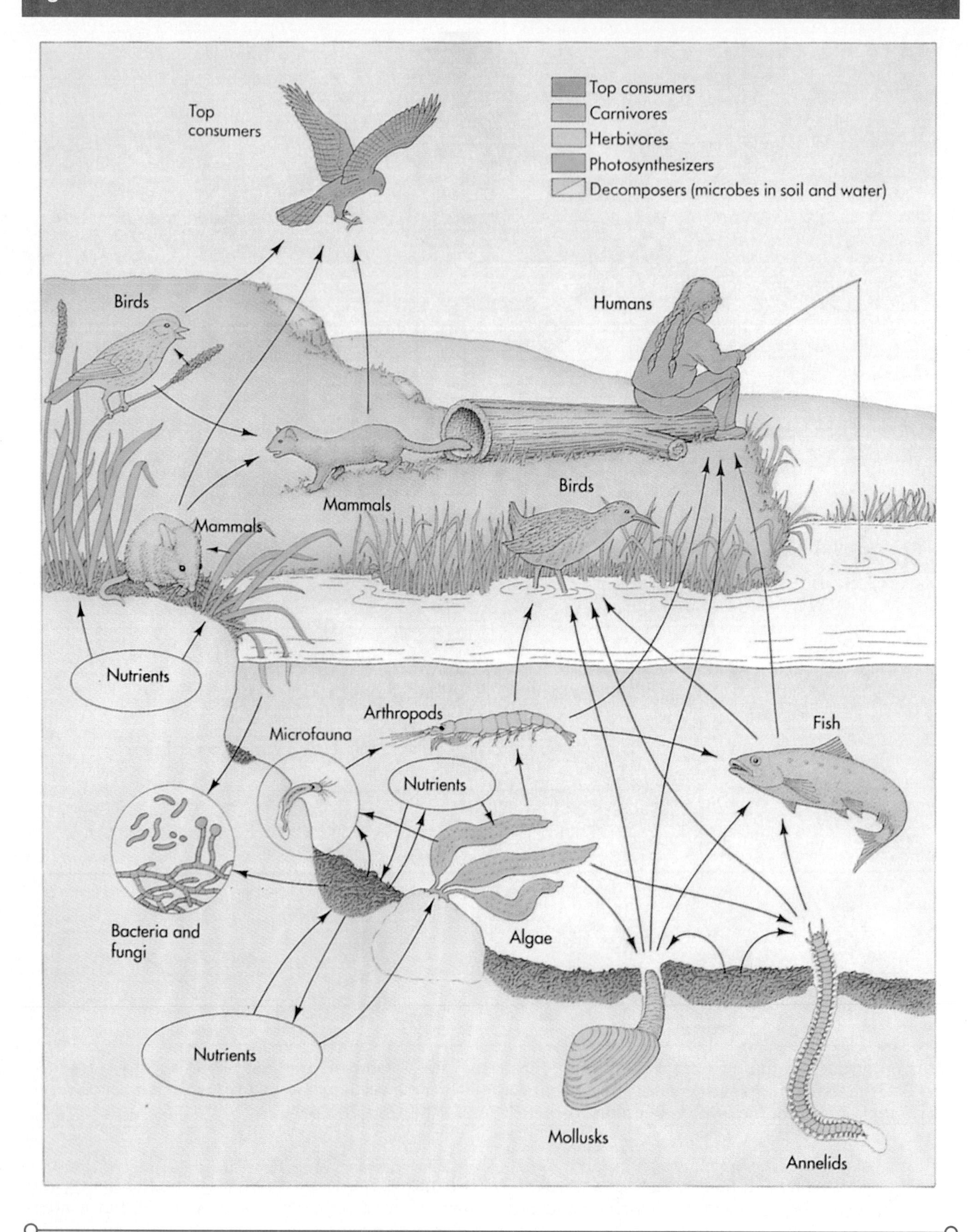

Food webs incorporate the trophic level hierarchies of food chains but also include actual feeding relationships between populations in the ecosystem.

gy of movement, about 75% of the energy is immediately lost to the environment as heat. This is also true in organisms. Anytime you exercise, some of the energy of muscular contraction is lost as heat. Some energy is also lost during a seed's germination or a sperm's whiplike movement.

The transfer of energy from one trophic level to the next is also inefficient or incomplete. When a grasshopper feeds on leaves, only some of the solar energy originally captured by the plant is available to the insect. Some energy was used by the plant to grow and maintain itself. Some was converted into the chemical bonds of molecules, such as cellulose, which the grasshopper cannot break down. Some was released as heat. Thus only a fraction of the energy captured by the first trophic level, the producer, is ever available to the primary consumer in the second trophic level. The energy consumed by the grasshopper is partially used to power its hopping from leaf to leaf, mouth movements, and head turning. Some energy is used to construct its indigestible exoskeleton, and a great deal is liberated as heat. All of this energy is unavailable to the bird, occupying the third trophic level, that eats the grasshopper. The bird uses much of the energy it gets from the grasshopper in maintaining itself, such as manufacturing indigestible feathers and bone. A great deal of energy is also expended as body heat, flight, and other bird behaviors. All of this energy will be unavailable to the hawk that catches the bird.

The average **net** or **actual energy transfer** between trophic levels is approximately 10%. This means the energy stored in primary consumers (for example, the grasshopper) represents only about 10% of the energy stored in the primary producers (for example, the leaves). The bird's body possesses about 10% of the energy stored in the grasshopper, the primary consumer. Generally, for every 100 Kcal of solar energy captured by producers, only about 10 Kcal is converted into primary consumers. As a result only 1 Kcal is converted into secondary consumers. The most numerous animals will be those feeding on plants, whereas there will always be relatively fewer animals such as carnivores.

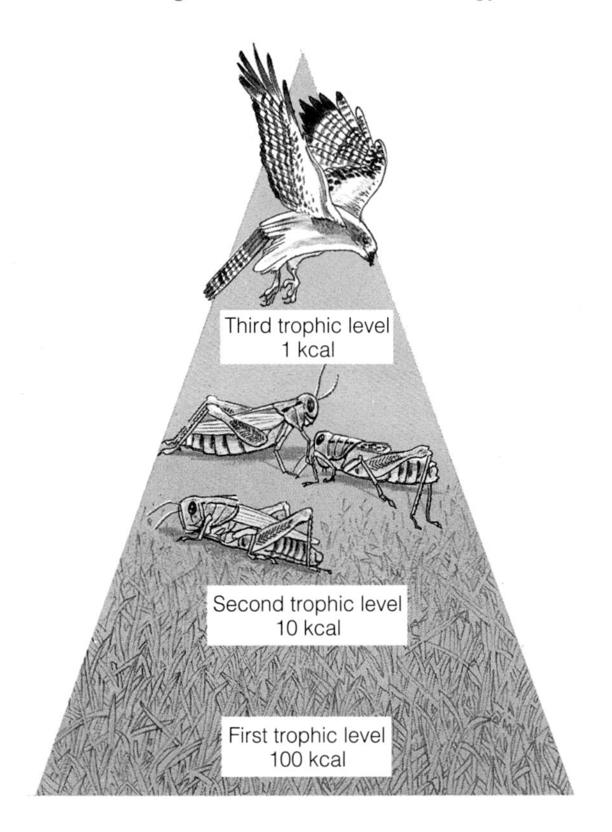

A side effect of energy transfer through an ecosystem is the concentration of toxic substances in the food chain, **biological magnification** or **bioaccumulation.** Some toxic substances cannot be broken down by bacteria and other decomposers. Therefore when other organisms incorporate them, they are not excreted. Once they enter the food chain they become more concentrated at each trophic level. For example, for many years the insecticide DDT was sprayed on the marshes of the mid-Atlantic state coastlines to control mosquitoes. Instead of being washed out to sea and diluted, the water-insoluble pesticide accumulated on the decomposing organic matter in the marshes. It was picked up by microscopically small life-forms, such as single-celled diatoms and small shrimp, collectively called **plankton.** Because fish such as minnows must eat many shrimp each day, they accumulated all the DDT from thousands of shrimp they ate during their lifetimes. Predatory fish such as pickerel consume many minnows daily, thus acquiring all the DDT each minnow had accumulated. Finally many water birds feed on the fish and receive large doses of DDT, causing effects ranging from death to producing eggs with thin shells that are usually crushed by the weight of the parent during the nesting season (Table 21-1).

Table 21-1 Biological Magnification of DDT on Long Island

FOUND IN:	DDT (PARTS PER MILLION)
Water	0.00005
Plankton	0.04
Sheepshead minnow	0.94
Pickerel	1.33
Needlefish	2.07
Merganser duck	22.8
Cormorant	26.4

Figure 21-3 The carbon cycle

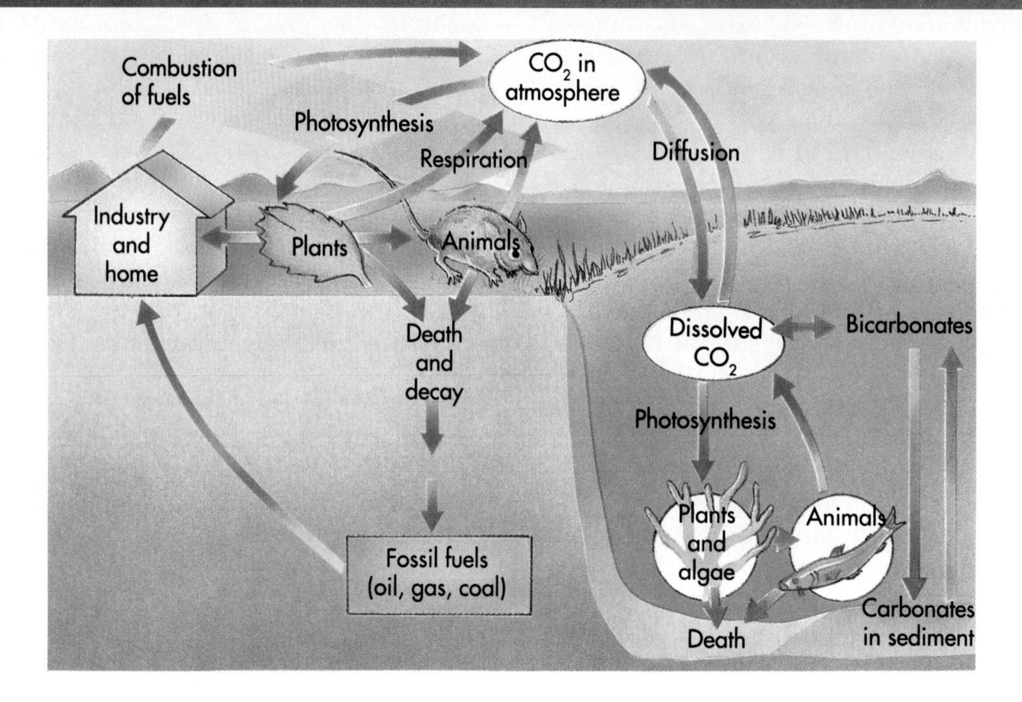

Carbon cycles back and forth from the abiotic and biotic portions of the environment. Two reservoirs for carbon are the atmosphere, where it combines with oxygen to form CO_2, and the oceans, where it exists as bicarbonate ions (HCO_3^-).

Chemical Cycling

In contrast to sunlight energy, chemicals do not flow down to the earth in a continuous stream from above. **Chemical cycles,** also called **nutrient cycles,** describe the pathways chemicals take as they move back and forth from the biotic to abiotic ecosystem portions. Generally the same pool of chemicals has been supporting life since its first appearance over 3 billion years ago. Some of these atoms and simple molecules used by life-forms, such as water, carbon, oxygen, nitrogen, phosphorus, and calcium, are needed in relatively large quantities. Others, primarily minerals, such as zinc, lead, and iodine, are required only in small amounts. In the next section we describe the nutrient cycles of carbon, nitrogen, and water.

Carbon Cycle

The two major reservoirs for carbon dioxide are in the atmosphere, where it comprises about 0.03% of the total gases, and the oceans, where it is dissolved in water. Recall from Chapter 2 that carbon atoms form the backbones or skeletons of all organic molecules. The carbon available to the first trophic level of living organisms is in combination with oxygen as carbon dioxide (CO_2). Carbon enters food webs through producers, such as plants, which take in CO_2 from the atmosphere during photosynthesis. Some oxygen molecules and CO_2 are returned to the atmosphere through the cellular respiration of producers at night, when photosynthesis is not possible (Figure 21-3).

Primary consumers take in plant material, respire some CO_2 back into the environment, and incorporate some carbon into their bodies. Ultimately, all living things are consumed by decomposers. Most of the carbon thus returns to the atmosphere as CO_2, where it can be recycled. However, some carbon cycles much more slowly. Fossil fuels are the remains of ancient plants and animals. The carbon in their organic molecules remains in these deposits. Under the high temperature and pressure created by geological forces over time, these deposits are transformed into coal, oil, and natural gas. When these substances are burned, the prehistoric sunlight of the chemical bonds is released as energy. Some of this trapped carbon is returned to the atmosphere as CO_2.

The carbon cycle also occurs in aquatic environments. Carbon dioxide dissolved in water produces bicarbonate ions (HCO_3^-). Bicarbonate ions are the source of carbon for algae and many other aquatic producers. The producers eventually become food for others. Also, when aquatic organisms respire, the CO_2 they release into the water becomes HCO_3^- in the water.

Nitrogen Cycle

Nitrogen is necessary for the production of amino acids, nucleic acids, and some vitamins. The atmosphere is approximately 79% nitrogen gas (N_2), but most organisms cannot use this gas directly. Plants, for example, must be supplied with nitrogen in the form of **nitrates** (NO_3^-) or **ammonia** (NH_3). How is atmospheric nitrogen converted into these usable molecules? Ammonia is synthesized by two types of bacteria. One lives in both soil and water, whereas the other lives in special swellings, called nodules, on the roots of plants called **legumes** (li′-gyums). These include soybeans, peas, and clover. Certain types of bacteria can also convert ammonia to nitrates.

Plants incorporate the nitrogen from ammonia and nitrates into amino acids, nucleic acids, and vitamins. These nitrogen-containing molecules from plants are eventually eaten, either by primary consumers or by decomposers. As the nitrogen is passed through the food web, some of it is released as waste products and is also bound up in dead bodies. Decomposing bacteria convert it back to ammonia and nitrates, which will be recycled via plants (Figure 21-4).

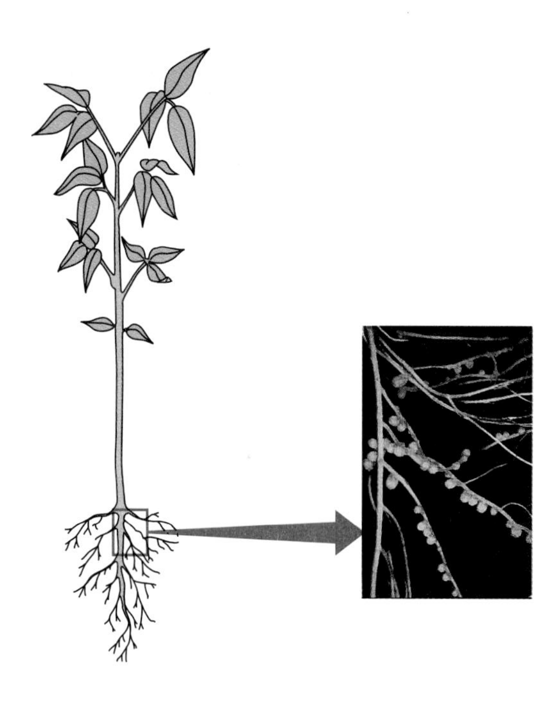

Figure 21-4 Nitrogen cycle

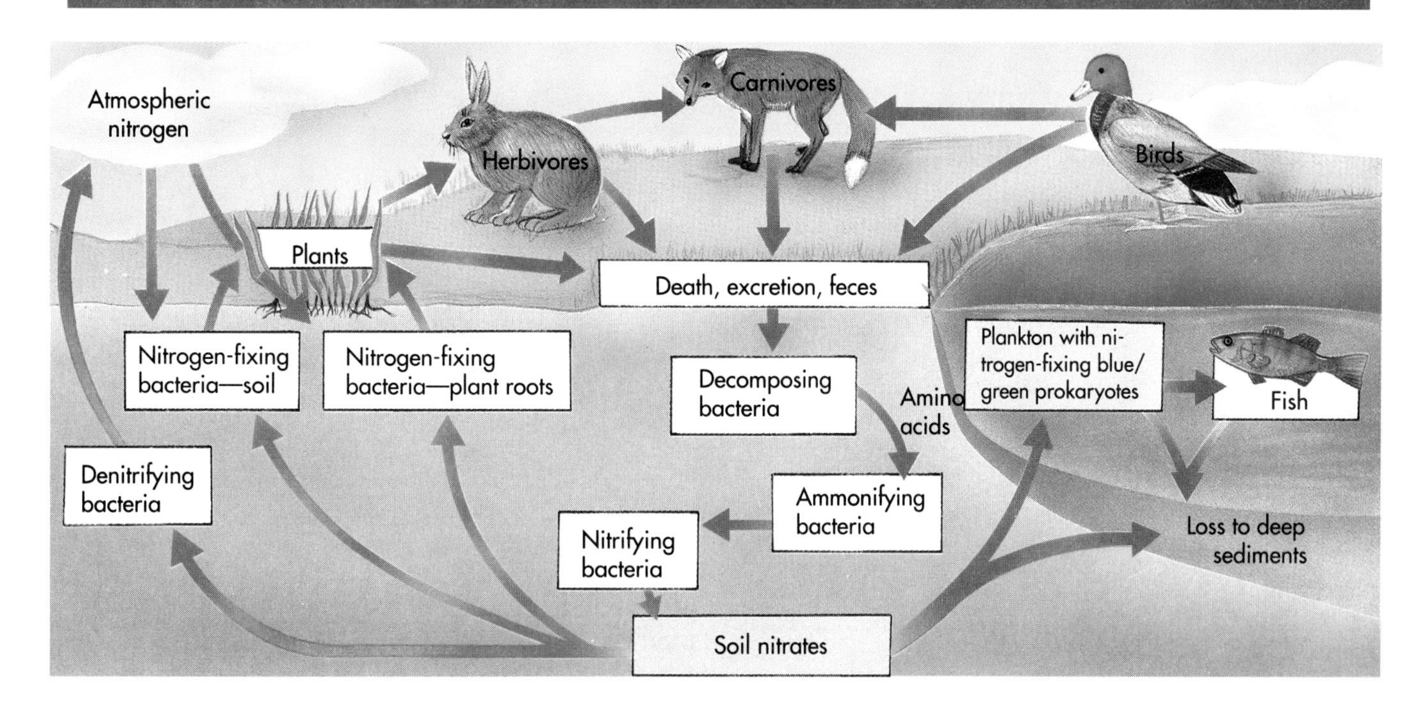

Nitrogen must be fixed in molecules of ammonia or nitrate compounds by bacteria before it can become available to the food web.

Figure 21-5 Water cycle

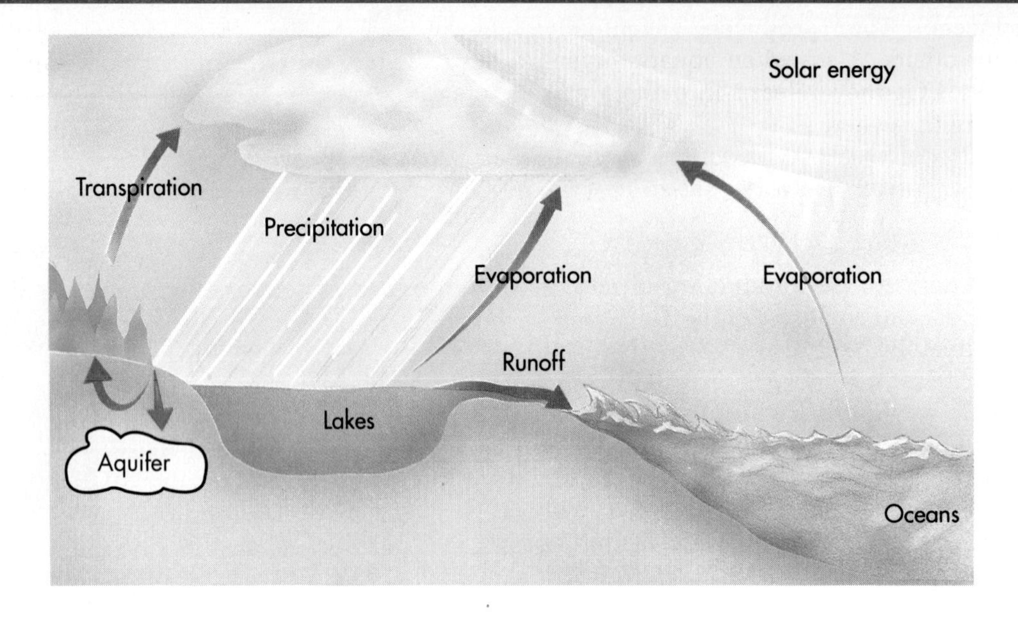

The oceans of the world are the primary reservoirs of water. Through the processes of evaporation and transpiration, water vapor rises into the atmosphere, condenses, and falls back to the earth as precipitation.

Water Cycle

The main pool of available water is the world's oceans. Water evaporates from the surface of the ocean and other bodies of water, from the surface of the soil, and from vegetation. Through the process of transpiration, water vapor is given off by vegetation. As the air carrying this water vapor rises, it begins to cool, causing the vapor to condense out of the air and form clouds. The water eventually returns to the earth's surface as precipitation. Some water runs off the land and back into streams, lakes, and oceans. Some water returning to the earth seeps into the ground to replenish vast underground reservoirs of water, called **aquifers,** found in highly porous rock. This temporarily stored water eventually rises to the earth's surface, where it again becomes available to living organisms (Figure 21-5).

Having had this description of ecosystem organization, we can now address how and in what ways populations are able to change through time.

Population Dynamics

Populations of organisms that constitute an undisturbed ecosystem tend to remain relatively stable over time. Yet, under certain conditions, populations can rapidly increase. Here we first examine how and why populations grow, then we look at the factors normally keeping their numbers in check.

Population growth can be summarized by a sweeping line on a graph called a **sigmoid** or **S-shaped curve.** For example, we could put a few

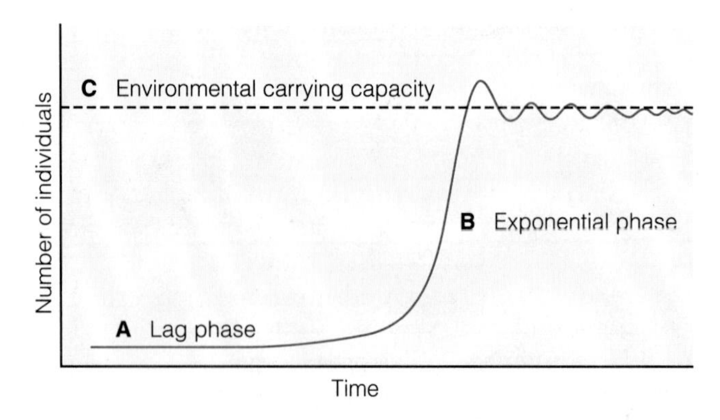

goats on an isolated island containing all the necessities of life. If the sigmoid curve represents population growth, the first, flat portion of the curve, labeled A and called the **lag phase,** represents a period of slow growth. This is followed by a portion of the curve, labeled B, that shoots up over a relatively short time or distance. The second portion of the curve represents a rapid increase in individuals and is called **exponential growth.** For example, if each goat in the original group leaves two offspring, the population will remain fairly small for the first few generations. But after a few more generations the population begins to increase rapidly. Exponential growth means the *rate* of population increase is doubling from generation to generation. Thus we see the ever-accelerating, upward-projecting second portion of the curve. In the last section of the graph, labeled C, population growth slows down and stabilizes, fluctuating around the maximum number of individuals the environment will hold. This is called the **carrying capacity.**

Two general factors determine the carrying capacity of an environment. These are **nonrenewable resources,** such as space, and **renewable resources,** primarily sunlight, water, and chemicals or nutrients. If, for example, space requirements are exceeded, many individuals may leave the area, but go to less desirable areas where their numbers decrease. Because these individuals may not find adequate breeding sites, reproduction declines. Similarly, if demands on renewable resources, such as food or water, are too high, organisms starve. In either case the population does not grow.

Now we can apply the concepts of population dynamics to world human population growth and examine the nature of human interaction and its impact on our ecosystems.

Human Population Growth

Humans did not cultivate plants until about 10,000 to 12,000 years ago. Until then, humans lived by hunting and foraging, using naturally occurring resources as they appeared in the environment. For example, if particular fruits or nuts were in season, they were gathered. If animals were migrating through a local area, they were hunted. Using what we know about the efficiency of such technology, and assuming that the earth's approximate land area able to sustain humans is about 58 million square miles, we can calculate that 12,000 years ago the earth would have supported no more than about 5 million people.

To put this into perspective, about 99% of our human evolution has been spent as hunters and foragers. The last 1% has been spent as cultivators and later as industrialized people. But during this later period the exponential increase in humans has had a tremendous impact on the environment. Currently there are over 5.5 billion humans on the planet.

The overall growth history of the world's human population shows two phases typical of living organisms. First is an early phase up through the Middle Ages, when population growth was slow and gradual. Second is a phase of exponential growth, from the end of the Middle Ages to the present. To explain this pattern of growth, we need to look at factors affecting birth and death rates.

The gradual increase in humans during the early phase of world population growth had two main causes. First, humans developed tools. Tool use increased hunting and foraging efficiency, resulting in an improved diet. This decreased death rates somewhat. The second cause came between about 10,000 and 12,000 years ago, with plant and animal cultivation. Cultivating plants and storing grain, instead of relying on wild foods as they occur in hunting and foraging economies, also decreased death rates. However, overall population growth was still kept in check by high infant mortality rates, disease, and periodic famine.

The exponential growth phase in the human population that began toward the end of the Middle Ages became apparent by the mid-eighteenth century. Several factors contributed to this accelerated growth. First were technological improvements in agriculture. These included improved seed strains resulting in higher yields, more efficient irrigation techniques, and crop rotation, which slowed down nutrient depletion of the soil or actually improved the soil. Second were improvements in sanitary conditions and gains in medical, scientific, and technological knowledge. In combination these two factors dramatically reduced death rates, including infant mortality rates, and modestly increased birth rates. This pushed world human population growth rates into an exponential phase (Figure 21-6). When the carrying capacity will be reached and human population growth will stabilize is the topic of an ongoing debate. However, given our present rates of growth and the impact of human activities on the environment, it is probably going to be within the next several human generations. In the following section we describe the impact our increasing numbers and activities have on the ecosystems we inhabit.

Human Impact on the Environment

A **pollutant** is any substance that enters the environment and interferes with the natural cycles in communities and ecosystems. Substances can act as pollutants in one of three ways. First are substances found naturally in ecosystems, but produced in such quantities they flood or overwhelm an ecosystem's ability to respond. For example, CO_2 is a naturally occurring by-product of cellular respiration. However, burning gasoline, oil, and natural gas produces CO_2 in such vast quantities that it is significantly increasing in the atmosphere and becoming a serious ecological problem. Other natural pollutants include sulfur dioxide, animal waste products, and soil sediments washed from farmland into natural waterways. Second are naturally occurring substances that do not exist in significant amounts in

Figure 21-6 World human population growth curve

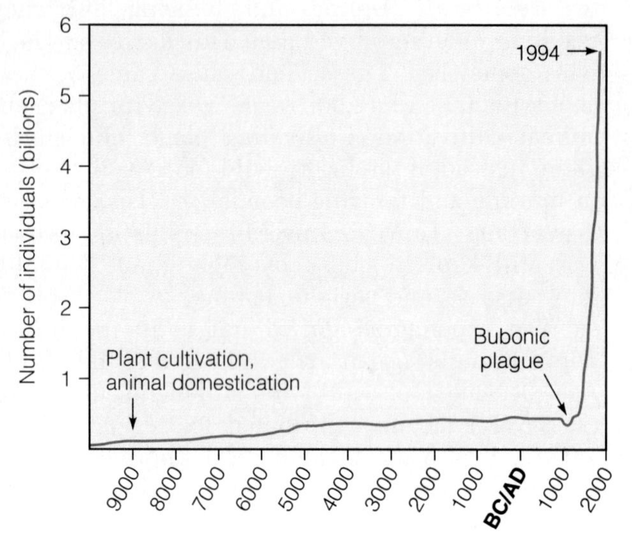

Time	Estimated world population	Doubling time (years)
8000 BC	5,000,000	1,500
AD 1650	500,000,000	200
1850	1,000,000,000	80
1930	2,000,000,000	45
1975	4,000,000,000	35
2025	8,000,000,000?	?

The vertical axis represents world population in billions. The slight dip between 1347 and 1351 shows when approximately 75 million people died in Europe as a result of the bubonic plague, or black death. At present growth rates, another billion will be added over the next 12-year period, between 1987 and 1999.

the environment but are normally bound up with other substances or confined in ecosystems. Human activities have freed or separated these substances, in some cases upsetting ecosystems. Arsenic, oil, lead, and mercury are some examples. Third are synthetic compounds such as pesticides, created by human activities, that accumulate in the ecosystems because organisms have not had sufficient time to evolve ways to metabolize or use them. Today most serious pollutants in ecosystems are naturally occurring substances produced in excess by human activities. We describe three of these pollutants—CO_2, sulfur dioxide, and nutrient runoff—and some of their environmental effects in the next section.

From approximately 345 to 280 million years ago, due to a unique series of environmental conditions, vast quantities of carbon were diverted from the carbon cycle when bodies of plants and animals were buried in sediments, temporarily escaping further decomposition. Through heat and pressure created by geological processes, these organic bodies were converted to fossil fuels, such as coal, oil, and natural gas. With the advent of industrialized technologies during the eighteenth century, human cultures have increasingly relied on the energy stored in these fuels. As these fuels are burned in power plants, factories, and automobiles, many chemicals are released into the atmosphere, one of which is CO_2. Another indirect source of added atmospheric CO_2 is through global deforestation, the cutting and processing of millions of acres of forests each year. This is occurring mainly in the tropical rain forests in attempts to increase agricultural land for growing populations. Recall that plants take in CO_2 and give off oxygen. In addition to this, much of the carbon stored in trees eventually returns to the atmosphere after they are cut, through either burning or decomposition. Both the burning of fossil fuels and global deforestation have caused about a 13% increase in the CO_2 content of our atmosphere since 1860.

Carbon dioxide has a potentially important environmental effect that makes its increasing atmospheric levels a cause for concern. It can increase global atmospheric temperatures. Atmospheric warming caused by CO_2 is called the **greenhouse effect** because it acts in a way similar to the glass in a greenhouse. Solar radiation passes through the greenhouse windows. Some of this energy is reflected off of the interior surfaces as longer infrared waves, or heat. These longer waves do not pass back through the glass, but are trapped inside the greenhouse, raising the interior temperature significantly. Similarly, solar radiation passes through the atmosphere and eventually hits the

Figure 21-7 The greenhouse effect

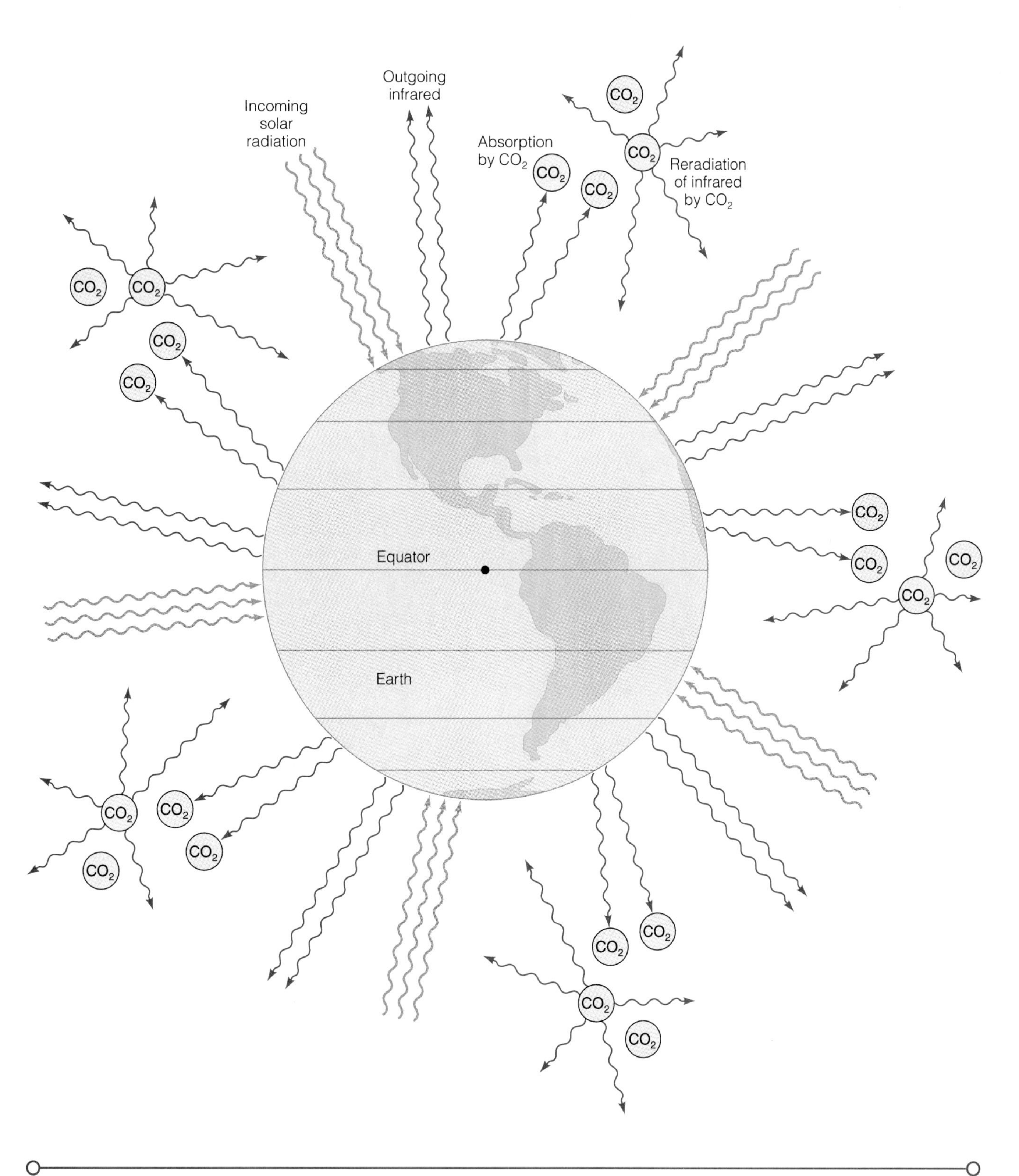

Incoming solar radiation passes through the atmosphere to reach the earth's surface. Some radiation is reflected back as longer waves, or heat. Some outgoing infrared waves hit CO_2 molecules in the atmosphere and are absorbed. The CO_2 reradiates the infrared waves, but on average, about one half of the waves are reradiated back into the atmosphere. The net effect is the more CO_2 in the atmosphere, the more heat kept in the atmosphere, resulting in global warming.

earth's surface. Some of this solar energy is reradiated as longer infrared waves that normally go into space. However, if an infrared wave hits a CO_2 molecule, it is absorbed and later reradiated away. The probability is about 50% that the released infrared wave will travel out toward space. However, it may be radiated back toward the earth. The overall effect is that CO_2 keeps heat from escaping the atmosphere (Figure 21-7).

A potentially serious consequence of this warming trend will be a shift in the global distribution of temperature range and rainfall amount. In the 1980s we have already had the four hottest years on record. Although it is difficult to predict the exact timing and magnitude of these changes, it is certain that your grandchildren may experience firsthand the consequences scientists are generally predicting today.

Coal burning, primarily by power plants to produce electricity, has released large quantities of sulfur dioxide into the atmosphere. Sulfur dioxide combines with water and oxygen in the atmosphere to produce sulfuric acid as follows:

$$\underset{\text{sulfur dioxide}}{2SO_2} + \underset{\text{water}}{2H_2O} + \underset{\text{oxygen}}{O_2} = \underset{\text{sulfuric acid}}{2H_2SO_4}$$

This in turn produces **acid rain,** defined as rainwater with a pH below 5.6. When acid rain falls into rivers and streams, it also lowers their pH. This kills fish and amphibian eggs, eventually eliminating those organisms from their aquatic habitat. Acid rain also removes minerals from the soil, which then run into waterways and can be highly toxic to mature fish.

We can minimize acid rain. Low-sulfur coal could be substituted for the high-sulfur coal presently being used. Also, devices called scrubbers could be installed in tall smoke stacks to remove most of the sulfur dioxide before it enters the air.

Among the greatest environmental problems facing humans today is water pollution. Both organic and inorganic sources contribute to water contamination. Normally, waterways are low in dissolved nutrients, which limits the growth of microscopic organisms that act as decomposers. Although water discharged from a sewage treatment plant is free of solid wastes, it is rich in dissolved nutrients such as nitrates and phosphates. Other sources of nutrient-rich water also wash off fertilized farm fields and manure-covered feedlots into lakes and streams. Adding these nutrients to the waterways eventually causes a sudden, explosive growth of algae and other photosynthetic organisms. The process of adding nutrients to a body of water is called **eutrophication** (u-trof'i-ka'shun), meaning "feeding well." The algae cloud up the water, preventing sunlight from reaching aquatic vegetation below. The plants and the organisms that feed on them die faster than they are consumed, and their bodies are decomposed by bacteria. The bacterial decomposition process requires oxygen, thus depleting the oxygen dissolved in the water. Deprived of oxygen, fish and other organisms also die. This fuels more of decomposer bacteria growth.

Possibly the most dangerous water pollutants are inorganic sources because many are so new that almost nothing is known of their long-term effects. For example, in 1950 we discovered that nitrates from commercial fertilizers in water supplies of many American cities could alter human hemoglobin, impairing its oxygen-carrying capacity. More recently we learned that certain bacteria found in the human digestive tract can also convert these ingested nitrates to carcinogenic nitrites.

A Possible Solution: Sustainable Development

Although serious damage is occurring to every ecosystem on the earth as a result, either directly or indirectly, of human activity, there are still ways to slow down and even reverse the damage. In recent years scientists have found ways that human populations around the world can maintain and even increase their quality of life without necessarily damaging the environment.

Sustainable development means using renewable natural resources in a manner that does not eliminate or degrade them or diminish their renewable usefulness for future generations. However, to implement sustainable development the United Nations Human Development Report of 1991 states that economic, human, environmental, and technological changes *must* also occur.

Economic Change

Currently, on a *per capita* basis, inhabitants of industrialized countries, such as the United States, use many times more of the world's natural resources than do inhabitants of less industrialized countries. For example, energy consumption from fossil fuels is 33 times higher per person in the United States than in India. For industrialized countries, sustainable development means steady reductions in wasteful energy and resource consumption through increased efficiency and behavioral changes. Economic assumptions based on the increasing consumption of goods and services must be changed to those based on steady-state consump-

Biology in Action CARS OF THE FUTURE

Stricter clean air laws and regulations in the U.S. are spurring a renewed push for the commercial development of electric-powered vehicles.

In California the Los Angeles Department of Water and Power and Southern California Edison are providing $17 million in an effort to put 10,000 electric vehicles on Los Angeles roads by 1995. From over 200 responses to its invitation for proposals, two companies were selected to produce electric vehicles. Clean Air Transport, based in Great Britain and Sweden, will produce four-passenger cars with a range of 150 miles. Unique Mobility, Inc., of Englewood, Colorado, will produce a midsized van with a range of 120 miles. As many as 15 other U.S. states are also considering rules that would require the availability of zero-emission (electric) vehicles in the 1990s.

Many automobile manufacturers are also moving rapidly to develop electric vehicles. Eight Japanese companies have developed intensive electric vehicle programs and have introduced working models. Many European companies are also developing electric vehicles. For example, BMW introduced a new four-passenger electric car with a range of 150 miles at the 1991 Frankfurt Motor Show.

Although 120 to 150 miles between battery recharges may not sound as if owning and operating an electric vehicle would be convenient, let us consider some aspects of how we currently use our automobiles.

About 85% of all automobile trips taken by U.S. drivers are within short distances of the home. Even more revealing is that 97% of the miles traveled by Americans in their cars are for trips of 12 miles or less. Considering the distances we actually travel, a 150-mile range between battery rechargings is adequate for most travel requirements.

The principal technical problem in developing electric vehicles is producing inexpensive batteries with lower weight and faster recharge times. As a result of its own research, General Motors announced during the summer of 1992 that these problems should be solved by 1997. With these problems well on their way to being solved, gasoline-powered automobiles may soon be history.

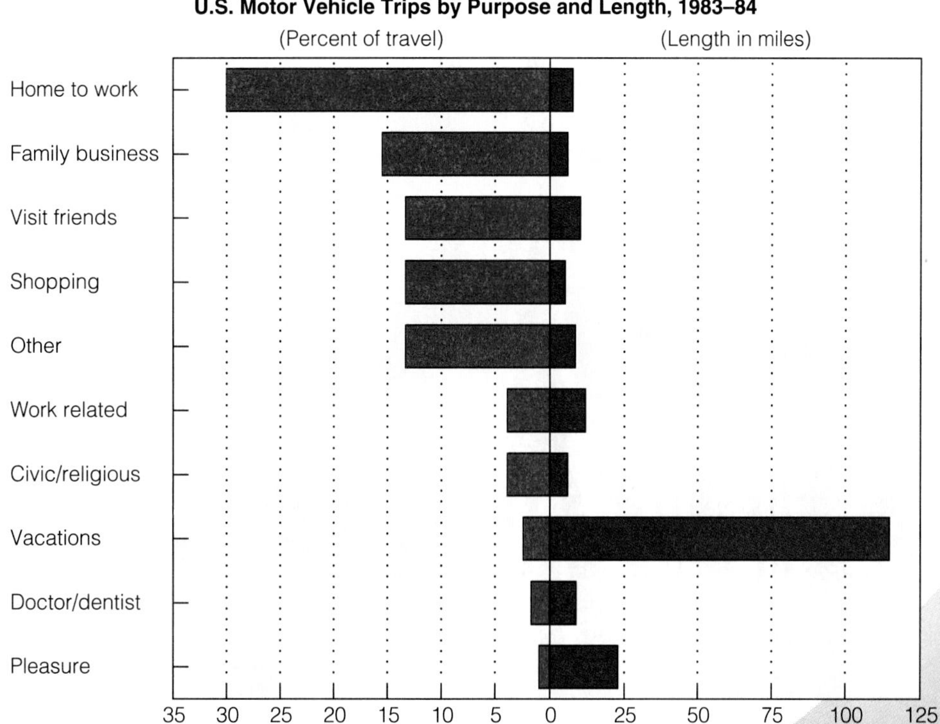

Source: James J. MacKenzie, "Toward a Sustainable Energy Future: The Critical Role of Rational Energy Pricing," in *WRI Issues and Ideas* (World Resources Institute, Washington, D.C., May 1991), p. 8.

tion of goods. For less industrialized countries this means adopting the most energy-efficient technologies and means of goods and services distribution available.

Changes in Human Population Numbers

Sustainable development means significant progress toward stabilizing human population growth. This means that continued growth at anything like current rates cannot continue for much longer. How much longer is still in debate, but most agree that time is running out.

Environment

Natural resources needed for food production and cooking fuels, from soil to forest to fisheries, must be protected, even though demand is increasing because of human population growth. These are potentially conflicting goals, but failing to conserve the natural resources on which food production depends would guarantee future food shortages.

Technological Change

Industry and manufacturing have often polluted the atmosphere, water, and land. In industrialized countries pollution control is accomplished at great expense. In many less industrialized countries pollutants are not controlled at all. However, pollution is not an inevitable consequence of industrial activity. Pollution results from using inefficient technologies or wasteful processes. Sustainable development means changing to technologies that are more efficient and less polluting, minimizing energy and natural resource consumption.

What is demanded of all people is a fundamental change in the way we view ourselves in relation to nature. We can no longer consider our natural environment an unlimited reservoir of raw materials that can be used without regard to potential consequences. Reaching the goal of sustainable development requires simultaneous progress along all of the four dimensions just discussed. These dimensions are closely linked and influence one another. For example, populations that are stabilized or reduced ease the pressure on using natural resources. Technological changes in industrialized countries could reduce pollution and resource use. Rapid technological change would also be needed in industrializing countries, so they can avoid repeating the mistakes and multiplying the environmental damage of industrialized countries.

In this chapter you have learned more about the complexities of the earth's living fabric. Each of our actions has consequences. Some of these are not immediately predictable or obvious. With the advantage and perspective of our knowledge comes greater responsibility. What past generations did to the fragile fabric of life out of ignorance, we can no longer continue to do. However, with the world's population continuing to increase at an exponential rate, humane and rational long-term decisions regarding the environment need to be made. No matter what the decisions might be, it is clear we can no longer maintain a way of life that mortgages the future of this planet.

CHAPTER REVIEW

SUMMARY

1. The study of the interrelationships among living organisms and their environments is called ecology. The complex interrelated network of living things and their nonliving surroundings is called an ecosystem. Ecosystems consist of communities, which in turn consist of local populations of different species.
2. Two properties organize ecosystems. First, the energy that powers them enters through the process of photosynthesis. Second, the basic chemicals or nutrients supporting ecosystems cycle through them to be used over and over again. Generally, energy flows through an ecosystem in one direction while chemicals cycle through them.
3. Living things are classified into trophic levels according to their role in the energy flow through an ecosystem. Energy from photosynthesizing organisms moves through trophic levels because all organisms either directly or indirectly consume them. The feeding web describes the actual feeding relationships between individuals in an ecosystem.
4. The actual net energy transfer between trophic levels is only about 10%. One by-product of energy transfer is biological magnification, or increased concentrations of toxic substances as they move through trophic levels.

5. Ultimately most of the carbon atoms forming the backbones of all organic molecules come from the atmosphere as carbon dioxide and enter the ecosystem by means of photosynthesis. Chemicals cycle between the biotic portion of an ecosystem and the abiotic portion, which acts as a reservoir.
6. Population growth of all organisms is described by a sigmoid curve. This includes a relatively flat portion representing slow growth, a steep upward curve indicating exponential growth, followed by either a flat or a slightly oscillating curve representing a steady state of no growth. Population growth is controlled by the availability of both renewable and nonrenewable environmental resources.
7. A pollutant is any substance that enters the environment and interferes with an ecosystem's natural cycles. At present most of the serious pollutants are also found naturally in ecosystems but are produced in excess by human activities. The potential disruptive effects of most inorganic substances on the environment remain unknown.

FILL-IN-THE-BLANK QUESTIONS

1. The living component of an environment is called the ______ portion, whereas the ______ portion of the environment consists of nonliving matter.
2. The populations of all species in a particular area make up a(n) ______.
3. The specific locale of a population is called a ______.
4. Energy enters the food chain through ______ organisms, sometimes called producers.
5. Because of inefficient energy transfer, the actual energy transferred between trophic levels is only ______.
6. The primary reservoir for carbon is the ______, where it exists in combination with oxygen as ______ ______.
7. Atmospheric nitrogen is fixed by ______ in both soil and water.
8. The population growth of all biological organisms can be described by a(n) ______ ______ or an S-shaped curve.
9. The maximum number of individuals that can be sustained over time is called the ______ ______ of the ecosystem.
10. A(n) ______ is any substance that enters the environment and interferes with its natural cycles.

SHORT-ANSWER QUESTIONS

1. Compare and contrast a food chain and a food web.
2. What is biological magnification?
3. Describe the changes nitrogen undergoes, beginning with its fixation and ending with its release by decomposers.
4. List the three main parts of the sigmoid growth curve and explain what each represents in describing population growth.
5. List three different renewable and nonrenewable resources in your environment.
6. Describe the greenhouse effect.
7. Define eutrophication and describe its impact on natural waterways.

VOCABULARY REVIEW

carnivore—*p. 388*
carrying capacity—*p. 395*
decomposer—*p. 388*
ecology—*p. 387*
heterotroph—*p. 388*
niche—*p. 387*
omnivorous—*p. 388*
pollutant—*p. 395*
producer—*p. 388*
sustainable development—*p. 398*
tertiary consumer—*p. 388*
trophic level—*p. 388*

Chapter Twenty-Two

HUMAN VARIABILITY

For Review

Here are some important terms and concepts that you will encounter in this chapter. If you are not familiar with them, you should review them before proceeding.

ABO and Rh blood system (page 90)
Alleles (page 73)
Continuous variation (page 96)
Genotype and phenotype (page 73)
Breeding population (page 109)
Polygenic inheritance (page 96)

OBJECTIVES

After reading this chapter you should be able to:

1. *Differentiate between traits of simple inheritance and traits of complex inheritance and give examples of each.*
2. *Define clinal distribution and its relationship to human skin-color variation.*
3. *Discuss the vitamin D hypothesis, explaining its importance in skin-color variation and natural selection.*
4. *Define race in terms of human variability.*

Humans differ from one another in many easily observed characteristics. Height, weight, hair color and form, and eye and skin color are a few examples. Human populations from different geographic areas of the world also differ. We need to be able to document the range of human variability and interpret differences among individuals and populations objectively.

Human Variability

Biological variation exists in every living species, including humans. Without variation, adaptation to environmental change is not possible. Without adaptation, a species cannot persist over time. We are clearly aware of individual human variability. It is as plain as the nose on your face, and her face, and his face, too. We differ from one another in many easily observed characteristics. Most of us have no difficulty picking out the face of a friend or a relative from a crowd. "Invisible variability" is also common. Most of us can recognize a friend's voice on the other end of the telephone or calling to us from the next room.

We are also aware that human populations from different geographic areas of the world also differ. Most people would not confuse an Eskimo from Greenland with a Pygmy from Zaire. However, our awareness of human variation sometimes confuses biological variations with cultural variations. For example, a Norwegian child may differ from a Japanese child not only in hair color, eye color, and superficial facial features, but also in facial expression, posture, dress, and language. The way the mouth is held or the smile and frown lines that seem to be biological features are superimposed on our faces by what we see and learn in our particular culture to be acceptable and normal. For example, during the nineteenth century Europeans believed Asians to be "inscrutable." Their emotional states were not readily apparent or easily recognized through facial expressions. Based on their cultural beliefs, the Japanese, for example, do not think it proper or polite to express emotions in public through obvious facial expressions. British explorers did not consider this to be a function of different *cultural values* but a function of the *physical differences* in the actual facial features.

Other human traits not as easily observable as physical differences are also variable. For example, in Chapter 13 you learned that traits such as blood type and Rh factor also display variability.

How do we interpret and understand human variability? For example, is type B blood an advatageous or disadvantageous variant in a particular environment? Is it, or was it in the recent past, selectively advantageous to have darkly or lightly pigmented skin? In this chapter we describe how human variability is measured, both among individuals and populations, and what evolutionary forces of change have caused their existence and distribution. We also discuss human variability and human races.

The Population

Whether a height of 5 feet 8 inches is considered tall, short, or average depends on the population a person is compared with. Similarly, a common group characteristic is considered a variation only in comparison with other groups with a different variety of that characteristic. If, for example, all the people in a particular group had type B blood, the only way to determine if there are other blood type variants is to compare this group with other groups. What then is the appropriate unit of comparison in which to consider human variability?

Biological variation must be understood in the context of a community of interbreeding individuals in a given locality, called a **breeding** or **local** population. For example, a local population of field mice may be confined to a meadow of a certain size, a population of skunks to the surrounding forest, and mosquitoes to a nearby pond. In that geographically determined population, individuals find their mates and reproduce. These geographical limitations sometimes influence human populations. For example, if individuals are isolated into groups in villages scattered throughout a rugged mountain range or on an island in the southern Pacific Ocean, there is not much chance of finding a mate outside of the immediate vicinity. But humans, in many ways, are unique. Our culturally directed behavior and our mobility often make human local or breeding populations a result of the interaction of both geographic and social factors.

In rural England during the nineteenth century, the average male found his mate within 600 yards of his birthplace. After the bicycle was introduced, the distance increased to 1600 yards. Extending this idea to the present, we might ask, has more convenient and faster modes of transportation turned the human species into one large local breeding population? Not necessarily. Even with today's rapid forms of transportation, among the five boroughs of New York City the average distance between birth residences of mating partners is only 25 miles. There must then be other factors at work that divide people into smaller groups.

In many cases social and cultural factors help maintain much smaller breeding populations. In these populations such diverse cultural factors as economic, religious, and ethnic affiliation play roles similar to that of geographic barriers or distance. In Franklin County, Pennsylvania, there are approximately 350 people belonging to a religious group known as Dunkers. They are descendants of a group that originally emigrated from Germany in the early 1700s. Since 1850, most of their marriages have been with other members of the group. Although living close to other groups of people, the Dunkers form a closed breeding population because they do not marry and mate with individuals outside of their religious community. Social and cultural factors also affect people who do not belong to such specialized groups like the Dunkers. Throughout the United States, including many metropolitan areas, social and cultural factors have kept marriages between diverse cultures to a minimum, effectively maintaining separate breeding populations.

Origin and Maintenance of Variability

Mutation generates new allelic variation and is the ultimate source of variability. However, individual variation in a population is also subject to processes that rearrange it and further mix this variation into new genotypes. Recall from Chapter 4 that recombination during meiosis rearranges the combination of individual alleles. With sexual reproduction, gametes from two individuals, each with different alleles for many characteristics, are united to form a unique individual. Recall from Chapter 5 that in populations, natural selection may alter allele frequencies from one generation to the next, increasing selectively advantageous allele frequencies and decreasing allele frequencies that are disadvantageous.

Measuring Human Variability

Mendelian patterns of inheritance indicate that the underlying genetic patterns of human variability consist primarily of two types, traits of simple inheritance and traits of complex inheritance.

Traits of Simple Inheritance

The first type of human variability, called **traits of simple inheritance,** is characteristics whose phenotypic variation falls into discrete, or discontinuous, nonoverlapping categories. All traits of simple inheritance share three characteristics. First, the trait is determined by one gene having two or more different alleles. For example, the ABO blood type system is determined by a single gene with three alleles, A, B, and O. Because the A and B allele are dominant over the O allele, four different phenotypes are produced, A, B, AB, and O. When a human population is blood typed, individuals fit into one of these four categories. No intermediate forms are possible. You are either type A or AB or B or O. Second, traits of simple inheritance are unaffected by changes either in the environment or during development. A person born with type A blood will have it throughout life and in any environment. Third, we can predict the proportion of offspring expressing each type. This trait-inheritance pattern is based on Mendelian principles. For example, if one parent is homozygous for type A blood and the other parent is type O, all the offspring will, phenotypically, be type A. If one parent is heterozygous for type B and the other parent is type O, one half of the offspring on average will be phenotypically type B and one half will be type O.

ABO Blood System

Alleles	*Genotypes*	*Phenotypes*
A	AA/AO	A
B	BB/BO	B
O	AB	AB
	OO	O

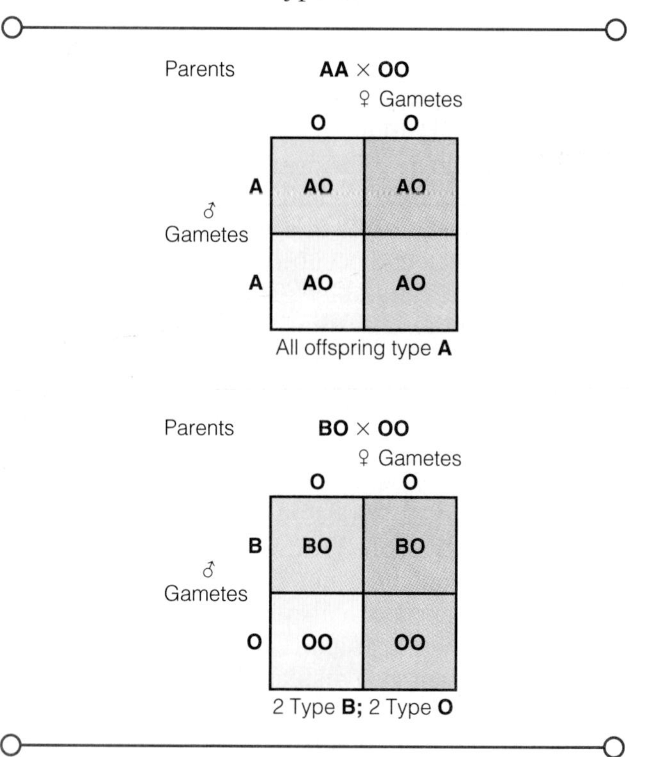

Traits of Complex Inheritance

People vary in their height, weight, and skin color. They also vary extensively, and often subtly, in their nose and ear shape, and even eye color. You recall from Chapter 5 that traits determined by two or more genes are called traits of polygenic inheritance. They are also called **traits of complex inheritance,** meaning their phenotypic expression is continuous throughout its range of variation, with no distinct, nonoverlapping categories. Traits of complex inheritance share two distinguishing characteristics. First, many different genes are involved in determining phenotypes. Second, environmental influences can affect phenotypic trait expression. Here we use a hypothetical example to illustrate the primary characteristics of traits of complex inheritance.

If the heights of a sample of male college students are measured to the nearest 5 cm, we find that a few of them are short, between 156 cm and 160 cm (5 feet 1 inch and 5 feet 3 inches) tall, and a few are tall, between 186 cm and 190 cm (6 feet 1 inch to 6 feet 3 inches) tall. Most, however, fall in the middle, from 171 cm to 175 cm (5 feet 7 inches to 5 feet 9 inches) tall. The number of people in each height class can be illustrated as a bar graph (Figure 22-1, *A*). This is a **frequency distribution,** or a listing of the numbers or proportion of people in a population in each designated class. The exact shape of the distribution depends on how fine or specific the measurements are. If the population were measured to the nearest centimeter, the distribution would change (Figure 22-1, *B*). If we created five times as many classes as before, the number of individuals in each class would also be smaller than in the first frequency distribution. However, you could measure five times as many students. Theoretically we could continue to refine our measurements to millimeters or even smaller units of measurement, each time increasing the number of students measured. In the end we would obtain a virtually continuous curve, representing a continuous frequency distribution of height in a large population of male college students (Figure 22-1, *C*).

Continuing with our example, if we compare each student's height with his father's height, we might note that the student is shorter than, taller than, or the same height as his father. This variation between generations suggests that environment also influences height expression. The fathers were reared in a different environment than their sons. For example, they may not have had access to the same type of nutrition as their sons, they may have been exposed to different diseases, or their medical care may not have been of the same quality.

Figure 22-1 Frequency distribution of a trait of complex inheritance

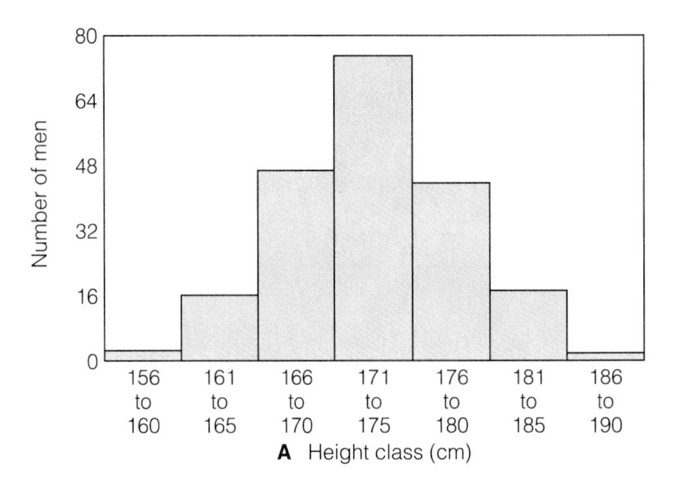

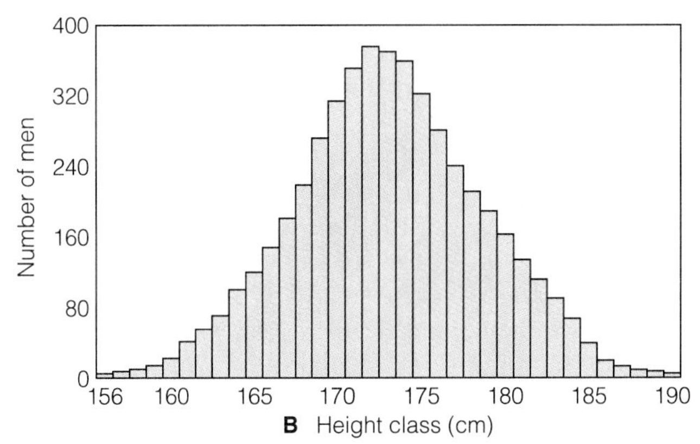

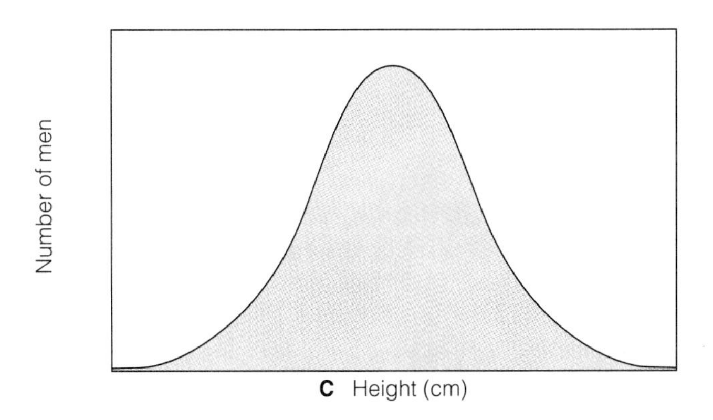

Frequency distribution of heights of a hypothetical population of male college students. A, Frequency distribution when measured to the nearest 5 cm. B, Frequency distribution when measured to the nearest centimeter. Five times more people were measured here than in A. C, Increasing the number of categories and individuals results in a continuous frequency distribution for height, which is common for traits of complex inheritance.

Traits of Simple Inheritance and Their Distribution

By analyzing traits of simple inheritance we can begin to document the range of variability in and among human populations and attempt to reconstruct the evolutionary forces that shaped them. We describe two well-known traits of simple inheritance, the ABO blood type and lactase deficiency.

Distribution of ABO Blood Type

Most of the world's populations contain two and usually all three of the ABO alleles. Worldwide the O allele is the most common, the A allele is next, and the B allele is the least common (Figure 22-2). The population allele frequencies of ABO blood type also fall within certain limits. The frequency of the O allele ranges from 40% to 100%, the A allele from 0% to 55%, and the B allele ranges from 0% to 30%

Within these limits the allele frequencies vary from population to population. For example, most Native American populations have a high frequency of type O blood, a low frequency of A, and almost no B. But the Blackfoot Indians have the world's highest frequency of A, little O, and no B allele. Eskimo populations from Alaska to Greenland show frequencies of B ranging from 1% to 5%. Why do frequencies of the ABO alleles vary so much among different populations?

TABLE 22-1 ABO Blood Types, Their Antigens, and Circulating Antibodies

TYPE	CELL ANTIGEN	ANTIBODY
A	A	Anti-B
B	B	Anti-A
AB	AB	None
O	None	Anti-A and anti-B

ABO Blood Type and Evolution

Since the initial discovery of the ABO system in 1900, millions of people have been typed in an attempt to discover associations between blood type and many different environmental factors. The distribution of ABO alleles in and among human populations strongly suggests that natural selection helps to produce these distributions. We describe three hypotheses in the next section.

One clue suggesting that natural selection may influence ABO system variability is that certain antibodies are associated with different blood groups (Table 22-1). Recall from Chapter 5 that if, for example, you have blood type A, you also have B antibodies circulating in your bloodstream. Information about **fertility rates** (the number of children born to an individual) of females with different blood types has shown that depending on the father's blood type, there is a significant difference in the rate of children born to type O mothers. Type O women give birth to significantly fewer children when the fathers are type A or B as compared with when the father is type O. If the father is heterozygous AO or BO, then there are more type O children than either type A or B children produced by the type O mother. This differential fertility may be acting before fertilization occurs. Type O females have a greater chance of fertilization by the sperm carrying the type O gene. Thus the resulting genotypes differ from the expected frequency of 50% type O, and 25% each for type A and type B. This selection of one male gamete over the other may be a result of the antibodies in the mucous secretions of the vagina, which react with sperm with type A or B antigens.

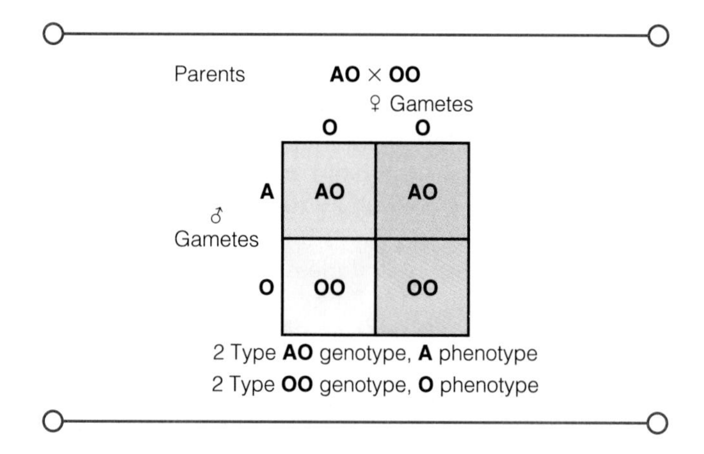

Many microorganisms causing both noninfectious and infectious diseases in humans also have ABO-like antigens on their cell-membrane surfaces. The general hypothesis is that if you have blood type A, and therefore circulating B antibodies, your immune system should be able to quickly detect and fight off any microorganisms carrying B-like antigens. On the other hand, if you are type B, you will not have antibodies to the B antigen in circulation. Consequently, your immune system may take significantly longer to react to an invasion of a

Figure 22-2 Worldwide distribution of A and B alleles

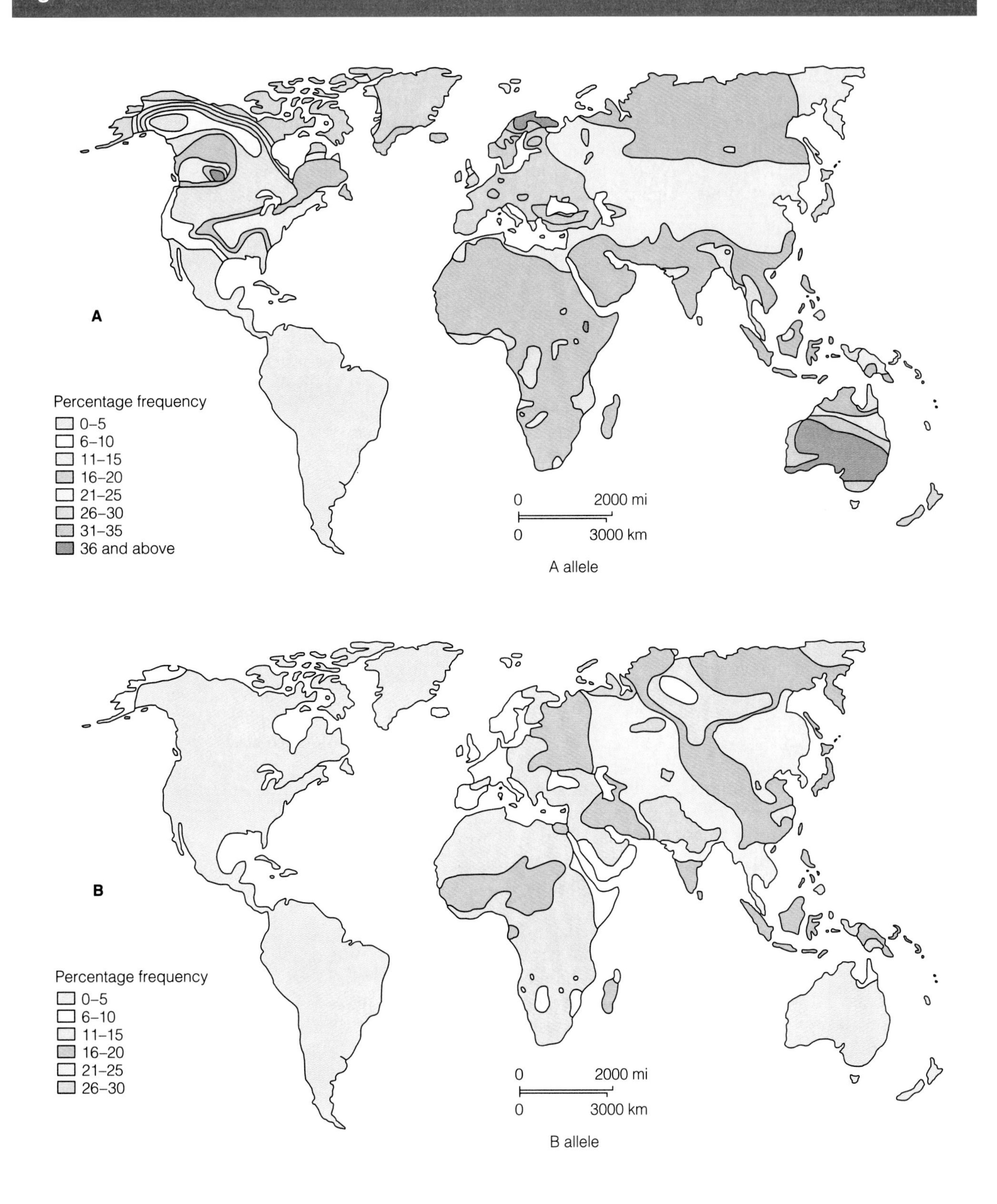

A and B allele distribution among indigenous populations.

microorganism with a B-like antigen on its surface. Is there evidence to support this hypothesis?

The smallpox virus has a type A–like antigen on its outer membrane. Therefore a person with type A blood might be more susceptible to smallpox than would a person with blood type B or O. A survey of rural village populations in northern India with several outbreaks of smallpox revealed a mortality rate of 50% in blood type A persons, which was approximately four times higher than with blood type B or O persons.

More evidence for this hypothesis involves the bubonic plague. It is caused by a bacterium with an O-like antigen on its membrane surface. The effectiveness of this disease as a "lethal" selective agent can be quite dramatic. The mortality rate during the Middle Ages was as high as 90% in the major urban centers of Europe and the Near East.

Areas where the plague has the longest history are also where type O is at its lowest frequency today. For example, central Asia and India were major centers for plague historically. These areas also have the lowest frequency of the O allele. If in the same population there was also selection against the type A allele because of smallpox, then both O and A would have low frequencies and the B allele would have a high frequency. This is exactly the situation in India where both diseases have had a long history. India's B allele frequency is among the highest in the world.

Lactase Deficiency

Infants usually receive nourishment from their mothers' breast milk. Infants have an enzyme, called lactase, that allows them to digest lactose (milk sugar). However, in many human populations lactase-enzyme production is switched off in children around 4 or 5 years of age.

Whether we continue to produce this enzyme after infancy is genetically variable. A dominant allele keeps lactase synthesis going, but the homozygous recessive genotype stops producing the enzyme sometime during childhood. A lactase-deficient child or adult ingesting lactose would then have severe diarrhea and cramps.

In most populations adults are lactase-deficient, but in several populations most adults are lactase-*sufficient* (Table 22-2). Populations with adults who can digest lactose have long depended on dairy herding and drinking fresh milk. Conversely, populations with lactase-deficient adults have not been dairy herders or fresh milk drinkers. Thus a hypothesis has been proposed stating that at one time almost all adults in every population were lactase-deficient. Several thousand years ago, when some populations started to herd cattle, lactose entered the adult diet. Selection then worked toward the relatively rare individual who was best able to use this new and highly nutritious food source. Over time lactase-sufficient adults became more numerous. Natural selection for the ability to digest fresh milk may have increased the frequency of lactase-sufficient adults, whereas it has been absent in populations that did not herd dairy animals or consume fresh milk.

Some evidence supports this hypothesis. Lactase-sufficient adults are numerous in European and European-derived American populations, some central Asian populations, and in some populations in southern and northeastern Africa. These populations have long practiced dairy herding. In contrast, west African populations and their American descendants, Native Americans, and many Asian populations are lactase-deficient.

Traits of Complex Inheritance and Their Distribution

As mentioned earlier, many human characteristics, such as height, weight, body shape, and skin-color variation, are traits of complex inheritance. Because we cannot consider all these traits in detail, we focus on one in particular, skin-color variation. We choose to discuss skin color for two important reasons. First, it is an excellent example of a trait of complex inheritance. Second, it is also a trait about which many cultural beliefs have been constructed. These beliefs are so strong as to override biological evidence invalidating them.

Skin-Color Variation

Historically the description and analysis of skin-color variation used crude measurement techniques. For example, one technique relied on a set of 30 colored tiles that the investigator could match to an individual's skin color. The population of individuals could then be described in terms of the number or proportion of individuals in a population matching one or another of these tiles. Because this method required matching a tile with a patch of skin, there was a wide variation in interpretation among researchers as a result of such diverse factors as differences in researchers' perceptual abilities and differences in light sources under which the comparison was made.

However, skin color is a continuously variable

TABLE 22-2 Frequencies of Lactase Deficiency in Some Human Populations

POPULATION	PERCENTAGE OF LACTASE DEFICIENCY
African ancestry	
African Americans	70–77
Ibos	99
Bantus	90
Fulani	22
Yoruba	99
Baganda	94
Asian ancestry	
Asian Americans	95–100
Thailand	97–100
Eskimos	72–88
Native Americans	58–67
European ancestry	
European Americans	2–19
Finland	18
Switzerland	12
Sweden	4

Sources: Lerner and Libby (1976:327); Molnar (1992:124).

trait. Attempting to match a person's skin with one or another colored tile is bound to be imprecise. Imagine that you had only 11 nonmixable colors. You could not accurately represent the whole visible color spectrum with only these 11 colors. Much of the actual "color variation" of the continuum would fall between these 11 colors.

These inaccurate techniques of skin-color analysis have long since been replaced by an instrument called a **reflectance spectrophotometer.** It measures the amount of light reflected by the skin compared with a pure white light standard. Recall from Chapter 9 that an object gets its color by absorbing some light waves and reflecting others. A green apple, for example, absorbs almost all waves of the visible light spectrum, reflecting only those light waves the eye perceives as the color green. So too with skin color.

The graphs produced by the reflectance spectrophotometer represent the specific wavelengths reflected by a person's skin. For example, the amount of light reflected from lightly pigmented skin is greater than that reflected from darkly pigmented skin. The slight troughs, or dips, in both lines at 550 nm are the result of the light absorption

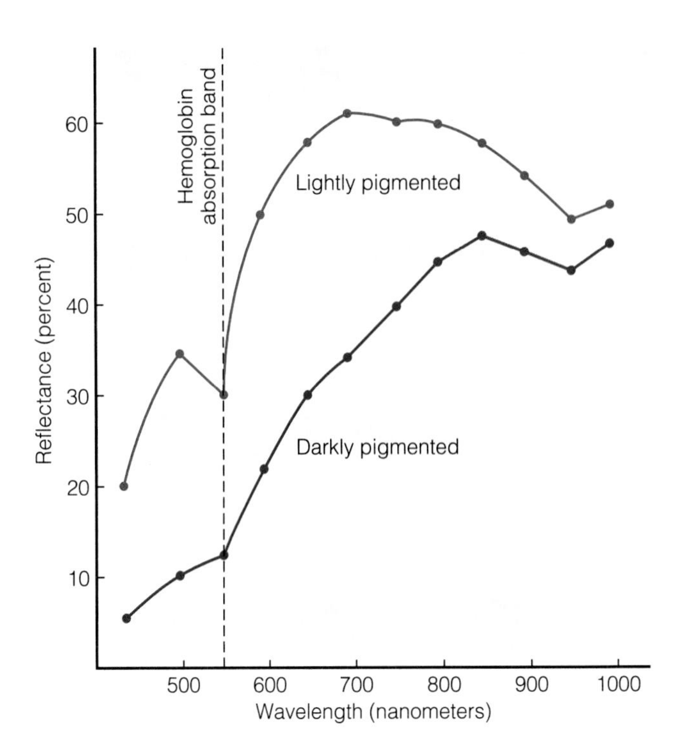

at these wavelengths by red blood cell hemoglobin in the capillaries below the skin's surface.

Human skin gets its color primarily from a brown pigment called **melanin.** To a lesser extent it obtains color from the hemoglobin of the red blood cells in the blood vessels directly beneath the epidermis. Melanin is formed in specialized cells called **melanocytes** (mel'a-no-sits) in the lower level of the epidermis. Each melanocyte produces melanin granules. These are deposited in organelles called **melanosomes** (mel'a-no-somz). The melanosomes then move through cellular extensions of the melanocytes to the surrounding developing epidermal cells. There they are incorporated into their cytoplasm. Once inside the developing epidermal cell, the melanosome breaks down, releasing the melanin granules into the cell's cytoplasm (Figure 22-3). Differences in the amount of melanin deposited in the epidermal cells account for skin-color differences among individuals. The average number of melanocytes per unit area of epidermis is the *same* between darkly pigmented and lightly pigmented individuals. The difference in skin color comes primarily from the varying number and size of the melanin granules formed in the melanocytes transferred to the surrounding epidermal cells.

In people whose epidermis produces little melanin, most of the coloration is produced by hemoglobin, which gives the skin a pinkish tone. The color of melanin itself is also affected by the transparency and thickness of the epidermis that overlies it. For example, oil increases the transparency of the epidermis. Thus a darkly pigmented

Figure 22-3 Melanocytes and melanin

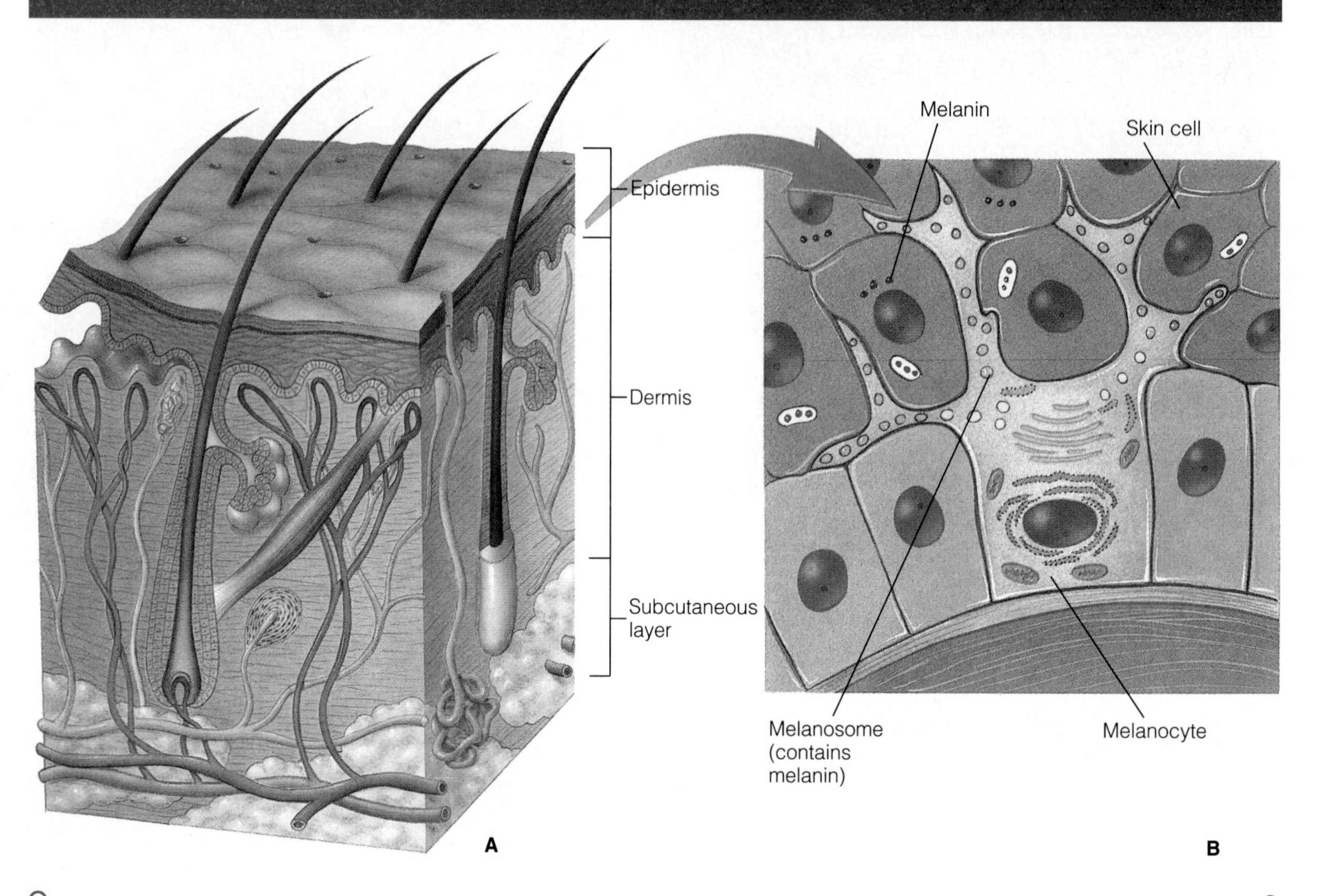

Melanin is produced by melanocytes in the deepest layers of the epidermis (A). Melanosomes transport melanin granules from melanocytes to surrounding developing epidermal cells, where they are taken up into cytoplasm (B).

epidermis will appear brown in tone if oily, whereas the same epidermis will, if dry, take on a grayer tone. Similarly, an increase in the thickness of the epidermis adds a yellowish tone to the skin regardless of the amount of melanin. The yellowish skin tone of many Asian populations thus is not caused by a different pigment. They actually have the same range of melanin as pinkish-toned populations referred to as "white." Differences result from a thickening of the epidermis that is common to them.

Melanin absorbs the ultraviolet radiation in sunlight. An immediate response of the epidermis to ultraviolet radiation exposure is to increase melanin granule production. This causes the skin to darken or become tan. All human populations tan in response to ultraviolet radiation exposure. The effects are simply more noticeable in those who are more lightly pigmented. The heat, reddening (most apparent in lightly pigmented skin), and peeling associated with sunburn occurs because the skin is actually burned. Severe tissue damage can result from allowing the skin to be exposed to ultraviolet radiation in amounts and rates exceeding the skin's capacity to produce an effective defense against potential damage.

Apart from the variations in skin color that occur as a result of an individual's response to ultraviolet light exposure, other differences in individual skin color result from underlying genetic variation. Skin color is determined by many different genes, making it a trait of complex inheritance.

When we look at skin-color variation as measured by reflectance spectrophotometry among human populations, we find that the amount of melanin decreases in populations the farther they are from the equator (Figure 22-4). Human skin-color variation can be expressed in terms of **clinal distribution.** Clinal distribution is a gradual variation in a trait over space shown by the alteration in its genotypic or phenotypic frequency from population to population. What would cause human skin-color variation to be distributed in this pattern?

Figure 22-4 Global distribution of human skin color

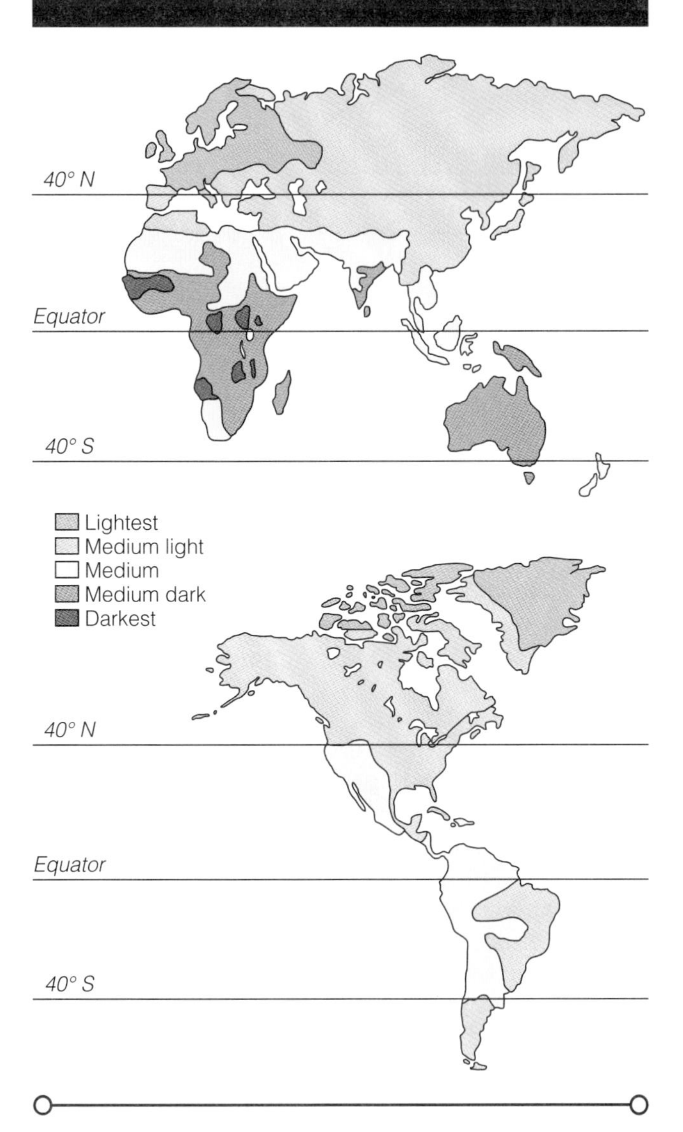

Skin Color and Natural Selection

By noting changing environmental conditions over geographic distance, it becomes possible to develop and test hypotheses about the cause of the clinal distribution of human skin color. Although many different hypotheses have been proposed to explain population differences in skin color, all of them consider sunlight, specifically ultraviolet radiation, as the ultimate cause. This is because the average amount of ultraviolet radiation in sunlight and the average amount of melanin decreases with distance from the equator. Here we must first describe the important role of vitamin D in humans and then discuss sunlight, ultraviolet radiation, and latitude.

Ultraviolet light stimulates vitamin D production in the skin's dermal layers. Vitamin D is essential for calcium absorption by the small intestine. If vitamin D is not present in sufficient quantities, either because of lack of exposure to sunlight or because of inadequate dietary intake, not enough calcium is absorbed in the small intestine. Recall from Chapter 11 that calcium is essential for proper CNS and muscle functioning. It is also a structural element in bone. When there is not enough calcium to maintain normal CNS activity, the body draws the mineral from the bones.

Prolonged calcium deficiency in children leads to insufficient mineralization of the growing bone. This insufficiency results in bone deformation, a condition called **rickets.** In rickets the bones remain soft because, although the collagen framework is laid down at a normal rate, bone mineralization, which gives bone its stiffness and rigidity, does not keep pace with the growing bone. As a consequence the softened bones bend and distort under the stress of body weight and movement and ultimately harden in their deformed state. The pelvis is also subject to distortion, and the size and shape of the birth canal of the female pelvis may be altered enough to make normal births impossible. Removing rickets-prone females from the reproductive segment of a population would be a powerful selective force acting directly against genotypes unable to adjust to reduced amounts of ultraviolet radiation sufficient to maintain physiological levels of vitamin D. Too much vitamin D, on the other hand, causes calcium deposits to form in soft tissues and can eventually lead to such problems as fatal kidney dysfunction.

The amount of sunlight reaching the earth during the course of a year is identical at all latitudes. However, more ultraviolet radiation penetrates the atmosphere and reaches the earth at the equator compared with more northern or southern latitudes because of less atmospheric filtering (Figure 22-5). The *distribution* of daylight hours also varies widely from the equator to the poles. For example, day and night are of equal length at the equator, and the midday sun and ultraviolet radiation are uniformly intense throughout the year. Away from the equator, the distribution and seasonal variation become greater. In the winter, days are shorter and there is less ultraviolet radiation, whereas during the summer days are longer and ultraviolet radiation levels may be as high as at the equator.

The vitamin D hypothesis states that in areas where sunlight and ultraviolet radiation are constant throughout the year, skin containing high levels of melanin is protected from damage while still allowing enough ultraviolet radiation to penetrate the skin to stimulate vitamin D synthesis. In contrast, at higher latitudes, because of seasonal variability and reduced intensity of solar radiation and

Figure 22-5 Solar radiation and Earth's atmosphere

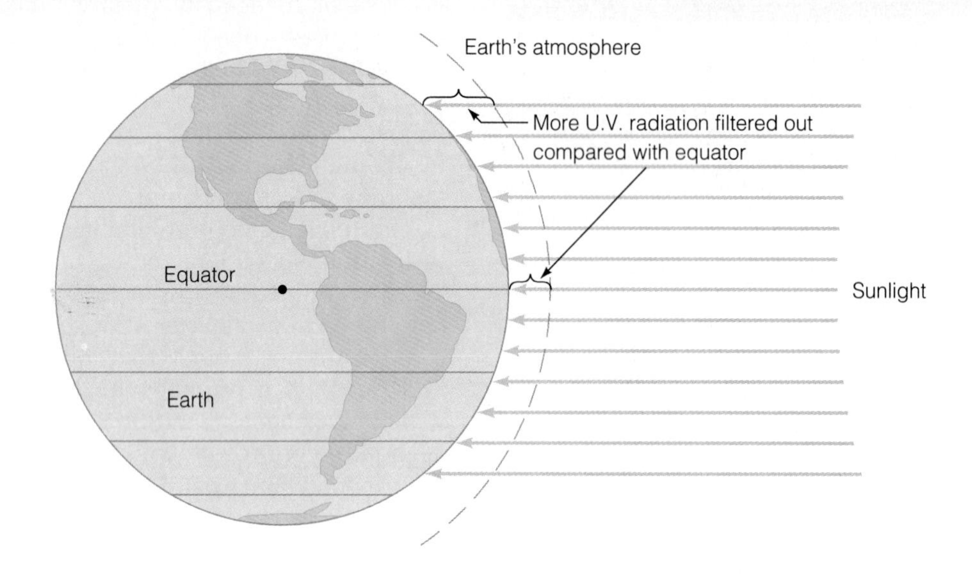

More ultraviolet radiation reaches Earth at the equator compared with more northern or southern latitudes. There is less atmospheric filtering at the equator because solar radiation hits the atmosphere more directly at the equator compared with other latitudes.

ultraviolet radiation, lightly pigmented skin is selectively advantageous because it allows most of the available ultraviolet radiation to penetrate the skin to stimulate vitamin D synthesis. The ability to tan during the months when solar radiation is high would, in those populations, provide a temporary adaptation against the damaging effects of higher levels of ultraviolet radiation.

We may never know exactly when human skin color began to vary. However, it could not have occurred until early humans moved out of the tropics and into higher latitudes. Based on dated fossil evidence of *Homo erectus,* this may have been about 1 million years ago. One implication of this hypothesis is that the early hominids, the australopithecines, were darkly pigmented.

Human Variability and the Problem of Race

We are usually acutely aware of the various categories into which human variability is cast, called "races." For many people, the notion of race is tied to the obvious physical or behavioral distinctions gained from a simple visual appraisal of an individual. But consider the different ways we use the term "race." Sometimes we speak of the "human race," or the "German race," "Jewish race," or "black race." In the first example we use "race" to distinguish the biological species *Homo sapiens.* In the second we use it to define a group of people by nationality. In the third we use it to distinguish religious affiliation. In the fourth example "race" refers to the single variable biological characteristic of skin color. So, what does the term mean?

Although the contexts in which we use the term "race" differ widely, they do share something in common. They reduce all the ways people differ to one easily recognizable, *socially* important trait. It might be skin color, religion, cultural identity, or something else. But even if the cultural or social definitions of human races are set aside, differences still persist. After all, the human eye and brain perceive a part of the visible color spectrum as "blue," regardless of whether it is called *blue* in English or *bleu* in French. Is it not the same thing for racial classification schemes based on skin-color variation?

We begin by considering the cultural definitions of race in the context of what we know about the biological variation of human populations. We then discuss whether these cultural definitions of race accurately reflect and describe known human population variations.

Do Human Races Exist?

Local populations of a species are dispersed throughout a particular habitat. Because habitats are not entirely homogeneous in their range, it is likely that local populations will often diverge in allele frequencies of various genetic traits. For example, two human populations, each living in slightly different environments, may have different allelic frequencies of ABO blood types because of slightly different selective factors in each environment. When allelic differences for many traits become detectable, biologists may classify these populations as **races.** Races are local populations that differ from other local populations in their frequency of several genetic traits.

Decisions about what traits to emphasize in describing differences among local populations are relatively arbitrary. Depending on which traits or set of traits are emphasized, local populations may be separated differently. For example, we might describe human populations based on blood type differences. If we used lactase deficiency or hemoglobin variants as our criteria, we would get different groupings. If we used blood type and lactase deficiency, or blood type, hemoglobin variants, and lactase deficiency as our sorting criteria, we would create entirely different groupings, or races.

Is skin-color variation a useful and accurate criterion for dividing human groups into races? If race, defined as local populations that differ from other local populations in their frequencies of *several* genetic traits, is to be a useful biological concept in describing human variability, the classification would have to work for many independent genetic traits. A classification based on skin-color variation would also need to show the same pattern of distribution in other traits, such as hair color and form, blood type, or facial characteristics. However, if each trait produces a different set of races, then the concept is not useful. Richard Lewontin, a population geneticist at Harvard University, tested this hypothesis in 1971.

Over the years biologists and geneticists have examined the gene frequencies of many genetic traits. Lewontin selected 17 different genetic traits with known worldwide distributions and gene frequencies. Included in Lewontin's analysis were genetic traits such as Rh factor and the ABO blood group. He then chose a traditional racial classification system of seven "major geographic human races," based primarily on skin color. These were Africans, Caucasians, Mongoloids, Amerindians, South Asians, Oceanians (people of the South Pacific Islands), and Australian Aborigines (Table 22-3). Using these categories, Lewontin calculated gene frequencies in local populations, among local populations in a major racial group, and finally, among racial groups themselves. For example, he would first determine the selected traits of a local population of Navajo Indians and then compare

TABLE 22-3 Traditional Racial Classifications

GEOGRAPHIC GROUPING (Based on skin color variation)	EXAMPLES OF POPULATIONS INCLUDED*
Caucasians	Arabs, Armenians, Tristan da Cunhans
Black Africans	Bantu, San, African Americans
Mongoloids	Ainu, Chinese, Turks
South Asian Aborigines	Andamanese, Tamils
Amerinds	Aleuts, Navajo, Yanomama
Oceanians	Easter Islanders, "Micronesians"
Australian Aborigines	All treated as a single group

*Not inclusive.

with them the gene frequencies of Cherokee Indians. Then he compared Amerindian gene frequencies with African gene frequencies (Figure 22-6).

Of all the genetic variations among the 17 traits he chose, Lewontin discovered that 85% of the total genetic variation was between people in local populations. The remaining variation was roughly split between populations within a racial group and variation between one race and another. In other words, about 94% of the total genetic variation occurred *within* racial groups and approximately 6% of the variation *between* races. We can interpret what this means by using the following extreme example. If, after a great disaster, only west Africans survived, the human species would still retain 94% of its total genetic variation, although humans as a whole would be darker skinned. If the disaster were even more extreme and only the Yanomama Indians of the Amazon rain forest survived, the human species would still retain over 80% of its genetic variation. Lewontin did not find "African only" alleles, nor did he discover any other alleles exclusively in any of the racial categories. It appears as if racial classifications based on skin color do not reveal much about the actual distribution of global human variability. Since 1972, many more genetic traits and their distributions have been discovered. This new information has generally strengthened Lewontin's original findings.

If you walk down the street of any large U.S. city, you could observe quite distinct differences in

Figure 22-6 Genetic variation and racial classifications

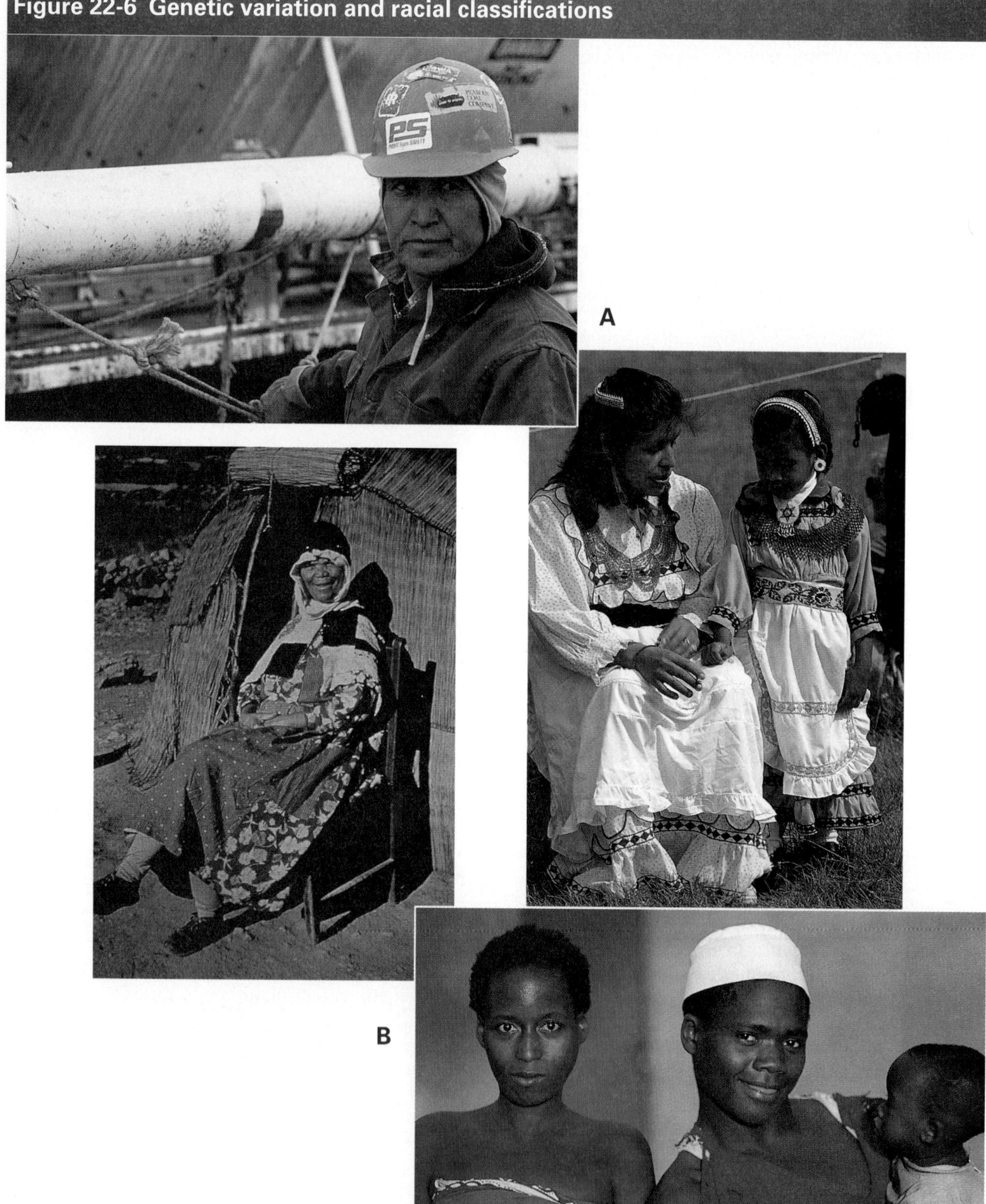

Lewontin first determined genetic variability of selected traits within populations. He then compared genetic variability among populations within a traditional racial category, such as between the Navaho and Cherokee Indians (A). Finally he compared the genetic variability among racial categories, such as between Amerindians and Africans (B).

the skin color of the people you pass. If skin-color variation shows clinal variation, a *gradual* variation over space shown by the alteration in its genotypic or phenotypic frequency among populations, why can we observe such abrupt and distinct differences among the people we see?

Most of the earliest immigrants to the United States were from the extreme ends of the range of skin-color variation. They were primarily darkly pigmented west Africans and lightly pigmented Europeans. As you walk down the street, you are in effect comparing the ends of the range of skin-color variation without benefit of the intervening geographic populations that at one time linked them.

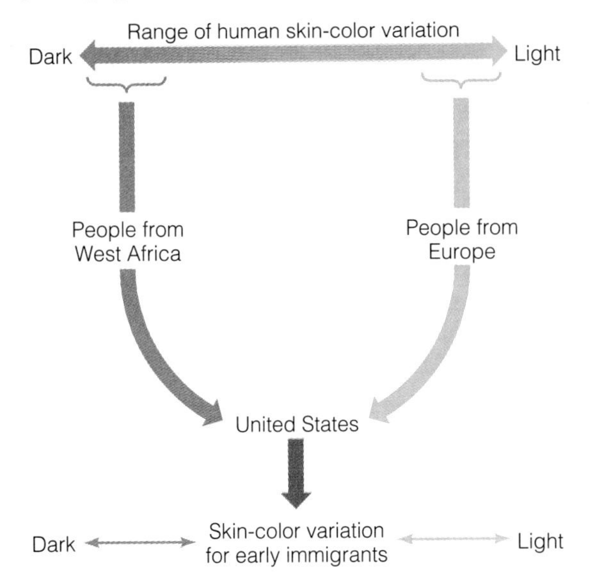

This is similar to comparing the color red from one end of the visible light spectrum with the color blue at the other end. They are, of course, distinctly different, but when the intermediate light waves between them are present, we see there is a *gradual* and *continuous* change from one color to the other.

Therefore, racial classifications based on skin color variation use a trait of complex inheritance with *continuous variation,* and break it up into largely arbitrary, discrete categories, such as "black," "white," "yellow," and "red." However, one may argue, along with the observed differences in skin color among individuals and groups, are there not also differences in other physical characteristics, such as hair form and color and facial characteristics? To answer this argument, we consider the distributions of two other human traits and compare them with the distribution of skin-color variation.

Skin Color Variation and Other Human Traits

One obvious characteristic of human head hair is its degree of curvature, which varies from completely straight to tightly coiled or curled. Hair form is categorized according to its degree of curvature from straight through wavy and helical to spiral. This curvature can be roughly determined by laying single hairs on a sheet of paper and measuring the diameter of the circle completely or partially formed.

Tightly spiraled hair, for example, is common in the San ("Bushmen") of southwest Africa. Helical hair is common throughout most of the rest of sub-saharan Africa and in Melanesia. Straight hair characterizes Asians and Native Americans, whereas wavy hair is most frequent in Europe, the Middle East, India, Polynesia, and among Australian Aborigines. However, hair-form distribution does not correspond in the same way as skin-color variation. Darkly pigmented Australian Aborigines have straight to wavy hair, as do equally darkly pigmented southern Indian populations (Figure 22-7). In contrast, lightly pigmented southern European and Middle Eastern populations display hair forms that vary from wavy to tightly spiraled.

What about facial characteristics? Consider lip shape as a distinctive facial feature. The degree to which the transition between facial skin and the mucous membrane lining the mouth is exposed and turned outward is called **lip eversion.** Lip eversion varies greatly among individuals within any human population (Figure 22-8). It is so variable *within* local populations that accurate generalizations of any sort regarding geographical populations are not possible.

The trait used to establish racial categories, skin color, is continuously variable throughout its distribution range. Then how can it be used to also describe culturally determined discrete categories such as "black" and "white" or "red" and "yellow"? Where does one category end and another begin? Other human traits do not correspond in the same ways as does the global distribution of skin-color variation. Racial classifications based on skin-color variation arbitrarily divide human variation into discrete categories, emphasizing differences among populations and de-emphasizing differences within populations.

This discussion about whether there is any biological evidence for culturally defined racial classifications may at first appear a bit abstract. However, as we stated earlier in the chapter, a common characteristic of culturally defined races is that they reduce all the ways people differ to an easily recognizable and usually *socially important* trait. The realization, however, that a culturally determined racial classification system has no biological validity does not diminish its influence nor, unfortunately, its destructive power.

Figure 22-7 Degree of hair curvature among Australian Aborigines

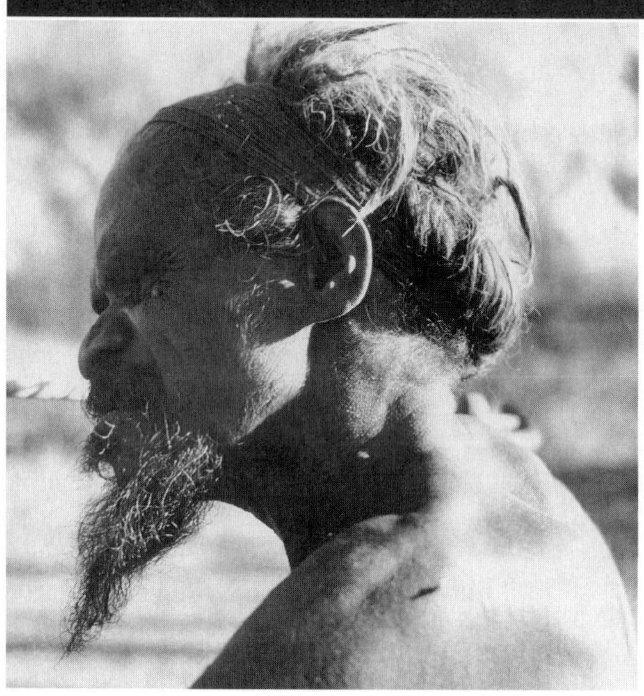

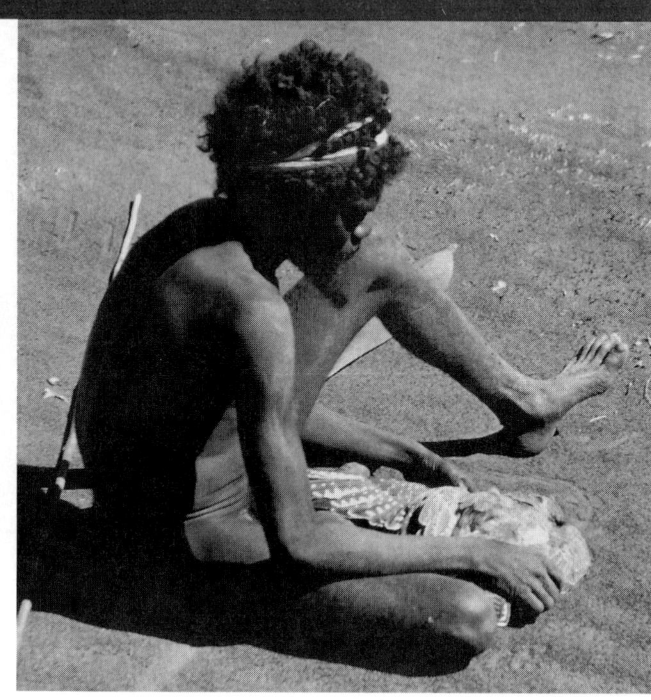

The subject of race is discussed almost daily in our media. Statistics on race are gathered by local, state, federal, and international organizations, with economic and political consequences of often immense proportions. It is both historically and currently true that cultural notions of race are used to justify discrimination and persecution of some people while granting favored or privileged status to others. In addition, some actually attempt to rationalize and justify racial classifications based on *evidence from biology.* It is distressingly common, for example, to read articles and even see television documentaries that discuss the dominance of "black athletes" in sports such as basketball, football, and track as resulting from particular biological traits associated with that "race." Similar debates continue regarding the biological bases for differences in levels of intelligence, creativity, criminality, or other behaviors believed to be displayed by different "races." In each instance, people continue to

Figure 22-8 Range of lip eversion among human populations

interpret human biological variation in terms of their cultural notions and beliefs about race, even though these biological facts contradict and invalidate these racial categories. It is similar to the fairy tale "The Emperor's New Clothes." If we *believe* the emperor has new clothes, does that mean he *actually* does have new clothes?

Perhaps by looking at the actual facts regarding human variation we can begin to change these powerful beliefs about human differences. These beliefs about people have succeeded only in producing large-scale and widespread human suffering and diminish not only the victims but also the perpetrators of all racist schemes. Admittedly, we have described only briefly the variability of known human traits. However, if we begin by first attacking what we believe most strongly to be true, that the differences we observe between people are based primarily on cultural beliefs, perhaps one day we shall truly see the emperor is not wearing any clothes at all.

Biology in Action: THE MYTH OF THE STRONGER SEX

In many cultures, strength and wellness are traits traditionally attributed to men, while women are often perceived as the weaker sex. Contrary to these views there is biological evidence indicating that males may be more vulnerable to a variety of developmental and environmental problems throughout their lives than females. From the very beginning, male development appears to be a more complex process than female development and consequently, more error prone at many of the critical prenatal stages. Recall from Chapter 5 that females have a pair of homologous sex chromosomes designated as XX, while the male's sex chromosomes are nonhomologous and designated X and Y. In females a recessive allele linked to the X chromosome, such as color blindness or hemophilia, can be masked by its dominant allele on the corresponding X chromosome. Recessive alleles on the single X chromosome of males have no dominant, masking counterpart on the Y chromosome and will therefore be expressed more frequently.

The catalog of male biological vulnerability is extensive. For example, approximately 140 males are conceived for every 100 females, but because of spontaneous abortions, the actual birth ratio is 105 males to 100 females. Boys display a higher percentage of congenital defects compared with girls and are more susceptible to childhood diseases. Boys have greater difficulties during their elementary school years. The incidence of reading problems, hyperactivity, mental retardation, and emotional and behavioral problems are all greater for boys than for girls.

Although biological evidence for male vulnerability is strong, many problems experienced by males may in fact be due to a harsher socialization of the male gender role than the female role. In other words it is not simply a question of differences in male and female biology but also the way in which culture defines the social roles assigned to each sex.

CHAPTER REVIEW

SUMMARY

1. Individual variation exists in all biological organisms, including humans. Without variation, adaptation to environmental change is not possible. Without adaptation, a species cannot persist over time.
2. Biological variation is interpreted in the context of a community of interbreeding individuals in a given locality, called a breeding or local population. Social and cultural factors also define and maintain human local populations.
3. The variations of human characteristics are of two types. The phenotypic expression of traits of simple inheritance fall into discrete categories with no overlapping or intermediate forms. In traits of complex inheritance variation is continuous throughout its range, with no distinct, nonoverlapping categories. Traits of simple inheritance remain unchanged throughout an individual's lifetime. A trait of

complex inheritance is subject to both external and internal environmental influences, and the resulting phenotype may be modified during an individual's lifetime.

4. Human skin-color variation, determined by the amount of a brown pigment called melanin in the epidermal skin cells, is a trait of complex inheritance that can be expressed in terms of clinal distribution. Clinal distribution is a gradual variation in a trait over space shown by the alteration in its genotypic or phenotypic frequency among populations.
5. The proposed selective force for skin-color variation is called the vitamin D hypothesis. It states that in areas where sunlight and ultraviolet radiation are constant throughout the year, skin containing high levels of melanin is protected from damage while still allowing enough ultraviolet radiation to penetrate the skin to stimulate vitamin D synthesis. Vitamin D is essential for calcium absorption by the small intestine. Too little or too much calcium disrupts the body's homeostatic balance.
6. In biology, race is defined as a local population that differs from other populations in their frequencies of several genetic traits. Depending on which trait or set of traits are emphasized, local populations may be separated and grouped differently.
7. Racial classification based on skin-color variation explains little about the actual distribution of human variation. About 94% of the total genetic variation of human traits occurs within any traditional racial grouping. Racial classifications based on skin color variation do not reveal much about the actual distribution of global human variability.
8. Other observable traits such as hair form and facial features do not correspond to the clinal distribution of human skin-color variation. For example, lip eversion is so variable within local populations that generalizations regarding geographic populations are not possible.

FILL-IN-THE-BLANK QUESTIONS

1. The phenotypic variation of traits of simple inheritance is discrete and ______.
2. The phenotypic variation of traits of complex inheritance is ______ throughout its range.
3. Some human populations are unable to produce the enzyme ______ and therefore cannot digest milk.
4. Skin color is a trait of ______ inheritance.
5. A pigment called ______ is responsible for human skin color.
6. A gradual variation in a trait over space shown by the alteration in its genotypic or phenotypic frequency among populations is called ______ ______.
7. Insufficient mineralization of the growing bone resulting in its deformation is called ______.
8. The ______ ______ hypothesis attempts to explain the selective advantage of skin-color variation.

SHORT-ANSWER QUESTIONS

1. What one problem is often associated with defining human breeding or local populations?
2. Compare and contrast one trait of simple inheritance and complex inheritance.
3. Describe two types of evidence supporting the conclusion that natural selection may influence the distribution of ABO blood types.
4. What is the selective advantage of being able to manufacture the enzyme lactase?
5. What evidence is there to support the vitamin D hypothesis?

VOCABULARY REVIEW

clinal distribution—*p. 410*
complex inheritance—*p. 405*
frequency distribution—*p. 405*
melanin—*p. 409*
simple inheritance—*p. 404*
race—*p. 413*

ANSWERS TO FILL-IN-THE-BLANK QUESTIONS

Chapter 1
1. cell
2. ecosystem; biome
3. homeostasis
4. natural causality
5. observation
6. independent
7. hypothesis
8. spontaneous generation
9. variation
10. photosynthesis

Chapter 2
1. neutrons; protons; electrons
2. electrons; negatively; electrons; positively
3. electrons
4. chemical bonds
5. polar covalent bond
6. hydrogen bond
7. heat of vaporization
8. pH
9. ionic

Chapter 3
1. organelles; cytoplasm
2. phospholipid; proteins
3. solvent; solute
4. solute; osmosis
5. passive transport
6. mitochondrion
7. ATP
8. cellular respiration
9. electron transport chain
10. anaerobic metabolic pathway
11. pyruvate
12. oxygen

Chapter 4
1. hybrid
2. dominate
3. test cross
4. 3:1
5. 9:3:3:1
6. sister chromatids
7. 23
8. spindle fibers
9. one half

Chapter 5
1. double; helical; single
2. restriction enzymes
3. uracil
4. two; A; P
5. mutation
6. karyotype
7. Down syndrome
8. recessive
9. sex-linked
10. pleiotropic
11. continuous
12. division; growth

Chapter 6
1. Thomas Malthus
2. Descent; natural selection
3. capacity; offspring
4. allele
5. reproduction; genotypes
6. bipedalism
7. erectus
8. fossils
9. homologous
10. biological species

Chapter 7
1. interstitial fluid
2. receptor
3. decreases
4. epithelial
5. tight junctions
6. collagen; elastin
7. tendons
8. tissues

Chapter 8
1. dendrites; axon
2. neuroglia
3. membrane potential
4. negatively; positively
5. synaptic; terminal knob
6. inhibitory
7. afferent; efferent
8. sensory; motor
9. postganglionic fiber; ganglion
10. reticular activating system
11. dorsal root; ventral root
12. cerebrum
13. primary motor area
14. medulla oblongata
15. electrical; structural

Chapter 9
1. stimulus
2. adaptation
3. free nerve endings; bradykinin
4. photopigments
5. cornea
6. cones; color
7. cochlea
8. crista ampullaris; semicircular canals

Chapter 10
1. hypothalamus; anterior
2. hypothalamus; posterior
3. cellular
4. cholesterol; gland; gonads
5. beta cells; glucose
6. adrenal cortex; stress
7. positive
8. synthesis; mitosis
9. anterior
10. gluconeogenesis

Chapter 11
1. axial
2. pectoral; pelvic
3. carpals
4. sutures
5. osteon; Haversian system
6. antagonistic pairs
7. Z
8. myosin; actin
9. calcium ions
10. creatine phosphate

Chapter 12
1. opposition
2. endothelium; smooth muscle; collagen
3. vasodilation; vasoconstriction
4. away from; back to
5. one-way valves
6. four; atria; ventricles
7. pulmonary valve
8. action potentials
9. S.A.; pacemaker
10. lymphatic system

Chapter 13
1. plasma; formed elements
2. plasma protein
3. hemoglobin
4. organelles
5. extracellular fluid
6. blood pressure
7. leukocytes
8. basophils; eosinophils
9. macrophages
10. thrombin

Chapter 14
1. antigen
2. B
3. immunoglobulins
4. 4; 2; heavy; 2; light
5. antigen-antibody
6. macrophage; self-antiself
7. lymphokines; cell
8. specific
9. immunological memory
10. allergies

Chapter 15
1. respiration
2. partial pressure
3. epiglottis
4. right; left primary bronchi
5. alveoli
6. diffusion
7. central; peripheral
8. bicarbonate
9. hemoglobin; oxyhemoglobin

Chapter 16
1. macromolecules; simple organic
2. alimentary; glands
3. involuntary
4. peristalsis
5. hydrochloric acid; pepsinogen
6. duodenum
7. villi
8. essential nutrients; diet

Chapter 17
1. renal artery
2. renal pelvis
3. 4
4. glomerular capsule
5. water; solute
6. distal convoluted tubule
7. urea
8. juxtaglomerular
9. renal failure

Chapter 18
1. spermatozoa; oocytes; ova
2. testes; ovaries
3. seminiferous tubules; testes
4. ovarian follicles; ovaries
5. ovarian; uterine; uterus; zygote
6. 14
7. menstruation
8. menopause
9. bacteria; viruses

Chapter 19
1. cell membrane
2. morula
3. monozygotic
4. trophoblast; human chorionic gonadotropin hormone
5. chorionic
6. endoderm; mesoderm; ectoderm; organogenesis
7. teratogens
8. oxytocin; posterior; prostaglandins
9. morning sickness
10. dilated; 10

Chapter 20
1. life span
2. postmitotic
3. collagen; cross-links
4. osteoporosis
5. left ventricle
6. growth

Chapter 21
1. biotic; abiotic
2. community
3. habitat
4. photosynthetic
5. 10%
6. atmosphere; carbon dioxide
7. bacteria
8. sigmoid curve
9. carrying capacity
10. pollutant

Chapter 22
1. discontinuous
2. continuous
3. lactase
4. complex
5. melanin
6. clinal variation
7. rickets
8. vitamin D

TABLE OF MEASUREMENTS

UNIT	METRIC EQUIVALENT	SYMBOL	U.S. EQUIVALENT
MEASURES OF LENGTH			
1 kilometer	= 1000 meters	km	0.62137 mile
1 meter	= 10 decimeters or 100 centimeters	m	39.37 inches
1 decimeter	= 10 centimeters	dm	3.937 inches
1 centimeter	= 10 millimeters	cm	0.3937 inch
1 millimeter	= 1000 micrometers	mm	
1 micrometer	= 1/1000 millimeter or 1000 nanometers	µm	
1 nanometer	= 10 angstroms or 1000 picometers	nm	No U.S. equivalent
1 angstrom	= 1/10,000,000 millimeter	Å	
1 picometer	= 1/1,000,000,000 millimeter	pm	
MEASURES OF VOLUME			
1 cubic meter	= 1000 cubic decimeters	m^3	1.308 cubic yards
1 cubic decimeter	= 1000 cubic centimeters	dm^3	0.03531 cubic foot
1 cubic centimeter	= 1000 cubic millimeters or 1 milliliter	cm^3 (cc)	0.06102 cubic inch
MEASURES OF CAPACITY			
1 kiloliter	= 1000 liters	kl	264.18 gallons
1 liter	= 10 deciliters	l	1.0567 quarts
1 deciliter	= 100 milliliters	dl	0.4227 cup
1 milliliter	= volume of 1 gram of water at standard temperature and pressure	ml	0.3381 ounces
MEASURES OF MASS			
1 kilogram	= 1000 grams	kg	2.2046 pounds
1 gram	= 100 centigrams or 1000 milligrams	g	0.0353 ounces
1 centigram	= 10 milligrams	cg	0.1543 grain
1 milligram	= 1/1000 gram	mg	

Note that a micrometer was formerly called a micron (µ), and a nanometer was formerly called a millimicron (mµ).

Comparative Temperature Scales

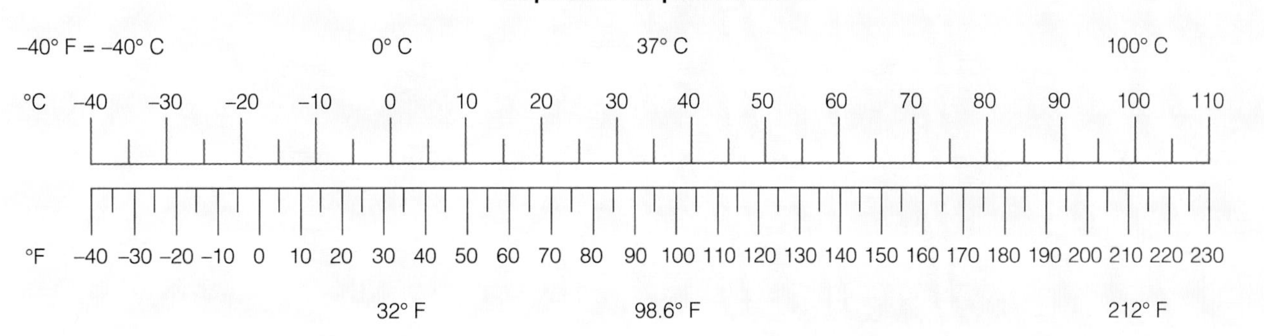

A

abiotic Pertaining to the nonliving portion of any environment, such as water, temperature, and soil. p. 387

absorption Transportation of simple molecules from the digestive system to the circulatory system for distribution to the rest of the body. p. 305

accommodation The ability of the eye to adjust the lens so that objects can be focused on the retina. p. 182

acetyl-CoA (acetyl coenzyme A) Formed in the metabolic processes by the combination of the two-carbon acetyl group with coenzyme A. p. 52

acetylcholine (Ach) (a'-sih-til-ko-len) A neurotransmitter released by neurons that stimulate skeletal muscles. Ach is also released by some neurons in the central nervous system. p. 149

Acheulian tradition Stone tools, directly associated with Homo erectus, worked on both sides of the stone producing a pointed, bi-face cutting edge. p. 118

acid Any molecule or compound that dissociates to form hydrogen ions in solution; pH of acids is less than 7. p. 23

acid rain Rain that, because of acid forming pollutants introduced into the atmosphere, has a pH lower than normal. p. 398

acrosome (ak'ro-som) The cap-like structure surrounding the anterior end of the nucleus of the sperm. It contains enzymes that function in the penetration of the ovum during fertilization. p. 339

actin A protein found in muscle fibers and the cytoskeleton that acts with myosin to bring about contraction and relaxation. Also called thin filament. p. 131

action potential The brief reversal of membrane potential brought about by rapid changes in the neuron's cell membrane permeability. This results in the reduction of a neuron's membrane potential to a less negative value. p. 144

active site The place on an enzyme where substrate molecules temporarily bond during catalysis. p. 31

active transport Carrier protein–mediated transport of molecules and ions across a cell membrane against their concentration gradients. This process requires the expenditure of energy by the cell. p. 43

adaptation A change or response to stress; in the neuron, receptors decrease the frequency of action potentials to the CNS over time; in evolution, any variety of structure or behavior that increases the likelihood of an organism's survival and the number of offspring it is able to produce relative to other organisms. pp. 7, 169

adaptive radiation The rise of many new species from a single species; new species are specialized for survival in diverse environments. p. 112

adenosine diphosphate (ADP) (ah-den'o-seen) The molecule remaining from the breaking off of a phosphate group when adenosine triphosphate is used to drive synthetic reactions. p. 49

adenosine triphosphate (ATP) (ah-den'o-seen) The energy transfer molecule. ATP is composed of adenine, a ribose sugar, and three phosphate groups held together by high-energy chemical bonds. p. 49

adipose tissue (ad'i-pos) Loose connective tissue specialized for fat storage. p. 129

adrenal glands (a-dre'nal) Paired glands that sit on top of the kidneys; they consist of the medulla and the cortex and secrete hormones. p. 200

adrenocorticotropic hormone (ACTH) (a-dre'no-kor'ti-ko-tro'pik) A tropic hormone secreted by the anterior lobe of the pituitary gland that triggers the adrenal cortex to produce and release corticosteroids. p. 202

afterbirth The placenta, the amnion, the chorion, and some amniotic fluid and blood are expelled from the uterus after childbirth. p. 365

agglutination (uh-gloo'-tih-na'-shun) The clumping together of cells as a result of interaction with specific antibodies. p. 267

aging The sum total of all the changes that lead to functional impairment and an increased probability of death. p. 375

agnosia Total or partial loss of the ability to recognize familiar objects or persons through sensory stimuli as a result of brain damage. The condition may affect any of the senses and is classified accordingly as auditory, visual, olfactory, and so on. p. 177

agranulocytes (mononuclear agranulocytes) White blood cells that lack cytoplasmic granules, such as monocytes and lymphocytes. p. 255

AIDS (acquired immune deficiency syndrome) Destruction of the T helper cells leading to the breakdown of the human immune system. AIDS is caused by the human immunodeficiency virus (HIV). p. 274

aldosterone (al'dah'-ster-on) A mineralocorticoid secreted by the adrenal cortex that regulates potassium and sodium ion concentration in extracellular fluids of the body. p. 202

alimentary canal (gastrointestinal or GI tract) A muscular membranous tube extending from the mouth to the anus and lined with mucous membrane. Its various portions include the mouth, pharynx, esophagus, stomach, small intestine, and large intestine. p. 305

alleles Alternative forms of a gene. p. 73

all-or-none response If the stimulus is sufficient to bring the neuron to threshold potential, an action potential will be generated. p. 145

alveoli (sing. alveolus) (al-ve-o-li) Small outpouchings of walls of alveolar space through which gas exchange takes place between alveolar air and pulmonary capillary blood. p. 285

amino acid (ah-me'no as'id) A simple organic molecule containing an amino group (—NH_2), a carboxyl group (—COOH), a hydrogen atom, and a variable group (R). Each of the four components is covalently bonded to a carbon atom. All the body's proteins are made from various combinations of twenty amino acids. p. 30

amniocentesis (am'-ne-o-sen-te'-sis) A procedure in which a small amount of amniotic fluid is removed for analysis. It is usually performed between the sixteenth and twentieth week of gestation to aid in the diagnosis of fetal abnormalities. p. 357

amnion (am'-ne-on) The membranous sac that is filled with fluid and contains the developing embryo. p. 359

amniotic cavity (am'-ne-ot-ik) The fluid-filled cavity of the amniotic sac surrounding the fetus. p. 359

amygdala (a-mig'da-lah) Part of the limbic system consisting of a group of nuclei located in front of the hippocampus. It receives impulses from all portions of the limbic system. p. 160

anabolic androgenic steroid Any one of several compounds derived from testosterone or prepared synthetically to promote general body growth or to promote masculinizing effects. p. 227

analogous structures Superficially similar structures that evolve in dissimilar and distantly related organisms. p. 110

androgens Any steroid hormone that increases male characteristics. p. 335

anemia A decrease in hemoglobin in the blood below normal range. p. 252

angiotensin (an'-je-o-ten'-sin) The active form of blood protein, angiotensinogen, that stimulates the release of aldosterone from the adrenal cortex. p. 203

antagonistic pairs Two muscles that exert opposite actions to each other. p. 221

antibody An immunoglobulin produced by B lymphocytes in response to bacteria, viruses, or other antigenic substances. An antibody is specific to an antigen. p. 265

antibody-mediated immune response The form of specific immune response in which antibody molecules are secreted in response to antigenic stimulation. p. 265

anticodon A three-nucleotide sequence complementary to the codon of mRNA that specifies the amino acid attached to that tRNA. p. 82

antidiuretic hormone (ADH) (an'ti-di'-u-ret-ik) A hormone produced by the hypothalamus but stored and released by the posterior lobe of the pituitary gland that helps control blood volume by regulating the amount of water reabsorbed by the kidneys. p. 195

antigen A substance, usually a protein, that causes the formation of an antibody and reacts specifically with that antibody. p. 264

aorta The main artery of the body that begins at the left ventricle of the heart. p. 235

appendix A pouch that hangs from the beginning of the large intestine and serves no essential purpose in humans. p. 314

appendicular skeleton (ap'pen-dik'u-lar) Bones of the arms, hands, legs, feet, pelvic girdle, and pectoral girdle. p. 209

aqueous humor A watery fluid filling a chamber behind the cornea that nourishes the cornea and the lens. p. 178

aquifers Underground rock strata that hold water. p. 394

arteries Large vessels that carry blood away from the heart. p. 232

arteriole The smallest blood vessel of arterial circulation. p. 232

artificial selection Controlled breeding in plants and animals to select for or against particular traits. p. 105

association neuron A neuron that combines the information from sensory neurons and then influences other neurons; found mainly in the body's areas of integration such as the brain and spinal cord. p. 142

atherosclerosis (ath'er-o-skle-ro'-sis) An arterial disorder characterized by plaques of cholesterol, lipids, and cellular debris in the inner layers of the wall of large and medium-sized arteries, narrowing the lumen. p. 242

atom The smallest particle of an element that retains the properties of that element. p. 4

atomic mass number The total number of protons and neutrons in the nucleus of an atom. p. 18

atomic nucleus The central mass of an atom that is composed of two types of subatomic particles, protons and neutrons. p. 17

atomic number The total number of protons in the nucleus of an atom. p. 18

atrioventricular (AV) node An area of specialized cardiac muscle that receives the impulses from the sinoatrial node and conducts them to the atrioventricular bundle and then to the walls of the ventricles. The AV node is located in the septal wall of the right atrium. p. 238

atrioventricular (AV) valves One-way valves located between the atria and ventricles. The AV valve on the right side of the heart is the tricuspid, and the AV valve on the left side is the mitral valve. p. 235

atrium (pl. atria) Either of the upper two chambers of the heart. The right atrium receives deoxygenated blood from the superior vena cava and the inferior vena cava. The left atrium receives oxygenated blood from the pulmonary veins. p. 235

Australopithecus (os-tral-oh-pith'-e-kus) A genus of fossil hominid that lived in Africa from approximately 1 to 4 million years ago, characterized by bipedal locomotion, relatively small brain size, large face, and large teeth. p. 115

autonomic nervous system (ANS) Subdivision of the PNS. This system carries sensory information from the internal body organs to the CNS. Efferent pathways conduct action potentials to the internal organs enabling the CNS to control involuntary or automatic activities of internal organs, non-skeletal muscle, and glands. p. 150

autosomal chromosomes The paired chromosomes other than the sex chromosomes. p. 86

autotroph A photosynthetic organism. They are capable of producing their own food from inorganic materials and an environmental energy source, such as sunlight. p. 388

axial skeleton The skull, vertebrae, ribs, and sternum. p. 209

axon A cellular projection extending from the cell body of the neuron that carries action potentials away from the cell body. p. 142

B

B lymphocyte (B cell) A type of lymphocyte that originates and matures in the bone marrow. When exposed to an antigen, they divide rapidly and produce large numbers of new B cells that produce antibodies specific to the antigen. p. 265

basal metabolic rate (BMR) The amount of energy the body requires to perform only the most essential activities of breathing, blood circulation, and cellular metabolism. p. 303

base Any molecule or compound that dissociates in solution to form hydroxide ions. p. 23

basement membrane Membrane that binds epithelium to underlying tissue. p. 125

basophil A granulocytic white blood cell. Its function is poorly understood. Basophils that take up residence in the body's connective tissues are called mast cells. p. 254

benign tumor A mass of cells forming a tumor that reaches a certain size, then stops growing. p. 97

bile A collection of molecules secreted by the liver that helps in the digestion of lipids. p. 312

bilirubin The orange yellow pigment of bile, formed by the breakdown of hemoglobin in red blood cells after termination of their normal life span. p. 312

biological magnification (bioaccumulation) The process by which substances such as pollutants come to be found in increasing concentrations in the tissues of organisms at higher trophic levels. p. 391

biological species Individuals capable of potentially interbreeding and producing viable and fertile offspring. p. 112

biome A group of ecosystems that share a similar climate and support the same types of plants. p. 388

biosphere The Earth's surface and all living organisms on it. p. 388

biotic Pertaining to the living component of any environment. p. 387

birth control pill Oral hormone medication for contraception. The two major hormones used are progesterone and estrogen. The hormones act by inhibiting the production of gonadotropin releasing hormone by the hypothalamus. p. 346

blastocyst (bla'-sto-sist) The embryonic form that follows the morula. It is a spherical mass of cells having a central, fluid-filled cavity surrounded by a single layer of cells. p. 355

blood Connective tissue whose extracellular matrix is liquid called plasma. p. 131

blood-brain barrier Consists of capillaries in the CNS joined by tight junctions, preventing free exchange between blood contents and interstitial fluid of the CNS. p. 154

blood clotting (coagulation) (ko-ag-u-la'shun) The conversion of blood from a free-flowing liquid to a semisolid gel. p. 255

blood pressure The pressure exerted by the circulating volume of blood on the walls of the arteries, veins, and the chambers of the heart. p. 241

bone Specialized connective tissue consisting of collagen fibers that contribute to flexibility and modified calcium that imparts stiffness. p. 131

bradykinin (brad'yki'nin) A peptide that is a vasodilator and also stimulates free nerve endings, resulting in pain sensation. p. 173

brain stem Portions of the brain continuous with the spinal cord. It consists of the medulla oblongata, pons, and midbrain. p. 157

buffer A chemical substance that aids in maintaining neutral pH conditions of a solution. p. 25

C

calorie The energy required to raise the temperature of 1g of water 1 degree C. p. 22

capillary The smallest blood vessel of the circulatory system that carries blood from the arterioles to venules. p. 123

carbohydrate A molecule that contains carbon, hydrogen and oxygen atoms. The ratio of hydrogen to oxygen atoms is 2 : 1. p. 27

carcinogen Any substance that causes cancer. p. 97

carnivore An animal that subsists on other animals. p. 388

carrier One whose chromosomes carry a recessive gene. p. 91

carrying capacity The maximum population of a species that a given environment can sustain over time. p. 395

cartilage (kar'ti-lij) Consists of chondrocytes separated by collagen, elastin fibers, and water. Cartilage is both tough and flexible. p. 129

catalyst A substance that increases the rate of chemical reaction without itself becoming a part of the reaction. A catalyst works by lowering the energy of activation required to initiate a chemical reaction. p. 27

cell A structure bounded by a semipermeable membrane containing cytoplasm, hereditary information and subcellular structures called organelles. p. 4

cell body The body of a nerve cell or neuron containing the nucleus, the sites of other cellular functions, and two types of cellular projections called dendrites and axons. p. 141

cell-mediated immune response A form of specific immune response provided by T lymphocytes, which directly destroy the antigen to which they are sensitized. p. 265

cell (plasma) membrane A membrane that consists primarily of two layers of phospholipid molecules stabilized by specifically oriented proteins. p. 37

cellular garbage theory Theory of aging stating that the gradual accumulation of lipofuscins interferes with normal cellular functioning, eventually causing cells to malfunction and die. p. 382

cellular respiration The conversion of the chemical bonds of nutrients to obtain energy needed for the cells' metabolic activity. It consists of energy-releasing metabolic pathways that hydrolyze glucose and other organic molecules, synthesizing ATP molecules. p. 51

central nervous system (CNS) Consists of the brain and spinal cord and is the site of information processing in the nervous system. p. 150

centriole An organelle surrounded by microtubule-organizing centers that play a key role in mitosis and meiosis. p. 47

centromere A specialized region of a chromosome where its duplicated copy temporarily attaches. p. 57

cerebellum (seh'-rih'beh'lum) Located behind the pons and medulla and protruding dorsally from the brain stem. Processes information received from higher brain centers, various brain stem areas, and sensory receptors from the body to provide precise timing and appropriate patterns of skeletal muscle contraction needed for smooth, coordinated movements. p. 158

cerebral dominance Designates which cerebral hemisphere is dominant for language. p. 163

cerebral hemispheres Two halves of the cerebrum connected by the corpus callosum. p. 160

cerebrospinal fluid A fluid that cushions the brain and spinal cord. p. 154

cerebrum (cerebral cortex) (ser-re'-brum) The outermost tissue and the largest part of the brain. It is the center for association, integration, and learning in the brain. p. 160

cervical cap A contraceptive device consisting of a small rubber cap fitted over the uterine cervix to prevent spermatozoa from entering the cervical canal. p. 348

cervix The part of the uterus that protrudes into the cavity of the vagina. p.341

chancre (shank'-er) A skin lesion, usually of syphilis, that begins at the site of infection. p. 348

chemical bond The attractive force that holds two atoms together in a molecule. p. 20

chemical cycle (nutrient cycle) The pathways chemicals take as they move back and forth from the biotic to abiotic portions of an ecosystem. p. 392

chlamydia (kluh-midh'-de-uh) A sexually transmitted disease caused by the bacterium *chlamydia trachomatous* that usually infects the epithelium of the urethra and cervix in females. p. 349

cholesterol (kuh-les'ter-ol) Steroid lipid found in many body tissues and in animal fats. p. 253

chondrocyte (kon'dro-sits) A cell that produces cartilage, a specialized connective tissue that is hard and strong. p. 129

chorion (kor'ri-on) The outermost embryonic membrane composed of trophoblast cells. It gives rise to the placenta and persists until birth. p. 357

choroid (ko'-royd) A thin, dark brown membrane that lines the sclera, containing blood vessels that nourish the retina and a dark pigment that absorbs light rays so that they will not be reflected within the eyeball. p. 176

chromosome The hereditary material of the cell composed primarily of DNA. p. 57

cilia (sih'-le-uh) Tiny, hairlike protrusions of the cell membrane surface that allow a cell to move materials across its surface. p. 48

ciliary muscle A tiny circular muscle that slightly changes the shape of the lens by contracting or relaxing. p. 176

circumcision A surgical procedure in which the prepuce, the foreskin of the penis, is removed. p. 337

citric acid cycle (Krebs cycle) The second metabolic pathway of cellular respiration. It is an aerobic metabolic pathway. p. 52

cleavage The series of repeated mitotic cell divisions occurring in the ovum immediately after fertilization to form a mass of cells that transforms the single-celled zygote into a multicellular embryo capable of growth and differentiation. p. 354

clinal distribution The gradual variation in a trait over space shown by the alteration in its genotypic or phenotypic frequency from population to population. p. 410

clitoris (klit'-o-ris) A small mass of erectile and nervous tissue in the female that responds to sexual stimulation; it is homologous to the tip of the penis in males. p. 341

clone The replication of identical copies of a cell. p. 100

cochlea A tightly coiled bony structure of the inner ear that houses the receptor organ for hearing. p. 186

codominance The equal degree of dominance of two alleles fully expressed in the heterozygous phenotype. p. 90

codon A sequence of three nucleotide bases in transcribed mRNA that code for a specific amino acid. p. 80

coenzyme A non-protein cofactor. p. 51

coenzyme A (CoA) A carrier molecule that accepts an acetyl group and a hydrogen atom during glycolysis to become acetyl-CoA. p. 52

cofactor A substance that assists an enzyme. p. 50

collagen (kol'la'jen) Large protein fibers that are strong and flexible but do not stretch. Collagen is found in connective tissue. p. 30

collecting tubule Any one of the many relatively large straight tubules of the kidney that funnel urine into the renal pelvis. p. 323

colon The portion of the large intestine extending from the cecum to the rectum. p. 314

compact bone Dense and highly organized bone. Osteocytes are organized into Haversian systems around canals containing blood vessels and nerves. p. 131

complementary base pairs The nucleotide bases that pair via hydrogen bonds. Adenine pairs with thymine and cytosine pairs with guanine. p. 79

complete protein Any protein that provides all the essential amino acids the body cannot manufacture. p. 300

compounds Molecules containing two or more different types of atoms. p. 19

concentration gradient A gradient from an area of higher concentration to lower concentration. p. 40

condom A soft, flexible sheath that covers the penis and prevents semen from entering the vagina in sexual intercourse, and is used to avoid transmission of an infection and to prevent conception. p. 347

connective tissue Binds structures together, provides support and protection, fills spaces, stores fat, and forms blood cells. p. 129

continuous variation Small degrees of phenotypic variation occurring over the entire range of a trait. p. 96

contraceptive sponge A sponge placed at the opening of the cervix and containing a spermicide. p. 348

control group A group identical to the experimental group, except for the single variable being tested. p. 10

convergence The relationship between a single neuron and other neurons synapsing on it. p. 181

cornea The transparent portion of the eye's outer layer, the sclera, that permits light to enter the eye. p. 176

corona radiata An aggregate of cells that surrounds the zona pellucida of the ovum. p. 342

corpus callosum A large tract of nerve fibers connecting the cerebral hemispheres. p. 160

corpus luteum A structure that emerges from a ruptured follicle in the ovary after ovulation. It secretes increased estrogen and progesterone, preparing the endometrium for the implantation of the fertilized ovum. p. 342

covalent bond A chemical bond formed when two atoms share one or more pairs of electrons. p. 20

Cowper's glands A set of tiny accessory glands lying beneath the prostate gland that secrete an alkaline fluid into the semen. p. 337

crista ampullaris A small elevation in the vestibule of the semicircular canal, containing hair cells. p. 187

cross-pollination Process of fertilization of one plant by another plant. p. 64

crossing-over When non-sister chromatids of homologous chromosome pairs exchange segments of DNA. p. 71

cytokinesis (si-to-kin-e'sis) The division of the cytoplasm into equal halves that occurs during mitosis. p. 59

cytoplasm (si'-to-plaz-zim) A semi-fluid material in the cell that contains chemical substances such as sugars, amino acids, proteins, and ions, that are necessary for cell maintenance, growth, and reproduction. p. 37

cytoskeleton Composed of several different protein fibers, it is responsible for maintaining cell shape and anchoring the organelles in the cytoplasm. p. 48

D

deamination The removal, usually by hydrolysis, of the NH_2 group from an amino acid. p. 313

decomposers Organisms that break down organic molecules of dead organisms and their byproducts and contribute to recycling both organic and inorganic molecules back to the environment. p. 388

deletion A type of chromosomal abnormality in which a portion of a chromosome is lost. p. 85

dendrite A cellular projection that extends from the cell body of the neuron, receives action potentials, and conducts them toward the cell body. p. 142

dense connective tissue Connective tissue containing a large amount of closely packed collagen oriented in the same direction. p. 129

deoxyribonucleic acid (DNA) (de-ox'si-ri'bo-nu-kle'ik) A nucleic acid present in the chromosomes that is the chemical basis of heredity and the carrier of genetic information. DNA is arranged in two long strands that twist around each other to form a double helix. p. 33

dependent variable The variable that is influenced by or dependent upon the independent variable. p. 10

dermis Layer underlying the epidermis and composed of connective tissue. p. 170

desmosome (dez-mo-som) A small circular area comprised of protein that forms the sites of adhesion between epithelial cells. p. 127

diabetes mellitus (di-uh-be'tez muh-li'tus) The hyposecretion of insulin by the pancreas. p. 205

dialysis (di-al'i-sis) A medical procedure for the removal of certain elements from the blood by virtue of the difference in their rates of diffusion through an external semipermeable membrane. Dialysis involves diffusion of particles from an area of higher concentration to an area of lower concentration. p. 331

diaphragm (di'uh-fram) In the respiratory system, a dome-shaped skeletal muscle sheet at the bottom of the thoracic cavity; upon contraction, it expands the thoracic cavity for inhalation. In contraception, a circular dome-shaped structure made of latex rubber inserted into the vagina to prevent sperm from entering the uterus. pp. 288, 348

diastole (di-as'-to-le) The period of time between contractions of the atria or the ventricles during which blood enters the relaxed chambers from systemic circulation and the lungs. p. 241

diffusion The random movement of molecules or ions down their concentration gradients from an area of higher concentration to an area of lower concentration. Also called simple diffusion. p. 40

digestion The mechanical and chemical breakdown of food and its absorption into the circulatory system for distribution to tissue cells. p. 304

dihybrid cross Breeding for two characteristics at a time. p. 67

diploid (dip'loyd) A cell with pairs of homoiogous chromosomes. There are 23 homologous pairs of chromosomes in human somatic cells. p. 57

dizygotic (fraternal) twins Two offspring born of the same pregnancy and developed from two ova that were released from the ovary simultaneously and fertilized at the same time. Such twins are as genetically similar as any two siblings born at different times. p. 355

DNA polymerase (pol'i-mer-as) An enzyme that controls the synthesis of a new complementary strand of DNA. p. 79

dominant The form or variety of a trait that will be expressed in a hybrid. p. 65

dorsal root ganglion A swelling consisting of sensory cell bodies located in the dorsal root of the spinal cord. p. 155

Down syndrome (trisomy 21) Condition in which an individual has three chromosomes #21. p. 85

E

ecology The science concerned with understanding the interrelationships among living things and their environments. p. 387

ecosystem The interrelated network of living organisms and their nonliving surroundings. p. 387

ectoderm The outermost of the three primary germ layers of an embryo. p. 359

edema (e-de'mah) The abnormal accumulation of fluid in interstitial spaces of tissues that causes swelling. p. 333

elastin A protein that will recoil back to its original shape if stretched. p. 129

electrocardiogram (ECG or EKG) A graphic record produced by an electrocardiograph. p. 238

electroencephalogram (EEG) (ih-lek-tro-en-seh'-fuh-lo-gram) A recording of the overall electrical activity of the brain. p. 165

electron A negatively charged subatomic particle that orbits the atomic nucleus. p. 17

electron transport chain The third metabolic pathway of cellular respiration. An aerobic metabolic pathway where a series of carrier proteins pass hydrogen ions and electrons to progressively lower energy states, releasing energy to produce ATP molecules. p. 52

element A substance made up of one type of atom and that cannot be divided or decomposed by chemical change into simpler substances. p. 17

embolus A foreign object, a quantity of air or gas, a bit of tissue or tumor or a piece of a thrombus that circulates in the bloodstream until it becomes lodged in a vessel. p. 242

embryo (em'bri-o) Human developmental stage between the third and eighth weeks. p. 353
embryonic disk The thickened plate from which the embryo develops in the second week of pregnancy. The disk develops from the endoderm and ectoderm. p. 359
encapsulated nerve ending Sensory neuron having one or more terminal dendrites enclosed in a connective tissue capsule. p. 171
endocrine gland A ductless gland that secretes chemical substances, called hormones, directly into the blood where they are distributed throughout the body. p. 126
endocytosis The process in which cells engulf large molecules or particles and bring these substances into the cell packaged within vesicles. p. 43
endoderm The innermost of the cell layers that develop from the embryonic disk of the inner cell mass of the blastocyst during early embryonic development. p. 359
endometrium The mucous membrane lining the uterus, consisting of two layers of cells, the functionalis and the deeper basalis. p. 341
endoplasmic reticulum (ER) (re-tik'u-lum) A membranous system of interconnected fluid-filled tubules and flattened sacs in the cytoplasm. It is continuous with the nuclear envelope. p. 45
endorphin A neuropeptide hormone acting on the CNS to reduce pain perception. p. 206
enkephalins (en-kef'a-linz) A class of peptide hormones responsible for blocking afferent signals from the body's pain receptors, thereby preventing the brain from perceiving pain. p. 206
enzyme A protein produced by living cells that catalyzes, or speeds up, specific chemical reactions. p. 27
eosinophil (e'-o-sin-o-fil) A granulocytic white blood cell. It increases in number with allergy and some parasitic infections. p. 254
epidermis Outer layer of the skin consisting of layered epithelial cells. p. 170
epididymis (ep-i-did'i-mis) A large, coiled tube that lies alongside the testes where sperm undergo further development after their formation within the seminiferous tubules of the testes. p. 335
epithelial tissue (ep-i-the-le-ul) A type of tissue that forms the outer covering of the internal and external organs of the body, as well as the lining of vessels, body cavities, glands, and organs. p. 125
erectile tissue A spongy network of tissue that fills with blood causing the penis to become stiff and erect. p. 337
erythrocyte (red blood cell) A biconcave cell that contains hemoglobin. Its principal function is to transport oxygen. p. 250
erythroblastosis fetalis (hemolytic disease of the newborn) (e-rith'ro-blasto'-sis feta'lis) Destruction of an unborn infant's red blood cells by the mother's immune system; usually caused by Rh factor incompatibilities between the blood of the mother and fetus. p. 273
esophagus (e-sof'-uh-gus) The muscular food tube about 24 cm long that extends from the pharynx to the stomach. p. 309
essential amino acids Amino acids that must be directly acquired from the diet because they cannot be manufactured. There are nine essential amino acids. p. 300
estrogen A steroid hormone secreted by the female ovaries that is responsible for the maturation of the female reproductive organs and the appearance of the secondary sexual characteristics in the female. p. 203
eustachian tube (u-sta'she-an) A tube, lined with mucous membrane, that joins the upper portion of the pharynx and the middle ear cavity, allowing equalization of the air pressure in the middle ear with atmospheric pressure. p. 185
evolution The permanent heritable changes accumulated in a population of organisms through time. p. 7
exocrine gland A gland that secretes its substance through a duct onto body surfaces (such as sweat glands) or into body cavities (such as salivary glands). p. 126
exocytosis A process of cellular secretion in which organelles create a membrane-bound vesicle around the material to be removed. The vesicle then fuses with the cell membrane, releasing its contents to the exterior of the cell. p. 43
experimental group The group in which the variable of interest is to be altered or manipulated to test the stated hypothesis. p. 10
exponential growth In ecology, rate of population growth doubles from generation to generation. p. 394
external respiration The exchange of oxygen and carbon dioxide between the blood and air in the alveoli of the lungs. p. 283

F

fallopian (uterine) tube (yoo'-ter-in) One of a pair of ducts opening at one end into the uterus and at the other over the ovary; serves as a passageway for the ovum to the uterus. p. 340
fat-soluble vitamins Vitamins that bind to ingested lipids and are absorbed with them from the digestive system and can be stored in the body. Fat-soluble vitamins include vitamins A, D, E, and K. p. 301
fatty acid A long hydrocarbon chain ending in a carboxyl group (COOH). A fatty acid can be saturated, unsaturated, or polyunsaturated. p. 28
feces (fe-sez) Body waste discharged by way of the anus. p. 318
fertilization The fusion of the sperm and ovum resulting in a zygote. p. 353
fetal alcohol syndrome (FAS) A set of congenital psychological, behavioral, and physical abnormalities that tend to appear in infants whose mothers consumed alcoholic beverages during pregnancy. p. 362
fetus The developing human from the beginning of the ninth week of gestation to birth. p. 353
fever An abnormal elevation of the temperature of the body above 37 degrees C. (98.6 degrees F.) because of disease. p. 264
fibroblast A branching, connective tissue cell that produces collagen. p. 129
fibrocartilage Specialized connective tissue consisting of fewer collagen fibers than hyaline cartilage. Fibrocartilage is found in weight-bearing areas of the body, such as knee joints and pads between the vertebrae of the spinal column. p. 130
fimbriae (fim-bre-e) The branched, finger-like projections at the distal end of each fallopian tube. The projections are connected to the ovary. p. 340
flagellum (pl. flagella) (fluh-jeh'-lum) A single, long, whip-like appendage that allows a cell to propel itself through its environment. p. 48
flavin adenine dinucleotide (FAD) A coenzyme in the citric acid cycle. Each FAD accepts a hydrogen ion to become FADH. FADH is used in the electron transport chain to produce ATP. p. 52
fluid mosaic model The most widely accepted theory of cell membrane structure, which states that the membrane consists of a fluid phospholipid bilayer in which individual proteins are embedded. p. 39
follicle A cavity or recess in an ovary containing a fluid that divides the follicular cells into layers that surround the ovum. Also called ovarian follicle. p. 340
follicle-stimulating hormone (FSH) A hormone secreted by the anterior pituitary gland that triggers the maturation of one oocyte each month in females. In males it triggers sperm production. p. 339
follicular phase The first part of the menstrual cycle, when ovarian follicles grow to prepare for ovulation. p. 343
food chain A sequence of consumption among organisms; the movement of energy and nutrients from one trophic level to the next. p. 388
food web The complex set of interrelationships among members of different trophic levels in an ecosystem. p. 388
formed elements The solid portion of blood plasma composed of red blood cells, white blood cells, and platelets. p. 249
fossil (fos-el) The mineralized parts of an organism's skeleton or impressions of its anatomy left in sedimentary rock. p. 109
fovea centralis The area of sharpest vision within the retina due to its high concentration of cones. p. 180
frameshift mutation A mutation that results in one nucleotide base pair in a gene being added or deleted. p. 85

G

Gaia hypothesis (Ga'uh hi-poth'e-sis) A hypothesis that states that the physical and chemical condition of the Earth's surface, the atmosphere, and oceans has been and is actively made fit by the presence of life itself. p. 14

gallbladder A pear-shaped sac that serves as a reservoir for bile, which it receives from the liver via the hepatic duct. During digestion of fats, the gallbladder contracts, ejecting bile through the common bile duct into the duodenum. p. 313

gamete (gam'-et) A mature haploid male or female sex cell that is capable of functioning in fertilization and contains the haploid number of chromosomes of the somatic cells. p. 70

ganglia (gang'-gli-ah) Groups or clusters of cell bodies in the PNS. p. 150

gap junction Site where clusters of small channels across the membrane of one cell match up with similar channels of an adjacent cell. p. 128

gastric glands Glands in the stomach mucosa that secrete hydrochloric acid and pepsinogen. p. 311

gene A sequence of DNA nucleotide triplets on a chromosome that codes for a polypeptide. p. 81

gene mutation theory Theory stating that with advancing age, mutations become more numerous, leading to cellular, tissue, and organ malfunction, and the increased probability of death. p. 382

gene theory of aging Theory stating that aging is the result of normal processes controlling earlier development and differentiation. With advancing age, these processes begin to interfere with cell function, leading to cellular dysfunction and death. p. 382

genotype (je'no-tip) The total set of genes of an individual. May also refer to the gene or set of genes for a particular characteristic. p. 73

geographical isolation When a single population is split up into two separate populations and mating between them is no longer possible. p. 112

germ layer One of the three primary cell layers (ectoderm, mesoderm, endoderm) formed during the early stages of embryonic development from which the entire range of body tissue is derived. p. 359

gerontology (jer-on-tol'o-je) The study of all aspects of the aging process, including the physical, psychological, economic, and sociologic problems encountered by the elderly. p. 375

gestation period (jes-ta'-shun) The time span between fertilization and labor. The period is approximately 40 weeks. p. 353

gland One or more cells that produce and secrete a product. p. 126

glans penis The conical tip of the penis. The urethral opening is normally located at the center of the distal tip of the glans penis. p. 337

glomerulus (glo-mer'u-lus) A structure composed of a cluster of capillaries that fits into the glomerular capsule of the nephron. p. 323

glucagon (gloo'-kuh-gon) A peptide hormone secreted by the alpha cells of the pancreas that increases blood glucose levels by enhancing the conversion of liver glycogen into glucose. p. 204

glycerol The three-carbon molecule to which fatty acids are attached in fats. p. 28

glycogen A polysaccharide that is the major form of glucose storage in the body. p. 27

glycolysis The first metabolic pathway in cellular respiration. It is an anaerobic pathway that breaks down each glucose molecule into two molecules of pyruvate. Glycolysis releases enough energy to produce a net gain of two ATP molecules. p. 51

Golgi bodies A set of flattened, slightly curved membrane-bound sacs in the cytoplasm of a cell. Golgi bodies collect, package, and distribute molecules synthesized elsewhere in the cell. p. 45

gonad (go'-nad) A male or female reproductive sex organ that produces gametes. p. 335

gonadotropin-releasing hormone (GnRH) A hormone secreted by the hypothalamus that stimulates the release of LH and FSH by the anterior pituitary gland. p. 339

gonorrhea (gah'-nor-e'uh) A sexually transmitted disease caused by the bacterium *Neisseria gonorrhoea* that causes a primary infection of the urethra in men and women and of the vagina and cervix in women. p. 348

granulocytes (polymorphonuclear granulocytes) White blood cell characterized by the presence of cytoplasmic granules. Kinds of granulocytes are basophil, eosinophil, and neutrophil. p. 254

greenhouse effect The build up in the Earth's atmosphere of carbon dioxide and other gases, allowing heat to enter the atmosphere but preventing it from escaping; may be contributing to global warming. p. 396

growth hormone (GH) (somatotrophic hormone) A hormone that stimulates most body cells to increase in size by promoting the cellular synthesis of proteins. It also increases the rate of cell mitosis. p. 198

H

habitat The location within an environment in which a local population lives. p. 387

haploid (N) (hap'loyd) A cell containing half of the hereditary material of normal body cells. p. 70

hard palate Located toward the front of the mouth, it consists of tissue-covered bone and forms a rigid surface against which the tongue can force food during chewing. p. 307

haversian canal (nutrient canal) One of the many tiny longitudinal canals in bone tissue. Each contains blood vessels, connective tissue, and nerves. p. 218

haversian system (osteon) The basic structural unit of compact bone, consisting of the haversian canal and its concentric rings of 4 to 20 lamellae. The units run with the long axis of the bone. p. 218

heart The muscular, cone-shaped organ, about the size of a clenched fist, that pumps blood throughout the body and beats normally about 70 time per minute, coordinating action potentials and muscular contractions. p. 235

hemoglobin A complex protein-iron compound in the red blood cell that carries oxygen to the cells from the lungs. p. 250

hemophilia (he'mo-fil'i-ah) A hereditary condition in which the blood is slow to clot or does not clot. Several forms of hemophilia are sex-linked. p. 91

hepatic duct The duct that carries bile produced by the liver to the gallbladder where it is temporarily stored. p. 313

herbivore An animal that subsists entirely on plants. p. 388

heredity (he-red'i-ti) The transmission of characteristics from parent to offspring. p. 5

herpes, genital (her'-pez) Sexually transmitted disease caused by herpes simplex type II virus. p. 350

heterotroph An organism that directly or indirectly consumes autotrophs and is incapable of producing its own food from inorganic materials. p. 388

heterozygous (het-er-o-zi'-gus) Having two different alleles for a trait. p. 73

high-density lipoprotein (HDL) A plasma protein made in the liver and containing about 50% protein with cholesterol and triglycerides. It is involved in transporting cholesterol and other lipids from the body to the liver. p. 253

hippocampus A part of the limbic system. It is responsible for generating particular limbic system responses based on current incoming sensory information compared with previous encounters with the same information found in memory. p. 159

histamine A compound found in all cells. It is released in allergic, inflammatory reactions and causes dilation of capillaries as well as constriction of smooth muscle of the bronchi and uterus. p. 254

homeostasis (ho'-me-o-sta'sis) The process of maintaining relatively stable internal conditions under changing external environmental conditions. p. 5

hominids Human-like primates. p. 115

Homo erectus The second genus that evolved from one of the Australopithecine groups. H. erectus is characterized by a larger brain and a relatively small face. This species lived from approximately 1.6 million years to 200,000 years ago. p. 117

Homo habilis The first known species of the genus Homo, dating between 1.5 and

2 million years ago and found in Africa. In overall appearance, this species is similar to the australopithecines but has a larger cranial capacity of, on average, 660 ml. p. 117

Homo sapiens Living, modern humans. A hominid whose fossil record dates back to approximately 35,000 years ago. p. 114

homologous structures Characteristics with similar internal structures. p. 110

homozygous (ho-mo-zi'gus) Having two identical alleles for a trait. p. 73

hormonal theory of aging The theory that the aging process is primarily a by-product of reduced levels of various hormones that cause tissue and organ structure and function to eventually decline. p. 382

hormone A chemical substance produced and secreted in one part or organ of the body that initiates or regulates the activity of an organ or group of cells in another part of the body. p. 191

host An organism in which another, usually parasitic, organism is nourished and harbored. p. 8

Huntington's disease An autosomal dominant genetic disorder causing a slow progressive deterioration of parts of the brain. p. 89

hyaline cartilage (hi'a-lin kar'ti-lij) Cartilage containing large amounts of collagen giving it a very smooth appearance; found in the nose, at ends of many bones and ribs, and rings of airway passages. p. 129

hybrid The offspring produced from cross-pollination. p. 64

hydrogen bond A weak chemical bond that occurs when a partially positive hydrogen atom in a polar molecule is attracted to a partial negative atom of another polar molecule. p. 21

hydrolysis reaction A process of breaking covalent bonds between two organic molecules. A hydrogen ion is added to one molecule and a hydroxide ion is added to the other molecule. The energy held in the chemical bond is released. p. 26

hyperopia (farsightedness) (hi-per-ah'-pe-uh) When a close object is focused behind the retina. p. 183

hypertension Disorder characterized by elevated blood pressure persistently exceeding 140/90 mm Hg. Essential hypertension, the most frequent kind, has no single identifiable cause. p. 241

hypertonic Solute concentration is greater on one side of the membrane than the other. p. 41

hypothalamus (hi'po-thal'a-mus) A mass of nervous tissue lying at the base of the brain that produces two hormones and controls the secretion of hormones by the pituitary gland. It controls homeostasis, and activates, controls, and integrates the ANS and endocrine system. p. 159

hypothesis (hi-poth'e-sis) A statement that a particular preceding event is the cause of a particular subsequent observation; that A causes event B. p. 10

hypotonic Solute concentration is less on one side of the membrane than on the other. p. 41

I

immunity The quality of being insusceptible to or unaffected by a particular disease or condition. p. 254

immunodeficiencies Any of a group of health conditions caused by a defect in the immune system and generally characterized by susceptibility to infection and chronic diseases. p. 273

implantation The process involving the attachment, penetration, and embedding of the blastocyst in the lining of the uterine wall. The process occurs over a period of a few days, beginning about the seventh or eighth day after fertilization. p. 356

incomplete dominance Two alleles, neither of which is dominant. p. 90

independent variable The causative agent controlled and manipulated by the experimenter. p. 10

induction The process of stimulating and determining differentiation in a developing embryo through the action of chemical substances transmitted from one to another of the embryonic parts. p. 361

inert element An atom whose outermost shell is maximally filled with electrons. p. 18

inflammatory response A tissue reaction to injury or an antigen. The response may include pain, swelling, itching, redness, and heat. There may be dilation of blood vessels, leakage of fluid from the vessels causing swelling, and the release of histamine. p. 261

inner cell mass A cluster of cells localized at one end of the blastocyst, from which the embryo develops. p. 355

insertion The place of attachment, such as of a muscle to the bone it moves. p. 220

inspiration Drawing air into the lungs; it occurs when the volume of the thoracic cavity is increased and the resulting negative pressure causes air to be drawn into the lungs. p. 283

insulin A peptide hormone secreted by the beta cells of the pancreas that increases blood glucose levels by enhancing membrane transport of glucose into muscle cells, fat cells and liver cells. p. 204

interferon A cellular protein formed when cells are exposed to a virus. It induces the production of interferon in non-infected cells. The induced interferon blocks translation of viral RNA, thus giving other cell protection against both the original and other viruses. p. 263

internal respiration The movement of oxygen from the blood into the body tissues and the removal of carbon dioxide from the tissues back into the blood. p. 283

interphase Phase of cell cycle preceding mitosis. p. 57

interstitial cells (in'-ter-stih'-shul) Cells of the interstitial tissue of the testes that secrete testosterone. p. 335

interstitial fluid (extracellular fluid) The medium through which substances are exchanged between the cells and the body's circulatory system. p. 123

intrapulmonary pressure The gas pressure within the lungs. When the thoracic cavity expands, a negative intrapulmonary pressure is created, drawing atmospheric air into the lungs. p. 289

intrauterine device (IUD) A plastic or metal device inserted into the uterus that prevents the developing embryo from being implanted in the endometrium. p. 348

inversion A type of chromosomal abnormality in which the orientation of a portion of a chromosome becomes reversed. p. 85

ion An atom that carries a positive or negative electrical charge. p. 18

ionic bond A chemical bond in which electrons are transferred from one atom to another. p. 20

iris A diaphragm lying between the cornea and the lens that controls the amount of light entering the eye. p. 178

islets of Langerhans Clusters of pancreatic cells that produce the hormones glucagon and insulin. p. 204

isotonic Solute concentration is equal on both sides of the membrane. p. 41

isotope An atom with the same number of protons but different numbers of neutrons. p. 18

J

joints Any one of the connections between bones. Each is classified according to structure and movability as immovable, slightly movable, and freely movable. p. 216

juxtaglomerular cells (JG cells) (juks'ta-glo-mer'u-lar) Specialized cells lining the glomerular end of the afferent arterioles that are in opposition to the distal convoluted tubule. These cells synthesize and store renin and release it in response to decreased renal blood pressure or decreased sodium concentration in fluid in the distal tubule. p. 328

K

karyotype The particular array of chromosomes that belongs to an individual. p. 86

keloid An overgrowth of collagenous scar tissue at the site of a wound of the skin. p. 133

kidneys Paired organs located at the back of the abdominal cavity. The kidneys function to excrete cellular waste products from the body and to maintain body fluid composition. p. 321

kilocalorie (Kcal) One kilocalorie equals 1,000 calories. It is also referred to as Calorie spelled with a capital C. p. 48

kingdom One of the two primary taxonomic divisions into which organisms are grouped. p. 114

Kleinfelter's syndrome The syndrome produced in a male by the presence of an extra X chromosome. The genotype is XXY. p. 94

L

labia majora (la'-be-uh mah-jor'-uh) Two long lips of skin, one on each side of the vaginal opening outside the labia minora. p. 341

labia minora (la'-be-uh mih-nor'-uh) Two folds of skin between the labia majora, extending from the clitoris backwards on both sides of the vaginal opening, ending between it and the labia majora. p. 341

lactase An enzyme that breaks down the disaccharide lactose into glucose and galactose. p. 315

lacteal (lak'-te-ul) One of the many lymphatic capillaries in the villi of the small intestine. p. 311

lag phase In ecology, the flat portion of a sigmoid curve representing a period of slow growth. p. 394

large intestine The portion of the digestive tract comprising the cecum; appendix; the ascending, transverse, and descending colons; and the rectum. The ileocecal valve separates the cecum from the small intestine. p. 314

law of independent assortment The distribution of hereditary factors for one characteristic into the reproductive cells does not affect the distribution of hereditary factors for other characteristics. p. 69

law of segregation Pairs of hereditary factors segregate or separate from one another during the formation of reproductive cells, and each reproductive cell receives only one of the pair of hereditary factors. p. 67

lens The convex, transparent structure located behind the iris in the eye that can change shape to allow focusing of light on the retina. p. 182

let-down reflex A normal reflex in a lactating woman elicited by tactile stimulation of the nipple, resulting in release of milk from the glands of the breast. p. 366

leukocyte (white blood cell) (lu'ko-sit) Any of several kinds of white blood cells including macrophages and lymphocytes, all functioning to defend the body against invading microorganisms and foreign substances. p. 249

limbic system A ring of structures surrounding the brain stem and interconnected by nerve pathways. It is responsible for many motivational states. p. 159

linked genes Genes located on the same chromosome that tend to be inherited together. p. 75

lipase An enzyme that breaks down the triglycerides in lipids to fatty acids and glycerol. p. 315

lipid Macromolecules made up of glycerol and fatty acids that are insoluble in polar solvents such as water. Examples include oils, fats, phospholipids, cholesterol, and steroids. p. 300

lipofuscins (lip-o-fus'inz) Modified fat molecules that are found in abundance in cells of adults. Lipofuscins may contribute to the aging process in a cell. p. 383

liver A large, complex organ weighing over three pounds, lying just under the diaphragm. The liver performs over 500 functions in the body, including aiding in the digestion of lipids. p. 311

locus The location of a gene on a chromosome. p. 73

long-term memory Memory storage involving the formation of new permanent synaptic connections between specific neurons, or the strengthening of existing but weak synaptic connections. p. 165

loop of Henle The U-shaped portion of the renal tubule, consisting of a thin descending limb and a thick ascending limb. p. 323

loose connective tissue Connective tissue that supports most epithelial tissue and organs. p. 129

low-density lipoprotein (LDL) A plasma protein containing relatively more cholesterol and triglycerides than protein. It delivers lipids to body tissues. p. 253

lumen A cavity or the channel within any organ or structure of the body. p. 231

luteal phase Second phase of the ovarian cycle beginning after ovulation and ending with the beginning of menstruation. p. 343

luteinizing hormone (LH) A hormone released by the anterior lobe of the pituitary gland that causes the gonads to secrete steroid hormones. p. 339

lymph A thin, watery fluid originating in organs and tissues of the body that circulates through the lymphatic vessels and is filtered by the lymph nodes. p. 244

lymph node One of the many small oval structures that filter the lymph and fight infection, and in which there are formed lymphocytes, monocytes, and plasma cells. p. 244

lymphatic system Network of vessels, duct, nodes, and organs that helps to protect, maintain, and circulate extracellular fluid of the body. It also returns some of the extracellular fluid to the blood. p. 135

lysosome A membrane-bound organelle containing digestive enzymes. Lysosomes are the primary organelles of intracellular digestion. p. 46

M

macromolecule Large organic molecules made up of smaller linked molecules. p. 26

major histocompatibility complex (MHC) A group of linked genes responsible for encoding protein self-recognition markers found on all cell membrane surfaces. p. 95

malignant tumor A mass of cells that continue to grow and often spread to other areas of the body. p. 97

mammary gland One of two hemispheric glands on the chest of mature females. Glandular tissue forms a radius of lobes containing sacs, each lobe having a system of ducts for the passage of milk from the sacs to the nipple. p. 366

mast cells Basophils that reside in the connective tissues of the body. Mast cells contain a number of substances, including bradykinin and histamine, which are released in response to injury and infection. p. 254

mechanoreceptor Sensory neuron sensitive to mechanical energy, such as touch. p. 170

medulla oblongata The lowest portion of the brain stem, continuous with the spinal cord. The site of neuron tracts that decussate. p. 158

meiosis (mi-o'-sis) Cell division that produces haploid gametes. p. 70

Meissner's corpuscle Egg-shaped mechanoreceptor found in the upper part of the dermis that detects texture. p. 172

melanin A brown pigment that occurs naturally in the skin. Melanin is produced by melanocytes, giving hair, skin, and the iris and choroid of the eye their color. p. 409

membrane A thin sheet or layer of tissue that covers a structure or lines a cavity. p. 133

membrane potential The separation of oppositely charged substances across a cell membrane. p. 143

memory cell Memory cells, derived from activated B and T cells, do not participate in the specific immune response, but do on re-exposure to the specific antigen. p. 265

menarche (me-nar'ke) The first menstruation and the beginning of cyclic menstrual function. p. 371

meninges Any of the three layers of membranes covering both the brain and the spinal cord. p. 154

menopause The cessation of menstrual activity and ovulation in a woman, usually between the ages 45 to 55. p. 345

menstrual phase Beginning of the uterine cycle with the shedding of the functionalis layer of the endometrium. p. 344

Merkel disc Sensory neuron located in the lower regions of the epidermis that is a light touch and superficial pressure receptor. p. 172

mesoderm The middle of the three primary germ layers of the developing embryo; helps form bone, blood, muscle, and reproductive organs. p. 359

messenger RNA (mRNA) The newly formed single stranded RNA produced from the process of transcription. p. 81

metabolic pathway The orderly sequence of all cellular metabolic chemical reactions. Each step of a metabolic pathway is controlled by a specific enzyme. p. 50

metabolism (me-tab'o-lizm) The process used by an organism of acquiring energy and material from the environment, converting them to new forms to grow and maintain itself. p. 5

metastases (me-tas'ta-sez) Processes in which cells of a malignant tumor can separate from the tumor, travel to a different area, and divide to produce a tumor. p. 97

midbrain Primary function is to pass action potentials between the spinal cord and the higher brain centers. p. 158

mineral Inorganic substances ingested as compounds that play a vital role in regulating many body functions. p. 302

mitochondrion (pl. mitochondria) (mi-to-kon-dre-uh) The energy organelle of the cell. The mitochondrion is able to extract usable energy from nutrients ingested by the cell. p. 46

mitosis Somatic cell division producing two identical daughter cells and divided into four phases: prophase, metaphase, anaphase, and telophase. p. 57

molecule A structure composed of two or more atoms that are chemically combined. p. 4

monohybrid cross Breeding for only one trait or characteristic at a time. p. 64

monosaccharide A simple sugar molecule. p. 27

monozygotic (identical) twins Two offspring born of the same pregnancy and developed from a single fertilized ovum that splits into equal halves during an early cleavage phase in embryonic development, giving rise to separate fetuses. Such twin are genetically identical. p. 355

mons pubis (monz pyoo'-bis) A pad of fatty tissue and skin that overlies the symphysis pubis in the female. After puberty it is covered with pubic hair. p. 341

morning sickness A common condition of early pregnancy, characterized by recurrent nausea, often in the morning, that may result in vomiting. It usually does not begin before the sixth week after the last menstrual period and ends by the twelfth to the fourteenth week of pregnancy. p. 366

morula (mor'uh-luh) A solid, spherical mass of cells resulting from the cleavage of the zygote in the early stages of development. p. 354

motor (efferent) neuron A neuron that carries nerve action potentials away from the spinal cord and brain to muscles or gland cells. p. 142

mucous membrane Membrane composed of epithelial cells, basement membrane, which rests on a thick layer of loose connective tissue. Also contains goblet cells that secrete mucus to protect the membrane. p. 133

muscle spindle A mechanoreceptor found throughout all skeletal muscle tissue. It is able to detect a muscle distention or stretch. p. 173

muscle tissue Any of three kinds of muscle fibers including skeletal, cardiac, and smooth. Muscle tissue contains two types of protein called myosin and actin. It is able to contract and is responsible for movement. p. 131

mutagen Any external factor or agent capable of damaging DNA. p. 86

mutation A change in the genetic message of a cell. p. 83

myelin sheath The lipid wrapping created by multiple layers of the Schwann cell membrane. The myelin sheath insulates the axon. p. 143

myocardial infarction Death of a portion of cardiac muscle caused by obstruction in a coronary artery from either atherosclerosis or an embolus. p. 243

myocardium (mi'o-kar'di-um) The thick middle layer of the heart wall composed of cardiac muscle tissue. p. 235

myofibril A cylindrical, organized arrangement of special thick and thin microfilaments capable of shortening a muscle fiber. p. 221

myopia (nearsightedness) (mi-o'-pe-uh) When distant objects are focused in front of the retina. p. 183

myosin A protein found in muscle fibers that acts with actin to bring about contraction and relaxation. Also called thick filament. p. 131

N

natural selection The process through which organisms exhibit variations that make them best suited to survive in the environment and produce more offspring than other organisms of the same population. In light of genetics, it is defined as differential reproduction of alternative genotypes. p. 8

Neanderthal A subspecies of Homo sapiens that existed from approximately 125,000 to 35,000 years ago. p. 118

negative feedback A decrease in function in response to a stimulus, occuring when an increase in the output of the system feeds back and decreases input into the system. p. 124

nephron A structural and functional unit of the kidney where urine is formed. p. 322

nerve A bundle of nonmyelinated and myelinated fibers or a bundle of axons surrounded by supporting neuroglial cells in the peripheral nervous system. p. 143

nervous tissue Tissue characterized by the ability to produce and conduct electrical impulses called action potentials. p. 132

neural plate A thick layer of ectodermal cells that gives rise to the neural tube and subsequently to the brain, spinal cord, and other tissues of the central nervous system. p. 360

neuroglia (nu-rog'li-ah) Support cells of the nervous tissue that nourish, protect, and insulate neurons. p. 132

neuromuscular junction A synapse between a neuron and a skeletal muscle fiber. p. 224

neuron A cell capable of producing and conducting electrical signals called action potentials. p. 132

neurotransmitter substance A chemical that is released when an action potential reaches the terminal knob of the axon. p. 147

neutron A subatomic particle with no electrical charge. p. 17

neutrophil A granulocytic white blood cell. Neutrophils are circulating white blood cells essential for phagocytosis by which bacteria, cellular debris, and solid particles are removed and destroyed. p. 254

nicotinamide adenine dinucleotide (NAD) A coenzyme carrier that accepts hydrogen ions, becoming $NADH^+$, stripped from the organic molecules being broken down during cellular respiration. $NADH^+$ is used in the electron transport chain to produce ATP. p. 51

node of Ranvier An noninsulated gap between two Schwann cells. p. 143

nondisjunction When two homologous chromosomes do not separate during meiosis, resulting in gametes with abnormal numbers of chromosomes. p. 85

nonspecific immune response Nonselective immune response to any foreign or abnormal material introduced into the body. p. 261

Norplant A trademark for a method of implanting capsules of a synthetic progesterone-like hormone beneath the skin of a woman's upper arm. p. 346

notochord (no-to-kord) An elongated strip of mesodermal tissue of the developing embryo that forms beneath the primitive streak that eventually becomes the vertebral column. p. 360

nuclear envelope The phospholipid bilayer which separates the nucleus from the rest of the cell. p. 43

nuclei Groups or clusters of cell bodies inside the CNS. p. 152

nucleic acid A macromolecule composed of nucleotides. p. 31

nucleoli (noo-kle-uh-li) Small spherical bodies found in the nucleus that produce ribosomes. p. 45

nucleotide An organic molecule composed of a five-carbon ring sugar, a phosphate group, and one or two nitrogen-containing carbon rings. p. 31

nucleus An organelle that contains DNA, the cell's genetic material. p. 43

nutrients Substances found in food that are used by the body to promote normal growth, maintenance, and repair. p. 299

O

Oldowan tradition Crude tools made by knocking several chips off a rounded stone to give it a rough cutting edge. This tool tradition is indirectly associated with Australopithecus. p. 116

omnivorous Referring to an organism that consumes both plants and animals. p. 388

oncogene A gene that causes cancer. p. 96

oocyte An immature ovum that has not undergone meiosis. p. 341

oogenesis (o-o-jen'ih-sis) The process of meiosis and development that produces mature sex cells, or ova. p. 337

oogonium (pl. oogonia) The precursor cell from which an oocyte develops in the fetus during intrauterine life. Near the time of birth it enters the prophase stage of meiosis I to form a primary oocyte. p. 341

opportunistic infection An infection caused by normally nonpathogenic organisms in a host whose resistance has been decreased. p. 11

organ A structure composed of two or more tissue types that perform one or more common functions. p. 125

organ of Corti The organ of hearing, a spiral structure within the cochlea containing hair cells that are stimulated by sound vibrations. The hair cells convert the vibrations into action potentials. p. 186

organ system A group of organs classified as a unit because of a common function or set of functions. p. 125

organelle A membrane-enclosed subcellular structure that performs a specialized task. p. 37

organogenesis (or'-gah-no-jeh'-nih-sis) The formation and differentiation of organs and organ systems during embryonic development. The period extends from approximately the end of the second week through the eighth week of gestation. p. 359

origin The fixed end of muscle attachment. p. 220

osmosis The diffusion of water across a selectively permeable membrane. p. 40

ossification (os'i-fi-ka'shun) The process of bone development. p. 219

osteoblasts Cells that synthesize the collagen to form the matrix and, with growth, develop into osteocytes. p. 218

osteoclast Bone-destroying cell that breaks down the inorganic component of bone and returns it to the bloodstream. p. 200

osteocyte (os'te-o-sit) A mature bone cell. p. 131

oval window An oval-shaped aperture in the wall of the middle ear leading to the inner ear; contains the stapes, which vibrate and transmit sound waves to the cochlea. p. 185

ovarian cycle The events occurring to the primary oocyte, to the follicle that contains it, and to other body regions in the female. p. 342

ovaries (o-ver-ez) A pair of female gonads found in the lower abdomen, beside the uterus. p. 340

ovulation The monthly process by which an ovum is produced and released by a Graafian follicle of the ovary. p. 342

ovum A female gamete released from the ovary at ovulation. p. 342

oxygen debt The quantity of oxygen taken up by the lungs during recovery from a period of exercise that is in excess of the quantity needed for resting metabolism during the pre-exercise period. p. 228

oxytocin (ok-se-to'sin) A hormone produced by the hypothalamus and stored in the posterior lobe of the pituitary gland that stimulates uterine smooth muscle contraction and milk secretion from the mammary glands. p. 196

P

Pacinian corpuscle Mechanoreceptor located deep in the skin's dermal layer and stimulated by deep pressure and vibration. p. 171

pancreas (pan'kre-as) Gland located in the abdominal cavity that secretes digestive enzymes and the hormones insulin and glucagon. p. 311

parasite An organism living in or on and obtaining nourishment from another organism. p. 8

parasympathetic system A division of the ANS that controls and regulates the normal activities of the body. p. 151

parathyroid hormone A peptide hormone secreted by the parathyroid glands important in the regulation of blood calcium levels. p. 200

parietal pleura That part of the pleural membrane that lines the inside of the thoracic wall. p. 287

parturition (par'-tu-rish'-un) The process of giving birth. p. 353

passive transport (facilitated diffusion) Transport across a cell membrane assisted by a carrier protein embedded in it. The molecule attaches to the carrier protein, moving down its concentration gradient across the cell membrane. p. 42

pectoral girdle (pek'-ter-ul) The part of the appendicular skeleton that consists of two bones, the clavicle and scapula, that attach the arms to the trunk. p. 213

pelvic girdle The part of the appendicular skeleton that consists of a bony ring formed by the hip bones, the sacrum, and the coccyx, that attaches the legs to the trunk. p. 213

pelvis The lower portion of the trunk of the body, composed of four bones, the two innominate bones laterally and ventrally and the sacrum and coccyx posteriorly. p. 211

penis A cylindrical organ that transfers sperm from the male reproductive tract to the female reproductive tract. It is also the male urinary organ. p. 337

pepsin An enzyme secreted by the gastric glands in the stomach that hydrolyzes protein. p. 315

peptide bond A covalent bond between two amino acids. p. 30

pericardial cavity Central compartment of the thoracic cavity containing the heart. p. 134

pericardium (per-i-kar'di-um) A double-walled fibrous sac that surrounds the heart. p. 235

peripheral nervous system (PNS) All the nerves of the body outside the brain and spinal cord. p. 150

peristalsis (payr'-ih-stal'-sis) The coordinated, rhythmic, serial contraction of smooth muscle that forces food through the digestive tract, bile through the bile duct, and urine through the ureters. p. 309

pH scale A measure of hydrogen ion concentration in a solution. The scale ranges from 0 to 14 with a pH of 7 being neutral. p. 24

phagocytosis (fag'o-si'to'sis) A type of endocytosis in which a cell ingests an organism or some other fragment of organic matter. Macrophages and neutrophils are phagocytes. p. 43

pharynx (fayr'inks) The throat; a tubular structure about 13 cm long that extends from the base of the skull to the esophagus and is the common passageway for food and air. p. 284

phenotype (fe'no-tip) The outward physical appearance or function of the genotype. May also refer to the form or function of a particular characteristic. p. 73

phospholipid A lipid consisting of one glycerol molecule, two fatty acid chains, and a phosphate-containing group of atoms with a nitrogen-containing group attached at the end. p. 28

pinna (outer ear) The visible, shell-shaped projection surrounding the opening of the external auditory canal. p. 185

pinocytosis (pin'o-si-to'sis) A type of endocytosis in which a cell engulfs liquid material. p. 43

pituitary gland (pit-u'i-ter'i) **(hypophysis)** (hi-pof'i-sis) A small, hormone-producing endocrine gland located in a bony cavity at the base of the brain. It is connected to the hypothalamus by a thin stalk of tissue that contains nerve cells and capillaries. p. 194

placebo (pla-se'bo) A substance that looks and tastes like the substance being tested but has no effect. p. 11

placenta (pluh-sen'-tuh) The highly vascular organ that grows into the uterine wall and through which the mother supplies the growing embryo and fetus with nutrients, water, and oxygen and through which carbon dioxide and other wastes are excreted. p. 357

plasma The fluid intercellular matrix within which blood cells float. Plasma contains practically every substance used and discarded by cells as well as nutrients, hormones, proteins, and ions. p. 131

plasma cells Cells derived from B lymphocytes that produce and secrete large amounts of antibodies; they are responsible for antibody-mediated immune response. p. 265

plasmid A circle of DNA found in bacteria. p. 99

platelets A cell fragment present in blood that plays an important role in clotting the blood. p. 249

pleura (pler'-uh) A delicate serous membrane enclosing the lung, composed of a single layer of cells resting on a delicate membrane of connective tissue. p. 287

point mutation The substitution or change in one pair of nucleotide bases in a gene. p. 83
polar body One of the small cells produced during the two meiotic divisions in the maturation process of female gametes. It is nonfunctional and incapable of being fertilized. p. 342
polar covalent bond A chemical bond in which a greater attraction for the electrons by one of the atoms results in an unequal sharing of the electrons involved. p. 21
polar molecule Produced by polar covalent bond, a molecule with weak opposite electrical charges. p. 21
pollutant Any substance that enters the environment and interferes with the natural cycles in communities and ecosystems. p. 395
polymer Similar organic molecules linked together. p. 26
polypeptide Three or more chemically bonded amino acids. p. 30
polysaccharide Covalently linked glucose molecules. p. 27
polyunsaturated fatty acid A fatty acid containing double covalent bonds between two or more pairs of carbon atoms. p. 28
pons The part of the brainstem that relays impulses from the spinal cord to higher brain centers. p. 158
population A group of organisms of the same kind living in the same area. p. 5
population coding Occurs when more receptors are stimulated from a larger area of a given tissue as the strength of the stimulus increases. p. 169
postganglionic neuron The second efferent neuron in both sympathetic and parasympathetic systems of the ANS. Its cell body is in a ganglion and its axon synapses with an organ or gland. p. 151
postmitotics Cells that actively divide during embryonic or postnatal development and no longer continue to divide afterwards. p. 376
potential energy Energy not actively engaged in change or work. p. 48
preganglionic neuron The first efferent neuron extending from the CNS in both sympathetic and parasympathetic systems of the ANS. The axon of the preganglionic neuron synapses in a ganglion. p. 151
prepuce (foreskin) A fold of skin that forms a retractable cover over the glans penis. p. 337
presbycusis (prez-bi-koo'sis) Loss of hearing sensitivity and speech intelligibility, associated with aging. p. 382
presbyopia (prez-be-o'pe-a) Farsightedness resulting from a loss of elasticity of the lens of the eye. The condition commonly develops with advancing age. p. 183
primary consumer An organism that eats plants; an herbivore. p. 388
primary immune response First exposure to a foreign antigen, in which there is a variable latent period before measurable amounts of specific antibodies appear in circulation. p. 269
primary spermatocyte The male sex cell that arises from a spermatogonium. Each primary spermatocyte gives rise to two haploid secondary spermatocytes. p. 338
primitive streak An area in the central region of the embryonic disk formed by the movement of a rapidly dividing mass of cells that spreads between the endoderm and ectoderm, giving rise to the mesoderm. p. 359
producers The first trophic level in a food chain, consisting of autotrophic organisms. p. 388
progesterone (pro-jes'-ter-on) A hormone secreted by the ovaries that affects the uterine endometrial lining to prepare it for implantation of the blastocyst. p. 203
prolactin (PRL) (lactogenic hormone) A hormone that promotes breast development and stimulates milk production in the mammary glands of the female breast. p. 198
proliferative phase The phase of the uterine cycle after menstruation. Under the influence of follicle-stimulating hormone the ovary produces increasing amounts of estrogen, causing the endometrium to become dense and vascular. The phase ends with ovulation. p. 344
promoter site A specific sequence of nucleotide bases recognized by RNA polymerase. The promoter site indicates where the RNA polymerase is to bind. p. 81
prostaglandins (pros'tah-glan'dinz) A class of lipid hormones that are found in most of the body's cells. Prostaglandins cause changes in blood vessel diameter and smooth muscle contractions. p. 205
prostate gland (prah'-stat) A gland surrounding the male urethra that adds a milky alkaline fluid to semen, neutralizing the acidity of the female vagina. p. 336
protein A chain of linked amino acids. p. 30
proto-oncogene A gene in which a mutation has occurred, altering its product so that it greatly increases the rate of cell division. p. 97
proton A positively charged subatomic particle. p. 17
proximal convoluted tubule The convoluted portion of the renal tubule nearest the glomerular capsule lying in the cortex of the kidney. p. 323
pseudostratified epithelium Consists of a single cell layer but appears multilayered. p. 126
puberty (pu'ber'ti) The period of life at which the ability to reproduce begins. p. 370
pulmonary circuit The blood flow through a network of vessels between the heart and lungs for the oxygenation of blood and removal of carbon dioxide. p. 235
pulmonary ventilation The process of inhaling and exhaling air through the lungs. p. 283
pulse The regular, recurrent expansion and contraction of an artery during each heartbeat. p. 232
pupil The opening in the center of the iris through which light passes. p. 178
Purkinje fibers (per-kin-ge') Myocardial fibers that are a continuation of the bundle branches and extend into the muscle walls of the ventricles. p. 238
pus A creamy whitish fluid found in wounds, composed of living and dead phagocytes, bacteria, and debris from damaged tissue cells. p. 262

R

races Local populations that differ from other populations in their frequency of several genetic traits. p. 413
radioisotope An unstable isotope that emits energy, subatomic particles, or both. p. 18
rapid eye movement sleep (REM sleep) Occurs about $1\frac{1}{2}$ to 2 hours after sleep begins. Brain waves become irregular, culminating in the reappearance of alpha waves. REM sleep occurs about 5 to 7 times during the sleep cycle. p. 166
reactant The organic and inorganic substances that enter into a metabolic pathway. p. 50
receptor A type of sensor that monitors the body's environment and responds to changes by sending information to the control center. Any sensory neuron. p. 123
recessive The form or variety of a trait that disappears entirely in a hybrid. p. 65
recombinant DNA The artificial association of DNA fragments or genes, usually from two different organisms, that are not found together naturally. p. 97
rectum The lower part of the large intestine, which terminates at the anus. p. 314
reflex Automatic, involuntary response to specific stimuli occurring inside or outside the body. p. 149
remodeling A type of bone growth involving the increase in overall size and diameter of the bone. p. 219
respiration The uptake of oxygen and the release of carbon dioxide by the body. Cellular, internal, and external respiration are all part of the general process of respiration. p. 283
resting potential Opposite charges separate across the neuron's cell membrane, and the neuron is said to be polarized. p. 144
restriction endonuclease (restriction enzyme) An enzyme capable of breaking the chemical bonds of DNA by cutting them within the DNA strand. p. 97
retina (reh'-tih-nuh) The innermost layer of the eyeball. p. 179
Rh factor A red blood cell antigen. Rh positive (Rh^+) individuals have the antigen present while Rh negative (Rh^-) individuals lack the antigen on their red blood cells. p. 273
rhodopsin A substance found in the rod photoreceptors of the eye. When exposed to light, rhodopsin breaks down into retinal and opsin, generating an action potential. p. 180

rhythm method Any one of several methods of family planning that does not rely on a medication or a device for avoiding pregnancy. p. 345

ribosomes (ri'bo-somz) Small, round structures found on endoplasmic reticulum; site where proteins are manufactured. p. 45

(RNA) ribonucleic acid (ri'bo-nu-kle'ik) A nucleic acid that contains ribose and the nucleotide bases, adenine, guanine, cytosine, and uracil. There are three types of RNA: messenger RNA, transfer RNA, and ribosomal RNA. p. 80

RNA polymerase An enzyme that controls the transcription of mRNA. p. 81

rough endoplasmic reticulum A portion of the ER on which ribosomes are attached to the membrane surface. p. 45

round window A round opening in the wall of the middle ear leading into the cochlea and covered by a thin membrane. p. 185

RU 486 A drug that can end pregnancy when administered as a one-dose pill within the first 6 weeks after conception. Two days after taking RU 486, the woman receives an injection of prostaglandins that causes the uterus to contract and expel the endometrium. p. 346

S

saliva The clear, viscous fluid secreted by the salivary glands of the mouth. Saliva contains water, mucin, and the digestive enzyme salivary amylase and serves to moisten the oral cavity, initiate the digestion of starches, and aid in the chewing and swallowing of food. p. 307

salivary gland One of three pairs of glands secreting into the mouth, thus aiding the digestive process. The salivary glands are the parotid, the submandibular, and the sublingual glands. p. 307

sarcomere (sar'-ko-mer) The smallest functional unit of a myofibril. Sarcomeres occur as repeating units along the length of a myofibril, occupying the region between Z lines of the myofibril. p. 221

sarcoplasmic reticulum A network of tubules and sacs in skeletal muscle that plays an important role in muscle contraction and relaxation by releasing and storing calcium ions. p. 221

Schwann cell The supporting cells associated with nerve fibers of the peripheral nervous system. p. 143

scientific method The experimental testing of a hypothesis. p. 9

sclera (skle'-ruh) Outermost layer of the eye composed of dense opaque connective tissue. p. 176

scrotum (skro'-tum) A sac of skin located outside the lower pelvic area of the male, which houses the testes. p. 335

secondary consumer A carnivore that eats herbivores. p. 388

secondary immune response Subsequent exposure of the same individual to the same antigen. Compared to the primary immune response, the secondary immune response is much more rapid. p. 269

secondary sex characteristics Any of the external physical characteristics of sexual maturity secondary to hormonal stimulation that develops in the maturing individual. These characteristics include adult distribution of hair and the development of the penis or of the breasts and labia. p. 339

secondary spermatocytes Haploid spermatocytes derived from meiosis of a primary spermatocyte. p. 338

secretory phase The phase of the uterine cycle after ovulation. The corpus luteum stimulated by luteinizing hormone develops from the ruptured follicle and secretes progesterone, which stimulates the development of the glands and arteries of the endometrium, causing it to become thick and spongy. p. 344

selective permeability The degree to which some molecules are allowed to pass through the cell membrane while others cannot. p. 38

self-pollination Process whereby the female gametes in a flower are fertilized by the male gametes of the same flower. p. 64

semen (se'-min) Fluid produced by the accessory glands of the male reproductive system combined with sperm. p. 336

semiconservative replication Process of DNA replication in which a complementary strand of DNA is formed from each of the original double DNA strands. This results in two new double strands of DNA, each made of one original single strand and one newly formed single strand. p. 79

semicircular canal Any of the three bony, fluid-filled loops of the inner ear, associated with the sense of balance. p. 187

seminal vesicles (seh'-mih-nul) One of three accessory glands that secrete a thick, clear fluid forming part of the semen. p. 336

seminiferous tubules (seh-mih-nih'-fer-us) A tightly coiled tube within the testis where sperm cells develop. p. 335

sensory (afferent) neurons Neurons that are receptors for specific stimuli from either the internal or external environment. These neurons carry action potentials toward the spinal cord and the brain. p. 142

Sertoli cells One of the supporting cells of the seminiferous tubules of the testes. p. 338

sex chromosomes The X and Y chromosomes. Females have two homologous X chromosomes and males have one X and one Y chromosome. p. 86

sex-linked inheritance Recessive alleles on the X chromosome that are always expressed in males. p. 91

short-term memory Memory storage involving the repeated activity of a particular neural circuit in the brain. p. 164

sickle-cell trait The heterozygous form of sickle-cell anemia, characterized by the presence of both hemoglobin S and normal hemoglobin in the red blood cells. p. 254

sigmoid curve Pertaining to an S-shaped curve. In ecology, used to describe population growth under particular conditions. p. 394

sinus A cavity or channel, such as a cavity within a bone. p. 210

sister chromatid A chromosome and its attached duplicated copy. p. 57

skeletal system Bones, cartilage, and other dense connective tissues of the body. p. 134

small intestine The longest portion of the digestive tract extending from the pylorus of the stomach to the ileocecal junction. It is divided into the duodenum, jejunum, and ileum. p. 311

smooth endoplasmic reticulum Portions of the ER without ribosomes attached to the membrane surface. p. 45

sodium-potassium pump A term given to the active transport of sodium ions and potassium ions across a cell membrane. p. 143

solute Any substance dissolved in a solution. p. 40

solution Atoms or molecules dissolved in a solvent. p. 40

solvent A liquid in which another substance can dissolve. p. 24

somatic cell Any nonsex cell of the body. Somatic cells undergo mitosis during cell replication. p. 57

somatic nervous system (SNS) Subdivision of the PNS. This system contains afferent pathways that relay sensory information to the CNS and efferent pathways carrying action potentials from the CNS to the body's skeletal muscles. p. 150

specific immune response A combined version of cellular and chemical (antibodies) reaction in which the defense is specifically tailored to deal with a certain agent. p. 264

spermatid (sper'-mah-tid) A male sex cell that arises from a secondary spermatocyte and that becomes a mature sperm in the last phase of spermatogenesis. p. 339

spermatogenesis (sper'-mah-to-jeh'nih-sis) The development of sperm cells within the coiled seminiferous tubules of the testis. Spermatogenesis is triggered by follicle-stimulating hormone. p. 337

spermatogonium (sper'-mah-togo-ne-um) A male sex cell that gives rise to a spermatocyte early in spermatogenesis. p. 337

spermatozoa (sperm) (sper'ma-tozo'ah) A mature male gamete that develops in the seminiferous tubules of the testes. p. 335

spermicide A chemical substance that kills spermatozoa by reducing their surface tension, causing the cell wall to break down. p. 347

spinal cord The part of the CNS that runs down the neck and back. A two-way conduction pathway to and from the brain to the body. p. 150

spindle fibers The formation of microtubules between centrioles. p. 57

spleen An organ situated between the stomach and the diaphragm. It is part of the lymphatic system because it contains lymph nodes. It performs various tasks, such as defense, hemopoiesis, blood storage, and the destruction of red blood cells and platelets. It also produces a variety of white blood cells. p. 251

spongy bone Composed of a latticework of trabeculae separated by irregular spaces. p. 131

spontaneous generation The production of living organisms from nonliving matter. p. 10

steroid A lipid composed of four connected carbon rings with hydrogen and other carbon atoms attached to the rings. p. 28

stimulus Anything that can elicit or evoke an action or response. Any environmental change that evokes an action potential in a sensory neuron. p. 169

stomach The major organ of digestion, located in the right upper quadrant of the abdomen. It receives partially processed food and drink funneled from the mouth through the esophagus and moves nutritional bulk into the intestines. p. 310

stratified epithelium Composed of more than one layer of cells with only one layer touching the basement membrane. p. 126

substrate A reactant in an enzyme-catalyzed reaction. p. 31

sustainable development The use of renewable natural resources in a manner that does not eliminate or degrade them or diminish their renewable usefulness for future generations. p. 398

sympathetic system A division of the ANS that, when stimulated, prepares the entire body for stressful physical activity. p. 151

synapse The region that incorporates the terminal knob of the axon of one neuron and the membrane of another tissue cell. p. 147

synaptic cleft The space or gap between the terminal knob and the next cell. p. 147

synovial joints (si-no'vi-al) Joints that consist of a fluid-filled cavity separating the articulating bones. p. 216

synthesis reaction A process of covalently bonding organic molecules together. A water molecule is produced for each created bond and energy is required for each created bond. p. 26

syphilis (sih'-fih-lis) A sexually transmitted disease caused by the bacterium *Treponema pallidum* that produces stages of localized infection to widespread infection. p. 348

systemic circuit The general blood circulation of the body, not including the lungs. p. 235

systole (sis'-to-le) The contraction of the heart, driving blood into the aorta and pulmonary arteries. p. 241

T

T-cytotoxic (T-killer) cells Those daughter cells of an antigen-activated T cell that attack and kill foreign cells and cancer cells. p. 269

T-helper cells Those daughter cells of an antigen-activated T cell that releases lymphokines, which stimulate both the production and activity of phagocytes, T cells and B cells. p. 269

T lymphocytes (T cells) Lymphocytes that have circulated through the thymus gland and have differentiated to become T lymphocytes. When exposed to an antigen, they divide rapidly and produce large numbers of new T cells, sensitized to that antigen. p. 265

T-memory cells T lymphocytes bearing specific surface recognition markers that enable them to respond rapidly to future exposures of the antigens to which they have been previously sensitized. p. 269

T-suppressor cells T lymphocytes that act to inhibit the production of antibodies against specific antigens. p. 269

taste bud A microscopic receptor embedded in the papillae of the tongue that detects dissolved chemicals to produce the sensation of taste. p. 175

taxonomy The science by which organisms are classified and placed into categories based on their evolutionary relationships. p. 114

Tay-Sachs disease An autosomal recessive genetic disorder in which the brain cells fail to synthesize an enzyme necessary to break down a special class of lipids called gangliosides. p. 89

tectorial membrane The roof-like membrane in which the cilia of the hair cells of the Organ of Corti are embedded. p. 186

tendons Fibrous bands of connective tissue that attach muscle to bone. p. 220

teratogens (ter'-ah-to-jinz) Agents causing the death of the embryo or birth defects of varying degrees. p. 362

terminal knob A small swelling at the end of an axon that allows a neuron to chemically communicate with other neurons. p. 142

tertiary consumer A carnivore that eats other carnivores. p. 388

testcross When a hybrid plant is cross-pollinated with a true-breeding recessive plant. p. 67

testis (pl. testes) The male gonads where sperm production occurs. p. 335

testosterone (tes-tah'-ster-on) Male steroid sex hormone secreted by the interstitial cells of the testes that initiates the maturation of the male reproductive organs and the appearance of male secondary sexual characteristics. p. 203

tetrad Formed during prophase I of meiosis when homologous chromosomes line up side by side at the cell's center. Crossing-over occurs during tetrad formation. p. 71

thalamus One structure comprising the limbic system. It acts like a filter, sorting out sensory information coming from the lower brain centers. p. 159

theory A generalization based on the repeated support and confirmation of a hypothesis or group of hypotheses that explain a large number of events. p. 13

thoracic cavity (tho-ras'ik) Subdivided into three compartments separated by serous membranes. The central compartment, called the pericardial cavity, contains the heart. The compartments on either side of the pericardial cavity, called the pleural cavities, each contain a lung. p. 134

thrombus A blood clot attached to the interior wall of a vein or artery, sometimes obstructing the lumen of the vessel. p. 242

thymus gland A two-lobed structure located beneath the sternum producing a group of hormones known as thymosins. p. 203

thyroid-stimulating hormone (TSH) A tropic hormone secreted by the anterior lobe of the pituitary gland that stimulates normal development and secretory activity of the thyroid gland. p. 199

thyroxin (thi-roks'in) A hormone made from iodine and secreted by the thyroid gland. It is responsible for the overall acceleration of the body's rate of cellular metabolism. p. 199

tight junction Where the cell membranes of adjacent cells are in direct contact and where there is no intercellular space between them. p. 128

tissue A group of cells with similar structure and function plus the extracellular substances located between them. p. 125

trabeculae (tra-bek'ule) The latticework of tiny bars comprising spongy bone. p. 131

trachea (tra'-ke-uh) A tube in the neck, composed of cartilage and membrane, that extends from the larynx and divides into two bronchi. The trachea conveys air to the lungs. p. 285

traits of complex inheritance Characteristics whose phenotypic expressions are continuous throughout their range of variation, with no distinct, non-overlapping categories. p. 405

traits of simple inheritance Characteristics whose phenotypic variation falls into discrete, or discontinuous, nonoverlapping categories. p. 404

transcription The cellular process that produces a single strand of RNA complementary to a segment of DNA. p. 81

transfer RNA (tRNA) RNA that attaches to an amino acid in the cytoplasm and brings it to the mRNA for assembly into a polypeptide. p. 82

translation The cellular process involving mRNA, ribosomes, and tRNA, which results in the synthesis of a polypeptide chain having an amino acid sequence specified by the sequence of codons in mRNA. p. 82

translocation A type of chromosomal abnormality in which a portion of one chromosome is moved to another non-homologous chromosome. p. 85
triglyceride A lipid consisting of one glycerol molecule to which are attached three fatty acid chains. p. 28
trophic level (feeding level) A level in a food chain. p. 388
trophoblast cells (tro-fo-blast) The layer of cells that forms the wall of the blastocyst. It functions in the implantation of the blastocyst in the uterine wall and in supplying nutrients to the embryo. p. 355
tubal ligation A sterilization procedure in which both fallopian tubes are blocked to prevent conception from occurring. p. 346
tumor A mass of cells produced by uncontrolled cell growth. p. 97
Turner's syndrome The syndrome produced in a female by the absence of one X chromosome. The genotype is XO. p. 94
tympanic membrane A thin membrane in the middle ear that transmits sound vibrations to the inner ear by means of the auditory ossicles. p. 185

U

umbilical cord A flexible structure connecting the fetus with the placenta. p. 357
unsaturated fatty acid Fatty acid containing double covalent bonds between one or more pairs of carbon atoms. p. 28
urea The final product of the urea cycle, produced from the deamination of amino acids, consisting of carbon and oxygen atoms and two molecules of ammonia. p. 313
ureter One of a pair of tubes that carries urine from the kidney into the bladder. p. 321
urethra A small tubular structure that drains urine from the bladder. p. 321
uric acid A product of the metabolism of protein present in the blood and excreted in the urine. p. 327
urine The fluid secreted by the kidneys, transported by the ureters, stored in the bladder, and voided through the urethra. Its normal components consist of water, urea, sodium chloride, phosphates, and uric acid. p. 321
uterine (menstrual) cycle (men'-stroo-ul) Monthly growth and shedding of an endometrium and maturation and release of ova. p. 342
uterus The hollow pear-shaped internal female organ of reproduction in which the fertilized ovum is implanted and the fetus develops, and from which menses flow. p. 340
uvula The small cone-shaped process suspended in the mouth from the middle of the posterior border of the soft palate. p. 307

V

vaccine (vak'sen) An injected solution of antigens used to stimulate antibody development, to protect individuals that might later be exposed to a disease agent. p. 270
vagina An organ of the female genitalia whose muscular, tube-like passageway to the exterior has three functions; it accepts the penis during intercourse, it is the lower portion of the birth canal, and it provides an exit for the menstrual flow. p. 341
varicose vein A dilated vein with nonfunctional valves. p. 234
vas deferens (vas deh'-fer-enz) A long connecting tube that ascends from the epididymis into the pelvic cavity curving around the bladder and connecting to the urethra in the penis. p. 335
vasectomy A procedure for male sterilization involving the cutting of the vas deferens. p. 346
vasoconstriction A reduction in the diameter of lumen resulting from smooth muscle contraction in the vessel wall. p. 232
vasodilation An increase in the diameter of the lumen resulting from smooth muscle relaxation in the vessel wall. p. 232
veins Vessels that carry blood from the capillaries to the heart. p. 231
vena cava One of two large veins that returns deoxygenated blood to the heart from parts of the body below the diaphragm. p. 235
ventricle The two lower, larger, and thick-walled chambers of the heart. The right ventricle pumps blood through the lungs. The left ventricle pumps blood throughout the rest of the body. p. 235
venule Small blood vessel that gathers blood from capillary networks to return blood to the heart. p. 233
vertebrae The vertebrae are bones composed of a body, an arch, and spinous processes for muscle attachment. Vertebrae make up the vertebral column. p. 155
vertebral column The flexible structure that forms the longitudinal axis of the skeleton. It includes 26 vertebrae arranged in a line from the base of the skull to the coccyx. p. 155
vesicle (ves'i-kl) A membrane-bound vessel. p. 43
villus (pl. villi) Tiny, finger-like projection clustered over the entire mucous surface of the small intestine. The villi diffuse and transport fluids and nutrients. p. 311
visceral pleura That part of the pleural membrane that covers the surface of the lungs. p. 287
vitamin A group of organic compounds essential for normal physiologic and metabolic functioning of the body. p. 300
vitreous humor A transparent jelly-like substance filling the cavity behind the lens of the eye that helps to maintain the spherical shape of the eyeball. p. 178

W

water-soluble vitamins Vitamins that are absorbed with water from the digestive system and cannot be stored in the body. The B vitamins and vitamin C are water-soluble vitamins. p. 301
white matter The tissue surrounding the gray matter of the spinal cord, consisting mainly of myelinated nerve fibers embedded in a network of neuroglia. p. 155

Z

zona pellucida (peh-loo'-sih-duh) The thick, transparent, noncellular membrane that encloses the ovum. p. 342
zygote (zi'got) A single diploid cell resulting from fusion of a male and female gamete. p. 70

CREDITS

Chapter 1

1-A, D,E, 1-2, Stewart Halperin; 1-B, M. Abbey/Visuals Unlimited; 1-1A,B,E,F, William Ober; 1-1C, Triarch/Visuals Unlimited; 1-1D, Biophoto Associates/Photo Researchers; 1-1G-J, E.S. Ross; 1-1K, NASA; 1-F, Laura J. Edwards.

Chapter 2

2-C, D, E, George Klatt; 2-I, 2-N, 2-10, William Ober; 2-4, 2-P, Ronald J. Ervin; 2-5(photo), Stewart Halperin; 2-7(photo), Manfred Kage/Peter Arnold.

Chapter 3

3-A, Nadine Sokol; 3-1, 3-K, Kevin Sommerville; 3-2, 3-C, 3-6, 3-18, Barbara Cousins; 3-D, 3-I, Christine Oleksyk; 3-5, Christy Krames; 3-7, 3-F, 3-G, 3-H, 3-J, William Ober; 3-8, 3-L, Rolin Graphics; 3-M-Q, Carolina Biological Supply Company.

Chapter 4

4-A, 4-1, Nadine Sokol after Bill Ober; 4-2, 4-3, 4-5, Nadine Sokol; 4-B, Barbara Cousins; 4-6, Raychel Ciemma; 4-7, Kevin Sommerville.

Chapter 5

5-B, Pagecrafters; 5-C, 5-D, Nadine Sokol; 5-I, 5-K, 5-N, 5-O, Barbara Cousins; 5-J, The Children's Hospital, Denver, Cytogenics Lab.; 5-2, CNRI/Science Photo Library; 5-L, Stewart Halperin; 5-M, Murayama/Biological Photo Service; 5-3, 5-5, 5-6, 5-8, Pagecrafters; 5-3(photo), Field Museum of Chicago, neg. #118.

Chapter 6

6-1, Pagecrafters; 6-3, William Ober; 6-5, Louise Van der Meid; 6-A(art), Molly Babich; 6-A(photo), Ken Lucas/Biological Photo Service; 6-B, Russell A. Mittermeier; 6-6, Nadine Sokol after Bill Ober; 6-8, 6-9, Beck Kent/Animals Animals; 6-11, 6-C, John Reader; 6-E, National Museum of Kenya; 6-G, Erik Trinkaus.

Chapter 7

7-1, Yvonne Wylie Waltson; 7-A(art), Christine Oleksyk; 7-A(photo), Patrick Watson; 7-B(1-3), Joan Beck; 7-3(art), Joan Beck; 7-3(photos), 7-E-G, 7-7, 7-H, Ed Reschke; 7-4, 7-5, Michael Schenk; 7-C, St. Louis Globe Democrat; 7-D, 7-I, Trent Stephens; 7-6, William Ober; 7-8 through 7-14, Joan Beck.

Chapter 8

8-1, 8-17, 8-H, 8-J, 8-K, 8-18, Scott Bodell; 8-2, 8-A(art), 8-C, 8-20, Nadine Sokol; 8-A(photo), E.S. Ross; 8-3, 8-7, 8-8, 8-12, Barbara Cousins; 8-B, Pagecrafters; 8-9, Raychel Ciemma; 8-11, John Daughtery; 8-E, Rolin Graphics; 8-16, 8-19, B. Vidic and F.R. Suarez (from *Photographic Atlas of the Human Body,* 1984, The C. V. Mosby Co.); 8-I, E. W. Beck; 8-21, Medical Communication Services.

Chapter 9

9-1, 9-C, 9-D, Rolin Graphics; 9-2, 9-4, Raychel Ciemma; 9-F, 9-6, 9-9, William Ober; 9-5; Marsha Dohrman/Christine Oleksyk; 9-G, Stewart Halperin; 9-8A,B, 9-13, 9-14A, Christine Oleksyk; 9-8C, Scott Mittman; 9-H, Graphic Works; 9-11, 9-12, 9-14B-D, G. David Brown.

Chapter 10

10-1, Kate Sweeney; 10-2, 10-3, 10-E, Barbara Cousins; 10-A, Rolin Graphics; 10-4, 10-9, 10-10, Nadine Sokol; 10-7, Barbara Stackhouse; 10-8, Raychel Ciemma.

Chapter 11

11-1, 11-C, Nadine Sokol; 11-A, 11-5, E. Beck; 11-2, 11-3, 11-4, David Mascaro; 11-B, 11-6, 11-7, 11-13, 11-D, Barbara Cousins; 11-9, Scott Bodell; 11-10, 11-12, Joan Beck; 11-11, William Ober; 11-14(photo), Richard Rodewald; 11-15, Yvonne Walston; 11-16, Raychel Ciemma; 11-E, Rolin Graphics.

Chapter 12

12-A, B, C, Christy Krames; 12-1, G. David Brown; 12-2, William Ober; 12-F, 12-3, George Wassilchenko; 12-G, 12-4, 12-8, Barbara Cousins; 12-5, Joan Beck; 12-6, Lisa Shoemaker/Joan Beck; 12-7, Barbara Stackhouse; 12-I, Joan Beck/Donna Odle; 12-J, Marcia Williams (from *Mosby's Medical and Nursing Dictionary, ed. 2,* 1986); 12-K, 12-L, G. David Brown.

Chapter 13

13-A, Rolin Graphics; 13-B, 13-D, G. Bevelander/J.A. Ramaley; 13-1, 13-C, Christine Oleksyk.

Chapter 14

14-3, Barbara Cousins; 14-C, William Ober; 14-D, 14-7, Molly Babick/John Daughtery; 14-E, Rolin Graphics.

Chapter 15

15-2, Kate Sweeney; 15-3, Jody Fulks; 15-5, Christy Krames; 15-6, Raychel Ciemma.

Chapter 16

16-1, Kate Sweeney; 16-A, Christy Krames; 16-B, 16-G, 16-8, 16-9, Rolin Graphics; 16-2, 16-C, 16-4, Nadine Sokol; 16-3, 16-7A, Barbara Cousins; 16-6, John Daughtery; 16-7B, David Phillips/Visuals Unlimited; 16-F, G. David Brown.

Chapter 17

17-1, Joan Beck; 17-2, 17-5, Barbara Cousins; 17-3, Barbara Stackhouse.

Chapter 18

18-3, Barbara Cousins; 18-A, William Ober; 18-B, Christine Oleksyk; 18-4, Joan Beck; 18-5, David Mascaro & Associates; 18-C, Raychel Ciemma; 18-7, Kevin Sommerville; 18-D, Ortho-Novum; 18-E, Wyeth-Ayerst Laboratories; 18-F-L, Laura J. Edwards; 18-8, Scott Bodell.

Chapter 19

19-A, 19-3, Rolin Graphics; 19-2, Scott Bodell; 19-B, 19-4, Raychel Ciemma; 19-5, Barbara Stackhouse; 19-6, Kevin Sommerville; 19-7, Marcia Hartsock; 19-9, Louise Brown/Wide World Photos.

Chapter 20

20-1, 20-5, Stewart Halperin; 20-4, Nathan Benn/Woodfin Camp & Associates; 20-A in box, Harris, Louis & Associates. Aging in the eighties: America in transition. Washington, DC: National Council on the Aging, 1981.

Chapter 21

21-2, Raychel Ciemma; 21-3, 21-4, 21-5, Nadine Sokol; 21-B, Desh Pal Verma/USDA.

Chapter 22

22-3, Rolin Graphics; Table 22-2, from Lerner/Libby: Heredity, Evolution and Environment, 2/e, W.H. Freeman and Company, 1976; 22-6A, 22-6B1, Stewart Halperin; 22-6B2, neg. #K4823/Tom Larson; 22-7A, neg. #333680, 22-7B, neg. #35151/R.A. Gould; 22-8A, neg. #123606, 22-8B, neg. #281668/R. Boulton, 22-8C, neg. #2223/Begoras, courtesy Department Library Services American Museum of Natural History.

A

B

C

D

E

F

G

H

I

N

O

P

Q

R

S

T

U

V

W

X

Y

Z